CONSTRUCTION AND HOME REPAIR TECHNIQUES SIMPLY EXPLAINED

WOODWORKING, SCAFFOLDING, LEVELING AND GRADING, CONCRETE, MASONRY, FLOORING AND ROOFING, EXTERIOR AND INTERIOR FINISHING, PLASTERING, STUCCOING, TILING, ETC.

Naval Education and Training Command

DOVER PUBLICATIONS, INC.
Mineola, New York

Bibliographical Note

This Dover edition, first published in 1999, is an unabridged republication of the revised edition of this training manual, as published by the Naval Education and Training Program Management Support Activity, United States Navy, in 1993–94, under the title *Builder 3 & 2,* volumes 1 and 2. For ready reference in this Dover book, the pages have been numbered consecutively and the indexes and glossaries from both volumes have been combined and placed at the back of the book.

Library of Congress Cataloging-in-Publication Data

Builder 3 & 2

Construction and home repair techniques simply explained : woodworking, scaffolding, leveling and grading, concrete, masonry, flooring and roofing, exterior and interior finishing, plastering, stuccoing, tiling, etc. / Naval Education and Training Command.

p. cm.

" . . . an unabridged republication of the revised edition of this training manual as published . . . under the title Builder 3 & 2, volumes 1 and 2 . . . pages have been numbered consecutively and the indexes and glossaries from both volumes have been combined and placed at the back of the book"—CIP t.p. verso.

Includes index.

ISBN 0-486-40481-1 (pbk.)

1. Building—United States Handbooks, manuals, etc. 2. United States. Navy—Builders Handbooks, manuals, etc. 3. United States. Navy—Military construction operations Handbooks, manuals, etc. I. United States. Naval Education and Training Command. II. Title.

VG593.B85 1999

690—dc21 99–30931

CIP

Manufactured in the United States of America

Dover Publications, Inc., 31 East 2nd Street, Mineola, N.Y. 11501

CONTENTS
VOLUME ONE

VOLUME TWO

PREFACE

This training manual (TRAMAN) and its associated nonresident training course (NRTC) are two units of a self-study program that will enable you, the Builder, to fulfill the requirements of your rating.

Designed for individual study and not formal classroom instruction, this TRAMAN provides subject matter that relates directly to the occupational standards of the Builder rating. The NRTC provides a way of satisfying the requirements for completing the TRAMAN. The assignments in the NRTC are intended to emphasize the key points in the TRAMAN.

Scope of revision—This TRAMAN contains new information on embarkation, K-span buildings, and tile, expands discussion of roof truss systems and coverings, and reorganizes sections dealing with interior and exterior finishing. The entire TRAMAN was reviewed for currency and updated as required.

This training manual and its separate nonresident training course were prepared by the Naval Education and Training Program Management Support Activity, Pensacola, Florida, for the Chief of Naval Education and Training. Technical assistance was provided by the Third Naval Construction Brigade, Pearl Harbor, Hawaii, and the Naval Construction Training Center, Gulfport, Mississippi.

SUMMARY OF THE BUILDER 3&2 RATE TRAINING MANUALS

VOLUME 1

Builder 3&2, Volume 1, NAVEDTRA 12520, is a basic book that should be mastered by those seeking advancement to Builder Third Class and Builder Second Class. The major topics addressed in this book include construction administration and safety; drawings and specifications; woodworking tools, materials and methods of woodworking; fiber line, wire rope, and scaffolding; leveling and grading; concrete; placing concrete; masonry; and planning, estimating, and scheduling.

VOLUME 2

Builder 3&2, Volume 2, NAVEDTRA 12521, continues where Volume 1 ends. The topics covered in this volume include floor and wall construction; roof framing; exterior and interior finishing; plastering, stuccoing, and ceramic tile; paints and preservatives; advanced base field structures; and heavy construction.

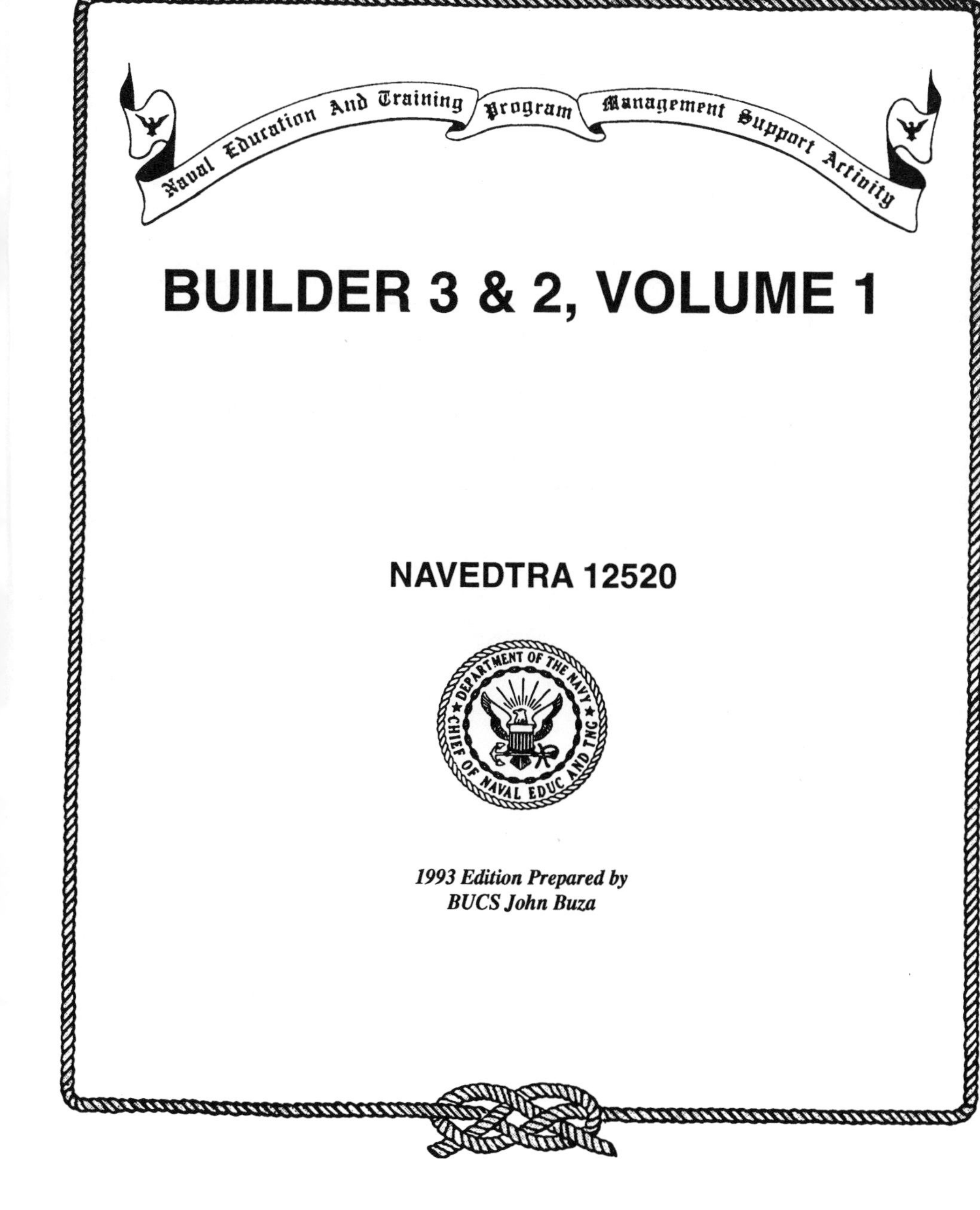

BUILDER 3 & 2, VOLUME 1

NAVEDTRA 12520

1993 Edition Prepared by
BUCS John Buza

Although the words "he," "him," and "his" are used sparingly in this manual to enhance communication, they are not intended to be gender driven nor to affront or discriminate against anyone reading this text.

CHAPTER 1

CONSTRUCTION ADMINISTRATION AND SAFETY

Being a petty officer carries many inherent responsibilities. These include your personal obligation to be a leader, an instructor, and an administrator in all the areas of your rating—military, technical, and safety.

As a petty officer, you need to develop an ability to control the work performed by your workers, as well as to lead them. As you gain experience as a petty officer and increase your technical competence as a Builder, you begin to accept a certain amount of responsibility for the work of others. With each advancement, you accept an increasing responsibility in military matters and in matters relating to the professional work of your rate. As you advance to third class and then to second class petty officer, you not only will have increased privileges but also increased responsibilities. You begin to assume greater supervisory and administrative positions.

The proper administration of any project, large or small, is as important as the actual construction. This chapter will provide you with information to help you to use and prepare the administrative paperwork that you encounter as a crew leader or as a crewmember.

ADMINISTRATION

LEARNING OBJECTIVE: Upon completing this section, you should be able to identify crew leader responsibilities in preparing tool kit inventories, preparing supply requisitions, and submitting labor time cards.

Administration is the means a person or an organization uses to keep track of what's happening. It provides a means of telling others what's been done and planned, who's doing it, and what's needed. Administration ranges from a simple notebook kept in your pocket to filling out a variety of reports and forms. As a growing leader in the Navy, you must learn about and become effective in the use of both the tools of your trade and administrative tools. Once you become comfortable with these, you can be a successful administrator.

PLANNING WORK ASSIGNMENTS

For our purposes here, planning means the process of determining requirements and developing methods and schemes of action for performing a task. Proper planning saves time and money and ensures a project is completed in a professional manner. Here, we'll look at some, but not all, of the factors you need to consider.

When you get a project, whether in writing or orally, make sure you clearly understand what is to be done. Study the plans and specifications carefully. If you have any questions, find the answers from those in a position to supply the information you need. Also, make sure you understand the priority of the project, expected time of completion, and any special instructions.

Consider the capabilities of your crew. Determine who is to do what and how long it should take. Also, consider the tools and equipment you will need. Arrange to have them available at the jobsite at the time the work is to get under way. Determine who will use the tools and make sure they know how to use them properly and safely.

To help ensure that the project is completed properly and on time, determine the best method of getting it done. If there is more than one way of doing a particular assignment, you should analyze the methods and select the one most suited to the job conditions. Listen to suggestions from others. If you can simplify a method and save time and effort, do it.

Establish goals for each workday and encourage your crew to work as a team in meeting these goals. Set goals that keep your crew busy, but make sure they are realistic. Discuss the project with the crew so they know what you expect from them. During an emergency, most crewmembers will make an all-out effort to meet a deadline. But when there is no emergency, don't expect them to work continuously at an excessively high rate. Again, set realistic goals. Daily briefings of this type cannot be over-emphasized.

DIRECTING WORK TEAMS

After a job has been properly planned, it is necessary to carefully direct the job. This ensures it is completed on time and with the quality that satisfies both the customer and the crew.

Before starting a project, make sure the crew knows what is expected. Give instructions and urge the crew to ask questions on **all** points that are not clear. Be honest in your answers. If you don't have an answer, say so; then find the answer and get back to the crew. Don't delay in getting solutions to the questions asked. Timely answers keep projects moving forward. They also show the crew your concern for the project is as genuine as theirs.

While a job is under way, spot check to ensure that the work is progressing satisfactorily. Determine whether the proper methods, materials, tools, and equipment are being used. When determining the initial requirements, do so early enough so there are no delays. If crewmembers are incorrectly performing a task, stop them and point out the correct procedures. When you check crewmembers' work, make them feel the purpose of checking is to teach, guide, or direct—not to criticize or find fault.

Make sure the crew complies with applicable safety precautions and wear safety apparel when required. Watch for hazardous conditions, improper use of tools and equipment, and unsafe work practices. These can cause mishaps and possibly result in injury to personnel. **There are no excuses for unsafe practices**. Proper safety instructions and training eliminate the desire to work carelessly. When directing construction crews, practice what you preach.

When time permits, rotate crewmembers on various jobs. Rotation gives you the opportunity to teach. It also gives each crewmember an opportunity to increase personal skill levels.

As a crew leader, you need to ensure that your crew work together in getting the job done. Develop an environment where each crewmember feels free to seek your advice when in doubt about any phase of the work. Emotional balance is especially important. Don't panic in view of your crew or be unsure of yourself when faced with a conflict.

Be tactful and courteous in dealing with your crew. It sounds obvious, but don't show any partiality. Keep every crewmember informed on both work and personal matters that affect his or her performance. Also, try to maintain a high level of morale. Low morale has a definite effect on the quantity and quality of a crew's work.

As you advance in rate, you spend more and more time supervising others. You have to learn as much as you can about supervision. Study books on both supervision and leadership. Also, watch how other supervisors—both good and bad—operate. Don't be afraid to ask questions.

TOOL KIT INVENTORY

Tool kits contain all the craft hand tools required by one, four-member construction crew or fire team of a given rating to pursue their trade. The kits may contain additional items required by a particular assignment. However, they should not be reduced in type of item and should be maintained at 100 percent of kit assembly allowance at all times.

As a crew leader, you can order and are responsible for all the tools required by the crew. This incurs the following responsibilities:

- Maintaining complete tools kits at all times;
- Assigning tools within the crew;
- Ensuring proper use and care of assigned tools by the crew;
- Preserving tools not in use;
- Securing assigned tools; and
- Ensuring that all electrical tools and cords are inspected on a regular basis.

To make sure tools are maintained properly, the operations officer and the supply officer establish a formal tool kit inventory and inspection program. As a crew leader, you perform a tool kit inventory at least every 2 weeks. Tools requiring routine maintenance are turned in to the central tool room (CTR) for repair and reissue. Damaged or worn tools should be returned to the CTR for replacement. You must submit requisitions for replacement items.

Tool management is further specified in instructions issued by Commander, Construction Battalion, Pacific (COMCBPAC) and Commander, Construction Battalion, Atlantic (COMCBLANT).

PREPARING REQUISITIONS

As a crew leader, you must become familiar with the forms used to request material or services through the Navy Supply System. Printed forms are available that provide all the information necessary for the physical transfer of the material and accounting requirements. The form you will use most often is NAVSUP Form 1250, shown in figure 1-1.

Crew leaders are not usually required to complete the entire form. However, you must list the stock number of the item, when available, the quantity required, and the name or description of each item needed. Turn this form in to the expediter, who checks it, fills in the remaining information, and signs it. The form then goes to the material liaison officer (MLO) or supply department for processing.

In ordering material, you need to know about the national stock number (NSN) system. Information on the NSN system and other topics about supply is given in *Military Requirements for Petty Officer Third Class*, NAVEDTRA 12044.

TIMEKEEPING

In both battalion and shore-based activities, you will be posting entries on time cards for military personnel. You need to know the type of information called for on the cards and understand the importance of accuracy in labor reporting. The reporting systems used primarily in naval mobile construction battalions (NMCBs) and the system employed at shore-based activities are similar.

A labor accounting system is used to record and measure the number of man-hours a unit spends on various functions. Labor utilization information is collected every day in sufficient detail and manner to allow the operations department to readily compile the data. This helps the operations officer to both manage manpower resources and prepare reports for higher authority. Although labor accounting systems may vary slightly from one command to another, the system described here is typical.

Each work unit accounts for all labor used to carry out its assignment. This lets management determine the amount of labor used on the project. Labor costs are figured, and actual man-hours are compared with other similar jobs. When completed, unit managers and higher commands use this information to develop planning standards.

The type of labor performed must be broken down and reported by category to show how labor has been used. For timekeeping and labor reporting

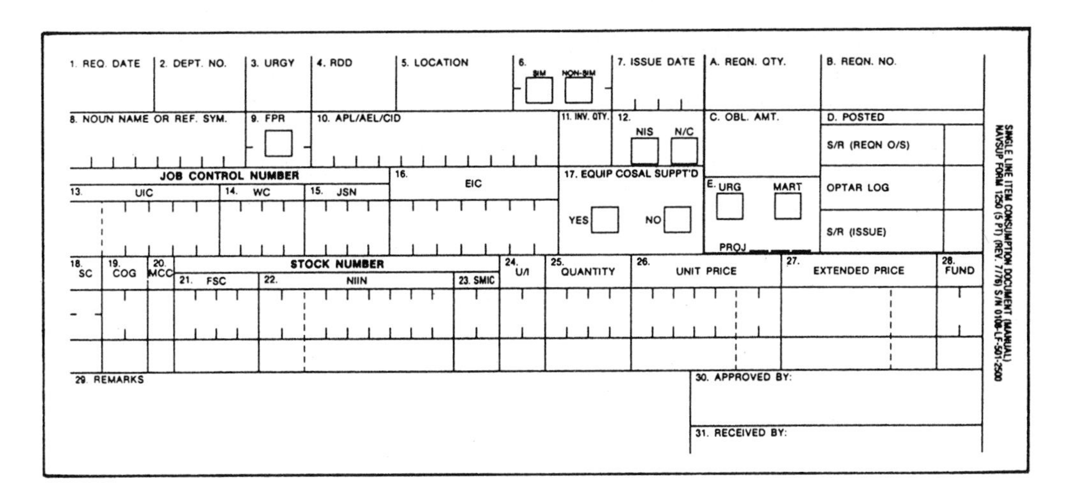

1. REQ. DATE
2. DEPT. NO.
3. URGY
4. RDD
5. LOCATION
6. SIM NON-SIM
7. ISSUE DATE
A. REQN. QTY.
B. REQN. NO.
8. NOUN NAME OR REF. SYM.
9. FPR
10. APL/AEL/CID
11. INV. QTY.
12. NIS N/C
C. OBL. AMT.
D. POSTED
S/R (REQN O/S)
JOB CONTROL NUMBER
13. UIC
14. WC
15. JSN
16. EIC
17. EQUIP COSAL SUPPT'D YES NO
E. URG MART PROJ
OPTAR LOG
S/R (ISSUE)
18. SC
19. COG
20. MCC
STOCK NUMBER
21. FSC
22. NIIN
23. SMIC
24. U/I
25. QUANTITY
26. UNIT PRICE
27. EXTENDED PRICE
28. FUND
29. REMARKS
30. APPROVED BY:
31. RECEIVED BY:
SINGLE LINE ITEM CONSUMPTION DOCUMENT (MANUAL)
NAVSUP FORM 1250 (5 PT) (REV. 7/78) S/N 0108-LF-501-2500

Figure 1-1.—NAVSUP 1250.

purposes, all labor is classified as either productive or overhead. Labor codes are shown in figure 1-2.

Productive labor either directly or indirectly contributes to the completion of the unit's mission, including construction operations and training. It is broken down into four categories: direct labor, indirect labor, military operations and readiness, and training.

Direct labor includes labor expended directly on assigned construction tasks contributing directly to the completion of an end product. It can be either in the field or in the shop. Direct labor must be reported separately for each assigned construction task. Indirect labor is labor required to support construction operations but not producing an end product itself.

Military operations and readiness includes work necessary to ensure the unit's military and mobility readiness. It consists of all manpower expended in actual military operations, unit embarkation, and planning and preparations.

Training includes attendance at service schools, factory and industrial courses, fleet-level training and short courses, military training, and organized training conducted within the battalion or unit.

Overhead labor, compared to productive labor, does not contribute directly or indirectly to the completion of an end product. It includes labor that must be performed regardless of the assigned mission.

During project planning and scheduling, each direct labor phase of the project is given an identifying code. For example, excavating and setting forms may be assigned code R-15; laying block, code R-16; and installing bond beams, code R-17. (Since there are many types of construction

PRODUCTIVE LABOR. Productive labor includes all labor that directly contributes to the accomplishment of the Naval Mobile Construction Battalion (NMCB), including construction operations and readiness, disaster recovery operations, and training.

DIRECT LABOR. This category includes all labor expended directly on assigned construction tasks, either in the field or in the shop, and which contributes directly to the completion of the end product.

INDIRECT LABOR. This category comprises labor required to support construction operations, but which does not produce in itself. Indirect labor reporting codes are as follows:

X01 Construction Equipment Maintenance, Repair and Records
X02 Operation and Engineering
X03 Project Supervision
X04 Project Expediting (Shop Planners)
X05 Location Moving
X06 Project Material Support
X07 Tool and Spare Parts Issue
X08 Other

MILITARY OPERATIONS AND READINESS. This category comprises all manpower expended in actual military operations, unit embarkation, and planning and preparations necessary to insure unit military and mobility readiness. Reporting codes are as follows:

M01 Military Operations
M02 Military Security
M03 Embarkation
M04 Unit Movement
M05 Mobility Preparation
M06 Contingency
M07 Military Administrative Functions
M08 Mobility & Defense Exercise
M09 Other

DISASTER CONTROL OPERATIONS

D01 Disaster Control Operations
D02 Disaster Control Exercise

TRAINING. This category includes attendance at service schools, factory, and industrial training courses, fleet type training, and short courses, military training, and organized training conducted within the battalion. Reporting codes are as follows:

T01 Technical Training
T02 Military Training
T03 Disaster Control Training
T04 Leadership Training
T05 Safety Training
T06 Training Administration

OVERHEAD LABOR. This category includes labor that must be performed regardless of whether a mission is assigned, and which does not contribute to the assigned mission. Reporting codes are as follows:

Y01 Administrative & Personnel
Y02 Medical & Dental Department
Y03 Navy Exchange and Special Services
Y04 Supply & Disbursing
Y05 Commissary
Y06 Camp Upkeep & Repairs
Y07 Security
Y08 Leave & Liberty
Y09 Sickcall, Dental & Hospitalization
Y10 Personal Affairs
Y11 Lost Time
Y12 TAD not for unit
Y13 Other

Figure 1-2.—Labor codes.

projects involving different operations, codes for direct labor may vary from one activity to another.) Use direct labor codes in reporting each hour spent by each of your crewmembers during each workday on an assigned activity code.

Submit your reports on a daily labor distribution report form (timekeeping card). Views A and B of figure 1-3 show typical timekeeping cards. The form provides a breakdown, by man-hours, of the activities in the various labor codes for each crewmember for each day on any given project. The form is reviewed at the company level by the staff and platoon commander. The company commander then initials the report and sends it to the operations department. The management division of the operations department tabulates the report, along with those received from all other companies and departments in the unit. This consolidated report is the means by which the operations office analyzes the labor

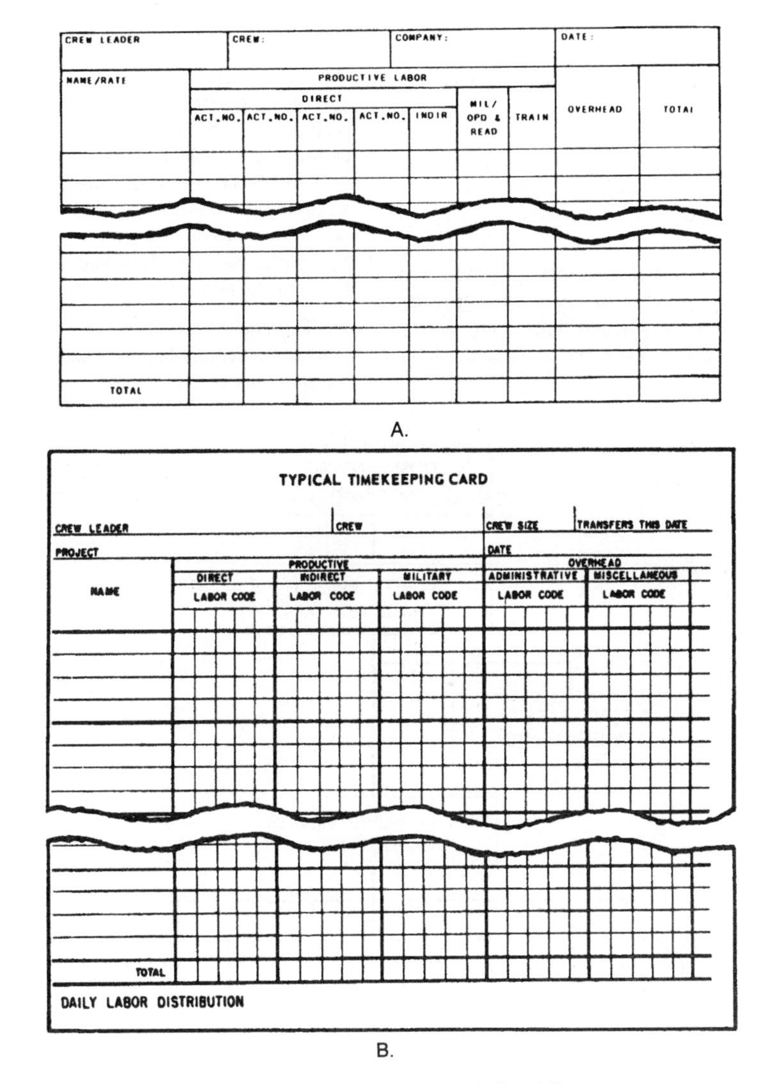

CREW LEADER		CREW:			COMPANY:			DATE:	
NAME/RATE	PRODUCTIVE LABOR							OVERHEAD	TOTAL
	DIRECT					MIL/ OPD & READ	TRAIN		
	ACT.NO.	ACT.NO.	ACT.NO.	ACT.NO.	INDIR				
TOTAL									

A.

TYPICAL TIMEKEEPING CARD

CREW LEADER	CREW	CREW SIZE	TRANSFERS THIS DATE
PROJECT		DATE	

NAME	PRODUCTIVE			OVERHEAD	
	DIRECT	INDIRECT	MILITARY	ADMINISTRATIVE	MISCELLANEOUS
	LABOR CODE	LABOR CODE	LABOR CODE	LABOR CODE	LABOR CODE
TOTAL					

DAILY LABOR DISTRIBUTION

B.

Figure 1-3.—Typical timekeeping cards (A and B).

distribution of total manpower resources for each day. It also serves as feeder information for preparing the monthly operations report, and any other source reports required of the unit. The information must be accurate and timely. Each level in the organization should review the report for an analysis of its own internal construction management and performance.

SAFETY PROGRAM

LEARNING OBJECTIVE: Upon completing this section, you should be able to describe the safety organization, function of the battalion or unit safety program, and the responsibilities of key personnel.

You must be familiar with the safety program at your activity. You cannot function effectively as a petty officer unless you are aware of how safety fits into your organization. You need to know who establishes and arbitrates safety policies and procedures. You should also know who provides guidelines for safety training and supervision. Every NMCB and shore command has a formal safety organization.

SAFETY ORGANIZATION

The NMCB's safety organization provides for the establishment of safety policy and control and reporting. As illustrated in figure 1-4, the battalion safety policy organization contains several committees: policy; supervisors'; and equipment, shop, and crew.

The executive officer presides over the safety policy committee. Its primary purpose is to develop safety rules and policy for the battalion. This committee reports to the commanding officer, who approves all changes in safety policy.

The battalion safety officer presides over the safety supervisors' committee. This committee includes safety supervisors assigned by company commanders, project officers, or officers in charge of a detail. Basically, it helps the safety officer manage an effective overall safety and health program. The committee provides a convenient forum for work procedures, safe practices, and safety suggestions. Its recommendations are sent to the policy committee.

The equipment, shop, and crew committees are assigned as required and are usually presided over by the company or project safety supervisor. The main

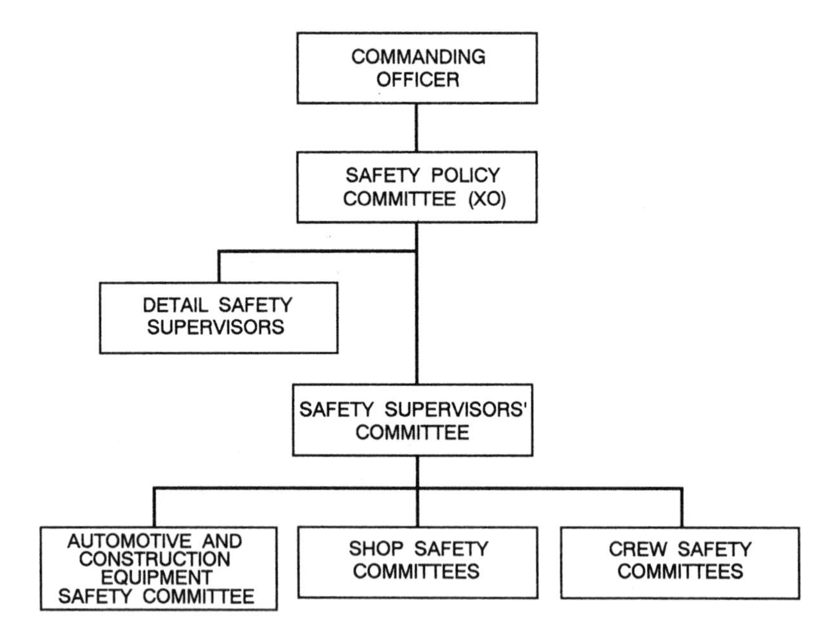

Figure 1-4.—NMCB safety policy organization.

objective of this committee is to propose changes in the battalion's safety policy to eliminate unsafe working conditions or prevent unsafe acts. It is **your** contact for recommending changes in safety matters. In particular, the equipment committee reviews all vehicle mishap reports, determines the cause of each mishap, and recommends corrective action. As a crew leader, you can expect to serve as a member. Each committee forwards reports and recommendations to the safety supervisors' committee.

SAFETY DUTIES

As a crew leader, you will report to the safety supervisor, who directs the safety program of a project. The safety supervisor is inherently responsible for all personnel assigned to that shop or project. Some of the duties include indoctrinating new crewmembers, compiling mishap statistics for the project, reviewing mishap reports submitted to the safety office, and comparing safety performances of all crews.

The crew leader is responsible for carrying out safe working practices. This is done under the direction of the safety supervisor or others in positions of authority (project chief, project officer, or safety officer). You, as the crew leader, ensure that each crewmember is thoroughly familiar with these working practices, has a general understanding of pertinent safety regulations, and makes proper use of protective clothing and safety equipment. Furthermore, you should be ready at all times to correct every unsafe working practice you observe, and report it immediately to the safety supervisor or the person in charge. When an unsafe condition exists, any crew or shop member can stop work until the condition is corrected.

In case of a **mishap**, make sure injured personnel get proper medical care as quickly as possible. Investigate each mishap involving crewmembers to determine its cause. Remove or permanently correct defective tools, materials, and machines. Do the same for environmental conditions contributing to a mishap. Afterward, submit required reports.

SAFETY TRAINING

New methods and procedures for safely maintaining and operating equipment are always coming out. You must keep up to date on the latest techniques in maintenance and operation safety and pass them on to your crewmembers. One method of keeping your crewmembers informed is by holding stand-up safety meetings before the day's work starts. As crew leader, you are responsible for conducting each meeting and passing on material from the safety supervisor. Information (such as the type of safety equipment to use, where to obtain it, and how to use it) is often the result of safety suggestions received by the safety supervisors' committee. Encourage your crew to submit ideas or suggestions. Don't limit yourself to just the safety lecture in the morning. Discuss minor safety infractions when they occur or at appropriate break times during the day. As the crew leader, you must impress safe working habits upon your crewmembers through proper instructions, constant drills, and continuous supervision.

You may hold group discussions on **specific mishaps** to guard against or that may happen on the job. Be sure to give plenty of thought to what you are going to say beforehand. Make the discussion interesting and urge the crew to participate. The final result should be a group conclusion as to how the specific mishap can be prevented.

Your stand-up safety meetings also give you the chance to discuss prestart checks, and the operation or maintenance of automotive vehicles assigned to a project. Vehicles are used for transporting crewmembers as well as cargo. It is important to emphasize how the prestart checks are to be made and how to care for the vehicles.

You can use a stand-up safety meeting to solve safety problems arising from a new procedure. An example might be starting a particular piece of equipment just being introduced. In this case, show the safe starting procedure for the equipment. Then, have your crewmembers practice the procedure.

Because of the variety of vehicles that may be assigned to a project, there is too much information and too many operating procedures for one person to remember. You need to know where to look for these facts and procedures. For specific information on prestart checks, operation, and maintenance of each vehicle assigned, refer to the manufacturer's operator/maintenance manuals. In addition, personnel from Alfa Company (equipment experts) will instruct all personnel in the proper start-up procedures for new equipment.

In addition to stand-up safety meetings, conduct day-to-day instruction and on-the-job training. Although it is beyond the scope of this chapter to describe teaching methods, a few words on your approach to safety and safety training at the crew level are appropriate. Getting your crew to work safely, like most other crew leader functions, is essentially a matter of leadership. Therefore, don't overlook the power of **personal example** in leading and teaching your crewmembers. They are quick to detect differences between what you say and what you do. Don't expect them to measure up to a standard of safe conduct that you, yourself, do not. Make your genuine concern for the safety of your crew visible at all times. Leadership by example is one of the most effective techniques you can use.

RECOMMENDED READING LIST

NOTE

Although the following references were current when this TRAMAN was published, their continued currency cannot be assured. You therefore need to ensure that you are studying the latest revision.

Naval Construction Force Manual, P-315, Naval Facilities Engineering Command, Washington, D.C., 1985.

Naval Construction Force Occupational Safety and Health Program Manual, COMCBPAC/COMCBLANTINST 5100.1F, Commander, Naval Construction Battalions, U.S. Pacific Fleet, Pearl Harbor, Hawaii, and Commander, Naval Construction Battalions, U.S. Atlantic Fleet, Naval Amphibious Base, Little Creek, Norfolk, Va., 1991.

Seabee Planner's Estimator's Handbook, NAVFAC P-405, Naval Facilities Engineering Command, Alexandria, Va., 1983.

CHAPTER 2

DRAWINGS AND SPECIFICATIONS

By this time in your Navy career, you have probably worked as a crewmember on various building projects. You probably did your tasks without thinking much about what it takes to lay out structures so they will conform to their location, size, shape, and other building features. In this chapter, you will learn how to extract these types of information from drawings and specifications. You will also be shown how to draw, read, and work from simple shop drawings and sketches.

We provide helpful references throughout the chapter. You are encouraged to study these, as required, for additional information on the topics discussed.

DESIGN OF STRUCTURAL MEMBERS

LEARNING OBJECTIVE: Upon completing this section, you should be able to identify the different types of structural members.

From the Builder's standpoint, building designs and construction methods depend on many factors. No two building projects can be treated alike. However, the factors usually considered before a structure is designed are its geographical location and the availability of construction materials.

It is easy to see why geographical location is important to the design of a structure, especially its main parts. When located in a temperate zone, for example, the roof of a structure must be sturdy enough not to collapse under the weight of snow and ice. Also, the foundation walls have to extend below the frost line to guard against the effects of freezing and thawing. In the tropics, a structure should have a low-pitch roof and be built on a concrete slab or have shallow foundation walls.

Likewise, the availability of construction materials can influence the design of a structure. This happens when certain building materials are scarce in a geographical location and the cost of shipping them is prohibitive. In such a case, particularly overseas, the structure is likely to be built with materials purchased locally. In turn, this can affect the way construction materials are used—it means working with foreign drawings and metric units of weights and measures.

By comparing the designs of the two structures shown in figures 2-1 and 2-2, you can see that each is designed according to its function. For example, light-frame construction is usually found in residential buildings where a number of small rooms are desired. Concrete masonry and steel construction is used for warehouse-type facilities where large open spaces are needed. You should study these figures carefully and learn the terminology. Depending on the use of the structure, you may use any combination of structural members.

DEAD AND LIVE LOADS

The main parts of a structure are the load-bearing members. These support and transfer the loads on the structure while remaining equal to each other. The places where members are connected to other members are called joints. The sum total of the load supported by the structural members at a particular instant is equal to the total dead load plus the total live load.

The total dead load is the total weight of the structure, which gradually increases as the structure rises and remains constant once it is completed. The total live load is the total weight of movable objects (such as people, furniture, and bridge traffic) the structure happens to be supporting at a particular instant.

The live loads in a structure are transmitted through the various load-bearing structural members to the ultimate support of the earth. Immediate or direct support for the live loads is first provided by horizontal members. The horizontal members are, in turn, supported by vertical members. Finally, the vertical members are supported by foundations or footings, which are supported by the earth. Look at figure 2-1, which illustrates both horizontal and vertical members of a typical light-frame structure. The weight of the roof material is distributed over the top supporting members and transferred through all joining members to the soil.

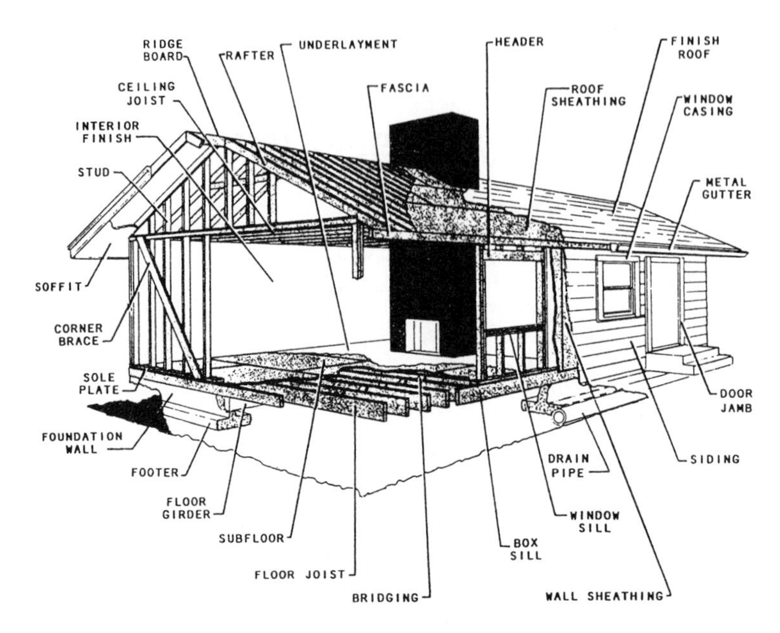

Figure 2-1.—Typical light-frame construction.

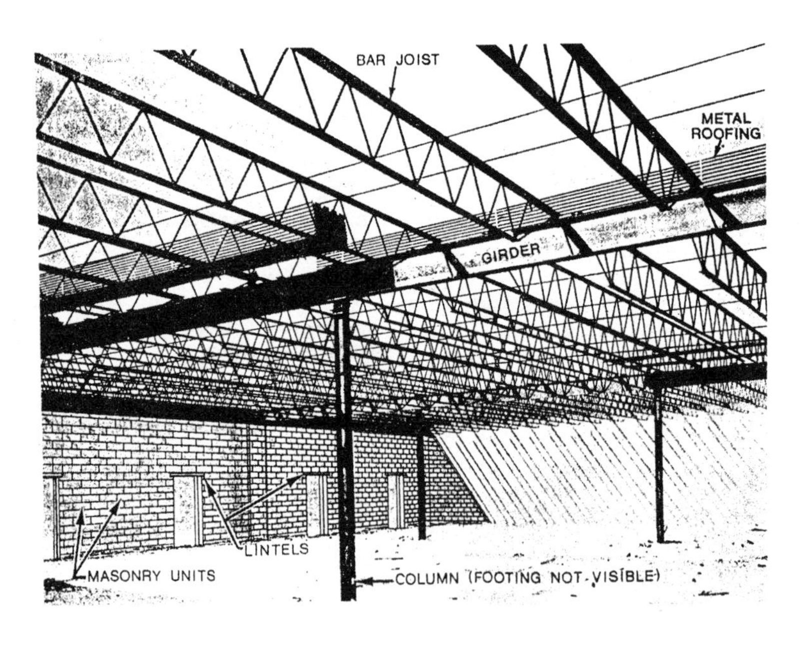

Figure 2-2.—Typical concrete masonry and steel structure.

The ability of the earth to support a load is called its soil-bearing capacity. This varies considerably with different types of soil. A soil of a given bearing capacity bears a heavier load on a wide foundation or footing than on a narrow one.

VERTICAL STRUCTURAL MEMBERS

In heavy construction, vertical structural members are high-strength columns. (In large buildings, these are called pillars.) Outside wall columns and inside bottom-floor columns usually rest directly on footings. Outside wall columns usually extend from the footing or foundation to the roof line. Inside bottom-floor columns extend upward from footings or foundations to the horizontal members, which, in turn, support the first floor or roof, as shown in figure 2-2. Upper floor columns are usually located directly over lower floor columns.

In building construction, a pier, sometimes called a short column, rests either directly on a footing, as shown in the lower center of figure 2-3, or is simply set or driven into the ground. Building piers usually support the lowermost horizontal structural members.

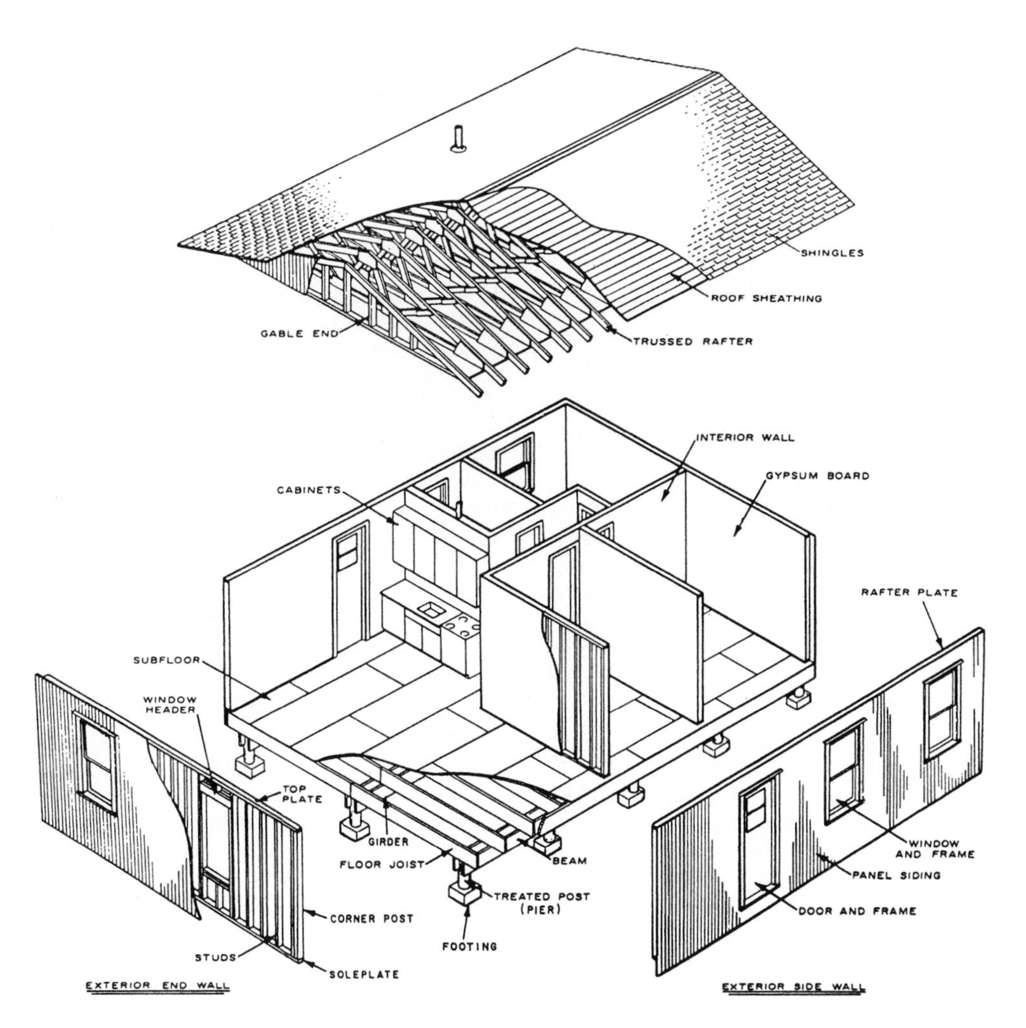

Figure 2-3.—Exploded view of a typical light-frame house.

In bridge construction, a pier is a vertical member that provides intermediate support for the bridge superstructure.

The chief vertical structural members in light-frame construction are called studs (see figures 2-1 and 2-3). They are supported by horizontal members called sills or soleplates, as shown in figure 2-3. Corner posts are enlarged studs located at the building corners. Formerly, in full-frame construction, a corner post was usually a solid piece of larger timber. In most modern construction, though, built-up corner posts are used. These consist of various members of ordinary studs nailed together in various ways.

HORIZONTAL STRUCTURAL MEMBERS

Technically, any horizontal load-bearing structural member that spans a space and is supported at both ends is considered a beam. A member fixed at one end only is called a cantilever. Steel members that consist of solid pieces of regular structural steel are referred to as "structural shapes." A girder (shown in figure 2-2) is a structural shape. Other prefabricated, open-web, structural-steel shapes are called bar joists (also shown in figure 2-2).

Horizontal structural members that support the ends of floor beams or joists in wood-frame construction are called sills or girders see figures 2-1 and 2-3). The name used depends on the type of framing and the location of the member in the structure. Horizontal members that support studs are called soleplates, depending on the type of framing. Horizontal members that support the wall ends of rafters are called rafter plates. Horizontal members that assume the weight of concrete or masonry walls above door and window openings are called lintels (figure 2-2).

The horizontal or inclined members that provide support to a roof are called rafters (figure 2-1). The lengthwise (right angle to the rafters) member, which supports the peak ends of the rafters in a roof, is called the ridge. The ridge may be called a ridge board, the ridge piece, or the ridge pole. Lengthwise members other than ridges are called purlins. In wood-frame construction, the wall ends of rafters are supported on horizontal members called rafter plates, which are, in turn, supported by the outside wall studs. In concrete or masonry wall construction, the wall ends of rafters may be anchored directly on the walls or on plates bolted to the walls.

A beam of given strength, without intermediate supports below, can support a given load over only a specific maximum span. When the span is wider than this maximum space, intermediate supports, such as columns, must be provided for the beam. Sometimes it is either not feasible or impossible to increase the beam size or to install intermediate supports. In such cases, a truss is used. A truss is a combination of members, such as beams, bars, and ties, usually arranged in triangular units to form a rigid framework for supporting loads over a span.

The basic components of a roof truss are the top and bottom chords and the web members. The top chords serve as roof rafters. The bottom chords act as ceiling joists. The web members run between the top and bottom chords. The truss parts are usually made of 2- by 4-inch or 2- by 6-inch material and are tied together with metal or plywood gusset plates, as shown in figure 2-4.

Roof trusses come in a variety of shapes and sizes. The most commonly used roof trusses, shown in figure 2-5, for light-frame construction are the king-post, the W-type, and the scissors trusses. The simplest type of truss used in frame construction is the king-post truss. It is mainly used for spans up to 22 feet. The most widely used truss in light-frame construction is the W-type truss. The W-type truss can be placed over spans up to 50 feet. The scissors truss is used for buildings with sloping ceilings. Generally, the slope of the bottom chord equals one-half the slope of the top chord. It can be placed over spans up to 50 feet.

DRAWINGS

LEARNING OBJECTIVE: Upon completing this section, you should be able to recognize the different types of drawings and their uses.

The building of any structure is described by a set of related drawings that give the Builder a complete, sequential, graphic description of each phase of the construction process. In most cases, a set of drawings begins by showing the location, boundaries, contours, and outstanding physical features of the construction site and its adjoining areas. Succeeding drawings give instructions for the excavation and disposition of existing ground; construction of the foundations and superstructure; installation of utilities, such as plumbing, heating, lighting, air conditioning, interior and exterior finishes; and whatever else is required to complete the structure.

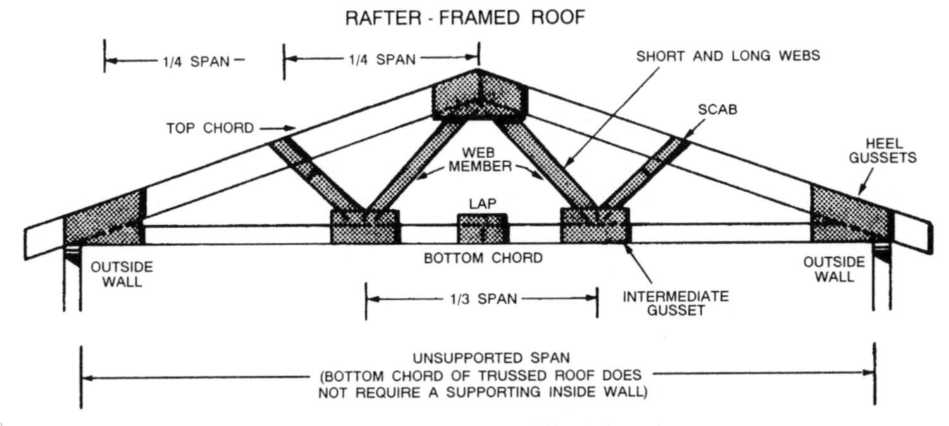

Figure 2-4.—A truss rafter.

The engineer works with the architect to decide what materials to use in the structure and the construction methods to follow. The engineer determines the loads that supporting members will carry and the strength qualities the members must have to bear the loads. The engineer also designs the mechanical systems of the structure, such as the lighting, heating, and plumbing systems. The end result is the architectural and engineering design sketches. These sketches guide draftsmen in preparing the construction drawings.

CONSTRUCTION DRAWINGS

Generally, construction or "working" drawings furnish enough information for the Builder to complete an entire project and incorporate all three main groups of drawings—architectural, electrical, and mechanical. In drawings for simple structures, this grouping may be hard to discern because the same single drawing may contain both the electrical and mechanical layouts. In complicated structures, however, a combination of layouts is not possible because of overcrowding. In this case, the floor plan may be traced over and over for drawings for the electrical and mechanical layouts.

All or any one of the three types of drawings gives you enough information to complete a project. The specific one to use depends on the nature of construction involved. The construction drawing furnishes enough information for the particular tradesman to complete a project, whether architectural, electrical, or mechanical. Normally, construction drawings include the detail drawings, assembly drawings, bill of materials, and the specifications.

A detail drawing shows a particular item on a larger scale than that of the general drawing in which the item appears. Or, it may show an item too small to appear at all on a general drawing.

An assembly drawing is either an exterior or sectional view of an object showing the details in the proper relationship to one another. Assembly drawings are usually drawn to a smaller scale from the dimensions of the detail drawings. This provides a check on the accuracy of the design drawings and often discloses errors.

Construction drawings consist mostly of right-angle and perpendicular views prepared by draftsmen

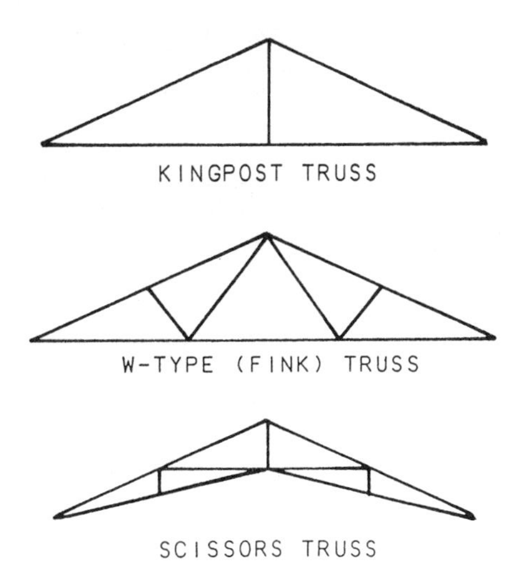

Figure 2-5.—The most commonly used roof trusses.

using standard technical drawing techniques, symbols, and other designations as stated in military standards (MIL-STDs). The first section of the construction drawings consists of the site plan, plot plan, foundation plans, floor plans, and framing plans. General drawings consist of plans (views from above) and elevations (side or front views) drawn on a relatively small scale. Both types of drawings use a standard set of architectural symbols. Figure 2-6 illustrates the conventional symbols for the more common types of material used on structures. Figure 2-7 shows the more common symbols used for doors and windows. Study these symbols thoroughly before proceeding further in this chapter.

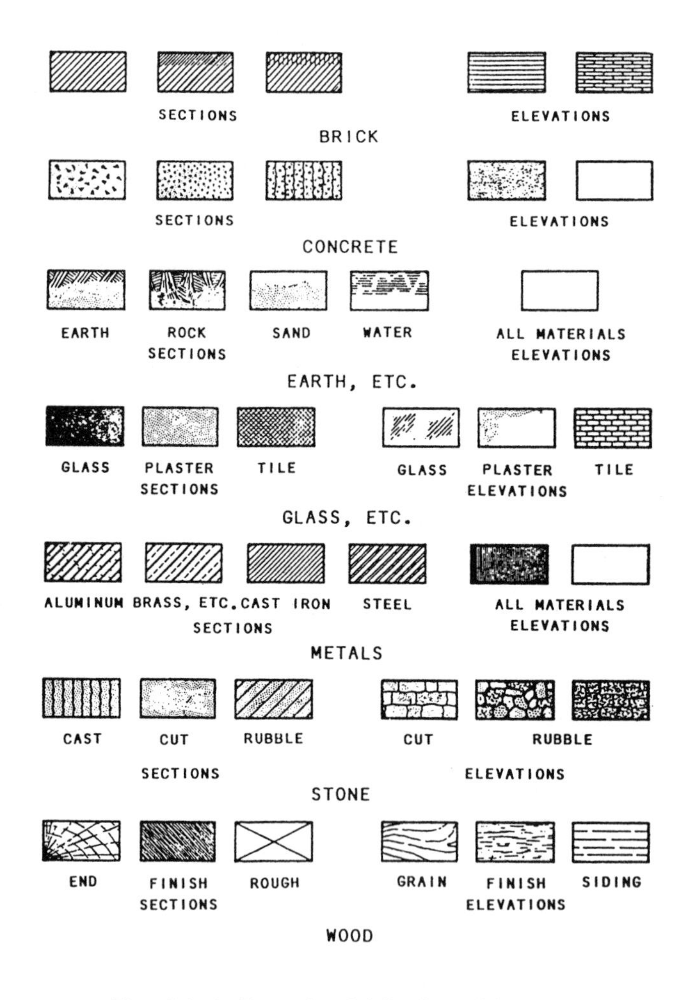

Figure 2-6.—Architectural symbols for plans and elevations.

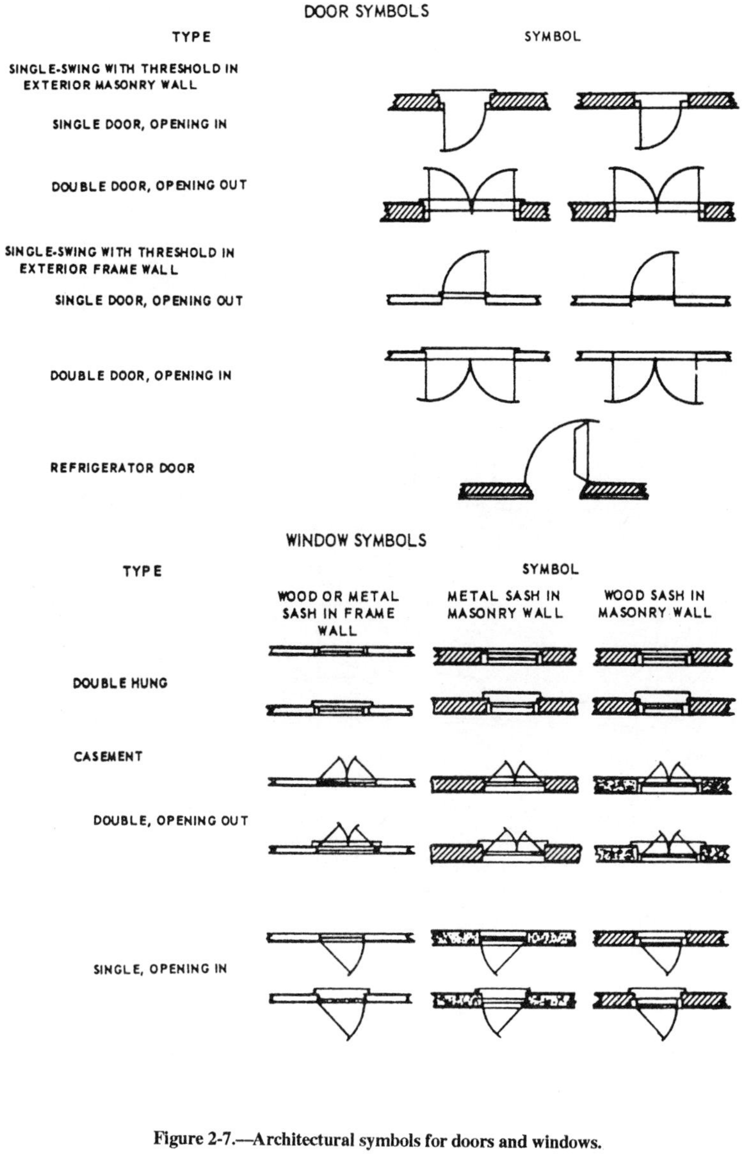

DOOR SYMBOLS

TYPE

SYMBOL

SINGLE-SWING WITH THRESHOLD IN EXTERIOR MASONRY WALL

SINGLE DOOR, OPENING IN

DOUBLE DOOR, OPENING OUT

SINGLE-SWING WITH THRESHOLD IN EXTERIOR FRAME WALL

SINGLE DOOR, OPENING OUT

DOUBLE DOOR, OPENING IN

REFRIGERATOR DOOR

WINDOW SYMBOLS

TYPE

SYMBOL

WOOD OR METAL SASH IN FRAME WALL

METAL SASH IN MASONRY WALL

WOOD SASH IN MASONRY WALL

DOUBLE HUNG

CASEMENT

DOUBLE, OPENING OUT

SINGLE, OPENING IN

Figure 2-7.—Architectural symbols for doors and windows.

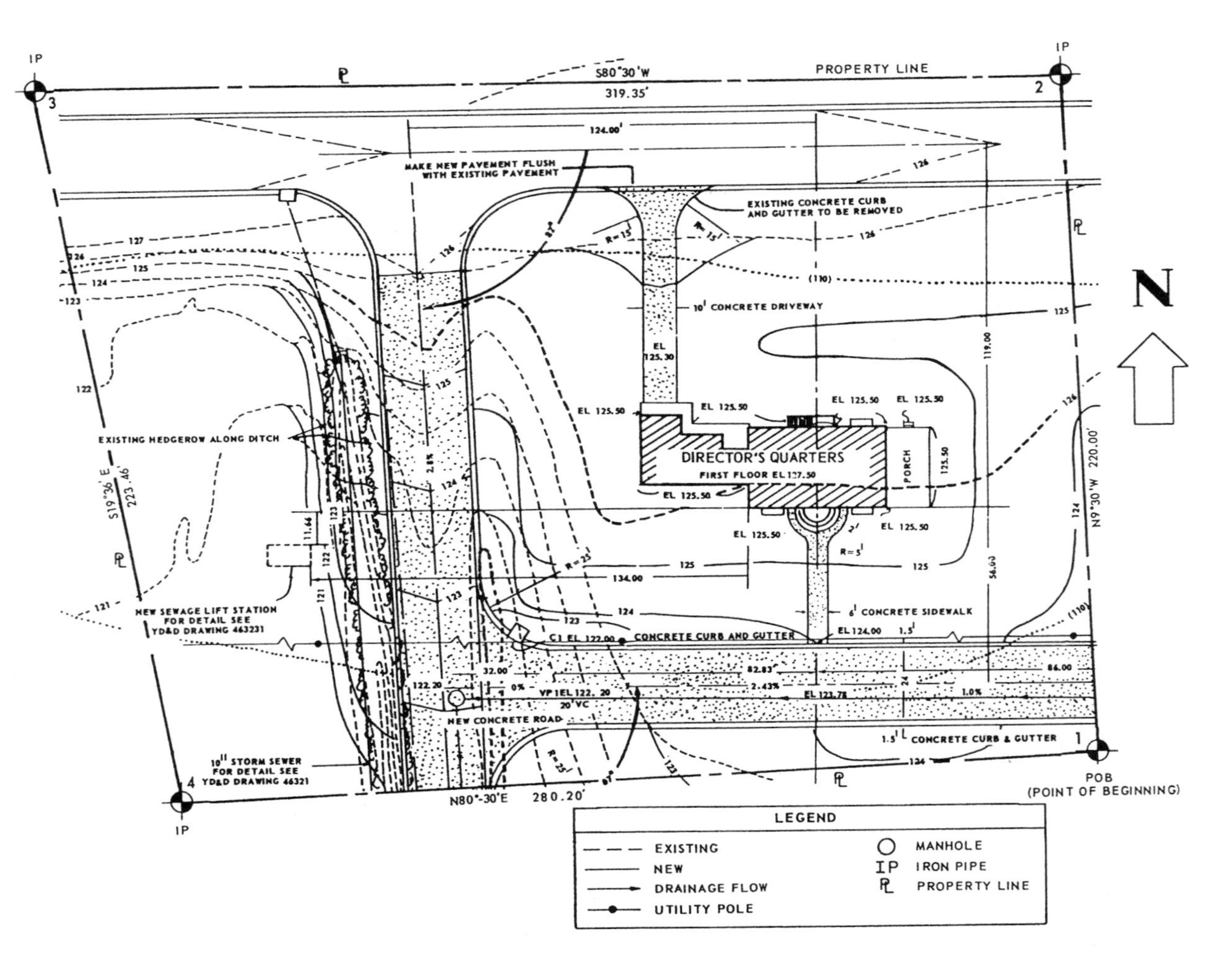

Figure 2-8.—Site plan.

Site Plan

A site plan (figure 2-8) shows the contours, boundaries, roads, utilities, trees, structures, and any other significant physical features on or near the construction site. The locations of proposed structures are shown in outline. This plan shows corner locations with reference to reference lines shown on the plot that can be located at the site. By showing both existing and finished contours, the site plan furnishes essential data for the graders.

Plot Plan

The plot plan shows the survey marks with the elevations and the grading requirements. The plot plan is used by the Engineering Aids to set up the corners and perimeter of the building using batter boards and line stakes, as shown in figure 2-9. Thus, the plot plan furnishes the essential data for laying out the building.

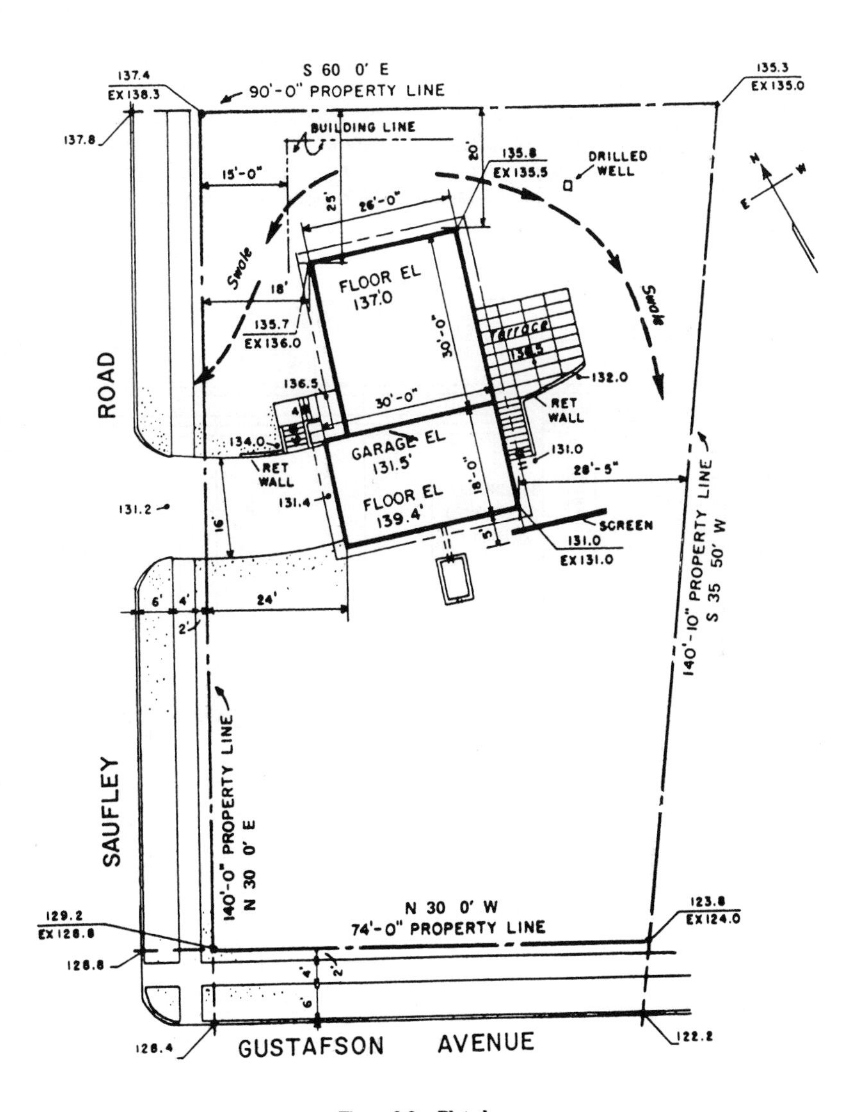

Figure 2-9.—Plot plan.

Foundation Plan

A foundation plan is a **plane** view of a structure. That is, it looks as if it were projected onto a horizontal plane and passed through the structure. In the case of the foundation plan, the plane is slightly below the level of the top of the foundation wall. The plan in figure 2-10 shows that the main foundation consists of 12-inch and 8-inch concrete masonry unit (CMU) walls measuring 28 feet lengthwise and 22 feet crosswise. The lower portion of each lengthwise section of wall is to be 12 inches thick to provide a concrete ledge 4 inches wide.

A girder running through the center of the building will be supported at the ends by two 4-by-12-inch concrete pilasters butting against the end foundation walls. Intermediate support for the girder will be provided by two 12-by-12-inch concrete piers, each supported on 18-by-18-inch spread footings, which are 10 inches deep. The dotted lines around the foundation walls indicate that these walls will also rest on spread footings.

Floor Plan

Figure 2-11 shows the way a floor plan is developed: from elevation, to cutting plane, to floor plan. An architectural or structural floor plan shows the structural characteristics of the building at the level of the plane of projection. A mechanical floor plan shows the plumbing and heating systems and any other mechanical components other than those that are electrical. An electrical floor plan shows the lighting system and any other electrical systems.

Figure 2-12 is a floor plan showing the lengths, thicknesses, and character of the outside walls and

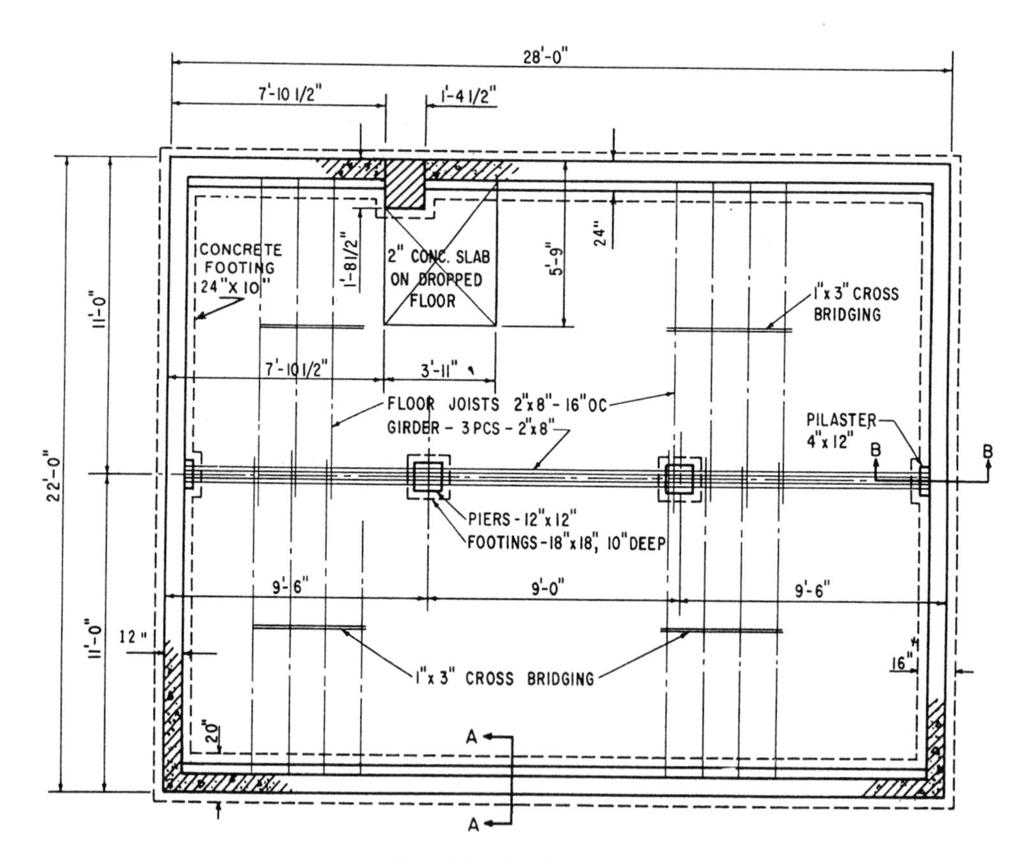

Figure 2-10.—Foundation plan.

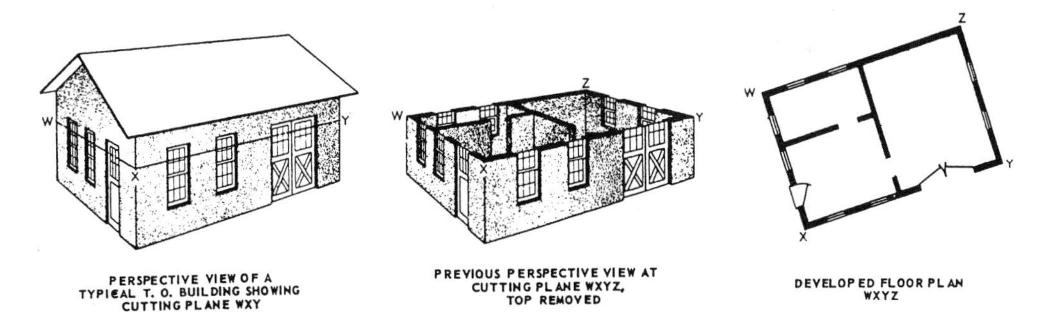

Figure 2-11.—Floor plan development.

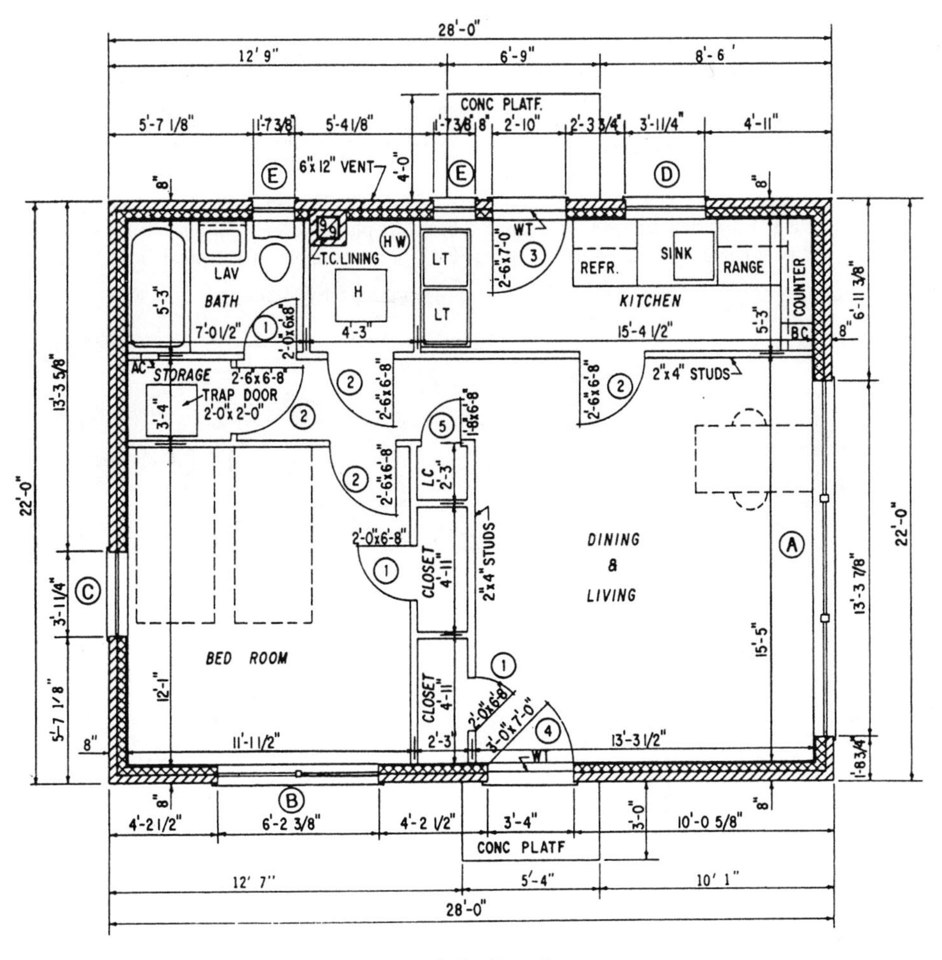

Figure 2-12.—Floor plan.

partitions at the particular floor level. It also shows the number, dimensions, and arrangement of the rooms, the widths and locations of doors and windows, and the locations and character of bathroom, kitchen, and other utility features. You should carefully study figure 2-12. In dimensioning floor plans, it is very important to check the overall dimension against the sum of the partial dimensions of each part of the structure.

Elevations

The front, rear, and sides of a structure, as they would appear projected on vertical planes, are shown in elevations. Studying the elevation drawing gives you a working idea of the appearance and layout of the structure.

Elevations for a small building are shown in figure 2-13. Note that the wall surfaces of this house will consist of brick and the roof covering of composition shingles. The top of the rafter plate will be 8 feet 2 1/4 inches above the level of the finished first floor, and the tops of the finished door and window openings 7 feet 1 3/4 inches above the same level. The roof will be a gable roof with 4 inches of rise for every 12 inches length. Each window shown in the elevations is identified by a capital letter that goes with the window schedule (which we'll discuss later in this chapter).

Framing Plans

Framing plans show the size, number, and location of the structural members (steel or wood) that make up the building framework. Separate framing plans may be drawn for the floors, walls, and roof. The floor framing plan must specify the sizes and spacing of joists, girders, and columns used to support the floor. When detail drawings are needed, the methods of anchoring joists and girders to the columns and foundation walls or footings must be shown. Wall framing plans show the location and method of framing openings and ceiling heights so that studs and posts can be cut. Roof framing plans show the construction of the rafters used to span the building and support the roof. Size, spacing, roof slope, and all details are shown.

FLOOR PLANS.—Framing plans for floors are basically plane views of the girders and joists. Figure 2-14 is an example of a typical floor framing plan.

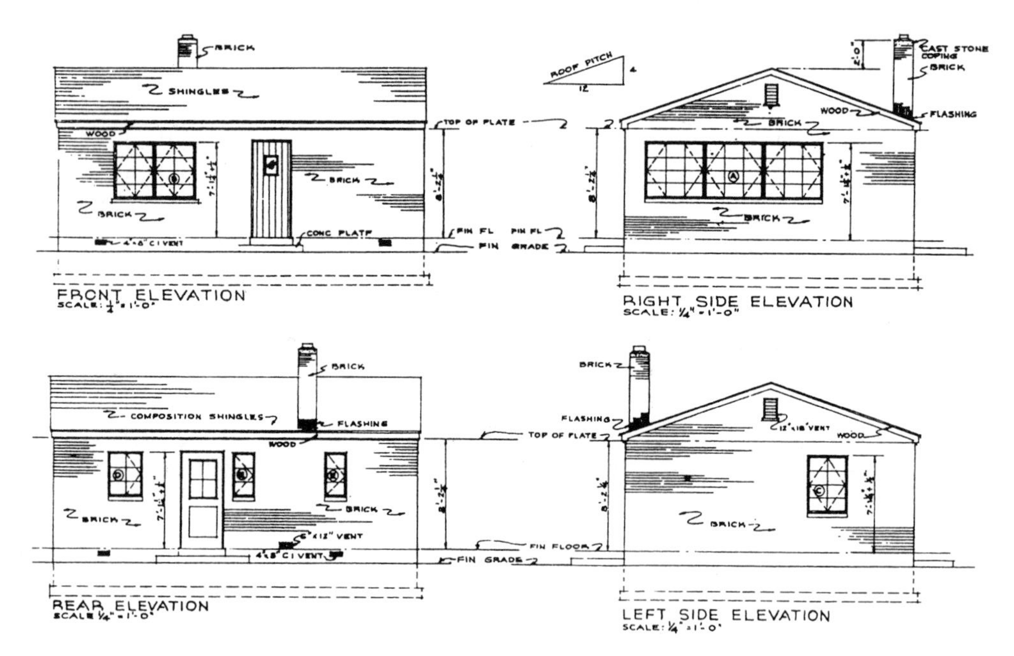

Figure 2-13.—Elevations.

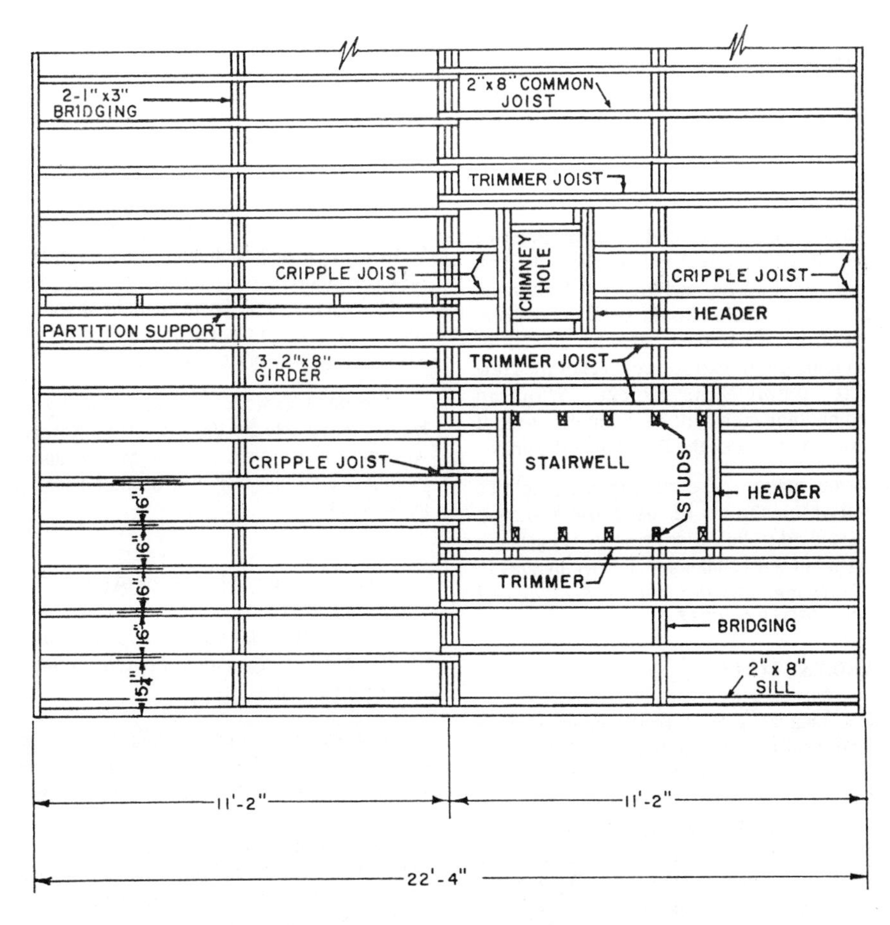

Figure 2-14.—Floor framing plan.

The unbroken, double-line symbol is used to indicate joists, which are drawn in the positions they will occupy in the completed building. Double framing around openings and beneath bathroom fixtures is shown where used. Bridging is shown by a double-line symbol that runs perpendicular to the joists. The number of rows of cross bridging is controlled by the span of the joists; they should not be placed more than 7 or 8 feet apart. A 14-foot span needs only one row of bridging, but a 16-foot span needs two rows.

Notes are used to identify floor openings, bridging, and girts or plates. Nominal sizes are used in specifying lumber. Dimensions need not be given between joists. Such information is given along with notes. For example, **1" x 6" joists @ 2'-0" cc** indicates that the joists are to be spaced at intervals of 2 feet 0 inches from center to center. Lengths might not be indicated in framing plans. If you find this to be the case, the overall building dimensions and the dimensions for each bay or distances between columns or posts provide such information.

ROOF PLANS.—Framing plans for roofs are drawn in the same manner as floor framing plans. A Builder should visualize the plan as looking down on the roof before any of the roofing material (sheathing) has been added. Rafters are shown in the same manner as joists.

SHOP DRAWINGS

Shop drawings are sketches, schedules, diagrams, and other information prepared by the contractor

(Builder) to illustrate some portion of the work. As a Builder, you will have to make shop drawings for minor shop and field projects. These may include shop items—such as doors, cabinets, and small portable buildings, prefabricated berthing quarters, and modifications of existing structures.

Shop drawings are prepared from portions of design drawings, or from freehand sketches based on the Builder's past building experience. They must include enough information for the crew to complete the job. Normally, the Builder bases the amount of required detailing on the experience level of the crew expected to complete the project. When an experienced building crew will be doing the work, it is not necessary to show all the fine standard details.

When you make actual drawings, templates (when available) should be used for standard symbols. Standard technical drawing techniques are recommended but not mandatory. For techniques in the skill of drawing, refer to *Blueprint Reading and Sketching*, NAVEDTRA 10077.

FREEHAND SKETCHES

Builders must be able to read and work from drawings and specifications and make quick, accurate sketches when conveying technical information or ideas. Sketches that you will prepare may be for your own use or for use by other crewmembers. One of the main advantages of sketching is that few materials are required. Basically, pencil and paper are all you need. The type of sketch prepared and personal preference determine the materials used.

Most of your sketches will be done on some type of scratch paper. The advantage of sketching on tracing paper is the ease with which sketches can be modified or redeveloped simply by placing transparent paper over previous sketches or existing drawings. Cross-sectional or graph paper may be used to save time when you need to draw sketches to scale. For making dimensional sketches in the field, you will need a measuring tape or pocket rule, depending on the extent of the measurements taken. In freehand pencil sketching, draw each line with a series of short strokes instead of with one stroke. Strive for a free and easy movement of your wrist and fingers. You don't need to be a draftsman or an artist to prepare good working sketches.

Freehand sketches are prepared by the crew leader responsible for the job. Any information that will make the project more understandable may be included, although sketches needn't be prepared in great detail.

SECTIONAL VIEWS

LEARNING OBJECTIVE: Upon completing this section, you should be able to interpret sectional views.

Sectional views, or sections, provide important information about the height, materials, fastening and support systems, and concealed features of a structure. Figure 2-15 shows the initial development of a section and how a structure looks when cut vertically by a cutting plane. The cutting plane is not necessarily continuous, but, as with the horizontal cutting plane in building plans, may be staggered to include as much construction information as possible. Like elevations, sectional views are vertical projections. They are also detail drawings drawn to large scale. This aids in reading, and provides

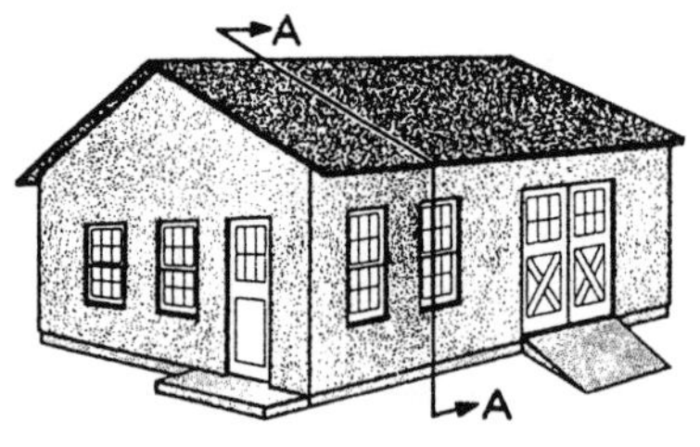

PERSPECTIVE VIEW

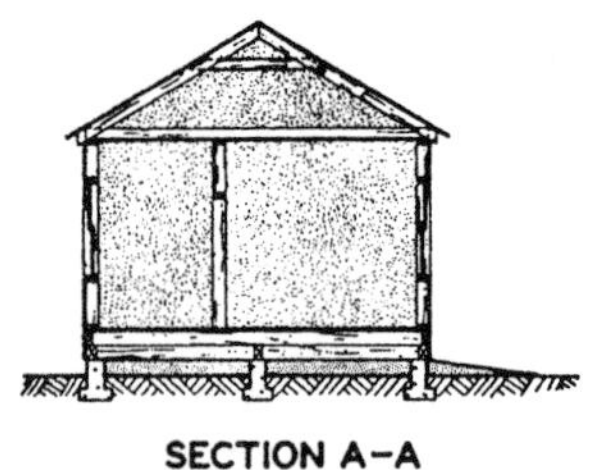
SECTION A–A

TYPICAL SMALL BUILDING SHOWING CUTTING PLANE A–A AND SECTION DEVELOPED FROM THE CUTTING PLANE

Figure 2-15.—Development of a sectional view.

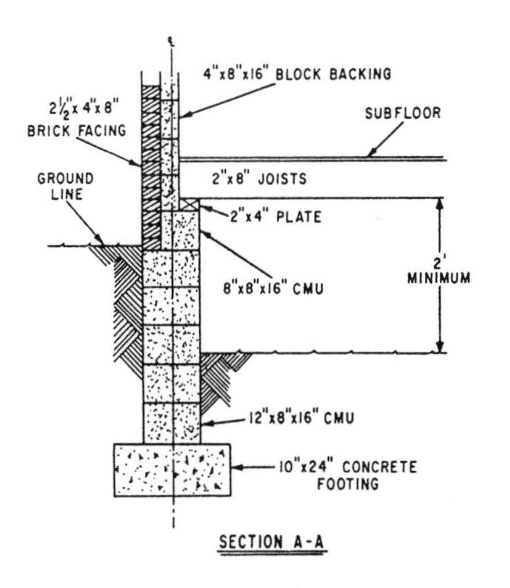

Figure 2-16.—A typical section of a masonry building.

information that cannot be given on elevation or plan views. Sections are classified as typical and specific.

Typical sections represent the average condition throughout a structure and are used when construction features are repeated many times. Figure 2-16 shows typical wall section A-A of the foundation plan in figure 2-10. You can see that it gives a great deal of information necessary for those constructing the building. Let's look at these a little more closely.

The foundation plan shown in figure 2-10 specifies that the main foundation of this structure will consist of a 22- by 28-foot concrete block rectangle. Figure 2-16, which is section A-A of the foundation plan, shows that the front and rear portions of the foundation (28-foot measurements) are made of 12-by-8-by-16-inch CMUs centered on a 10-by-24-inch concrete footing to an unspecified height. These are followed by 8-inch CMUs, which form a 4-inch ledger for floor joist support on top of the 12-inch units. In this arrangement, the 8-inch CMUs serve to form a 4-inch support for the brick. The main wall is then laid with standard 2 1/2-by-4-by-8-inch face brick backed by 4-by-8-by-16-inch CMUs.

Section B-B (figure 2-17) of the foundation plan shows that both side walls (22-foot measurements) are 8 inches thick centered on a 24-inch concrete footing to an unspecified height. It also illustrates the pilaster, a **specific** section of the wall to be constructed for support of the girder. It shows that the pilaster is constructed of 12-by-8-by-16-inch CMUs alternated with 4-by-8-by-16-inch and 8-by-8-by-16-inch CMUs. The hidden lines (dashed

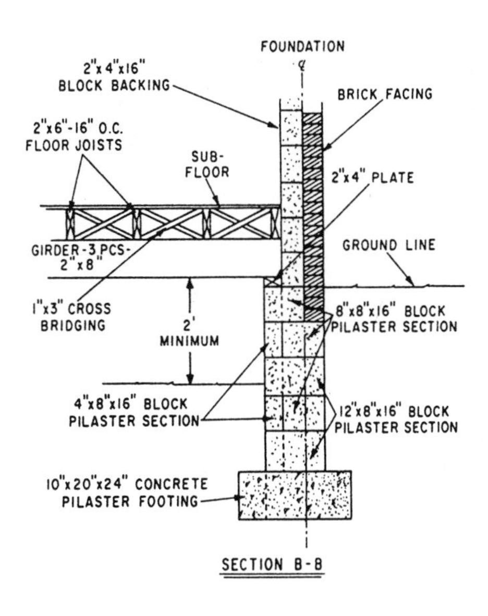

Figure 2-17.—A specific section of a concrete masonry wall.

lines) on the 12-inch-wide units indicate that the thickness of the wall beyond the pilaster is 8 inches. Note how the extra 4-inch thickness of the pilaster provides a center support for the girder, which, in turn, will support the floor joists.

Details are large-scale drawings that show the builders of a structure how its various parts are to be connected and placed. Although details do not use the cutting plane indication, they are closely related to sections. The construction of doors, windows, and eaves is customarily shown in detail drawings of buildings. Typical door and window details are shown in figure 2-18. Detail drawings are used whenever the information provided in elevations, plans, and sections is not clear enough for the constructors on the job. These drawings are usually grouped so that references may be made easily from the general drawing.

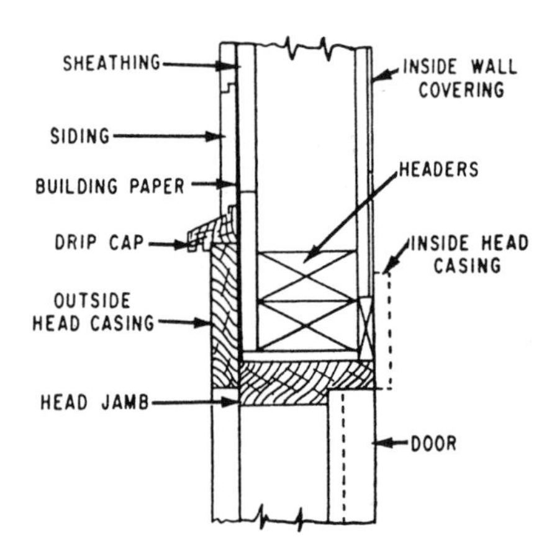

SECTION THROUGH HEAD JAMB

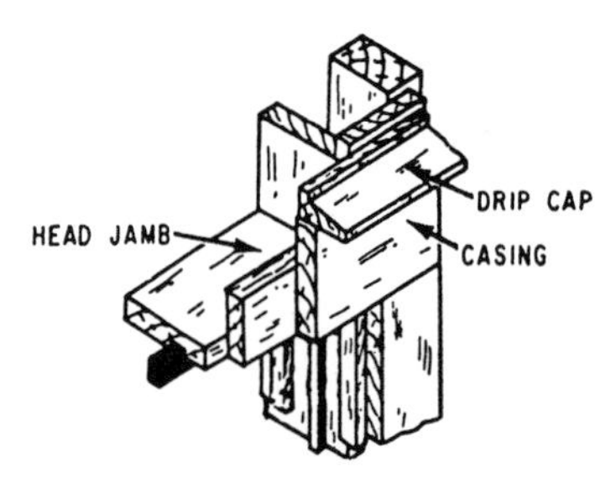

UPPER AND LOWER CORNER DETAILS WINDOW FRAME

Figure 2-18.—Door and window details.

SCHEDULES

LEARNING OBJECTIVE: Upon completing this section, you should be able to interpret building schedules.

A schedule is a group of general notes, usually grouped in a tabular form according to materials of construction. General notes refer to all notes on the drawing not accompanied by a leader and an arrowhead. Item schedules for doors, rooms, footings, and so on, are more detailed. Typical door and window finish schedule formats are presented in the next section.

DOOR SCHEDULE

Doors may be identified as to size, type, and style with code numbers placed next to each symbol in a plan view. This code number, or mark, is then entered on a line in a door schedule, and the principal characteristics of the door are entered in successive columns along the line. The "Amount Required" column allows a quantity check on doors of the same design as well as the total number of doors required. By using a number with a letter, you will find that the mark serves a double purpose: the number identifies the floor on which the door is located, and the letter identifies the door design. The "Remarks" column allows identification by type (panel or flush), style, and material. The schedule is a convenient way of presenting pertinent data without making the Builder refer to the specification. A typical door schedule is shown in table 2-1.

WINDOW SCHEDULE

A window schedule is similar to a door schedule in that it provides an organized presentation of the significant window characteristics. The mark used in the schedule is placed next to the window symbol that applies on the plan view of the elevation view (figure 2-13). A similar window schedule is shown in table 2-2.

FINISH SCHEDULE

A finish schedule specifies the interior finish material for each room and floor in the building. The finish schedule provides information for the walls, floors, ceilings, baseboards, doors, and window trim. An example of a finish schedule is shown in table 2-3.

Table 2-1.—Door Schedule

MARK	SIZE	AMOUNT REQUIRED	REMARKS
1	2′0" × 6′8" × 1 3/8"	3	Flush door
2	2′6" × 6′8" × 1 3/8"	4	Flush door
3	2′6" × 6′8" × 1 3/8"	1	Ext flush door, 1 light
4	3′0" × 7′0" × 1 3/4"	1	Ext flush door, 4 lights
5	1′8" × 6′8" × 1 3/8"	1	Flush door

Table 2-2.—Window Schedule

MARK	SIZE	AMOUNT REQUIRED	REMARKS
A	4′5 1/8" × 4′2 5/8"	3	Metal frame
B	3′1 1/8" × 4′2 5/8"	2	Metal frame
C	3′1" × 4′2 5/8"	1	Metal frame
D	3′1" × 4′2 5/8"	1	Metal frame
E	1′7 5/8" × 4′2 5/8"	2	Metal frame

Table 2-3.—Finish Schedule

ROOM	FLOOR	WALLS	CEILING	BASEBOARD	TRIM
Dining and living	1" × 3" oak	1/2" Drywall paint white	1/2" Drywall paint white	Wood	Wood
Bedroom	1" × 3" oak	1/2" Drywall paint white	1/2" Drywall paint white	Wood	Wood
Bathroom	Linoleum-tan	1/2" Drywall paint white	1/2" Drywall paint white	Lino-cove	Wood
Kitchen	Linoleum-tan	1/2" Drywall paint white	1/2" Drywall paint white	Lino-cove	Wood
Utility room	Linoleum-tan	1/2" Drywall paint white	1/2" Drywall paint white	Lino-cove	Wood
Hall	1" × 3" oak	1/2" Drywall paint white	1/2" Drywall paint white	Wood	Wood

NOTES ON SCHEDULES

Notes are generally placed a minimum of 3 inches below the "Revision" block in the right-hand side of the first sheet. The purpose of these notes is to give additional information that clarifies a detail or explains how a certain phase of construction is to be performed. You should read all notes, along with the specifications, while you are planning a project.

WRITTEN SPECIFICATIONS

LEARNING OBJECTIVE: Upon completing this section, you should be able to interpret written construction specifications.

Because many aspects of construction cannot be shown graphically, even the best prepared construction drawings often inadequately show some portions of a project. For example, how can anyone show on a drawing the quality of workmanship required for the installation of doors and windows? Or, who is responsible for supplying the materials? These are things that can be conveyed only by hand-lettered notes. The standard procedure is to supplement construction drawings with detailed written instructions. These written instructions, called specifications (or more commonly **specs**), define and limit materials and fabrication to the intent of the engineer or designer.

Usually, it is the responsibility of the design engineer to prepare project specifications. As a Builder, you will be required to read, interpret, and use these in your work as a crew leader or supervisor. You must be familiar with the various types of federal, military, and nongovernmental reference specifications used in preparing project specs. When assisting the engineer in preparing or using specifications, you also need to be familiar with the general format and terminology used.

NAVFAC SPECIFICATIONS

NAVFAC specifications are prepared by the Naval Facilities Engineering Command (NAVFAC-ENGCOM), which sets standards for all construction work performed under its jurisdiction. This includes work performed by the Seabees. There are three types of NAVFAC specifications.

NAVFACENGCOM Guide Specifications

NAVFACENGCOM guide specifications (NFGS) are the primary basis for preparing specifications for construction projects. These specifications define and establish minimum criteria for construction, materials, and workmanship and must be used as guidance in the preparation of project specifications. Each of these guide specifications (of which there are more than 300) has been written to encompass a wide variety of different materials, construction methods, and circumstances. They must also be tailored to suit the work actually required by the specific project.

To better explain this, let's look at figure 2-19, which is a page taken from a NAVFACENGCOM guide specification. In this figure, you can see that there are two paragraphs numbered 3.2.1. This indicates that the spec writer must choose the paragraph that best suits the particular project for which he is writing the specification. The capital letters *I* and *J* in the right-hand margin next to those paragraphs refer to footnotes (contained elsewhere in the same guide specification) that the spec writer must follow when selecting the best paragraph. Additionally, you can see that some of the information in figure 2-19 is enclosed in brackets ([]). This indicates other choices that the spec writer must make. Guide specifications should be modified and edited to reflect the latest proven technology, materials, and methods.

EFD Regional Guide Specifications

Engineering Field Division regional guide specifications are used in the same way as the NAVFACENGCOM guide specifications but only in areas under the jurisdiction of an EFD of the Naval Facilities Engineering Field Command. When the spec writer is given a choice between using an EFD regional guide specification or a NAVFACENGCOM guide specification with the same identification number, the writer must use the one that has the most recent date. This is because there can only be one valid guide specification for a particular area at any one time.

NFGS-07310 (August 1983)

PART 3 - EXECUTION

3.1 SURFACES AND CONDITIONS: Do not apply shingle roofing on surfaces that are unsuitable or that prevent a satisfactory application. Ensure that roof is smooth, clean, dry, and without loose knots. Cover knotholes and cracks with sheet metal nailed securely to the sheathing. Properly flash and secure vents and other roof projections and drive projecting nails firmly home.

3.2 APPLICATION: The manufacturer's written instructions shall be followed for applications not listed in this specification and in cases of conflict with this specification.

3.2.1 Underlayment (for Roof Slopes 4 Inches Per Foot and Greater): Apply underlayment consisting of one layer of No. 15 asphalt-saturated felt to the roof deck. Lay felt parallel to roof eaves continuing from eaves to ridge, using 2-inch head laps, 6-inch laps from both sides over all hips and ridges, and 4-inch end laps in the field of the roof. Nail felt sufficiently to hold until shingles are applied. Turn underlayment up vertical surfaces not less than 4 inches. (I)

OR

3.2.1 Underlayment (for Roof Slopes (Between 2 Inches and 4 Inches Per Foot) (4 Inches Per Foot and Greater)): Apply underlayment consisting of two layers of No. 15 asphalt-saturated felt to the roof deck. Provide a 19-inch wide strip of felt as a starter sheet to maintain the specified number of layers throughout the roof. Lay felt parallel to roof eaves continuing from eaves to ridge, using 19-inch head laps for 6-inch laps from both sides over all hips and ridges, and 12-inch end laps in the field of the roof. Nail felt sufficiently to hold until shingles are applied. Confine nailing to the upper 17 inches of each felt. Turn underlayment up vertical surfaces not less than 4 inches. (I) (J)

3.2.2 Metal Drip Edges: Provide metal drip edges as specified in Section 07600, "Flashing and Sheet Metal," applied directly on the wood deck at the eaves and over the underlayment at the rakes. Extend back from the edge of the deck not more than 3 inches and secure with fasteners spaced not more than 10 inches on center along the inner edge.

(3.2.3 Eaves Flashing (for Roof Slopes 4 Inches Per Foot and Greater): Provide eaves flashing strips consisting of 55-pound or heavier smooth-surface roll roofing. The flashing strips shall overhang the metal drip edge 1/4 to 3/8 inch and extend up the slope far enough to cover a point 12 inches inside the interior face of the exterior wall. Where overhangs require flashings wider than 36 inches, locate the laps outside the exterior wall face. The laps shall be at least 2 inches wide and cemented. End laps shall be 12 inches and cemented.) (K)

Figure 2-19.—Sample page from a NAVFACENGCOM guide specification.

Standard Specifications

Standard specifications are written for a small group of specialized structures that must have uniform construction to meet rigid operational requirements. NAVFAC standard specifications contain references to federal, military, other command and bureau, and association specifications. NAVFAC standard specifications are referenced or copied in project specifications, and can be modified with the modification noted and referenced. An example of a standard specification with modification is shown below:

"The magazine shall be Arch, Type I, conforming to specifications S-M8E, except that all concrete shall be class F-1."

OTHER SPECIFICATIONS

The following specifications establish requirements mainly in terms of performance. Referencing these documents in project specifications assures the procurement of economical facility components and services while considerably reducing the number of words required to state such requirements.

Federal and Military Specifications

Federal specifications cover the characteristics of materials and supplies used jointly by the Navy and other government agencies. These specifications do not cover installation or workmanship for a particular project, but specify the technical requirements and tests for materials, products, or services. The engineering technical library should have all the commonly used federal specifications pertinent to Seabee construction.

Military specifications are those specifications that have been developed by the Department of Defense. Like federal specifications, they also cover the characteristics of materials. They are identified by **DOD** or **MIL** preceding the first letter and serial number.

Technical Society and Trade Association Specifications

Technical society specifications should be referenced in project specifications when applicable. The organizations publishing these specifications include, but are not limited to, the American National Standards Institute (ANSI), the American Society for Testing and Materials (ASTM), the Underwriters Laboratories (UL), and the American Iron and Steel Institute (AISI). Trade association specifications contain requirements common to many companies within a given industry.

Manufacturer's Specifications

Manufacturer's specifications contain the precise description for the manner and process for making, constructing, compounding, and using any items the manufacturer produces. They should not be referenced or copied verbatim in project specifications but may be used to aid in preparing project specifications.

PROJECT SPECIFICATIONS

Construction drawings are supplemented by written project specifications. Project specifications give detailed information regarding materials and methods of work for a particular construction project. They cover various factors relating to the project, such as general conditions, scope of work, quality of materials, standards of workmanship, and protection of finished work.

The drawings, together with the project specifications, define the project in detail and show exactly how it is to be constructed. Usually, drawings for an important project are accompanied by a set of project specifications. The drawings and project specifications are inseparable. Drawings indicate what the project specifications do not cover. Project specifications indicate what the drawings do not portray, or they further clarify details that are not covered amply by the drawings and notes on the drawings. When you are preparing project specifications, it is important that the specifications and drawings be closely coordinated so that discrepancies and ambiguities are minimized. Whenever there is conflicting information between the drawings and project specs, the specifications take precedence over the drawings.

ORGANIZATION OF SPECIFICATIONS

For consistency, the Construction Standards Institute (CSI) has organized the format of specifications into 16 basic divisions. These divisions, used throughout the military and civilian construction industry, are listed in order as follows:

1. **General Requirements** include information that is of a general nature to the project, such as inspection requirements and environmental protection.

2. **Site Work** includes work performed on the site, such as grading, excavation, compaction, drainage, site utilities, and paving.

3. **Concrete** includes precast and cast-in-place concrete, formwork, and concrete reinforcing.

4. **Masonry** includes concrete masonry units, brick, stone, and mortar.
5. **Metals** include such items as structural steel, open-web steel joists, metal stud and joist systems, ornamental metal work, grills, and louvers. (Sheet-metal work is usually included in Division 7.)
6. **Wood and Plastics** include wood and wood framing, rough and finish carpentry, foamed plastics, fiberglass-reinforced plastics, and laminated plastics.
7. **Thermal and Moisture Protection** includes such items as waterproofing, dampproofing, insulation, roofing materials, sheet metal and flashing, caulking, and sealants.
8. **Doors and Windows** include doors, windows, finish hardware, glass and glazing, storefront systems, and similar items.
9. **Finishes** include such items as floor and wall coverings, painting, lathe, plaster, and tile.
10. **Specialties** include prefabricated products and devices, such as chalkboards, moveable partitions, fire-fighting devices, flagpoles, signs, and toilet accessories.
11. **Equipment** includes such items as medical equipment, laboratory equipment, food service equipment, kitchen and bath cabinetwork, and counter tops.
12. **Furnishings** include prefabricated cabinets, blinds, drapery, carpeting, furniture, and seating.
13. **Special Construction** includes such items as prefabricated structures, integrated ceiling systems, and swimming pools.
14. **Conveying Systems** include dumbwaiters, elevators, moving stairs, material-handling systems, and other similar conveying systems.
15. **Mechanical Systems** include plumbing, heating, air conditioning, fire-protection systems, and refrigeration systems.
16. **Electrical Systems** include electrical service and distribution systems, electrical power equipment, electric heating and cooling systems, lighting, and other electrical items.

Each of the above divisions is further divided into sections. You can find a discussion of the required sections of Division 1 in *Policy and Procedures for Project Drawing and Specification Preparation*, MIL-HDBK-1006/1. The Division 1 sections, sometimes referred to as "boilerplate," are generally common to all projects accomplished under a construction contract. Divisions 2 through 16 contain the technical sections that pertain to the specific project for which the spec writer has prepared the specification. These technical sections follow the CSI-recommended three-part section format. The first part, General, includes requirements of a general nature. Part 2, Products, addresses the products or quality of materials and equipment to be included in the work. The third part, Execution, provides detailed requirements for performance of the work.

GUIDANCE

Usually, the engineer or spec writer prepares each section of a specification based on the appropriate guide specification listed in the *Engineering and Design Criteria for Navy Facilities*, MIL-BUL-34. This military bulletin (issued quarterly by the Naval Construction Battalion Center, Port Hueneme, California) lists current NAVFACENGCOM guide specifications, standard specifications and drawings, definitive drawings, NAVFAC design manuals, and military handbooks that are used as design criteria.

As discussed earlier, when writing the specifications for a project, you must modify the guide specification you are using to fit the project. Portions of guide specifications that concern work not included in the project should be deleted. When portions of the required work are not included in a guide specification, then you must prepare a suitable section to cover the work, using language and form similar to the guide specification. Do not combine work covered by various guide specifications into one section unless the work is minor in nature. Do not reference the guide specification in the project specifications. You must use the guide spec only as a manuscript that can be edited and incorporated into the project specs.

The preceding discussion provides only a brief overview of construction specifications. For additional guidance regarding specification preparation, you should refer to *Policy and Procedures for Project Drawing and Specification Preparation*, MIL-HDBK-1006/1.

RECOMMENDED READING LIST

NOTE

Although the following reference was current when this TRAMAN was published, its continued currency cannot be assured. You therefore need to ensure that you are studying the latest revision.

Engineering Aid 3 & 2, Vol. 3, NAVEDTRA 10629-1, Naval Education and Training Program Management Support Activity, Pensacola, Fla., 1987.

CHAPTER 3

WOODWORKING TOOLS, MATERIALS, AND METHODS

As a Builder, hand and power woodworking tools are essential parts of your trade. To be a proficient woodworking craftsman, you must be able to use and maintain a large variety of field and shop tools effectively. To perform your work quickly, accurately, and safely, you must select and use the correct tool for the job at hand. Without the proper tools and the knowledge to use them, you waste time, reduce efficiency, and may injure yourself or others.

Power tools not only are essential in performing specific jobs, but also play an important role in your daily work activities. Keep in mind that you are responsible for knowing and observing all safety precautions applicable to the tools and equipment you operate. For additional information on the topics discussed in this chapter, you are encouraged to study *Tools and Their Uses*, NAVEDTRA 10085-B2. Because that publication contains a detailed discussion of common tools used by Builders, we will not repeat that information in this chapter.

In this chapter, several of the most common power tools used by Builders are briefly described. Their uses, general characteristics, attachments, and safety and operating features are outlined. To become skilled with these power tools and hand tools, you must use them. You should also study the manufacturer's operator and maintenance guides for each tool you use for additional guidance. We will also be covering materials and methods of woodworking.

POWER TOOLS

LEARNING OBJECTIVE: Upon completing this section, you should be able to determine the proper use and maintenance requirements of portable power tools.

Your duties as a Builder include developing and improving your skills and techniques when working with different power tools. In this section, we'll identify and discuss the most common power tools that are in the Builder's workshop or used on the jobsite. We'll also discuss safety precautions as they relate to the particular power tool under discussion. You must keep in mind and continually stress to your crew that woodworking power tools can be dangerous, and that safety is everyone's responsibility.

SHOP TOOLS

As a Builder, you might be assigned to a shop. Therefore, you will need to know some of the common power tools and equipment found there.

Shop Radial Arm Saw

Figure 3-1 illustrates a typical shop radial arm saw. The procedures used in the operation, maintenance, and lubrication of any shop radial arm saw are found in the manufacturers' operator and maintenance manuals. The safety precautions to be observed for this saw are found in these same manuals. The primary difference between this saw and other saws of this type (field saws) is the location of controls.

Tilt-Arbor Table Bench Saw

A tilt-arbor table bench saw (figure 3-2) is so named because the saw blade can be tilted for cutting bevels by tilting the arbor. The arbor, located beneath the table, is controlled by the tilt handwheel. In earlier types of bench saws, the saw blade remained stationary and the table was tilted. A canted (tilted) saw table is hazardous in many ways; most modern table saws are of the tilt-arbor type.

To rip stock, remove the cutoff gauges and set the rip fence away from the saw by a distance equal to the desired width of the piece to be ripped off. The piece is placed with one edge against the fence and fed through with the fence as a guide.

To cut stock square, set the cutoff gauge at 90° to the line of the saw and set the ripping fence to the outside edge of the table, away from the stock to be cut. The piece is then placed with one edge against

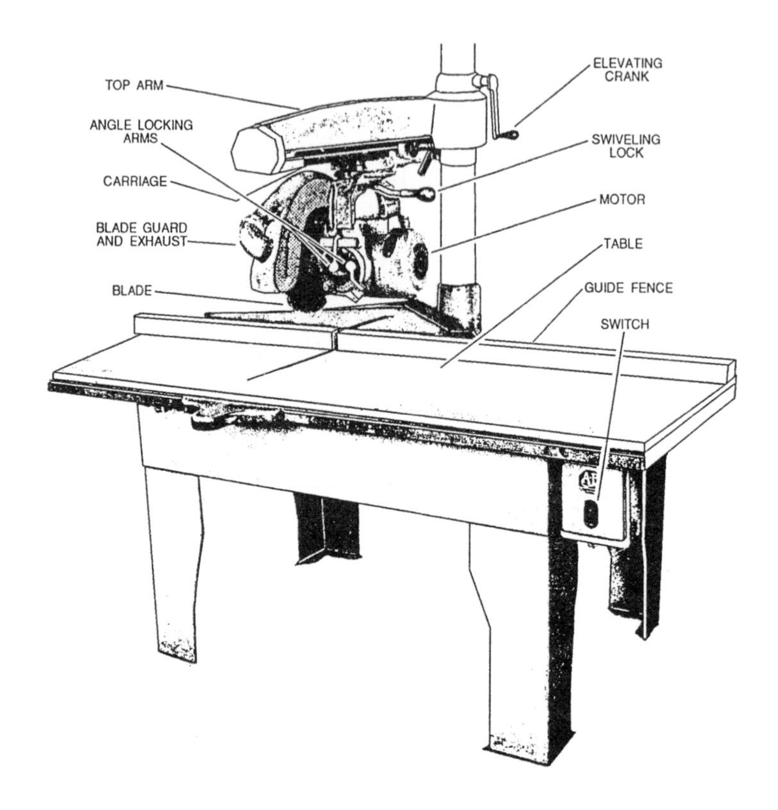

Figure 3-1.—A shop radial arm saw.

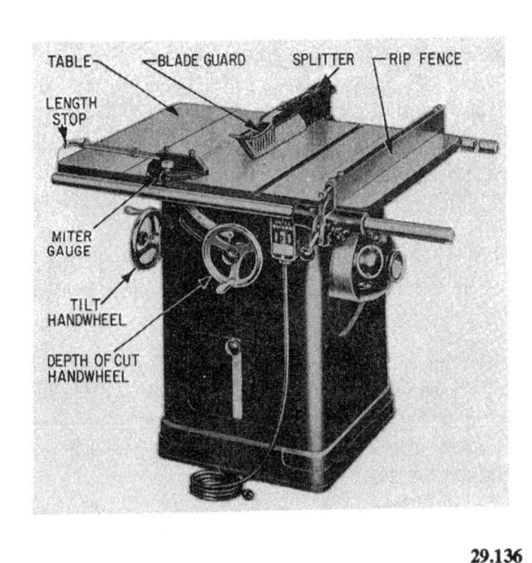

29.136

Figure 3-2.—Tilt-arbor bench saw.

the cutoff gauge, held firmly, and fed through by pushing the gauge along its slot.

The procedure for cutting stock at an angle other than 90° (called miter cutting) is similar, except that the cutoff gauge is set to bring the piece to the desired angle with the line of the saw.

For ordinary ripping or cutting, the saw blade should extend above the table top 1/8 to 1/4 inch plus the thickness of the piece to be sawed. The vertical position of the saw is controlled by the depth of cut handwheel, shown in figure 3-2. The angle of the saw blade is controlled by the tilt handwheel. Except when its removal is absolutely unavoidable, the guard must be kept in place.

The slot in the table through which the saw blade extends is called the throat. The throat is contained in a small, removable section of the table called the throat plate. The throat plate is removed when it is necessary to insert a wrench to remove the saw blade.

The blade is held on the arbor by the arbor nut. A saw is usually equipped with several throat plates, containing throats of various widths. A wider throat is required when a dado head is used on the saw. A dado head consists of two outside grooving saws (which are much like combination saws) and as many intermediate chisel-type cutters (called chippers) as are required to make up the designated width of the groove or dado. Grooving saws are usually 1/8-inch thick; consequently, one grooving saw will cut a 1/8-inch groove, and the two, used together, will cut a 1/4-inch groove. Intermediate cutters come in various thicknesses.

Observe the following safety precautions when operating the tilt-arbor table bench saw:

- Do not use a ripsaw blade for crosscutting or a crosscut saw blade for ripping. When ripping and crosscutting frequently, you should install a combination blade to eliminate constantly changing the blade. Make sure the saw blade is sharp, unbroken, and free from cracks before using. The blade should be changed if it becomes dull, cracked, chipped, or warped.

- Be sure the saw blade is set at proper height above the table to cut through the wood.

- Avoid the hazard of being hit by materials caused by kickbacks by standing to one side of the saw.

- Always use a push stick to push short, narrow pieces between the saw blade and the gauge.

- Keep stock and scraps from accumulating on the saw table and in the immediate working area.

- Never reach over the saw to obtain material from the other side.

- When cutting, do not feed wood into the saw blade faster than it will cut freely and cleanly.

- Never leave the saw unattended with the power on.

Band Saw

Although the band saw (figure 3-3) is designed primarily for making curved cuts, it can also be used for straight cutting. Unlike the circular saw, the band saw is frequently used for freehand cutting.

The band saw has two large wheels on which a continuous narrow saw blade, or band, turns, just as a belt is turned on pulleys. The lower wheel, located below the working table, is connected to the motor directly or by means of pulleys or gears and serves as the driver pulley. The upper wheel is the driven pulley.

The saw blade is guided and kept in line by two sets of blade guides, one fixed set below the table and one set above with a vertical sliding adjustment. The alignment of the blade is adjusted by a mechanism on the backside of the upper wheel. Tensioning of the blade—tightening and loosening—is provided by another adjustment located just back of the upper wheel.

Cutoff gauges and ripping fences are sometimes provided for use with band saws, but you'll do most of your work freehand with the table clear. With this type of saw, it is difficult to make accurate cuts when gauges or fences are used.

The size of a band saw is designated by the diameter of the wheels. Common sizes are 14-, 16-, 18-, 20-, 30-, 36-, 42-, and 48-inch-diameter wheel machines. The 14-inch size is the smallest practical band saw. With the exception of capacity, all band

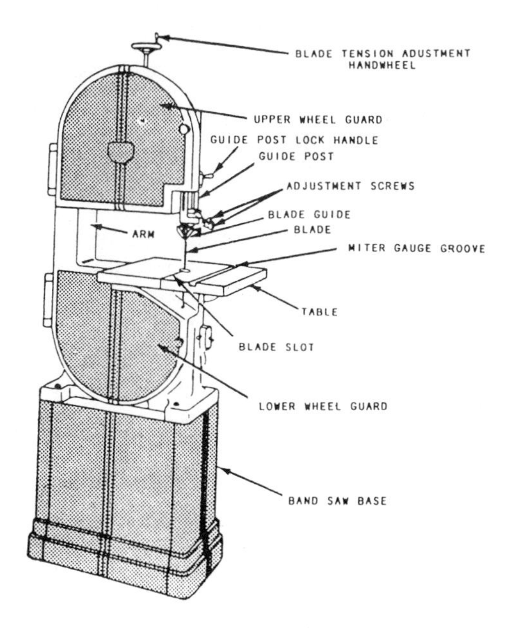

Figure 3-3.—Band saw.

saws are much the same with regard to maintenance, operation, and adjustment.

A rule of thumb used by many Seabees is that the width of the blade should be one-eighth the minimum radius to be cut. Therefore, if the piece on hand has a 4-inch radius, the operator should select a 1/2-inch blade. Don't construe this to mean that the minimum radius that can be cut is eight times the width of the blade; rather, the ratio indicates the practical limit for high-speed band saw work.

Blades, or bands, for band saws are designated by points (tooth points per inch), thickness (gauge), and width. The required length of a blade is found by adding the circumference of one wheel to twice the distance between the wheel centers. Length can vary within a limit of twice the tension adjustment range.

Band saw teeth are shaped like the teeth in a hand ripsaw blade, which means that their fronts are filed at 90° to the line of the saw. Reconditioning procedures are the same as those for a hand ripsaw, except that very narrow band saws with very small teeth must usually be set and sharpened by special machines.

Observe the following safety precautions when operating a band saw:

- Keep your fingers away from the moving blade.
- Keep the table clear of stock and scraps so your work will not catch as you push it along.
- Keep the upper guide just above the work, not excessively high.
- Don't use cracked blades. If a blade develops a click as it passes through the work, the operator should shut off the power because the click is a danger signal that the blade is cracked and may be ready to break. After the saw blade has stopped moving, it should be replaced with one in proper condition.
- If the saw blade breaks, the operator should shut off the power immediately and not attempt to remove any part of the saw blade until the machine is completely stopped.
- If the work binds or pinches on the blade, the operator should never attempt to back the work away from the blade while the saw is in motion since this may break the blade. The operator should always see that the blade is working freely through the cut.
- A band saw should not be operated in a location where the temperature is below 45°F. The blade may break from the coldness.
- Using a small saw blade for large work or forcing a wide saw on a small radius is bad practice. The saw blade should, in all cases, be as wide as the nature of the work will permit.
- Band saws should not be stopped by thrusting a piece of wood against the cutting edge or side of the band saw blade immediately after the power has been shut off; doing so may cause the blade to break. Band saws with 36-inch-wheel diameters and larger should have a hand or foot brake.
- Particular care should be taken when sharpening or brazing a band saw blade to ensure the blade is not overheated and the brazed joints are thoroughly united and finished to the same thickness as the rest of the blade. It is recommended that all band saw blades be butt welded where possible; this method is much superior to the old style of brazing.

Drill Press

Figure 3-4 shows a drill press. (The numbers in the figure correspond to those in the following text.) The drill press is an electrically operated power machine that was originally designed as a metal-working tool; as such, its use would be limited in the average woodworking shop. However, accessories, such as a router bit or shaper heads, jigs, and special techniques, now make it a versatile woodworking tool as well.

The motor (10) is mounted to a bracket at the rear of the head assembly (1) and designed to permit V-belt changing for desired spindle speed without removing the motor from its mounting bracket. Four spindle speeds are obtained by locating the V-belt on any one of the four steps of the spindle-driven and motor-driven pulleys. The belt tensioning rod (16) keeps proper tension on the belt so it doesn't slip.

The controls of all drill presses are similar. The terms "right" and "left" are relative to the operator's position standing in front of and facing the drill press. "Forward" applies to movement toward the operator. "Rearward" applies to movement away from the operator.

The on/off switch (11) is located in the front of the drill press for easy access.

The spindle and quill feed handles (2) radiate from the spindle and quill pinion feed (3) hub, which is located on the lower right-front side of the head assembly (1). Pulling forward and down on any one of the three spindle and quill feed handles, which point upward at the time, moves the spindle and quill assembly downward. Release the feed handle (2) and the spindle and quill assembly return to the retracted or upper position by spring action.

The quill lock handle (4) is located at the lower left-front side of the head assembly. Turn the quill lock handle clockwise to lock the quill at a desired operating position. Release the quill by turning the quill lock handle counterclockwise. However, in most cases, the quill lock handle will be in the released position.

The head lock handle (5) is located at the left-rear side of the head assembly. Turn the head lock handle clockwise to lock the head assembly at a desired vertical height on the bench column. Turn the head lock handle counterclockwise to release the head assembly. When operating the drill press, you must ensure that the head lock handle is tight at all times.

The head support collar handle (6) is located at the right side of the head support collar and below the head assembly. The handle locks the head support collar, which secures the head vertically on the bench column, and prevents the head from dropping when the head lock handle is released. Turn the head support collar lock handle clockwise to lock the support to the bench column and counterclockwise to

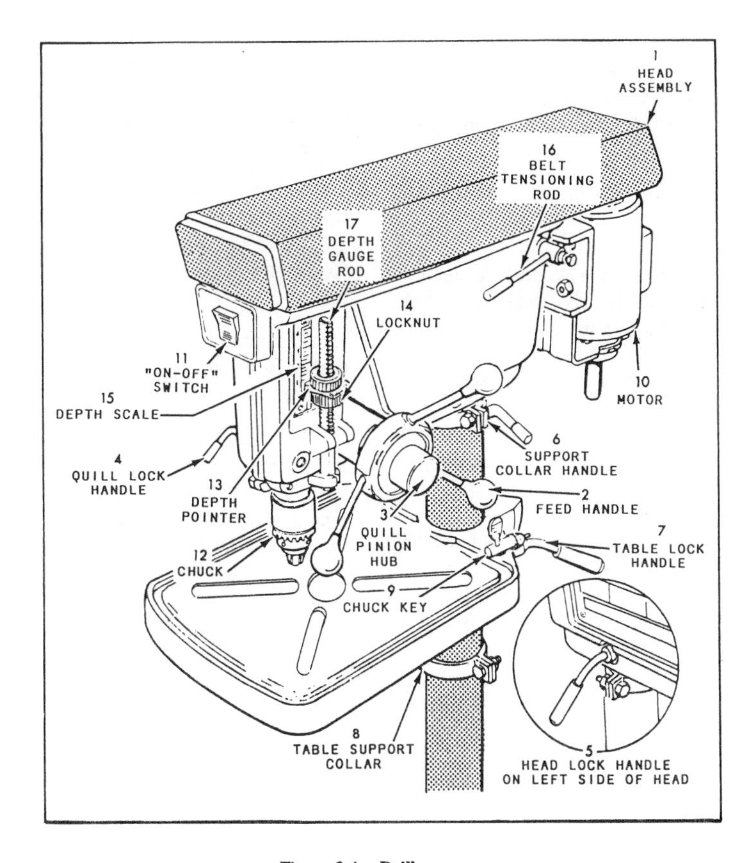

Figure 3-4.—Drill press.

release the support. When operating the drill press, ensure that the head support collar lock handle is tight at all times.

As you face the drill press, the tilting table lock handle is located at the right-rear side of the tilting table bracket. The lockpin secures the table at a horizontal or 45° angle. This allows you to move the table to the side, out of the way for long pieces of wood. The table support collar (8) allows you to raise or lower the table. Turn the tilting table lock handle counterclockwise to release the tilting table bracket so it can be moved up and down or around the bench column. Lock the tilting table assembly at the desired height by turning the lock handle clockwise. When operating the drill press, ensure that the tilting table lock handle is tight at all times.

The adjustable locknut (14) is located on the depth gauge rod (17). The purpose of the adjustable locknut is to regulate depth drilling. Turn the adjustable locknut clockwise to decrease the downward travel of the spindle. The locknut must be secured against the depth pointer (13) when operating the drill press. The depth of the hole is shown on the depth scale (15).

Observe the following safety precautions when operating a drill press:

- Make sure that the drill is properly secured in the chuck (12) and that the chuck key (9) is removed before starting the drill press.
- Make sure your material is properly secured.
- Operate the feed handle with a slow, steady pressure to make sure you don't break the drill bit or cause the V-belt to slip.
- Make sure all locking handles are tight and that the V-belt is not slipping.
- Make sure the electric cord is securely connected and in good shape.
- Make sure you are not wearing hanging or loose clothing.
- Listen for any sounds that may be signs of trouble.
- After you have finished operating the drill press, make sure the area is clean.

Woodworking Lathe

The woodworking lathe is, without question, the oldest of all woodworking machines. In its early form, it consisted of two holding centers with the suspended stock being rotated by an endless rope belt. It was operated by having one person pull on the rope hand over hand while the cutting was done by a second person holding crude hand lathe tools on an improvised beam rest.

The actual operations of woodturning performed on a modern lathe are still done to a great degree with woodturner's hand tools. However, machine lathe work is coming more and more into use with the introduction of newly designed lathes for that purpose.

The lathe is used in turning or shaping round drums, disks, and any object that requires a true diameter. The size of a lathe is determined by the maximum diameter of the work it can swing over its bed. There are various sizes and types of wood lathes, ranging from very small sizes for delicate work to large surface or bull lathes that can swing jobs 15 feet in diameter.

Figure 3-5 illustrates a type of lathe that you may find in your shop. It is made in three sizes to swing 16-, 20-, and 24-inch diameter stock. The lathe has four major parts: bed, headstock, tailstock, and tool rest.

The lathe shown in figure 3-5 has an iron bed and comes in assorted lengths. The bed is a broad, flat surface that supports the other parts of the machine.

The headstock is mounted on the left end of the lathe bed. All power for the lathe is transmitted through the headstock. It has a fully enclosed motor that gives variable spindle speed. The spindle is threaded at the front end to receive the faceplates. A faceplate attachment to the motor spindle is furnished to hold or mount small jobs having large diameters. There is also a flange on the rear end of the spindle to receive large faceplates, which are held securely by four stud bolts.

The tailstock is located on the right end of the lathe and is movable along the length of the bed. It supports one end of the work while the other end is being turned by the headstock spur. The tail center can be removed from the stock by simply backing the screw. The shank is tapered to center the point automatically.

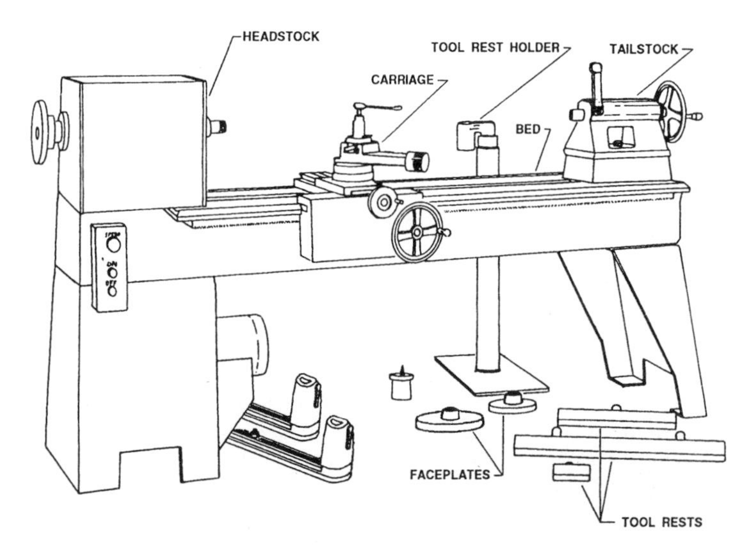

Figure 3-5.—A woodworking lathe with accessories.

Most large sizes of lathes are provided with a power-feeding carriage. A cone-pulley belt arrangement provides power from the motor, and trackways are cast to the inside of the bed for sliding the carriage back and forth. All machines have a metal bar that can be attached to the bed of the lathe between the operator and the work. This serves as a hand tool rest and provides support for the operator in guiding tools along the work. It may be of any size and is adjustable to any desired position.

In lathe work, wood is rotated against the special cutting tools (illustrated in figure 3-6). These tools include turning gouges (view A); skew chisels (view B); parting tools (view C); round-nose (view D); square-nose (view E); and spear-point (view F)

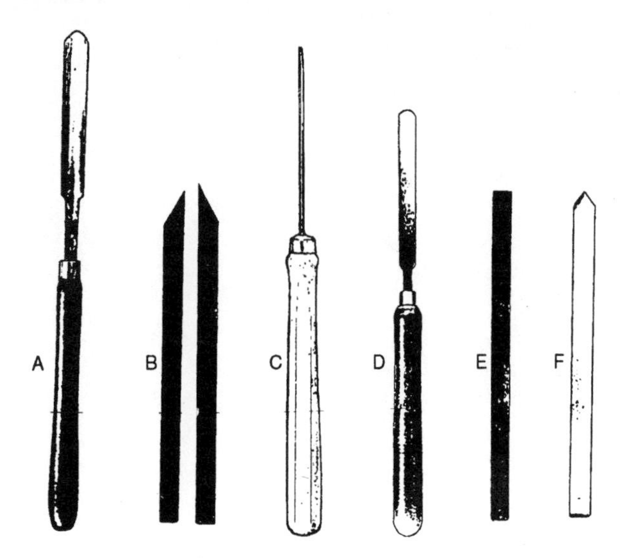

Figure 3-6.—Lathe cutting tools.

chisels. Other cutting tools are toothing irons and auxiliary aids, such as calipers, dividers, and templates.

Turning gouges are used chiefly to rough out nearly all shapes in spindle turning. The gouge sizes vary from 1/8 to 2 or more inches, with 1/4-, 3/4-, and 1-inch sizes being most common.

Skew chisels are used for smoothing cuts to finish a surface, turning beads, trimming ends or shoulders, and for making V-cuts. They are made in sizes from 1/8 to 2 1/2 inches in width and in right-handed and left-handed pairs.

Parting tools are used to cut recesses or grooves with straight sides and a flat bottom, and also to cut off finished work from the faceplate. These tools are available in sizes ranging from 1/8 to 3/4 inch.

Scraping tools of various shapes are used for the most accurate turning work, especially for most faceplate turning. A few of the more commonly used shapes are illustrated in views D, E, and F of figure 3-6. The chisels shown in views B, E, and F are actually old jointer blades that have been ground to the required shape; the wood handles for these homemade chisels are not shown in the illustration.

A toothing iron (figure 3-7) is basically a square-nose turning chisel with a series of parallel grooves cut into the top surface of the iron. These turning tools are used for rough turning of segment work mounted on the face plate. The points of the toothing iron created by the parallel grooves serve as a series of spear point chisels (detail A); therefore, the tool is not likely to catch and dig into the work like a square-nose turning chisel. The toothing iron is made with course, medium, and fine parallel grooves and varies from 1/2 to 2 inches in width.

Lathe turning can be extremely dangerous. You therefore must use particular care in this work. Observe the following safety precautions:

- When starting the lathe motor, stand to one side. This helps you avoid the hazard of flying debris in the event of defective material.

Figure 3-7.—Toothing iron lathe tool.

- The tool rest must be used when milling stock.
- Adjust and set the compound or tool rest for the start of the cut before turning the switch on.
- Take very light cuts, especially when using hand tools.
- Never attempt to use calipers on interrupted surfaces while the work is in motion.

Jointer

The jointer is a machine for power planing stock on faces, edges, and ends. The planing is done by a revolving cutterhead equipped with two or more knives, as shown in figure 3-8. Tightening the set screws forces the throat piece against the knife for holding the knife in position. Loosening the set screws releases the knife for removal. The size of a jointer is designated by the width, in inches, of the cutterhead; sizes range from 4 to 36 inches. A 6-inch jointer is shown in figure 3-9.

The principle on which the jointer functions is illustrated in figure 3-10. The table consists of two parts on either side of the cutterhead. The stock is started on the infeed table and fed past the cutterhead onto the outfeed table. The surface of the outfeed table must be exactly level with the highest point reached by the knife edges. The surface of the infeed table is depressed below the surface of the outfeed

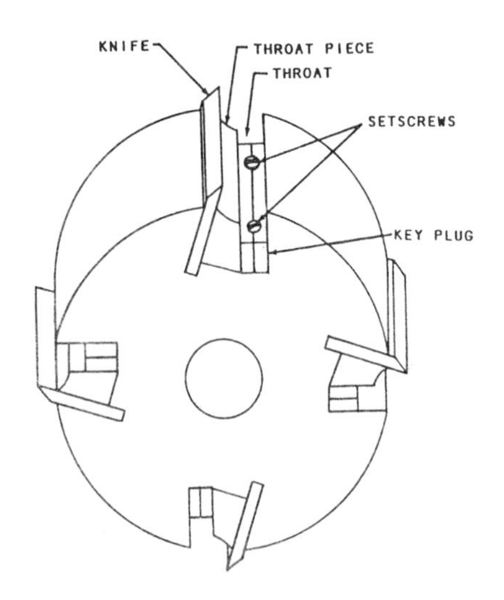

Figure 3-8.—Four-knife cutterhead for a jointer.

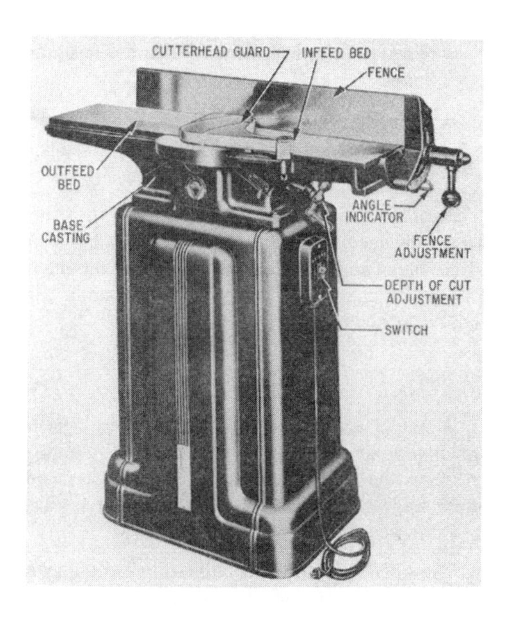

29.138

Figure 3-9.—Six-inch jointer.

table an amount equal to the desired depth of cut. The usual depth of cut is about 1/16 to 1/8 inch.

The level of the outfeed table must be frequently checked to ensure the surface is exactly even with the highest point reached by the knife edges. If the outfeed table is too high, the cut will become progressively more shallow as the piece is fed through. If the outfeed table is too low, the piece will drop downward as its end leaves the infeed table, and the cut for the last inch or so will be too deep.

To set the outfeed table to the correct height, first feed a piece of waste stock past the cutterhead until a few inches of it lie on the outfeed table. Then, stop the machine and look under the outfeed end of the piece. If the outfeed table is too low, there will be a space between the surface of the table and the lower face of the piece. Raise the outfeed table until this space is eliminated. If no space appears, lower the outfeed table until a space does appear. Now, run the stock back through the machine. If there is still a space, raise the table just enough to eliminate it.

Note that the cutterhead cuts toward the infeed table; therefore, to cut with the grain, you must place the piece with the grain running toward the infeed table. A piece is edged by feeding it through on edge with one of the faces held against the fence. A piece is surfaced by feeding it through flat with one of the edges against the fence. However, this operation should, if possible, be limited to straightening the face of the stock. The fence can be set at 90° to produce squared faces and edges, or at any desired angle to produce beveled edges or ends.

Only sharp and evenly balanced knives should be used in a jointer cutting head. The knives must not be set to take too heavy a cut because a kickback is almost certain to result, especially if there is a knot or change of grain in the stock. The knives must be securely refastened after the machine has been standing in a cold building over the weekend.

Each hand-fed jointer should be equipped with a cylindrical cutting head, the throat of which should not exceed 7/16 inch in depth or 5/8 inch in width. It is strongly recommended that no cylinder be used in which the throat exceeds 3/8 inch in depth or 1/2 inch in width.

Each hand-fed jointer should have an automatic guard that covers all the sections of the head on the working side of the fence or gauge. The guard should automatically adjust horizontally for edge jointing and vertically for surface work, and it should remain in contact with the material at all times.

When operating the jointer, observe the following safety precautions:

- Always plane with the grain. A piece of wood planed against the grain on a jointer may be kicked back.
- Never place your hands directly over the inner cutterhead. Should the piece of wood kick

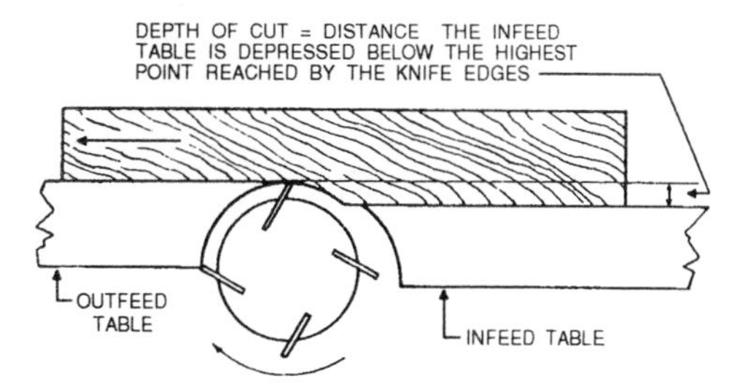

Figure 3-10.—Operating principle of a jointer.

back, your hands will drop on the blades. Start with your hands on the infeed bed. When the piece of wood is halfway through, reach around with your left hand and steady the piece of wood on the outfeed bed. Finish with both your hands on the outfeed bed.

- Never feed a piece of wood with your thumb or finger against the end of the piece of wood being fed into the jointer. Keep your hands on top of the wood at all times.

- Avoid jointing short pieces of wood whenever possible. Joint a longer piece of wood and then cut it to the desired size. If you must joint a piece of wood shorter than 18 inches, use a push stick to feed it through the jointer.

- Never use a jointer with dull cutter blades. Dull blades have a tendency to kick the piece, and a kickback is always dangerous.

- Keep the jointer table and the floor around the jointer clear of scraps, chips, and shavings. Always stop the jointer before brushing off and cleaning up those scraps, chips, and shavings.

- Never joint a piece of wood that contains loose knots.

- Keep your eyes and undivided attention on the jointer as you are working. Do not talk to anyone while operating the jointer.

Remember, the jointer is one of the most dangerous machines in the woodworking shop. Only experienced and responsible personnel should be allowed to operate it using the basic safety precautions provided above.

Surfacer

A single surfacer (also called a single planer) is shown in figure 3-11. This machine surfaces stock on one face (the upper face) only. (Double surfacers, which surface both faces at the same time, are used only in large planing mills.)

The single surfacer cuts with a cutterhead like the one on the jointer, but, on the single surfacer, the cutterhead is located above instead of below the drive rollers. The part adjacent to the cutterhead is pressed down against the feed bed by the chip breakers (just ahead of the cutterhead) and the pressure bar (just behind the cutterhead). The pressure bar temporarily

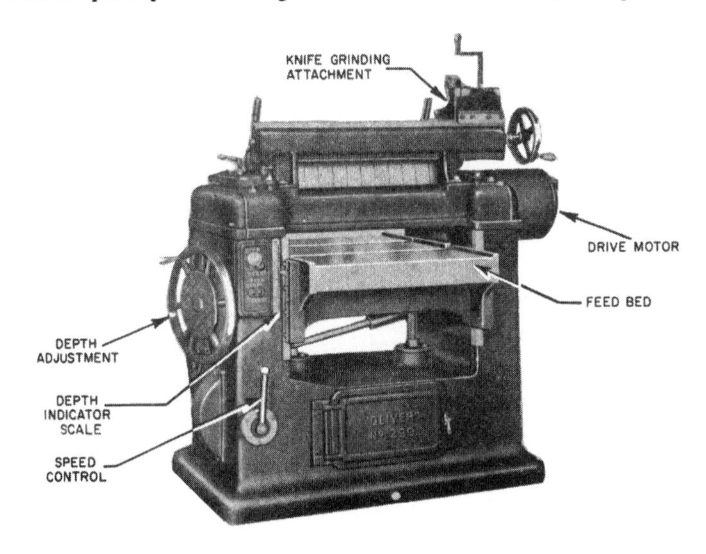

29.135

Figure 3-11.—Single surfacer.

straightens out any warp a piece may have; a piece that goes into the surfacer warped will come out still warped. This is not a defect in the machine; the surfacer is designed for surfacing only, not for truing warped stock. If true plane surfaces are desired, one face of the stock (the face that goes down in the surfacer) must be trued on the jointer before the piece is feed through the surfacer. If the face that goes down in the surfacer is true, the surfacer will plane the other face true.

Observe the following safety precautions when operating a surfacer:

- The cutting head should be covered by metal guards.
- Feed rolls should be guarded by a hood or a semicylindrical guard.
- Never force wood through the machine.
- If a piece of wood gets stuck, turn off the surfacer and lower the feed bed.

Figure 3-12.—Three-wing cutter for a shaper.

Shaper

The shaper is designed primarily for edging curved stock and for cutting ornamental edges, as on moldings. It can also be used for rabbeting, grooving, fluting, and beading.

The flat cutter on a shaper is mounted on a vertical spindle and held in place by a hexagonal spindle nut. A grooved collar is placed below and above the cutter to receive the edges of the knives. Ball bearing collars are available for use as guides on irregular work where the fence is not used. The part of the edge that is to remain uncut runs against a ball bearing collar underneath the cutter, as shown in the bottom view of figure 3-12. A three-wing cutter (top view of figure 3-12) fits over the spindle. Cutters come with cutting edges in a great variety of shapes.

For shaping the side edges on a rectangular piece, a light-duty shaper has an adjustable fence, like the one shown on the shaper in figure 3-13. For shaping the end edges on a rectangular piece, a machine of this type has a sliding fence similar to the cutoff gauge on a circular saw. The sliding fence slides in the groove shown in the table top.

On larger machines, the fence consists of a board straightedge, clamped to the table with a hand screw,

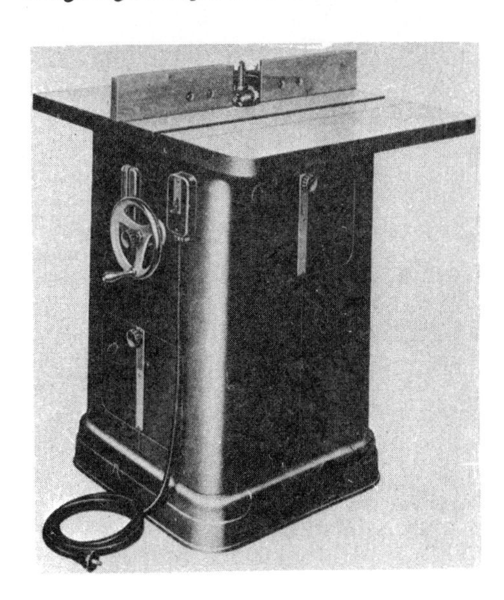

68.27

Figure 3-13.—Light-duty shaper with adjustable fence.

as shown in figure 3-14. A semicircular opening is sawed in the edge of the straightedge to accommodate the spindle and the cutters. Whenever possible, a guard of the type shown in the figure should be placed over the spindle.

For shaping curved edges, there are usually a couple of holes in the table, one on either side of the spindle, in which vertical starter pins can be inserted. When a curved edge is being shaped, the piece is guided by and steadied against the starter pin and the ball bearing collar on the spindle.

When operating a shaper, observe the following safety precautions:

- Like the jointer and surfacer, the shaper cuts toward the infeed side of the spindle, which is against the rotation of the spindle. Therefore, stock should be placed with the grain running toward the infeed side.
- Make sure the cutters are sharp and well secured.
- If curved or irregularly shaped edges are to be shaped, place the stock in position and make sure the collar will rub against the part of the edge, which should not be removed.
- Whenever the straight fence cannot be used, always use a starting pin in the table top.
- Never make extremely deep cuts.
- Make sure the shaper cutters rotate toward the work.
- Whenever possible, always use a guard, pressure bar, hold-down, or holding jig.
- If possible, place the cutter on the shaper spindle so that the cutting will be done on the lower side of the stock.
- Do not attempt to shape small pieces of wood.
- Check all adjustments before turning on the power.

SAFETY NOTE

The spindle shaper is one of the most dangerous machines used in the shop. Use extreme caution at all times.

PORTABLE HAND TOOLS

In addition to using power shop tools, you will be required to operate different types of portable hand tools in the field. You therefore need to understand the safety precautions associated with these.

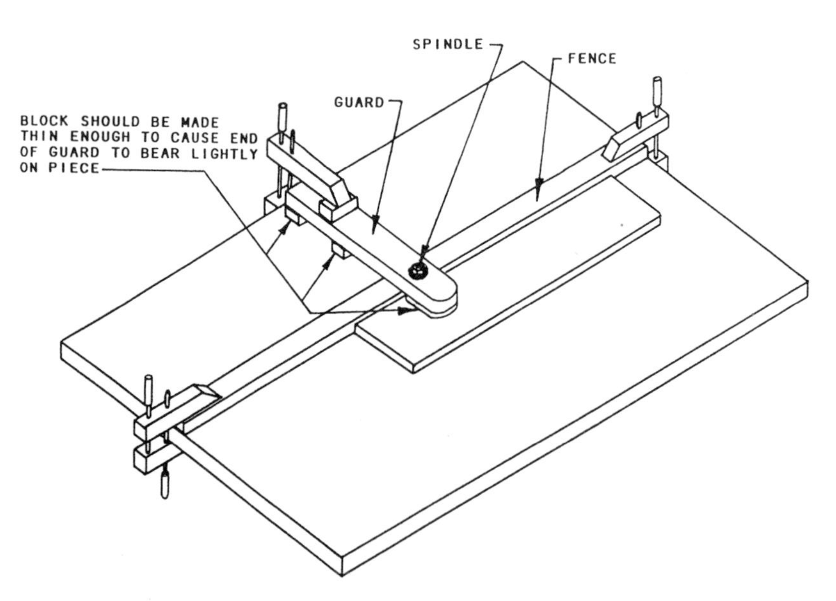

Figure 3-14.—Shaper table showing straightedge fence and guard.

Portable Electric Circular Saw

The portable electric circular saw is used chiefly as a great labor-saving device in sawing wood framing members on the job. The size of a circular saw is determined by the diameter of the largest blade it can use. The most commonly used circular saws are the 7 1/4- and 8 1/4-inch saws. There are two different types of electric saws, as shown in figure 3-15: the side-drive (view A) and the

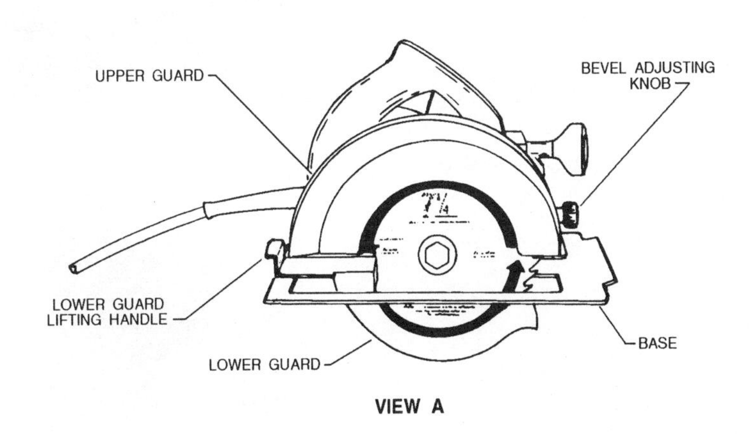

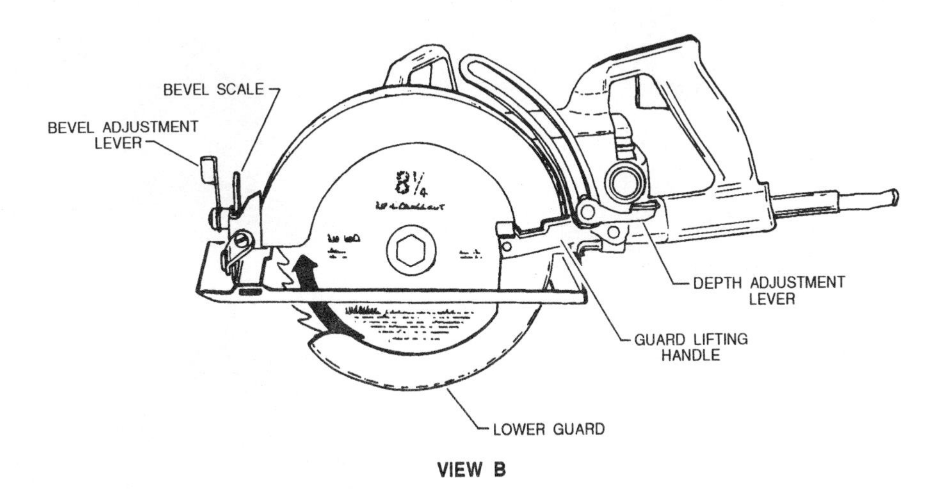

Figure 3-15.—Side-drive (view A) and worm-drive (view B) circular saws.

worm-drive (view B). Circular saws can use many different types of cutting blades, some of which are shown in figure 3-16.

COMBINATION CROSSCUT AND RIP BLADES.—Combination blades are all-purpose blades for cutting thick and thin hardwoods and softwoods, both with or across the grain. They can also be used to cut plywood and hardboard.

CROSSCUT BLADES.—Crosscut blades have fine teeth that cut smoothly across the grain of both hardwood and softwood. These blades can be used for plywood, veneers, and hardboard.

RIP BLADES.—Rip blades have bigger teeth than combination blades, and should be used only to cut with the grain. A rip fence or guide will help you make an accurate cut with this type of blade.

HOLLOW-GROUND BLADES.—Hollow-ground blades have no set. They make the smoothest cuts on thick or thin stock. Wood cut with these blades requires little or no sanding.

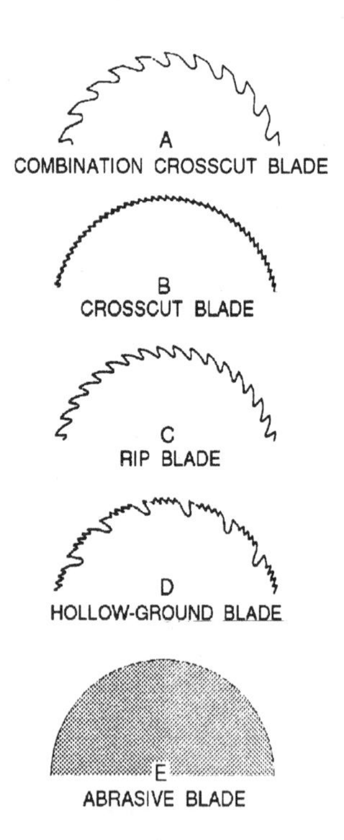

Figure 3-16.—Circular saw blades.

ABRASIVE BLADES.—Abrasive blades are used for cutting metal, masonry, and plastics. These blades are particularly useful for scoring bricks so they can be easily split.

Figure 3-17 shows how versatile the circular saw can be. To make an accurate ripping cut (view A), the

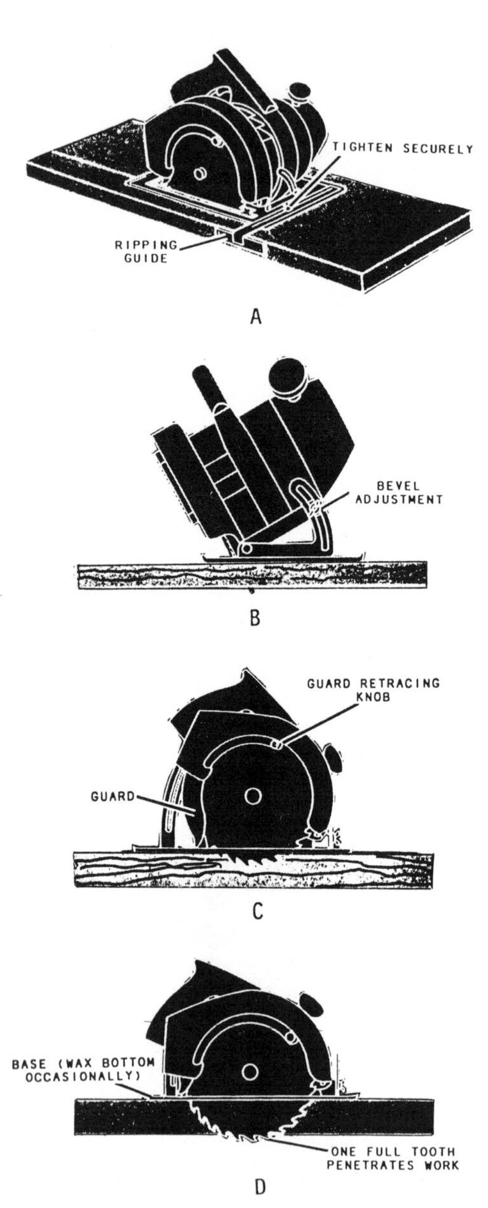

Figure 3-17.—Different ways to use a circular saw.

ripping guide is set a distance away from the saw equal to the width of the strip to be ripped off. It is then placed against the edge of the piece as a guide for the saw. To make a bevel angle cut up to 45° (view B), you just set the bevel adjustment knob to the angle you want and cut down the line. To make a pocket cut (views C and D), a square cut in the middle of a piece of material, you retract the guard back and tilt the saw so that it rests on the front of the base. Then, lowering the rear of the saw into the material, hold it there until it goes all the way through the wood. Then, follow your line.

Observe the following safety precautions when operating a circular saw:

- Don't force the saw through heavy cutting stock. If you do, you may overload the motor and damage it.
- Before using the saw, carefully examine the material to be cut and free it of nails or other metal objects. Cutting into or through knots should be avoided, if possible.
- Disconnect the saw from its power source before making any adjustments or repairs to the saw. This includes changing the blade.
- Make sure all circular saws are equipped with guards that automatically adjust themselves to the work when in use so that none of the teeth protrude above the work. Adjust the guard over the blade so that it slides out of its recess and covers the blade to the depth of the teeth when you lift the saw off the work.
- Wear goggles or face shields while using the saw and while cleaning up debris afterward.
- Grasp the saw with both hands and hold it firmly against the work. Take care to prevent the saw from breaking away from the work and thereby causing injury.
- Inspect the blade at frequent intervals and always after it has locked, pinched, or burned the work. Disconnect the saw from the power source before performing this inspection.
- Inspect daily the electric cords that you use for cuts or breaks. Before cutting boards, make sure the cord is not in the way of the blade.

Saber Saw

The saber saw (figure 3-18) is a power-driven jigsaw that cuts smooth and decorative curves in wood and light metal. Most saber saws are light-duty machines and not designed for extremely fast cutting.

There are several different, easily interchangeable blades (figure 3-19) designed to operate in the saber saw. Some blades are designed for cutting wood and some for cutting metal.

The best way to learn how to handle this type of tool is to use it. Before trying to do a finished job with the saber saw, clamp down a piece of scrap plywood and draw some curved as well as straight lines to follow. You will develop your own way of

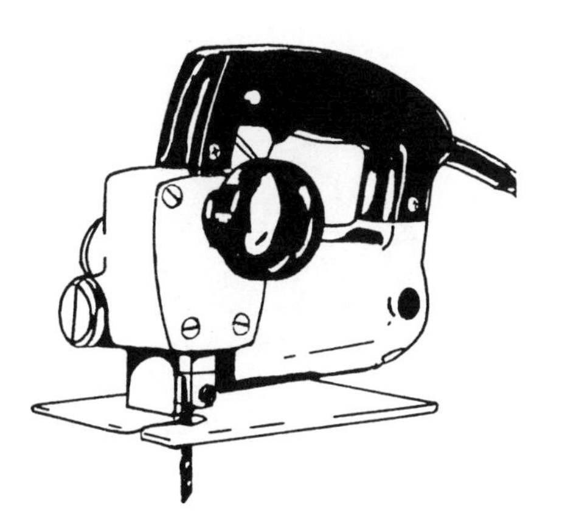

Figure 3-18.—Saber saw.

Figure 3-19.—Saber saw blades.

gripping the tool, which will be affected somewhat by the particular tool you are using. On some tools, for example, you will find guiding easier if you apply some downward pressure on the tool as you move it forward. If you don't use a firm grip, the tool will tend to vibrate excessively and roughen the cut. Do not force the cutting faster than the design of the blade allows or you will break the blade.

You can make a pocket cut with a saber saw just like you can with a circular saw, although you need to drill a starter hole to begin work. A saber saw can also make bevel-angle and curve cuts.

Observe the following safety precautions when operating the saber saw:

- Before working with the saber saw, be sure to remove your rings, watches, bracelets, and other jewelry.
- If you are wearing long sleeves, roll them up.
- Be sure the saber saw is properly grounded.
- Use the proper saw blade for the work to be done, and ensure the blade is securely locked in place.
- Be sure the material to be cut is free of any obstructions.
- Keep your full attention focused on the work being performed.
- Grip the handle of the saw firmly. Control the forward and turning movements with your free hand on the front guide.
- To start a cut, place the forward edge of the saw base on the edge of the material being worked, start the motor, and move the blade into the material.

Portable Reciprocating Saw

The portable reciprocating saw (saw-all) (figure 3-20) is a heavy-duty power tool that you can use for a variety of woodworking maintenance work, remodeling, and roughing-in jobs. You can use it to cut rectangular openings, curved openings, along straight or curved lines, and flush.

Blades for reciprocating saws are made in a great variety of sizes and shapes. They vary in length from 2 1/2 to 12 inches and are made of high-speed steel or carbon steel. They have cutting edges similar to those shown in figure 3-19.

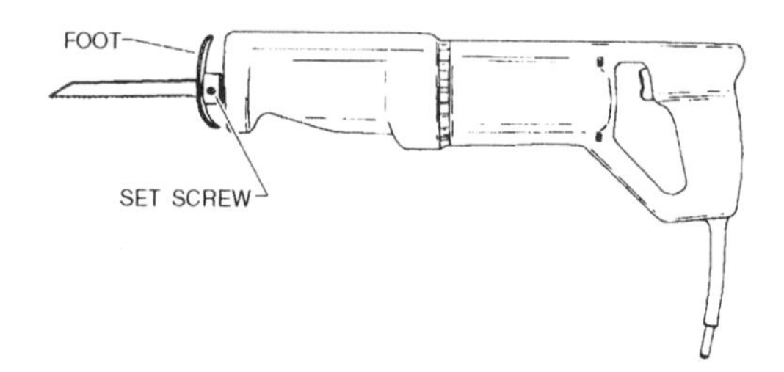

Figure 3-20.—Reciprocating saw.

Before operating this saw, be sure you are using a blade that is right for the job. The manufacturer's instruction manual shows the proper saw blade to use for a particular material. The blade must be pushed securely into the opening provided. Rock it slightly to ensure a correct fit, then tighten the setscrew.

To start a cut, place the saw blade near the material to be cut. Then, start the motor and move the blade into the material. Keep the cutting pressure constant, but do not overload the saw motor. Never reach underneath the material being cut.

Observe the following safety precautions when operating a reciprocating saw:

- Disconnect the saw when changing blades or making adjustments.

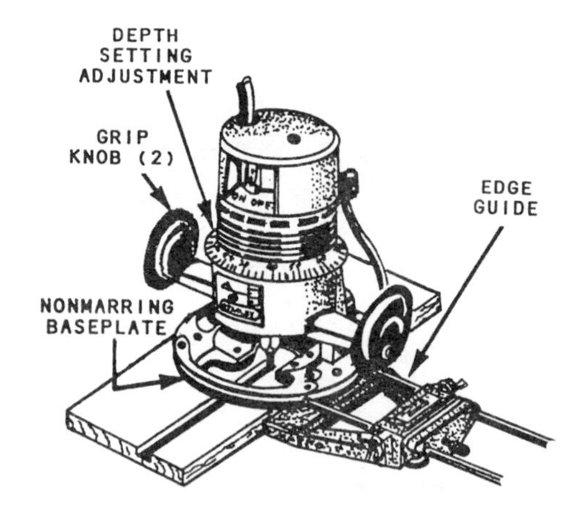

Figure 3-21.—Portable router with edge guide.

- Place the foot of the saw firmly on the stock before starting to cut.
- Don't cut curves shaper than the blade can handle.
- When cutting through a wall, make sure you don't cut electrical wires.

Router

The router is a versatile portable power tool that can be used free hand or with jigs and attachments. Figure 3-21 shows a router typical of most models. It consists of a motor containing a chuck into which the router bits are attached. The motor slides into the base in a vertical position. By means of the depth adjustment ring, easy regulation of the depth of a cut is possible. Routers vary in size from 1/4 to 2 1/2 horsepower, and the motor speed varies from 18,000 to 27,000 rpm.

One of the most practical accessories for the router is the edge guide. It is used to guide the router in a straight line along the edge of the board. The edge guide is particularly useful for cutting grooves on long pieces of lumber. The two rods on the edge guide slip into the two holes provided on the router base. The edge guide can be adjusted to move in or out along the two rods to obtain the desired lateral depth cut.

There are two classifications of router bits. Built-in, shank-type bits fit into the chuck of the router. Screw-type bits have a threaded hole through the center of the cutting head, which allows the cutting head to be screwed to the shank. Figure 3-22 shows a few of the most common router bits.

Observe the following safety precautions when operating a router:

- Before operating a router, be sure the work piece is well secured and free of obstruction.
- Make sure the router is disconnected from the power source before making any adjustment or changing bits.
- Don't overload the router when cutting the material.
- Use both hands to hold the router when cutting material.

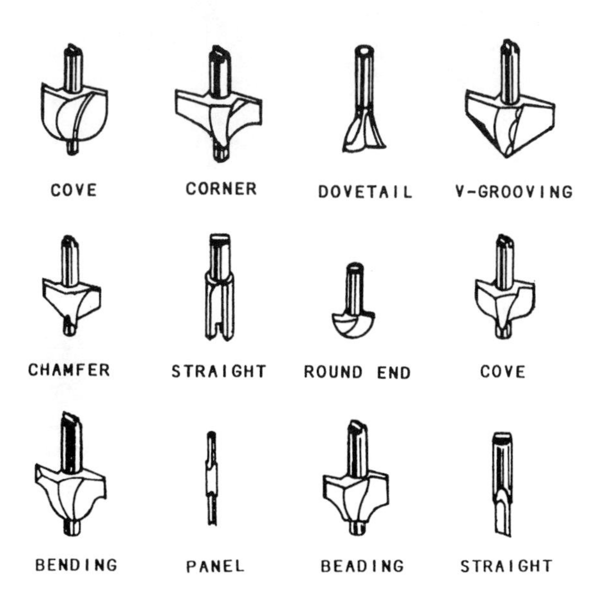

Figure 3-22.—Router bits.

Portable Power Plane

The portable electric power plane (figure 3-23) is widely used for trimming panels, doors, frames, and so forth. It is a precision tool capable of exact depth of cut up to 3/16 inch on some of the heavier models. However, the maximum safe depth of cut on any model is 3/32 inch in any one pass.

The power plane is essentially a high-speed motor that drives a cutter bar, containing either straight or spiral blades, at high speed.

Operating the power plane is simply a matter of setting the depth of cut and passing the plane over the work. First, make careful measurements of the piece, where it is to fit, and determine how much material has to be removed. Then, the stock being planed should be held in a vise, clamped to the edge of a bench, or otherwise firmly held. Check the smoothness and straightness of all the edges.

If a smoothing cut is desired, make that cut first and then recheck the dimensions. Make as many passes as necessary with the plane to reach the desired dimensions, checking frequently so as not to remove too much material. The greater the depth of the cut, the slower you must feed the tool into the work. Feed pressure should be enough to keep the tool cutting, but not so much as to slow it down excessively. Keep wood chips off the work because they can mar the surface of the stock as the tool passes over them. Keep your hands away from the cutterhead or blades when a cut is finished.

The L-shaped base, or fence, of the plane should be pressed snugly against the work when planing, assuring that the edge will be cut square. For bevel cuts, loosen the setscrew on the base, set the base at the desired bevel, and then retighten the setscrew.

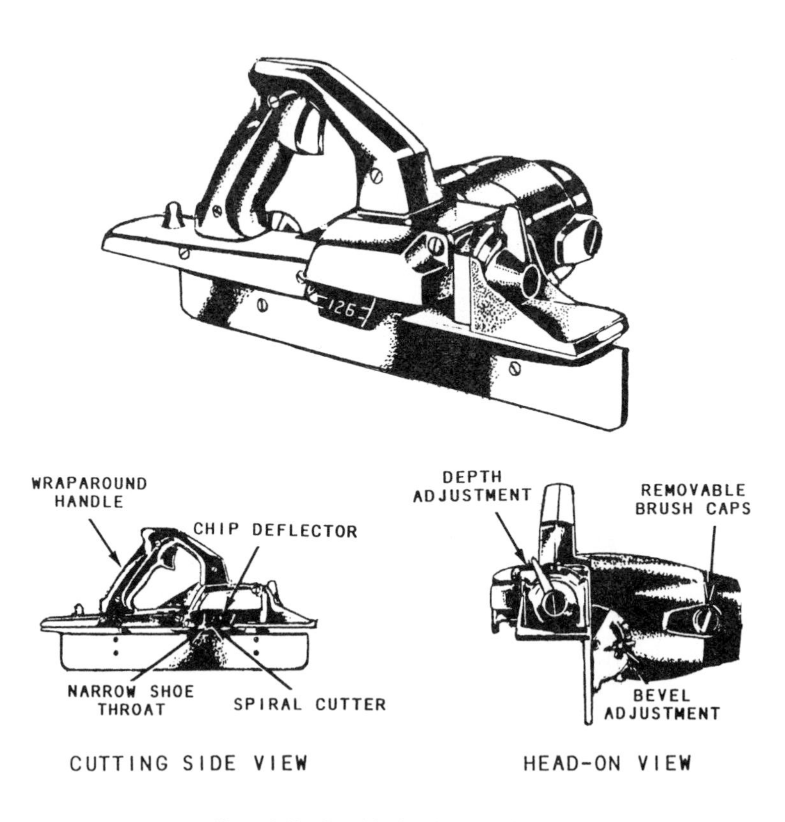

Figure 3-23.—Portable electric power plane.

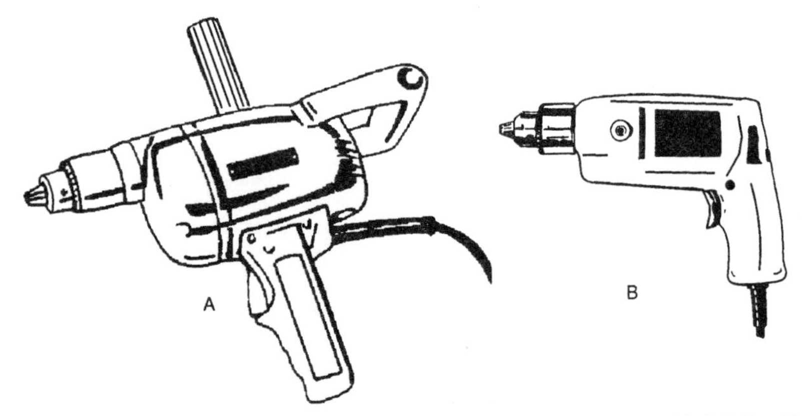

Figure 3-24.—Heavy-duty 1/2-inch portable drill (view A) and light-duty 1/2-inch portable drill (view B).

Observe the following safety precautions when operating a portable power plane:

- Make sure that the plane is turned off before plugging it in.
- Make sure you disconnect the plug before making any adjustment.
- Don't attempt to power plane with one hand—you need two.
- Always clamp your work securely in the best position to perform the planing.
- When finished planing, make sure you disconnect the power cord.

Portable Power Drills

Portable power drills have generally replaced hand tools for drilling holes because they are faster and more accurate. With variable-speed controls and special clutch-drive chucks, they can also be used as electric screwdrivers. More specialized power-driven screwdrivers are also available; these have greatly increased the efficiency of many fastening operations in construction work.

The two basic designs for portable electric drills (figure 3-24) are the spade design for heavy-duty construction (view A) and the pistol-grip design for lighter work (view B). Sizes of power drills are based on the diameter of the largest drill shank that will fit into the chuck of the drill.

The right-angle drill is a specialty drill used in plumbing and electrical work. It allows you to drill holes at a right angle to the drill body.

Observe the following safety precautions when operating a portable drill:

- Make sure that the drill or bit is securely mounted in the chuck.
- Hold the drill firmly as prescribed by the manufacturer of the drill.
- When feeding the drill into the material, vary the pressure you apply to accommodate the different kinds of stock. Be careful not to bind the drill or bit.
- When drilling a deep hole, withdraw the drill several times to clean the drill bit.

Portable Sanders

There are three types of portable sanders: belt, disk, and finish sanders. When using a belt sander (figure 3-25), be careful not to gouge the wood. The size of a belt sander is usually identified by the width of its sanding belt. Belt widths on heavier duty

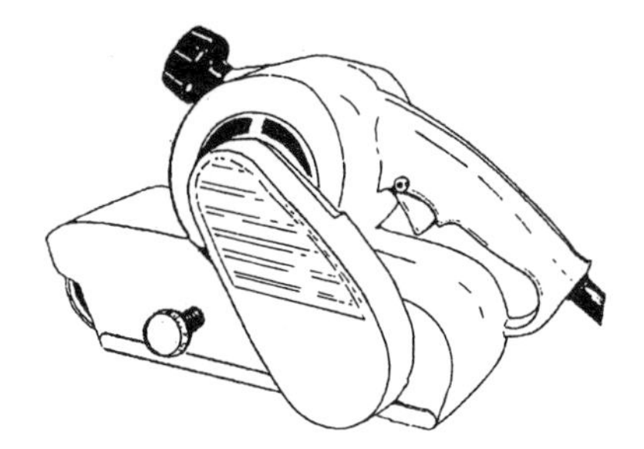

Figure 3-25.—Belt sander.

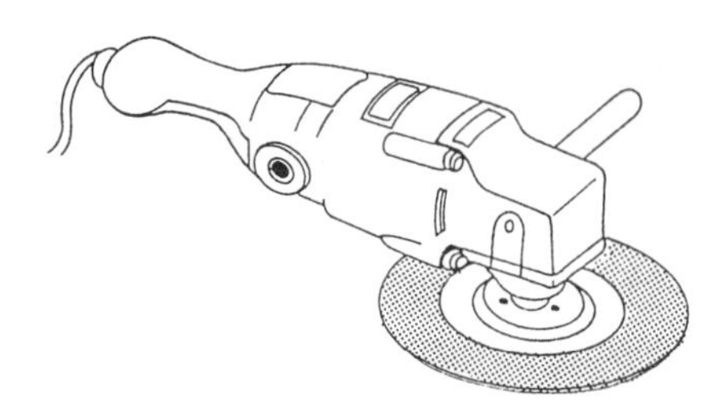

Figure 3-26.—Portable disk sander.

models are usually 3 or 4 inches. Depending on the make and model, belt lengths vary from 21 to 27 inches. Different grades of abrasives are available.

The disk sander (figure 3-26) is a useful tool for removing old finish, paint, and varnish from siding, wood flooring, and concrete. For best results with a disk sander, tip the machine lightly with just enough pressure to bend the disk. Use a long, sweeping motion, back and forth, advancing along the surface. When using a disk sander, always operate it with both hands.

The finish sander (figure 3-27) is used for light and fine sanding. Two kinds of finish sanders are available. One operates with an orbital (circular) motion (view A), and the other has an oscillating (back and forth) movement (view B). Finish sanders use regular abrasive paper (sandpaper) cut to size from full sheets.

Observe the following safety tips when operating portable sanders:

- Make sure the sander is off before plugging it in.
- Make sure that you use two hands if using the belt sander.
- Don't press down on the sander. The weight of the sander is enough to sand the material.
- Make sure the sander is disconnected when changing sandpaper.
- Keep the electrical cord away from the area being sanded.

Power Nailers and Staplers

There is a wide variety of power nailers and staplers available. A typical example of each is shown in figure 3-28. A heavy-duty nailer is used for

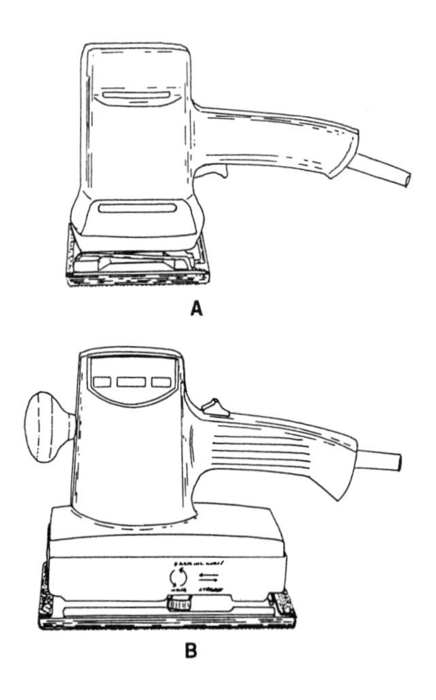

Figure 3-27.—Two types of finish sanders: orbital (view A) and oscillating (view B).

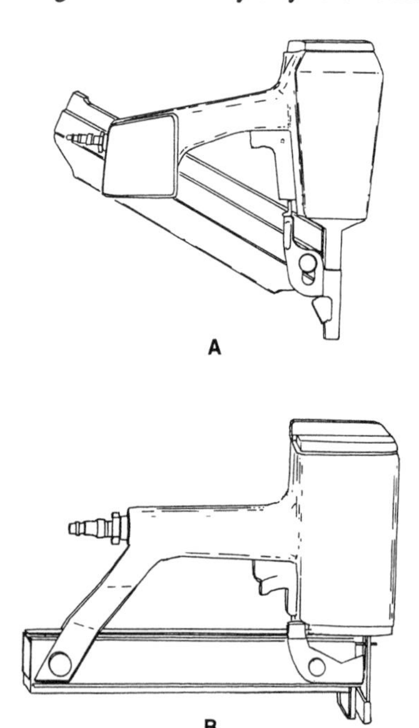

Figure 3-28.—Heavy-duty pneumatic nailer (view A) and pneumatic stapler (view B).

framing or sheathing work; finish nailers are used for paneling or trimming. There is also a wide variety of staplers that you can use for jobs, such as fastening sheeting, decking, or roofing. These tools are often driven by compressed air. The amount of pneumatic, or air, pressure required to operate the tool depends on the size of the tool and the type of operation you are performing. Check the manufacturer's manual for the proper air pressure to operate the tool.

The power nailer and power stapler are great timesaving tools, but they are also **very dangerous** tools. Observe the following safety precautions when using them:

- Use the correct air pressure for the particular tool and job.
- Use the right nailer or stapler for the job and also the correct nails and staples.
- Keep the nose of the tool pointed away from your body.
- When you are not using a nailer or stapler or if you are loading one, **disconnect the air supply.**

MATERIALS

LEARNING OBJECTIVE: Upon completing this section, you should be able to identify the types, sources, uses, and characteristics of the common woods used on various construction projects.

Of all the different construction materials, wood is probably the most often used and perhaps the most important. The variety of uses of wood is practically unlimited. Few Seabee construction projects are accomplished without using some type of wood. It is used for permanent structures as well as concrete forms, scaffolding, shoring, and bracing, which may be used again and again. The types, sources, uses, and characteristics of common woods are given in table 3-1. The types of classifications of wood for a large project are usually designated in the project specifications and included in the project drawings.

Table 3-1.—Common Woods

TYPES	SOURCES	USES	CHARACTERISTICS
ASH	East of Rockies	Oars, boat thwarts, benches, gratings, hammer handles, cabinets, ball bats, wagon construction, farm implements	Strong, heavy, hard, tough, elastic, close straight grain, shrinks very little, takes excellent finish, lasts well
BEECH	East of Mississippi and southeastern Canada	Cabinetwork, imitation mahogany furniture, wood dowels, capping, boat trim, interior finish, tool handles, turnery, shoe lasts, carving, flooring	Similar to birch but not so durable when exposed to weather, shrinks and checks considerably, close grain, light or dark red color
BIRCH	East of Mississippi River and north of gulf coast states, southeast Canada, and Newfoundland	Cabinetwork, imitation mahogany furniture, wood dowels, capping, boat trim, interior finish, tool handles, turnery, carving	Hard, durable, fine grain, even texture, heavy, stiff, strong, tough, takes high polish, works easily, forms excellent base for white enamel finish, but not durable when exposed. Heartwood is light to dark reddish brown in color

Table 3-1.—Common Woods—Continued

TYPES	SOURCES	USES	CHARACTERISTICS
BUTTERNUT	Southern Canada, Minnesota, eastern U.S. as far south as Alabama and Florida	Toys, altars, woodenware, millwork, interior trim, furniture, boats, scientific instruments	Very much like walnut in color but softer, not so soft as white pine and basswood, easy to work, coarse grained, fairly strong
DOUGLAS FIR	Pacific coast and British Columbia	Deck planking on large ships, shores, strongbacks, plugs, filling pieces and bulkheads of small boats, building construction, dimension timber, plywood	Excellent structural lumber, strong, easy to work, clear straight grained, soft but brittle. Heartwood is durable in contact with ground, best structural timber of northwest
ELM	States east of Colorado	Agricultural implements, wheel-stock, boats, furniture, crossties, posts, poles	Slippery, heavy, hard, tough, durable, difficult to split, not resistant to decay
HICKORY	Arkansas, Tennessee, Ohio, and Kentucky	Tools, handles, wagon stock, hoops, baskets, vehicles, wagon spokes	Very heavy, hard, stronger and tougher than other native woods, but checks, shrinks, difficult to work, subject to decay and insect attack
MAPLE	All states east of Colorado and Southern Canada	Excellent furniture, high-grade floors, tool handles, ship construction, crossties, counter tops, bowling pins	Fine grained, grain often curly or "Birds's Eyes," heavy, tough, hard, strong, rather easy to work, but not durable. Heartwood is light brown, sap wood is nearly white
LIVE OAK	Southern Atlantic and gulf coasts of U.S., Oregon, and California	Implements, wagons, shipbuilding	Very heavy, hard, tough, strong, durable, difficult to work, light brown or yellow sap wood nearly white
MAHOGANY	Honduras, Mexico, Central America, Florida, West Indies, Central Africa, and other tropical sections	Furniture, boats, decks, fixtures, interior trim in expensive homes, musical instruments	Brown to red color, one of most useful of cabinet woods, hard, durable, does not split badly, open grained, takes beautiful finish when grain is filled but checks, swells, shrinks, warps slightly
NORWAY PINE	States bordering Great Lakes	Dimension timber, masts, spars, piling, interior trim	Light, fairly hard, strong, not durable in contact with ground

Table 3-1.—Common Woods—Continued

TYPES	SOURCES	USES	CHARACTERISTICS
PHILIPPINE MAHOGANY	Philippine Islands	Pleasure boats, medium-grade furniture, interior trim	Not a true mahogany, shrinks, expands, splits, warps, but available in long, wide, clear boards
POPLAR	Virginias, Tennessee, Kentucky, and Mississippi Valley	Low-grade furniture, cheaply constructed buildings, interior finish, shelving drawers, boxes	Soft, cheap, obtainable in wide boards, warps, shrinks, rots easily, light, brittle, weak, but works easily and holds nails well, fine-textured
RED CEDAR	East of Colorado and north of Florida	Mothproof chests, lining for linen closets, sills, and other uses similar to white cedar	Very light, soft, weak, brittle, low shrinkage, great durability, fragrant scent, generally knotty, beautiful when finished in natural color, easily worked
RED OAK	Virginias, Tennessee, Arkansas, Kentucky, Ohio, Missouri, Maryland	Interior finish, furniture, cabinets, millwork, crossties when preserved	Tends to warp, coarse grain, does not last well when exposed to weather, porous, easily impregnated with preservative, heavy, tough, strong
REDWOOD	California	General construction, tanks, paneling	Inferior to yellow pine and fir in strength, shrinks and splits little, extremely soft, light, straight grained, very durable, exceptionally resistant to decay
SPRUCE	New York, New England, West Virginia, central Canada, Great Lakes states, Idaho, Washington, Oregon	Railway ties, resonance wood, piles, airplanes, oars, masts, spars, baskets	Light, soft, low strength, fair durability, close grain, yellowish, sap wood indistinct
SUGAR PINE	California and Oregon	Same as white pine	Very light, soft, resembles white pine
TEAK	India, Burma, Thailand, and Java	Deck planking, shaft logs for small boats	Light brown color, strong, easily worked, durable, resistant to moisture damage

Table 3-1.—Common Woods—Continued

TYPES	SOURCES	USES	CHARACTERISTICS
WALNUT	Eastern half of U.S. except southern Atlantic and gulf coasts, some in New Mexico, Arizona, California	Expensive furniture, cabinets, interior woodwork, gun stocks, tool handles, airplane propellers, fine boats, musical instruments	Fine cabinet wood, coarse grained but takes beautiful finish when pores closed with woodfiller, medium weight, hard, strong, easily worked, dark chocolate color, does not warp or check, brittle
WHITE CEDAR	Eastern coast of U.S., and around Great Lakes	Boat planking, railroad ties, shingles, siding, posts, poles	Soft, lightweight, close grained, exceptionally durable when exposed to water, not strong enough for building construction, brittle, low shrinkage, fragment, generally knotty
WHITE OAK	The Virginias, Tennessee, Arkansas, Kentucky, Ohio, Missouri, Maryland, and Indiana	Boat and ship stems, stern-posts, knees, sheer strakes, fenders, capping, transoms, shaft logs, framing for buildings, strong furniture, tool handles, crossties, agricultural implements, fence posts	Heavy, hard, strong, medium coarse grain, tough, dense, most durable of hardwoods, elastic, rather easy to work, but shrinks and likely to check. Light brownish grey in color with reddish tinge, medullary rays are large and outstanding and present beautiful figures when quarter sawed, receives high polish
WHITE PINE	Minnesota, Wisconsin, Maine, Michigan, Idaho, Montana, Washington, Oregon, and California	Patterns, any interior job or exterior job that doesn't require maximum strength, window sash, interior trim, millwork, cabinets, cornices	Easy to work, fine grain, free of knots, takes excellent finish, durable when exposed to water, expands when wet, shrinks when dry, soft, white, nails without splitting, not very strong, straight grained
YELLOW PINE	Virginia to Texas	Most important lumber for heavy construction and exterior work, keelsons, risings, filling pieces, clamps, floors, bulkheads of small boats, shores, wedges, plugs, strongbacks, staging, joists, posts, piling, ties, paving blocks	Hard, strong, heartwood is durable in the ground, grain varies, heavy, tough, reddish brown in color, resinous, medullary rays well marked

LUMBER

The terms "wood," "lumber," and "timber" are often spoken of or written in ways to suggest that their meanings are alike or nearly so. But in the Builder's language, the terms have distinct, separate meanings. Wood is the hard, fibrous substance that forms the major part of the trunk and branches of a tree. Lumber is wood that has been cut and surfaced for use in construction work. Timber is lumber that is 5 inches or more in both thickness and width.

SEASONING OF LUMBER

Seasoning of lumber is the result of removing moisture from the small and large cells of wood—drying. The advantages of seasoning lumber are to reduce its weight; increase its strength and resistance to decay; and decrease shrinkage, which tends to avoid checking and warping after lumber is placed. A seldom used and rather slow method of seasoning lumber is air-drying in a shed or stacking in the open until dry. A faster method, known as kiln drying, has lumber placed in a large oven or kiln and dried with heat, supplied by gas- or oil-fired burners. Lumber is considered dry enough for most uses when its moisture content has been reduced to about 12 or 15 percent. As a Builder, you will learn to judge the dryness of lumber by its color, weight, smell, and feel. Also, after the lumber is cut, you will be able to judge the moisture content by looking at the shavings and chips.

DEFECTS AND BLEMISHES

A defect in lumber is any flaw that tends to affect the strength, durability, or utility value of the lumber. A blemish is a flaw that mars only the appearance of lumber. However, a blemish that affects the utility value of lumber is also considered to be a defect; for example, a tight knot that mars the appearance of lumber intended for fine cabinet work.

Various flaws apparent in lumber are listed in table 3-2.

Table 3-2.—Wood Defects and Blemishes

COMMON NAME	DESCRIPTION
Bark Pocket	Patch of bark over which the tree has grown, and has entirely or almost entirely enclosed
Check	Separation along the lengthwise grain, caused by too rapid or nonuniform drying
Cross Grain	Grain does not run parallel to or spiral around the lengthwise axis
Decay	Deterioration caused by various fungi
Knot	Root section of a branch that may appear on a surface in cross section or lengthwise. A cross-sectional knot may be loose or tight. A lengthwise knot is called a spike knot
Pitch Pocket	Deposit of solid or liquid pitch enclosed in the wood
Shake	Separation along the lengthwise grain that exists before the tree is cut. A heart shake moves outward from the center of the tree and is caused by decay at the center of the trunk. A wind shake follows the circular lines of the annual rings; its cause is not definitely known
Wane	Flaw in an edge or corner of a board or timber. It is caused by the presence of bark or lack of wood in that part
Warp	Twist or curve caused by shrinkage that develops in a once flat or straight board
Blue Stain	A blemish caused by a mold fungus; it does not weaken the wood

CLASSIFICATION OF LUMBER

Trees are classified as either softwood or hardwood (table 3-3). Therefore, all lumber is referred to as either "softwood" or "hardwood." The terms "softwood" and "hardwood" can be confusing since some softwood lumber is harder than some hardwood lumber. Generally, however, hardwoods are more dense and harder than softwoods. In addition, lumber can be further classified by the name of the tree from which it comes. For example, Douglas fir lumber comes from a Douglas fir tree; walnut lumber comes from a walnut tree, and so forth.

The quality of softwood lumber is classified according to its intended use as being yard, structural, factory, or shop lumber. Yard lumber consists of those grades, sizes, and patterns generally intended for ordinary building purposes. Structural lumber is 2 or more inches in nominal thickness and width and is used where strength is required. Factory and shop lumber are used primarily for building cabinets and interior finish work.

Lumber manufacturing classifications consist of rough dressed (surfaced) and worked lumber. Rough lumber has not been dressed but has been sawed, edged, and trimmed. Dressed lumber is rough lumber that has been planed on one or more sides to attain smoothness and uniformity. Worked lumber, in addition to being dressed, has also been matched, shiplapped, or patterned. Matched lumber is tongue and groove, either sides or ends or both. Shiplapped lumber has been rabbeted on both edges to provide a close-lapped joint. Patterned lumber is designed to a pattern or molded form.

Softwood Grading

The grade of a piece of lumber is based on its strength, stiffness, and appearance. A high grade of lumber has very few knots or other blemishes. A low grade of lumber may have knotholes and many loose knots. The lowest grades are apt to have splits, checks, honeycombs, and some warpage. The grade of lumber to be used on any construction job is usually stated in the specifications for a set of blueprints. Basic classifications of softwood grading include boards, dimension, and timbers. The grades within these classifications are shown in table 3-4.

Lumber is graded for quality in accordance with American Lumber Standards set by the National Bureau of Standards for the U.S. Department of Commerce. The major quality grades, in descending order of quality, are select lumber and common

Table 3-3.—Different Types of Softwoods and Hardwoods

SOFTWOODS	HARDWOODS
Douglas fir Southern pine Western larch	Basswood Willow American elm
Hemlock White fir Spruce	Mahogany* Sweet gum White ash*
Ponderosa pine Western red cedar Redwood	Beech Birch Cherry
Cypress White pine Sugar pine	Maple Oak* Walnut*

*Open-grained wood

Table 3-4.—Softwood Lumber Grades

BOARDS

		Grade	
APPEARANCE GRADES	SELECTS	B & BETTER C SELECT D SELECT	(IWP—SUPREME) (IWP—CHOICE) (IWP—QUALITY)
	FINISH	SUPERIOR PRIME E	
	PANELING	CLEAR (ANY SELECT OR FINISH GRADE) NO. 2, 3 COMMON SELECTED FOR KNOTTY PANELING	
	SIDING (BEVEL, BUNGALOW)	SUPERIOR PRIME	

		ALTERNATE BOARD GRADES
BOARDS SHEATHING	NO. 1 COMMON (IWP—COLONIAL) NO. 2 COMMON (IWP—STERLING) NO. 3 COMMON (IWP—STANDARD) NO. 4 COMMON (IWP—UTILITY)	SELECT MERCHANTABLE CONSTRUCTION STANDARD UTILITY

DIMENSION

LIGHT FRAMING 2 in. to 4 in. Thick 2 in. to 4 in. Wide	CONSTRUCTION STANDARD UTILITY ECONOMY	This category for use where high strength values are **NOT** required; such as studs, plates, sills, cripples, blocking, etc.
	STUD ECONOMY STUD	An optional all-purpose grade limited to 10 feet and shorter. Characteristics affecting strength and stiffness values are limited so that the Stud grade is suitable for all stud uses, including load bearing walls.
STRUCTURAL LIGHT FRAMING 2 in. to 4 in. Thick 2 in. to 4 in. Wide	SELECT STRUCTURAL NO. 1 NO. 2 NO. 3 ECONOMY	These grades are designed to fit those engineering applications where higher bending strength ratios are needed in light framing sizes. Typical uses would be for trusses, concrete pier wall forms, etc.
STRUCTURAL JOISTS & PLANKS 2 in. to 4 in. Thick 6 in. and Wider	SELECT STRUCTURAL NO. 1 NO. 2 NO. 3 ECONOMY	These grades are designed especially to fit in engineering applications for lumber 6 inches and wider, such as joists, rafters and general-framing uses.

TIMBERS

BEAMS & STRINGERS	SELECT STRUCTURAL NO. 1 NO. 2 (NO. 1 MINING) NO. 3 (NO. 2 MINING)	POSTS & TIMBERS	SELECT STRUCTURAL NO. 1 NO. 2 (NO. 1 MINING NO. 3 (NO. 2 MINING)

lumber. Table 3-5 lists the subdivisions for each grade in descending order of quality.

Hardwood Grades

Grades of hardwood lumber are established by the National Hardwood Lumber Association. FAS (firsts and seconds) is the best grade. It specifies that pieces be no less than 6-inches wide by 8-feet long and yield at least 83 1/3 percent clear cuttings. The next lower grade is selects, which permits pieces 4-inches wide by 6-feet long. A still lower grade is No. 1 common. Lumber in this group is expected to yield 66 2/3 percent clear cuttings.

Lumber Sizes

Standard lumber sizes have been established in the United States for uniformity in planning structures and in ordering materials. Lumber is identified by nominal sizes. The nominal size of a piece of lumber is larger than the actual dressed dimensions. Referring to table 3-6, you can determine the common widths and thicknesses of lumber in their nominal and dressed dimensions.

Table 3-5.—Grades and Subdivisions of Lumber

SELECT LUMBER	
Grade A	This lumber is practically free of defects and blemishes
Grade B	This lumber contains a few minor blemishes
Grade C	This lumber contains more numerous and more significant blemishes than grade B. It must be capable of being easily and thoroughly concealed with paint
Grade D	This lumber contains more numerous and more significant blemishes than grade C, but it is still capable of presenting a satisfactory appearance when painted
COMMON LUMBER	
No. 1	Sound, tight-knotted stock containing only a few minor defects. Must be suitable for use as watertight lumber
No. 2	Contains a limited number of significant defects but no knotholes or other serious defects. Must be suitable for use as grain-tight lumber
No. 3	Contains a few defects that are larger and coarser than those in No. 2 common; for example, occasional knotholes
No. 4	Low-quality material containing serious defects like knotholes, checks, shakes, and decay
No. 5	Capable only of holding together under ordinary handling

Table 3-6.—Nominal and Dressed Sizes of Lumber

NOMINAL SIZE (INCHES)	DRESSED SIZE (INCHES)
1 × 3	3/4 × 2 1/2
1 × 4	3/4 × 3 1/2
1 × 6	3/4 × 5 1/2
1 × 8	3/4 × 7 1/4
1 × 10	3/4 × 9 1/4
1 × 12	3/4 × 11 1/4
2 × 4	1 1/2 × 3 1/2
2 × 6	1 1/2 × 5 1/2
2 × 8	1 1/2 × 7 1/4
2 × 10	1 1/2 × 9 1/4
2 × 12	1 1/2 × 11 1/4
3 × 8	2 1/2 × 7 1/4
3 × 12	2 1/2 × 11 1/4
4 × 12	3 1/2 × 11 1/4
4 × 16	3 1/2 × 15 1/4
6 × 12	5 1/2 × 11 1/2
6 × 16	5 1/2 × 15 1/2
6 × 18	5 1/2 × 17 1/2
8 × 16	7 1/2 × 15 1/2
8 × 20	7 1/2 × 19 1/2
8 × 24	7 1/2 × 23 1/2

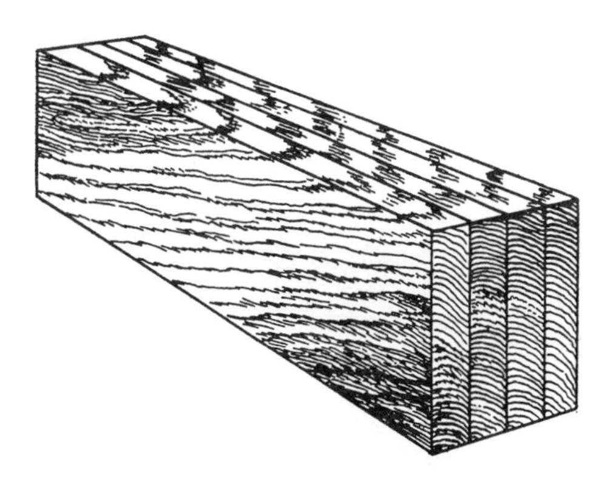

Figure 3-29.—Laminated lumber.

LAMINATED LUMBER

Laminated lumber (figure 3-29) is made of several pieces of lumber held together as a single unit, a process called lamination. Usually 1 1/2-inches thick, the pieces are nailed, bolted, or glued together with the grain of all pieces running parallel. Laminating greatly increases the load-carrying capacity and rigidity of the wood. When extra length is needed, the pieces are spliced—with the splices staggered so that no two adjacent laminations are spliced at the same point. Built-up beams and girders are examples. They are built as shown in figure 3-30, usually nailed or bolted together, and spliced.

Lamination can be used independently or with other materials in the construction of a structural unit. Trusses can be made with lamination for the chords and sawed

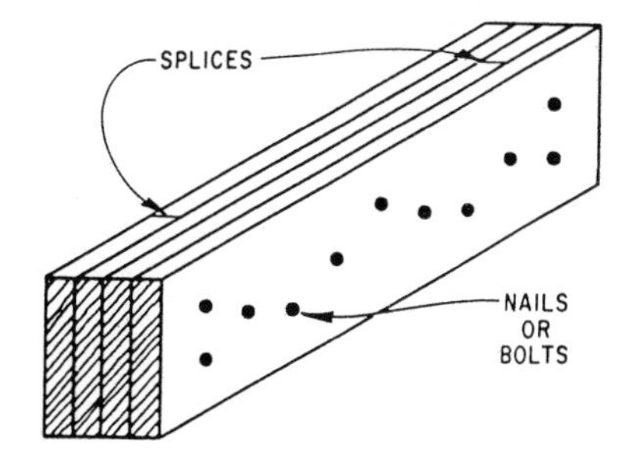

Figure 3-30.—Built-up beam.

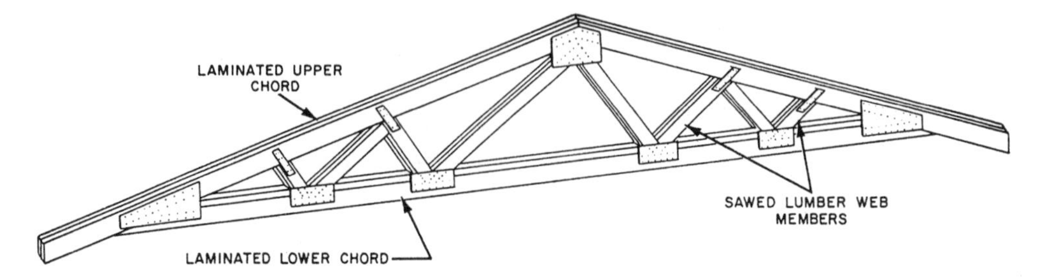

Figure 3-31.—Truss using laminated and sawed lumber.

lumber, or for the web members (figure 3-31). Special beams can be constructed with lamination for the flanges and plywood or sawed lumber, for the web, as shown in figure 3-32. Units, such as plywood box beams and stressed skin panels, can contain both plywood and lamination (figure 3-33).

Probably the greatest use of lamination is in the fabrication of large beams and arches. Beams with spans in excess of 100 feet and depths of 8 1/2 feet have been constructed using 2-inch boards. Laminations this large are factory produced. They are glued together under pressure. Most laminations are spliced using scarf joints (figure 3-34), and the entire piece is dressed to ensure uniform thickness and width. The depth of the lamination is placed in a horizontal position and is usually the full width of the beam (figure 3-35).

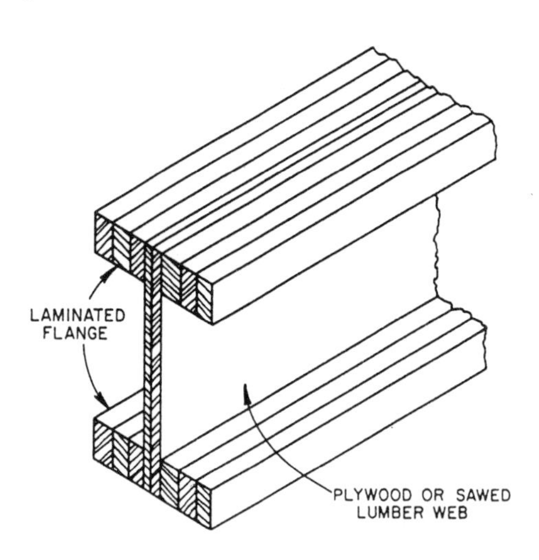

Figure 3-32.—Laminated and sawed lumber or plywood beam.

PLYWOOD

Plywood is constructed by gluing together a number of layers (plies) of wood with the grain direction turned at right angles in each successive layer. This design feature makes plywood highly resistant to splitting. It is one of the strongest building materials available to Seabees. An odd number (3, 5, 7) of plies is used so that they will be balanced on either side of a center core and so that the grain of the outside layers runs in the same direction. The outer plies are called faces or face and back. The next layers under these are called crossbands, and the other inside layer or layers are called the core (figure 3-36). A plywood panel made of three layers would consist of two faces and a core.

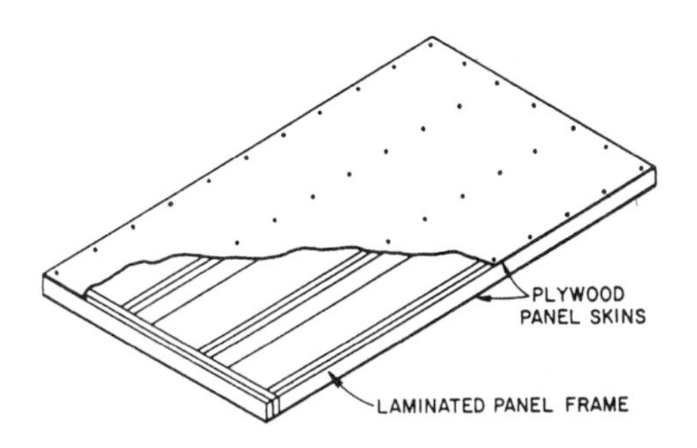

Figure 3-33.—Stressed skin panel.

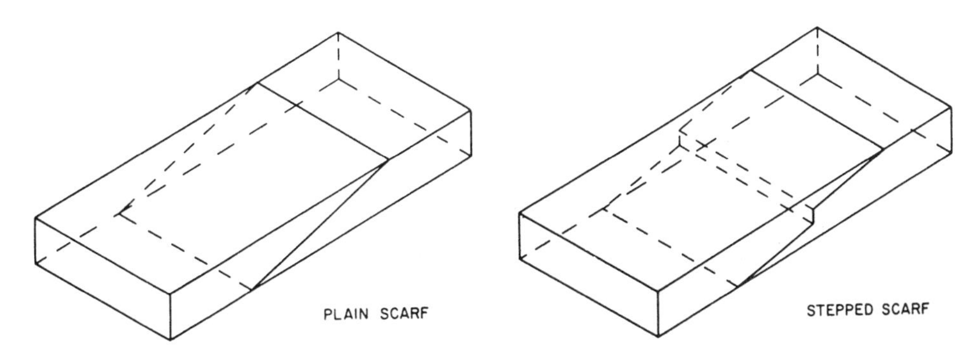

Figure 3-34.—Scarf joints.

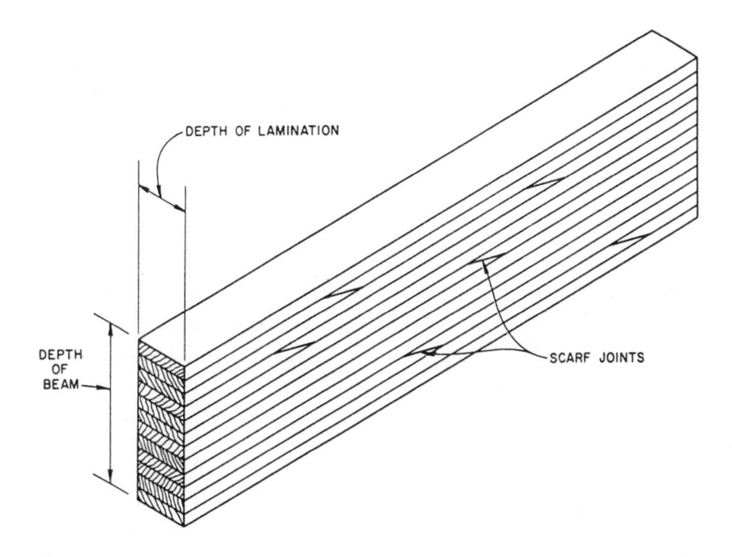

Figure 3-35.—Laminated beam.

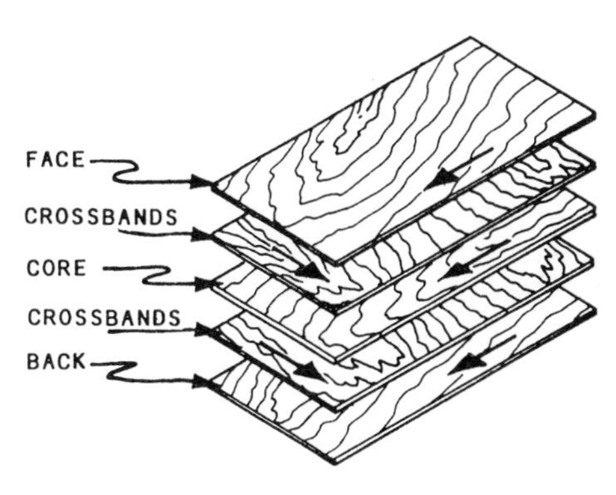

Figure 3-36.—Grain direction in a sheet of plywood.

There are two basic types of plywood: exterior and interior. Exterior plywood is bonded with waterproof glues. It can be used for siding, concrete forms, and other constructions where it will be exposed to the weather or excessive moisture. Interior plywood is bonded with glues that are not waterproof. It is used for cabinets and other inside construction where the moisture content of the panels will not exceed 20 percent.

Plywood is made in thicknesses of 1/8 inch to more than 1 inch, with the common sizes being 1/4, 3/8, 1/2, 5/8, and 3/4 inch. A standard panel size is 4-feet wide by 8-feet long. Smaller size panels are available in the hardwoods.

Table 3-7.—Plywood Veneer Grades

	DESCRIPTION
N	Special order "natural-finish" veneer. Select all heartwood or all sapwood. Free of open defects. Allows some repairs
A	Smooth and paintable. Neatly made repairs permissible. Also used for natural finish in less demanding applications
B	Solid surface veneer. Circular repair plugs and tight knots permitted
C	Knotholes to 1 inch. Occasional knotholes 1/2 inch larger permitted providing total width of all knots and knotholes within a specified section does not exceed certain limits. Limited splits permitted. Minimum veneer permitted in exterior-type plywood
C Plgd	Improved C veneer with splits limited to 1/8 inch in width and knotholes and borer holes limited to 1/4 inch by 1/2 inch
D	Permits knots and knotholes to 2 1/2 inches in width and 1/2 inch larger under certain specified limits. Limited splits permitted

Table 3-8.—Classification of Softwood Plywood Rates Species for Strength and Stiffness

GROUP 1	GROUP 2		GROUP 3	GROUP 4	GROUP 5
Apitong	Cedar,	Maple,	Alder,	Aspen,	Basswood
Beech,	Port	Black	Port	Bigtooth	Fir,
American	Orford	Mengkulang	Birch,	Quaking	Balsam
Birch,	Cypress	Meranti,	Paper	Cativo	Poplar,
Sweet	Douglas fir	Red	Cedar,	Cedar,	Balsam
Yellow	Fir,	Mersawa	Alaska	Incense	
Douglas fir	California	Pine,	Fir,	Western	
Kapur	Red	Pond	Subalpine	Red	
Keruing	Grand	Red	Hemlock,	Cottonwood,	
Larch,	Noble	Virginia	Eastern	Eastern	
Western	Pacific	Western	Maple,	Black	
Maple,	Silver	White	Bigleaf	Western popular	
Sugar	Hemcock,	Sruce,	Pine,	Pine,	
Pine,	White	Red	Jack	Eastern	
Caribbean	Western	Sitka	Lodgepole	White	
Ocote	Lauan,	Sweetgum	Ponderosa	Sugar	
Pine,	Almon	Tamarack	Spruce		
South	Bagtikan	Yellow poplar	Redwood		
Loblolly	Mayapis		Spruce,		
Longleaf	Red lauan		Black		
Shortleaf	Tangile		Engelmann		
Slash	White lauan		White		

Plywood can be worked quickly and easily with common carpentry tools. It holds nails well and normally does not split when nails are driven close to the edges. Finishing plywood presents no unusual problems; it can be sanded or texture coated with a permanent finish or left to weather naturally.

There is probably no other building material as versatile as plywood. It is used for concrete forms, wall and roof sheathing, flooring, box beams, soffits, stressed-skin panels, paneling, shelving, doors, furniture, cabinets, crates, signs, and many other items.

Softwood Plywood Grades

All plywood panels are quality graded based on products standards (currently PS 1/74). The grade of each type of plywood is determined by the kind of veneer (N, A, B, C, or D) used for the face and back of the panel and by the type of glue used in construction. The plywood veneer grades are shown in table 3-7.

Many species of softwood are used in making plywood. There are five separate plywood groups based on stiffness and strength. Group 1 includes the stiffest and strongest; group 5 includes the weakest woods. A listing of groupings and associated woods is shown in table 3-8.

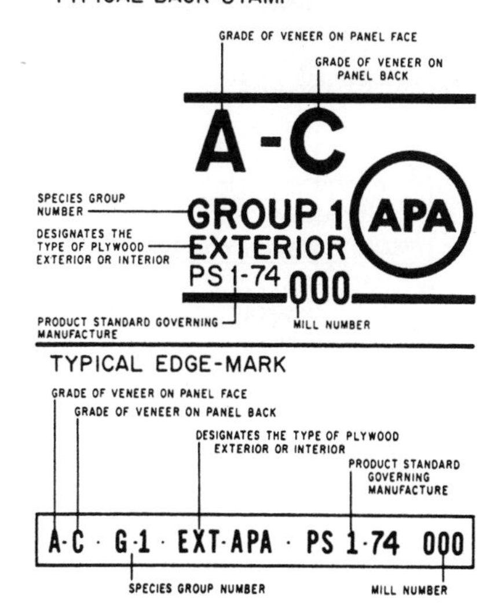

Figure 3-37.—Standard plywood identification symbols.

GRADE/TRADEMARK STAMP.—Construction and industrial plywood panels are marked with different stamps.

Construction Panels.—Grading identification stamps (such as those shown in figure 3-37) indicate the kind and type of plywood. The stamps are placed on the back and sometimes on the edges of each sheet of plywood.

For example, a sheet of plywood having the designation "A-C" would have A-grade veneer on the face and C-grade veneer on the back. Grading is also based on the number of defects, such as knotholes, pitch pockets, splits, discolorations, and patches in the face of each panel. Each panel or sheet of plywood has a stamp on the back that gives all the information you need. Table 3-9 lists some uses for construction-grade plywood.

Industrial Panels.—Structural and sheeting panels have a stamp found on the back. A typical example for an industrial panel grade of plywood is shown in figure 3-38.

The span rating shows a pair of numbers separated by a slash mark (/). The number on the left indicates the maximum recommended span in inches when the plywood is used as roof decking (sheeting). The right-hand number applies to span when the plywood is used as subflooring. The rating applies only when the sheet is placed the long dimension across three or more supports. Generally, the larger the span rating, the greater the stiffness of the panel.

Figure 3-39 lists some typical engineered grades of plywood. Included are descriptions and most common uses.

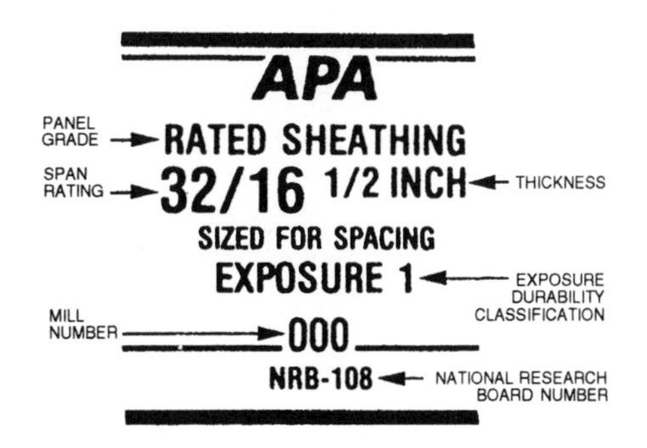

Figure 3-38.—Structural stamp.

Table 3-9.—Plywood Uses

SOFTWOOD PLYWOOD GRADES FOR EXTERIOR USES				
Grade (Exterior)	**Face**	**Back**	**Inner Plies**	**Uses**
A-A	A	A	C	Outdoor where appearance of both sides is important
A-B	A	B	C	Alternate for A-A where appearance of one side is less important
A-C	A	C	C	Siding, soffits, fences. Face is finish grade
B-C	B	C	C	For utility uses, such as farm buildings, some kinds of fences, etc.
C-C (Plugged)	C (Plugged)	C	C	Excellent base for tile and linoleum, backing for wall coverings
C-C	C	C	C	Unsanded, for backing and rough construction exposed to weather
B-B Concrete Forms	B	B	C	Concrete forms. Reuse until wood literally wears out
MDO	B	B or C	C or C-Plugged	Medium density overlay. Ideal base for paint; for siding, built-ins, signs, displays
HDO	A or B	A or B	C-Plugged	High density overlay. Hard surface; no paint needed. For concrete forms, cabinets, counter tops, tanks
SOFTWOOD PLYWOOD GRADES FOR INTERIOR USES				
Grade (Interior)	**Face**	**Back**	**Inner Plies**	**Uses**
A-A	A	A	D	Cabinet doors, built-ins, furniture where both sides will show
A-B	A	B	D	Alternate of A-A. Face is finish grade, back is solid and smooth
A-D	A	D	D	Finish grade face for paneling, built-ins, backing
B-D	B	D	D	Utility grade. One paintable side. For backing, cabinet sides, etc
Standard	C	D	D	Sheathing and structural uses such as temporary enclosures, subfloor. Unsanded

Typical Trademarks	Description and Common Uses	Grade Designation	
APA RATED SHEATHING 48/24 3/4 INCH SIZED FOR SPACING EXTERIOR 000 NRB-108	Exterior sheathing panel for subflooring and wall and roof sheathing, siding on service and farm buildings, crating, pallets, pallet bins, cable reels, etc. Manufactured as conventional veneered plywood. Common thicknesses: 5/16, 3/8, 1/2, 5/8, 3/4.	APA RATED SHEATHING EXT	EXTERIOR USE
APA RATED SHEATHING STRUCTURAL I 42/20 5/8 INCH SIZED FOR SPACING EXTERIOR 000 PS 1-74 C-C NRB-108	For engineered applications in construction and industry where resistance to permanent exposure to weather or moisture is required. Manufactured only as conventional veneered PS 1 plywood. Unsanded. STRUCTURAL I more commonly available. Common thicknesses: 5/16, 3/8, 1/2, 5/8, 3/4. (3)	APA STRUCTURAL I & II RATED SHEATHING EXT	EXTERIOR USE
APA RATED STURD-I-FLOOR 20 oc 19/32 INCH SIZED FOR SPACING EXTERIOR 000 NRB-108	For combination subfloor-underlayment under resilient floor coverings where severe moisture conditions may be present, as in balcony decks. Possesses high concentrated and impact load resistance. Manufactured only as conventional veneered plywood. Available square edge or tongue-and-groove. Common thicknesses: 5/8 (19/32), 3/4 (23/32).	APA RATED STURD-I-FLOOR EXT	EXTERIOR USE
APA RATED SHEATHING 32/16 1/2 INCH SIZED FOR SPACING EXPOSURE 1 000 NRB-108	Specially designed for subflooring and wall and roof sheathing, but can also be used for a broad range of other construction and industrial applications. Can be manufactured as conventional veneered plywood, as a composite, or as a nonveneered panel. For special engineered applications, including high load requirements and certain industrial uses, veneered panels conforming to PS 1 may be required. Specify Exposure 1 when long construction delays are anticipated. Common thicknesses: 5/16, 3/8, 7/16, 1/2, 5/8, 3/4.	APA RATED SHEATHING EXP 1 or 2	PROTECTED OR INTERIOR USE
APA RATED SHEATHING STRUCTURAL I 24/0 3/8 INCH SIZED FOR SPACING EXPOSURE 1 000 PS 1-74 C-D INT/EXT GLUE NRB-108	Unsanded all-veneer PS 1 plywood grades for use where strength properties are of maximum importance: structural diaphragms, box beams, gusset plates, stressed-skin panels, containers, pallet bins. Made only with exterior glue (Exposure 1). STRUCTURAL I more commonly available. Common thicknesses: 5/16, 3/8, 1/2, 5/8, 3/4. (3)	APA STRUCTURAL I & II RATED SHEATHING EXP 1	PROTECTED OR INTERIOR USE
APA RATED STURD-I-FLOOR 24 oc 23/32 INCH SIZED FOR SPACING T&G NET WIDTH 47-1/2 EXPOSURE 1 000 NRB-108	For combination subfloor-underlayment. Provides smooth surface for application of resilient floor covering and possesses high concentrated and impact load resistance. Can be manufactured as conventional veneered plywood, as a composite, or as a nonveneered panel. Available square edge or tongue-and-groove. Specify Exposure 1 when long construction delays are anticipated. Common thicknesses: 5/8 (19/32), 3/4 (23/32).	APA RATED STURD-I-FLOOR EXP 1 or 2	PROTECTED OR INTERIOR USE
APA RATED STURD-I-FLOOR 48 oc 1-1/8 INCH (2-4-1) SIZED FOR SPACING EXPOSURE 1 T&G 000 INT/EXT GLUE NRB-108 FHA-UM-66	For combination subfloor-underlayment on 32- and 48-inch spans and for heavy timber roof construction. Provides smooth surface for application of resilient floor coverings and possesses high concentrated and impact load resistance. Manufactured only as conventional veneered plywood and only with exterior glue (Exposure 1). Available square edge or tongue-and-groove. Thickness: 1-1/8.	APA RATED STURD-I-FLOOR 48 oc (2-4-1) EXP 1	PROTECTED OR INTERIOR USE

(1) Specific grades, thicknesses, constructions and exposure durability classifications may be in limited supply in some areas. Check with your supplier before specifying.

(2) Specify Performance-Rated Panels by thickness and Span Rating.

(3) All plies in STRUCTURAL I panels are special improved grades and limited to Group 1 species. All plies in STRUCTURAL II panels are special improved grades and limited to Group 1, 2, or 3 species.

Figure 3-39.—List of engineered grade of softwood plywood.

Exposure Ratings.—The grade/trademark stamp lists the exposure durability classification for plywood. There are two basic types or ratings: exterior and interior. The exterior type has a 100-percent waterproof glue line, and the interior type has a highly moisture-resistant glue line. However, panels can be manufactured in three exposure durability classifications: Exterior, Exposure 1, and Exposure 2.

Panels marked "Exterior" can be used where there is continual exposure to weather and moisture. Panels marked "Exposure 1" can withstand moisture during extended periods, but they should be used only indoors. Panels marked "Exposure 2" can be used in protected locations. They may be subjected to some water leakage or high humidity but generally should be protected from weather.

Most plywood is made with waterproof exterior glue. However, interior panels may be made with intermediate or interior glue.

Hardwood Plywood Grades

Hardwood plywood panels are primarily used for door skins, cabinets, and wall paneling. The Hardwood Plywood Manufacturers' Association has established a grading system with the following grades: premium (A), good grade (1), sound grade (2), utility grade (3), and backing grade (4). For example, an A-3 grade hardwood plywood would have a premium face and a utility back. A 1-1 grade would have a good face and a good back.

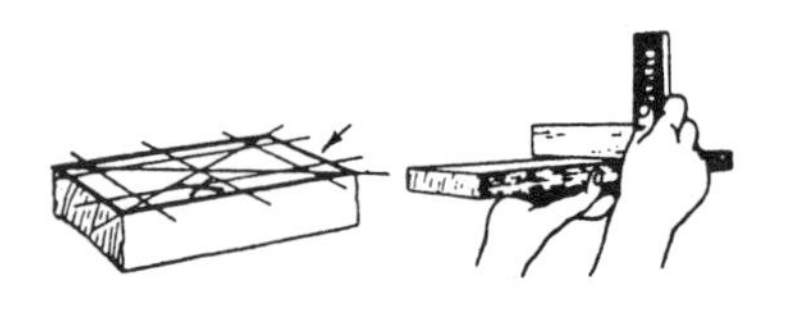

1. WORK FACE

PLANE ONE BROAD SURFACE SMOOTH AND STRAIGHT. TEST IT CROSSWISE, LENGTHWISE, AND FROM CORNER TO CORNER. MARK THE WORK FACE X.

2. WORK EDGE

PLANE ONE EDGE SMOOTH, STRAIGHT AND SQUARE TO THE WORK FACE. TEST IT FROM THE WORK FACE. MARK THE WORK EDGE X.

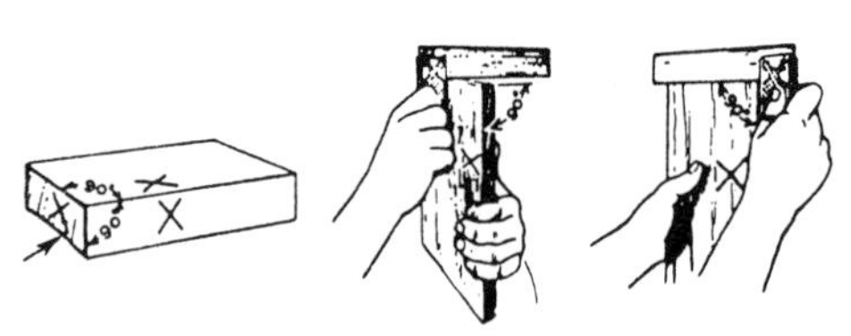

3. WORK END

PLANE ONE END SMOOTH AND SQUARE. TEST IT FROM THE WORK FACE AND WORK EDGE. MARK THE WORK END X.

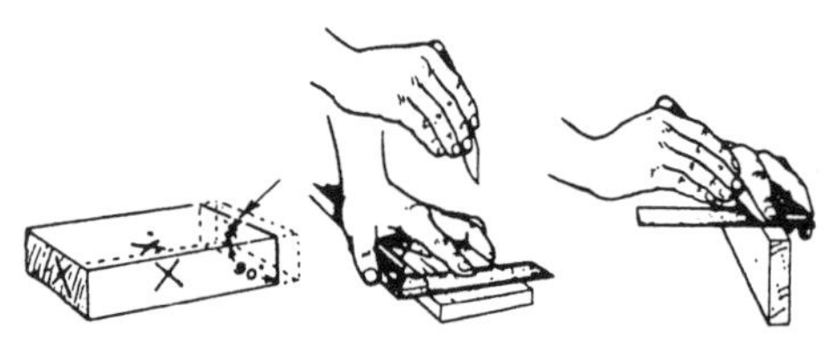

4. SECOND END

MEASURE LENGTH AND SCRIBE AROUND THE STOCK A LINE SQUARE TO THE WORK EDGE AND WORK FACE. SAW OFF EXCESS STOCK NEAR THE LINE AND PLANE SMOOTH TO THE SCRIBED LINE. TEST THE SECOND END FROM BOTH THE WORK FACE AND THE WORK EDGE.

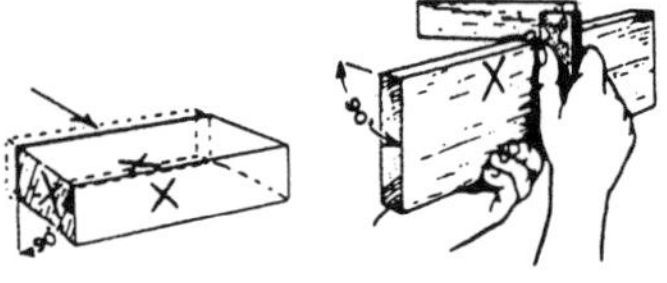

5. SECOND EDGE

FROM THE WORK EDGE GAUGE A LINE FOR WIDTH ON BOTH FACES. PLANE SMOOTH, STRAIGHT, SQUARE AND TO THE GAUGE LINE. TEST THE SECOND EDGE FROM THE WORK FACE.

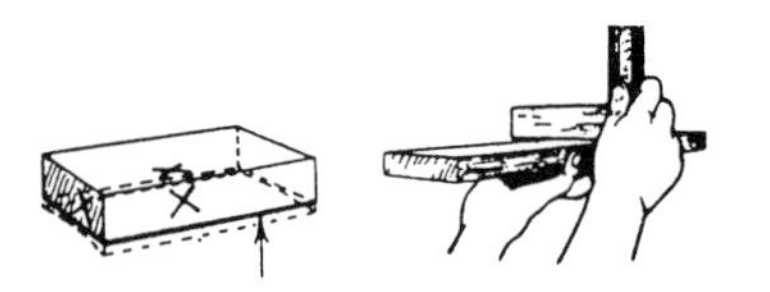

6. SECOND FACE

FROM THE WORK FACE GAUGE A LINE FOR THICKNESS AROUND THE STOCK. PLANE THE STOCK TO THE GAUGE LINE. TEST THE SECOND FACE AS THE WORK FACE IS TESTED.

Figure 3-40.—Planing and squaring to dimensions.

Figure 3-41.—90° plain butt joints.

WOODWORKING METHODS

LEARNING OBJECTIVE: Upon completing this section, you should be able to identify the various methods and joints associated with woodworking.

In the following section, we will cover some of the methods used by Builders in joining wood.

PLANING AND SQUARING TO DIMENSIONS

Planing and squaring a small piece of board to dimensions is what you might call the first lesson in woodworking. Like many other things you may have tried to do, it looks easy until you try it. The six major steps in this process are illustrated and described in figure 3-40. You should practice these steps until you can get a smooth, square board with a minimum of planing.

JOINTS AND JOINING

One basic skill of woodworking is the art of joining pieces of wood to form tight, strong, well-made joints. The two pieces that are to be joined together are called members. The two major steps in making joints are (1) laying out the joint on the ends, edges, or faces and (2) cutting the members to the required shapes for joining.

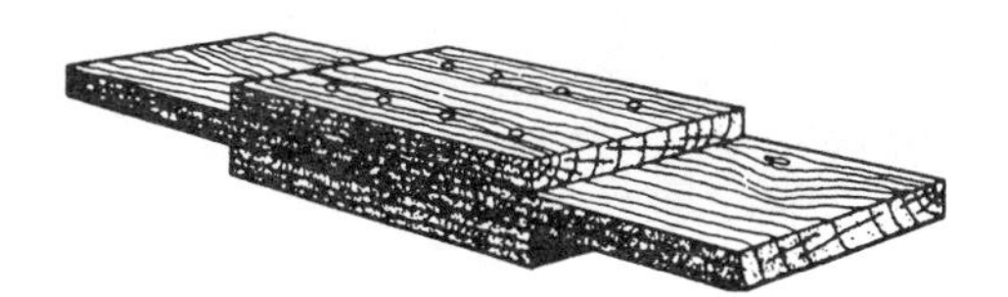

Figure 3-42.—End butt joints with fishplates.

The instruments normally used for laying out joints are the try square, miter square, combination square, the sliding T-bevel, the marking or mortising gauge, a scratch awl, and a sharp pencil or knife for scoring lines. For cutting the more complex joints by hand, the backsaw dovetail saw and various chisels are essential. The rabbet-and-fillister plane (for rabbet joints) and the router plane (for smoothing the bottoms of dadoes and gains) are also helpful.

Simple joints, like the butt (figures 3-41 and 3-42), the lap (figure 3-43), and the miter joints

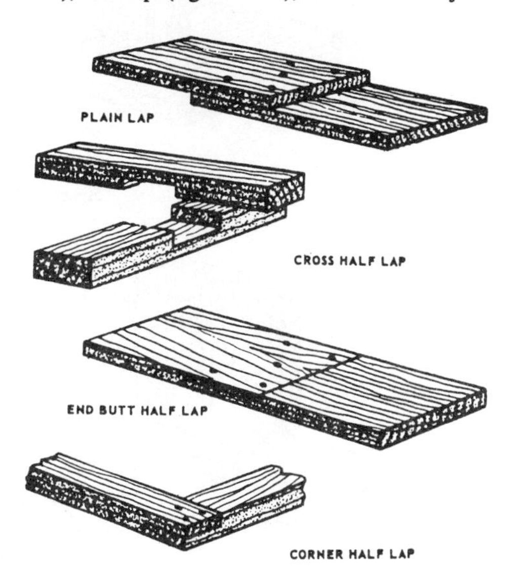

Figure 3-43.—Lap joints.

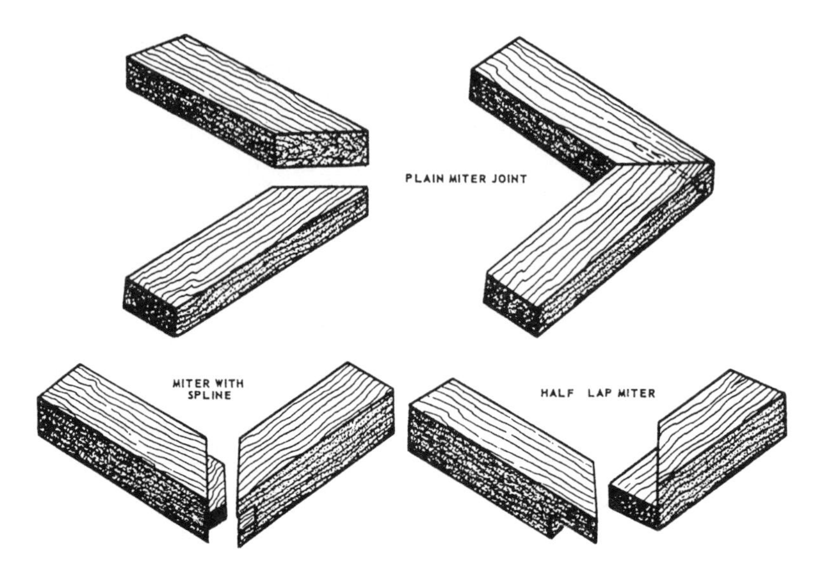

Figure 3-44.—Miter joints.

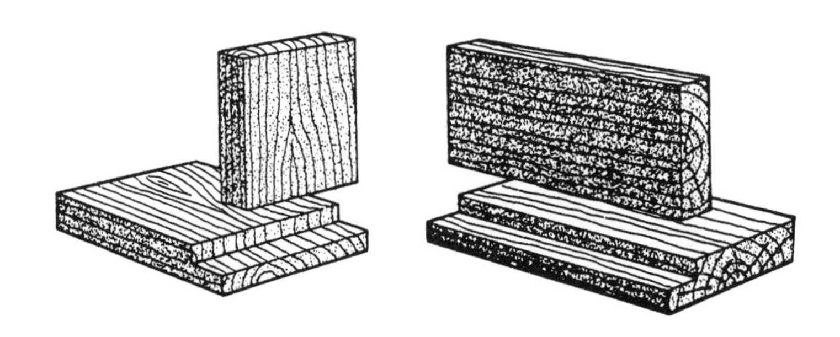

Figure 3-45.—Rabbet joints.

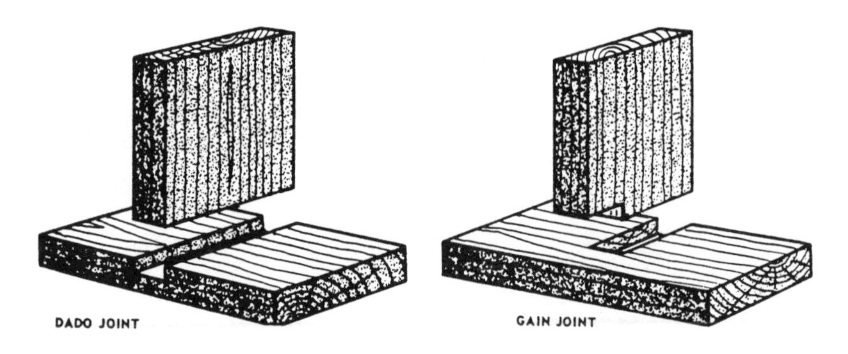

Figure 3-46.—Dado and gain joints.

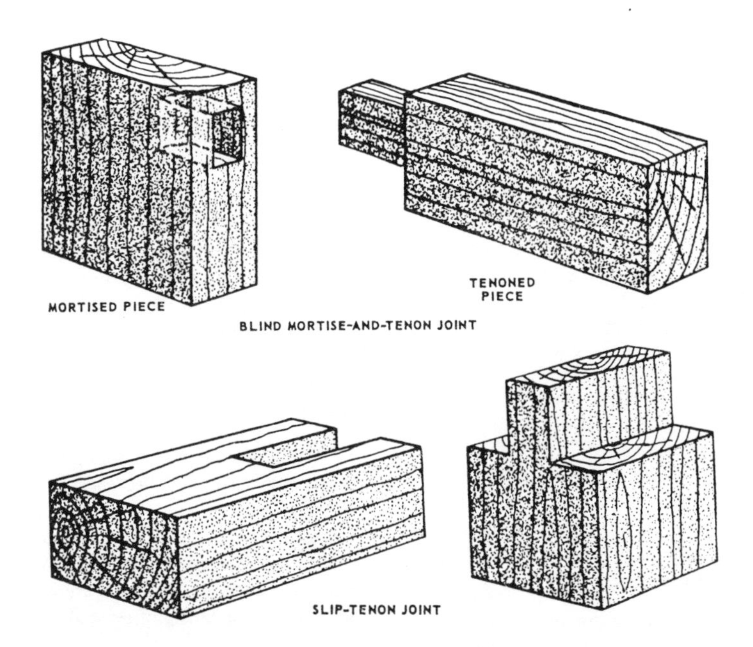

Figure 3-47.—Tenon joints.

(figure 3-44), are used mostly in rough or finish carpentry though they may be used occasionally in millwork and furniture making. More complex joints, like the rabbet joints (figure 3-45), the dado and gain joints (figure 3-46), the blind mortise-and-tenon and slip-tenon joints (figure 3-47), the box corner joint (figure 3-48), and the dovetail joints (figure 3-49), are used mostly in making furniture and cabinets and in

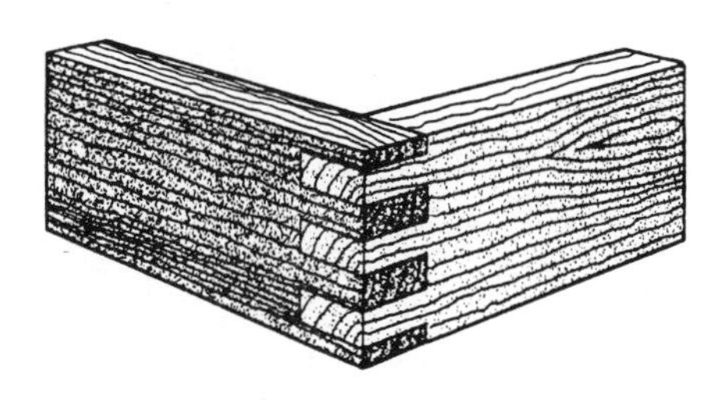

Figure 3-48.—Box corner joint.

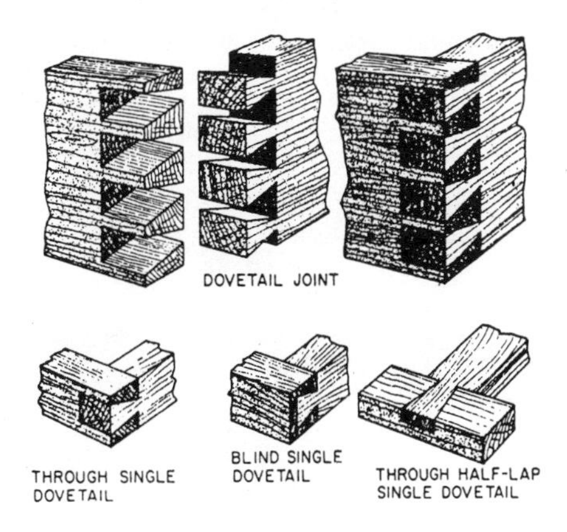

Figure 3-49.—Dovetail joints.

millwork. Of the edge joints shown in figure 3-50, the dowel and spline joints are used mainly in furniture and cabinet work, whereas the plain butt and the tongue-and-groove joints are used in practically all types of woodworking.

The joints used in rough and finished carpentry are, for the most part, simply nailed together. Nails in a 90° plain butt joint can be driven through the member abutted against and into the end of the abutting member. The joints can also be toenailed at an angle through the faces of the abutting member into the face of the member abutted against, as shown in figure 3-51. Studs and joists are usually toenailed to soleplates and sills.

The more complex furniture and cabinet-making joints are usually fastened with glue. Additional strength can be provided by dowels, splines, corrugated fasteners, keys, and other types of joint fasteners. In the dado joint, the gain joint, the mortise-and-tenon joint, the box corner joint, and the dovetail joint, the interlocking character of the joint is an additional factor in fastening.

All the joints we have been mentioned can be cut either by hand or by machine. Whatever the method used and whatever the type of joint, remember: To ensure a tight joint, always cut on the waste side of the line; never on the line itself. Preliminary grooving on the waste side of the line with a knife or chisel will help a backsaw start smoothly.

Half-Lap Joints

For half-lap joints, the members to be jointed are usually of the same thickness, as shown in figure 3-43.

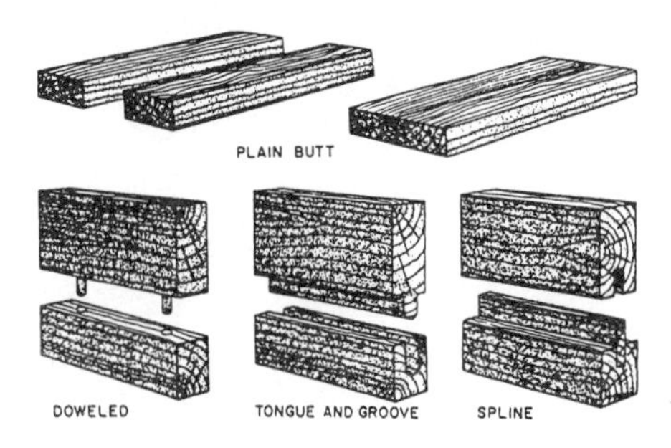

Figure 3-50.—Edge joints.

The method of laying out and cutting an end butt half lap (figure 3-43) is to measure off the desired amount of lap from each end of each member and square a line all the way around at this point. For a corner half lap (figure 3-43), measure off the width of the member from the end of each member and square a line all the way around. These lines are called shoulder lines.

Next, select the best surface for the face and set a marking gauge to one-half the thickness and score a line (called the cheek line) on the edges and end of each member from the shoulder line on one edge to the shoulder line on the other edge. Be sure to gauge the cheek line from the face of each member. This ensures that the faces of each member will be flush after the joints are cut.

Next, make the shoulder cuts by cutting along the waste side of the shoulder lines down to the waste side of the cheek line. Then, make the cheek cuts along the waste side of the cheek lines. When all cuts have been made, the members should fit together with faces, ends, and edges flush or near enough to be made flush with the slight paring of a wood chisel.

Other half-lap joints are laid out in a similar manner. The main difference is in the method of cutting. A cross half-lap joint may best be cut with a dado head or wood chisel rather than a handsaw. Others may easily be cut on a bandsaw, being certain

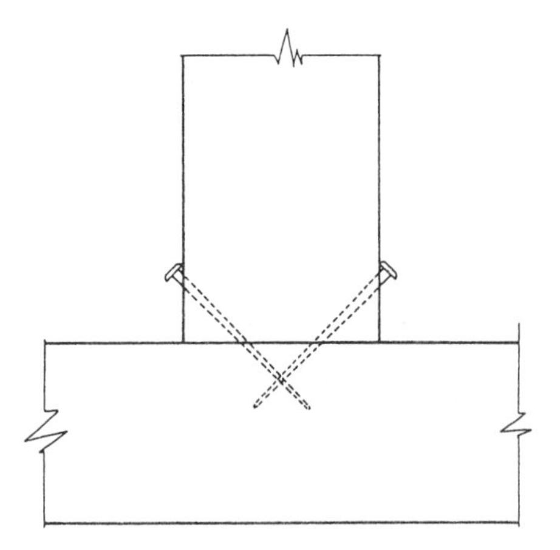

Figure 3-51.—Toenailing.

to cut on the waste side of the lines and making all lines from the face of the material.

Miter Joints

A miter joint is made by mitering (cutting at an angle) the ends or edges of the members that are to be joined together (figure 3-44). The angle of the miter cut is one-half of the angle formed by the joined members. In rectangular mirror frames, windows, door casing boxes, and the like, adjacent members form a 90° angle, and, consequently, the correct angle for mitering is one-half of 90°, or 45°. For members forming an equal-sided figure with other than four sides (such as an octagon or a pentagon), the correct mitering angle can be found by dividing the number of sides the figure will have into 180° and subtracting the result from 90°. For an octagon (an eight-sided figure), determine the mitering angle by subtracting from 90° 180° divided by 8 or 90° minus 22.5° equals 67.5°. For a pentagon (a five-sided figure), the angle is

$$90° - (180° \div 5) \; or \; 90° - 36° = 54°$$

Members can be end mitered to 45° in the wooden miter box and to any angle in the steel miter box by setting the saw to the desired angle, or on the circular saw, by setting the miter gauge to the desired angle. Members can be edge mitered to any angle on the circular saw by tilting the saw to the required angle.

Sawed edges are sometimes unsuitable for gluing. However, if the joint is to be glued, the edges can be mitered on a jointer, as shown in figure 3-52.

Figure 3-52.—Beveling on a jointer for a mitered edge joint.

SAFETY NOTE

This is a dangerous operation and caution should be taken.

Since abutting surfaces of end-mitered members do not hold well when they are merely glued, they should be reinforced. One type of reinforcement is the corrugated fastener. This is a corrugated strip of metal with one edge sharpened for driving into the joint. The fastener is placed at a right angle to the line between the members, half on one member and half on the other, and driven down flush with the member. The corrugated fastener mars the appearance of the surface into which it is driven; therefore, it is used only on the backs of picture frames and the like.

A more satisfactory type of fastener for a joint between end-mitered members is the slip feather. This is a thin piece of wood or veneer that is glued into a kerf cut in the thickest dimension of the joint. First, saw about halfway through the wood from the outer to the inner corner, then apply glue to both sides of the slip feather, pushing the slip feather into the kerf. Clamp it tightly and allow the glue to dry. After it has dried, remove the clamp and chisel off the protruding portion of the slip feather.

A joint between edge-mitered members can also be reinforced with a spline. This is a thick piece of wood that extends across the joint into grooves cut in the abutting surfaces. A spline for a plain miter joint is shown in figure 3-44. The groove for a spline can be cut either by hand or by a circular saw.

Grooved Joints

A three-sided recess running with the grain is called a groove, and a recess running across the grain is called a dado. A groove or dado that does not extend all the way across the wood is called a stopped groove or a stopped dado. A stopped dado is also known as a gain (figure 3-46). A two-sided recess running along an edge is called a rabbet T (figure 3-45). Dadoes, gains, and rabbets are not, strictly speaking, grooves; but joints that include them are generally called grooved joints.

A groove or dado can be cut with a circular saw as follows: Lay out the groove or dado on the end wood (for a groove) or edge wood (for a dado) that will first come in contact with the saw. Set the saw to the desired depth of the groove above the table, and set the fence at a distance from the saw that will cause the first cut to run on the waste side of the line that indicates the left side of the groove. Start the saw and bring the wood into light contact with it; then stop the saw and examine the layout to ensure the cut will be on the waste side of the line. Readjust the fence, if necessary. When the position of the fence is right, make the cut. Then, reverse the wood and proceed to set and test as before for the cut on the opposite side of the groove. Make as many recuts as necessary to remove the waste stock between the side kerfs.

The procedure for grooving or dadoing with the dado head is about the same, except that, in many cases, the dado head can be built up to take out all the waste in a single cut. The two outside cutters alone will cut a groove 1/4 inch wide. Inside cutters vary in thickness from 1/16 to 1/4 inch.

A stopped groove or stopped dado can be cut on the circular saw, using either a saw blade or a dado head, as follows: If the groove or dado is stopped at only one end, clamp a stop block to the rear of the table in a position that will stop the wood from being fed any farther when the saw has reached the place where the groove or dado is supposed to stop. If the groove or dado is stopped at both ends, clamp a stop block to the rear of the table and a starting block to the front. The starting block should be placed so the saw will contact the place where the groove is supposed to start when the infeed end of the piece is against the block. Start the cut by holding the wood above the saw, with the infeed end against the starting block and the edge against the fence. Then, lower the wood gradually onto the saw, and feed it through to the stop block.

A rabbet can be cut on the circular saw as follows: The cut into the face of the wood is called the shoulder cut, and the cut into the edge or end, the cheek cut. To make the shoulder cut (which should be made first), set the saw to extend above the table a distance equal to the desired depth of the cheek. Be sure to measure this distance from a sawtooth set to the left, or away from the ripping fence. If you measure it from a tooth set to the right or toward the fence, the cheek will be too deep by an amount equal to the width of the saw kerf.

By using the dado head, you can cut most ordinary rabbets in a single cut. First, build up a dado head equal in thickness to the desired width of the cheek. Next, set the head to protrude above the table a distance equal to the desired depth of the shoulder. Clamp a 1-inch board to the fence to serve as a guide for the piece, and set the fence so the edge of the board barely contacts the right side of the dado head. Set the piece against the miter gauge (set at 90°), hold the edge or end to be rabbeted against the 1-inch board, and make the cut.

On some jointers, a rabbeting ledge attached to the outer edge of the infeed table can be depressed for rabbeting, as shown in figure 3-53. The ledge is located on the outer end of the cutterhead. To rabbet on a jointer of this type, you depress the infeed table and the rabbeting ledge the depth of the rabbet below the outfeed table, and set the fence the width of the rabbet away from the outer end of the cutterhead. When the piece is fed through, the unrabbeted part feeds onto the rabbeting ledge. The rabbeted portion feeds onto the outfeed table.

Various combinations of the grooved joints are used in woodworking. The tongue-and-groove joint is a combination of the groove and the rabbet, with the tongued member rabbeted on both faces. In some types of paneling, the tongue is made by rabbeting only one face. A tongue of this kind is called a barefaced tongue. A joint often used in making boxes, drawers, and cabinets is the dado and rabbet joint, shown in figure 3-54. As you can see, one of the members is rabbeted on one face to form a barefaced tongue.

Mortise-and-Tenon Joints

The mortise-and-tenon joint is most frequently used in furniture and cabinet work. In the blind mortise-and-tenon joint, the tenon does not penetrate

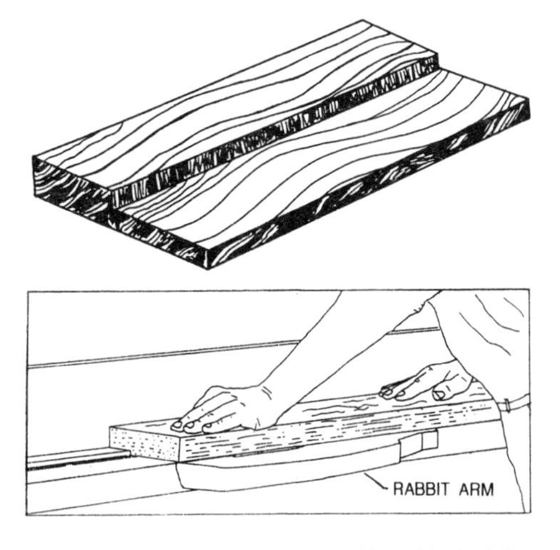

Figure 3-53.—Rabbeting on a jointer with a rabbeting ledge.

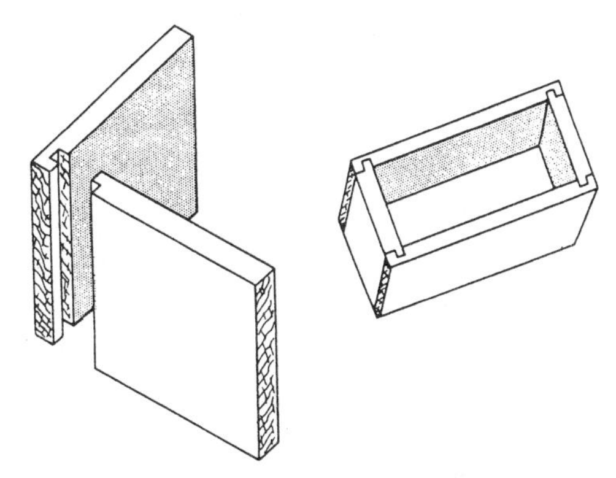

Figure 3-54.—Dado and rabbet joint.

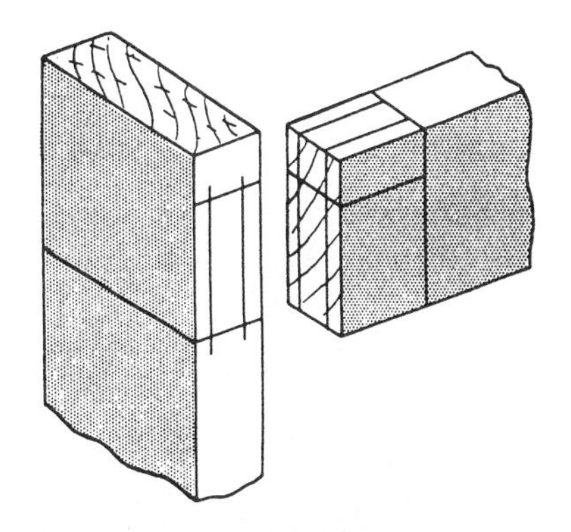

Figure 3-56.—Layout of stub mortise-and-tenon joint.

all the way through the mortised member (figure 3-47).

A joint in which the tenon does penetrate all the way through is a through mortise-and-tenon joint (figure 3-55). Besides the ordinary stub joint (view A), there are haunched joints (view B) and table-haunched joints (view C). Haunching and table-haunching increase the strength and rigidity of the joint.

The layout procedure for an ordinary stub mortise-and-tenon joint is shown in figure 3-56. The shoulder and cheek cuts of the tenon are shown in figures 3-57 and 3-58. To maintain the stock upright while making the cheek cuts, use a push board similar to the one shown in figure 3-58. Tenons can also be cut with a dado head by the same method previously described for cutting end half-lap joints.

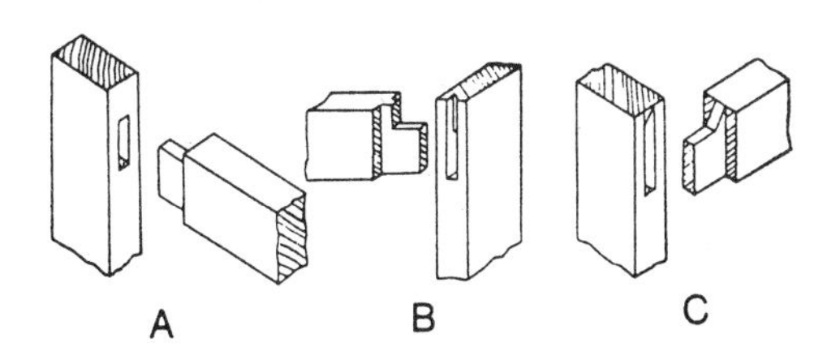

Figure 3-55.—Stub (view A), haunched (view B), and table-haunched (view C) mortise-and-tenon joints.

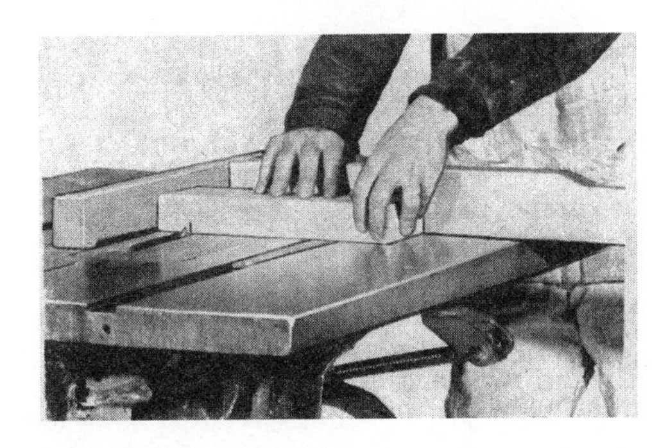

103.22
Figure 3-57.—Making tenon shoulder cut on a table saw.

103.23
Figure 3-58.—Making tenon cheek cut on a table saw using a push board.

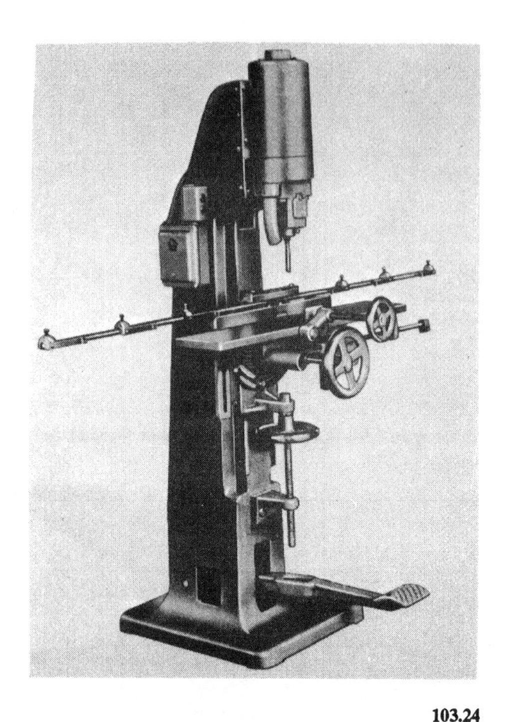

103.24

Figure 3-59.—Hollow-chisel mortising machine.

Mortises are cut mechanically on a hollow-chisel mortising machine like the one shown in figure 3-59. The cutting mechanism on this machine consists of a boring bit encased in a square, hollow, steel chisel. As the mechanism is pressed into the wood, the bit takes out most of the waste while the chisel pares the sides of the mortise square. Chisels come in various sizes, with corresponding sizes of bits to match. If a mortising machine is not available, the same results can be attained by using a simple drill press to take out most of the waste and a hand chisel, for paring the sides square.

In some mortise-and-tenon joints, such as those between rails and legs in tables, the tenon member is much thinner than the mortise member. Sometimes a member of this kind is too thin to shape in the customary manner, with shoulder cuts on both faces. When this is the case, a barefaced mortise-and-tenon joint can be used. In a barefaced joint, the tenon member is shoulder cut on one side only. The cheek on the opposite side is simply a continuation of the face of the member.

Mortise-and-tenon joints are fastened with glue and with additional fasteners, as required.

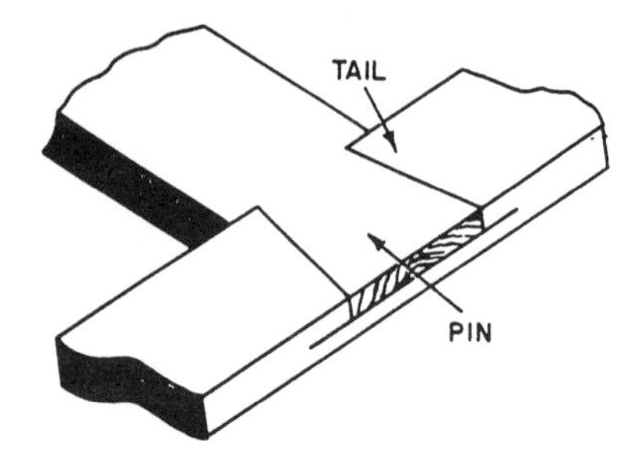

Figure 3-60.—Dovetail half-lap joint.

Dovetail Joints

The dovetail joint (figure 3-49) is the strongest of all the woodworking joints. It is used principally for joining the sides and ends of drawers in fine grades of furniture and cabinets. In the Seabee units, you will seldom use dovetail joints since they are laborious and time-consuming to make.

A through dovetail joint is a joint in which the pins pass all the way through the tail member. Where the pins pass only part way through, the member is known as a blind dovetail joint.

The simplest of the dovetail joints is the dovetail half-lap joint, shown in figure 3-60. Figure 3-61 shows how this type of joint is laid out, and figure 3-62 shows the completed joint.

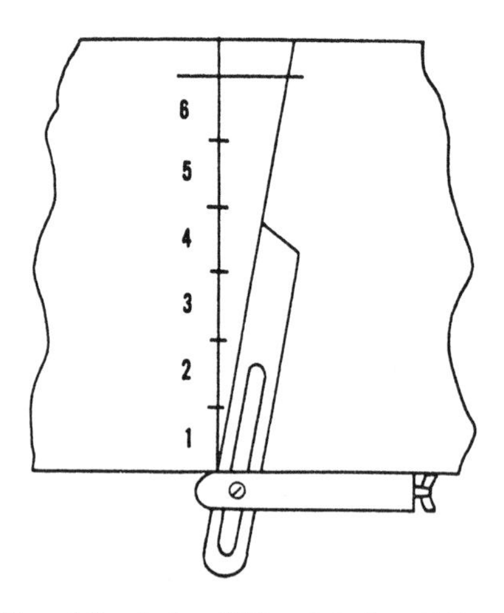

Figure 3-61.—Laying off 10° angle for dovetail joint.

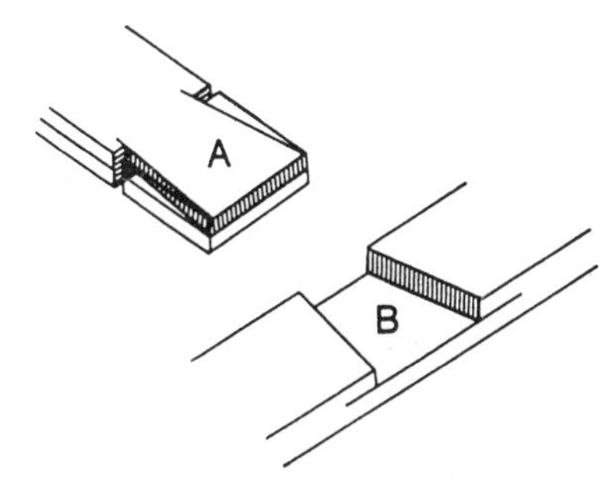

Figure 3-62.—Making a dovetail half-lap joint.

A multiple dovetail joint is shown in figure 3-63; figure 3-64 indicates how the waste is chiseled from the multiple joint.

Box Corner Joints

With the exception of the obvious difference in the layout, the box corner joint (figure 3-48) is made in a similar manner as the through-multiple-dovetail joint.

Coping Joints

Inside corner joints between molding trim members are usually made by butting the end of one member against the face of the other. Figure 3-65

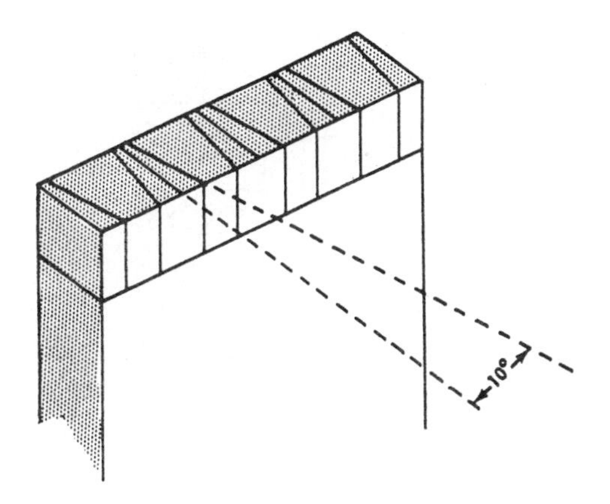

Figure 3-63.—Laying out a pin member for a through-multiple-dovetail joint.

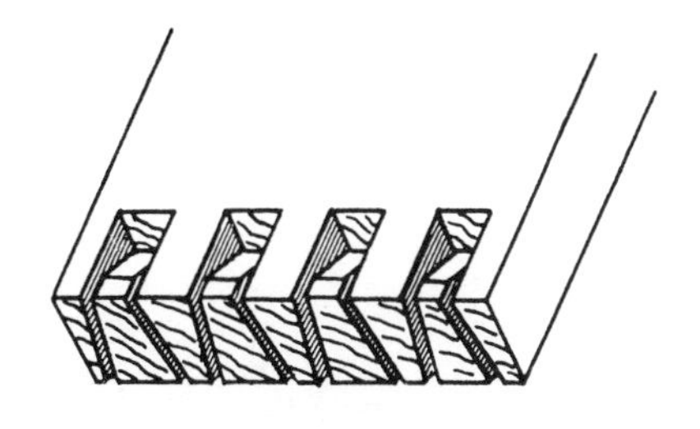

Figure 3-64.—Chiseling out waste in a through-multiple-dovetail joint.

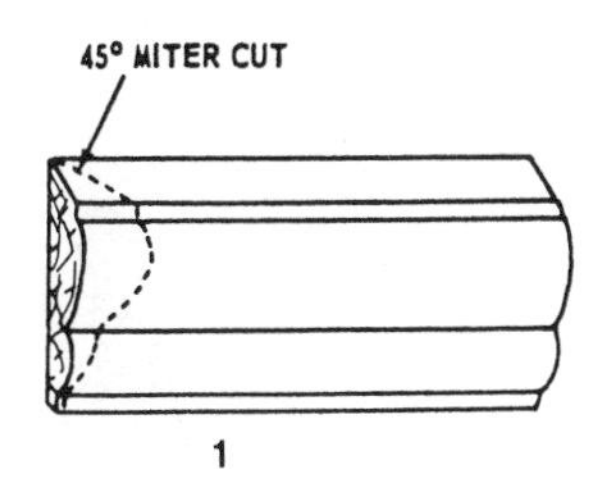

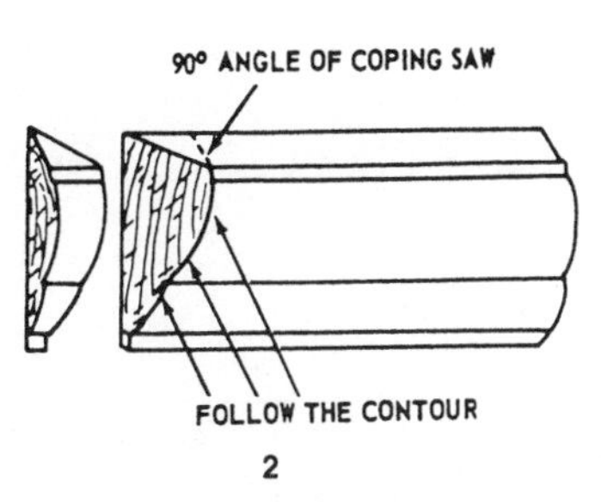

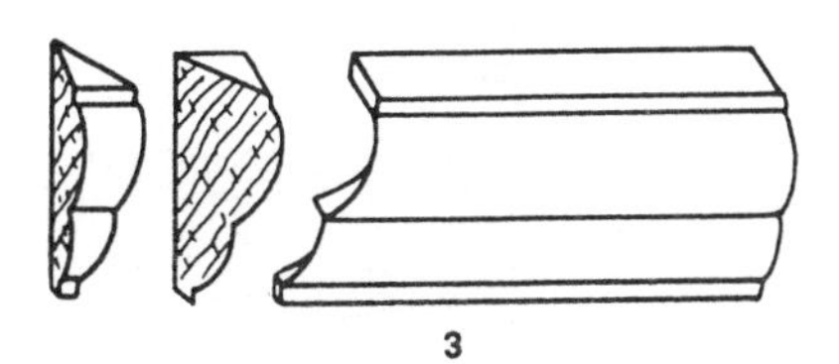

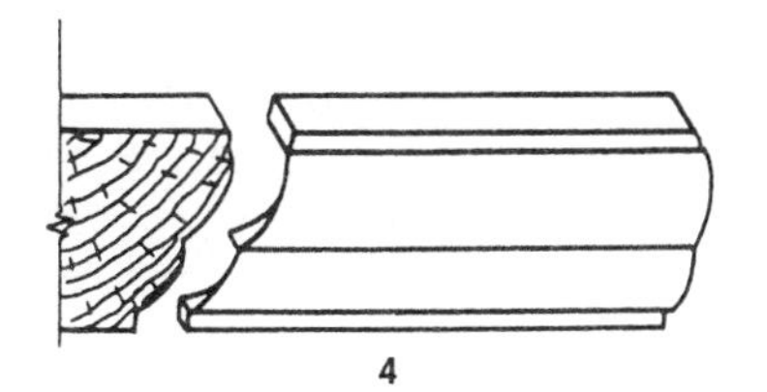

Figure 3-65.—Making a coping joint.

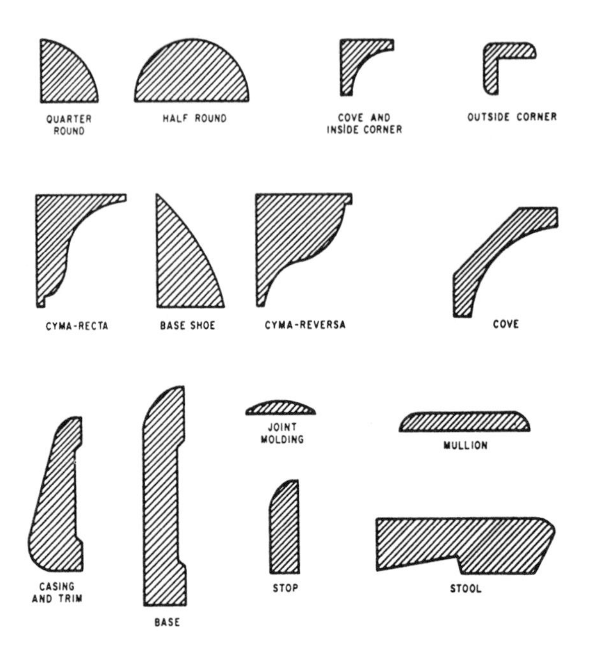

Figure 3-66.—Simple molding and trim shapes.

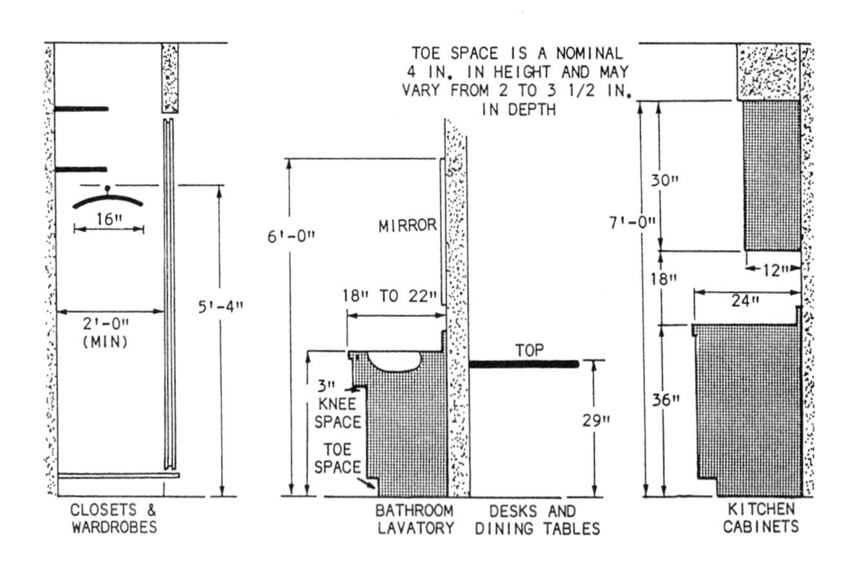

Figure 3-67.—Typical dimensions for cabinetwork.

shows the method of shaping the end of the abutting member to fit the face of the other member. First, saw off the end of the abutting member square, as you would for an ordinary butt joint between ordinary flat-faced members. Then, miter the end to 45°, as shown in the first and second views of figure 3-65. Set the coping saw at the top of the line of the miter cut, hold the saw at 90° to the lengthwise axis of the piece, and saw off the segment shown in the third view, following closely the face line left by the 45° miter cut. The end of the abutting member will then match the face of the other member, as shown in the third view. A joint made in this manner is called a coping joint. You will have to cut coping joints on a large variety of moldings. Figure 3-66 shows the simplest and most common moldings and trims used in woodworking.

MILLWORK

LEARNING OBJECTIVE: Upon completing this section, you should be able to recognize the various types of millwork products and procedures.

As a general term, millwork usually embraces most wood products and components that require manufacturing. It not only includes the interior trim and doors, but also kitchen cabinets and similar units. Most of these units are produced in a millwork manufacturing plant and are ready to install. Figure 3-67 is an example of the dimensions you might be working with.

BUILDING CABINETS IN PLACE

One of the most common ways of building cabinets, such as those shown in figure 3-68, is to cut

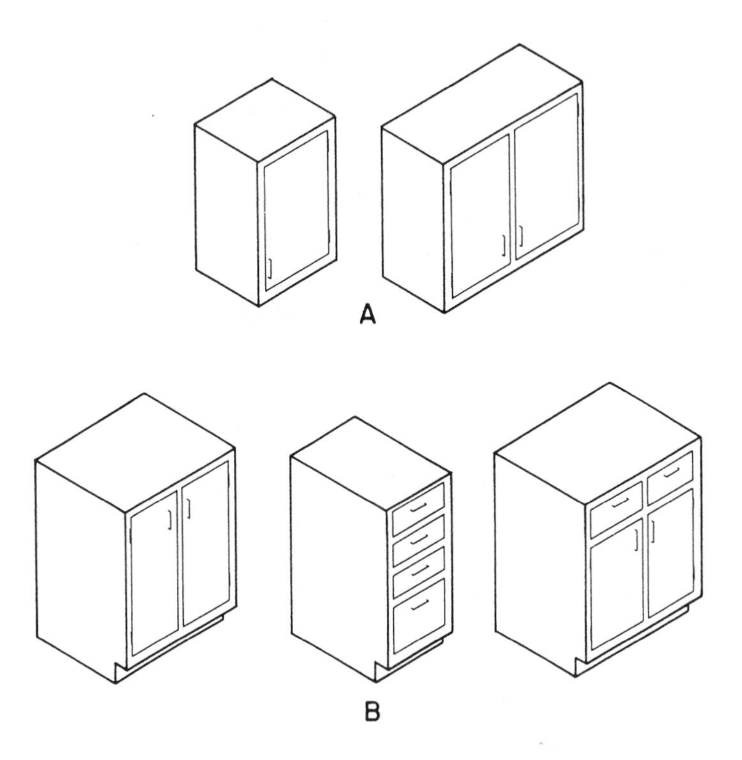

Figure 3-68.—Typical kitchen cabinets: wall (view A) and base (view B).

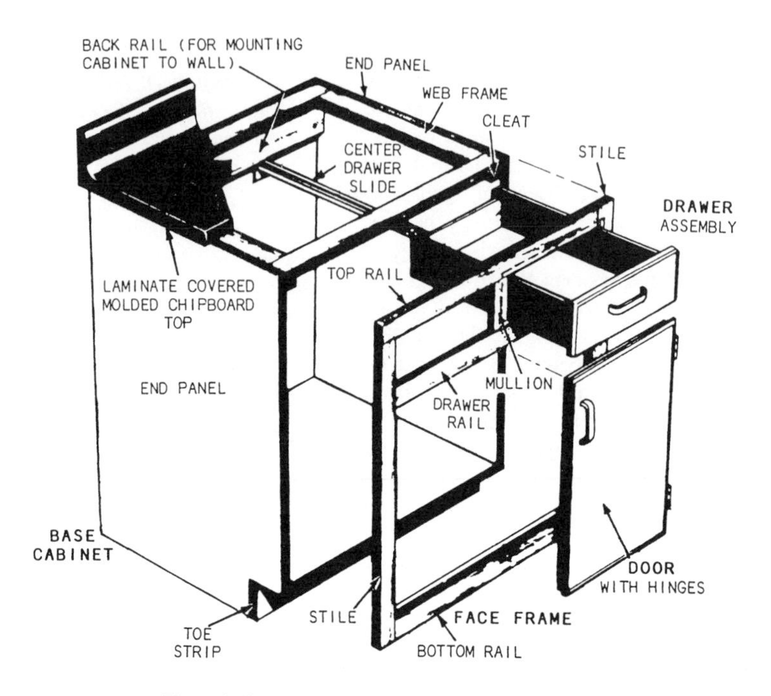

Figure 3-69.—Typical frame construction of a cabinet.

the pieces (figure 3-69) and assemble them in place. Think of building in-place cabinets in four steps.

1. Construct the base first. Use straight 2-by-4 lumber for the base. Nail the lumber to the floor and to a strip attached to the wall. If the floor is not level, place shims under the various members of the base. Later, you can face any exposed 2-by-4 surfaces with a finished material, or the front edge can be made of a finished piece, such as base molding.

2. Next, cut and install the end panels. Attach a strip along the wall between the end panels and level with the top edge. Be sure the strip is level throughout its length. Nail it securely to the wall studs.

3. Cut the bottom panels and nail them in place on the base. Follow this with the installation of the partitions, which are notched at the back corner of the top edge so they will fit over the wall strip.

4. Finally, plumb the front edge of the partitions and end panels. Secure them with temporary strips nailed along the top.

Wall units are made using the same basic steps as the base units. You should make your layout lines directly on the ceiling and wall. Nail the mounting strips through the wall into the studs. At the inside corners, end panels can be attached directly to the wall.

Remember to make your measurements for both base and wall units carefully, especially for openings for built-in appliances. Refer frequently to your drawings and specifications to ensure accuracy.

Shelves

Shelves are an integral part of cabinetmaking, especially for wall units. Cutting dadoes into cabinet walls to fit in shelves may actually strengthen the cabinet (figure 3-70.) When adding shelves, try to make them adjustable so the storage space can be altered as needed. Figure 3-71 shows two methods of installing adjustable shelves.

Whatever method of shelf support you use, make sure that your measurements are accurate and the shelves are level. Most of the time, you will find it easier to do your cutting and drilling before you start assembling the cabinets. If the shelf standards are the type that are set in a groove, you must cut the groove

Figure 3-70.—End panels of a wall cabinet in place (view A) and completed framing with facing partially applied (view B).

before assembly. Some adjustable shelf supports can be mounted on the surface.

Shelving supports for 3/4-inch shelves should be placed no more than 42-inches apart. Shelves designed to hold heavy loads should have closer supports. To improve the appearance of plywood shelving, cover the laminated edge with a strip of wood that matches the stock used for the cabinet.

Cabinet Facing

After completing the frame construction and shelving, apply finished facing strips to the front of the cabinet frame. These strips are sometimes assembled into a framework (called a faceplate or face frame) by commercial sources before they are attached to the basic cabinet structure. The vertical members of the facing are called stiles, and the horizontal members are known as rails.

As previously mentioned for built-in-place cabinets, you cut each piece and install it separately. The size of each piece is laid out by positioning the facing stock on the cabinet and marking it. Then, the finished cuts are made. A cut piece can be used to lay out duplicate pieces.

Cabinet stiles are generally attached first, then the rails (figure 3-72). Sometimes a Builder will attach a

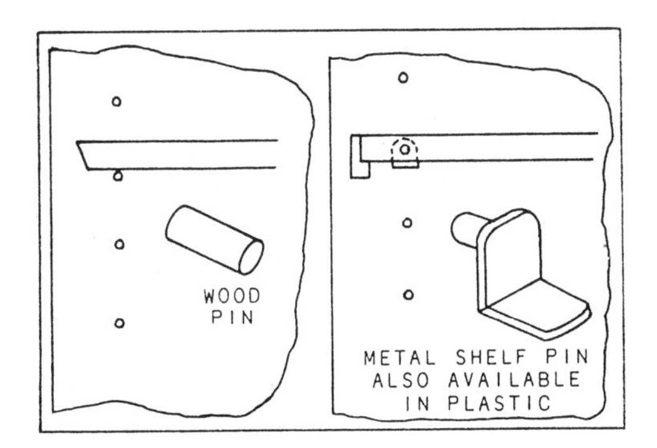

Figure 3-71.—Two methods of supporting shelves.

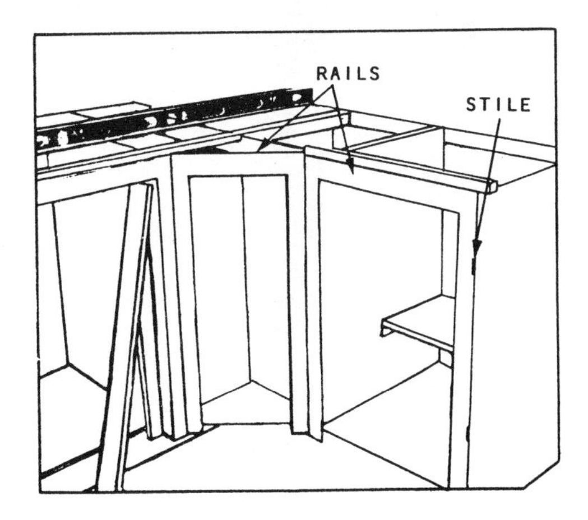

Figure 3-72.—Facing being placed on a cabinet.

plumb end stile first, and then attach rails to determine the position of the next stile.

Use finishing nails and glue to install facing. When nailing hardwoods, drill nail holes where you think splitting might occur.

Drawers

Seabees use many methods of building drawers. The three most common are the multiple dovetail, lock-shouldered, and square-shouldered methods (figure 3-73).

There are several types of drawer guides available. The three most commonly used are the side guide, the corner guide, and the center guide (shown in figure 3-74, view A).

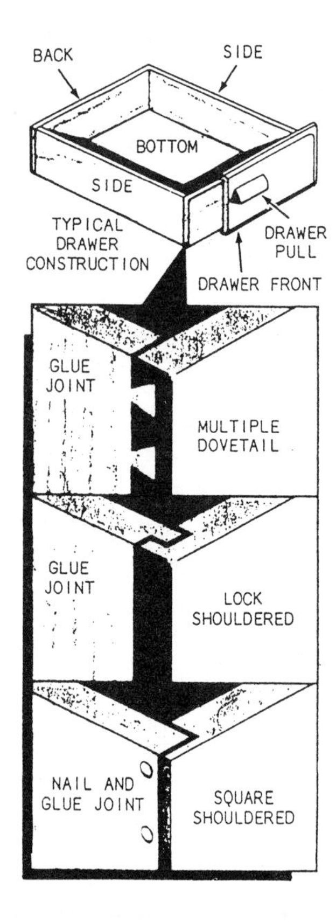

Figure 3-73.—Three common types of joints used in drawer construction.

The two general types of drawer faces are the lip and flush faces (shown in figure 3-74, view B). A flush drawer must be carefully fitted. A lip drawer must have a rabbet along the top and sides of the front. The lip style overlaps the opening and is much easier to construct.

Cabinet Doors

The four types of doors commonly used on cabinets are the flush, lipped, overlay, and sliding doors. A flush door, like the flush drawer, is the most difficult to construct. For a finished look, each type of door must be fitted in the cabinet opening within 1/16-inch clearance around all four edges. A lipped door is simpler to install than a flush door since the lip, or overlap, feature allows you a certain amount of adjustment and greater tolerances. The lip is formed by cutting a rabbet along the edge.

Overlay doors are designed to cover the edges of the face frame. There are several types of sliding doors used on cabinets. One type of sliding door is rabbeted to fit into grooves at the top and bottom of the cabinet. The top groove is always made to allow the door to be removed by lifting it up and pulling the bottom out.

INSTALLING PREMADE CABINETS

To install premade cabinets, you can begin with either the wall or base cabinets. The general procedures for each are similar.

Installing the Wall Cabinets First

When layouts are made and wall studs located, the wall units are lifted into position. They are held with a padded T-brace that allows the worker to stand close to the wall while making the installation. After the wall cabinets are securely attached and checked, the base cabinets are moved into place, leveled, and secured.

Installing the Base Cabinets First

When base cabinets are installed first, the tops of the base cabinets can be used to support braces that hold the wall units in place while they are fastened to the wall.

Procedures

The following procedures are a simple way of installing premade cabinets:

1. First, locate and mark the location of all wall studs where the cabinets are to be hung. Find and mark the highest point in the floor. This will ensure the base cabinet is level on uneven floor surfaces. (Shims should be used to maintain the cabinet at its designated leveled height.)

2. Start the installation of a base cabinet with a corner or end unit. After all base cabinets are in position, fasten the cabinets together. To get maximum holding power from screws, place one hole close to the top and one close to the bottom.

3. Starting at the highest point in the floor, level the leading edges of the cabinets. After leveling all the leading edges, fasten them to the wall at the studs to obtain maximum holding power.

4. Next, install the countertop on the base cabinets making sure to drill or screw through the top.

5. Then, make a brace to help support the wall cabinets while they are being fastened. Start the wall cabinet installation with a corner or end cabinet. Make sure you check for plumb and level as you install these cabinets.

6. After installing the cabinets and checking for plumb and level, join the wall cabinets through the sides as you did with the base cabinets.

7. Finally, after they are plumb and level, secure the cabinets to the wall at the studs for maximum holding power.

Here are some helpful hints for the general construction of cabinets:

- Cabinet parts are fastened together with screws or nails. They are set below the surface, and the holes are filled with putty. Glue is used at all joints. Clamps should be used to produce better fitting, glued joints.

- A better quality cabinet is rabbeted where the top, bottom, back, and side pieces come together. However, butt joints are also used. If panels are less than 3/4-inch thick, a reinforcing block should be used with the butt joint. Fixed shelves are dadoed into the sides.

- Screws should go through the hanging strips and into the stud framing. Never use nails. Toggle bolts are required when studs are inaccessible. Join units by first clamping them together and then, while aligned, install bolts and T-nuts.

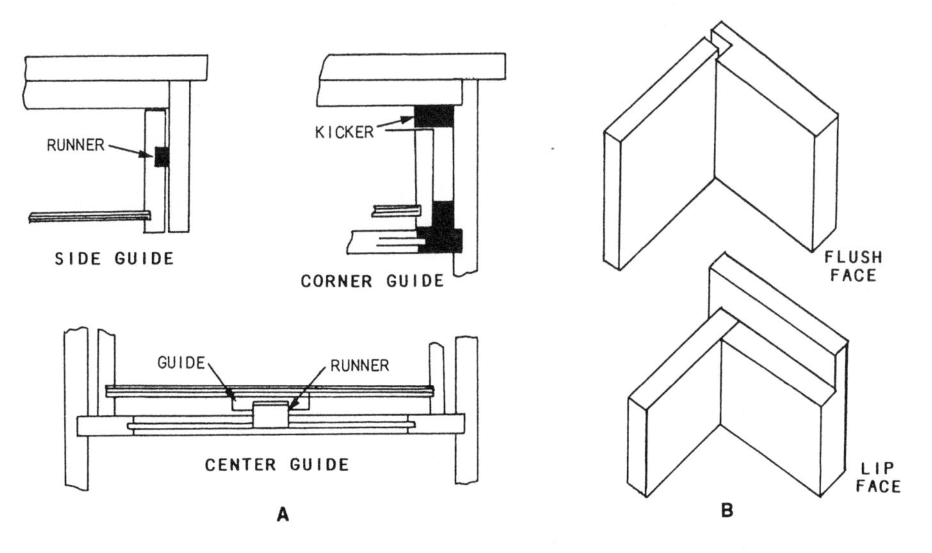

Figure 3-74.—Types of drawer guides (view A) and faces (view B).

COUNTERS AND TOPS

In cabinetwork, the counters and tops are covered with a 1/16-inch layer of high-pressure plastic laminate. Although this material is very hard, it does not possess great strength and is serviceable only when it is bonded to plywood, particle board, or wafer wood. This base, or core material, must be smooth and is usually 3/4-inch thick.

Working Laminates

Plastic laminates can be cut to rough size with a table saw, portable saw, or saber saw. Use a fine-tooth blade, and support the material close to the cut. If no electrical power is available, you can use a finish handsaw or a hacksaw. When cutting laminates with a saw, place masking tape over the cutting area to help prevent chipping the laminate. Make cut markings on the masking tape.

Measure and cut a piece of laminate to the desired size. Allow at least 1/4-inch extra to project past the edge of the countertop surface. Next, mix and apply the contact bond cement to the underside of the laminate and to the topside of the countertop surface. **Be sure to follow the manufacturer's recommended directions for application.**

Adhering Laminates

Allow the contact bond cement to set or dry. To check for bonding, press a piece of waxed brown paper on the cement-coated surface. When no adhesive residue shows, it is ready to be bonded. Be sure to lay a full sheet of waxed brown paper across the countertop. This allows you to adjust the laminate into the desired position without permanent bonding. Now, you can gradually slide the paper out from under the laminate, and the laminate becomes bonded to the countertop surface.

Be sure to roll the laminate flat by hand, removing any air bubbles and getting a good firm bond. After sealing the laminate to the countertop surface, trim the edges by using either a router with a special guide or a small block plane. If you want to bevel the countertop edge, use a mill file.

METHODS OF FASTENING

LEARNING OBJECTIVE: Upon completing this section, you should be able to identify the different types of fastening devices.

A variety of metal fastening devices are used by Seabees in construction. Although nails are the most commonly used fastener, the use of staples to attach wood structural members is growing. For certain operations, screws and bolts are required. In addition, various metal devices exist for anchoring materials into concrete, masonry, and steel.

The increasing use of adhesives (glues and mastics) is an important development in the building industry. Adhesives are used in combination with, or in place of, nails and screws.

NAILS

Nails, the most common type of metal fasteners, are available in a wide range of types and sizes.

Basic Nail Types

Some basic types are shown in figure 3-75. The common nail is designed for rough framing. The box nail is used for toenailing and light work in frame construction. The casing nail is used in finished carpentry work to fasten doors and window casings and other wood trim. The finishing nail and brad are used for light, wood-trim material and are easy to drive below the surface of lumber with a nail set.

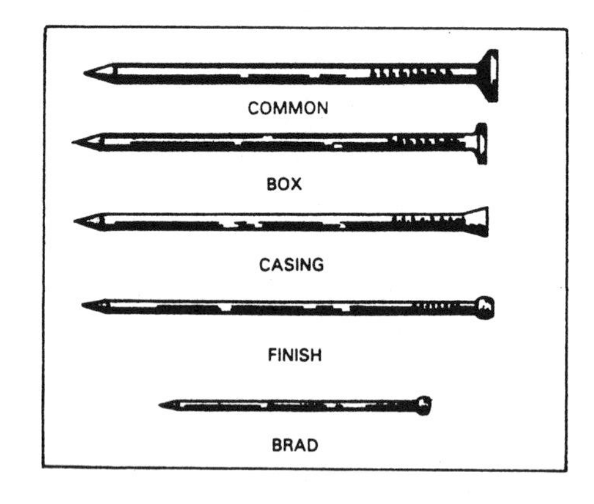

Figure 3-75.—Basic types of nails.

The size of a nail is measured in a unit known as a penny. Penny is abbreviated with the lowercase letter **d**. It indicates the length of the nail. A 6d (6-penny) nail is 2-inches long. A 10d (10-penny) nail is 3-inches long (figure 3-76). These measurements apply to common, box, casing, and finish nails only. Brads and small box nails are identified by their actual length and gauge number.

A nail, whatever the type, should be at least three times as long as the thickness of the wood it is intended to hold. Two-thirds of the length of the nail is driven into the other piece of wood for proper anchorage. The other one-third of the length provides the necessary anchorage of the piecebeingfastened. Protrudingnailsshouldbebent over to prevent damage to materials and injury to personnel.

There are a few general rules to be followed in the use of nails in building. Nails should be driven at an angle slightly toward each other to improve their holding power. You should be careful in placing nails to provide the greatest holding power. Nails driven with the grain do not hold as well as nails driven across the grain. A few nails of proper type and size, properly placed and properly driven, will hold better than a great many driven close together. Nails can generally be considered the cheapest and easiest fasteners to be applied.

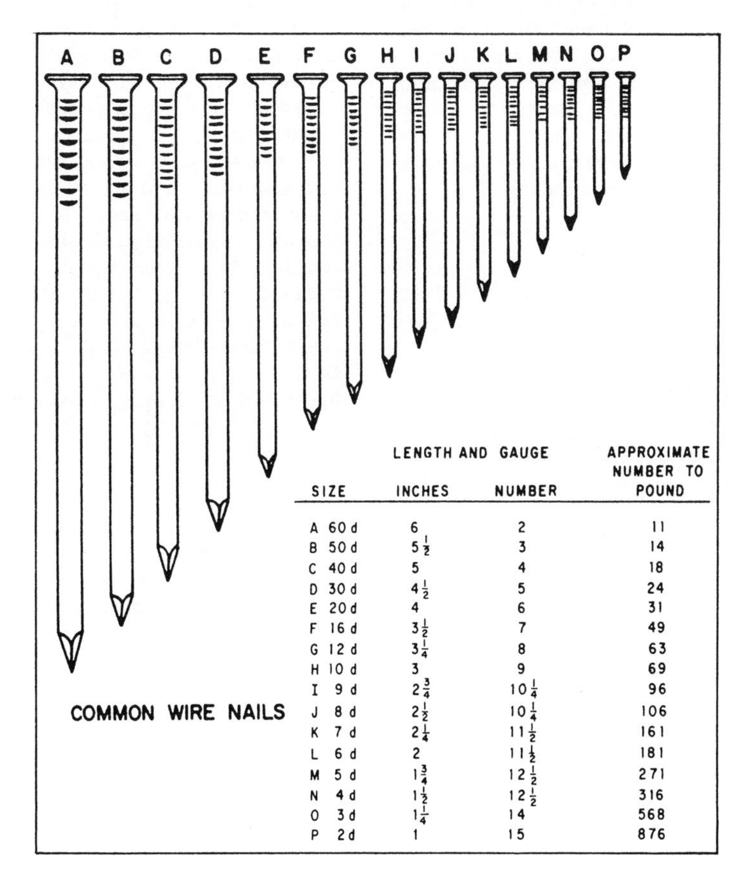

SIZE	LENGTH AND GAUGE INCHES	NUMBER	APPROXIMATE NUMBER TO POUND
A 60 d	6	2	11
B 50 d	$5\frac{1}{2}$	3	14
C 40 d	5	4	18
D 30 d	$4\frac{1}{2}$	5	24
E 20 d	4	6	31
F 16 d	$3\frac{1}{2}$	7	49
G 12 d	$3\frac{1}{4}$	8	63
H 10 d	3	9	69
I 9 d	$2\frac{3}{4}$	$10\frac{1}{4}$	96
J 8 d	$2\frac{1}{2}$	$10\frac{1}{4}$	106
K 7 d	$2\frac{1}{4}$	$11\frac{1}{2}$	161
L 6 d	2	$11\frac{1}{2}$	181
M 5 d	$1\frac{3}{4}$	$12\frac{1}{2}$	271
N 4 d	$1\frac{1}{2}$	$12\frac{1}{2}$	316
O 3 d	$1\frac{1}{4}$	14	568
P 2 d	1	15	876

Figure 3-76.—Nail sizes given in "penny" (d) units.

Specialty Nails

Figure 3-77 shows a few of the many specialized nails. Some nails are specially coated with zinc, cement, or resin materials. Some have threading for increased holding power of the nails. Nails are made from many materials, such as iron, steel, copper, bronze, aluminum, and stainless steel.

Annular and spiral nails are threaded for greater holding power. They are good for fastening paneling or plywood flooring. The drywall nail is used for hanging drywall and has a special coating to prevent rust. Roofing nails are not specified by the penny system; rather, they are referred to by length. They are available in lengths from 3/4 inch to 2 inches and have large heads. The double-headed nail, or duplex-head nail, is used for temporary construction, such as form work or scaffolding. The double head on this nail makes it easy to pull out when forms or scaffolding are torn down. Nails for power nailing come in rolls or clips for easy loading into a nailer. They are coated for easier driving and greater holding power. Table 3-10 gives the general size and type of nails preferable for specific applications.

STAPLES

Staples are available in a wide variety of shapes and sizes, some of which are shown in figure 3-78. Heavy-duty staples are used to fasten plywood sheeting and subflooring. Heavy-duty staples are driven by electrically or pneumatically operated tools. Light-duty and medium-duty staples are used for attaching molding and other interior trim. Staples are sometimes driven in by hand-operated tools.

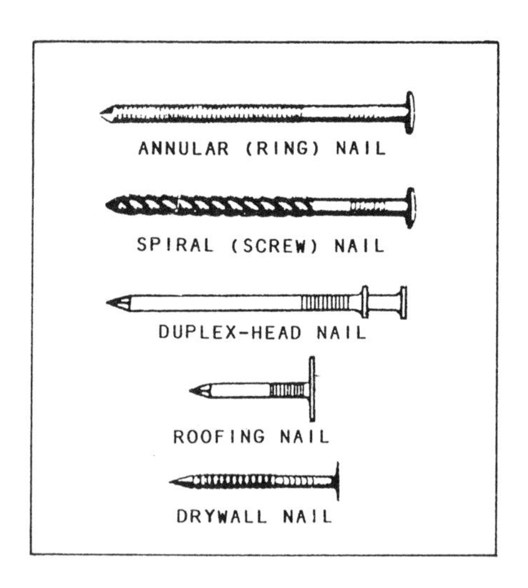

Figure 3-77.—Specialized nails.

SCREWS

The use of screws, rather than nails, as fasteners may be dictated by a number of factors. These may include the type of material to be fastened, the requirement for greater holding power than can be obtained by the use of nails, the finished appearance desired, and the fact that the number of fasteners that can be used is limited. Using screws, rather than nails, is more expensive in terms of time and money, but it is often necessary to meet requirements for superior results. The main advantages of screws are that they provide more holding power, can be easily tightened to draw the items being fastened securely together, are neater in appearance if properly driven, and can be withdrawn without damaging the material. The common wood screw is usually made of unhardened steel, stainless steel, aluminum, or brass. The steel may be bright finished or blued, or zinc, cadmium, or chrome plated. Wood screws are threaded from a gimlet point for approximately two-thirds of the length of the screw and are provided with a slotted head designed to be driven by an inserted driver. Wood screws, as shown in figure 3-79, are designated according to head style. The most common types are flathead, oval head, and

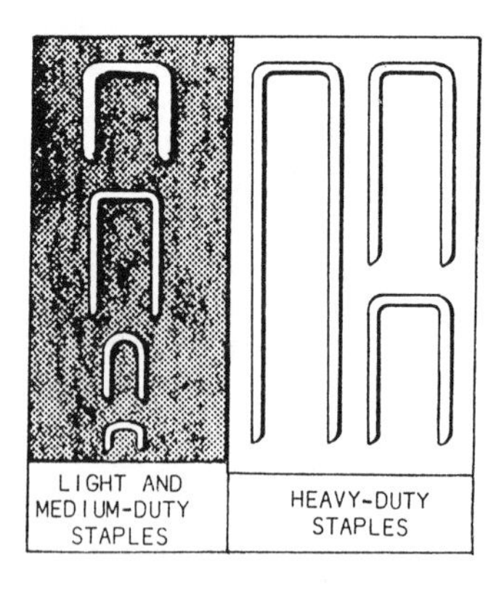

Figure 3-78.—Types of staples.

Table 3-10.—Size, Type, and Use of Nails

SIZE	LENGTH (IN.)[1]	DIAMETER (IN.)	REMARKS	WHERE USED
2d	1	.072	Small head	Finish work, shop work
2d	1	.072	Large flathead	Small timber, wood shingles, lathes
3d	1 1/4	.08	Small head	Finish work, shop work
3d	1 1/4	.08	Large flathead	Small timber, wood shingles, lathes
4d	1 1/2	.098	Small head	Finish work, shop work
4d	1 1/2	.098	Large flathead	Small timber, lathes, shop work
5d	1 3/4	.098	Small head	Finish work, shop work
5d	1 3/4	.098	Large flathead	Small timber, lathes, shop work
6d	2	.113	Small head	Finish work, casing, stops, etc., shop work
6d	2	.113	Large flathead	Small timber, siding, sheathing, etc., shop work
7d	2 1/4	.113	Small head	Casing, base, ceiling, stops, etc.
7d	2 1/4	.113	Large flathead	Sheathing, siding, subflooring, light framing
8d	2 1/2	.131	Small head	Casing, base, ceiling, wainscot, etc., shop work
8d	2 1/2	.131	Large flathead	Sheathing, siding, subflooring, light framing, shop work
8d	1 1/4	.131	Extra-large flathead	Roll roofing, composition shingles
9d	2 3/4	.131	Small head	Casing, base, ceiling, etc.
9d	2 3/4	.131	Large flathead	Sheathing, siding, subflooring, framing, shop work
10d	3	.148	Small head	Casing, base, ceiling, etc., shop work
10d	3	.148	Large flathead	Sheathing, siding, subflooring, framing, shop work
12d	3 1/4	.148	Large flathead	Sheathing, subflooring, framing
16d	3 1/2	.162	Large flathead	Framing, bridges, etc.
20d	4	.192	Large flathead	Framing, bridges, etc.
30d	4 1/2	.207	Large flathead	Heavy framing, bridges, etc.
40d	5	.225	Large flathead	Heavy framing, bridges, etc.
50d	5 1/2	.244	Large flathead	Extra-heavy framing, bridges, etc.
60d	6	.262	Large flathead	Extra-heavy framing, bridges, etc.

1 This chart applies to wire nails, although it may be used to determine the length of cut nails.

roundhead, as illustrated in that order in figure 3-79. All of these screws can have slotted or Phillips heads.

To prepare wood for receiving the screws, bore a body hole the diameter of the screw to be used in the piece of wood that is to be fastened (figure 3-80). You should then bore a starter hole in the base wood with a diameter less than that of the screw threads and a depth of one-half or two-thirds the length of the threads to be anchored. The purpose of this careful preparation is to assure accuracy in the placement of the screws, to reduce the possibility of splitting the wood, and to reduce the time and effort required to drive the screw. Properly set slotted and Phillips flathead and oval head screws are countersunk sufficiently to permit a covering material to be used to cover the head. Slotted roundhead and Phillips roundhead screws are not countersunk, but they are driven so that the head is firmly flush with the surface of the wood. The slot of the roundhead screw is left parallel with the grain of the wood.

The proper name for a lag screw (shown in figure 3-79) is lag bolt or wood screw. These screws are often required in constructing large projects, such as a building. They are longer and much heavier than the common wood screw and have coarser threads that extend from a cone, or gimlet point, slightly more than half the length of the screw. Square-head and hexagonal-head lag screws are always externally driven, usually by means of a wrench. They are used when ordinary wood screws would be too short or too light and spikes would not be strong enough. Sizes of

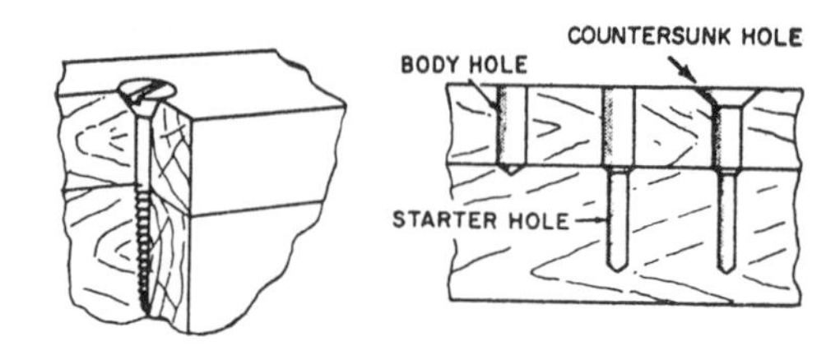

Figure 3-80.—Proper way to sink a screw.

lag screws are shown in table 3-11. Combined with expansion anchors, they are used to frame timbers to existing masonry.

Expansion shields, or expansion anchors as they are sometimes called, are used for inserting a predrilled hole, usually in masonry, to provide a gripping base or anchor for a screw, bolt, or nail intended to fasten an item to the surface in which the hole was bored. The shield can be obtained separately, or it may include the screw, bolt, or nail. After the expansion shield is inserted in the predrilled hole, the fastener is driven into the hole in the shield, expanding the shield and wedging it firmly against the surface of the hole.

For the assembly of metal parts, sheet metal screws are used. These screws are made regularly in steel and brass with four types of heads: flat, round, oval, and fillister, as shown in that order in figure 3-79.

Wood screws come in sizes that vary from 1/4 inch to 6 inches. Screws up to 1-inch in length increase by

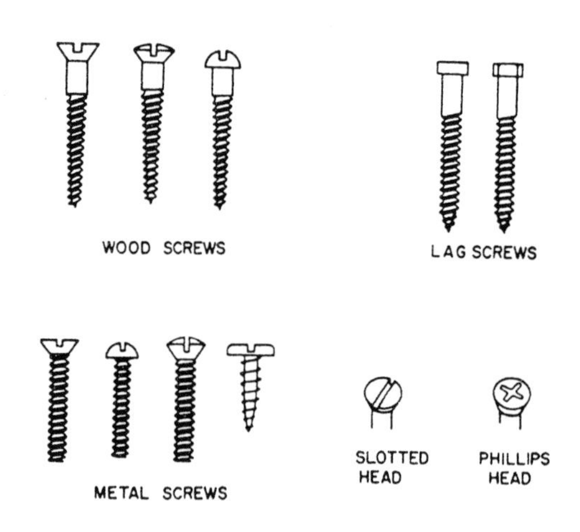

Figure 3-79.—Types of screws.

Table 3-11.—Lag Screw Sizes

LENGTH (INCHES)	DIAMETER (INCHES)			
	1/4	3/8 7/16 1/2	5/8 3/4	7/8 1
1	×	×		
1 1/2	×	×	×	
2, 2 1/2, 3 3 1/2, etc., 7 1/2 8 to 10	×	×	×	×
11 to 12		×	×	×
13 to 16			×	×

eighths, screws from 1 to 3 inches increase by quarters, and screws from 3 to 6 inches increase by half inches. Screws vary in length and size of shaft. Each length is made in a number of shaft sizes specified by an arbitrary number that represents no particular measurement but indicates relative differences in the diameter of the screws. Proper nomenclature of a screw, as shown in figure 3-81, includes the type, material, finish, length, and screw size number, which indicates the wire gauge of the body, drill or bit size for the body hole, and drill or bit size for the starter hole. Tables 3-12 and 3-13 provide size, length, gauge, and applicable drill and auger bit sizes for screws. Table 3-11 gives lengths and diameters of lag screws.

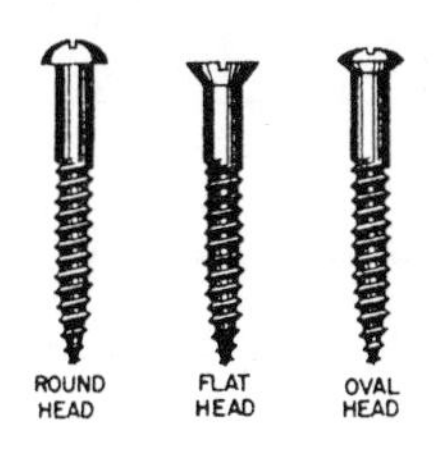

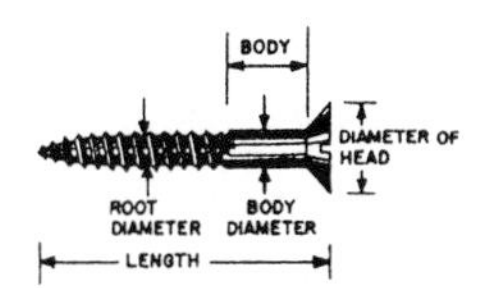

Figure 3-81.—Types and nomenclature of wood screws.

BOLTS

Bolts are used in construction when great strength is required or when the work under construction must be frequently disassembled. Their use usually implies the use of nuts for fastening and, sometimes, the use of washers to protect the surface of the material they are used to fasten. Bolts are selected for application to specific requirements in terms of length, diameter, threads, style of head, and type. Proper selection of head style and type of bolt results in good appearance as well as good construction. The use of washers between the nut and a wood surface or between both the nut and the head and their opposing surfaces helps you avoid marring the surfaces and permits additional torque in tightening.

Carriage Bolts

Carriage bolts fall into three categories: square neck finned neck and ribbed neck (figure 3-82). These bolts have round heads that are not designed to

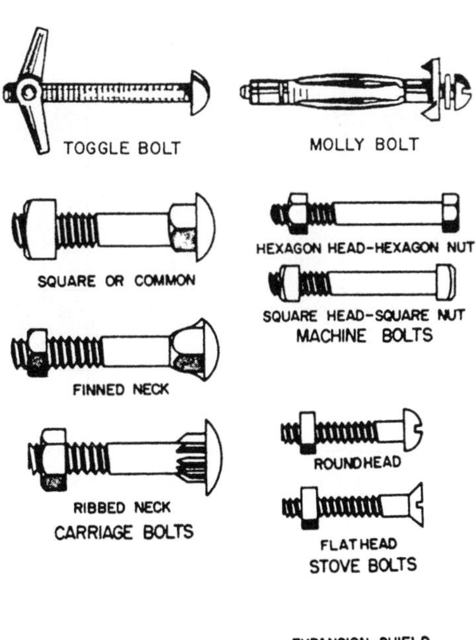

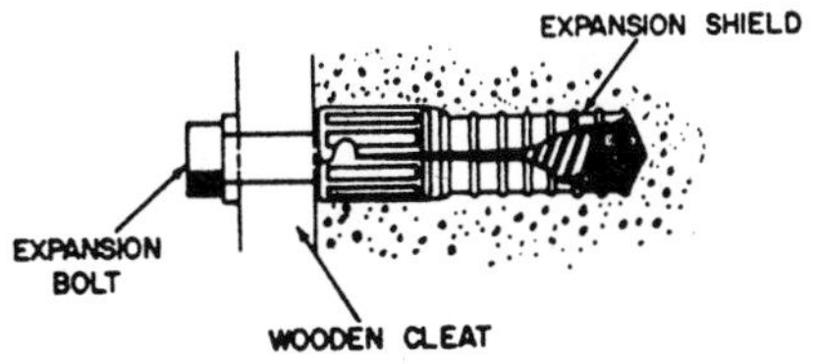

Figure 3-82.—Types of bolts.

Table 3-12.—Screw Sizes and Dimensions

LENGTH (in.)	SIZE NUMBERS																				
	0	1	2	3	4	5	6	7	8	9	10	11	12	13	14	15	16	17	18	20	22	24
1/4	×	×	×	×																		
3/8	×	×	×	×	×	×	×	×	×	×												
1/2		×	×	×	×	×	×	×	×	×	×	×	×									
5/8		×	×	×	×	×	×	×	×	×	×	×	×		×							
3/4			×	×	×	×	×	×	×	×	×	×	×		×		×					
7/8			×	×	×	×	×	×	×	×	×	×	×		×		×					
1				×	×	×	×	×	×	×	×	×	×		×		×		×	×		
1 1/4					×	×	×	×	×	×	×	×	×		×		×		×	×		×
1 1/2					×	×	×	×	×	×	×	×	×		×		×		×	×		×
1 3/4						×	×	×	×	×	×	×	×		×		×		×	×		×
2						×	×	×	×	×	×	×	×		×		×		×	×		×
2 1/4						×	×	×	×	×	×	×	×		×		×		×	×		×
2 1/2						×	×	×	×	×	×	×	×		×		×		×	×		×
2 3/4							×	×	×	×	×	×	×		×		×		×	×		×
3							×	×	×	×	×	×	×		×		×		×	×		×
3 1/2									×	×	×	×	×		×		×		×	×		×
4									×	×	×	×	×		×		×		×	×		×
4 1/2													×		×		×		×	×		×
5													×		×		×		×	×		×
6															×		×		×	×		×
THREADS PER INCH	32	28	26	24	22	20	18	16	15	14	13	12	11		10		9		8	8		7
DIA OF SCREW (in.)	.060	.073	.086	.099	.112	.125	.138	.151	.164	.177	.190	.203	.216		.242		.268		.294	.320		.372

Table 3-13.—Drill and Auger Bit Sizes for Wood Screws

SCREW SIZE NO.		1	2	3	4	5	6	7	8	9	10	12	14	16	18
NOMINAL SCREW		.073	.086	.099	.112	.125	.138	.151	.164	.177	.190	.216	.242	.268	.294
BODY DIAMETER		5/64	3/32	3/32	7/64	1/8	9/64	5/32	11/64	11/64	3/16	7/32	15/64	17/64	19/64
PILOT HOLE	Drill size	5/64	3/32	7/64	7/64	1/8	9/64	5/32	11/64	3/16	3/16	7/32	1/4	17/64	19/64
	Bit size	—	—	—	—	—	—	—	—	—		4	4	5	5
STARTER HOLE	Drill size	—	1/16	1/16	5/64	5/64	3/32	7/64	7/64	1/8	1/8	9/64	5/32	3/16	13/64
	Bit size	—	—	—	—	—	—	—	—	—	—	—	—		4

Table 3-14.—Carriage Bolt Sizes

LENGTH (INCHES)	DIAMETER (INCHES)			
	3/16, 1/4, 5/16, 3/8	7/16, 1/2	9/16, 5/8	3/4
3/4	×	—	—	—
1	×	×	—	—
1 1/4	×	×	×	—
1 1/2, 2, 2 1/2, etc., 9 1/2, 10 to 20	×	×	×	×

be driven. They are threaded only part of the way up the shaft. Usually, the threads are two to four times the diameter of the bolt in length. In each type of carriage bolt, the upper part of the shank, immediately below the head, is designed to grip the material in which the bolt is inserted and keep the bolt from turning when a nut is tightened down on it or removed. The finned type is designed with two or more fins extending from the head to the shank. The ribbed type is designed with longitudinal ribs, splines, or serrations on all or part of a shoulder located immediately beneath the head. Holes bored to receive carriage bolts are bored to be a tight fit for the body of the bolt and counterbored to permit the head of the bolt to fit flush with, or below the surface of, the material being fastened. The bolt is then driven through the hole with a hammer. Carriage bolts are chiefly for wood-to-wood application, but they can also be used for wood-to-metal applications. If used for wood-to-metal application, the head should be fitted to the wood item. Metal surfaces are sometimes predrilled and countersunk to permit the use of carriage bolts metal to metal. Carriage bolts can be obtained from 1/4 inch to 1 inch in diameter and from 3/4 inch to 20 inches long (table 3-14). A common flat washer should be used with carriage bolts between the nut and the surface.

Machine Bolts

Machine bolts (figure 3-82) are made with cut national fine and national coarse threads extending in length from twice the diameter of the bolt plus 1/4 inch (for bolts less than 6 inches in length) to twice the diameter of the bolt plus 1/2 inch (for bolts over 6 inches in length). They are precision made and generally applied metal to metal where close tolerance is desirable. The head may be square, hexagonal, rounded, or flat countersunk. The nut usually corresponds in shape to the head of the bolt with which it is used. Machine bolts are externally driven only. Selection of the proper machine bolt is made on the basis of head style, length, diameter, number of threads per inch, and coarseness of thread. The hole through which the bolt is to pass is bored to the same diameter as the bolt. Machine bolts are made in diameters from 1/4 inch to 3 inches and may be obtained in any length desired (table 3-15).

Table 3-15.—Machine Bolt Sizes

LENGTH (INCHES)	DIAMETER (INCHES)				
	1/4, 3/8	7/16	1/2, 9/16, 5/8	1/2, 7/8, 1	1 1/8, 1 1/4
3/4	×	—	—	—	—
1 1/4	×	×	×	—	—
1 1/2, 2, 2 1/2	×	×	×	×	—
3, 3 1/2, 4, 4 1/2, etc., 9 1/2, 10 to 20	×	×	×	×	×
21 to 25	—	—	×	×	×
26 to 39	—	—	—	×	×

Stove Bolts

Stove bolts (figure 3-82) are less precisely made than machine bolts. They are made with either flat or round slotted heads and may have threads extending over the full length of the body, over part of the body, or over most of the body. They are generally used with square nuts and applied metal to metal, wood to wood, or wood to metal. If flatheaded, they are countersunk. If roundheaded, they are drawn flush to the surface.

Expansion Bolt

An expansion bolt (figure 3-82) is a bolt used in conjunction with an expansion shield to provide anchorage in substances in which a threaded fastener alone is useless. The shield, or expansion anchor, is inserted in a predrilled hole and expands when the bolt is driven into it. It becomes wedged firmly in the hole, providing a secure base for the grip of the fastener.

Toggle Bolts

A toggle bolt (figure 3-82) is a machine screw with a spring-action, wing-head nut that folds back as the entire assembly is pushed through a prepared hole in a hollow wall. The wing head then springs open inside the wall cavity. As the screw is tightened, the wing head is drawn against the inside surface of the finished wall material. Spring-action, wing-head toggle bolts are available in a variety of machine screw combinations. Common sizes range from 1/8 inch to 3/8 inch in diameter and 2 inches to 6 inches in length. They are particularly useful with sheetrock wall surfaces.

Molly Bolt

The molly bolt or molly expansion anchor (figure 3-82) is used to fasten small cabinets, towel bars, drapery hangers, mirrors, electrical fixtures, and other lightweight items to hollow walls. It is inserted in a prepared hole. Prongs on the outside of the shield grip the wall surfaces to prevent the shield from turning as the anchor screw is being driven. As the screw is tightened, the shield spreads and flattens against the interior of the wall. Various sizes of screw anchors can be used in hollow walls 1/8 inch to 1 3/4 inches thick.

Figure 3-83.—Driftpin (driftbolt).

Driftpins

Driftpins are long, heavy, threadless bolts used to hold heavy pieces of timber together (figure 3-83). They have heads that vary in diameter from 1/2 to 1 inch and in length from 18 to 26 inches. The term "driftpin" is almost universally used in practice. However, for supply purposes, the correct designation is driftbolt.

To use the driftpin, you make a hole slightly smaller than the diameter of the pin in the timber. The pin is driven into the hole and is held in place by the compression action of the wood fibers.

CORRUGATED FASTENERS

The corrugated fastener is one of the many means by which joints and splices are fastened in small timber and boards. It is used particularly in the miter joint. Corrugated fasteners are made of 18- to 22-gauge sheet metal with alternate ridges and grooves; the ridges vary from 3/16 to 5/16 inch, center to center. One end is cut square; the other end is sharpened with beveled edges. There are two types of corrugated fasteners: one with the ridges running parallel (figure 3-84, view A); the other with ridges running at a slight angle to one another (figure 3-84, view B). The latter type has a tendency to compress the material since the ridges and grooves are closer at the top than at the bottom. These fasteners are made in several different lengths and widths. The width varies from 5/8 to 1 1/8 inches; the length varies from 1/4 to 3/4 inch. The fasteners also are made with different numbers of ridges, ranging from three to six ridges per fastener. Corrugated fasteners are used in a number of ways—to fasten parallel boards together, as in fastening tabletops; to make any type of joint; and as a substitute for nails where nails may split the timber. In small timber, corrugated fasteners have greater holding power than nails. The proper method of using the fasteners is shown in figure 3-84.

ADHESIVES

Seabees use many different types of adhesives in various phases of their construction projects. Glues

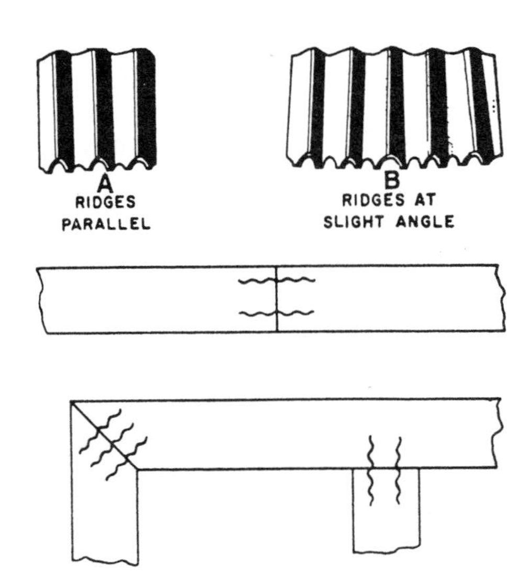

Figure 3-84.—Corrugated fasteners and their uses.

(which have a plastic base) and mastics (which have an asphalt, rubber, or resin base) are the two major categories of adhesives.

The method of applying adhesives, their drying time, and their bonding characteristics vary. Some adhesives are more resistant to moisture and to hot and cold temperatures than others.

SAFETY NOTE

Some adhesives are highly flammable; they should be used only in a well-ventilated work area. Others are highly irritating to the skin and eyes. **ALWAYS FOLLOW MANUFACTURER'S INSTRUCTIONS WHEN USING ADHESIVES.**

Glues

The primary function of glue is to hold together joints in mill and cabinet work. Most modern glues have a plastic base. Glues are sold as a powder to which water must be added or in liquid form. Many types of glue are available under brand names. A brief description of some of the more popular types of glue is listed below.

Polyvinyl resin, or white glue, is a liquid that comes in ready-to-use plastic squeeze bottles. It does a good job of bonding wood together, and it sets up (dries) quickly after being applied. Because white glue is not waterproof, it should not be used on work that will be subjected to constant moisture or high humidity.

Urea resin is a plastic based glue that is sold in a powder form. The required amount is mixed with water when the glue is needed. Urea resin makes an excellent bond for wood and has fair water resistance.

Phenolic resin glue is highly resistant to temperature extremes and water. It is often used for bonding the veneer layers of exterior grade plywood.

Resorcinol glue has excellent water resistance and temperature resistance, and it makes a very strong bond. Resorcinol resin is often used for bonding the wood layers of laminated timbers.

Contact cement is used to bond plastic laminates to wood surfaces. This glue has a neoprene rubber base. Because contact cement bonds very rapidly, it is useful for joining parts that cannot be clamped together.

Mastics

Mastics are widely used throughout the construction industry. The asphalt, rubber, or resin base of mastics gives them a thicker consistency. Mastics are sold in cans, tubes, or canisters that fit into hand-operated or air-operated caulking guns.

These adhesives can be used to bond materials directly to masonry or concrete walls. If furring strips are required on a wavy concrete wall, the strips can be applied with mastic rather than by the more difficult procedure of driving in concrete nails. You can also fasten insulation materials to masonry and concrete walls with a mastic adhesive. Mastics can also be used to bond drywall (gypsum board) directly to wall studs. They can also be used to bond gypsum board to furring strips or directly to concrete or masonry walls. Because you don't use nails, there are no nail indentations to fill.

By using mastic adhesives, you can apply paneling with very few or no nails at all. Wall panels can be bonded to studs, furring strips, or directly against concrete or masonry walls. Mastic adhesives can be used with nails or staples to fasten plywood panels to floor joists. The mastic adhesive helps eliminate squeaks, bounce, and nail popping. It also increases the stiffness and strength of the floor unit.

RECOMMENDED READING LIST

NOTE

Although the following references were current when this TRAMAN was published, their continued currency cannot be assured. You therefore need to ensure that you are studying the latest revision.

Carpentry I, Headquarters, EN5155, U.S. Army Engineering School, Fort Belvoir, Va., 1988.

Carpentry III, Headquarters, EN0533, U.S. Army Engineering School, Fort Belvoir, Va., 1987.

CHAPTER 4

FIBER LINE, WIRE ROPE, AND SCAFFOLDING

This chapter presents information on how to use fiber line, wire rope, and timber in rigging and erecting hoisting devices (such as shear legs, tripods, blocks and tackles), and different types of scaffolds and ladders. Formulas are given on how to determine or find the safe working load of these materials.

FIBER LINE

LEARNING OBJECTIVE: Upon completing this section, you should be able to determine the use, breaking strength, and care of fiber lines and rope used for rigging.

Fiber line is made from either natural or synthetic fiber. Natural fibers, which come from plants, include manila, sisal, and hemp. The synthetic fibers include nylon, polyester, and polypropylene.

NATURAL FIBER ROPES

The two most commonly used natural fiber ropes are manila and sisal, but the only type suitable for construction rigging is a good grade of manila. High-quality manila is light cream in color, smooth, clean, and pliable. The quality of the line can be distinguished by varying shades of brown: Number 1 grade is very light in color; Number 2 grade is slightly darker; Number 3 grade is considerably darker. The next best line-making fiber is sisal. The sisal fiber is similar to manila, but it is lighter in color. This type of fiber is only about 80 percent as strong as manila fiber.

SYNTHETIC FIBER ROPES

Synthetic fiber rope, such as nylon and polyester, has rapidly gained wide use by the Navy. It is lighter in weight, more flexible, less bulky, and easier to handle and store than manila line. It is also highly resistant to mildew, rot, and fungus. Synthetic rope is stronger than natural fiber rope. For example, nylon is about three times stronger than manila. When nylon line is wet or frozen, the loss of strength is relatively small. Nylon rope will hold a load even though several stands may be frayed. Ordinarily, the line can be made reusable by cutting away the chafed or frayed section and splicing the good line together.

FABRICATION OF LINE

The fabrication of line consists essentially of three twisting operations. First, the fibers are twisted to the right to form the yarns. Next, the yarns are twisted to the left to form the strands. Finally, the strands are twisted to the right to form the line. Figure 4-1 shows you how the fibers are grouped to form a three-strand line.

The operation just described is the standard procedure, and the resulting product is known as a right-laid line. When the process is reversed, the result is a left-laid line. In either instance, the principle of opposite twists must always be observed. The two main reasons for the principle of opposite twists are to keep the line tight to prevent the fibers from unlaying with a load suspended on it and to prevent moisture penetration.

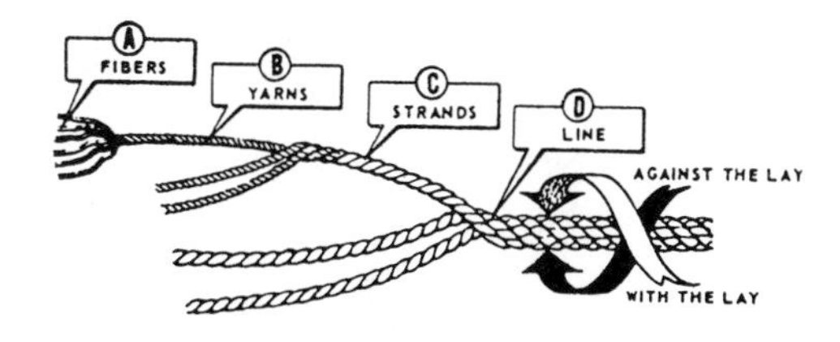

Figure 4-1.—Fiber groupings in a three-strand line.

Types of Line Lays

There are three types of fiber line lays: hawser-laid, shroud-laid, and cable-laid lines. Each type is illustrated in figure 4-2.

Hawser-laid line generally consists of three strands twisted together, usually in a right-hand direction. A shroud-laid line ordinarily is composed of four strands twisted together in a right-hand direction around a center strand, or core, which usually is of the same material, but smaller in diameter than the four strands. You will find that shroud-laid line is more pliable and stronger than hawser-laid line, but it has a strong tendency toward kinking. In most instances, it is used on sheaves and drums. This not only prevents kinking, but also makes use of its pliability and strength. Cable-laid line usually consists of three right-hand, hawser-laid lines twisted together in a left-hand direction. It is especially safe to use in heavy construction work; if cable laid line untwists, it will tend to tighten any regular right-hand screw connection to which it is attached.

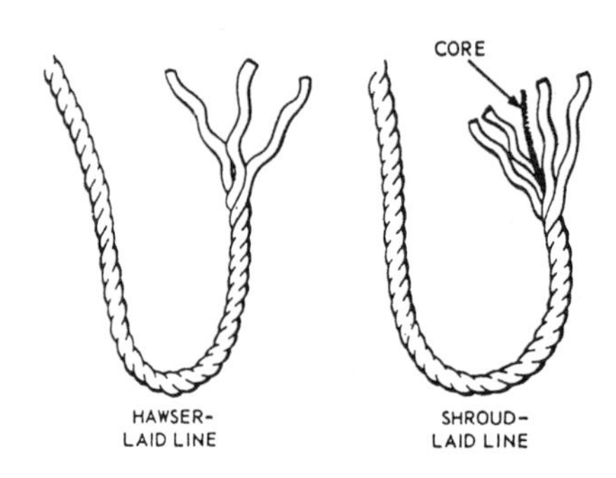

Figure 4-2.—Three types of fiber line.

Size Designation

Line that is 1 3/4 inches or less in circumference is called small stuff; this size is usually designated by the number of threads (or yarns) that make up each strand. You may use from 6- to 24-thread strands, but the most commonly used are 9- to 21-thread strands (figure 4-3). You may hear some small stuff designated by name without reference to size. One such type is marline—a tarred, two-strand, left-laid hemp. Marline is the small stuff you will use most for seizing. When you need something stronger than marline, you will use a tarred, three-strand, left-laid hemp called houseline.

Line larger than 1 3/4 inches in circumference is generally size designated by its circumference in inches. A 6-inch manila line, for instance, is constructed of manila fibers and measures 6 inches in circumference. Line is available in sizes ranging up to 16 inches in circumference, but 12 inches is about the largest carried in stock. Anything larger is used only on special jobs.

If you have occasion to order line, you may find that in the catalogs, it is designated and ordered by diameter. The catalog may also use the term "rope" rather than "line."

Rope yarns for temporary seizing, whippings, and lashings are pulled from large strands of old line that

MANILA LINE

SOME COMMONLY USED SIZES	* CIRCUMFERENCE		THREAD
	INCHES	MILLIMETERS	
	3/4	19.05	6
	1	25.40	9
	1 1/8	28.58	12
	1 1/4	31.76	15
	1 1/2	38.10	21
	1 3/4	44.45	24
	2	50.80	
	3	76.20	
	4	101.6	
	5	127.0	
	6	152.4	

* SIZE IS DESIGNATED BY THE CIRCUMFERENCE

Figure 4-3.—Some commonly used sizes of manila line.

has outlived its usefulness. Pull your yarn from the middle, away from the ends, or it will get fouled.

STRENGTH OF FIBER LINE

Overloading a line poses a serious threat to the safety of personnel, not to mention the heavy losses likely to result through damage to material. To avoid overloading, you must know the strength of the line with which you are working. This involves three factors: breaking strength, safe working load (swl), and safety factor.

Breaking strength refers to the tension at which the line will part when a load is applied. Breaking strength has been determined through tests made by rope manufacturers, who provide tables with this information. In the absence of manufacturers' tables, a rule of thumb for finding the breaking strength of manila line using the formula: $C^2 \times 900 = BS$. C equals the circumference in inches, and BS equals the breaking strength in pounds. To find BS, first square the circumference; you then multiply the value obtained by 900. With a 3-inch line, for example, you will get a BS of 8,100, or $3 \times 3 \times 900 = 8{,}100$ pounds.

The breaking strength of manila line is higher than that of sisal line. This is caused by the difference in strength of the two fibers. The fiber from which a particular line is constructed has a definite bearing on its breaking strength. The breaking strength of nylon line is almost three times that of manila line of the same size.

The best rule of thumb for the breaking strength of nylon is $BS = C^2 \times 2{,}400$. The symbols in the rule are the same as those for fiber line. For 2 1/2-inch nylon line, $BS = 2.5 \times 2.5 \times 2{,}400 = 15{,}000$ pounds.

Briefly defined, the safe working load of a line is the load that can be applied without damaging the line. Note that the safe working load is considerably less than the breaking strength. A wide margin of difference between breaking strength and safe working load is necessary. This difference allows for such factors as additional strain imposed on the line by jerky movements in hoisting or bending over sheaves in a pulley block.

You may not always have a chart available to tell you the safe working load for a particular size line. Here is a rule of thumb that will adequately serve your needs on such an occasion: $swl = C^2 \times 150$. In this equation, swl equals the safe working load in pounds, and C equals the circumference of the line in inches. Simply take the circumference of the line, square it, then multiply by 150. For a 3-inch line, $3 \times 3 \times 150 = 1{,}350$ pounds. Thus, the safe working load of a 3-inch line is equal to 1,350 pounds.

If line is in good shape, add 30 percent to the swl arrived at by means of the preceding rule; if it is in bad shape, subtract 30 percent from the swl. In the example given above for the 3-inch line, adding 30 percent to the 1,350 pounds gives you a safe working load of 1,755 pounds. On the other hand, subtracting 30 percent from the 1,350 pounds leaves you with a safe working load of 945 pounds.

Remember that the strength of a line decreases with age, use, and exposure to excessive heat, boiling water, or sharp bends. Especially with used line, these and other factors affecting strength should be given careful consideration and proper adjustment made in determining the breaking strength and safe working load capacity of the line. Manufacturers of line provide tables that show the breaking strength and safe working load capacity of line. You will find such tables very useful in your work. You must remember, however, that the values given in manufacturers' tables only apply to new line being used under favorable conditions. For that reason, you must progressively reduce the values given in manufacturers' tables as the line ages or deteriorates with use.

Keep in mind that a strong strain on a kinked or twisted line will put a permanent distortion in the line. Figure 4-4 shows what frequently happens when pressure is applied to a line with a kink in it. The kink that could have been worked out is now permanent, and the line is ruined.

The safety factor of a line is the ratio between the breaking strength and the safe working load. Usually, a safety factor of 4 is acceptable, but this is not always the case. In other words, the safety factor varies depending on such things as the condition of the line and circumstances under which it is to be used. Although the safety factor should never be less than 3, it often must be well above 4 (possibly as high as 8 or

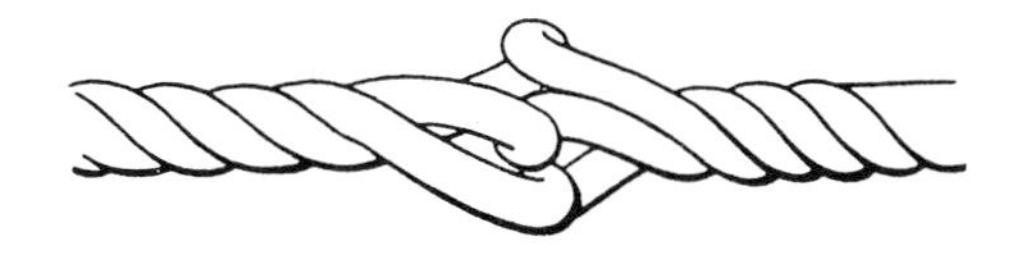

Figure 4-4.—Results of a strong strain on a line with a kink in it.

10). For best, average, or unfavorable conditions, the following safety factors may often be suitable:

- Best conditions (new line): 4;
- Average conditions (line used, but in good condition): 6; and
- Unfavorable conditions (frequently used line, such as running rigging): 8.

HANDLING AND CARE OF LINES

If you expect the fiber line you work with to give safe and dependable service, make sure it is handled and cared for properly. Study the precautions and procedures given here and carry them out properly.

Cleanliness is part of the care of fiber line. Never drag a line over the deck or ground, or over rough or dirty surfaces. The line can easily pick up sand and grit, which will work into the strands and wear the fibers. If a line does get dirty, use only water to clean it. Do not use soap because it will remove oil from the line, thereby weakening it.

Avoid pulling a line over sharp edges because the strands may break. When you encounter a sharp edge, place chafing gear, such as a board, folded cardboard or canvas, or part of a rubber tire between the line and the sharp edge to prevent damaging the line.

Never cut a line unless you have to. When possible, always use knots that can be untied easily.

Fiber line contracts, or shrinks, when it gets wet. If there is not enough slack in a wet line to permit shrinkage, the line is likely to become overstrained and weakened. If a taut line is exposed to rain or dampness, make sure the line, while still dry, is slackened to allow for the shrinkage.

Line should be inspected carefully at regular intervals to determine whether it is safe. The outside of a line does not show the condition of the line on the inside. Untwisting the strands slightly allows you to check the condition of the line on the inside. Mildewed line gives off a musty odor. Broken strands or yarns usually can be spotted immediately by a trained observer. You will want to look carefully to ensure there is not dirt or sawdust-like material inside the line. Dirt or other foreign matter inside reveals possible damage to the internal structure of the line. A smaller circumference of the line is usually a sure sign that too much strain has been applied to the line.

For a thorough inspection, a line should be examined at several places along its length. **Only one weak spot—anywhere in a line—makes the entire line weak.** As a final check, pull out a couple of fibers from the line and try to break them. Sound fibers show a strong resistance to breakage.

If an inspection discloses any unsatisfactory conditions in a line, make sure the line is destroyed or cut up in small pieces as soon as possible. This precaution prevents the defective line from being used for hoisting.

WIRE ROPE

LEARNING OBJECTIVE: Upon completing this section, you should be able to determine the use, breaking strength, and care of wire rope used for rigging.

During the course of a project, Seabees often need to hoist or move heavy objects. Wire rope is used for heavy-duty work. The characteristics, construction, and usage of many types of wire rope are discussed in the following paragraphs. We will also discuss the safe working load, use of attachments and fittings, and procedures for the care and handling of wire rope.

CONSTRUCTION

Wire rope consists of three parts: wires, strands, and core (figure 4-5). In the manufacture of rope, a number of wires are laid together to form the strand. Then a number of strands are laid together around a core to form the rope.

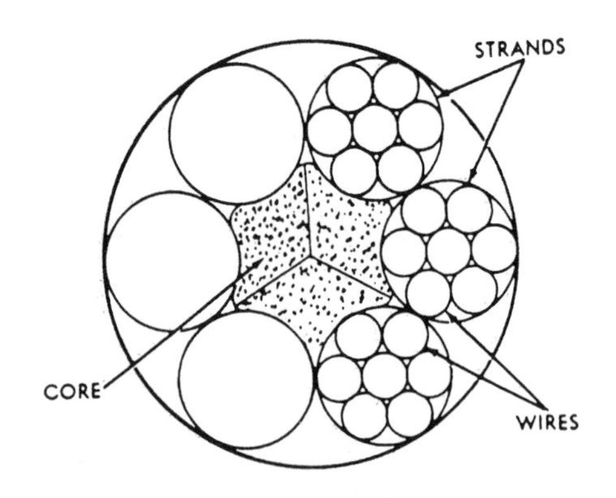

Figure 4-5.—Parts of wire rope.

The basic unit of wire rope construction is the individual wire, which may be made of steel, iron, or other metal in various sizes. The number of wires to a strand varies, depending on the purpose for which the rope is intended. Wire rope is designated by the number of strands per rope and the number of wires per strand. Thus, a 1/2-inch 6-by-19 rope will have 6 strands with 19 wires per strand; but it will have the same outside diameter as a 1/2-inch 6-by-37 wire rope, which will have 6 strands with 37 wires of much smaller size per strand. Wire rope made up of a large number of small wires is flexible, but the small wires are easily broken, so the wire rope does not resist external abrasion. Wire rope made up of a smaller number of larger wires is more resistant to external abrasion but is less flexible.

The core is the element around which the strands are laid to form the rope. It may be a hard fiber (such as manila, hemp, plastic, paper, asbestos, or sisal), a wire strand, or an independent wire rope. Each type of core serves the same basic purpose—to support the strands laid around it.

A fiber core offers the advantage of increased flexibility. Also, it serves as a cushion to reduce the effects of sudden strain and acts as a reservoir for the oil to lubricate the wires and strands to reduce friction between them. Wire rope with a fiber core is used in places where flexibility of the rope is important.

A wire strand core not only resists heat more than a fiber core, but also adds about 15 percent to the strength of the rope. On the other hand, the wire strand makes the rope less flexible than a fiber core.

An independent wire rope core is a separate wire rope over which the main strands of the rope are laid. It usually consists of six, seven-wire strands laid around either a fiber core or a wire strand core. This core strengthens the rope more, provides support against crushing, and supplies maximum resistance to heat.

Wire rope may be made by either of two methods. If the strands or wires are shaped to conform to the curvature of the finished rope before laying up, the rope is termed "preformed." If they are not shaped before fabrication, the rope is termed "nonpreformed." When cut, preformed wire rope tends not to unlay, and it is more flexible than nonpreformed wire rope. With nonpreformed wire rope, twisting produces a stress in the wires; and, when it is cut or broken, the stress causes the strands to unlay. **In nonpreformed wire, unlaying is rapid and almost instantaneous, which could cause serious injury to someone not familiar with it.**

The main types of wire rope used by the Navy consist of 6, 7, 12, 19, 24, or 37 wires in each strand. Usually, the rope has six strands laid around a fiber or steel center. Two common types of wire rope, 6-by-19 and 6-by-37 rope, are illustrated in views A and B of figure 4-6, respectively. The 6-by-19 type of rope, having 6 strands with 19 wires in each strand, is commonly used for rough hoisting and skidding work where abrasion is likely to occur. The 6-by-37 wire rope, having 6 strands with 37 wires in each strand, is the most flexible of the standard 6-strand ropes. For that reason, it is particularly suitable when small sheaves and drums are to be used, such as on cranes and similar machinery.

GRADES OF WIRE ROPE

Wire rope is made in a number of different grades. Three of the most common are mild plow steel, plow steel, and improved plow steel.

Mild plow steel rope is tough and pliable. It can stand up under repeated strain and stress, and it has a tensile strength of from 200,000 to 220,000 pounds per square inch (psi). Plow steel wire rope is unusually tough and strong. It has a tensile strength (resistance to lengthwise stress) of 220,000 to 240,000 psi. This rope is suitable for hauling, hoisting, and logging. Improved plow steel rope is one of the best grades of rope available, and most, if not all, of the wire rope in your work will probably be made of this material. It is stronger, tougher, and more resistant to wear than either plow steel or mild plow steel. Each square inch of improved plow steel can withstand a strain of 240,000 to 260,000 psi.

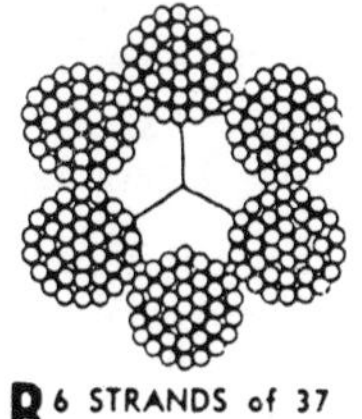

Figure 4-6.—Two common types of wire rope.

MEASURING WIRE ROPE

The size of wire rope is designated by its diameter. The true diameter of a wire rope is the diameter of a circle that will just enclose all of its strands. Correct and incorrect methods of measuring wire rope are illustrated in figure 4-7. In particular, note that the correct way is to measure from the top of one strand to the top of the strand directly opposite it. The wrong way is to measure across two strands side by side. Use calipers to take the measurement. If calipers are not available, an adjustable wrench will do.

To ensure an accurate measurement of the diameter of a wire rope, always measure the rope at three places, at least 5 feet apart. Use the average of the three measurements as the diameter of the rope.

SAFE WORKING LOAD

The term "safe working load" (swl), as used in reference to wire rope, means the load that can be applied and still obtain the most efficient service and also prolong the life of the rope. Most manufacturers provide tables that show the safe working load for their rope under various conditions. In the absence of these tables, you must apply a thumb rule formula to obtain the swl. There are rules of thumb that may be used to compute the strength of wire rope. The one recommended by the Naval Facilities Engineering Command (NAVFAC) is swl (in tons) = $D^2 \times 8$. This particular formula provides an ample safety margin to account for such variables as the number, size, and location of sheaves and drums on which the rope runs. Also included are dynamic stresses, such as the speed of operation and the acceleration and deceleration of the load. All can affect the endurance and breaking strength of the rope.

Let's work an example. In the above formula, D represents the diameter of the rope in inches. Suppose you want to find the swl of a 2-inch rope. Using the formula above, your figures would be: swl = $2^2 \times 8$, or $4 \times 8 = 32$. The answer is 32, meaning that the rope has a swl of 32 tons.

It is very important to remember that any formula for determining swl is only a rule of thumb. In computing the swl of old rope, worn rope, or rope that is otherwise in poor condition, you should reduce the swl as much as 50 percent, depending on the condition of the rope. The manufacturer's data concerning the breaking strength (BS) of wire rope should be used if available. But if you do not have that information, one rule of thumb recommended is BS = $C^2 \times 8{,}000$ pounds.

As you recall, wire rope is measured by the diameter (D). To obtain the circumference (C) required in the formula, multiply D by pi (usually shown by the Greek letter π), which is approximately 3.1416. Thus, the formula to find the circumference is $C = D\pi$.

WIRE ROPE FAILURE

Wire can fail due to any number of causes. Here is a list of some of the common causes of wire rope failure.

- Using the incorrect size, construction, or grade of wire rope;
- Dragging rope over obstacles;
- Having improper lubrication;
- Operating over sheaves and drums of inadequate size;
- Overriding or crosswinding on drums;
- Operating over sheaves and drums with improperly fitted grooves or broken flanges;
- Jumping off sheaves;
- Subjecting it to acid fumes;
- Attaching fittings improperly;
- Promoting internal wear by allowing grit to penetrate between the strands; and
- Subjecting it to severe or continuing overload.

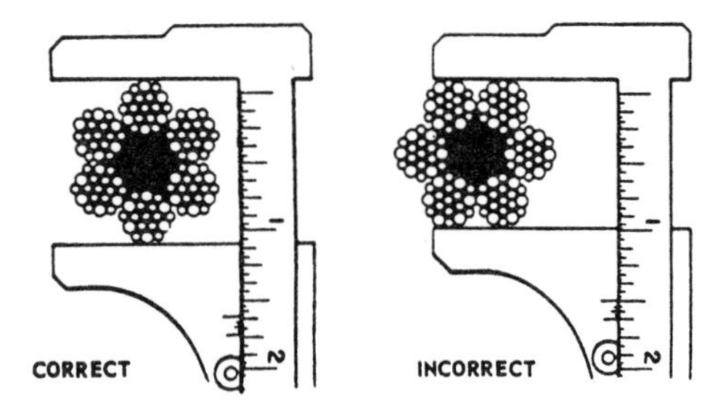

Figure 4-7.—Correct and incorrect methods of measuring wire rope.

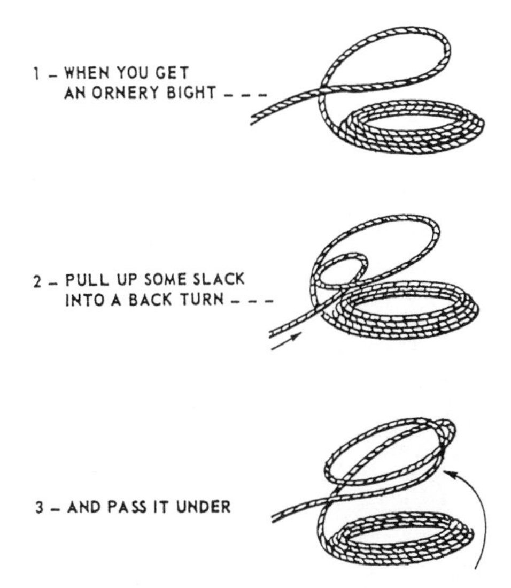

Figure 4-8.—Throwing a back turn to make wire lie down.

HANDLING AND CARE OF WIRE ROPE

To render safe, dependable service over a maximum period of time, wire rope must have the care and upkeep necessary to keep it in good condition. In this section, we'll discuss various ways of caring for and handling wire rope. Not only should you study these procedures carefully, you should also practice them on your job to help you do a better job now. In the long run, the life of the wire rope will be longer and more useful.

Coiling and Uncoiling

Once a new reel has been opened, it may be either coiled or faked down like line. The proper direction of coiling is counterclockwise for left-laid wire rope and clockwise for right-laid rope. Because of the general toughness and resilience of wire, however, it occasionally tends to resist being coiled down. When this occurs, it is useless to fight the wire by forcing down a stubborn turn; it will only spring up again. But if it is thrown in a back turn, as shown in figure 4-8, it will lie down properly. A wire rope, when faked down, will run right off like line; but when wound in a coil, it must always be unwound.

Wire rope tends to kink during uncoiling or unreeling, especially if it has been in service for a long time. A kink can cause a weak spot in the rope, which will wear out quicker than the rest of the rope. A good method for unreeling wire rope is to run a pipe or rod through the center and mount the reel on drum jacks or other supports so the reel is off the ground or deck (figure 4-9.) In this way, the reel will turn as the rope is unwound, and the rotation of the reel will help keep the rope straight. During unreeling, pull the rope straight forward, as shown in figure 4-9, and try to avoid hurrying the operation. As a safeguard against kinking, never unreel wire rope from a stationary reel.

To uncoil a small coil of wire rope, simply stand the coil on edge and roll it along the ground or deck like a wheel or hoop, as illustrated in figure 4-9. Never lay the coil flat on the deck or ground and uncoil it by pulling on the end because such practice can kink or twist the rope.

To rewind wire rope back onto a reel or a drum, you may have difficulty unless you remember that it tends to roll in the direction opposite the lay. For example, a right-laid wire rope tends to roll to the left.

Figure 4-9.—Unreeling wire rope (left) and uncoiling wire rope (right).

FOR RIGHT-LAY ROPE
(USE RIGHT HAND)

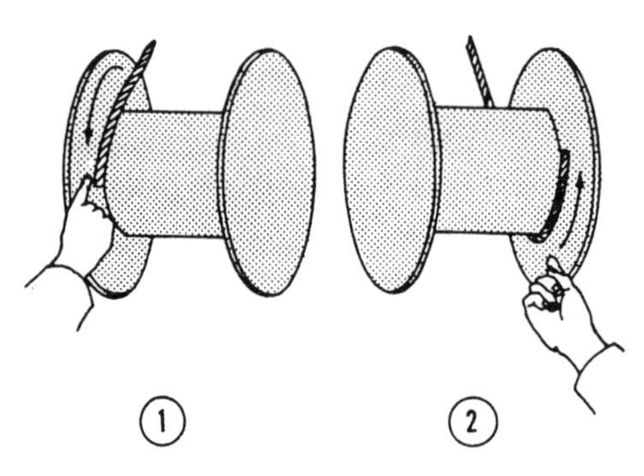

FOR OVERWIND
ON DRUM:

The palm is down, facing the drum.
The index finger points at on-winding rope.
The index finger must be closest to the left-side flange.
The wind of the rope must be from left to right along the drum.

FOR UNDERWIND
ON DRUM:

The palm is up, facing the drum.
The index finger points at on-winding rope.
The index finger must be closest to the right-side flange.
The wind of the rope must be from right to left along the drum.

FOR LEFT-LAY ROPE
(USE LEFT HAND)

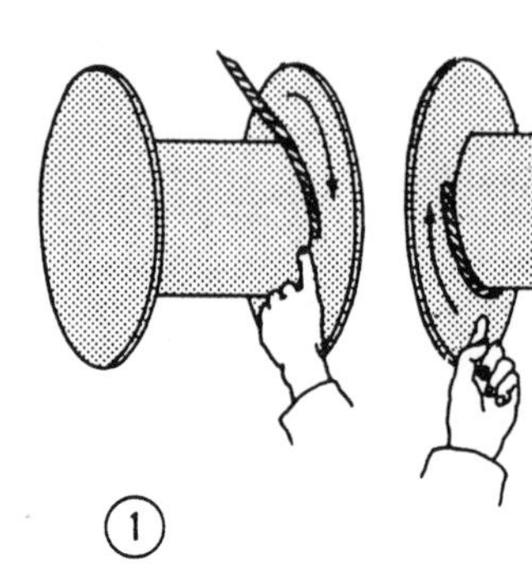

FOR OVERWIND
ON DRUM:

The palm is down, facing the drum.
The index finger points at on-winding rope.
The index finger must be closest to the right-side flange.
The wind of the rope must be from right to left along the drum.

FOR UNDERWIND
ON DRUM:

The palm is up, facing the drum.
The index finger points at on-winding rope.
The index finger must be closest to the left-side flange.
The wind of the rope must be from left to right along the drum.

If a smooth-face drum has been cut or scored by an old rope, the methods shown may not apply.

Figure 4-10.—Drum windings diagram for selecting the proper lay of rope.

Carefully study figure 4-10, which shows drum-winding diagrams selecting the proper lay of rope. When putting wire rope onto a drum, you should have no trouble if you know the methods of overwinding and underwinding shown in the illustration. When wire rope is run off one reel onto another, or onto a winch or drum, it should be run from top to top or from bottom to bottom, as shown in figure 4-11.

Kinks

If a wire rope should form a loop, never try to pull it out by putting strain on either part. As soon as a loop is noticed, uncross the ends by pushing them apart. (See steps 1 and 2 in figure 4-12.) This reverses the process that started the loop. Now, turn the bent portion over and place it on your knee or some firm object and push downward until the loop straightens out somewhat. (See step 3 in figure 4-12.) Then, lay the bent portion on a flat surface and pound it smooth with a wooden mallet. (See step 4 in figure 4-12.)

If a heavy strain has been put on a wire rope with a kink in it, the rope can no longer be trusted. Replace the wire rope altogether.

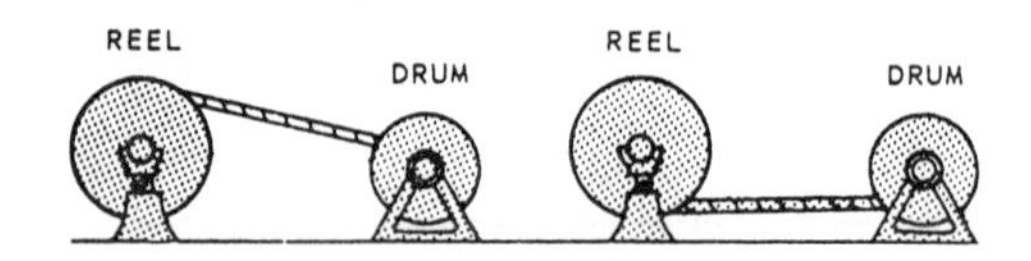

Figure 4-11.—Transferring wire from reel to drum.

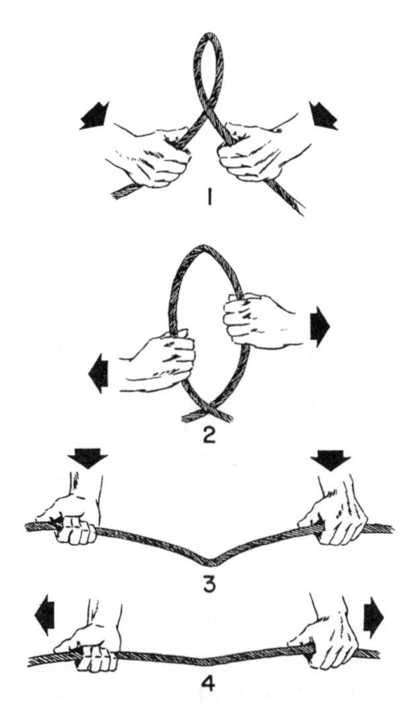

Figure 4-12.—The correct way to take out a loop in wire rope.

Lubrication

Used wire rope should be cleaned at frequent intervals to remove any accumulation of dirt, grit, rust, or other foreign matter. The frequency of cleaning depends on how much the rope is used. However, rope should always be well cleaned before lubrication. The rope can be cleaned by wire brushes, compressed air, or steam. **Do not use oxygen** in place of compressed air; it becomes very dangerous when it comes in contact with grease or oil. The purpose is to remove all old lubricant and foreign matter from the valleys between the strands and from the spaces between the outer wires. This gives newly applied lubricant ready entrance into the rope. Wire brushing affords a good opportunity to find any broken wires that may otherwise go unnoticed.

Wire rope is initially lubricated by the manufacturer, but this initial lubrication isn't permanent and periodic reapplications have to be made by the user. Each time a wire rope bends and straightens, the wires in the strands and the strands in the rope slide upon each other. To prevent the rope wearing out by this sliding action, a film of lubricant is needed between the surfaces in contact. The lubricant also helps prevent corrosion of the wires and deterioration of fiber centers. A rusty wire rope is a liability! With wire rope, the same as with any machine or piece of equipment, proper lubrication is essential to smooth, efficient performance.

The lubricant should be a good grade of lubricating oil, free from acids and corrosive substances. It must also be of a consistency that will penetrate to the center of the core, yet heavy enough to remain as a coating on the outer surfaces of the strands. Two good lubricants for this purpose are raw linseed oil and a medium graphite grease. Raw linseed oil dries and is not greasy to handle. Graphite grease is highly resistant to saltwater corrosion. Of course, other commercial lubricants may be obtained and used. One of the best is a semiplastic compound that is thinned by heating before being applied. It penetrates while hot, then cools to a plastic filler, preventing the entrance of water.

One method of applying the lubricant is by using a brush. In doing so, remember to apply the coating of fresh lubricant evenly and to work it in well. Another method involves passing the wire rope through a trough or box containing hot lubricant (figure 4-13). In this method, the heated lubricant is placed in the trough, and the rope passed over a sheave, through the lubricant, and under a second sheave. Hot oils or greases have very good penetrating qualities. Upon cooling, they have high adhesive and film strength around each wire.

As a safety precaution, always wipe off any excess when lubricating wire rope. This is especially important where heavy equipment is involved. Too much lubricant can get on brakes or clutches, causing them to fail. While in use, the motion of machinery

Figure 4-13.—Trough method of lubrication.

can throw excess oil onto crane cabs and catwalks, making them unsafe to work on.

Storage

Wire rope should not be stored in places where acid is or has been kept. The slightest trace of acid coming in contact with wire rope damages it at that particular spot. Many times, wire rope that has failed has been found to be acid damaged. The importance of keeping acid or acid fumes away from wire rope must be stressed to all hands.

It is especially important that wire rope be cleaned and lubricated properly before it is placed in storage. Fortunately, corrosion of wire rope during storage can be virtually eliminated if the lubricant film is applied properly beforehand and if adequate protection is provided from the weather. Bear in mind that rust, corrosion of wires, and deterioration of the fiber core greatly reduce the strength of wire rope. It is not possible to state exactly the loss of strength that results from these effects. It is certainly great enough to require close observance of those precautions prescribed for protection against such effects.

Inspection

Wire rope should be inspected at regular intervals, the same as fiber line. In determining the frequency of inspection, you need to carefully consider the amount of use of the rope and conditions under which it is used.

During an inspection, the rope should be examined carefully for fishhooks, kinks, and worn, corroded spots. Usually, breaks in individual wires are concentrated in those portions of the rope that consistently run over the sheaves or bend onto the drum. Abrasion or reverse and sharp bends cause individual wires to break and bend back. The breaks are known as fishhooks. When wires are only slightly worn, but have broken off squarely and stick out all over the rope, the condition is usually caused by overloading or rough handling. Even if the breaks are confined to only one or two strands, the strength of the rope may be seriously reduced. When 4 percent of the total number of wires in the rope are found to have breaks within the length of one lay of the rope, the wire rope is unsafe. Consider a rope unsafe when three broken wires are found in one strand of 6-by-7 rope, six broken wires in one strand of 6-by-19 rope, or nine broken wires in one strand of 6-by-37 rope.

Overloading a rope also causes its diameter to be reduced. Failure to lubricate the rope is another cause of reduced diameter since the fiber core will dry out and eventually collapse or shrink. The surrounding strands are thus deprived of support, and the rope's strength and dependability are correspondingly reduced. Rope that has its diameter reduced to less than 75 percent of its original diameter should be removed from service.

A wire rope should also be removed from service when an inspection reveals widespread corrosion and pitting of the wires. Particular attention should be given to signs of corrosion and rust in the valleys or small spaces between the strands. Since such corrosion is usually the result of improper or infrequent lubrication, the internal wires of the rope are then subject to extreme friction and wear. This form of internal, and often invisible, destruction of the wire is one of the most frequent causes of unexpected and sudden failure of wire rope. The best safeguard, of course, is to keep the rope well lubricated and to handle and store it properly.

WIRE ROPE ATTACHMENTS

Many attachments can be fitted to the ends of wire rope so that the rope can be connected to other wire ropes, pad eyes, or equipment. The attachment used most often to attach dead ends of wire ropes to pad eyes or like fittings on earthmoving rigs is the wedge socket shown in figure 4-14. The socket is applied to the bitter end of the wire rope, as shown in the figure.

Remove the pin and knock out the wedge first. Then, pass the wire rope up through the socket and

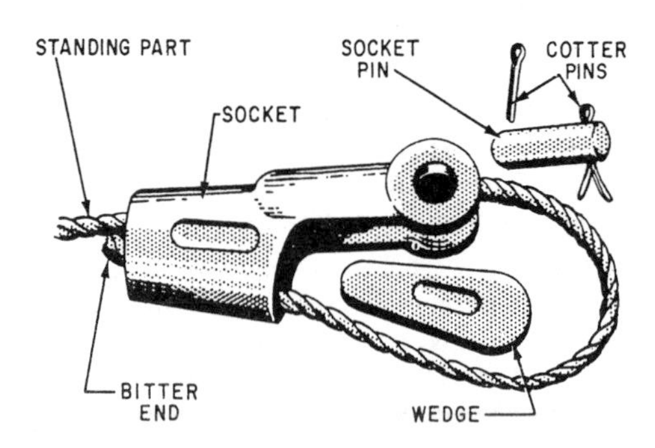

Figure 4-14.—Parts of a wedge socket.

lead enough of it back through the socket to allow a minimum of 6 to 9 inches of the bitter end to extend below the socket. Next, replace the wedge, and haul on the bitter end of the wire rope until the bight closes around the wedge, as shown in figure 4-15. A strain on the standing part will tighten the wedge. You need at least 6 to 9 inches on the dead end (the end of the line that doesn't carry the load). Finally, place one wire rope clip on the dead end to keep it from accidentally slipping back through the wedge socket. The clip should be approximately 3 inches from the socket. Use one size smaller clip than normal so that the threads on the U-bolt are only long enough to clamp tightly on one strand of wire rope. The other alternative is to use the normal size clip and loop the dead end back as shown in figure 4-15. Never attach the clip to the live end of the wire rope.

The advantage of the wedge socket is that it is easy to remove; just take off the wire clip and drive out the wedge. The disadvantage of the wedge socket is that it reduces the strength of wire rope by about 30 percent. Of course, reduced strength means less safe working load.

To make an eye in the end of a wire rope, use new wire rope clips, like those shown in figure 4-16. The U-shaped part of the clip with the threaded ends is called the U-bolt; the other part is called the saddle. The saddle is stamped with the diameter of the wire rope that the clip will fit. Always place a clip with the U-bolt on the bitter end, not on the standing part of the wire rope. If clips are attached incorrectly, the standing part (live end) of the wire rope will be distorted or have mashed spots. An easy way to remember is never saddle a dead horse.

You also need to determine the correct number of clips to use and the correct spacing. Here are two simple formulas.

3 × wire rope diameter + 1 = number of clips

6 × wire rope diameter = spacing between clips

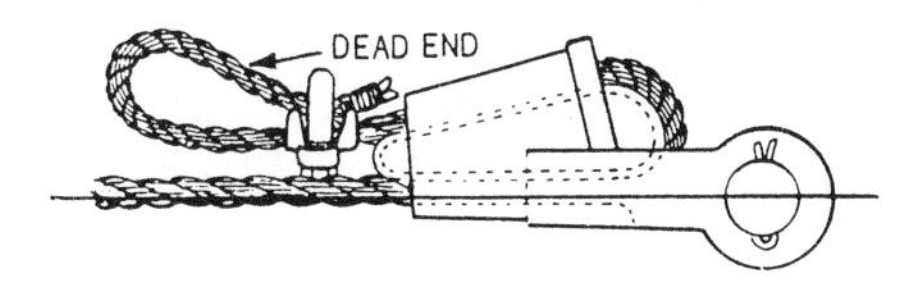

Figure 4-15.—Wedge socket attached properly.

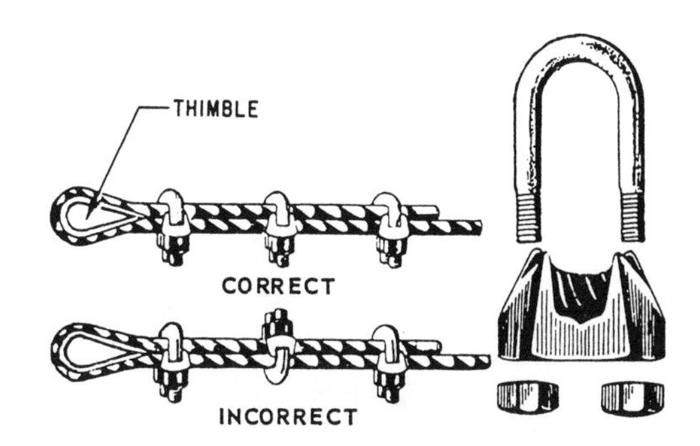

Figure 4-16.—Wire rope clips.

Another type of wire rope clip is the twin-base clip (sometimes referred to as the "universal" or "two-clamp") shown in figure 4-17. Since both parts of this clip are shaped to fit the wire rope, correct installation is almost certain. This considerably reduces potential damage to the rope. The twin-base clip also allows for a clean 360° swing with the wrench when the nuts are being tightened. When an eye is made in a wire rope, a metal fitting (called a thimble) is usually placed in the eye, as shown in figure 4-16, to protect the eye against wear. Clipped eyes with thimbles hold approximately 80 percent of the wire rope strength.

After the eye made with clips has been strained, the nuts on the clips must be retightened. Occasional checks should be made for tightness or damage to the rope caused by the clips.

Figure 4-17.—Twin-base wire clip.

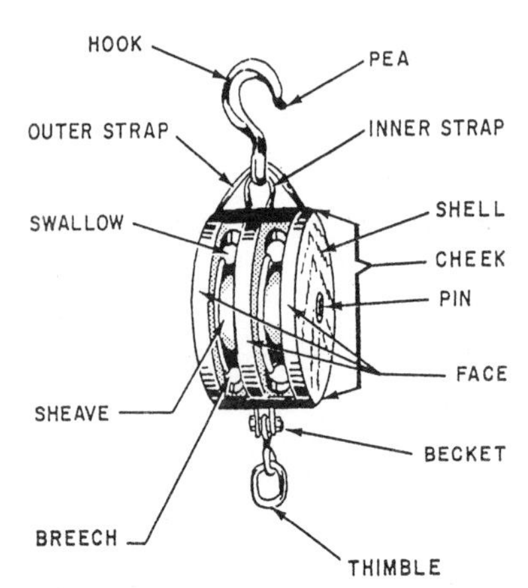

Figure 4-18.—Nomenclature of a fiber line block.

BLOCK AND TACKLE

LEARNING OBJECTIVE: Upon completing this section, you should be able to identify the components and operating characteristics of block and tackle units.

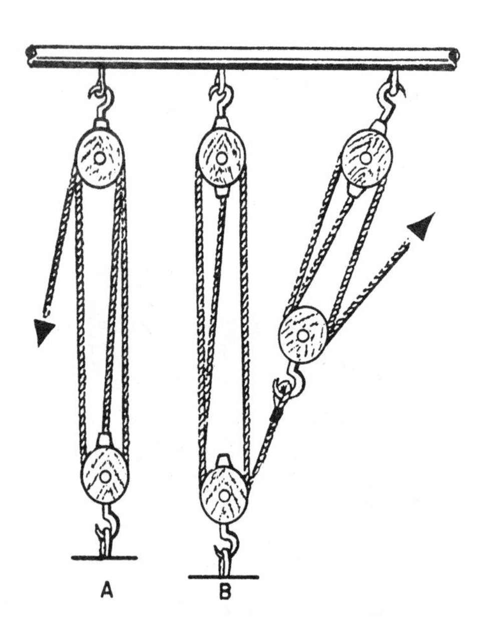

Figure 4-19.—Types of tackle: simple (view A) and compound (view B).

A block (figure 4-18) consists of one or more sheaves fitted in a wood or metal frame supported by a shackle inserted in the strap of the block. A tackle (figure 4-19) is an assembly of blocks and lines used to gain a mechanical advantage in lifting and pulling.

In a tackle assembly, the line is reeved over the sheave(s) of blocks. The two types of tackle systems are simple and compound. A simple tackle system is an assembly of blocks in which a single line is used (view A of figure 4-19). A compound tackle system is an assembly of blocks in which more than one line is used (view B of figure 4-19).

TACKLE TERMS

To help avoid confusion in working with tackle, you need a working knowledge of tackle vocabulary. Figure 4-20 will help you organize the various terms.

A fall is a line, either a fiber line or a wire rope, reeved through a pair of blocks to form a tackle. The hauling part is the part of the fall leading from one of the blocks upon which the power is exerted. The standing part is the end of the fall, which is attached to one of the beckets. The movable (or running) block of a tackle is the block attached to the object to be moved. The fixed (or standing) block is the block attached to a fixed object or support. When a tackle is being used, the movable block moves, and the fixed

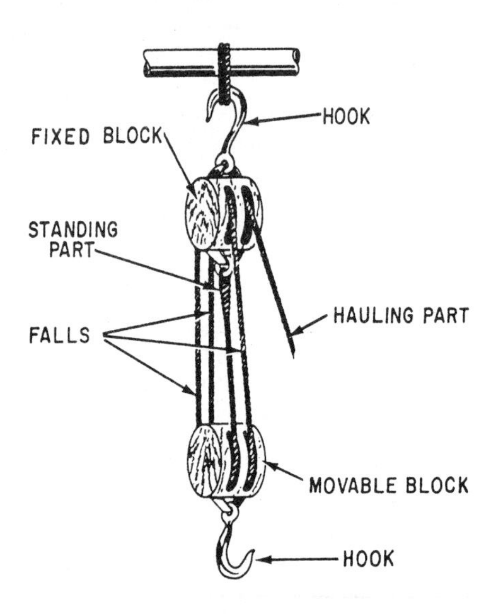

Figure 4-20.—Parts of a tackle.

block remains stationary. The term "two-blocked" means that both blocks of a tackle are as close together as they will go. You may also hear this term called block-and-block. To overhaul is to lengthen a tackle by pulling the two blocks apart. To round in means to bring the blocks of a tackle toward each other, usually without a load on the tackle (opposite of overhaul).

Don't be surprised if your coworkers use a number of different terms for a tackle. For example, line-and-blocks, purchase, and block-and-falls are typical of other names frequently used for tackle.

BLOCK NOMENCLATURE

The block (or blocks) in a tackle assembly changes (or change) the direction of pull or mechanical advantage, or both. The name and location of the key parts of a fiber line block are shown in figure 4-18.

The frame (or shell), made of wood or metal, houses the sheaves. The sheave is a round, grooved wheel over which the line runs. Ordinarily, blocks used in your work will have one, two, three, or four sheaves. Blocks come with more than this number of sheaves; some come with 11 sheaves. The cheeks are the solid sides of the frame, or shell. The pin is a metal axle that the sheave turns on. It runs from cheek to cheek through the middle of the sheave. The becket is a metal loop formed at one or both ends of a block; the standing part of the line is fastened to this part. The straps hold the block together and support the pin on which the sheaves rotate. The swallow is the opening in the block through which the line passes. The breech is the part of the block opposite the swallow.

CONSTRUCTION OF BLOCKS

Blocks are constructed for use with fiber line or wire rope. Wire rope blocks are heavily constructed and have a large sheave with a deep groove. Fiber line blocks are generally not as heavily constructed as wire rope blocks and have smaller sheaves with shallower wide grooves. A large sheave is needed with wire rope to prevent sharp bending. Since fiber line is more flexible and pliable than wire rope, it does not require a sheave as large as the same size of wire rope.

Blocks fitted with one, two, three, or four sheaves are often referred to as single, double, triple, and quadruple blocks, respectively. Blocks are fitted with a number of attachments, the number depending upon their use. Some of the most commonly used fittings are hooks, shackles, eyes, and rings. Figure 4-21 shows two metal frame, heavy-duty blocks. Block A is designed for manila line, and block B is for wire rope.

RATIO OF BLOCK SIZE TO LINE OR WIRE SIZE

The size of fiber line blocks is designated by the length in inches of the shell or cheek. The size of standard wire rope blocks is controlled by the diameter of the rope. With nonstandard and special-purpose wire rope blocks, the size is found by measuring the diameter of one of its sheaves in inches.

Use care in selecting the proper size line or wire for the block to be used. If a fiber line is reeved onto a tackle whose sheaves are below a certain minimum diameter, the line will be distorted and will soon wear badly. A wire rope too large for a sheave tends to be pinched and damages the sheave. The wire will also be damaged due to the too short a radius of the bend. A wire rope too small for a sheave lacks the necessary bearing surface, puts the strain on only a few strands, and shortens the life of the wire.

With fiber line, the length of the block used should be about three times the circumference of the

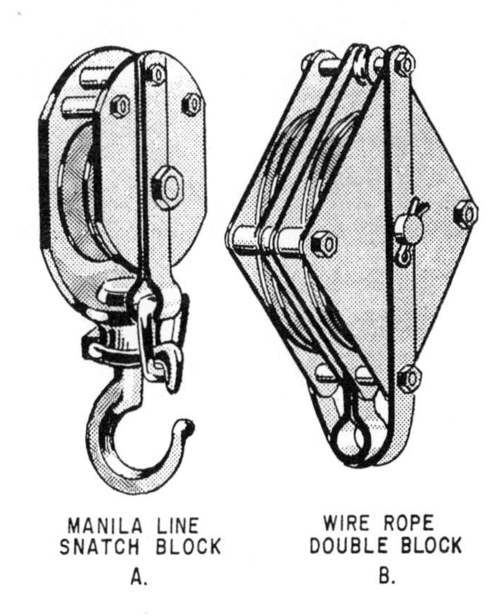

Figure 4-21.—Metal frame, heavy-duty blocks.

line. However, an inch or so either way doesn't matter too much; for example, a 3-inch line may be reeved onto an 8-inch block with no ill effects. As a rule, you are more likely to know the block size than the sheave diameter. However, the sheave diameter should be about twice the size of the circumference of the line used.

Wire rope manufacturers issue tables that give the proper sheave diameters used with the various types and sizes of wire rope they manufacture. In the absence of these, a rough rule of thumb is that the sheave diameter should be about 20 times the diameter of the wire. Remember that with wire rope, it is diameter rather than circumference that is important. Also, remember that this rule refers to the diameter of the sheave rather than to the size of the block.

SNATCH BLOCKS AND FAIRLEADS

A snatch block (figure 4-22) is a single-sheave block made so that the shell opens on one side at the base of the hook to permit a rope or line to be slipped over a sheave without threading the end of it through the block. Snatch blocks ordinarily are used where it is necessary to change the direction of the pull on a line.

Figure 4-23 shows a system of moving a heavy object horizontally away from the power source using snatch blocks. This is an ideal way to move objects in limited spaces. Note that the weight is pulled by a single luff tackle, which has a mechanical advantage of 3 (mechanical advantage is discussed below). Adding snatch blocks to a rigging changes the direction of pull, but the mechanical advantage is not affected. It is, therefore, wise to select the proper rigging system to be used based upon the weight of the object and the type and capacity of the power that is available.

The snatch block that is used as the last block in the direction of pull to the power source is called the leading block. This block can be placed in any convenient location provided it is within 20 drum widths of the power source. This is required because the fairlead angle, or fleet angle, cannot exceed 2° from the center line of the drum; therefore, the 20-drum width distance from the power source to the leading block will assure the fairlead angle. If the fairlead angle is not maintained, the line could jump the sheave of the leading block and cause the line on the reel to jump a riding turn.

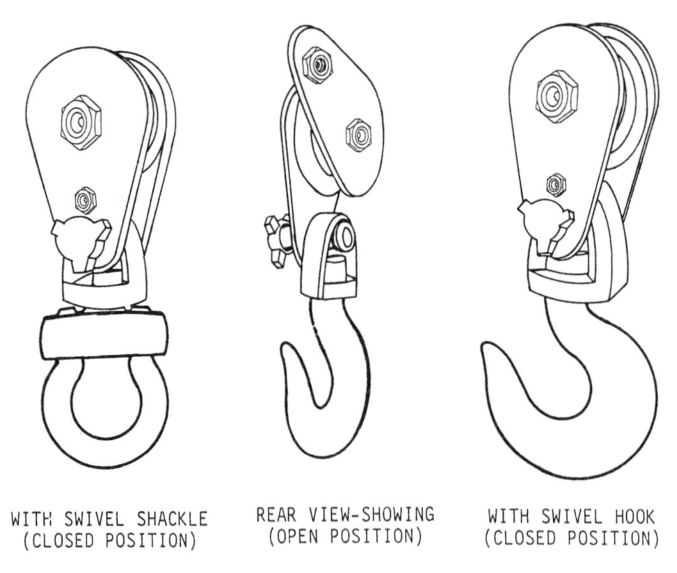

Figure 4-22.—Top dead end snatch blocks.

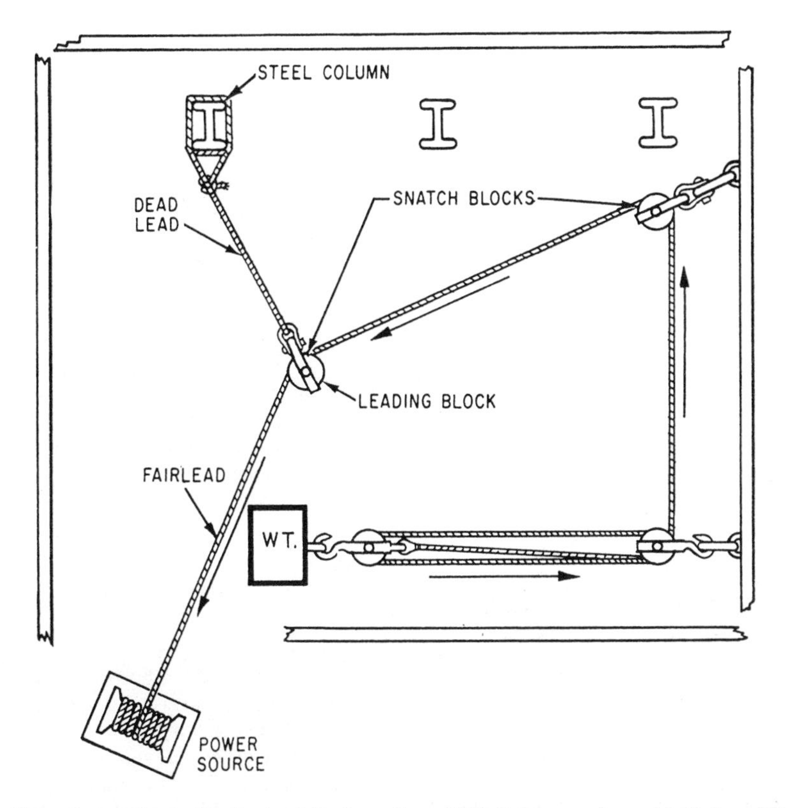

Figure 4-23.—Moving a heavy object horizontally along a floor with limited access using snatch blocks and fairleads.

MECHANICAL ADVANTAGE

The mechanical advantage of a tackle is the term applied to the relationship between the load being lifted and the power required to lift it. If the load and the power required to lift it are the same, the mechanical advantage is 1. However, if a load of 50 pounds requires only 10 pounds to lift it, then you have a mechanical advantage of 5 to 1, or 5 units of weight are lifted for each unit of power applied.

The easiest way to determine the mechanical advantage of a tackle is by counting the number of parts of the falls at the running block. If there are two parts, the mechanical advantage is two times the power applied (disregarding friction). A gun tackle, for instance, has a mechanical advantage of 2. Therefore, lifting a 200-pound load with a gun tackle requires 100 pounds of power, disregarding friction.

To determine the amount of power required to lift a given load by means of a tackle, determine the weight of the load to be lifted and divide that by the mechanical advantage. For example, if it is necessary to lift a 600-pound load by means of a single luff tackle, first determine the mechanical advantage gained by the tackle. By counting the parts of the falls at the movable block, you determine a mechanical advantage of 3. By dividing the weight to be lifted, 600 pounds, by the mechanical advantage in this tackle, 3, we find that 200 pounds of power is required to lift a weight of 600 pounds using a single luff tackle.

Remember though, a certain amount of the force applied to a tackle is lost through friction. Friction develops in a tackle by the lines rubbing against each other, or against the shell of a block. Therefore, an adequate allowance for the loss from friction must be added. Roughly, 10 percent of the load must be allowed for each sheave in the tackle.

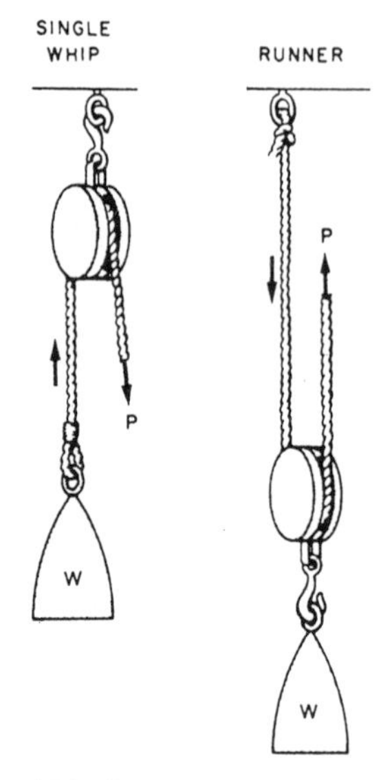

Figure 4-24.—Single-whip and runner tackle.

TYPES OF TACKLE

Tackles are designated in two ways: first, according to the number of sheaves in the blocks that are used to make the tackle, such as single whip or twofold purchase; and second, by the purpose for which the tackle is used, such as yard tackles or stay tackles. In this section, we'll discuss some of the different types of tackle in common use: namely, single whip, runner, gun tackle, single luff, twofold purchase, double luff, and threefold purchase. Before proceeding, we should point out that the purpose of the letters and arrows in figures 4-24 through 4-30 is to indicate the sequence and direction in which the standing part of the fall is led in reeving. You may want to refer to these illustrations when we discuss reeving of blocks in the next sections.

A single-whip tackle consists of one single-sheave block (tail block) fixed to a support with a rope passing over the sheave (figure 4-24.) It has a mechanical advantage of 1. If a 100-pound load is lifted, a pull of 100 pounds, plus an allowance for friction, is required.

A runner (figure 4-24) is a single-sheave movable block that is free to move along the line on which it is reeved. It has a mechanical advantage of 2.

A gun tackle is made up of two single-sheave blocks (figure 4-25). This tackle got its name in the old days because it was used to haul muzzle-loading guns back into the battery after the guns had been fired and reloaded. A gun tackle has a mechanical advantage of 2. To lift a 200-pound load with a gun tackle requires 100 pounds of power, disregarding friction.

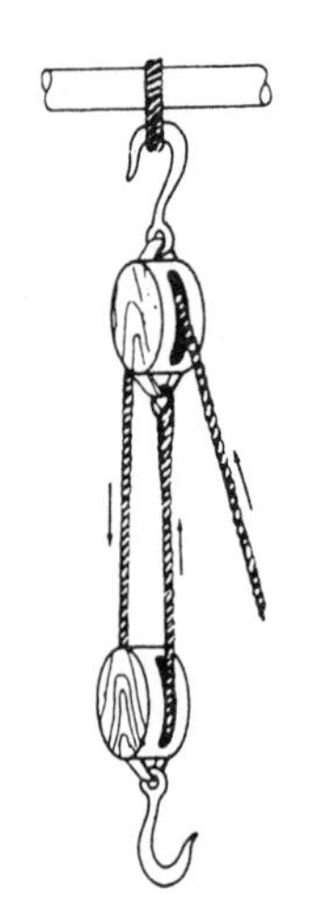
Figure 4-25.—Gun tackle.

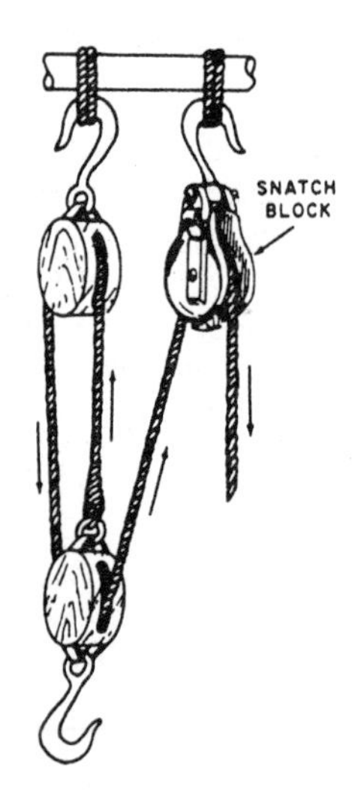

Figure 4-26.—Inverted gun tackle.

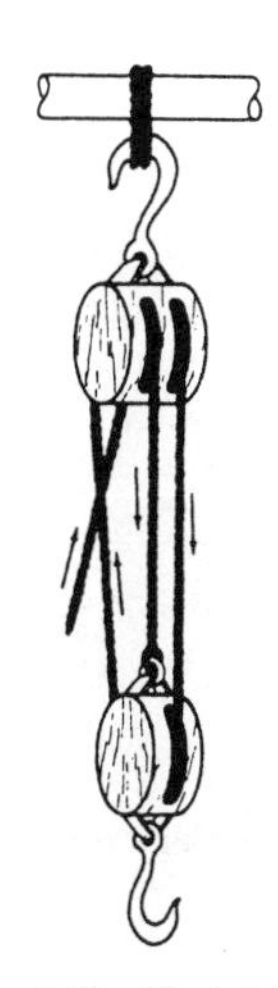

Figure 4-27.—Single-luff tackle.

By inverting any tackle, you always gain a mechanical advantage of 1 because the number of parts at the movable block is increased. By inverting a gun tackle, for example, you gain a mechanical advantage of 3 (figure 4-26). When a tackle is inverted, the direction of pull is difficult. This can easily be overcome by adding a snatch block, which changes the direction of the pull, but does not increase the mechanical advantage.

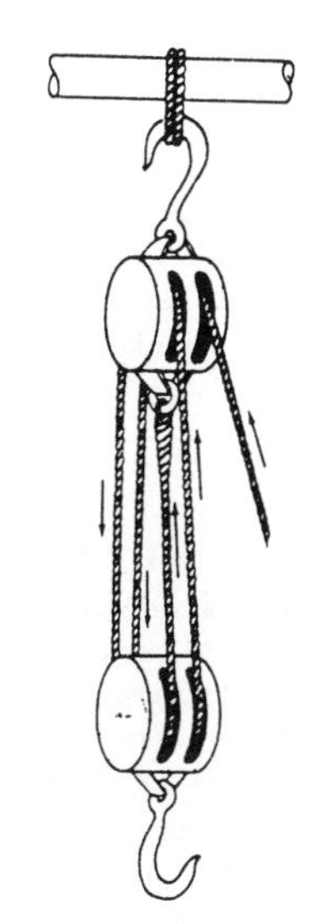

Figure 4-29.—Twofold purchase.

A single-luff tackle consists of a double and single block as indicated in figure 4-27, and the double-luff tackle has one triple and one double

Figure 4-28.—Double-luff tackle.

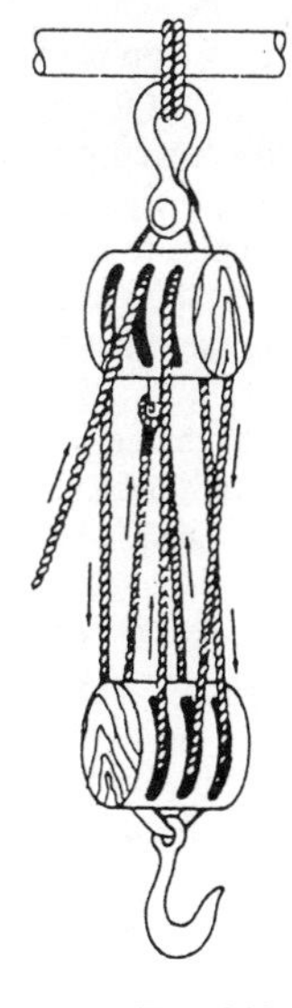

Figure 4-30.—Threefold purchase.

block, as shown in figure 4-28. The mechanical advantage of the single is 3, whereas the mechanical advantage of the double is 5.

A twofold purchase consists of two double blocks, as shown in figure 4-29, whereas a threefold purchase consists of two triple blocks, as shown in figure 4-30. The mechanical advantage of the twofold purchase is 4; the advantage of the threefold is 6.

REEVING TACKLE

In reeving a simple tackle, lay the blocks a few feet apart. The blocks should be placed down with the sheaves at right angles to each other and the becket ends pointing toward each other.

To begin reeving, lead the standing part of the falls through one sheave of the block that has the greatest number of sheaves. If both blocks have the same number of sheaves, begin at the block fitted with the becket. Then, pass the standing part around the sheaves from one block to the other, making sure no lines are crossed, until all sheaves have a line passing over them. Now, secure the standing part of the falls at the becket of the block containing the least number of sheaves, using a becket hitch for a temporary securing or an eye splice for a permanent securing.

With blocks of more than two sheaves, the standing part of the falls should be led through the sheave nearest the center of the block. This method places the strain on the center of the block and prevents the block from toppling and the lines from being cut by rubbing against the edges of the block.

Falls are generally reeved through 8- or 10-inch wood or metal blocks in such a manner as to have the lower block at right angles to the upper block. Two, three-sheave blocks are the usual arrangement, and the method of reeving these is shown in figure 4-31. The hauling part must go through the middle sheave of the upper block, or the block will tilt to the side and the falls jam when a strain is taken.

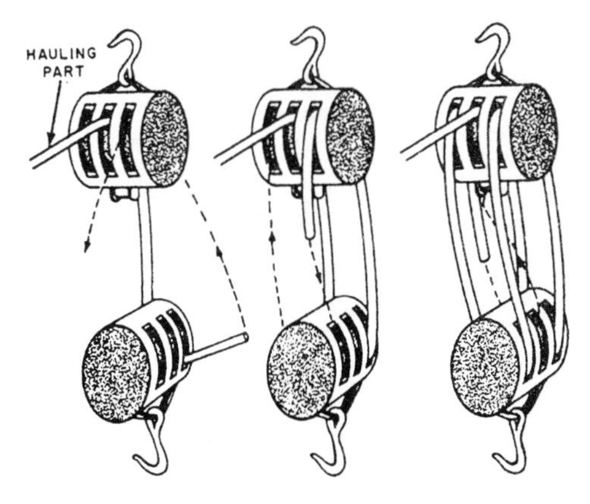

Figure 4-31.—Reeving a threefold purchase.

If a three- and two-sheave block rig is used, the method of reeving is about the same (figure 4-32), but, in this case, the becket for the dead end must be on the lower, rather than the upper, block.

Naturally, you must reeve the blocks before you splice in the becket thimble, or you will have to reeve the entire fall through from the opposite end.

SAFE WORKING LOAD OF A TACKLE

You know that the force applied at the hauling part of a tackle is multiplied as many times as there are parts of the fall on the movable block. Also, an allowance for friction must be made, which adds roughly 10 percent to the weight to be lifted for every sheave in the system. For example, if you are lifting a weight of 100 pounds with a tackle containing five sheaves, you must add 10 percent times 5, or 50 percent, of 100 pounds to the weight in your calculations. In other words, you determine that this tackle is going to lift 150 pounds instead of 100 pounds.

Disregarding friction, the safe working load of a tackle should be equal to the safe working load of the line or wire used, multiplied by the number of parts of the fall on the movable block. To make the necessary

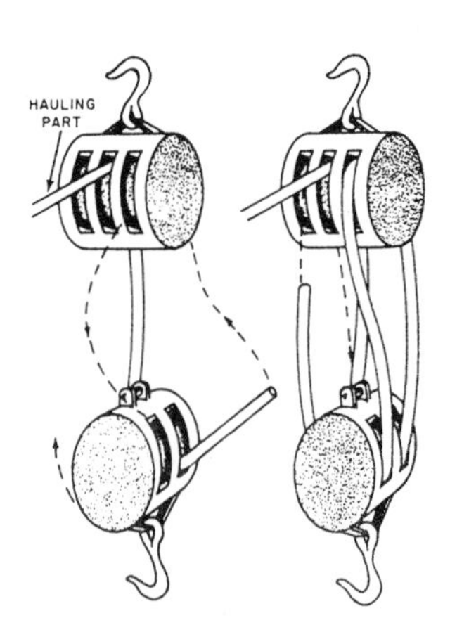

Figure 4-32.—Reeving a double-luff tackle.

allowance for friction, you multiply this result by 10, and then divide what you get by 10 plus the number of sheaves in the system.

Suppose you have a threefold purchase, a mechanical advantage of 6, reeved with a line that has a safe working load of 2 tons. Disregarding friction, 6 times 2, or 12 tons, should be the safe working load of this setup. To make the necessary allowance for friction, however, you first multiply 12 by 10, which gives you 120. This you divide by 10 plus 6 (number of sheaves in a threefold purchase), or 16. The answer is 7 1/2 tons safe working load.

Lifting a Given Weight

To find the size of **fiber** line required to lift a given load, use this formula:

$$C\ (in\ inches) = \sqrt{15 \times P\ (tons)}$$

C in the formula is the circumference, in inches, of the line that is safe to use. The number *15* is the conversion factor. *P* is the weight of the given load expressed in tons. The radical sign, or symbol, over $15 \times P$ indicates that you are to find the square root of that product.

To square a number means to multiply that number by itself. Finding the square root of a number simply means finding the number that, multiplied by itself, gives the number whose square root you are seeking. Most pocket calculators today have the square root function. Now, let's determine what size fiber line you need to hoist a 5-ton load. First, circumference equals 15 times five, or $C = 15 \times 5$, or 75. Next, the number that multiplied by itself comes nearest to 75 is 8.6. Therefore, a fiber line 8 1/2 inches in circumference will do the job.

The formula for finding the size of **wire** rope required to lift a given load is: $C\ (in\ inches) = 2.5 \times P\ (tons)$. You work this formula in the same manner explained above for fiber line. One point you should be careful not to overlook is that these formulas call for the circumference of the wire. You are used to talking about wire rope in terms of its diameter, so remember that circumference is about three times the diameter, roughly speaking. You can also determine circumference by the following formula, which is more accurate than the rule of thumb: circumference equals diameter times pi (π). In using this formula, remember that π equals approximately 3.14.

Size of Line to Use in a Tackle

To find the size of line to use in a tackle for a given load, add one-tenth (10 percent for friction) of its value to the weight to be hoisted for every sheave in the system. Divide the result you get by the number of parts of the fall at the movable block, and use this result as P in the formula

$$C = \sqrt{15 \times P}$$

For example, let's say you are trying to find the size of fiber line to reeve in a threefold block to lift 10 tons. There are six sheaves in a threefold block. Ten tons plus one-tenth for each of the six sheaves (a total of 6 tons) gives you a theoretical weight of 16 tons to be lifted. Divide 16 tons by 6 (number of parts on the movable block in a threefold block), and you get about 2 2/3. Using this as P in the formula you get

$$C = \sqrt{15 \times 2\frac{2}{3}}\ \ or \sqrt{40}\ \ or\ about\ 6.3$$

The square root of 40 is about 6.3, so it will take a line of about 6 1/2 inches in this purchase to hoist 10 tons safely. As you seldom find three-sheave blocks that will take a line as large as 6 1/2 inches, you will probably have to rig two threefold blocks with a continuous fall, as shown in figure 4-33. Each of

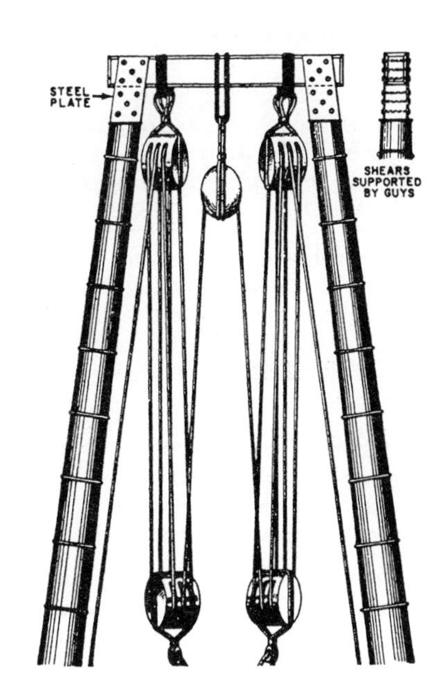

Figure 4-33.—Rigging two tackles with continuous fall.

these will have half of the load. To find the size of the line to use, calculate what size fiber line in a threefold block will lift 5 tons. It works out to about 4 1/2 inches.

TACKLE SAFETY PRECAUTIONS

In hoisting and moving heavy objects with blocks and tackle, stress safety for people and materials.

Always check the condition of blocks and sheaves before using them on a job to make sure they are in safe working order. See that the blocks are properly greased. Also, make sure that the line and sheave are the right size for the job.

Remember that sheaves or drums that have become worn, chipped, or corrugated must not be used because they will damage the line. Always find out whether you have enough mechanical advantage in the amount of blocks to make the load as easy to handle as possible.

Sheaves and blocks designed for use with fiber line must not be used for wire rope since they are not strong enough for that service, and the wire rope does not fit the sheave grooves. Also, sheaves and blocks built for wire rope should never be used for fiber line.

HOOKS AND SHACKLES

Hooks and shackles are handy for hauling or lifting loads without tying them directly to the object with a line or wire rope. They can be attached to wire rope, fiber line, or blocks. Shackles should be used for loads too heavy for hooks to handle.

Hooks should be inspected at the beginning of each workday and before lifting a full-rated load. Figure 4-34, view A, shows where to inspect a hook for wear and strain. Be especially careful during the inspection to look for cracks in the saddle section and at the neck of the hook.

When the load is too heavy for you to use a hook, use a shackle. Shackles, like hooks, should be inspected on a daily routine and before lifting heavy loads. Figure 4-34, view B, shows the area to look for wear.

You should never replace the shackle pin with a bolt. Never use a shackle with a bent pin, and never allow the shackle to be pulled at an angle; doing so will reduce its carrying capacity. Packing the pin with washers centralizes the shackle (figure 4-34, view B).

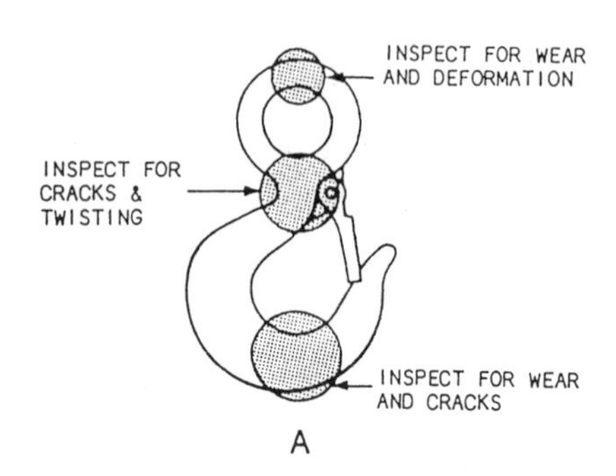

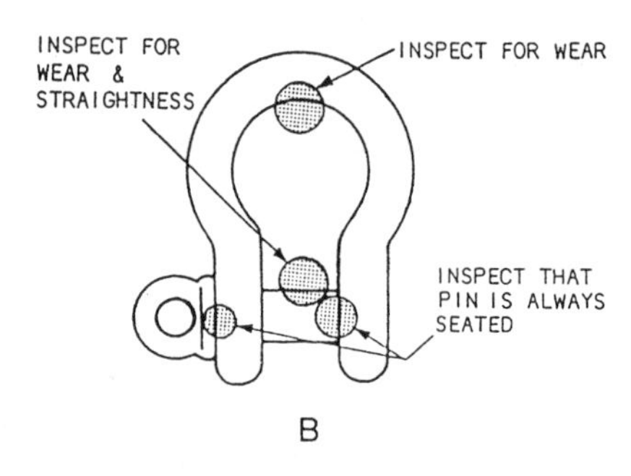

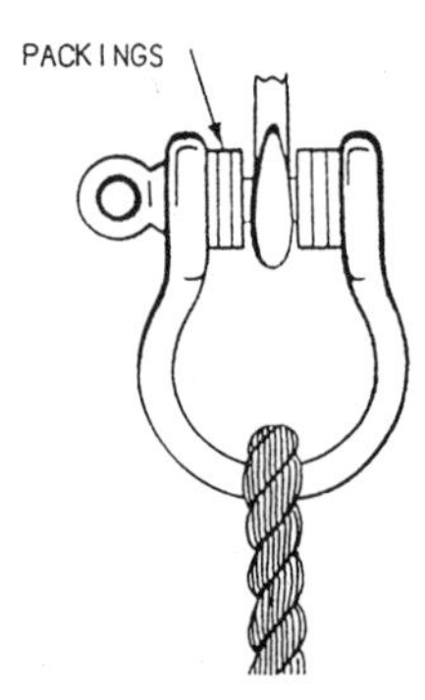

Figure 4-34.—Hook and shackle inspection (views A and B) and packing a shackle with washers.

If you need a hook or shackle for a job, always get it from Alfa Company. This way, you will know that it has been load tested.

Mousing is a technique often used to close the open section of a hook to keep slings, straps, and so on, from slipping off the hook (figure 4-35). To some extent, it also helps prevent straightening of the hook. Hooks may be moused with rope yarn, seizing wire, or a shackle. When using rope yarn or wire, make 8

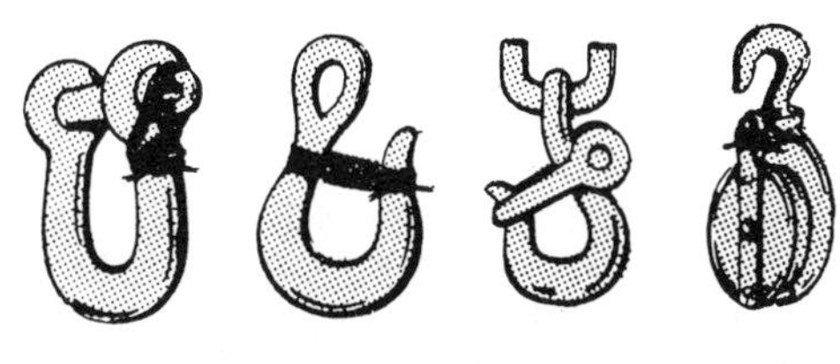

Figure 4-35.—Mousing.

or 10 wraps around both sides of the hook. To finish off, make several turns with the yarn or wire around the sides of the mousing, and then tie the ends securely (figure 4-35).

Shackles are moused when there is danger of the shackle pin working loose and coming out because of vibration. To mouse a shackle, simply take several turns with seizing wire through the eye of the pin and around the bow of the shackle. Figure 4-35 shows what a properly moused shackle looks like.

HOISTING

LEARNING OBJECTIVE: Upon completing this section, you should have a basic understanding of hoisting, handsignals used in lifting loads, and some of the safety rules of lifting.

In lifting any load, it takes two personnel to ensure a safe lift: an equipment operator and a signalman. In the following paragraphs, we will discuss the importance of the signalman and a few of the safety rules to be observed by all hands engaged in hooking on.

SIGNALMAN

One person, and one person only, should be designated as the official signalman for the operator of a piece of hoisting equipment, and both the signalman and the operator must be thoroughly familiar with the standard hand signals. When possible, the signalman should wear some distinctive article of dress, such as a bright-colored helmet. The signalman must maintain a position from which the load and the crew working on it can be seen, and also where he can be seen by the operator.

Appendix III at the end of this TRAMAN shows the standard hand signals for hoisting equipment. Some of the signals shown apply only to mobile equipment; others, to equipment with a boom that can be raised, lowered, and swung in a circle. The two-arm hoist and lower signals are used when the signalman desires to control the speed of hoisting or lowering. The one-arm **hoist** or **lower** signal allows the operator raise or lower the load. To dog off the load and boom means to set the brakes so as to lock both the hoisting mechanism and the boom hoist mechanism. The signal is given when circumstances require that the load be left hanging motionless.

With the exception of the emergency stop signal, which may be given by anyone who sees a necessity for it, and which must be obeyed instantly by the operator, only the official signalman gives the signals. The signalman is responsible for making sure that members of the crew remove their hands from slings, hooks, and loads before giving a signal. The signalman should also make sure that all persons are clear of bights and snatch block lines.

ATTACHING A LOAD

The most common way of attaching a load to a lifting hook is to put a sling around the load and hang the sling on the hook (figure 4-36). A sling can be made of line, wire, or wire rope with an eye in each

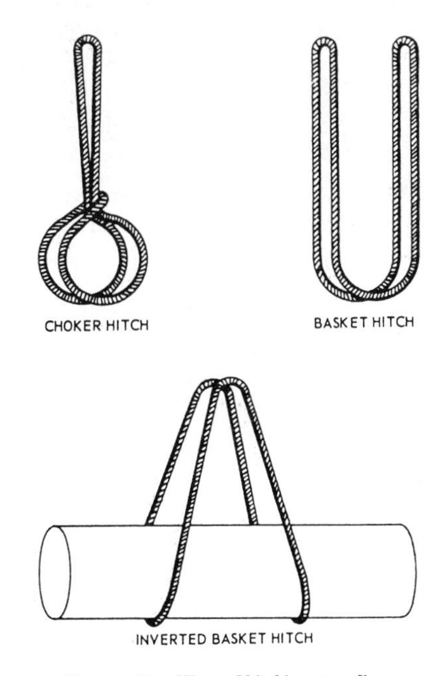

Figure 4-36.—Ways of hitching on a sling.

end (also called a strap) or an endless sling (figure 4-37). When a sling is passed through its own bight or eye, or shackled or hooked to its own standing part, so that it tightens around the load like a lasso when the load is lifted, the sling is said to be choked, or it may be called a choker, as shown in figures 4-36 and 4-37. A two-legged sling that supports the load at two points is called a bridle, as shown in figure 4-38.

SAFETY RULES

The following safety rules must be given to all hands engaged in hooking on. They must be strictly observed.

- The person in charge of hooking on must know the safe working load of the rig and the weight of every load to be hoisted. The hoisting of any load heavier than the safe working load of the rig is absolutely prohibited.
- When a cylindrical metal object, such as a length of pipe, a gas cylinder, or the like, is hoisted in a choker bridle, each leg of the bridle should be given a round turn around the load before it is hooked or shackled to its own part or have a spreader bar placed between the legs. The purpose of this is to ensure that the legs of the bridle will not slide together along the load, thereby upsetting the balance and possibly dumping the load.
- The point of strain on a hook must never be at or near the point of the hook.
- Before the hoist signal is given, the person in charge must be sure that the load will balance evenly in the sling.
- Before the hoist signal is given, the person in charge should be sure that the lead of the whip or falls is vertical. If it is not, the load will take a swing as it leaves the deck or ground.
- As the load leaves the deck or ground, the person in charge must watch carefully for kinked or fouled falls or slings. If any are observed, the load must be lowered at once for clearing.
- Tag lines must be used to guide and steady a load when there is a possibility that the load might get out of control.
- Before any load is hoisted, it must be inspected carefully for loose parts or objects that might drop as the load goes up.
- All personnel must be cleared from and kept out of any area that is under a suspended load, or over which a suspended load may pass.
- Never walk or run under a suspended load.
- Loads must not be placed and left at any point closer than 4 feet 8 inches from the nearest rail

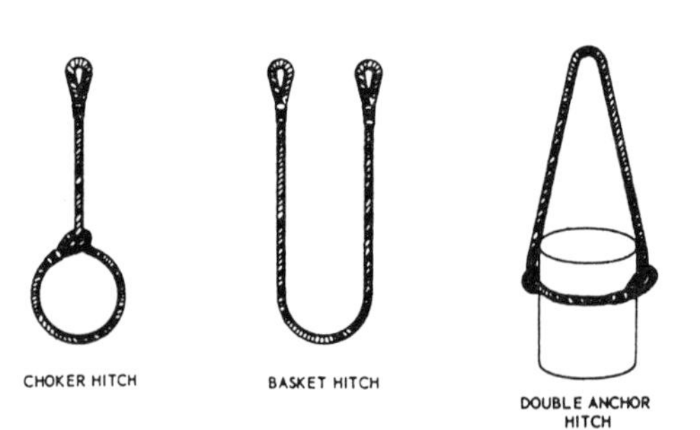

Figure 4-37.—Ways of hitching on straps.

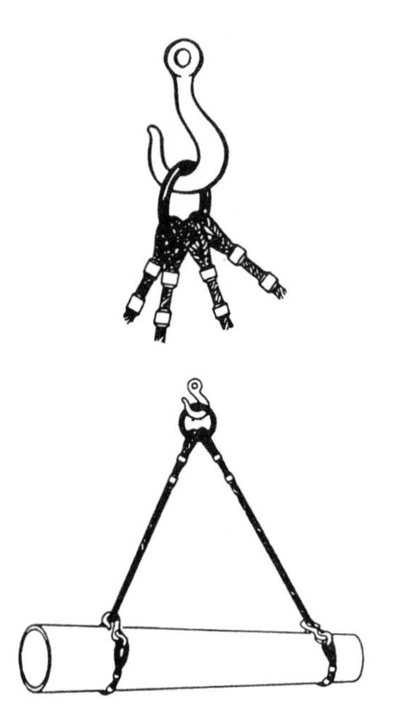

Figure 4-38.—Bridles.

of a railroad track or crane truck, or in any position where they would impede or prevent access to fire-fighting equipment.

- When materials are being loaded or unloaded from any vehicle by crane, the vehicle operators and all other persons, except the rigging crew, should stand clear.
- When materials are placed in work or storage areas, dunnage or shoring must be provided, as necessary, to prevent tipping of the load or shifting of the materials.
- All crew members must stand clear of loads that tend to spread out when landed.
- When slings are being heaved out from under a load, all crew members must stand clear to avoid a backlash, and also to avoid a toppling or a tip of the load, which might be caused by fouling of a sling.

SHEAR LEGS

The shear legs are formed by crossing two timbers, poles, planks, pipes, or steel bars and lashing or bolting them together near the top. A sling is suspended from the lashed intersection and is used as a means of supporting the load tackle system (figure 4-39). In addition to the name shear legs, this rig often is referred to simply as a "shears". (It has also been called an A-frame.)

The shear legs are used to lift heavy machinery and other bulky objects. They may also be used as end supports of a cableway and highline. The fact that the shears can be quickly assembled and erected is a major reason why they are used in field work.

A shears requires only two guy lines and can be used for working at a forward angle. The forward guy does not have much strain imposed on it during hoisting. This guy is used primarily as an aid in adjusting the drift of the shears and in keeping the top of the rig steady in hoisting or placing a load. The after guy is a very important part of the shears' rigging, as it is under considerable strain when hoisting. It should be designed for a strength equal to one-half the load to be lifted. The same principles for thrust on the spars or poles apply; that is, the thrust increases drastically as the shear legs go off the perpendicular.

In rigging the shears, place your two spars on the ground parallel to each other and with their butt ends even. Next, put a large block of wood under the tops of the legs just below the point of lashing, and place a small block of wood between the tops at the same point to facilitate handling of the lashing. Now, separate the poles a distance equal to about one-third the diameter of one pole.

As lashing material, use 18- or 21-thread small stuff. In applying the lashing, first make a clove hitch around one of the legs. Then, take about eight or nine turns around both legs above the hitch, working towards the top of the legs. Remember to wrap the turns tightly so that the finished lashing will be smooth and free of kinks. To apply the frapping (tight lashings), make two or three turns around the lashing between the legs; then, with a clove hitch, secure the end of the line to the other leg just below the lashing (figure 4-39).

Now, cross the legs of the shears at the top, and separate the butt ends of the two legs so that the spread between them is equal to one-half the height of the shears. Dig shallow holes, about 1 foot (30 cm) deep, at the butt end of each leg. The butts of the legs should be placed in these holes in erecting the shears. Placing the legs in the holes will keep them from kicking out in operations where the shears are at an angle other than vertical.

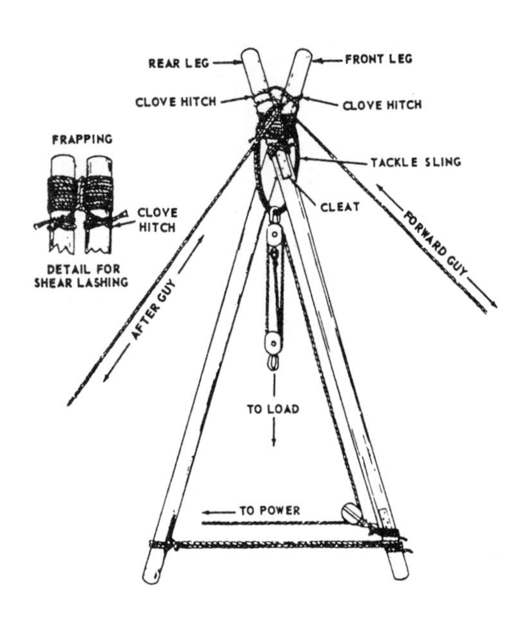

Figure 4-39.—Shear legs.

The next step is to form the sling for the hoisting falls. To do this, take a short length of line, pass it a sufficient number of times over the cross at the top of the shears, and tie the ends together. Then, reeve a set of blocks and place the hook of the upper block through the sling, and secure the hook by mousing the open section of the hook with rope yarn to keep it from slipping off the sling. Fasten a snatch block to the lower part of one of the legs, as indicated in figure 4-39.

The guys—one forward guy and one after guy—are secured next to the top of the shears. Secure the forward guy to the rear leg and the after guy to the front leg using a clove hitch in both instances. If you need to move the load horizontally by moving the head of the shears, you must rig a tackle in the after guy near its anchorage.

TRIPODS

A tripod consists of three legs of equal length that are lashed together at the top (figure 4-40). The legs are generally made of timber poles or pipes. Materials used for lashing include fiber line, wire rope, and chain. Metal rings joined with short chain sections are also available for insertion over the top of the tripod legs.

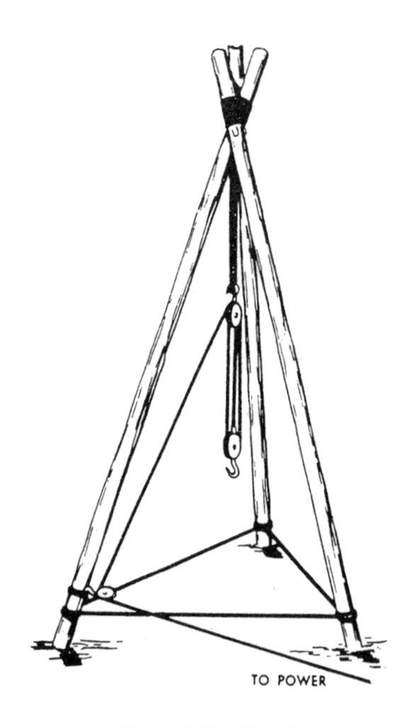

Figure 4-40.—Tripod.

When compared with other hoisting devices, the tripod has a distinct disadvantage: it is limited to hoisting loads only vertically. Its use will be limited primarily to jobs that involve hoisting over wells, mine shafts, or other such excavations. A major advantage of the tripod is its great stability. In addition, it requires no guys or anchorages, and its load capacity is approximately one-third greater than shears made of the same-size timbers. Table 4-1 gives the load-carrying capacities of shear legs and tripods for various pole sizes.

Rigging Tripods

The strength of a tripod depends largely on the strength of the material used for lashing, as well as the amount of lashing used. The following procedure for

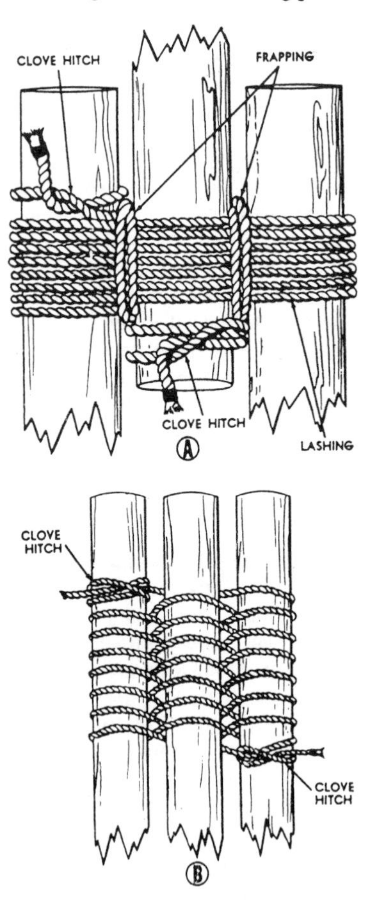

Figure 4-41.—Lashings for a tripod.

Table 4-1.—Load-Carrying Capacities of Shear Legs and Tripods

POLE SIZE (INCHES)	LENGTH (FEET)	WORKING CAPACITY (TONS) SHEAR LEGS (2) POLES	WORKING CAPACITY (TONS) TRIPODS (3) POLES
6 × 6	20	8	13
	25	5	7
	30	3	5
8 × 8	25	12	18
	30	8	13
	40	5	7
	50	3	5
10 × 10	20	35	52
	25	26	39
	30	17	26
	40	10	15
	50	7	10
12 × 12	30	35	52
	40	21	31
	50	14	21
	60	10	15

lashing applies to a line 3 inches in circumference or smaller. For extra heavy loads, use more turns than specified in the procedure given here. For light loads, use fewer turns than specified here.

As the first step of the procedure, take three spars of equal length and place a mark near the top of each to indicate the center of the lashing. Now, lay two of the spars parallel with their tops resting on a skid (or block). Place the third spar between the two, with the butt end resting on a skid. Position the spars so that the lashing marks on all three are in line. Leave an interval between the spars equal to about one-half the diameter of the spars. This will keep the lashing from being drawn too tightly when the tripod is erected.

With the 3-inch line, make a clove hitch around one of the outside spars; put it about 4 inches above the lashing mark. Then, make eight or nine turns with the line around all three spars. (See view A of figure 4-41.) In making the turns, remember to maintain the proper amount of space between the spars.

Now, make one or two close frapping turns around the lashing between each pair of spars. Do not draw the turns too tightly. Finally, secure the end of the line with a clove hitch on the center spar just above the lashing, as shown in view A of figure 4-41.

There is another method of lashing a tripod that you may find preferable to the method just given. It may be used in lashing slender poles up to 20 feet in length, or when some means other than hand power is available for erection.

First, place the three spars parallel to each other, leaving an interval between them slightly greater than twice the diameter of the line to be used. Rest the top of each pole on a skid so that the end projects about 2 feet over the skid. Then, line up the butts of the three spars, as indicated in view B of figure 4-41.

Next, make a clove hitch on one outside leg at the bottom of the position the lashing will occupy, which is about 2 feet from the end. Now, proceed to weave the line over the middle leg, under and around the other outside leg, under the middle leg, over and around the first leg, and so forth, until completing about eight or nine turns. Finish the lashing by forming a clove hitch on the other outside leg (view B of figure 4-41).

ERECTING TRIPODS

In the final position of an erected tripod, it is important that the legs be spread an equal distance apart. The spread between legs must be no more than two-thirds nor less than one-half the length of a leg. Small tripods, or those lashed according to the first procedure given in the preceding section, may be raised by hand. Here are the main steps.

Start by raising the top ends of the three legs about 4 feet, keeping the butt ends of the legs on the ground. Now, cross the tops of the two outer legs, and position the top of the third or center leg so that it rests on top of the cross.

A sling for the hoisting tackle can be attached readily by first passing the sling over the center leg, and then around the two outer legs at the cross. Place the hook of the upper block of a tackle on the sling, and secure the hook by mousing.

The raising operation can now be completed. To raise an ordinary tripod, a crew of about eight may be required. As the tripod is being lifted, spread the legs so that when it is in the upright position, the legs will be spread the proper distance apart. After getting the tripod in its final position, lash the legs near the bottom with line or chain to keep them from shifting (figure 4-40). Where desirable, a leading block for the hauling part of the tackle can be lashed to one of the tripod legs, as indicated in figure 4-40.

In erecting a large tripod, you may need a small gin pole to aid in raising the tripod into position. To erect a tripod lashed according to the first procedure described in the preceding section, you first raise the tops of the legs far enough from the ground to permit spreading them apart. Use guys or tag lines to help hold the legs steady while they are being raised. Now, with the legs clear of the ground, cross the two outer legs and place the center leg so that it rests on top of the cross. Then, attach the sling for the hoisting tackle. Here, as with a small tripod, simply pass the sling over the center leg and then around the two outer legs at the cross.

SCAFFOLDING

LEARNING OBJECTIVE: Upon completing this section, you should be able to determine the proper usage of wood and prefabricated metal scaffolding.

As the working level of a structure rises above the reach of crew members on the ground or deck, temporary elevated platforms, called scaffolding, are erected to support the crew members, their tools, and materials.

There are two types of scaffolding in use today—wood and prefabricated. The wood types include the swinging scaffold, which is suspended from above, and the pole scaffold, which is supported on the ground or deck. The prefabricated type is made of metal and is put together in sections, as needed.

SWINGING SCAFFOLD CONSTRUCTION

The simplest type of a swinging scaffold consists of an unspliced plank that is made from 2-by-8-inch (minimum) lumber. Hangers should be placed between 6 and 18 inches from the ends of the plank. The span between hangers should not exceed 10 feet. Make sure that the hangers are secured to the plank to stop them from slipping off. Figure 4-42 shows the construction of a hanger with a guardrail. The guardrail should be made of 2-by-4-inch material between 36- and 42-inches high. A midrail, if required, should be constructed of 1-by-4 lumber.

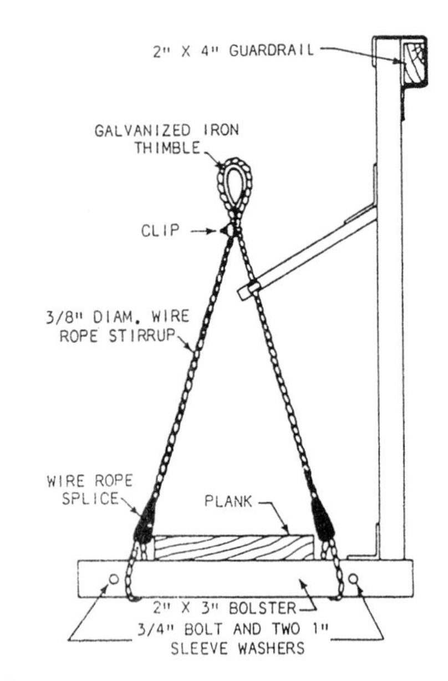

Figure 4-42.—Typical hanger to use with plank scaffold.

Swing scaffolds should be suspended by wire or fiber line secured to the outrigger beams. A minimum safety factor of 6 is required for suspension ropes. The blocks for fiber ropes should be the standard 6-inch size consisting of at least one double block and one single block. The sheaves of all blocks should fit the size of rope used.

The outrigger beams should be spaced no more than the hanger spacing and should be constructed of no less than 2-by-10 lumber. The beam should not extend more than 6 feet beyond the face of the building. The inboard side should be 9 feet beyond the edge of the building and should be securely fastened to the building.

Figure 4-43 shows a swinging scaffold that can be used for heavy work with block and tackle.

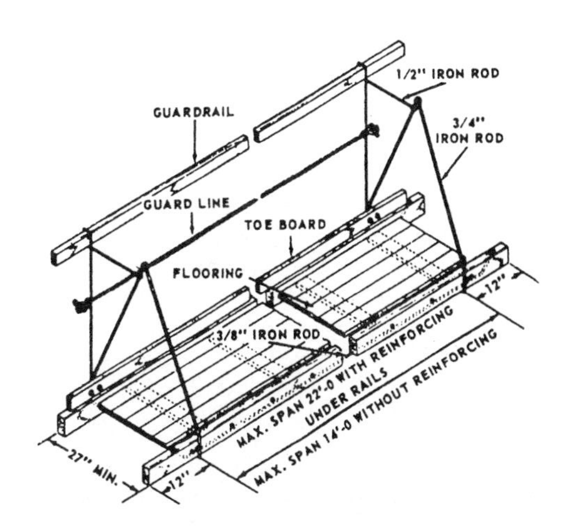

Figure 4-43.—Swinging scaffold.

POLE SCAFFOLD CONSTRUCTION

The poles on a job-built pole scaffold should not exceed 60 feet in height. If higher poles are required, the scaffolding must be designed by an engineer.

- All poles must be set up perfectly plumb.
- The lower ends of poles must not bear directly on a natural earth surface. If the surface is earth, a board footing 2-inches thick and 6- to 12-inches wide (depending on the softness of the earth) must be placed under the poles.
- If poles must be spliced, splice plates must not be less than 4-feet long, not less than the width of the pole wide, and each pair of plates must have a combined thickness not less than the thickness of the pole. Adjacent poles must not be spliced at the same level.
- A ledger must be long enough to extend over two pole spaces, and it must overlap the poles at the ends by at least 4 inches. Ledgers must be spliced by overlapping and nailing at poles—never between poles. If platform planks are raised as work progresses upward, the ledgers and logs on which the planks previously rested must be left in place to brace and stiffen the poles. For a heavy-duty scaffold, ledgers must be supported by cleats, nailed or bolted to the poles, as well as by being nailed themselves to the poles.
- A single log must be set with the longer section dimension vertical, and logs must be long enough to overlap the poles by at least 3 inches. They should be both face nailed to the poles and toenailed to the ledgers. When the inner end of the log butts against the wall (as it does in a single-pole scaffold), it must be supported by a 2-by-6-inch bearing block, not less than 12 inches long, notched out the width of the log and securely nailed to the wall. The inner end of the log should be nailed to both the bearing block and the wall. If the inner end of a log is located in a window opening, it must be supported on a stout plank nailed across the opening. If the inner end of a log is nailed to a building stud, it must be supported on a cleat, the same thickness as the log, and nailed to the stud.
- A platform plank must never be less than 2-inches thick. Edges of planks should be close enough together to prevent tools or materials from falling through the opening. A plank must be long enough to extend over three logs, with an overlap of at least 6 inches, but not more than 12 inches.

PREFABRICATED SCAFFOLD ERECTION

Several types of scaffolding are available for simple and rapid erection, one of which is shown in figure 4-44. The scaffold uprights are braced with diagonal members, and the working level is covered with a platform of planks. All bracing must form triangles, and the base of each column requires adequate footing plates for bearing area on the ground or deck. The steel scaffolding is usually erected by placing the two uprights on the ground or deck and inserting the diagonal members. The diagonal members have end fittings that permit rapid locking in position. In tiered scaffolding, figure 4-45, the first tier is set on steel bases on the ground, and a second tier is placed in the same manner on the first tier with the bottom of each upright locked to the top of the lower tier. A third and fourth upright can be placed on the ground level and locked to the first set with diagonal bracing. The scaffolding can be built as high as desired, but high scaffolding should be tied to the main structure. Where necessary, scaffolding can be mounted on casters for easy movement.

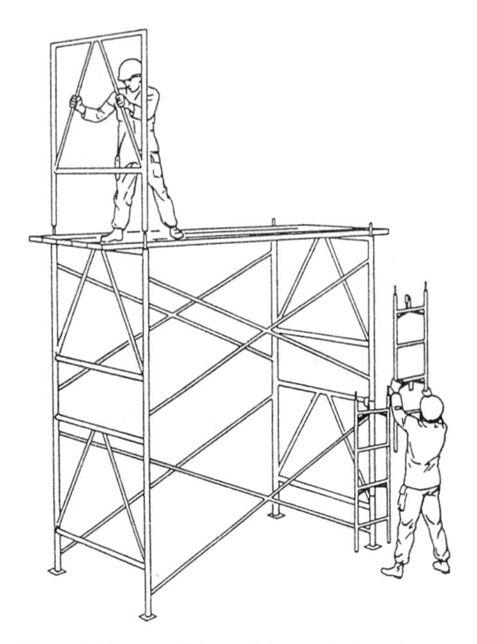

Figure 4-44.—Assembling prefabricated independent-pole scaffolding.

Prefabricated scaffolding comes in three categories: light, medium, and heavy duty. Light duty has nominal 2-inch-outside-diameter steel-tubing bearers. Posts are spaced no more than 6- to 10-feet apart. Light-duty scaffolding must be able to support 25-pound-per-square-foot loads.

Medium-duty scaffolding normally uses 2-inch-outside-diameter steel-tubing bearers. Posts should be spaced no more than 5- to 8-feet apart. If 2 1/2-inch-outside-diameter steel-tubing bearers are used, posts are be spaced 6- to 8-feet apart. Medium-duty scaffolding must be able to support 50-pound-per-square-foot loads.

Heavy-duty scaffolding should have bearers of 2-1/2-inch-outside-diameter steel tubing with the posts spaced not more than 6-feet to 6-feet 6-inches apart. This scaffolding must be able to support 75-pound-per-square-foot loads.

To find the load per square foot of a pile of materials on a platform, divide the total weight of the pile by the number of square feet of platform it covers.

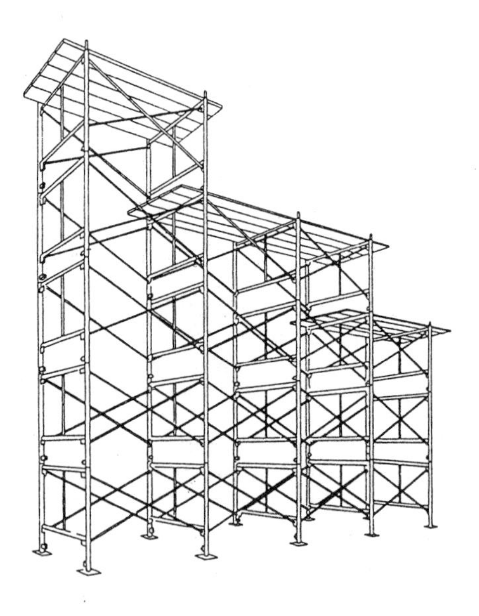

Figure 4-45.—Tiered scaffolding.

BRACKET SCAFFOLDING

The bracket, or carpenter's scaffold (figure 4-46), is built of a triangular wood frame not less than 2- by 3-inch lumber or metal of equivalent strength. Each bracket is attached to the structure in one of four ways: a bolt (at least 5/8 inch) that extends through to the inside of the building wall; a metal stud attachment device; welded to a steel tank; or hooked over a secured supporting member.

The brackets must be spaced no more than 8-feet apart. No more than two persons should be on any 8-foot section at one time. Tools and materials used on the scaffold should not exceed 75 pounds.

The platform is built of at least two 2- by 10-inch nominal size planks. The planks should extend between 6 and 12 inches beyond each support.

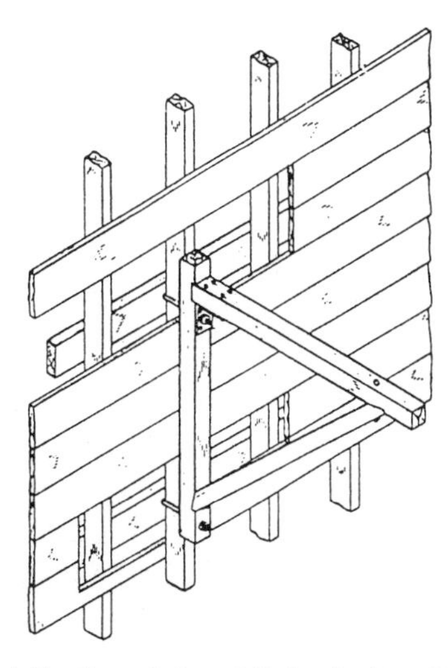

Figure 4-46.—Carpenter's portable bracket for scaffolding.

SCAFFOLD SAFETY

When working on scaffolding or tending others on scaffolding, you must observe all safety precautions. Builder petty officers must not only observe the safety precautions themselves, but they must also issue them to their crew and ensure that the crew observes them.

RECOMMENDED READING LIST

NOTE

Although the following references were current when this TRAMAN was published, their continued currency cannot be assured. You therefore need to ensure that you are studying the latest revisions.

Navy Occupational Safety and Health (NAVOSH) Program Manual for Forces Afloat, Volume I, Office of the Chief of Naval Operations (OP-45), Washington, D.C., 1989.

Safety and Health Requirements Manual, EM 385-1-1, U.S. Army Corps of Engineers, Washington, D.C. 1981.

CHAPTER 5

LEVELING AND GRADING

This chapter describes the common types of leveling instruments. It also describes their principles, uses, procedures of establishing elevations, and techniques of laying outbuilding lines. As a Builder, you will find the information especially useful in performing such duties as setting up a level, reading a leveling rod, interpreting and setting grade stakes, and setting batterboards. Also included in this chapter are practices and measures that help prevent slides and cave-ins at excavation sites, and the procedures for computing volume of land mass.

LEVELS

LEARNING OBJECTIVE: Upon completing this section, you should be able to describe the types of leveling instruments and their uses.

The engineer's level, often referred to as the "dumpy level," is the instrument most commonly used to attain the level line of sight required for differential leveling (defined later). The dumpy level and the self-leveling level can be mounted for use on a tripod, usually with adjustable legs (figure 5-1).

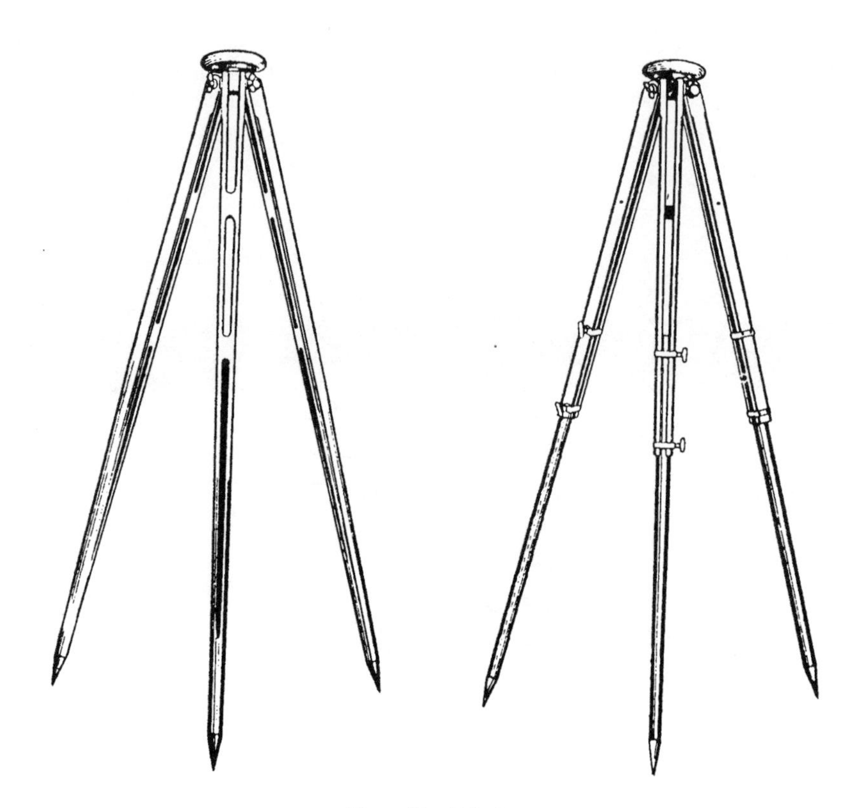

Figure 5-1.—Tripods.

Mounting is done by engaging threads at the base of the instrument (called the footplate) with the threaded head on the tripod. These levels are the ones most frequently used in ordinary leveling projects. For rough leveling, the hand level is used.

DUMPY LEVEL

Figure 5-2 shows a dumpy level and its nomenclature. Notice that the telescope is rigidly fixed to the supporting frame.

Inside the telescope there is a ring, or diaphragm, known as the reticle, which supports the cross hairs. The cross hairs are brought into exact focus by manipulating the knurled eyepiece focusing ring near the eyepiece, or the eyepiece itself on some models. If the cross hairs get out of horizontal adjustment, they can be made horizontal again by slackening the reticle adjusting screws and turning the screws in the appropriate direction. This adjustment should be performed only by trained personnel. The object to which you are sighting, regardless of shape, is called a target. The target is brought into clear focus by manipulating the focusing knob shown on top of the telescope. The telescope can be rotated only horizontally, but, before it can be rotated, the azimuth clamp must be released. After training the telescope as nearly on the target as you can, tighten the azimuth clamp. You then bring the vertical cross hair into exact alignment on the target by rotating the azimuth tangent screw.

The level vial, leveling head, leveling screws, and footplate are all used to adjust the instrument to a perfectly level line of sight once it is mounted on the tripod.

SELF-LEVELING LEVEL

You can save time using the self-leveling, or so-called "automatic," level in leveling operations. The self-leveling level (figure 5-3) has completely eliminated the use of the tubular spirit level, which required excessive time because it had to be reset quite often during operation.

The self-leveling level is equipped with a small bull's-eye level and three leveling screws. The leveling screws, which sit on a triangular footplate,

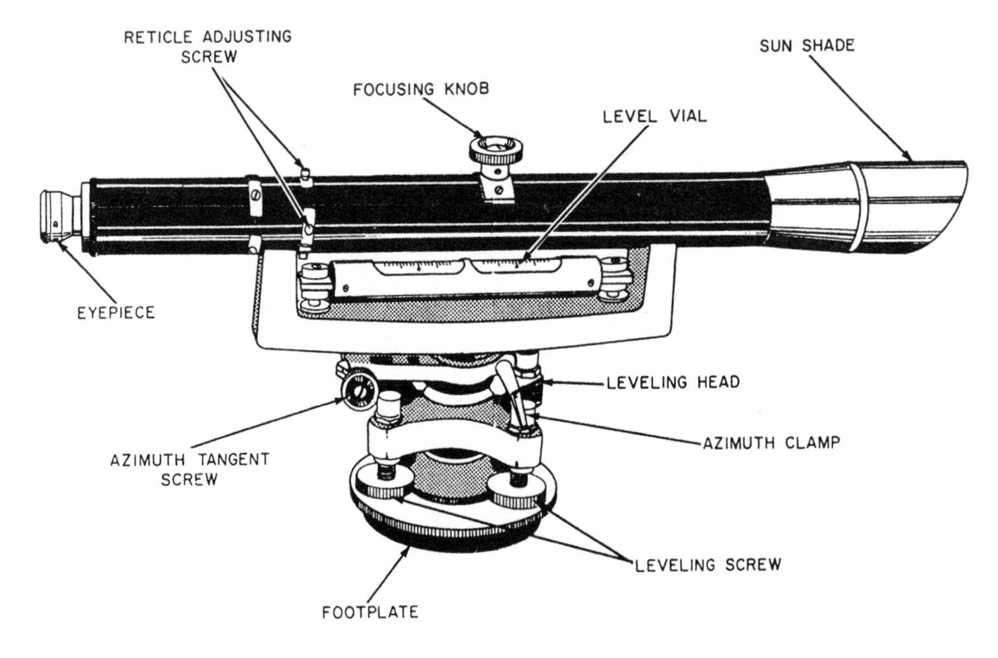

Figure 5-2.—Dumpy level.

are used to center, as much as possible, the bubble of the bull's-eye level. The line of sight automatically becomes horizontal and remains horizontal as long as the bubble remains approximately centered.

HAND LEVEL

The hand level, like all surveying levels, is an instrument that combines a level vial and a sighting device. Figure 5-4 shows the Locke level, a type of hand level. A horizontal line, called an index line, is provided in the sight tube as a reference line. The level vial is mounted atop a slot in the sighting tube in which a reflector is set at a 45° angle. This permits the observer, who is sighting through the tube, to see the object, the position of the level bubble in the vial, and the index line at the same time.

To get the correct sighting through the tube, you should stand straight, using the height of your eye (if known) above the ground to find the target. When your eye height is not known, you can find it by sighting the rod at eye height in front of your body. Since the distances over which you sight a hand level are rather short, no magnification is provided in the tube.

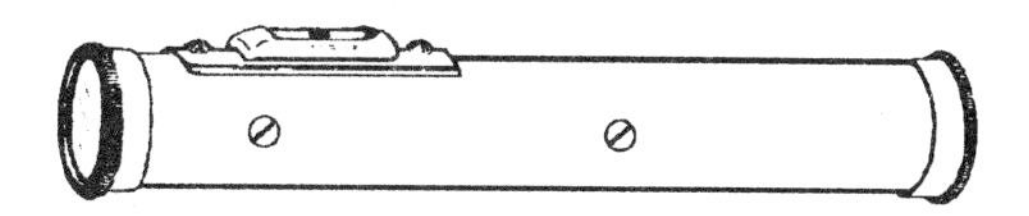

Figure 5-4.—Locke level.

SETTING UP A LEVEL

After you select the proper location for the level, your first step is to set up the tripod. This is done by spreading two of the legs a convenient distance apart and then bringing the third leg to a position that will bring the protector cap (which covers the tripod head threads) about level when the tripod stands on all three legs. Then, unscrew the protector cap, which exposes the threaded head, and place it in the carrying case where it will not get lost or dirty. The tripod protective cap should be in place when the tripod is not being used.

Lift the instrument out of the carrying case by the footplate—not by the telescope. Set it squarely and gently on the tripod head threads and engage the head nut threads under the footplate by rotating the footplate clockwise. If the threads will not engage

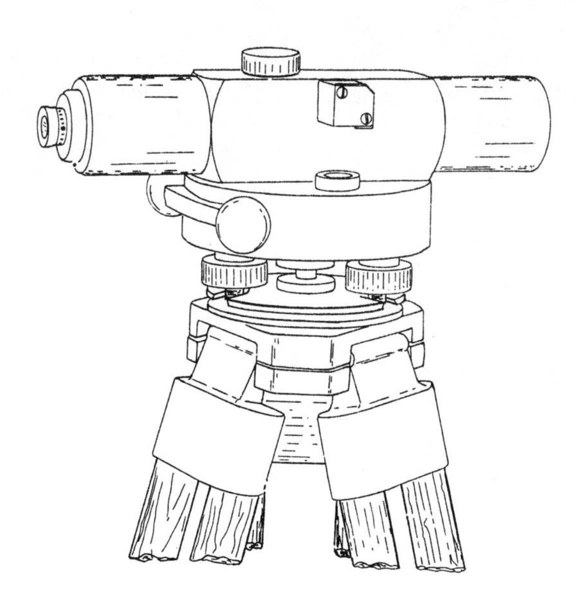

Figure 5-3.—Self-leveling level.

smoothly, they may be cross-threaded or dirty. Do not force them if you encounter resistance; instead, back off, and, after checking to see that they are clean, square up the instrument, and then try again gently. Screw the head nut up firmly, but not too tightly. Screwing it too tightly causes eventual wearing of the threads and makes unthreading difficult. After you have attached the instrument, thrust the leg tips into the ground far enough to ensure that each leg has stable support, taking care to maintain the footplate as near level as possible. With the instrument mounted and the legs securely positioned in the soil, the thumbscrews at the top of each leg should be firmly tightened to prevent any possible movement.

Quite frequently, the Builder must set up the instrument on a hard, smooth surface, such as a concrete pavement. Therefore, steps must be taken to prevent the legs from spreading. Figure 5-5 shows two good ways of doing this. In view A, the tips of the legs are inserted in joints in the pavement. In view B, the tips are held by a wooden floor triangle.

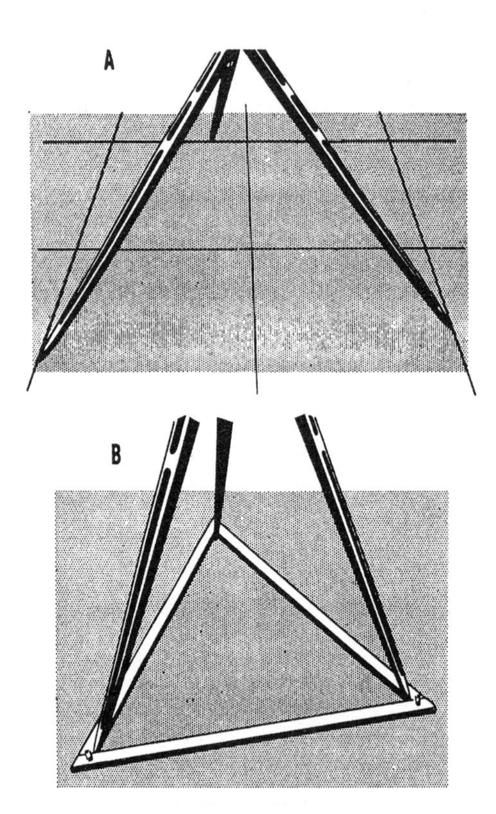

Figure 5-5.—Methods of preventing tripod legs from spreading.

LEVELING A LEVEL

To function accurately, the level must provide a line of sight that is perfectly horizontal in any direction the telescope is trained. To ensure this, you must level the instrument as discussed in the next paragraphs.

When the tripod and instrument are first set up, the footplate should be made as nearly level as possible. Next, train the telescope over a pair of diagonally opposite leveling screws, and clamp it in that position. Then, manipulate the leveling thumbscrews, as shown in figure 5-6, to bring the bubble in the level vial exactly into the marked center position.

The thumbscrews are manipulated by simultaneously turning them in opposite directions, which shortens one spider leg (threaded member running through the thumbscrew) while it lengthens the other. It is helpful to remember that the level vial bubble will move in the same direction that your left thumb moves while you rotate the thumbscrews. In other words, when your left thumb pushes the thumbscrew clockwise, the bubble will move towards your left hand; when you turn the left thumbscrew counterclockwise, the bubble moves toward your right hand.

After leveling the telescope over one pair of screws, train it over the other pair and repeat the process. As a check, set the telescope in all four possible positions and be sure that the bubble centers exactly in each.

Various techniques for using the level will develop with experience; however, in this section we will only discuss the techniques that we believe are essential to the Builder rating.

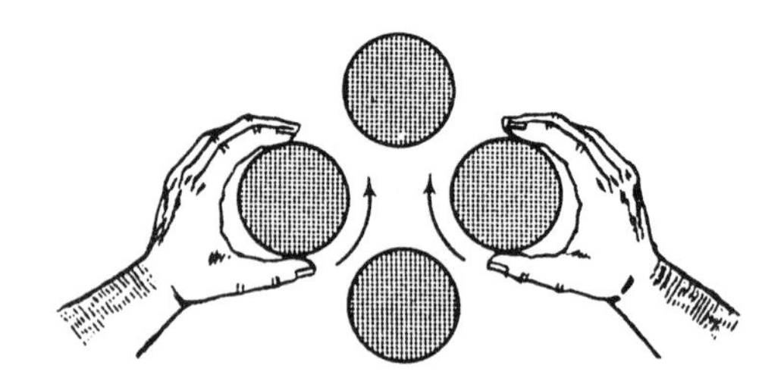

Figure 5-6.—Manipulating leveling thumbscrews.

CARE OF LEVELS

An engineer's level is a precision instrument containing many delicate and fragile parts. It must therefore be handled gently and with the greatest care at all times; it must never be subjected to shock or jar. Movable parts (if not locked or clamped in place) should work easily and smoothly. If a movable part resists normal pressure, there is something wrong. If you force the part to move, you will probably damage the instrument. You will also cause wear or damage if you excessively tighten clamps and screws.

The only proper place to stow the instrument when it is detached from the tripod is in its own carrying box or case. The carrying case is designed to reduce the effect of jarring to a minimum. It is strongly made and well padded to protect the instrument from damage. Before stowing, the azimuth clamp and leveling screws should be slightly tightened to prevent movement of parts inside the box. When it is being transported in a vehicle, the case containing the instrument should be placed as nearly as possible midway between the front and rear wheels. This is the point where jarring of the wheels has the least effect on the chassis.

You should never lift the instrument out of the case by grasping the telescope. Wrenching the telescope in this manner will damage a number of delicate parts. Instead, lift it out by reaching down and grasping the footplate or the level bar.

When the instrument is attached to the tripod and carried from one point to another, the azimuth clamp and level screws should be set up tight enough to prevent part motion during the transport but loose enough to allow a "give" in case of an accidental bump against some object. When you are carrying the instrument over terrain that is free of possible contacts (across an open field, for example), you may carry it over your shoulder like a rifle. When there are obstacles around, you should carry it as shown in figure 5-7. Carried in this manner, the instrument is always visible to you, and this makes it possible for you to avoid striking it against obstacles.

Figure 5-7.—Safest carrying position for instrument when obstacles may be encountered.

LEVELING RODS

LEARNING OBJECTIVE: Upon completing this section, you should be able to interpret the readings from a leveling rod.

A leveling rod, in essence, is a tape supported vertically that is used to measure vertical distance (difference in elevation) between a line of sight and a required point above or below it. Although there are several types of rods, the most popular and frequently used is the Philadelphia rod. Figure 5-8 shows the face and back of this rod.

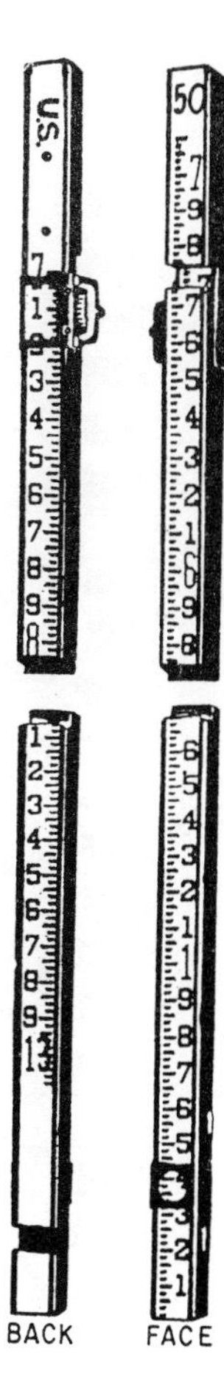

Figure 5-8.—Back and face of Philadelphia leveling rod.

The Philadelphia rod consists of two sliding sections, which can be fully extended to a total length of 13.10 feet. When the sections are entirely closed, the total length is 7.10 feet. For direct readings (that is, for readings on the face of the rod) of up to 7.10 and 13.10 feet, the rod is used extended and read on the back by the rodman. If you are in the field and don't have a Philadelphia rod, you can use a 1-by-4 with a mark or a 6-foot wooden ruler attached to a 2-by-4.

In direct readings, the person at the instrument reads the graduation on the rod intercepted by the cross hair through the telescope. In target readings, the rodman reads the graduation on the face of the rod intercepted by a target. In figure 5-8, the target does not appear; however, it is shown in figure 5-9. As you can see, it is a sliding, circular device that can be moved up or down the rod and clamped in position. It is placed by the rodman on signals given by the instrumentman.

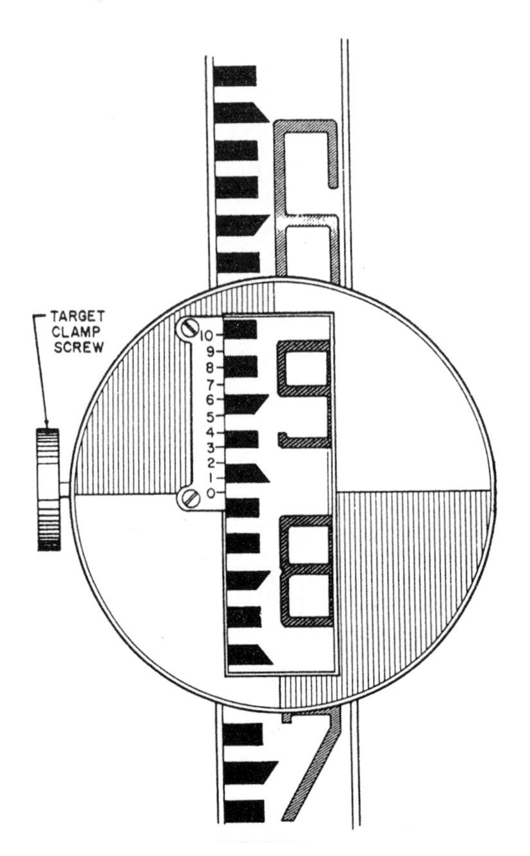

Figure 5-9.—Philadelphia rod set for target reading of less than 7.000 ft.

The rod shown in the figures is graduated in feet and hundredths of a foot. Each even foot is marked with a large red numeral, and, between each pair of adjacent red numerals, the intermediate tenths of a foot are marked with smaller black numerals. Each intermediate hundredth of a foot between each pair of adjacent tenths is indicated by the top or bottom of one of the short, black dash graduations.

DIRECT READINGS

As the levelman, you can make direct readings on a self-reading rod held plumb on the point by the rodman. If you are working to tenths of a foot, it is relatively simple to read the foot mark below the cross hair and the tenth mark that is closest to the cross hair. If greater precision is required, and you must work to hundredths, the reading is more complicated (see figure 5-10).

For example, suppose you are making a direct reading that should come out to 5.67 feet. If you are using a Philadelphia rod, the interval between the top and the bottom of each black graduation and the interval between the black graduations (figure 5-11) each represent 0.01 foot. For a reading of 5.76 feet, there are three black graduations between the 5.70-foot mark and the 5.76-foot mark. Since there are three graduations, a beginner may have a tendency to misread 5.76 feet as 5.73 feet.

As you can see, neither the 5-foot mark nor the 6-foot mark is shown in figure 5-11. Sighting through the telescope, you might not be able to see the foot marks to which you must refer for the reading. When you cannot see the next lower foot mark through the telescope, it is a good idea to order the rodman to **raise the red**. On the Philadelphia rod, whole feet numerals are in red. Upon hearing this order, the rodman slowly raises the rod until the next lower red figure comes into view.

TARGET READINGS

For more precise vertical measurements, level rods may be equipped with a rod target that can be set and clamped by the rodman at the directions of the instrumentman. When the engineer's level rod target and the vernier scale are being used, it is possible to make readings of 0.001 (one-thousandth of a foot), which is slightly smaller than one sixty-fourth of an

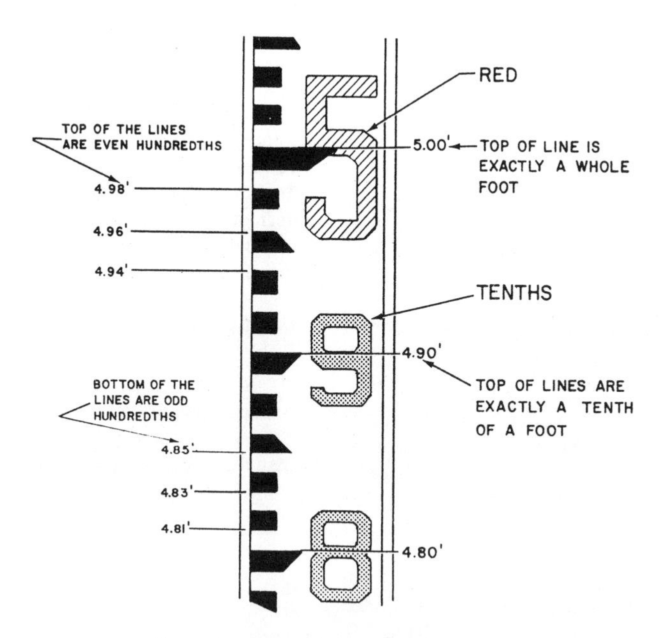

Figure 5-10.—Philadelphia rod markings.

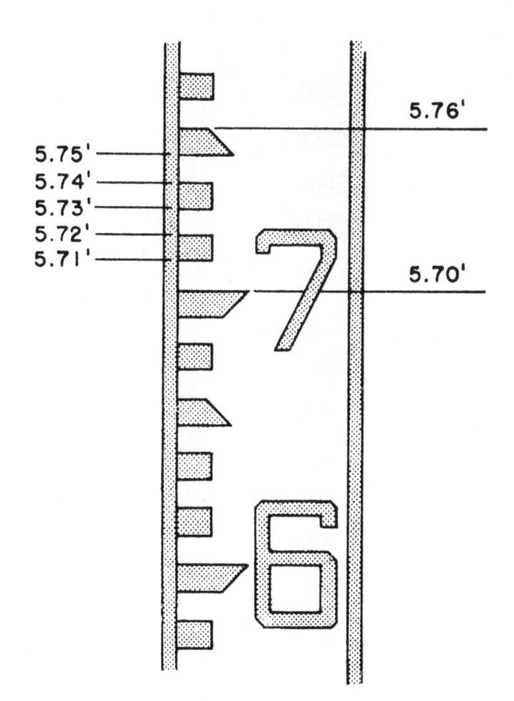

Figure 5-11.—Direct reading of 5.76 ft on Philadelphia rod.

inch. The indicated reading of the target can be read either by the rodman or the instrumentman. In figure 5-12, you can see that the 0 on the vernier scale is in exact alignment with the 4-foot mark. If the position of the 0 on the target is not in exact alignment with a line on the rod, go up the vernier scale on the target to the line that is in exact alignment with the hundredths line on the rod, and the number located will be the reading in thousandths.

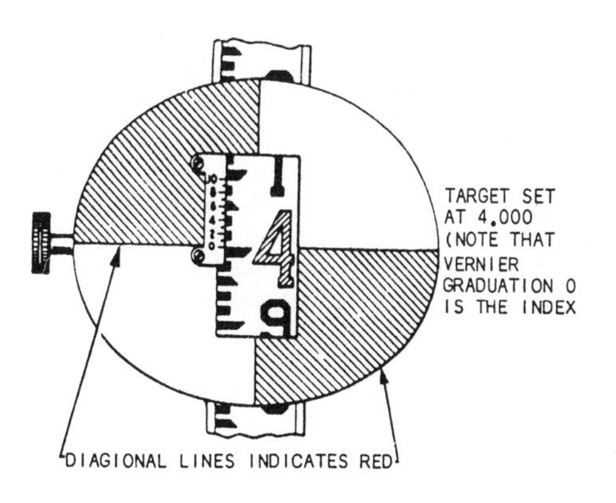

Figure 5-12.—Target.

There are three situations in which target reading, rather than direct reading, is done on the face of the rod:

- When the rod is too far from the level to be read directly through the telescope:
- When a reading to the nearest 0.001 foot, rather than to the nearest 0.01 foot, is desired (a vernier on the target or on the back of the rod makes this possible;
- When the instrumentman desires to ensure against the possibility of reading the wrong foot (large red letter) designation on the rod.

For target readings up to 7.000 feet, the rod is used fully closed, and the rodman, on signals from the instrumentman, sets the target at the point where its horizontal axis is intercepted by the cross hair, as seen through the telescope. When the target is located, it is clamped in place with the target screw clamp, as shown in figure 5-9. When a reading to only the nearest 0.01 foot is desired, the graduation indicated by the target's horizontal axis is read; in figure 5-9, this reading is 5.84 feet.

If reading to the nearest 0.001 foot is desired, the rodman reads the vernier (small scale running from 0 to 10) on the target. The 0 on the vernier indicates that the reading lies between 5.840 feet and 5.850 feet. To determine how many thousandths of a foot over 5.840 feet, you examine the graduations on the vernier to determine which one is most exactly in line with a graduation (top or bottom of a black dash) on the rod. In figure 5-9, this graduation on the vernier is the 3; therefore, the reading to the nearest 0.001 foot is 5.843 feet.

For target readings of more than 7.000 feet, the procedure is a little different. If you look at the left-hand view of figure 5-8 (showing the back of the rod), you will see that only the back of the upper section is graduated, and that it is graduated downward from 7.000 feet at the top to 13.09 feet at the bottom. You can also see there is a rod vernier fixed to the top of the lower section of the rod. This vernier is read against the graduations on the back of the upper section.

For a target reading of more than 7.000 feet, the rodman first clamps the target at the upper section of the rod. Then, on signals from the instrumentman, the rodman extends the rod upward to the point where the horizontal axis of the target is intercepted by the cross hair. The rodman then clamps the rod, using the rod clamp screw shown in figure 5-13, and reads the vernier on the back of the rod, also shown in that figure. In this case, the 0 on the vernier indicates a certain number of thousandths more than 7.100 feet. Remember, that in this case, you read the rod and the vernier down from the top, not up from the bottom. To determine the thousandths, determine which vernier graduation lines up most exactly with a graduation on the rod. In this case, it is the 7; therefore, the rod reading is 7.107 feet.

Rod Levels

A rod reading is accurate only if the rod is perfectly plumb (vertical) at the time of the reading. If the rod is out of plumb, the reading will be greater than the actual vertical distance between the height of

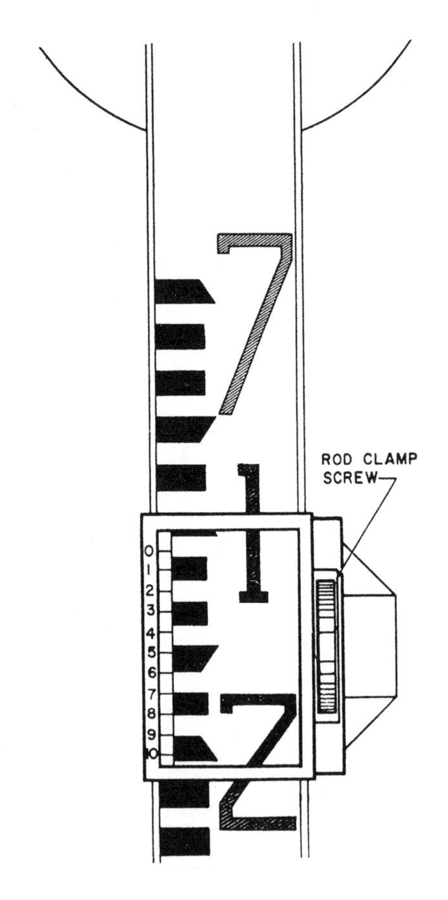

Figure 5-13.—Philadelphia rod target reading of more than 7.000 ft.

instrument (H.I.) and the base of the rod. On a windy day, the rodman may have difficulty holding the rod plumb. In this case, the levelman can have the rodman wave the rod back and forth, allowing the levelman to read the lowest reading touched on the engineer's level cross hairs.

The use of a rod level ensures a vertical rod. A bull's-eye rod level is shown in figure 5-14. When it is held as shown (on a part of the rod where readings are not being taken to avoid interference with the instrumentman's view of the scale) and the bubble is centered, the rod is plumb. A vial rod level has two spirit vials, each of which is mounted on the upper edge of one of a pair of hinged metal leaves. The vial level is used like the bull's-eye level, except that two bubbles must be watched instead of one.

Figure 5-14.—Bull's-eye rod level.

Care of Leveling Rods

A leveling rod is a precision instrument and must be treated as such. Most rods are made of carefully selected, kiln-dried, well-seasoned hardwood. Scale graduations and numerals on some are painted directly on the wood; however, on most rods they are painted on a metal strip attached to the wood. Unless a rod is handled at all times with great care, the painted scale will soon become scratched, dented, worn, or otherwise marked and obscured. Accurate readings on a scale in this condition are difficult.

Allowing an extended sliding-section rod to **close on the run**, by permitting the upper section to drop, may jar the vernier scale out of position or otherwise damage the rod. Always close an extended rod by easing the upper section down gradually.

A rod will read accurately only if it is perfectly straight. It follows, then, that anything that might bend or warp the rod must be avoided. Do not lay a rod down flat unless it is supported throughout, and never use a rod for a seat, a lever, or a pole vault. In short, never use a rod for any purpose except the one for which it is designed.

Store a rod not in use in a dry place to avoid warping and swelling caused by dampness. Always wipe off a wet rod before putting it away. If there is dirt on the rod, rinse it off, but do not scrub it off. If a soap solution must be used (to remove grease, for example), make it a very mild one. The use of a strong soap solution will soon cause the paint on the rod to degenerate.

Protect a rod as much as possible against prolonged exposure to strong sunlight. Such exposure causes paint to chalk (that is, degenerate into a chalk-like substance that flakes from the surface).

DIFFERENTIAL LEVELING

LEARNING OBJECTIVE: Upon completing this section, you should be able to determine elevations in the field to locate points at specified elevations.

The most common procedure for determining elevations in the field, or for locating points at specified elevations, is known as differential leveling. This procedure, as its name implies, is nothing more than finding the vertical difference between the known or assumed elevation of a bench mark and the elevation of the point in question. Once the difference is measured, it can (depending on the circumstances) be added to or subtracted from the bench mark elevation to determine the elevation of the new point.

ELEVATION AND REFERENCE

The elevation of any object is its vertical distance above or below an established height on the earth's surface. This established height is referred to as either a "reference plane" or "simple reference." The most commonly used reference plane for elevations is mean (or average) sea level, which has been assigned an assumed elevation of 000.0 feet. However, the reference plane for a construction project is usually the height of some permanent or semipermanent object in the immediate vicinity, such as the rim of a manhole cover, a rod, or the finish floor of an existing structure. This object may be given its relative sea level elevation (if it is known); or it may be given a convenient, arbitrarily assumed elevation, usually a whole number, such as 100.0 feet. An object of this type, with a given, known, or assumed elevation, which is to be used in determining the elevations of other points, is called a bench mark.

PRINCIPLES OF DIFFERENTIAL LEVELING

Figure 5-15 illustrates the principle of differential leveling. The instrument shown in the center represents an engineer's level. This optical instrument provides a perfectly level line of sight through a telescope, which can be trained in any direction. Point A in the figure is a bench mark (it could be a concrete monument, a wooden stake, a sidewalk curb, or any other object) having a known elevation of 365.01 feet. Point B is a ground surface point whose elevation is desired.

The first step in finding the elevation point of point B is to determine the elevation of the line of sight of the instrument. This is known as the height of instrument and is often written and referred to simply as "H.I." To determine the H.I., you take a backsight on a level rod held vertically on the bench mark (B.M.) by a rodman. A backsight (B.S.) is always taken after a new instrument position is set up by sighting back to a known elevation to get the new H.I. A leveling rod is graduated upward in feet, from 0 at its base, with appropriate subdivisions in feet.

In figure 5-15, the backsight reading is 11.56 feet. Thus, the elevation of the line of sight (that is, the H.I.) must be 11.56 feet greater than the bench mark elevation, point A. Therefore, the H.I. is 365.01 feet plus 11.56 feet, or 376.57 feet as indicated.

Next, you train the instrument ahead on another rod (or more likely, on the same rod carried ahead) held vertically on B. This is known as taking a foresight. After reading a foresight (F.S.) of 1.42 feet on the rod, it follows that the elevation at point B must be 1.42 feet lower than the H.I. Therefore, the elevation of point B is 376.57 feet minus 1.42 feet, or 375.15 feet.

GRADING

The term "grade" is used in several different senses in construction. In one sense, it refers to the steepness of a slope; for example, a slope that rises 3 vertical feet for every 100 horizontal feet has a grade of 3 percent. Although the term "grade" is commonly used in this sense, the more accurate term for indicating steepness of slope is "gradient."

In another sense, the term "grade" simply means surface. On a wall section, for example, the line that

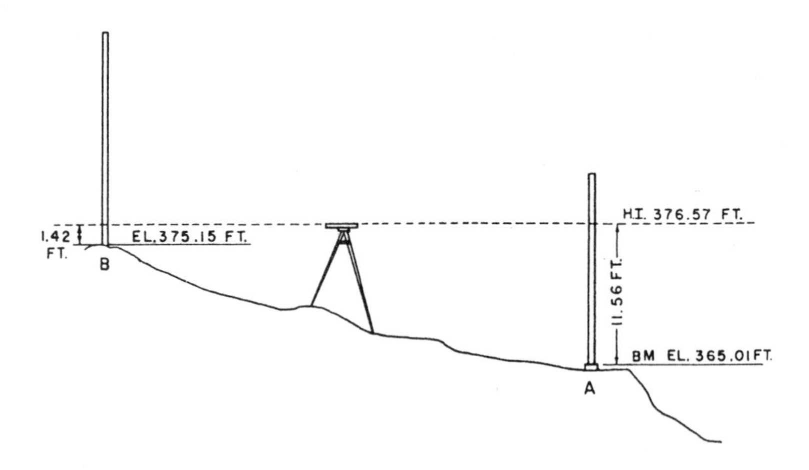

Figure 5-15.—Procedure for differential leveling.

indicates the ground surface level outside the building is marked "Grade" or "Grade Line."

The elevation of a surface at a particular point is a grade elevation. A grade elevation may refer to an existing, natural earth surface or to a hub or stake used as a reference point, in which case the elevation is that of existing grade or existing ground. It may also refer to a proposed surface to be created artificially, in which case the elevation is that of prescribed grade, plan grade, or finished grade.

Grade elevations of the surface area around a structure are indicated on the plot plan. Because a natural earth surface is usually irregular in contour, existing grade elevations on such a surface are indicated by contour lines on the plot plan; that is, by lines that indicate points of equal elevation on the ground. Contour lines that indicate existing grade are usually made dotted; however, existing contour lines on maps are sometimes represented by solid lines. If the prescribed surface to be created artificially will be other than a horizontal-plane surface, prescribed grade elevations will be indicated on the plot plan by solid contour lines.

On a level, horizontal-plane surface, the elevation is the same at all points. Grade elevation of a surface of this kind cannot be indicated by contour lines because each contour line indicates an elevation different from that of each other contour line. Therefore, a prescribed level surface area, to be artificially created, is indicated on the plot plan by outlining the area and inscribing inside the outline the prescribed elevation, such as "First floor elevation 127.50 feet."

BUILDING LAYOUT

LEARNING OBJECTIVE: Upon completing this section, you should be able to determine boundaries of building layout.

Before foundation and footing excavation for a building can begin, the building lines must be laid out to determine the boundaries of the excavations. Points shown on the plot plan, such as building corners, are located at the site from a system of horizontal control points established by the battalion engineering aids. This system consists of a framework of stakes, driven pipes, or other markers located at points of known horizontal location. A point in the structure, such as a building corner, is located on the ground by reference to one or more nearby horizontal control points.

We cannot describe here all the methods of locating a point with reference to a horizontal control point of a known horizontal location. We will take, as an illustrative example, the situation shown in figure 5-16. This figure shows two horizontal control points, consisting of monuments A and B. The term "monument," incidentally, doesn't necessarily mean an elaborate stone or concrete structure. In structural horizontal control, it simply means any permanently located object, either artificial (such as a driven length of pipe) or natural (such as a tree) of known horizontal location.

In figure 5-16, the straight line from A to B is a control base line from which the building corners of the structure can be located. Corner E, for example, can be located by first measuring 15 feet along the base line from A to locate point C; then measuring off 35 feet on CE, laid off at 90° to (that is, perpendicular to) AB. By extending CE another 20 feet, you can locate building corner F. Corners G and H can be similarly located along a perpendicular run from point D, which is itself located by measuring 55 feet along the base line from A.

PERPENDICULAR BY PYTHAGOREAN THEOREM

The easiest and most accurate way to locate points on a line or to turn a given angle, such as 90°,

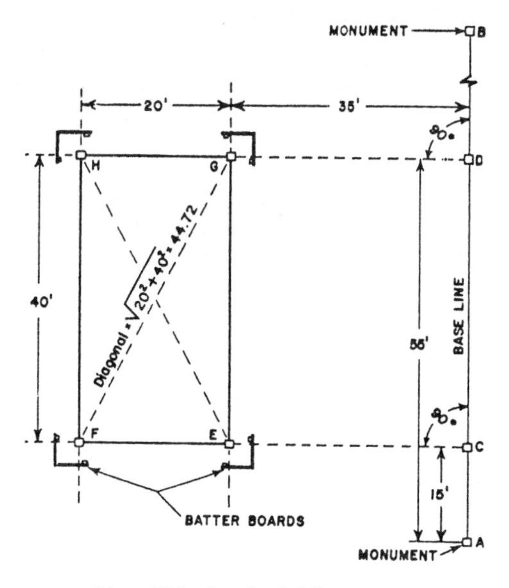

Figure 5-16.—Locating building corners.

from one line to another is to use a surveying instrument called a transit. However, if you do not have a transit, you can locate the corner points with tape measurements by applying the Pythagorean theorem. First, stretch a cord from monument A to monument B, and locate points C and D by tape measurements from A. Now, if you examine figure 5-16, you will observe that straight lines connecting points C, D, and E form a right triangle with one side 40 feet long and the adjacent side 35 feet long. By the Pythagorean theorem, the length of the hypotenuse of this triangle (the line ED) is equal the square root of $35^2 + 40^2$, which is approximately 53.1 feet. Because figure EG DC is a rectangle, the diagonals both ways (ED and CG) are equal. Therefore, the line from C to G should also measure 53.1 feet. If you have one person hold the 53.1-foot mark of a tape on D, have another hold the 35-foot mark of another tape on C, and have a third person walk away with the joined 0-foot ends, when the tapes come taut, the joined 0-foot ends will lie on the correct location for point E. The same procedure, but this time with the 53.1-foot length of tape running from C and the 35-foot length running from D, will locate corner point G. Corner points F and H can be located by the same process, or by extending CE and DG 20 feet.

PERPENDICULAR BY 3:4:5 TRIANGLE

If you would rather avoid the square root calculations required in the Pythagorean theorem method, you can apply the basic fact that any triangle with sides in the proportions of 3:4:5 is a right triangle. In locating point E, you know that this point lies 35 feet from C on a line perpendicular to the base line. You also know that a triangle with sides 30 and 40 feet long and a hypotenuse 50 feet long is a right triangle.

To get the 40-foot side, you measure off 40 feet from C along the base line; in figure 5-16, the segment from C to D happens to measure 40 feet. Now, if you run a 50-foot tape from D and a 30-foot tape from C, the joined ends will lie on a line perpendicular from the base line, 30 feet from C. Drive a hub at this point, and extend the line to E (5 more feet) by stretching a cord from C across the mark on the hub.

BATTER BOARDS

Hubs driven at the exact locations of building corners will be disturbed as soon as the excavation for the foundation begins. To preserve the corner locations, and also to provide a reference for measurement down to the prescribed elevations, batter boards are erected as shown in figure 5-17.

Each pair of boards is nailed to three 2-by-4 corner stakes as shown. The stakes are driven far enough outside the building lines so that they will not be disturbed during excavation. The top edges of the boards are located at a specific elevation, usually some convenient number of whole feet above a significant prescribed elevation, such as that at the top of the foundation. Cords located directly over the lines through corner hubs, placed by holding plumb bobs on the hubs, are nailed to the batter boards. Figure 5-17 shows how a corner point can be located in the excavation by dropping a plumb bob from the point of intersection between two cords.

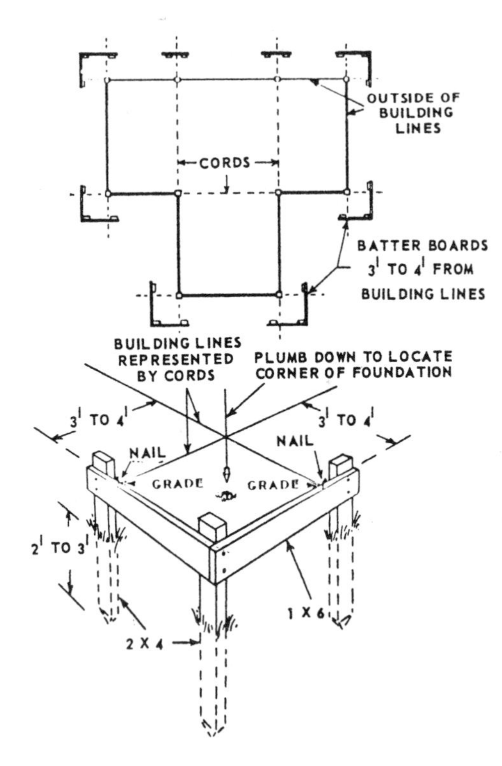

Figure 5-17.—Batter boards.

In addition to their function in horizontal control, batter boards are also used for vertical control. The top edge of a batter board is placed at a specific elevation. Elevations of features in the structure, such as foundations and floors, can be located by measuring downward or upward from the cords stretched between the batter boards.

You should always make sure that you have complete information as to exactly what lines and elevations are indicated by the batter boards. You should emphasize to your crewmembers that they exercise extreme caution while working around batter boards. If the boards are damaged or moved, additional work will be required to replace them and to relocate reference points.

RECOMMENDED READING LIST

NOTE

Although the following reference was current when this TRAMAN was published, its continued currency cannot be assured. You therefore need to ensure that you are studying the latest revision.

Engineering Aid 3 & 2, Vol. 3, NAVEDTRA 10629-1, Naval Education and Training Program Management Support Activity, Pensacola, Fla., 1987.

CHAPTER 6

CONCRETE

Concrete is one of the most important construction materials. It is comparatively economical, easy to make, offers continuity and solidity, and will bond with other materials. The keys to good-quality concrete are the raw materials required to make concrete and the mix design as specified in the project specifications. In this chapter, we'll discuss the characteristics of concrete, the ingredients of concrete, concrete mix designs, and mixing concrete. We'll conclude the chapter with a discussion of precast and tilt-up concrete. At the end of the discussion, we provide helpful references. You are encouraged to study these references, as required, for additional information on the topics discussed.

CONCRETE CHARACTERISTICS

LEARNING OBJECTIVE: Upon completing this section, you should be able to define the characteristics of concrete.

Concrete is a synthetic construction material made by mixing cement, fine aggregate (usually sand), coarse aggregate (usually gravel or crushed stone), and water in the proper proportions. The product is not concrete unless all four of these ingredients are present.

CONSTITUENTS OF CONCRETE

The fine and coarse aggregates in a concrete mix are the inert, or inactive, ingredients. Cement and water are the active ingredients. The inert ingredients and the cement are first thoroughly mixed together. As soon as the water is added, a chemical reaction begins between the water and the cement. The reaction, called hydration, causes the concrete to harden. This is an important point. The hardening process occurs through hydration of the cement by the water, not by drying out of the mix. Instead of being dried out, concrete must be kept as moist as possible during the initial hydration process. Drying out causes a drop in water content below that required for satisfactory hydration of the cement. The fact that the hardening process does not result from drying out is clearly shown by the fact that concrete hardens just as well underwater as it does in air.

CONCRETE AS BUILDING MATERIAL

Concrete may be cast into bricks, blocks, and other relatively small building units, which are used in concrete construction. Concrete has a great variety of applications because it meets structural demands and lends itself to architectural treatment. All important building elements, foundations, columns, walls, slabs, and roofs are made from concrete. Other concrete applications are in roads, runways, bridges, and dams.

STRENGTH OF CONCRETE

The compressive strength of concrete (meaning its ability to resist compression) is very high, but its tensile strength (ability to resist stretching, bending, or twisting) is relatively low. Consequently, concrete which must resist a good deal of stretching, bending, or twisting—such as concrete in beams, girders, walls, columns, and the like—must be reinforced with steel. Concrete that must resist only compression may not require reinforcement. As you will learn later, the most important factor controlling the strength of concrete is the water-cement ratio, or the proportion of water to cement in the mix.

DURABILITY OF CONCRETE

The durability of concrete refers to the extent to which the material is capable of resisting deterioration caused by exposure to service conditions. Concrete is also strong and fireproof. Ordinary structural concrete that is to be exposed to the elements must be watertight and weather-resistant. Concrete that is subject to wear, such as floor slabs and pavements, must be capable of resisting abrasion.

The major factor that controls the durability of concrete is its strength. The stronger the concrete, the more durable it is. As we just mentioned, the chief factor controlling the strength of concrete is the water-cement ratio. However, the character, size, and grading (distribution of particle sizes between the largest permissible coarse and the smallest permissible fine) of the aggregate also have important effects on both strength and durability. However, maximum strength and durability will still not be attained unless the sand and coarse aggregate you use consist of well-graded, clean, hard, and durable particles free of undesirable substances (figure 6-1).

WATERTIGHTNESS OF CONCRETE

The ideal concrete mix is one with just enough water required for complete hydration of the cement. However, this results in a mix too stiff to pour in forms. A mix fluid enough to be poured in forms always contains a certain amount of water over and above that which will combine with the cement. This water eventually evaporates, leaving voids, or pores, in the concrete. Penetration of the concrete by water is still impossible if these voids are not interconnected. They may be interconnected, however, as a result of slight sinking of solid particles in the mix during the hardening period. As these particles sink, they leave water-filled channels that become voids when the water evaporates. The larger and more numerous these voids are, the more the watertightness of the concrete is impaired. The size and number of the voids vary directly with the amount of water used in excess of the amount required to hydrate the cement. To keep the concrete as watertight as possible, you must not use more water than the minimum amount required to attain the necessary degree of workability.

GENERAL REQUIREMENTS FOR GOOD CONCRETE

The first requirement for good concrete is to use a cement type suitable for the work at hand and have a satisfactory supply of sand, coarse aggregate, and water. Everything else being equal, the mix with the best graded, strongest, best shaped, and cleanest aggregate makes the strongest and most durable concrete.

Second, the amount of cement, sand, coarse aggregate, and water required for each batch must be carefully weighed or measured according to project specifications.

Third, even the best designed, best graded, and highest quality mix does not make good concrete if it is not workable enough to fill the form spaces thoroughly. On the other hand, too much fluidity also results in defects. Also, improper handling during the overall concrete making process, from the initial aggregate handling to the final placement of the mix, causes segregation of aggregate particles by sizes, resulting in nonuniform, poor-quality concrete.

Finally, the best designed, best graded, highest quality, and best placed mix does not produce good concrete if it is not properly cured, that is, properly protected against loss of moisture during the earlier stages of setting.

CONCRETE INGREDIENTS

LEARNING OBJECTIVE: Upon completing this section, you should be able to identify the ingredients essential for good concrete.

The essential ingredients of concrete are cement, aggregate, and water. A mixture of only cement and water is called cement paste. In large quantities, however, cement paste is prohibitively expensive for most construction purposes.

PORTLAND CEMENT

Most cement used today is portland cement. This is a carefully proportioned and specially processed combination of lime, silica, iron oxide, and alumina. It is usually manufactured from limestone mixed with shale, clay, or marl. Properly proportioned raw materials are pulverized and fed into kilns where they are heated to a temperature of 2,700°F and maintained at that temperature for a specific time. The heat produces chemical changes in the mixture and transforms it into clinker—a hard mass of fused clay and limestone. The clinker is then ground to a fineness that will pass through a sieve containing 40,000 openings per square inch.

Types of Cement

There are five types of portland cement covered under "Standard Specifications for Portland Cement." These specifications are governed by the American Society for Testing and Material (ASTM) types. Separate specifications, such as those required for air-entraining portland cements, are found under a separate ASTM. The type of construction, chemical composition of the soil, economy, and requirements for use of the finished concrete are factors that influence the selection of the kind of cement to be used.

TYPE I.—Type I cement is a general-purpose cement for concrete that does not require any of the special properties of the other types. In general, type I cement is intended for concrete that is not subjected

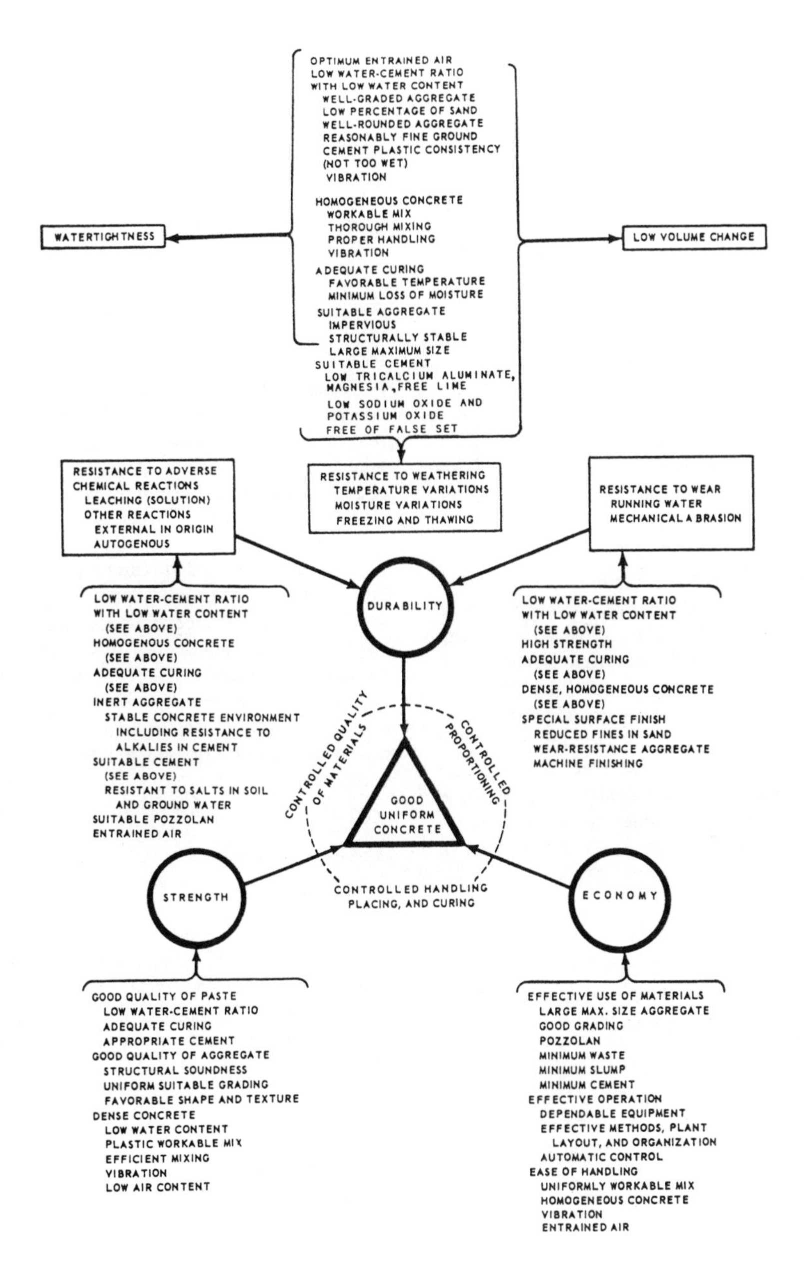

Figure 6-1.—The principal properties of good concrete.

to sulfate attack or damage by the heat of hydration. Type I portland cement is used in pavement and sidewalk construction, reinforced concrete buildings and bridges, railways, tanks, reservoirs, sewers, culverts, water pipes, masonry units, and soil-cement mixtures. Generally, it is more available than the other types. Type I cement reaches its design strength in about 28 days.

TYPE II.—Type II cement is modified to resist moderate sulfate attack. It also usually generates less heat of hydration and at a slower rate than type I. A typical application is for drainage structures where the sulfate concentrations in either the soil or groundwater are higher than normal but not severe. Type II cement is also used in large structures where its moderate heat of hydration produces only a slight temperature rise in the concrete. However, the temperature rise in type II cement can be a problem when concrete is placed during warm weather. Type II cement reaches its design strength in about 45 days.

TYPE III.—Type III cement is a high-early-strength cement that produces design strengths at an early age, usually 7 days or less. It has a higher heat of hydration and is more finely ground than type I. Type III permits fast form removal and, in cold weather construction, reduces the period of protection against low temperatures. Richer mixtures of type I can obtain high early strength, but type III produces it more satisfactorily and economically. However, use it cautiously in concrete structures having a minimum dimension of 2 1/2 feet or more. The high heat of hydration can cause shrinkage and cracking.

TYPE IV.—Type IV cement is a special cement. It has a low heat of hydration and is intended for applications requiring a minimal rate and amount of heat of hydration. Its strength also develops at a slower rate than the other types. Type IV is used primarily in very large concrete structures, such as gravity dams, where the temperature rise from the heat of hydration might damage the structure. Type IV cement reaches its design strength in about 90 days.

TYPE V.—Type V cement is sulfate-resistant and should be used where concrete is subjected to severe sulfate action, such as when the soil or groundwater contacting the concrete has a high sulfate content. Type V cement reaches its design strength in 60 about days.

Air-Entrained Cement

Air-entrained portland cement is a special cement that can be used with good results for a variety of conditions. It has been developed to produce concrete that is resistant to freeze-thaw action, and to scaling caused by chemicals applied for severe frost and ice removal. In this cement, very small quantities of air-entraining materials are added as the clinker is being ground during manufacturing. Concrete made with this cement contains tiny, well-distributed and completely separated air bubbles. The bubbles are so small that there may be millions of them in a cubic foot of concrete. The air bubbles provide space for freezing water to expand without damaging the concrete. Air-entrained concrete has been used in pavements in the northern states for about 25 years with excellent results. Air-entrained concrete also reduces both the amount of water loss and the capillary/water-channel structure.

An air-entrained admixture may also be added to types I, II, and III portland cement. The manufacturer specifies the percentage of air entrainment that can be expected in the concrete. An advantage of using air-entrained cement is that it can be used and batched like normal cement. The air-entrained admixture comes in a liquid form or mixed in the cement. To obtain the proper mix, you should add the admixture at the batch plant.

AGGREGATES

The material combined with cement and water to make concrete is called aggregate. Aggregate makes up 60 to 80 percent of concrete volume. It increases the strength of concrete, reduces the shrinking tendencies of the cement, and is used as an economical filler.

Types

Aggregates are divided into fine (usually consisting of sand) and coarse categories. For most building concrete, the coarse aggregate consists of gravel or crushed stone up to 1 1/2 inches in size. However, in massive structures, such as dams, the coarse aggregate may include natural stones or rocks ranging up to 6 inches or more in size.

Purpose of Aggregates

The large, solid coarse aggregate particles form the basic structural members of the concrete. The voids between the larger coarse aggregate particles are filled by smaller particles. The voids between the smaller particles are filled by still smaller particles. Finally, the voids between the smallest coarse aggregate particles are filled by the largest fine aggregate particles. In turn, the voids between the largest fine aggregate particles are filled by smaller fine aggregate particles, the voids between the smaller fine aggregate particles by still smaller particles, and so on. Finally, the voids between the finest grains are filled with cement. You can see from this that the better the aggregate is graded (that is, the better the distribution of particles sizes), the more solidly all voids will be filled, and the denser and stronger will be the concrete.

The cement and water form a paste that binds the aggregate particles solidly together when it hardens. In a well-graded, well-designed, and well-mixed batch, each aggregate particle is thoroughly coated with the cement-water paste. Each particle is solidly bound to adjacent particles when the cement-water paste hardens.

AGGREGATE SIEVES.—The size of an aggregate sieve is designated by the number of meshes to the linear inch in that sieve. The higher the number, the finer the sieve. Any material retained on the No. 4 sieve can be considered either coarse or fine. Aggregates larger than No. 4 are all course; those smaller are all fines. No. 4 aggregates are the dividing point. The finest coarse-aggregate sieve is the same No. 4 used as the coarsest fine-aggregate sieve. With this exception, a coarse-aggregate sieve is designated by the size of one of its openings. The sieves commonly used are 1 1/2 inches, 3/4 inch, 1/2 inch, 3/8 inch, and No. 4. Any material that passes through the No. 200 sieve is too fine to be used in making concrete.

PARTICLE DISTRIBUTION.—Experience and experiments show that for ordinary building concrete, certain particle distributions consistently seem to produce the best results. For fine aggregate, the recommended distribution of particle sizes from No. 4 to No. 100 is shown in table 6-1.

The distribution of particle sizes in aggregate is determined by extracting a representative sample of the material, screening the sample through a series of sieves ranging in size from coarse to fine, and determining the percentage of the sample retained on each sieve. This procedure is called making a sieve analysis. For example, suppose the total sample weighs 1 pound. Place this on the No. 4 sieve, and shake the sieve until nothing more goes through. If what is left on the sieve weighs 0.05 pound, then 5 percent of the total sample is retained on the No. 4 sieve. Place what passes through on the No. 8 sieve and shake it. Suppose you find that what stays on this sieve weighs 0.1 pound. Since 0.1 pound is 10 percent of 1 pound, 10 percent of the total sample was retained on the No. 8 sieve. The cumulative retained weight is 0.15 pound. By dividing 0.15 by 1.0 pound, you will find that the total retained weight is 15 percent.

The size of coarse aggregate is usually specified as a range between a minimum and a maximum size; for example, 2 inches to No. 4, 1 inch to No. 4,

Table 6-1.—Recommended Distribution of Particle Sizes

SIEVE NUMBER	PERCENT RETAINED ON SQUARE MESH LABORATORY SIEVES
3/8"	0
No. 4	18
No. 8	27
No. 16	20
No. 30	20
No. 50	10
No. 100	4

Table 6-2.—Recommended Maximum and Minimum Particle Sizes

SIZE OF COURSE AGGREGATE, IN INCHES	PERCENTAGES BY WEIGHT PASSING LABORATORY SIEVES HAVING SQUARE OPENINGS										
	4-in.	3 1/2 in.	3-in.	2 1/2 in.	2-in.	1 1/2 in.	1-in.	3/4 in.	1/2 in.	3/8 in.	No. 4
1.5	—	—	—	—	100	95-100	—	35-70	—	10-30	0-5
2	—	—	—	100	95-100	—	35-70	—	10-30	—	0-5
2.5	—	—	100	90-100	—	35-70	—	10-40	—	0-15	0-5
3.5	100	90-100	—	45-80	—	25-50	—	10-30	—	0-15	0-5

2 inches to 1 inch, and so on. The recommended particle size distributions vary with maximum and minimum nominal size limits, as shown in table 6-2.

A blank space in table 6-2 indicates a sieve that is not required in the analysis. For example, for the 2 inch to No. 4 nominal size, there are no values listed under the 4-inch, the 3 1/2-inch, the 3-inch, and the 2 1/2-inch sieves. Since 100 percent of this material should pass through a 2 1/2-inch sieve, there is no need to use a sieve coarser than that size. For the same size designation (that is, 2 inch size aggregate), there are no values listed under the 1 1/2-inch, the 3/4-inch, and the 3/8-inch sieves. Experience has shown that it is not necessary to use these sieves in making this particular analysis.

Quality Standards

Since 66 to 78 percent of the volume of the finished concrete consists of aggregate, it is imperative that the aggregate meet certain minimum quality standards. It should consist of clean, hard, strong, durable particles free of chemicals that might interfere with hydration. The aggregate should also be free of any superfine material, which might prevent a bond between the aggregate and the cement-water paste. The undesirable substances most frequently found in aggregate are dirt, silt, clay, coal, mica, salts, and organic matter. Most of these can be removed by washing. Aggregate can be field-tested for an excess of silt, clay, and the like, using the following procedure:

1. Fill a quart jar with the aggregate to a depth of 2 inches.
2. Add water until the jar is about three-fourths full.
3. Shake the jar for 1 minute, then allow it to stand for 1 hour.
4. If, at the end of 1 hour, more than 1/8 inch of sediment has settled on top of the aggregate, as shown in figure 6-2, the material should be washed.

An easily constructed rig for washing a small amount of aggregate is shown in figure 6-3.

Weak, friable (easily pulverized), or laminated (layered) aggregate particles are undesirable. Especially avoid shale, stones laminated with shale, and most varieties of chert (impure flint-like rock). For most ordinary concrete work, visual inspection is enough to reveal any weaknesses in the coarse

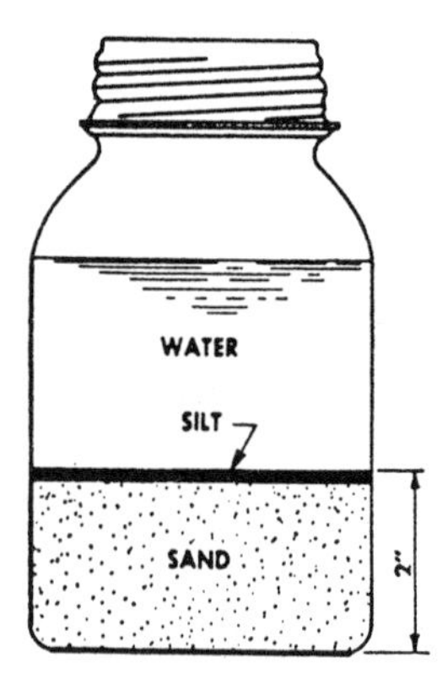

Figure 6-2.—Quart jar method of determining silt content of sand.

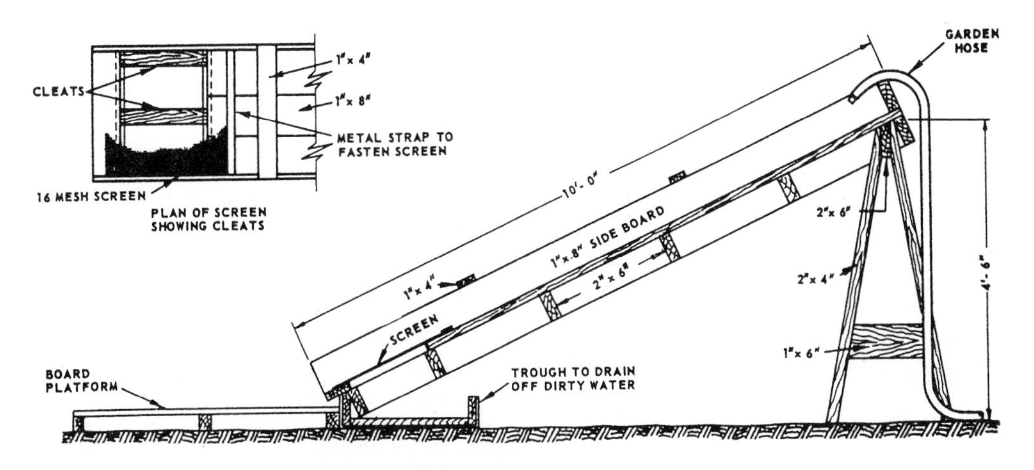

Figure 6-3.—Field-constructed rig for washing aggregate.

aggregate. For work in which aggregate strength and durability are of vital importance, such as paving concrete, aggregate must be laboratory tested.

Handling and Storage

A mass of aggregate containing particles of different sizes has a natural tendency toward segregation. "Segregation" refers to particles of the same size tending to gather together when the material is being loaded, transported, or otherwise disturbed. Aggregate should always be handled and stored by a method that minimizes segregation.

Stockpiles should not be built up in cone shapes, formed by dropping successive loads at the same spot. This procedure causes segregation. A pile should be built up in layers of uniform thickness, each made by dumping successive loads alongside each other.

If aggregate is dropped from a clamshell, bucket, or conveyor, some of the fine material may be blown aside, causing a segregation of fines on the lee side (that is, the side away from the wind) of the pile. Conveyors, clamshells, and buckets should be discharged in contact with the pile.

When a bin is being charged (filled), the material should be dropped from a point directly over the outlet. Material chuted in at an angle or material discharged against the side of a bin will segregate. Since a long drop will cause both segregation and the breakage of aggregate particles, the length of a drop into a bin should be minimized by keeping the bin as full as possible at all times. The bottom of a storage bin should always slope at least 50° toward the central outlet. If the slope is less than 50°, segregation will occur as material is discharged out of the bin.

WATER

The two principal functions of water in a concrete mix are to effect hydration and improve workability. For example, a mix to be poured in forms must contain more water than is required for complete hydration of the cement. Too much water, however, causes a loss of strength by upsetting the water-cement ratio. It also causes "water-gain" on the surface—a condition that leaves a surface layer of weak material, called laitance. As previously mentioned, an excess of water also impairs the watertightness of the concrete.

Water used in mixing concrete must be clean and free from acids, alkalis, oils, and organic materials. Most specifications recommend that the water used in mixing concrete be suitable for drinking, should such water be available.

Seawater can be used for mixing unreinforced concrete if there is a limited supply of fresh water. Tests show that the compressive strength of concrete made with seawater is 10 to 30 percent less than that obtained using fresh water. Seawater is not suitable for use in making steel-reinforced concrete because of the risk of corrosion of the reinforcement, particularly in warm and humid environments.

ADMIXTURES

Admixtures include all materials added to a mix other than portland cement, water, and aggregates.

Admixtures are sometimes used in concrete mixtures to improve certain qualities, such as workability, strength, durability, watertightness, and wear resistance. They may also be added to reduce segregation, reduce the heat of hydration, entrain air, and accelerate or retard setting and hardening.

We should note that the same results can often be obtained by changing the mix proportions or by selecting other suitable materials without resorting to the use of admixtures (except air-entraining admixtures when necessary). Whenever possible, comparison should be made between these alternatives to determine which is more economical or convenient. Any admixture should be added according to current specifications and under the direction of the crew leader.

Workability Agents

Materials, such as hydrated lime and bentonite, are used to improve workability. These materials increase the fines in a concrete mix when an aggregate is tested deficient in fines (that is, lacks sufficient fine material).

Air-Entraining Agents

The deliberate adding of millions of minute disconnected air bubbles to cement paste, if evenly diffused, changes the basic concrete mix and increases durability, workability, and strength. The acceptable amount of entrained air in a concrete mix, by volume, is 3 to 7 percent. Air-entraining agents, used with types I, II, or III cement, are derivatives of natural wood resins, animal or vegetable fats, oils, alkali salts of sulfated organic compounds, and water-soluble soaps. Most air-entraining agents are in liquid form for use in the mixing water.

Accelerator

The only accepted accelerator for general concrete work is calcium chloride with not more than 2 percent by weight of the cement being used. This accelerator is added as a solution to the mix water and is used to speed up the strength gain. Although the final strength is not affected, the strength gain for the first 7 days is greatly affected. The strength gain for the first 7 days can be as high as 1,000 pounds per square inch (psi) over that of normal concrete mixes.

Retarders

The accepted use for retarders is to reduce the rate of hydration. This permits the placement and consolidation of concrete before initial set. Agents normally used are fatty acids, sugar, and starches.

CEMENT STORAGE

Portland cement is packed in cloth or paper sacks, each weighing 94 pounds. A 94-pound sack of cement amounts to about 1 cubic foot by loose volume.

Cement will retain its quality indefinitely if it does not come in contact with moisture. If allowed to absorb appreciable moisture in storage, however, it sets more slowly and strength is reduced. Sacked cement should be stored in warehouses or sheds made as watertight and airtight as possible. All cracks in roofs and walls should be closed, and there should be no openings between walls and roof. The floor should be above ground to protect the cement against dampness. All doors and windows should be kept closed.

Sacks should be stacked against each other to prevent circulation of air between them, but they should not be stacked against outside walls. If stacks are to stand undisturbed for long intervals, they should be covered with tarpaulins.

When shed or warehouse storage cannot be provided, sacks that must be stored in the open should be stacked on raised platforms and covered with waterproof tarps. The tarps should extend beyond the edges of the platform to deflect water away from the platform and the cement.

Cement sacks stacked in storage for long periods sometimes acquire a hardness called warehouse pack. This can usually be loosened by rolling the sack around. However, cement that has lumps or is not free flowing should not be used.

CONCRETE MIX DESIGN

LEARNING OBJECTIVE: Upon completing this section, you should be able to calculate concrete mix designs.

Before proportioning a concrete mix, you need information concerning the job, such as size and shapes of structural members, required strength of the concrete, and exposure conditions. The end use of the concrete and conditions at time of placement are additional factors to consider.

INGREDIENT PROPORTIONS

The ingredient proportions for the concrete on a particular job are usually set forth in the specifications under "CONCRETE—General Requirements." See table 6-3 for examples of normal

Table 6-3.—Normal Concrete

CLASS OF CONCRETE (FIGURES DENOTE SIZE OF COARSE AGGREGATE IN INCHES)	ESTIMATED 28-DAY COMPRESSIVE STRENGTH (POUNDS PER SQUARE INCH)	CEMENT FACTOR BAGS (94 POUNDS) OF CEMENT PER CUBIC YARD OF CONCRETE FRESHLY MIXED	MAXIMUM WATER (GALLONS) PER BAG (94 POUNDS) OF CEMENT	FINE AGGREGATE RANGE IN PERCENT OF TOTAL AGGREGATE BY WEIGHT	APPROXIMATE WEIGHTS OF SATURATED SURFACE-DRY AGGREGATES PER BAG (94 POUNDS) OF CEMENT	
					FINE AGGREGATE POUNDS	COURSE AGGREGATE POUNDS
(1)	(2)	(3)	(4)	(5)	(6)	(7)
B-1	1,500	4.10	9.50	42-52	368	415
B1-1.5	1,500	3.80	9.50	38-48	376	498
B-2	1,500	3.60	9.50	35-45	378	567
B-2.5	1,500	3.50	9.50	33-43	373	609
B-3.5	1,500	3.25	9.50	30-40	378	702
C-1	2,000	4.45	8.75	41-51	329	387
C-1.5	2,000	4.10	8.75	37-47	338	467
C-2	2,000	3.90	8.75	34-44	338	529
C-2.5	2,000	3.80	8.75	32-42	332	565
C-3.5	2,000	3.55	8.75	29-39	334	648
D-0.5	2,500	5.70	7.75	50-60	282	231
D-0.75	2,500	5.30	7.75	45-55	288	288
D-1	2,500	5.05	7.75	40-50	279	341
D-1.5	2,500	4.65	7.75	36-46	287	413
D-2	2,500	4.40	7.75	34-42	288	471
D-2.5	2,500	4.25	7.75	32-40	287	509
D-3.5	2,500	4.00	7.75	29-37	285	578
E-0.5	3,000	6.50	6.75	50-58	238	203
E-0.75	3,000	6.10	6.75	45-53	240	249
E-1	3,000	5.80	6.75	40-48	233	297
E-1.5	3,000	5.35	6.75	36-44	239	359
E-2	3,000	5.05	6.75	33-41	241	410
E-2.5	3,000	4.90	6.75	31-39	238	441
E-3.5	3,000	4.60	6.75	28-36	237	503
F-0.375	3,500	7.70	6.00	56-64	210	140
F-0.5	3,500	7.35	6.00	48-56	198	183
F-0.75	3,500	6.85	6.00	43-51	201	226
F-1	3,500	6.50	6.00	38-46	195	270
F-1.5	3,500	6.00	6.00	34-42	200	325
F-2	3,500	5.70	6.00	31-29	199	369
F-2.5	3,500	5.50	6.00	29-37	197	400
F-3.5	3,500	5.20	6.00	27-35	200	444
G-0.25	4,000	8.95	5.50	100	281	—
G-0.375	4,000	8.40	5.50	55-63	186	129
G-0.5	4,000	8.00	5.50	47-55	175	168
G-0.75	4,000	7.45	5.50	42-50	178	209
G-1	4,000	7.10	5.50	37-45	172	247
G-1.5	4,000	6.55	5.50	33-41	175	299
G-2	4,000	6.20	5.50	30-38	175	340
G-2.5	4,000	6.00	5.50	28-36	173	368
G-3.5	4,000	5.65	5.50	26-34	176	411
H-0.25	5,000	11.50	4.25	100	202	—
H-0.375	5,000	10.80	4.25	53-61	130	98
H-0.5	5,000	10.35	4.25	45-53	121	126
H-0.75	5,000	9.65	4.25	40-48	123	157
H-1	5,000	9.20	4.25	35-43	119	186
H-1.5	5,000	8.45	4.25	31-39	122	228
H-2	5,000	8.00	4.25	28-36	122	260

concrete-mix design according to NAVFAC specifications.

In table 6-3, one of the formulas for 3,000 psi concrete is 5.80 bags of cement per cubic yard, 233 pounds of sand (per bag of cement), 297 pounds of coarse aggregate (per bag of cement), and a water-cement ratio of 6.75 gallons of water to each bag of cement. These proportions are based on the assumption that the inert ingredients are in a saturated surface-dry condition, meaning that they contain all the water they are capable of absorbing, but no additional free water over and above this amount.

We need to point out that a saturated surface-dry condition almost never exists in the field. The amount of free water in the coarse aggregate is usually small enough to be ignored, but the ingredient proportions set forth in the specs must almost always be adjusted to allow for the existence of free water in the fine aggregate. Furthermore, since free water in the fine aggregate increases its measured volume or weight over that of the sand itself, the specified volume or weight of sand must be increased to offset the volume or weight of the water in the sand. Finally, the number of gallons of water used per sack of cement must be reduced to allow for the free water in the sand. The amount of water actually added at the mixer must be the specified amount per sack, less the amount of free water that is already in the ingredients in the mixer.

Except as otherwise specified in the project specifications, concrete is proportioned by weighing and must conform to NAVFAC specifications. (See table 6-3 for normal concrete.)

MATERIAL ESTIMATES

When tables, such as table 6-3, are not available for determining quantities of material required for 1 cubic yard of concrete, a rule of thumb, known as rule 41 or 42, may be used for a rough estimation. According to this rule, it takes either 41 or 42 cubic feet of the combined dry amounts of cement, sand, and aggregates to produce 1 cubic yard of mixed concrete. Rule 41 is used to calculate the quantities of material for concrete when the size of the coarse aggregate is not over 1 inch. Rule 42 is used when the size of the coarse aggregate is not over 2 1/2 inches. Here is how it works.

As we mentioned earlier, a bag of cement contains 94 pounds by weight, or about 1 cubic foot by loose volume. A batch formula is usually based on the number of bags of cement used in the mixing machine.

For estimating the amount of dry materials needed to mix 1 cubic yard of concrete, rules 41 and 42 work in the same manner. The decision on which rule to use depends upon the size of the aggregate. Let's say your specifications call for a 1:2:4 mix with 2-inch coarse aggregates, which means you use rule 42. First, add 1:2:4, which gives you 7. Then compute your material requirements as follows:

$42 \div 7 = 6$ bags, or 6 cu ft of cement;

$6 \times 2 = 12$ cu ft of sand;

$6 \times 4 = 24$ cu ft of coarse aggregates.

Adding your total dry materials, $6 + 12 + 24 = 42$, so your calculations are correct.

Frequently, you will have to convert volumes in cubic feet to weights in pounds. In converting, multiply the required cubic feet of cement by 94 since 1 cubic foot, or 1 standard bag of cement, weighs 94 pounds. When using rule 41 for coarse aggregates, multiply the quantity of coarse gravel in cubic feet by 105 since the average weight of dry-compacted fine aggregate or gravel is 105 pounds per cubic feet. By rule 42, however, multiply the cubic feet of rock (1-inch-size coarse aggregate) by 100 since the average dry-compacted weight of this rock is 100 pounds per cubic foot.

A handling-loss factor is added in ordering materials for jobs. An additional 5 percent of materials is added for jobs requiring 200 or more cubic yards of concrete, and 10 percent is added for smaller jobs. This loss factor is based on material estimates after the requirements have been calculated. Additional loss factors may be added where conditions indicate the necessity for excessive handling of materials before batching.

Measuring Water

The water-measuring controls on a machine concrete mixer are described later in this chapter. Water measurement for hand mixing can be done with a 14-quart bucket, marked off on the inside in gallons, half-gallons, and quarter-gallons.

Never add water to the mix without carefully measuring the water, and always remember that the amount of water actually placed in the mix varies according to the amount of free water that is already in the aggregate. This means that if the aggregate is wet by a rainstorm, the proportion of water in the mix may have to be changed.

Measuring Aggregate

The accuracy of aggregate measurement by volume depends upon the accuracy with which the amount of "bulking," caused by moisture in the aggregate, can be determined. The amount of bulking varies not only with different moisture contents but also with different gradations. Fine sand, for example, is bulked more than coarse sand by the same moisture content. Furthermore, moisture content itself varies from time to time, and a small variation causes a large change in the amount of bulking. For these and other reasons, aggregate should be measured by weight rather than by volume whenever possible.

To make grading easier, to keep segregation low, and to ensure that each batch is uniform, you should store and measure coarse aggregate from separate piles or hoppers. The ratio of maximum to minimum particle size should not exceed 2:1 for a maximum nominal size larger than 1 inch. The ratio should not exceed 3:1 for a maximum nominal size smaller than 1 inch. A mass of aggregate with a nominal size of 1 1/2 inches to 1/4 inch, for example, should be separated into one pile or hopper containing 1 1/2-inch to 3/4-inch aggregate, and another pile or hopper containing 3/4-inch to 1/4-inch aggregate. A mass with a nominal size of 3 inches to 1/4 inch should be separated into one pile or hopper containing 3-inch to 1 1/2-inch aggregate, another containing 1 1/2-inch to 3/4-inch aggregate, and a third containing 3/4-inch to 1/4-inch aggregate.

Water-Cement Ratio

The major factor controlling strength, everything else being equal, is the amount of water used per bag of cement. Maximum strength is obtained by using just the amount of water, and no more, required for the complete hydration of the cement. As previously mentioned, however, a mix of this type may be too dry to be workable. Concrete mix always contains more water than the amount required to attain maximum strength. The point for you to remember is that the strength of concrete **decreases** as the amount of extra water **increases**.

The specified water-cement ratio is the happy medium between the maximum possible strength of the concrete and the necessary minimum workability requirements. The strength of building concrete is expressed in terms of the compressive strength in pounds per square inch (psi) reached after a 7- or 28-day set. This is usually referred to as "probable average 7-day strength" and "probable average 28-day strength."

SLUMP TESTING

Slump testing is a means of measuring the consistency of concrete using a "slump cone." The cone is made of galvanized metal with an 8-inch-diameter base, a 4-inch-diameter top, and a 12-inch height. The base and the top are open and parallel to each other and at right angles to the axis of the cone (figure 6-4). A tamping rod 5/8 inch in diameter and 24 inches long is also needed. The tamping rod should be smooth and bullet-pointed. **Do not** use a piece of reinforcing bar (rebar).

Samples of concrete for test specimens are taken at the mixer or, in the case of ready-mixed concrete, from the transportation vehicle during discharge. The sample of concrete from which test specimens are made should be representative of the entire batch. Such samples are obtained by repeatedly passing a scoop or pail through the discharging stream of concrete, starting the sampling operation at the beginning of discharge, and repeating the operation until the entire batch is discharged. To counteract segregation when a sample must be transported to a test site, the concrete should be remixed with a shovel until it is uniform in appearance. The job location from which the sample was taken should be noted for future reference. In the case of paving concrete,

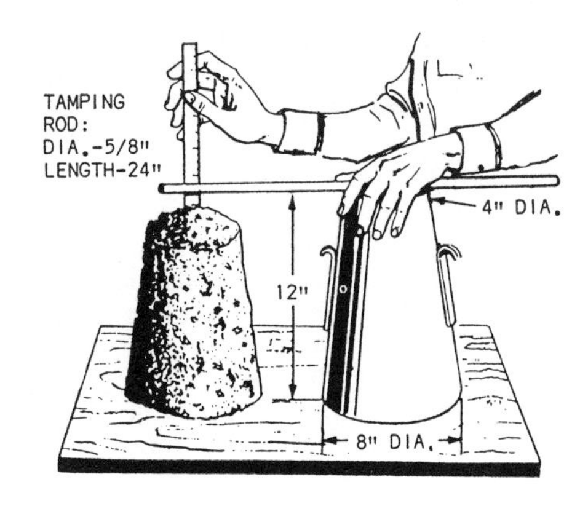

Figure 6-4.—Measurement of slump.

samples may be taken from the batch immediately after depositing it on the subgrade. At least five samples should be taken at different times, and these samples should be thoroughly mixed to form the test specimen.

When making a slump test, dampen the cone and place it on a flat, moist, nonabsorbent surface. From the sample of concrete obtained, immediately fill the cone in three layers, each approximately one-third the volume of the cone. In placing each scoop full of concrete in the cone, move the scoop around the edge of the cone as the concrete slides from the scoop. This ensures symmetrical distribution of concrete within the cone. Each layer is then "rodded in" with 25 strokes. The strokes should be distributed uniformly over the cross section of the cone and penetrate into the underlying layer. The bottom layer should be rodded throughout its depth.

If the cone becomes overfilled, use a straightedge to strike off the excess concrete flush with the top. The cone should be immediately removed from the concrete by raising it carefully in a vertical direction. The slump should be measured immediately after removing the cone. You determine the slump by measuring the difference between the height of the cone and the height of the specimen (figure 6-4). The slump should be recorded in terms of inches of subsidence of the specimen during the test.

After completing the slump measurement, gently tap the side of the mix with the tamping rod. The behavior of the concrete under this treatment is a valuable indication of the cohesiveness, workability, and placability of the mix. In a well-proportioned mix, tapping only causes it to slump lower. It doesn't crumble apart or segregate by the dropping of larger aggregate particles to a lower level in the mix. If the concrete crumbles apart, it is oversanded. If it segregates, it is undersanded.

WORKABILITY

A mix must be workable enough to fill the form spaces completely, with the assistance of a reasonable amount of shoveling, spading, and vibrating. Since a fluid or "runny" mix does this more readily than a dry or "stiff" mix, you can see that workability varies directly with fluidity. The workability of a mix is determined by the slump test. The amount of the slump, in inches, is the measure of the concrete's workability—the more the slump, the higher the workability.

The slump can be controlled by a change in any one or all of the following: gradation of aggregates, proportion of aggregates, or moisture content. If the moisture content is too high, you should add more cement to maintain the proper water-cement ratio.

The desired degree of workability is attained by running a series of trial batches, using various amounts of fine to coarse aggregate, until a batch is produced that has the desired slump. Once the amount of increase or decrease in fines required to produce the desired slump is determined, the aggregate proportions, not the water proportion, should be altered in the field mix to conform. If the water proportion were changed, the water-cement ratio would be upset.

Never yield to the temptation to add more water without making the corresponding adjustment in the cement content. Also, make sure that crewmembers who are spreading a stiff mix by hand do not ease their labors by this method without telling you.

As you gain experience, you will discover that adjustments in workability can be made by making very minor changes in the amount of fine or coarse aggregate. Generally, everything else remaining equal, an increase in the proportion of fines stiffens a mix, whereas an increase in the proportion of coarse loosens a mix.

NOTE

Before you alter the proportions set forth in a specification, you must find out from higher authority whether you are allowed to make any such alterations and, if you are, the permissible limits beyond which you must not go.

GROUT

As previously mentioned, concrete consists of four essential ingredients: water, cement, sand, and coarse aggregate. The same mixture without aggregate is mortar. Mortar, which is used chiefly for bonding masonry units together, is discussed in a later chapter. Grout refers to a water-cement mixture called neat-cement grout and to a water-sand-cement mixture called sand-cement grout. Both mixtures are used to plug holes or cracks in concrete, to seal joints, to fill spaces between machinery bedplates and concrete foundations, and for similar plugging or sealing purposes. The consistency of grout may range from stiff (about 4 gallons of water per sack of cement) to fluid (as many as 10 gallons of water per sack of cement), depending upon the nature of the grouting job at hand.

BATCHING

When bagged cement is used, the field mix proportions are usually given in terms of designated amounts of fine and coarse aggregate per bag (or per 94 pounds) of cement. The amount of material that is mixed at a time is called a batch. The size of a batch is usually designated by the number of bags of cement it contains, such as a four-bag batch, a six-bag batch, and so forth.

The process of weighing out or measuring out the ingredients for a batch of concrete is called batching. When mixing is to be done by hand, the size of the batch depends upon the number of persons available to turn it with hand tools. When mixing is to be done by machine, the size of the batch depends upon the rated capacity of the mixer. The rated capacity of a mixer is given in terms of cubic feet of mixed concrete, not of dry ingredients.

On large jobs, the aggregate is weighed out in an aggregate batching plant (usually shortened to "batch plant"), like the one shown in figure 6-5. Whenever possible, a batch plant is located near to and used in conjunction with a crushing and screening plant. In a crushing and screening plant, stone is crushed into various particle sizes, which are then screened into separate piles. In a screening plant, the aggregate in its natural state is screened by sizes into separate piles.

29.153

Figure 6-5.—Aggregate batching plant.

The batch plant, which is usually portable and can be taken apart and moved from site to site, is generally set up adjacent to the pile of screened aggregate. The plant may include separate hoppers for several sizes of fine and coarse aggregates, or only one hopper for fine aggregate and another for coarse aggregate. It may have one or more divided hoppers, each containing two or more separate compartments for different sizes of aggregates.

Each storage hopper or storage hopper compartment can be discharged into a weigh box, which can, in turn, be discharged into a mixer or a batch truck. When a specific weight of aggregate is called for, the operator sets the weight on a beam scale. The operator then opens the discharge chute on the storage hopper. When the desired weight is reached in the weigh box, the scale beam rises and the operator closes the storage hopper discharge chute. The operator then opens the weigh box discharge chute, and the aggregate discharges into the mixer or batch truck. Batch plant aggregate storage hoppers are usually loaded with clamshell-equipped cranes.

The following guidelines apply to the operation of batch plants:

- All personnel working in the batch plant area should wear hard hats at all times.
- While persons are working in conveyor line areas, the switches and controls should be secured and tagged so that no one can engage them until all personnel are clear.
- When hoppers are being loaded, personnel should stay away from the area of falling aggregate.
- The scale operator should be the only person on the scale platform during batching operations.
- Housekeeping of the charging area is important. Personnel should do everything possible to keep the area clean and free of spoiled material or overflow.
- Debris in aggregate causes much of the damage to conveyors. Keep the material clean at all times.

- When batch operations are conducted at night, good lighting is a must.

- Personnel working in batch plants should use good eye hygiene. Continual neglect of eye care can have serious consequences.

MIXING CONCRETE

LEARNING OBJECTIVE: Upon completing this section, you should be able to determine methods and mixing times of concrete.

Concrete is mixed either by hand or machine. No matter which method is used, you must follow well-established procedures if you expect finished concrete of good quality. An oversight in proper concrete mixing, whether through lack of competence or inattention to detail, cannot be corrected later.

MIXING BY HAND

A batch to be hand mixed by a couple of crewmembers should not be much larger than 1 cubic yard. The equipment required consists of a watertight metal or wooden platform, two shovels, a metal-lined measuring box, and a graduated bucket for measuring the water.

The mixing platform does not need to be made of expensive materials. It can be an abandoned concrete slab or concrete parking lot that can be cleaned after use. A wooden platform having tight joints to prevent the loss of paste may be used. Whichever surface is used, you should ensure that it is cleaned prior to use and level.

Let's say your batch consists of two bags of cement, 5.5 cubic feet of sand, and 6.4 cubic feet of coarse aggregate. Mix the sand and cement together first, using the following procedure:

1. Dump 3 cubic feet of sand on the platform first, spread it out in a layer, and dump a bag of cement over it.

2. Spread out the cement and dump the rest of the sand (2.5 cubic feet) over it.

3. Dump the second sack of cement on top of the lot.

This use of alternate layers of sand and cement reduces the amount of shoveling required for complete mixing.

Personnel doing the mixing should face each other from opposite sides of the pile and work from the outside to the center. They should turn the mixture as many times as is necessary to produce a uniform color throughout. When the cement and sand are completely mixed, the pile should be leveled off and the coarse material added and mixed by the same turning method.

The pile should next be troughed in the center. The mixing water, after being carefully measured, should be poured into the trough. The dry materials should then be turned into the water, with great care taken to ensure that none of the water escapes. When all the water has been absorbed, the mixing should continue until the mix is of a uniform consistency. Four complete turnings are usually required.

MIXING BY MACHINE

The size of a concrete mixer is designated by its rated capacity. As we mentioned earlier, the capacity is expressed in terms of the volume of mixed concrete, not of dry ingredients the machine can mix in a single batch. Rated capacities run from as small as 2 cubic feet to as large as 7 cubic yards (189 cubic feet). In the Naval Construction Forces (NCFs), the most commonly used mixer is the self-contained Model 16-S (figure 6-6) with a capacity of 16 cubic feet (plus a 10-percent overload).

29.151

Figure 6-6.—Model 16-S concrete mixer.

The production capacity of the 16-S mixer varies between 5 and 10 cubic yards per hour, depending on the efficiency of the personnel. Aggregate larger than 3 inches will damage the mixer. The mixer consists of a frame equipped with wheels and towing tongue (for easy movement), an engine, a power loader skip, mixing drum, water tank, and an auxiliary water pump. The mixer may be used as a central mixing plant.

Charging the Mixer

Concrete mixers may be charged by hand or with the mechanical skip. Before loading the mechanical skip, remove the towing tongue. Then cement, sand, and gravel are loaded and dumped into the mixer together while the water runs into the mixing drum on the side opposite the skip. A storage tank on top of the mixer measures the mixing water into the drum a few seconds before the skip dumps. This discharge also washes down the mixer between batches. The coarse aggregate is placed in the skip first, the cement next, and the sand is placed on top to prevent excessive loss of cement as the batch enters the mixer.

Mixing Time

It takes a mixing machine having a capacity of 27 cubic feet or larger 1 1/2 minutes to mix a 1-cubic yard batch. Another 15 seconds should be allowed for each additional 1/2 cubic yard or fraction thereof. The water should be started into the drum a few seconds before the skip begins to dump, so that the inside of the drum gets a washout before the batched ingredients go in. The mixing period should be measured from the time all the batched ingredients are in, provided that all the water is in before one-fourth of the mixing time has elapsed. The time elapsing between the introduction of the mixing water to the cement and aggregates and the placing of the concrete in the forms should not exceed 1 1/2 hours.

Discharging the Mixer

When the material is ready for discharge from the mixer, the discharge chute is moved into place to receive the concrete from the drum of the mixer. In some cases, stiff concrete has a tendency to carry up to the top of the drum and not drop down in time to be deposited on the chute. Very wet concrete may not carry up high enough to be caught by the chute. This condition can be corrected by adjusting the speed of the mixer. For very wet concrete, the speed of the drum should be increased. For stiff concrete, the drum speed should be slowed down.

Cleaning and Maintaining the Mixer

The mixer should be cleaned daily when it is in continuous operation or following each period of use if it is in operation less than a day. If the outside of the mixer is kept coated with oil, the cleaning process can be speeded up. The outside of the mixer should be washed with a hose, and all accumulated concrete should be knocked off. If the blades of the mixer become worn or coated with hardened concrete, the mixing action will be less efficient. Badly worn blades should be replaced. Hardened concrete should not be allowed to accumulate in the mixer drum. The mixer drum must be cleaned out whenever it is necessary to shut down for more than 1 1/2 hours. Place a volume of coarse aggregate in the drum equal to one-half of the capacity of the mixer and allow it to revolve for about 5 minutes. Discharge the aggregate and flush out the drum with water. Do not pound the discharge chute, drum shell, or the skip to remove aggregate or hardened concrete. Concrete will readily adhere to the dents and bumps created. For complete instructions on the operation, adjustment, and maintenance of the mixer, study the manufacturer's manual.

All gears, chains, and rollers of mixers should be properly guarded. All moving parts should be cleaned and properly serviced to permit safe performance of the equipment. When the mixer drum is being cleaned, the switches must be open, the throttles closed, and the control mechanism locked in the OFF position. The area around the mixer must be kept clear.

Skip loader cables and brakes must be inspected frequently to prevent injuries caused by falling skips. When work under an elevated skip is unavoidable, you must shore up the skip to prevent it from falling in the event that the brake fails or is accidentally released. The mixer operator must never lower the skip without first making sure that there is no one underneath.

Dust protection equipment must be issued to the crew engaged in handling cement, and the crew must wear the equipment when so engaged. Crewmembers should stand with their backs to the wind, whenever possible. This helps prevent cement and sand from being blown into their eyes and faces.

HANDLING AND TRANSPORTING CONCRETE

When ready-mixed concrete is carried by an ordinary type of carrier (such as a wheelbarrow or buggy), jolting of the carrier increases the natural tendency of the concrete to segregate. Carriers should therefore be equipped with pneumatic tires whenever possible, and the surface over which they travel should be as smooth as possible.

A long free fall also causes concrete to segregate. If the concrete must be discharged at a level more than 4 feet above the level of placement, it should be dumped into an "elephant trunk" similar to the one shown in figure 6-7.

Segregation also occurs when discharged concrete is allowed to glance off a surface, such as the side of a form or chute. Wheelbarrows, buggies, and conveyors should discharge so that the concrete falls clear.

Concrete should be transported by chute only for short distances. It tends to segregate and dry out when handled in this manner. For a mix of average workability, the best slope for a chute is about 1 foot of rise to 2 or 3 feet of run. A steeper slope causes segregation, whereas a flatter slope causes the concrete to run slowly or not at all. The stiffer the mix, the steeper the slope required. All chutes and spouting used in concrete pours should be clean and well-supported by proper bracing and guys.

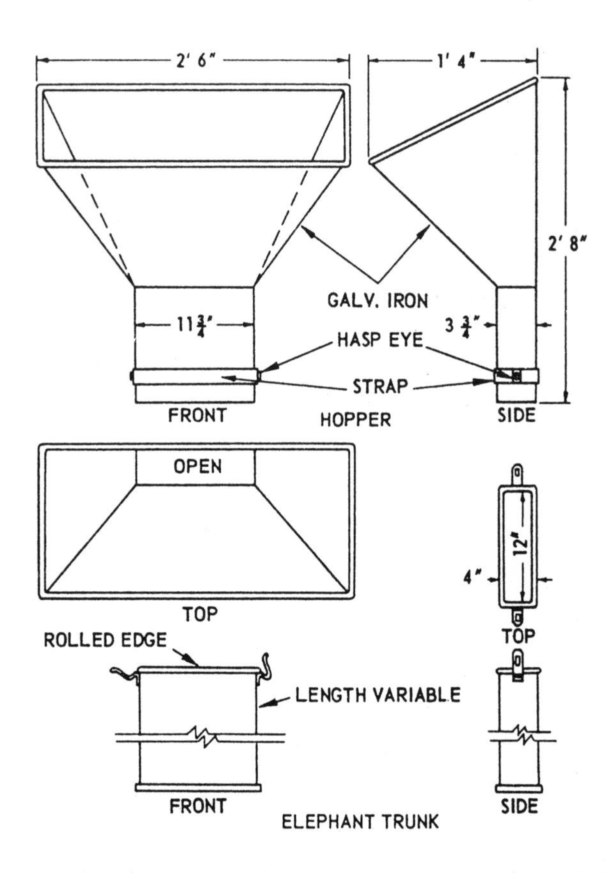

Figure 6-7.—Chute, or downpipe, used to check free fall of concrete.

133.499

Figure 6-8.—Precast wall panels in stacks of three each.

When spouting and chutes run overhead, the area beneath must be cleared and barricaded during placing. This eliminates the danger of falling concrete or possible collapse.

READY-MIXED CONCRETE

On some jobs, such as large highway jobs, it is possible to use a batch plant that contains its own mixer. A plant of this type discharges ready-mixed concrete into transit mixers, which haul it to the construction site. The truck carries the mix in a revolving chamber much like the one on a mixer. Keeping the mix agitated in route prevents segregation of aggregate particles. A ready-mix plant is usually portable so that it can follow the job along. It must be certain, of course, that a truck will be able to deliver the mix at the site before it starts to set. Discharge of the concrete from the drum should be completed within 1 1/2 hours.

TRANSIT-MIXED CONCRETE

By transit-mixing, we refer to concrete that is mixed, either wet or dry, en route to a job site. A transit-mix truck carries a mixer and a water tank from which the driver can, at the proper time, introduce the required amount of water into the mix. The truck picks up the dry ingredients at the batch plant, together with a slip which tells how much water is to be introduced to the mix upon arrival at the site. The mixer drum is kept revolving in route and at the job site so that the dry ingredients do not segregate. Transit-mix trucks are part of the battalion's equipment inventory and are widely used on all but the smallest concrete jobs assigned to a battalion.

PRECAST AND TILT-UP CONCRETE

LEARNING OBJECTIVE: Upon completing this section, you should be able to determine projects suitable for and lifting methods necessary for precast and tilt-up construction.

Concrete cast in the position it is to occupy in the finished structure is called cast-in-place concrete. Concrete cast and cured elsewhere is called precast concrete. Tilt-up concrete is a special type of precast concrete in which the units are tilted up and placed using cranes or other types of lifting devices.

Wall construction, for example, is frequently done with precast wall panels originally cast horizontally (sometimes one above the other) as slabs. This method has many advantages over the conventional method of casting in place in vertical wall forms. Since a slab form requires only edge forms and a single surface form, the amount of formwork and form materials required is greatly reduced. The labor involved in slab form concrete casting is much less than that involved in filling a high wall form. One side of a precast unit cast as a slab may be finished by hand to any desired quality of finishing. The placement of reinforcing steel is much easier in slab forms, and it is easier to attain thorough filling and vibrating. Precasting of wall panels as slabs may be expedited by mass production methods not available when casting in place.

Relatively light panels for concrete walls are precast as slabs (figure 6-8). The panels are set in

133.500

Figure 6-9.—Precast panels being erected by use of crane and spreader bars.

place by cranes, using spreader bars (figure 6-9). Figure 6-10 shows erected panels in final position.

CASTING

The casting surface is very important in making precast concrete panels. In this section, we will cover two common types: earth and concrete. Regardless of which method you use, however, a slab must be cast in a location that will permit easy removal and handling.

Castings can be made directly on the ground with cement poured into forms. These "earth" surfaces are

133.501

Figure 6-10.—Precast panels in position.

economical but only last for a couple of concrete pours. Concrete surfaces, since they can be reused repeatedly, are more practical.

When building casting surfaces, you should keep the following points in mind:

- The subbase should be level and properly compacted.
- The slab should be at least 6 inches thick and made of 3,000 psi or higher reinforced concrete. Large aggregate, 2 1/2 inches to 3 inches maximum, may be used in the casting slabs.
- If pipes or other utilities are to be extended up through the casting slab at a later date, they should be stopped below the surface and the openings temporarily closed. For wood, cork, or plastic plugs, fill almost to the surface with sand and top with a thin coat of mortar that is finished flush with the casting surface.
- It is important to remember that any imperfections in the surface of the casting slab will show up on the cast panels. When finishing the casting slab, you must ensure there is a flat, level, and smooth surface without humps, dips, cracks, or gouges. If possible, cure the casting surface keeping it covered with water (ponding). However, if a curing compound or surface hardener is used, make sure it will not conflict with the later use of bond-breaking agents.

FORMS

The material most commonly used for edge forms is 2-by lumber. The lumber must be occasionally replaced, but the steel or aluminum angles and channels may be reused many times. The tops of the forms must be in the same plane so that they may be used for screeds. They must also be well braced to remain in good alignment.

Edge forms should have holes in them for rebar or for expansion/contraction dowels to protrude. These holes should be 1/4 inch larger in diameter than the bars. At times, the forms are spliced at the line of these bars to make removal easier.

The forms, or rough bucks, for doors, windows, air-conditioning ducts, and so forth, are set before the steel is placed and should be on the same plane as the edge forms.

BOND-BREAKING AGENTS

Bond-breaking agents are one of the most important items of precast concrete construction. The most important requirement is that they must break the bond between the casting surface and the cast panel. Bond-breaking agents must also be economical, fast drying, easily applied, easily removed, or leave a paintable surface on the cast panel, if desired. They are broken into two general types: sheet materials and liquids.

There are many commercially available bond-breaking agents available. You should obtain the type best suited for the project and follow the manufacturer's application instructions. If commercial bond-breaking agents are not available, several alternatives can be used.

- Paper and felt effectively prevent a bond with a casting surface, but usually stick to the cast panels and may cause asphalt stains on the concrete.
- When oiled, plywood, fiberboard, and metal effectively prevent a bond and can be used many times. The initial cost, however, is high and joint marks are left on the cast panels.
- Canvas gives a very pleasing texture and is used where cast panels are lifted at an early stage. It should be either dusted with cement or sprinkled with water just before placing the concrete.
- Oil gives good results when properly used, but is expensive. The casting slab must be dry when the oil is applied, and the oil must be allowed to absorb before the concrete is placed. Oil should not be used if the surface is to be painted, and crankcase oil should never be used.
- Waxes, such as spirit wax (paraffin) and ordinary floor wax, give good-to-excellent results. One mixture that may be used is 5 pounds of paraffin mixed with 1 1/2 gallons of light oil or kerosene. The oil must be heated to dissolve the paraffin.
- Liquid soap requires special care to ensure that an excess amount is not used or the surface of the cast panel will be sandy.

Materials should be applied after the side forms are in place and the casting slab is clean but **before** any reinforcing steel is placed. To ensure proper adhesion of the concrete, keep all bond-breaking materials off the reinforcing steel.

REINFORCEMENTS AND INSERTS

Reinforcing bars (rebar) should be assembled into mats and placed into the forms as a unit. This allows for rapid assembly on a jig and reduces walking on the casting surface, which has been treated with the bond-breaking agent.

Extra rebars must be used at openings. They should be placed parallel to and about 2 inches from the sides of openings or placed diagonally across the corners of openings.

The bars may be suspended by conventional methods, such as with high chairs or from members laid across the edge forms. However, high chairs should not be used if the bottom of the cast panel is to be a finished surface. Another method is to first place half the thickness of concrete, place the rebar mat, and then complete the pour. However, this method must be done quickly to avoid a cold joint between the top and bottom layers.

When welded wire fabric (WWF) is used, dowels or bars must still be used between the panels and columns. WWF is usually placed in sheets covering the entire area and then clipped along the edges of the openings after erection.

If utilities are going to be flush-mounted or hidden, pipe, conduit, boxes, sleeves, and so forth should be put into the forms at the same time as the reinforcing steel. If the utilities pass from one cast panel to another, the connections must be made after the panels are erected but before the columns are poured. If small openings are to go through the panel, a greased pipe sleeve is the easiest method of placing an opening in the form. For larger openings, such as air-conditioning ducts, forms should be made in the same manner as doors or windows.

After rebar and utilities have been placed, all other inserts should be placed. These will include lifting and bracing inserts, anchor bolts, welding plates, and so forth. You need to make sure these items are firmly secured so they won't move during concrete placement or finishing.

POURING, FINISHING, AND CURING

With few exceptions, pouring cast panels can be done in the same manner as other pours. Since the panels are poured in a horizontal position, a stiffer mix can be used. A minimum of six sacks of cement per cubic yard with a maximum of 6 gallons of water per sack of cement should be used along with well-graded aggregate. As pointed out earlier, though, you will have to reduce the amount of water used per sack of cement to allow for the free water in the sand. Large aggregate, up to 1 1/2 inches in diameter, may be used effectively. The concrete should be worked into place by spading or vibration, and extra care must be taken to prevent honeycomb around outer edges of the panel.

Normal finishing methods should be used, but many finishing styles are available for horizontally cast panels. Some finishing methods include patterned, colored, exposed aggregate, broomed, floated, or steel-troweled. Regardless of the finish used, finishers must be cautioned to do the finishing of all panels in a uniform manner. Spots, defects, uneven brooming, or troweling, and so forth will be highly visible when the panels are erected.

Without marring the surface, curing should be started as soon as possible after finishing. Proper curing is important, so cast panels should be cured just like any other concrete to achieve proper strength. Curing compound, if used, prevents bonding with other concrete or paint.

LIFTING EQUIPMENT AND ATTACHMENTS

Tilt-up panels can be set up in many different ways and with various kinds of power equipment. The choice depends upon the size of the job. Besides the equipment, a number of attachments are used.

Equipment

The most popular power equipment is a crane. But other equipment used includes a winch and an A frame, used either on the ground or mounted on a truck. When a considerable number of panels are ready for tilting at one time, power equipment speeds up the job.

Attachments

Many types of lifting attachments are used to lift tilt-up panels. Some of these attachments are locally made and are called hairpins; other types are available commercially. Hairpin types are made on the job site from rebar. These are made by making 180° bends in the ends of two vertical reinforcing bars. The hairpins are then placed in the end of the panel before the concrete is poured. These lifting attachments must protrude from the top of the form for attaching the lifting chains or cables, but go deep enough in the panel form so they won't pull out.

Among the commercial types of lifting attachments, you will find many styles with greater lifting capacities that are more dependable than hairpins if properly installed. These are used with lifting plates. For proper placement of lifting inserts, refer to the plans or specs.

Spreader Bars

Spreader bars (shown in figure 6-9) may be permanent or adjustable, but must be designed and made according to the heaviest load they will carry plus a safety factor. They are used to distribute the lifting stresses evenly, reduce the lateral force applied by slings, and reduce the tendency of panels to bow.

POINT PICKUP METHODS

Once the concrete has reached the desired strength, the panels are ready to be lifted. The strength of the inserts is governed by the strength of the concrete.

CAUTION

An early lift may result in cracking the panel, pulling out the insert, or total concrete failure. The time taken to wait until the concrete has reached its full strength prevents problems and minimizes the risk of injury.

There are several different pickup methods. The following are just some of the basics. Before using these methods on a job, make sure that you check plans and specs to see if these are stated there. Figure 6-11 shows four different pickup methods: 2, 2-2, 4-4, and 2-2-2.

The 2-point pickup is the simplest method, particularly for smaller panels. The pickup cables or chains are fastened directly from the crane hook or spreader bar to two pickup points on or near the top of the precast panel.

The 2-2 point pickup is a better method and is more commonly used. Variations of the 2-2 are 4-4 and 2-2-2, or combinations of pickup points as designated in the job site specifications. These methods use a combination of spreader bars, sheaves, and equal-length cables. The main purpose is to distribute the lifting stresses throughout the panel during erection. Remember, the cables must be long enough to allow ample clearance between the top of the panel and the sheaves or spreader bar.

ERECTING, BRACING, AND JOINTING PANELS

Erecting is an important step in the construction phase of the project. Before you start the erecting phase and for increased safety, you should make sure that all your tools, equipment, and braces are in proper working order. All personnel must be well informed and the signalman and crane operator understand **and** agree on the signals to be used. During the erection of the panels, make sure that the signalman and line handler are not under the panel and that all unnecessary personnel and equipment are away from the lifting area. After the erection is done, make sure that all panels are properly braced and secured before unhooking the lifting cables.

Bracing is an especially important step. After all the work of casting and placing the panels, you want them to stay in place. The following are some steps to take before lifting the panels:

- Install the brace inserts into the panels during casting if possible.
- Install the brace inserts into the floor slab either during pouring or the day before erection.
- Install solid brace anchors before the day of erection.
- If brace anchors must be set during erection, use a method that is fast and accurate.

Although there are several types of bracing, pipe or tubular braces are the most common. They usually have a turnbuckle welded between sections for adjustment. Some braces are also made with telescoping sleeves for greater adaptability. Figure 6-10 shows tube-type braces used to hold up panels. Cable braces are normally used for temporary bracing and for very tall panels. Their flexibility and tendency to stretch, however, make them unsuitable for most projects. Wood bracing is seldom used except for low, small panels or for temporary bracing.

Jointing the panels is simple. Just tie all the panels together, covering the gap between them. You can weld, bolt, or pour concrete columns or beams. Steps used to tie the panels should be stated in the plans and specs.

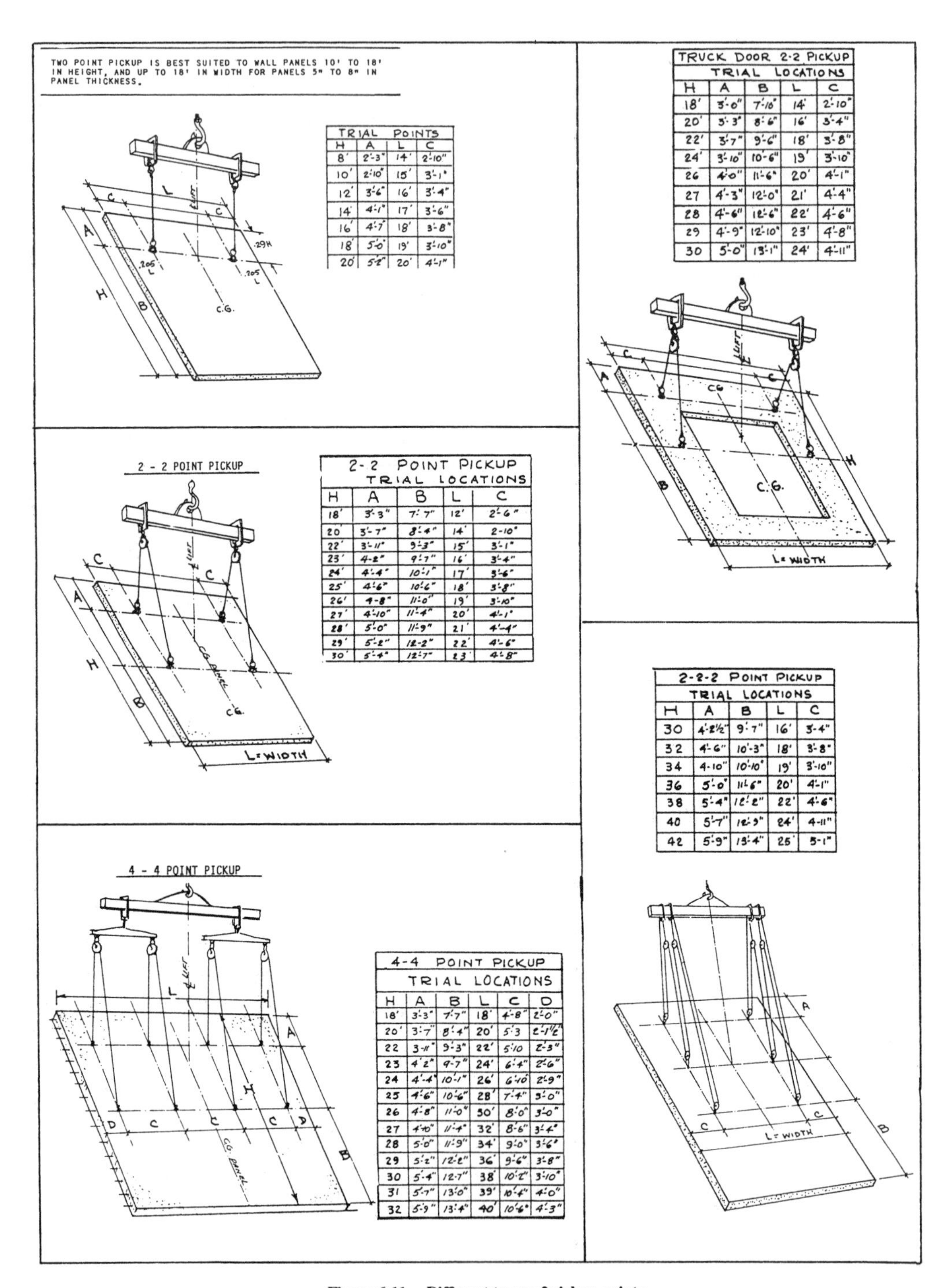

TRIAL POINTS			
H	A	L	C
8'	2'-3"	14'	2'-10"
10'	2'-10"	15'	3'-1"
12'	3'-6"	16'	3'-4"
14'	4'-1"	17'	3'-6"
16'	4'-7"	18'	3'-8"
18'	5'-0"	19'	3'-10"
20'	5'-2"	20'	4'-1"

TRUCK DOOR 2-2 PICKUP TRIAL LOCATIONS				
H	A	B	L	C
18'	3'-0"	7'-10"	14'	2'-10"
20'	3'-3"	8'-6"	16'	3'-4"
22'	3'-7"	9'-6"	18'	3'-8"
24'	3'-10"	10'-6"	19'	3'-10"
26	4'-0"	11'-6"	20'	4'-1"
27	4'-3"	12'-0"	21'	4'-4"
28	4'-6"	12'-6"	22'	4'-6"
29	4'-9"	12'-10"	23'	4'-8"
30	5'-0"	13'-1"	24'	4'-11"

2-2 POINT PICKUP TRIAL LOCATIONS				
H	A	B	L	C
18'	3'-3"	7'-7"	12'	2'-6"
20'	3'-7"	8'-4"	14'	2-10"
22'	3'-11"	9'-3"	15'	3'-1"
23'	4-2"	9'-7"	16'	3'-4"
24'	4'-4"	10'-1"	17'	3'-6"
25'	4'-6"	10'-6"	18'	3'-8"
26'	4-8"	11'-0"	19'	3'-10"
27'	4'-10"	11'-4"	20'	4'-1"
28'	5'-0"	11'-9"	21'	4'-4"
29'	5'-2"	12-2"	22'	4'-6"
30'	5'-4"	12'-7"	23'	4'-8"

2-2-2 POINT PICKUP TRIAL LOCATIONS				
H	A	B	L	C
30	4'-2½"	9'-7"	16'	3'-4"
32	4'-6"	10'-3"	18'	3'-8"
34	4-10"	10'-10"	19'	3'-10"
36	5'-0"	11'-6"	20'	4'-1"
38	5'-4"	12'-2"	22'	4'-6"
40	5'-7"	12'-9"	24'	4-11"
42	5'-9"	13'-4"	25'	5-1"

4-4 POINT PICKUP TRIAL LOCATIONS					
H	A	B	L	C	D
18'	3'-3"	7'-7"	18'	4'-8"	2'-0"
20'	3'-7"	8'-4"	20'	5'-3	2'-1½"
22	3-11"	9'-3"	22'	5'-10	2'-3"
23	4'-2"	9'-7"	24'	6'-4"	2'-6"
24	4'-4"	10'-1"	26'	6'-10"	2'-9"
25	4'-6"	10'-6"	28'	7'-4"	3'-0"
26	4'-8"	11'-0"	30'	8'-0"	3'-0"
27	4'-10"	11'-4"	32'	8'-6"	3'-4"
28	5'-0"	11'-9"	34'	9'-0"	3'-6"
29	5'-2"	12'-2"	36'	9'-6"	3'-8"
30	5'-4"	12'-7"	38'	10'-2"	3'-10"
31	5'-7"	13'-0"	39'	10'-4"	4'-0"
32	5'-9"	13'-4"	40'	10'-6"	4'-3"

Figure 6-11.—Different types of pickup points.

RECOMMENDED READING LIST

NOTE

Although the following references were current when this TRAMAN was published, their continued currency cannot be assured. You therefore need to ensure that you are studying the latest revision.

Concrete and Masonry, FM 5-742, Headquarters, Department of the Army, Washington, D.C., 1985.

Concrete Formwork, Koel, Leonard, American Technical Publishers, Inc., Homewood, Ill., 1988.

Design and Control of Concrete Mixtures, Portland Cement Association, Skokie, Ill., 1988.

CHAPTER 7

WORKING WITH CONCRETE

Concrete is the principal construction material used in most construction projects. The quality control of concrete and its placement are essential to ensure its final strength and appearance. Proper placement methods must be used to prevent segregation of the concrete.

This chapter provides information and guidance for you, the Builder, in the forming, placement, finishing, and curing of concrete. Information is also provided on the placement of reinforcing steel, and the types of ties required to ensure nonmovement of reinforcing once positioned. You will also be provided necessary information on concrete construction joints and the concrete saw. At the end of the chapter, you will find helpful references. You are encouraged to study these references, as required, for additional information on the topics discussed.

FORMWORK

LEARNING OBJECTIVE: Upon completing this section, you should be able to describe the types of concrete forms and their construction.

Most structural concrete is made by placing or "casting" plastic concrete into spaces enclosed by previously constructed forms. The plastic concrete hardens into the shape outlined by the forms. The size and shape of the formwork are always based on the project plans and specifications.

Forms for all concrete structures must be tight, rigid, and strong. If the forms are not tight, there will be excessive leakage at the time the concrete is placed. This leakage can result in unsightly surface ridges, honeycombing, and sand streaks after the concrete has set. The forms must be able to safely withstand the pressure of the concrete at the time of placement. **No shortcuts should be taken.** Proper form construction material and adequate bracing in place prevent the forms from collapsing or shifting during the placement of the concrete.

Forms or form parts are often omitted when a firm earth surface exists that is capable of supporting or molding the concrete. In most footings, the bottom of the footing is cast directly against the earth and only the sides are molded in forms. Many footings are cast with both the bottom and the sides against the natural earth. In these cases, however, the specifications usually call for larger footings. A foundation wall is often cast between a form on the inner side and the natural earth surface on the outer side.

FORM MATERIALS

Forms are generally constructed from either earth, metal, wood, fiber, or fabric.

Earth

Earthen forms are used in subsurface construction where the soil is stable enough to retain the desired shape of the concrete. The advantages of earthen forms are that less excavation is required and there is better settling resistance. The obvious disadvantage is a rough surface finish, so the use of earthen forms is generally restricted to footings and foundations. Precautions must be taken to avoid collapse of the sides of trenches.

Metal

Metal forms are used where high strength is required or where the construction is duplicated at more than one location. They are initially more expensive than wood forms, but may be more economical if they can be reused repeatedly. Originally, all prefabricated metal forms were made of steel. These forms were heavy and hard to handle. Currently, aluminum forms, which are lightweight and easier to handle, are replacing steel.

Prefabricated metal forms are easy to erect and strip. The frame on each panel is designed so that the panels can be easily and quickly fastened and unfastened. Metal forms provide a smooth surface finish so that little concrete finishing is required after the forms are stripped. They are easily cleaned, and maintenance is minimal.

Metal-wood forms are just like metal forms except for the face. It is made with a sheet of B-grade exterior plywood with waterproof glue.

Wood

Wooden forms are by far the most common type used in building construction. They have the advantage of economy, ease in handling, ease of production, and adaptability to many desired shapes. Added economy may result from reusing form lumber later for roofing, bracing, or similar purposes. Lumber should be straight, structurally sound, strong, and only partially seasoned. Kiln-dried timber has a tendency to swell when soaked with water from the concrete. If the boards are tight-jointed, the swelling will cause bulging and distortion. When green lumber is used, an allowance should be made for shrinkage, or the forms should be kept wet until the concrete is in place. Soft woods, such as pine, fir, and spruce, make the best and most economical form lumber since they are light, easy to work with, and available in almost every region.

Lumber that comes in contact with concrete should be surfaced at least on one side and both edges. The surfaced side is turned toward the concrete. The edges of the lumber may be square, shiplap, or tongue and groove. The latter makes a more watertight joint and tends to prevent warping.

Plywood can be used economically for wall and floor forms if it is made with waterproof glue and is identified for use in concrete forms. Plywood is more warp resistant and can be reused more often than lumber. Plywood is made in 1/4-, 3/8-, 1/2-, 9/16-, 5/8- and 3/4-inch thicknesses and in widths up to 48 inches. Although longer lengths are manufactured, 8-foot lengths are the most common. The 5/8- and 3/4-inch thicknesses are most economical; thinner sections require additional solid backing to prevent bulging. However, the 1/4-inch thickness is useful for forming curved surfaces.

Fiber

Fiber forms are prefabricated from impregnated waterproofed cardboard and other fiber materials. Successive layers of fiber are first glued together and then molded in the desired shape. Fiber forms are ideal for round concrete columns and other applications where preformed shapes are feasible since they require no form fabrication at the job site. This saves considerable time and money.

Fabric

Fabric forming is made of two layers of nylon fabric. These layers are woven together, forming an envelope. Structural mortar is injected into these envelopes, forming nylon-encased concrete "pillows." These are used to protect the shorelines of waterways, lakes and reservoirs, and as drainage channel linings.

Fabric forming offers exceptional advantages in the structural restoration of bearing piles under waterfront structures. A fabric sleeve with a zipper closure is suspended around the pile to be repaired, and mortar is pumped into the sleeve. This forms a strong concrete jacket.

FORM DESIGN

Forms for concrete construction must support the plastic concrete until it has hardened. Stiffness is an important feature in forms. Failure to provide form stiffness may cause unfortunate results. Forms must be designed for all the weight to which they are likely to be subjected. This includes the dead load of the forms, the plastic concrete in the forms, the weight of the workmen, the weight of equipment and materials, and the impact due to vibration. These factors vary with each project, but none should be ignored. The ease of erection and removal is also an important factor in the economical design of forms. Platform and ramp structures independent of formwork are sometimes preferred to avoid displacement of forms due to loading and impact shock from workmen and equipment.

When concrete is placed in forms, it is in a plastic state and exerts hydrostatic pressure on the forms. The basis of form design, therefore, is the maximum pressure developed by concrete during placing. The maximum pressure developed depends on the placing rate and the temperature. The rate at which concrete is placed affects the pressure because it determines how much hydrostatic head builds up in the form. The hydrostatic head continues to increase until the concrete takes its initial set, usually in about 90 minutes. At low temperatures, however, the initial set takes place much more slowly. This makes it necessary to consider the temperature at the time of placing. By knowing these two factors and the type of form material to be used, you can calculate a tentative design.

FORM CONSTRUCTION

Strictly speaking, it is only those parts of the formwork that directly mold the concrete that are correctly referred to as the "forms." The rest of the formwork consists of various bracing and tying members. In the following discussion on forms, illustrations are provided to help you understand the names of all the formwork members. You should study these illustrations carefully so that you will understand the material in the next section.

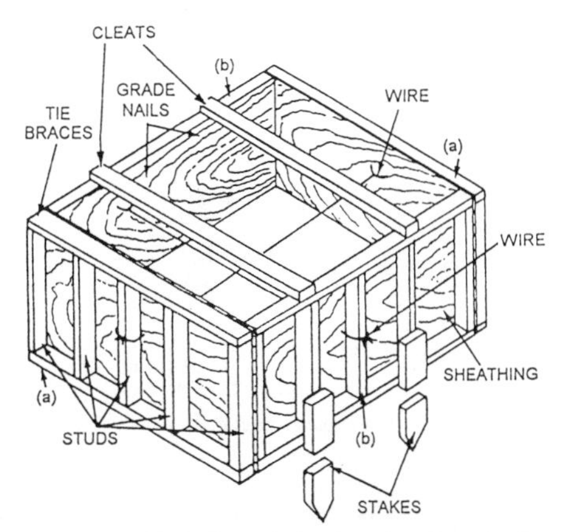

Figure 7-1.—Typical foundation form for a large footing.

Foundation Forms

The portion of a structure that extends above the ground level is called the superstructure. The portion below the ground level is called the substructure. The parts of the substructure that distribute building loads to the ground are called foundations. Footings are installed at the base of foundations to spread the loads over a larger ground area. This prevents the structure from sinking into the ground. It's important to remember that the footings of any foundation system should always be placed below the frost line. Forms for large footings, such as bearing wall footings, column footings, and pier footings, are called foundation forms. Footings, or foundations, are relatively low in height since their primary function is to distribute building loads. Because the concrete in a footing is shallow, pressure on the form is relatively low. Therefore, a form design based on high strength and rigidity considerations is generally not necessary.

SIMPLE FOUNDATION.—Whenever possible, excavate the earth and use it as a mold for concrete footings. You should thoroughly moisten the earth before placing the concrete. If this is not possible, you must construct a form. Because most footings are rectangular or square, you can build and erect the four sides of the form in panels.

Make the first pair of opposing panels (figure 7-1 (a)) to exact footing width. Then, nail vertical cleats to the exterior sides of the sheathing. Use at least 1-by-2-inch lumber for the cleats, and space them 2 1/2 inches from each end of the exterior sides of the panels (a), and on 2-foot centers between the ends. Next, nail two cleats to the ends of the interior sides of the second pair of panels (figure 7-1 (b)). The space between these panels should equal the footing length plus twice the sheathing thickness. Then, nail cleats on the exterior sides of the panels (b) spaced on 2-foot centers.

Erect the panels into either a rectangle or square, and hold them in place with form nails. Make sure that all reinforcing bars are in place. Now, drill small holes on each side of the center cleat on each panel. These holes should be less than 1/2 inch in diameter to prevent paste leakage. Pass No. 8 or No. 9 black annealed iron wire through these holes and wrap it around the center cleats of the opposing panels to hold them together (see figure 7-1). Mark the top of the footing on the interior side of the panels with grade nails.

For forms 4 feet square or larger, drive stakes against the sheathing, as shown in figure 7-1. Both the stakes and the 1 by 6 tie braces nailed across the top of the form keep it from spreading apart. If a footing is less than 1-foot deep and 2-feet square, you can construct the form from 1-inch sheathing without cleats. Simply make the side panels higher than the footing depth, and mark the top of the footing on the interior sides of the panels with grade nails. Cut and

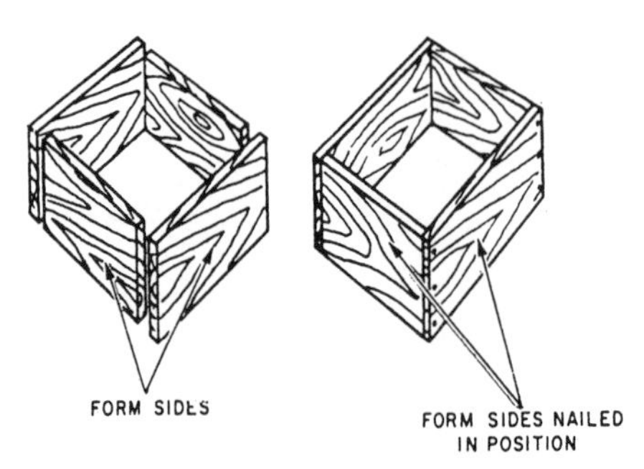

Figure 7-2.—Typical small footing form.

nail the lumber for the sides of the form, as shown in figure 7-2.

FOUNDATION AND PIER FORMS COMBINED.—You can often place a footing and a small pier at the same time. A pier is a vertical member that supports the concentrated loads of an arch or bridge superstructure. It can be either rectangular or round. You build a pier form as shown in figure 7-3. The footing form should look like the one in figure 7-1. You must provide support for the pier form while not interfering with concrete placement in the footing form. You can do this by first nailing 2-by-4s or 4-by-4s across the footing form, as shown in figure 7-3. These serve as both supports and tie braces. Then, nail the pier form to these support pieces.

BEARING WALL FOOTINGS.—Figure 7-4 shows a typical footing formwork for a bearing wall, and figure 7-5 shows bracing methods for a bearing

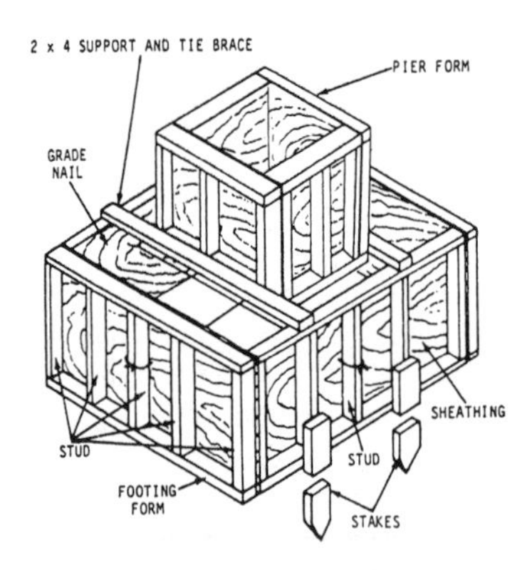

Figure 7-3.—Footing and pier form.

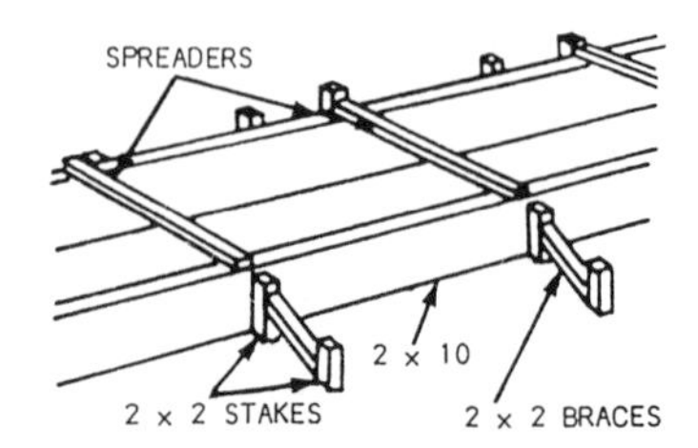

Figure 7-4.—Typical footing form.

wall footing. A bearing wall, also called a load-bearing wall, is an exterior wall that serves as an enclosure and also transmits structural loads to the foundation. The form sides are 2-inch lumber whose width equals the footing depth. Stakes hold the sides in place while spreaders maintain the correct distance between them. The short braces at each stake hold the form in line.

A keyway is made in the wet concrete by placing a 2-by-2-inch board along the center of the wall footing form. After the concrete is dry, the board is removed. This leaves an indentation, or key, in the concrete. When you pour the foundation wall, the key provides a tie between the footing and wall. Although not discussed in this training manual, there are several commercial keyway systems available for construction projects.

Columns

Square column forms are made of wood. Round column forms are made of steel, or cardboard

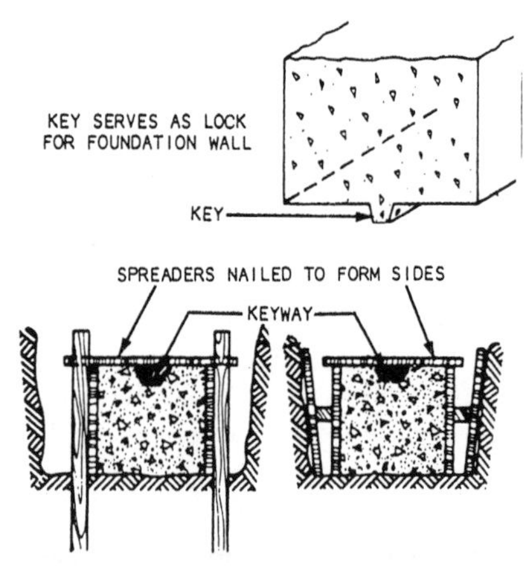

Figure 7-5.—Methods of bracing bearing wall footing forms and placing a keyway.

impregnated with waterproofing compound. Figure 7-6 shows an assembled column and footing form. After constructing the footing forms, build the column form sides, and then nail the yokes to them.

Figure 7-7 shows a column form with two styles of yokes. View A shows a commercial type, and view B shows yokes made of all-thread bolts and 2-by material. Since the rate of placing concrete in a column form is very high and the bursting pressure exerted on the form by the concrete increases directly with the rate of placing, a column form must be securely braced, as shown by the yokes in the figure. Because the bursting pressure is greater at the bottom of the form than it is at the top, yokes are placed closer together at the bottom.

The column form should have a clean-out hole cut in the bottom from which to remove construction debris. Be sure to nail the pieces that you cut to make the clean-out hole to the form. This way, you can replace them exactly before placing concrete in the column. The intention of the clean-out is to ensure that the surface which bonds with the new concrete is clear of all debris.

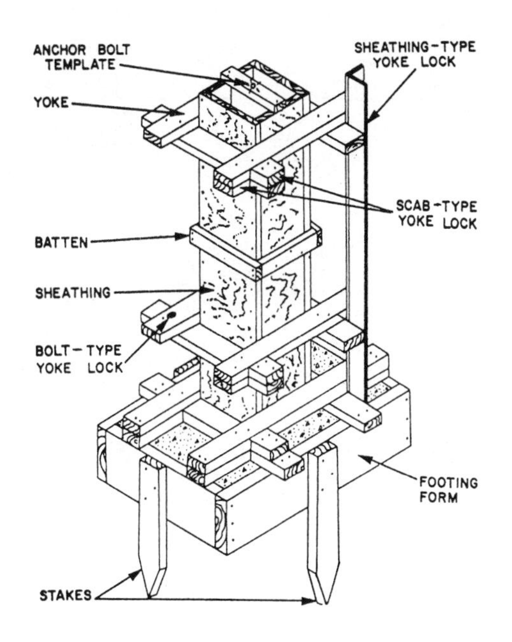

Figure 7-6.—Form for a concrete column.

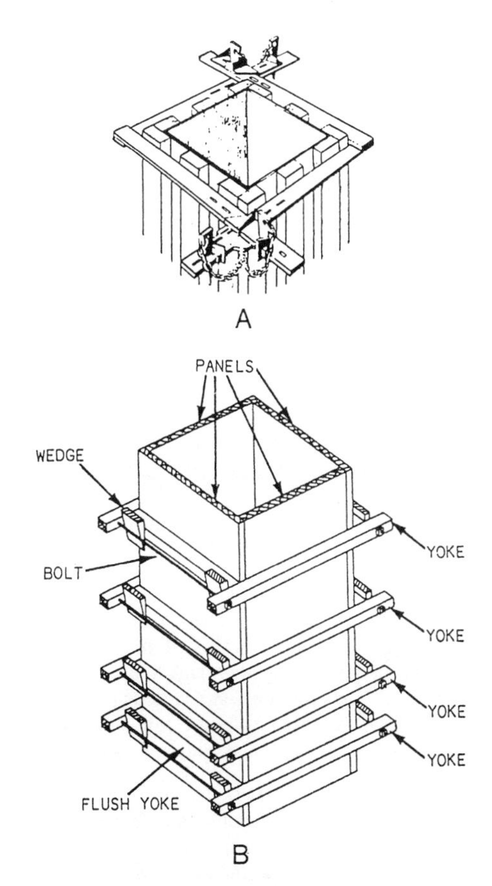

Figure 7-7.—Column form with scissor clamp (View A), and yolk and wedge (View B).

Walls

Wall forms (figure 7-8) may be built in place or prefabricated, depending on shape and desirability of

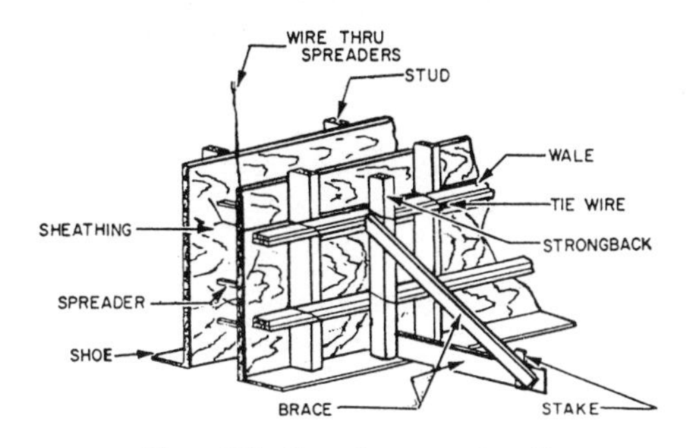

Figure 7-8.—Form for a concrete wall.

form reuse. Some of the elements that make up wooden forms are sheathing, studs, wales, braces, shoe plates, spreaders, and tie wires.

CONSTRUCTION.—Sheathing forms the surfaces of the concrete. It should be as smooth as possible, especially if the finished surfaces are to be exposed. Since the concrete is in a plastic state when placed in the form, the sheathing should be watertight. Tongue-and-groove sheathing gives a smooth, watertight surface. Plywood or hardboard can also be used and is the most widely accepted construction method.

The weight of the plastic concrete causes sheathing to bulge if it is not reinforced. As a result, studs are run vertically to add rigidity to the wall form. Studs are generally made from 2-by-4 or 3-by-6 material.

Studs also require reinforcing when they extend over 4 or 5 feet. This reinforcing is supplied by double wales. Double wales also serve to tie prefabricated panels together and keep them in a straight line. They run horizontally and are lapped at the corners of the forms to add rigidity. Wales are usually made of the same material as the studs.

The shoe plate is nailed into the foundation or footing. It is carefully placed to maintain the correct wall dimension and alignment. The studs are tied into the shoe and spaced according to the correct design.

Small pieces of wood are cut the same length as the thickness of the wall and are placed between the forms to maintain proper distance between forms. These pieces are known as spreaders. The spreaders are not nailed but are held in place by friction and must be removed before the concrete covers them. A wire should be securely attached to each spreader so that the spreaders can be pulled out after the concrete has exerted enough pressure on the walls to allow them to be easily removed.

Tie wire is designed to hold the forms securely against the lateral pressure of unhardened concrete. A double strand of tie wire is always used.

BRACING.—Many types of braces can be used to add stability and bracing to the forms. The most common type is a diagonal member and horizontal member nailed to a stake and to a stud or wale, as shown in figure 7-8. The diagonal member should make a 30° angle with the horizontal member. Additional bracing may be added to the form by placing vertical members (strongbacks) behind the wales or by placing vertical members in the corner formed by intersecting wales. Braces are not part of the form design and are not considered as providing any additional strength.

REINFORCEMENT.—Wall forms are usually reinforced against displacement by the use of ties. Two types of simple wire ties, used with wood spreaders, are shown in figure 7-9. The wire is passed around the studs, the wales, and through small holes bored in the sheathing. Each spreader is placed as close as possible to the studs, and the tie is set taut by the wedge, as shown in view A of figure 7-9, or by twisting with a small toggle, as shown in view B. As the concrete reaches the level of each spreader, the spreader is knocked out and removed. Figure 7-10 shows you an easy way to remove the spreaders by drilling holes and placing a wire through them. The parts of the wire that are inside the forms remain in the concrete; the outside surplus is cut off after the forms are removed.

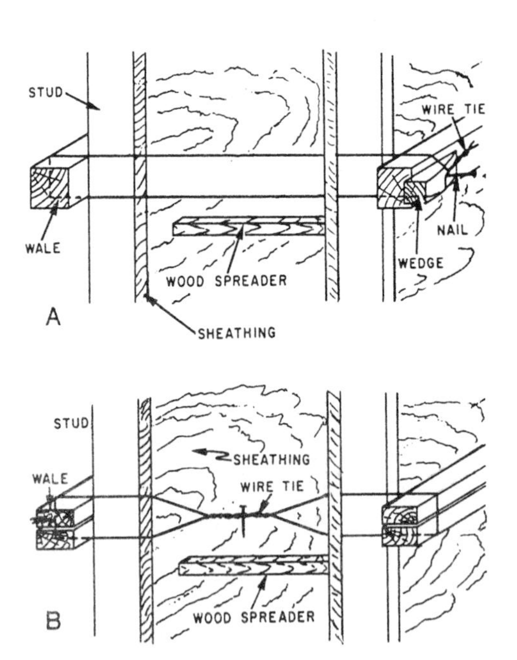

Figure 7-9.—Wire ties for wall forms.

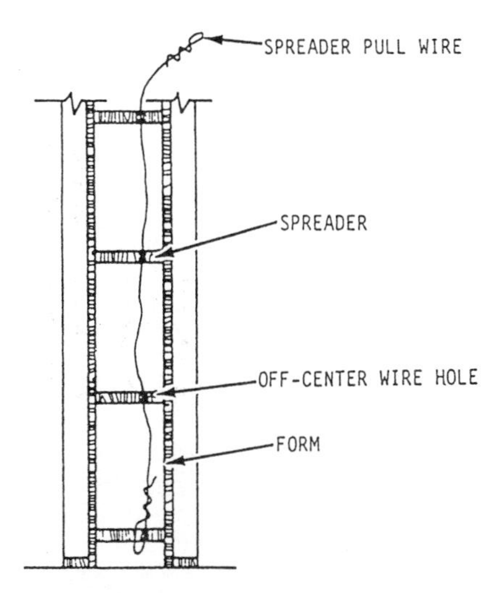

Figure 7-10.—Removing wood spreaders.

Wire ties and wooden spreaders have been largely replaced by various manufactured devices in which the function of the tie and the function of the spreader are combined. Figure 7-11 shows one of these. It is called a snap tie. These ties are made in various sizes to fit various wall thicknesses. The tie holders can be removed from the tie rod. The rod goes through small holes bored in the sheathing, and also through the wales, which are usually doubled for that purpose. Tapping the tie holders down on the ends of the rod brings the sheathing to bear solidly against the spreader washers. You can prevent the tie holder from coming loose by driving a duplex nail in the provided hole. After the concrete has hardened, the tie holders can be detached to strip the forms. After the forms are stripped, a special wrench is used to break off the outer sections of rods. The rods break off at the breaking points, located about 1-inch inside the surface of the concrete. Small surface holes remain, which can be plugged with grout if necessary.

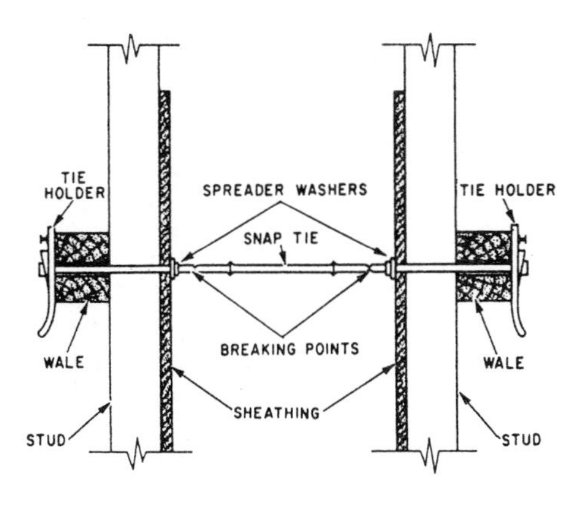

Figure 7-11.—Snap tie.

Another type of wall-form tie is the tie rod (figure 7-12). This rod consists of an inner section that is threaded on both ends and two threaded outer sections. The inner section with the cone nuts set to the thickness of the wall is placed between the forms, and the outer sections are passed through the wales and sheathing and threaded into the cone nuts. The clamps are then threaded on the outer sections to bring the forms to bear against the cone nuts. After the concrete hardens, the clamps are loosened, and the outer sections of rod are removed by threading them out of the cone nuts. After the forms are stripped, the cone nuts are removed from the concrete by threading them off the inner sections of the rod with a special wrench. The cone-shaped surface holes that remain can be plugged with grout. The inner sections of the rod remain in the concrete. The outer sections and the cone nuts may be reused indefinitely.

Wall forms are usually constructed as separate panels. Make the panels by first nailing sheathing to the studs. Next, connect the panels, as shown in

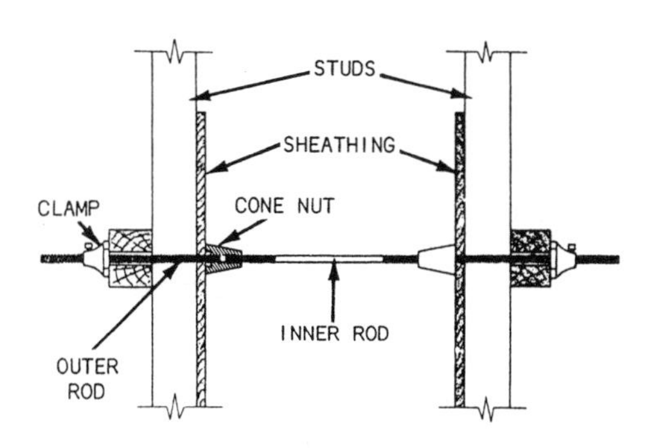

Figure 7-12.—Tie rod.

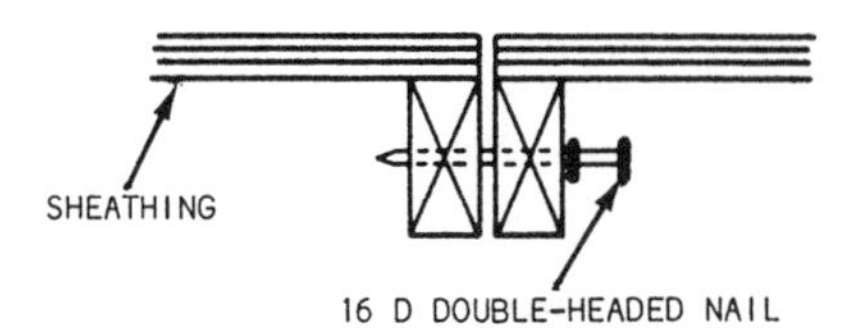

Figure 7-13.—Joining wall form panels together in line.

figure 7-13. Figure 7-14 shows the form details at the wall corner. When placing concrete panel walls and columns at the same time, construct the wall form, as shown in figure 7-15. Make the wall form shorter than the distance between the column forms to allow for a wood strip that acts as a wedge. When stripping the forms, remove the wedge first to aid in form removal.

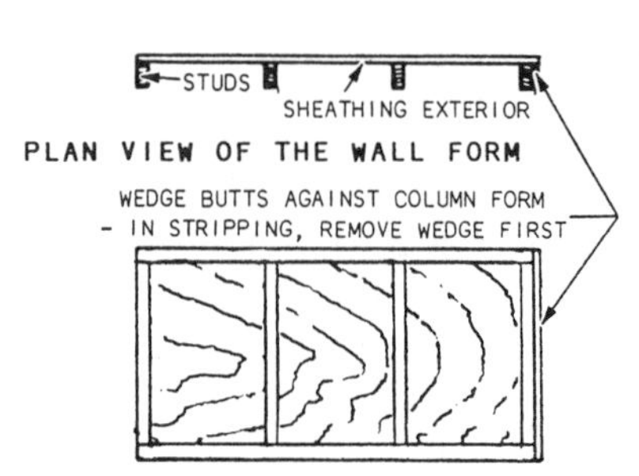

Figure 7-15.—Form for panel wall and columns.

Stair Forms

Concrete stairway forms require accurate layout to ensure accurate finish dimensions for the stairway. Stairways should always be reinforced with rebars (reinforcing bars) that tie into the floor and landing. They are formed monolithically or formed after the concrete for the floor slab has set. Stairways formed after the slab has set must be anchored to a wall or beam by tying the stairway rebars to rebars projecting from the walls or beams, or by providing a keyway in the beam or wall. You can use various stair forms, including prefabricated forms. For moderate-width stairs joining typical floors, a design based on strength considerations is generally not necessary. Figure 7-16 shows one way to construct forms for stair widths up to and including 3 feet. Make the sloping wood platform that serves as the form for the underside of the steps from 3/4-inch plywood. The platform should extend about 12 inches beyond each side of the stairs to support the stringer bracing blocks. Shore up the back of the platform with 4-by-4 supports, as shown in figure 7-16. The post supports should rest on wedges for easy adjustment and removal. Cut 2-by-12 planks for the side stringers to fit the treads and risers. Bevel the bottom of the 2-by-12 risers for easy form removal and finishing.

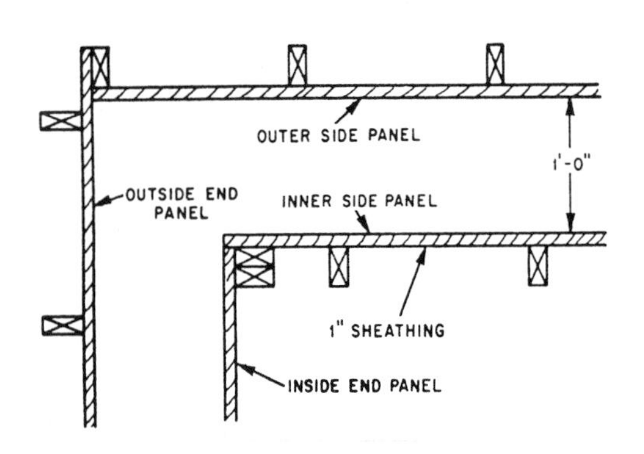

Figure 7-14.—Joining wall form panels at a corner.

Beams and Girders

The type of construction used for beam and girder forms depends upon whether the forms are to be removed in one piece or whether the sides are to be

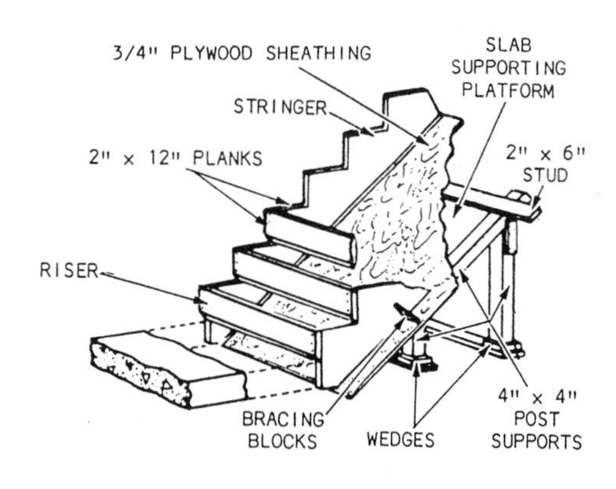

Figure 7-16.—Stairway form.

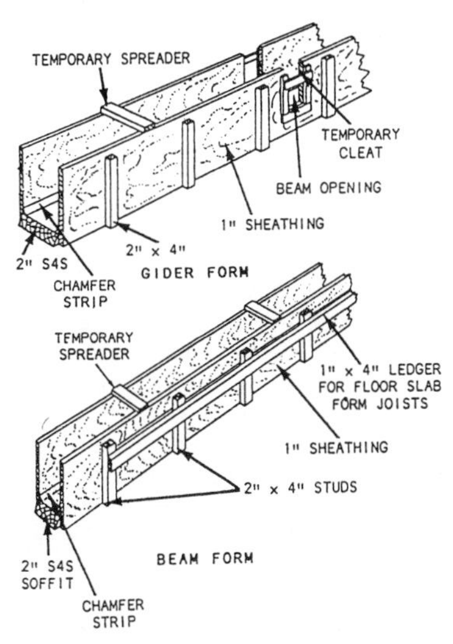

Figure 7-17.—Typical beam and girder form.

stripped and the bottom left in place until the concrete has hardened enough to permit removal of the shoring. The latter type of form is preferred, and details for this type are shown in figure 7-17. Although beam and girder forms are subjected to very little bursting pressure, they must be shored up at frequent intervals to prevent sagging under the weight of fresh concrete.

The bottom of the form should be the same width as the beam and should be in one piece for the full width. The sides of the form should be 1-inch-thick tongue-and-groove sheathing and should lap over the bottom as shown in figure 7-17. The sheathing is nailed to 2-by-4-inch studs placed on 3-foot centers. A 1-by-4-inch piece is nailed along the studs. These pieces support the joist for the floor panel, as shown in figure 7-18, detail E. The beam sides of the form are not nailed to the bottom. They are held in position by continuous strips, as shown in detail E. The

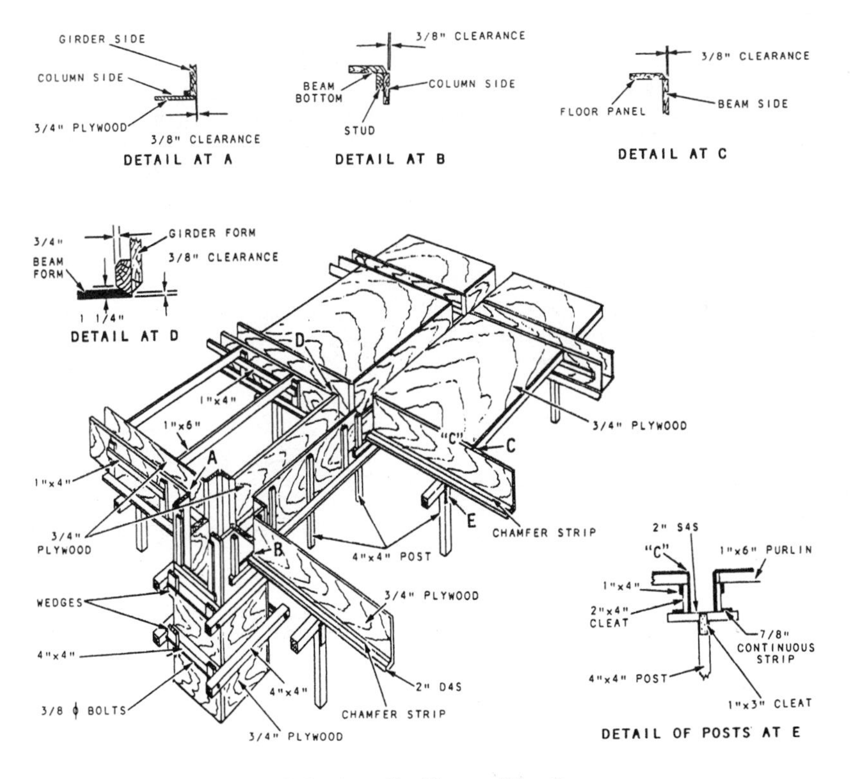

Figure 7-18.—Assembly of beam and floor forms.

crosspieces nailed on top serve as spreaders. After erection, the slab panel joists hold the beam sides in position. Girder forms (figure 7-17) are the same as beam forms except that the sides are notched to receive the beam forms. Temporary cleats should be nailed across the beam opening when the girder form is being handled.

The entire method of assembling beam and girder forms is illustrated in figure 7-18. The connection of the beam and girder is illustrated in detail D. The beam bottom butts up tightly against the side of the girder form and rests on a 2-by- 4-inch cleat nailed to the girder side. Detail C shows the joint between the beam and slab panel, and details A and B show the joint between the girder and column. The clearances given in these details are needed for stripping and also to allow for movement that occurs due to the weight of the fresh concrete. The 4-by-4 posts (detail E) used for shoring the beams and girders should be spaced to provide support for the concrete and forms. They should be wedged at the bottom to obtain proper elevation.

Figure 7-19 shows you how the same type of forming can be done by using quick beams, scaffolding, and I-beams—if they are available. This type of system can be set up and taken down in minimum time.

Oiling and Wetting Forms

You should never use oils or other form coatings that may soften or stain the concrete surface, prevent the wet surfaces from water curing, or hinder the proper functioning of sealing compounds used for curing. If you cannot obtain standard form oil or other form coating, you can wet the forms to prevent sticking in an emergency.

OIL FOR WOOD FORMS.—Before placing concrete in wood forms, treat the forms with a suitable form oil or other coating material to prevent the concrete from sticking to them. The oil should penetrate the wood and prevent water absorption. Almost any light-bodied petroleum oil meets these specifications. On plywood, shellac works better than oil in preventing moisture from raising the grain and detracting from the finished concrete surface. Several commercial lacquers and similar products are also available for this purpose. If you plan to reuse wood forms repeatedly, a coat of paint or sealing compound will help preserve the wood. Sometimes lumber contains enough tannin or other organic substance to soften the concrete surface. To prevent this, treat the form surfaces with whitewash or limewater before applying the form oil or other coating.

OIL FOR STEEL FORMS.—Oil wall and steel column forms before erecting them. You can oil all other steel forms when convenient, but they should be oiled before the reinforcing steel is placed. Use specially compounded petroleum oils, not oils intended for wood forms. Synthetic castor oil and some marine engine oils are examples of compound oils that give good results on steel forms.

APPLYING OIL.—The successful use of form oil depends on how you apply it and the condition of the forms. They should be clean and have smooth surfaces. Because of this, you should not clean forms with wire brushes, which can mar their surfaces and cause concrete to stick. Apply the oil or coating with a brush, spray, or swab. Cover the form surfaces evenly, but do not allow the oil or coating to contact construction joint surfaces or any reinforcing steel in the formwork. Remove all excess oil.

OTHER COATING MATERIALS.—Fuel oil, asphalt paint, varnish, and boiled linseed oil are also suitable coatings for forms. Plain fuel oil is too thin to use during warm weather, but mixing one part petroleum grease to three parts of fuel oil provides adequate thickness.

Form Failure

Even when all form work is adequately designed, many form failures occur because of human error, improper supervision, or using damaged materials. The following list highlights some, but not all, of the most common construction deficiencies that supervisory personnel should consider when working with concrete:

- Inadequately tightened or secured form ties;
- Inadequate diagonal bracing of shores;
- Use of old, damaged, or weathered form materials;
- Use of undersized form material;
- Shoring not plumb;
- Failure to allow for lateral pressures on form work; and

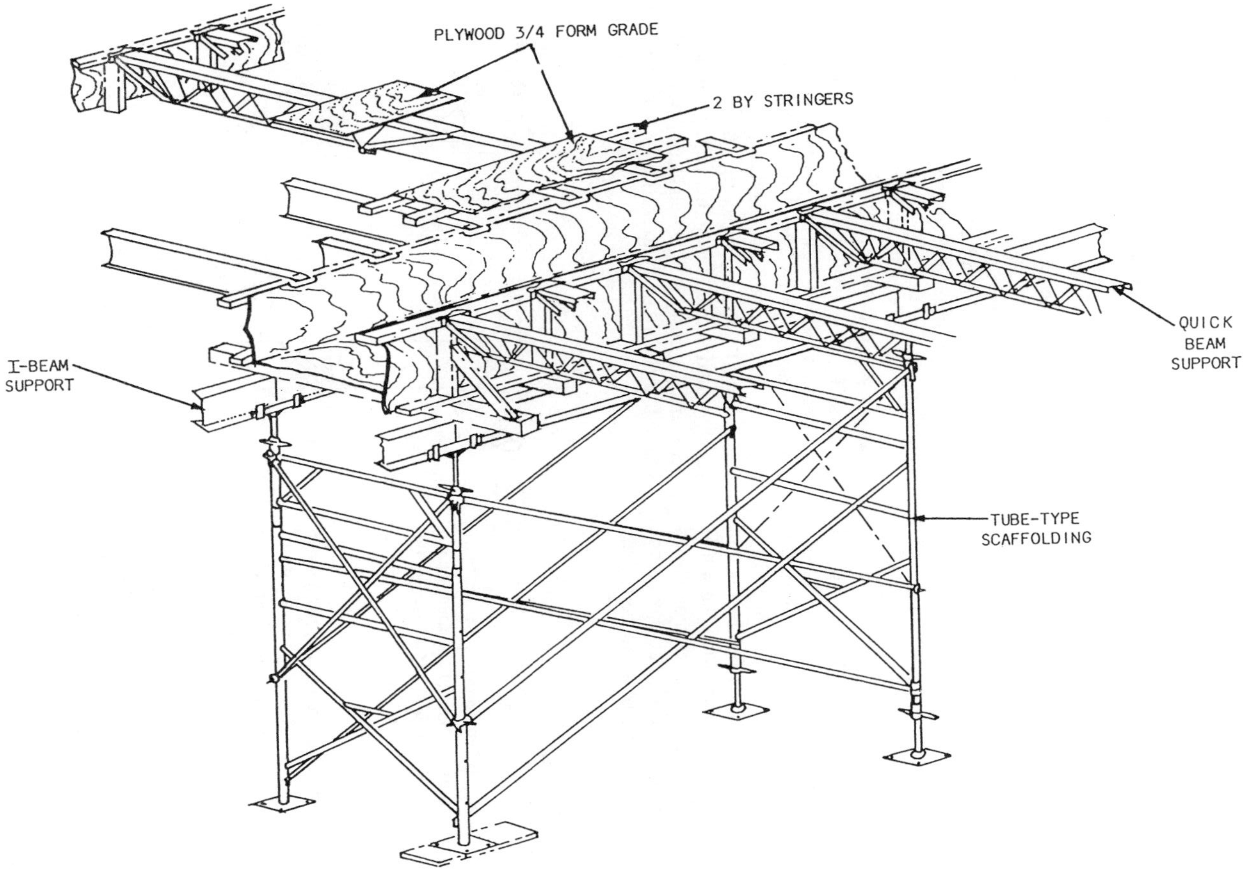

PLYWOOD 3/4 FORM GRADE

2 BY STRINGERS

I-BEAM SUPPORT

QUICK BEAM SUPPORT

TUBE-TYPE SCAFFOLDING

Figure 7-19.—Beam and floor forms.

- Failure to inspect form work during and after concrete placement to detect abnormal deflections or other signs of imminent failure.

There are many reasons why forms fail. It is the responsibility of the Builder to ensure that the forms are correctly constructed according to design, and that proper techniques are followed.

REINFORCED CONCRETE

LEARNING OBJECTIVE: Upon completing this section, you should be able to determine the types of ties for and placement of reinforcing steel.

Concrete is strong under compression, but relatively weak under tension. The reverse is true for steel. Therefore, when the two are combined, one makes up for the deficiency of the other. When steel is embedded in concrete in a manner that assists it in carrying imposed loads, the combination is known as reinforced concrete. The steel may consist of welded wire fabric or expanded metal mesh, but, more often, it consists of reinforcing bars, or more commonly "rebar."

WELDED WIRE FABRIC

Welded wire fabric, often referred to as "wire mesh," comes in rolls and sheets. These must be cut to fit your individual application. The individual sections of fabric must be tied together, or "lapped," to form a continuous sheet of fabric.

Specifications and designs are usually used when wire fabric is being lapped. However, as a rule of thumb, one complete lap is usually sufficient with a minimum of 2 inches between laps. Whenever the rule of thumb is not allowed, use the end lap or side lap method.

In the end lap method, the wire mesh is lapped by overlapping one full mesh measured from the end of the longitudinal wires in one piece to the end of longitudinal wires in the adjacent piece. The two pieces are then tied at 1 1/2-foot centers with a snap tie. In the side lap method, the two longitudinal side wires are placed one alongside and overlapping the other and then are tied with a snap tie every 3 feet.

REINFORCING STEEL

Before placing reinforcing steel in forms, all form oiling should be completed. As mentioned earlier, oil or other coating should not contact the reinforcing steel in the formwork. Oil on reinforcing bars reduces the bond between the bars and the concrete. Use a piece of burlap to clean the bars of rust, scale, grease, mud, or other foreign matter. A light film of rust or mill scale is not objectionable.

Rebars must be tied together for the bars to remain in a desired arrangement during pouring. Tying is also a means of keeping laps or splices in place. Laps allow bond stress to transfer the load from one bar, first into the concrete and then into the second bar.

Methods of Tying

Several types of ties can be used with rebar. Some are more effective than others. The views in figure 7-20 illustrate the six types used by the Seabees: (A) snap, or simple, tie, (B) wall tie, (C) double-strand tie, (D) saddle tie, (E) saddle tie with twist, and (F) cross, or figure-eight, tie. As a Builder, you will probably be concerned only with the snap

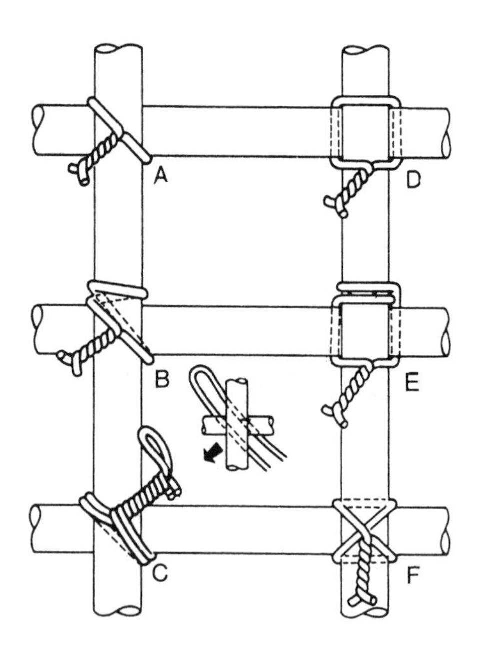

Figure 7-20.—Types of ties.

and saddle ties. However, as a professional, you should be familiar with all six types.

SNAP, OR SIMPLE, TIE.—The snap, or simple, tie (view A of figure 7-20) is simply wrapped once around the two crossing bars in a diagonal manner with the two ends on top. The ends are then twisted together with a pair of side cutters until they are very tight against the bars. Finally, the loose ends are cut off. This tie is used mostly on floor slabs.

WALL TIE.—The wall tie (view B of figure 7-20) is made by taking one and one-half turns around the vertical bar, then one turn diagonally around the intersection. The two ends are twisted together until the connection is tight, then the excess is cut off. The wall tie is used on light vertical mats of steel.

DOUBLE-STRAND SINGLE TIE.—The double-strand tie (view C) is a variation of the simple tie. It is favored in some localities and is especially used for heavy work.

SADDLE TIE.—The wires of the saddle tie (view D) pass half way around one of the bars on either side of the crossing bar and are brought squarely or diagonally around the crossing bar. The ends are then twisted together and cut off.

SADDLE TIE WITH TWIST.—The saddle tie with twist (view E) is a variation of the saddle tie. The tie wire is carried completely around one of the bars, then squarely across and halfway around the other, either side of the crossing bars, and finally brought together and twisted either squarely or diagonally across. The saddle tie with twist is used for heavy mats that are to be lifted by crane.

CROSS, OR FIGURE-EIGHT, TIE.—The cross, or figure-eight, tie (view F) has the advantage of causing little or no twist in the bars.

CARRYING WIRE.—When tying reinforcing bars, you must have a supply of tie wire available. There are several ways you can carry your tie wire. One way is to coil it to a diameter of 18 inches, then slip it around your neck and under one arm (figure 7-21). This leaves a free end for tying. Coil enough wire so it weighs about 9 pounds.

Another way to carry tie wire is to take pieces of wire about 9-inches long, fold them, and hook one end in your belt. Then, you can pull the wires out as needed. The tools you use in tying reinforcing bars include a 6-foot folding rule, side cutters, leather gloves, 50-foot tape measure, and a keel crayon, either yellow, red, or blue.

Figure 7-21.—Carrying tie wire.

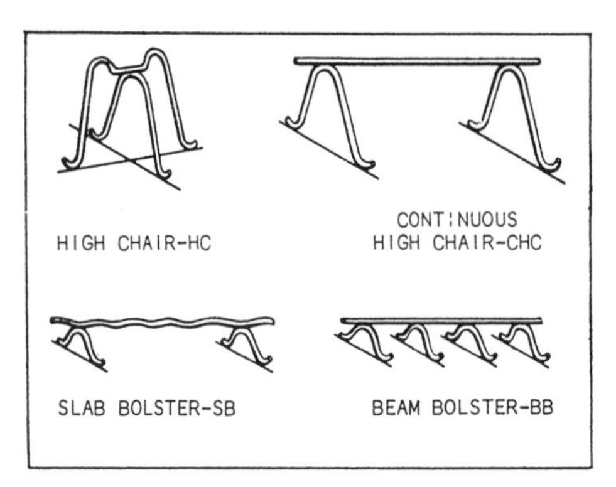

Figure 7-22.—Devices used to support horizontal reinforcing bars.

Location for Reinforcing Steel

The proper location for reinforcing bars is given on the drawings. To ensure that the structure can withstand the loads it must carry, place the steel in exactly the position shown. Secure the bars in position so that they will not move when the concrete is placed. This can be accomplished by using the reinforcing bar supports shown in figures 7-22, 7-23, and 7-24.

Footings and other principal structural members that are against the ground should have at least 3

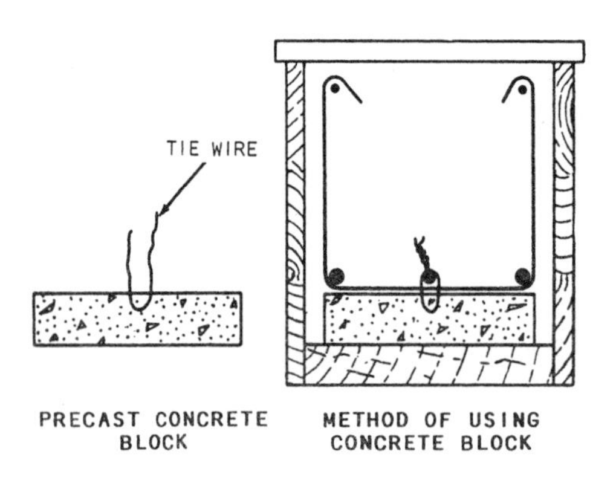

Figure 7-23.—Precast concrete block used for reinforcing steel support.

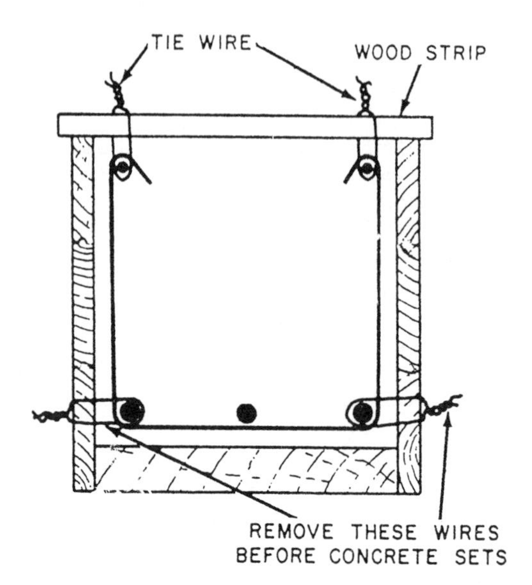

Figure 7-24.—Beam-reinforcing steel hung in place.

inches of concrete between steel and ground. If the concrete surface is to be in contact with the ground or exposed to the weather after removal of the forms, the protective covering of concrete over the steel should be 2 inches for bars larger than No. 5 and 1 1/2 inches for No. 5 or smaller. The protective covering may be reduced to 1 1/2 inches for beams and columns and 3/4 inch for slabs and interior wall surfaces, but it should be 2 inches for all exterior wall surfaces.

The clear distance between parallel bars in beams, footings, walls, and floor slabs should be a minimum of 1 inch, or one and one-third times the largest size aggregate particle in the concrete. In columns, the clear distance between parallel bars should be a minimum of one and one-half times the bar diameter, one and one-half times the maximum size of the coarse aggregate, or not less than 1 1/2 inches.

The support for reinforcing steel in floor slabs is shown in figure 7-25. The height of the slab bolster is determined by the concrete protective cover required. Concrete blocks made of sand-cement mortar can be used in place of the slab bolster. Wood blocks should never be used for this purpose if there is any possibility the concrete might become wet and if the construction is of a permanent type. Bar chairs, like those shown in figure 7-25, are available from commercial sources in heights up to 6 inches. If a height greater than 6 inches is required, make the

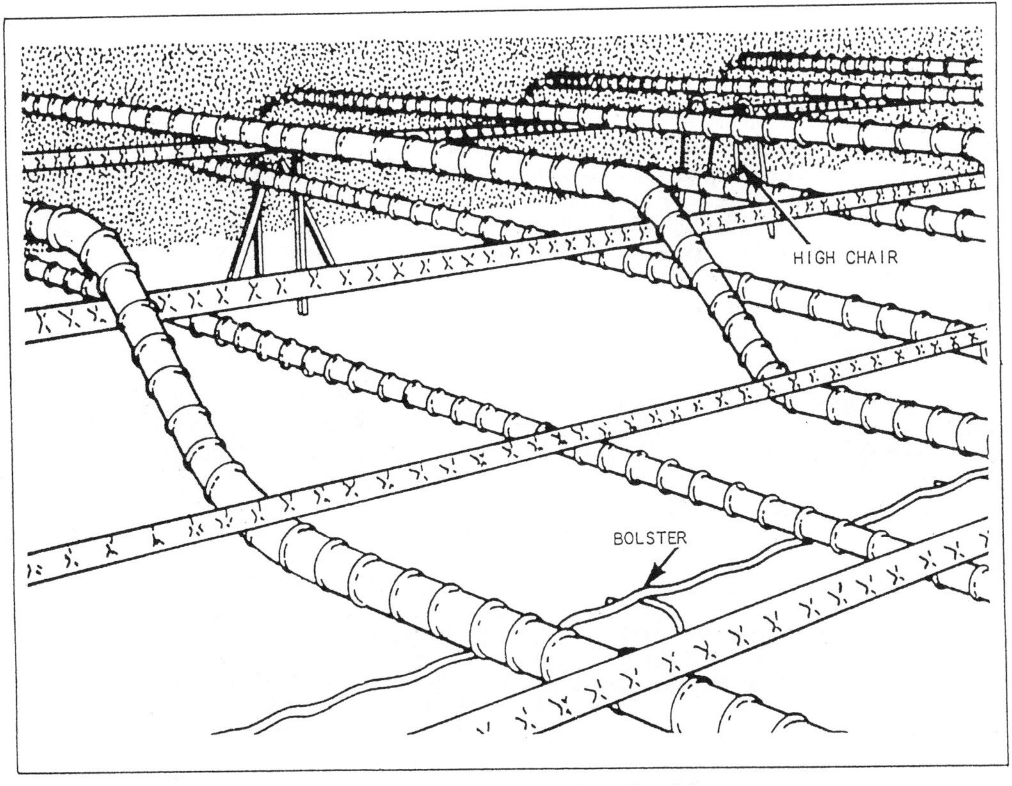

Figure 7-25.—Reinforcing steel for a floor slab.

chair of No. 0 soft annealed iron wire. Tie the bars together at frequent intervals with a snap tie to hold them firmly in position.

Steel for column ties can be assembled into cages by laying the vertical bars for one side of the column horizontally across a couple of sawhorses. The proper number of ties is slipped over the bars, the remaining vertical bars are added, and then the ties are spaced out as required by the placing plans. A sufficient number of intersections are wired together to make the assembly rigid. This allows it to be hoisted and set as a unit.

After the column form is raised, it is tied to the dowels or reinforcing steel carried up from below. This holds it firmly in position at the base. The column form is erected, and the reinforcing steel is tied to the column form at 5-foot intervals, as shown in figure 7-26.

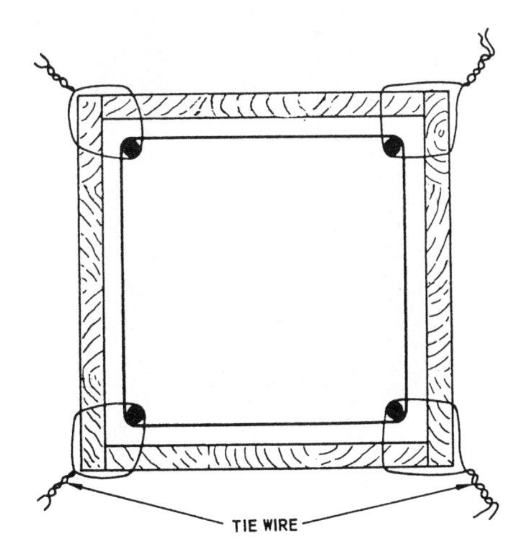

Figure 7-26.—Securing a column with reinforcing steel against displacement.

The use of metal supports to hold beam-reinforcing steel in position is shown in figure 7-27. Note the position of the beam bolster. The stirrups are tied to the main reinforcing steel with a snap tie. Whenever possible, you should assemble the stirrups and main reinforcing steel outside the form and then place the assembled unit in position. Wood blocks should be substituted for the metal supports only if there is no possibility of the concrete becoming wet or if the construction is known to be temporary. Precast concrete blocks, as shown in figure 7-23, may be substituted for metal supports or, if none of the types of bar supports described above seem suitable, the method shown in figure 7-24 may be used.

Placement of steel in walls is the same as for columns except that the steel is erected in place and not preassembled. Horizontal steel is tied to vertical steel at least three times in any bar length. Steel in place in a wall is shown in figure 7-28. The wood block is removed when the form has been filled up to the level of the block. For high walls, ties in between the top and bottom should be used.

Steel is placed in footings very much as it is placed in floor slabs. Stones, rather than steel supports, may be used to support the steel at the proper distance above the subgrade. Steel mats are generally preassembled and placed in small footings after the forms have been set. A typical arrangement is shown in figure 7-29. Steel mats in large footings are generally constructed in place.

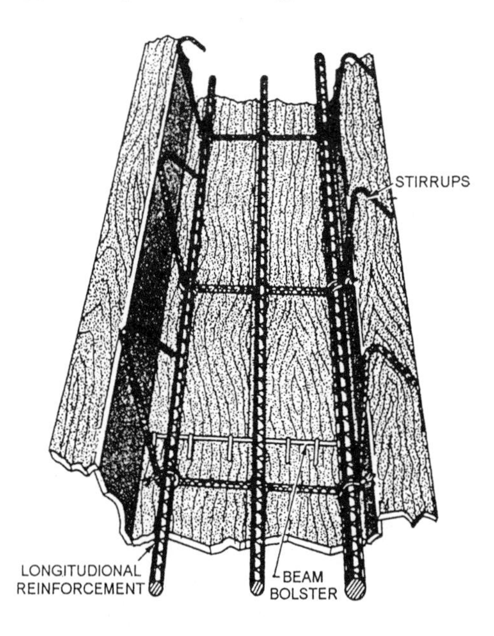

Figure 7-27.—Beam-reinforcing steel supported on beam bolsters.

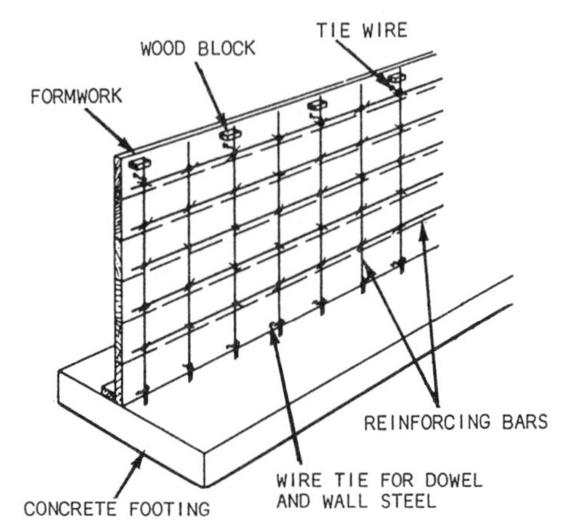

Figure 7-28.—Steel in place in a wall.

Welded wire fabric (figure 7-30) is also used as limited reinforcement for concrete footings, walls, and slabs, but its primary use is to control crack widths due to temperature changes.

Form construction for each job has its peculiarities. However, certain natural conditions prevail in all situations. Wet concrete always develops hydrostatic pressure and strain on forms. Therefore, all stakes,

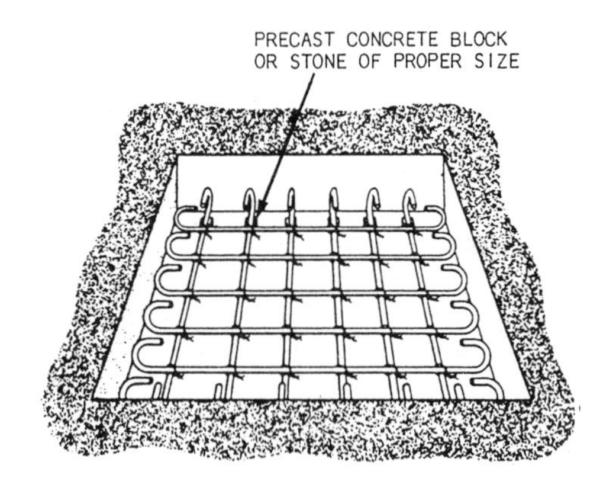

Figure 7-29.—Steel in place in a footing.

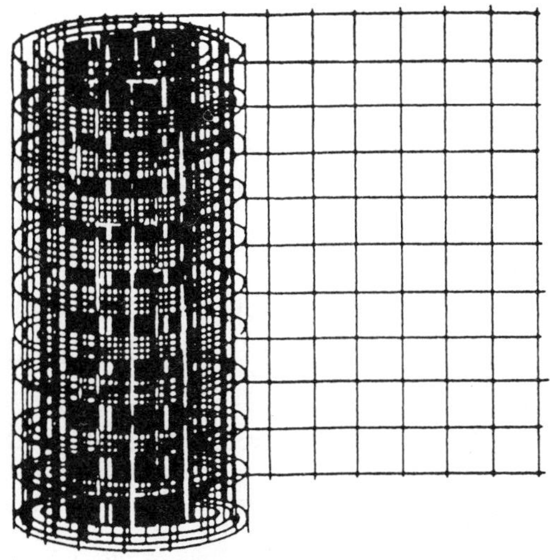

Figure 7-30.—Welded wire mesh fabric.

braces, walers, ties, and shebolts should be properly secured before placing concrete.

Splicing Reinforcing Bar

Because rebar is available only in certain lengths, it must be spliced together for longer runs. Where splices are not dimensioned on the drawings, the bars should be lapped not less than 30 times the bar diameter, or not less than 12 inches.

The stress in a tension bar can be transmitted through the concrete and into another adjoining bar by a lap splice of proper length. The lap is expressed as the number of bar diameters. If the bar is No. 2, make the lap at least 12 inches. Tie the bars together with a snap tie (figure 7-31).

CONCRETE CONSTRUCTION JOINTS

LEARNING OBJECTIVE: Upon completing this section, you should be able to determine the location of construction joints.

Concrete structures are subjected to a variety of stresses. These stresses are the result of shrinkage and differential movement. Shrinkage occurs during hydration, and differential movement is caused by temperature changes and different loading conditions. These stresses can cause cracking, spalling, and scaling of concrete surfaces and, in extreme cases, can result in failure of the structure.

TYPES OF JOINTS

Stresses in concrete can be controlled by the proper placement of joints in the structure. We'll discuss three basic types of joints: isolation joints, control joints, and construction joints.

Isolation Joints

Isolation joints are used to separate (isolate) adjacent structural members. An example is the joint that separates the floor slab from a column. An isolation joint allows for differential movement in the vertical plane due to loading conditions or uneven settlement. Isolation joints are sometimes called expansion or contraction joints. In this context, they allow for differential movement as a result of temperature changes (as in two adjacent slabs). All isolation joints (expansion or contraction) extend completely through the member and have no load

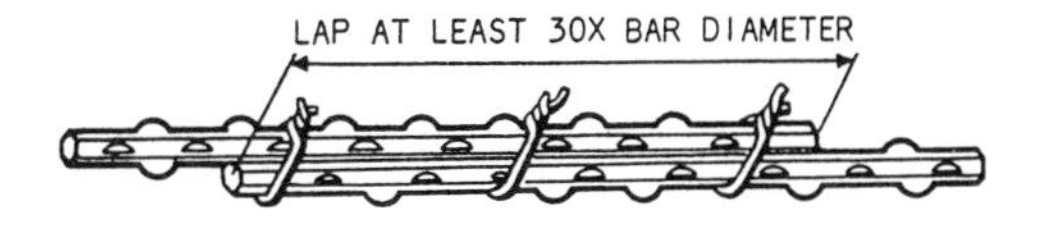

Figure 7-31.—Bars spliced by lapping.

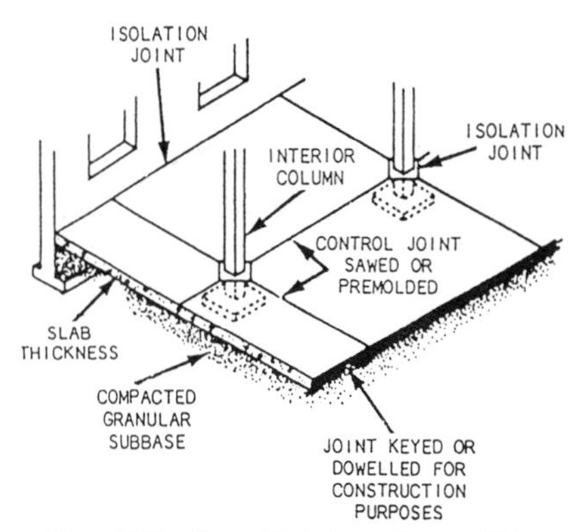

Figure 7-32.—Typical isolation and control joints.

transfer devices built into them. Examples of these are shown in figures 7-32, 7-33, and 7-34.

Control Joints

Movement in the plane of a concrete slab is caused by drying shrinkage and thermal contraction. Some shrinkage is expected and can be tolerated, depending on the design and exposure of the particular structural elements. In a slab, shrinkage occurs more rapidly at the exposed surfaces and causes upward curling at the edges. If the slab is restrained from curling, cracking will occur wherever the restraint imposes stress greater than the tensile strength. Control joints (figure 7-35) are cut into the concrete slab to create a plane of weakness, which forces cracking (if it happens) to occur at a designated place rather than randomly. These joints run in both directions at right angles to each other. Control joints in interior slabs are typically cut 1/3 to 1/4 of the slab thickness and then filled with joint filler. See table 7-1 for suggested control joint spacings. Temperature steel (welded wire fabric) can be used to restrict crack width. For sidewalks and driveways, tooled joints spaced at intervals equal to the width of the slab, but not more than 20 feet (6 meters) apart, should be used. The joint should be 3/4 to 1 inch deep. Surface irregularities along the plane of the

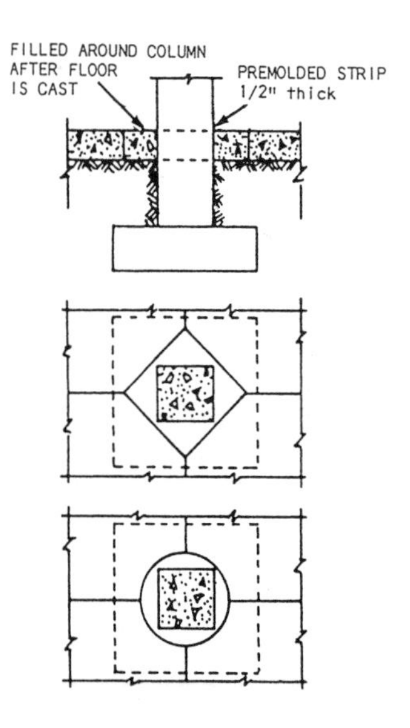

Figure 7-33.—Isolation joints at columns and walls.

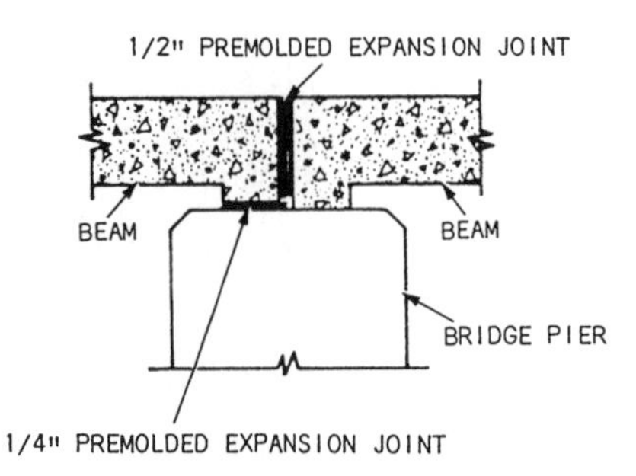

Figure 7-34.—Expansion/contraction joint for a bridge.

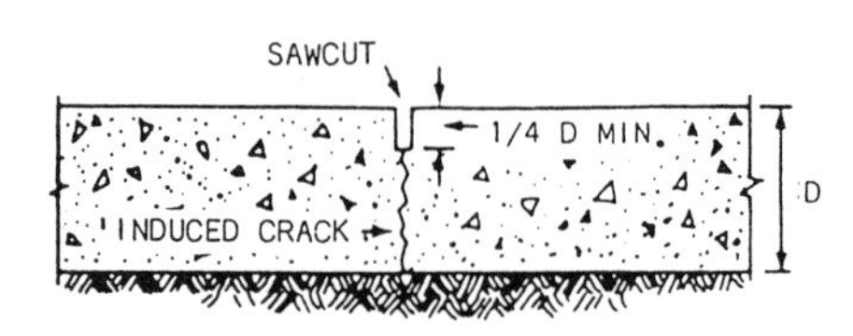

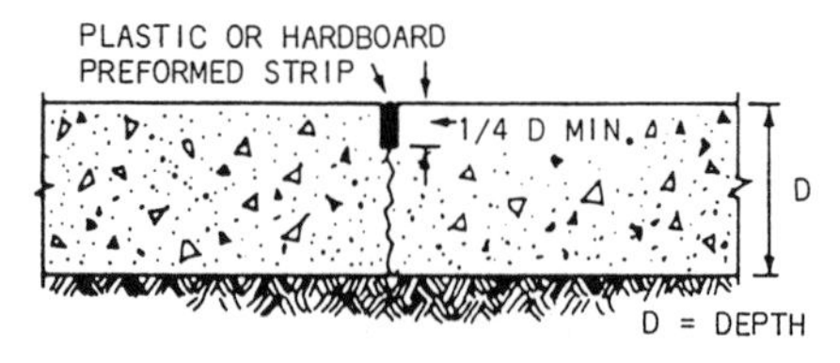

Figure 7-35.—Control joints.

Table 7-1.—Suggested Spacing of Control Joints

SLAB THICKNESS (IN)	LESS THAN 3/4 IN AGGREGATE: SPACING (FT)	LARGER THAN 3/4 IN AGGREGATE: SPACING (FT)	SLUMP LESS THAN 4 IN: SPACING (FT)
5	10*	13	15
6	12	15	18
7	14	18	21
8	16	20	24
9	18	23	27
10	20	25	30

*Given spacings also apply to the distance from control joints to parallel isolation joints or to parallel construction joints. Refer to text for other factors that may call for different spacing.

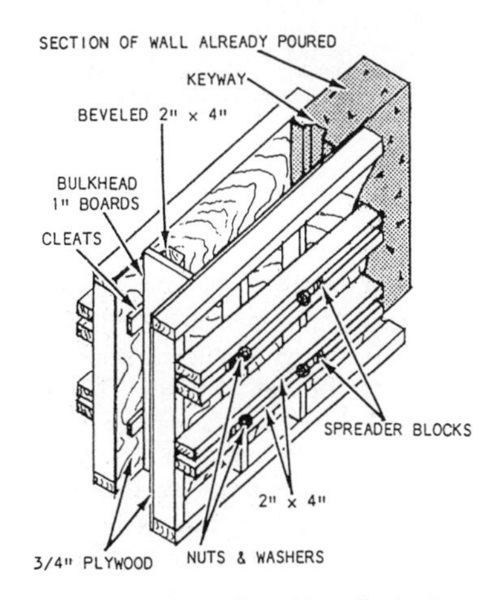

Figure 7-36.—Vertical bulkhead in wall using keyway.

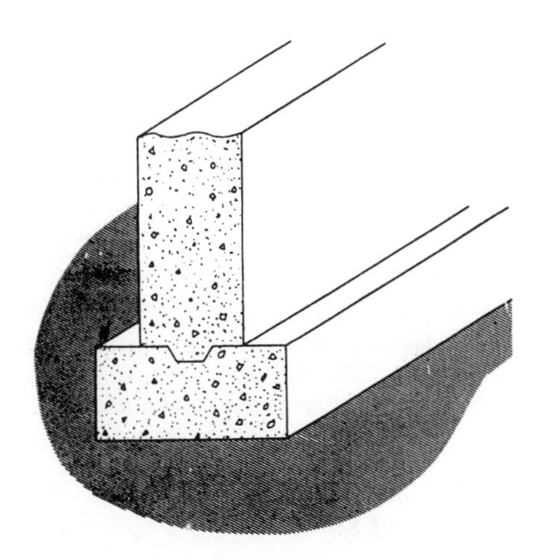

Figure 7-38.—Construction joint between wall and footing with a keyway.

crack are usually sufficient to transfer loads across the joint in slabs on grade.

Construction Joints

Construction joints (figures 7-36, 7-37, 7-38, and 7-39) are made where the concrete placement

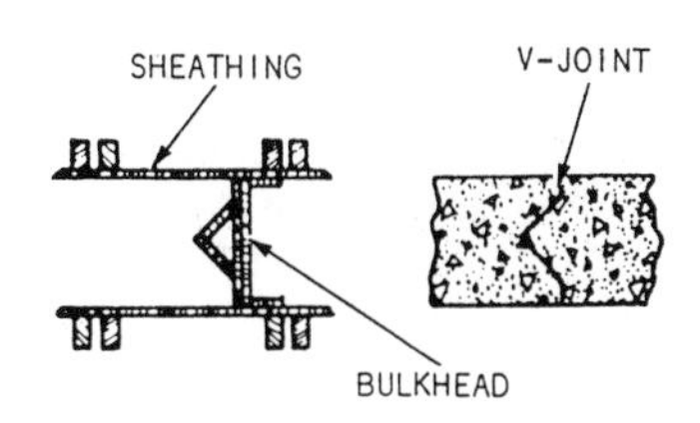

Figure 7-37.—Keyed wall construction joint.

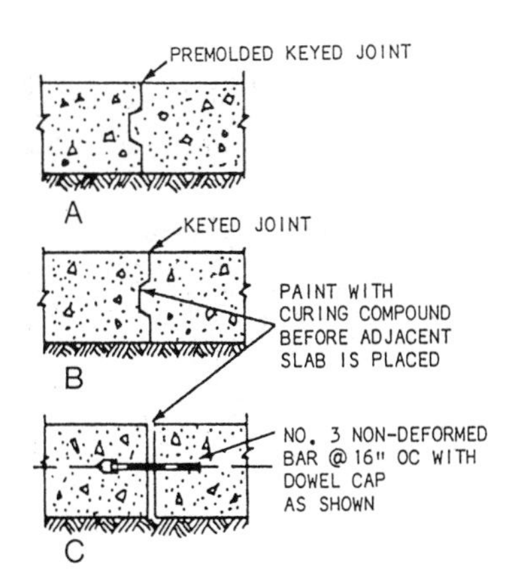

Figure 7-39.—Types of construction joints.

operations end for the day or where one structural element is cast against previously placed concrete. These joints allow some load to be transferred from one structural element to another through the use of keys or (for some slabs and pavement) dowels. Note that the construction joint extends entirely through the concrete element.

SAWING CONCRETE

LEARNING OBJECTIVE: Upon completing this section, you should be able to determine proper occasions for using the concrete saw.

THE CONCRETE SAW

The concrete saw is used to cut longitudinal and transverse joints in finished concrete pavements. The saw is small and can be operated by one person (figure 7-40). Once the cut has been started, the machine provides its own tractive power. A water spray is used to flush the saw cuttings from the cutting area and to cool the cutting blade.

133.18

Figure 7-40.—Concrete saw.

Several types of blades are available. The most common blades have either diamond or carborundum cutting surfaces. The diamond blade is used for cutting **hard** or old concrete; the carborundum blade is used for cutting **green** concrete (under 30 hours old). Let's take a closer look at these two blades.

DIAMOND BLADES

Diamond blades have segments made from a sintered mixture of industrial diamonds and metal powders, which are brazed to a steel disk. They are generally used to cut old concrete, asphalt, and green concrete containing the harder aggregates. Diamond blades must always be used wet. Many grades of diamond blades are available to suit the conditions of the job.

Twelve-inch-diameter diamond blades are the most popular size. This size makes a cut about 3 1/4-inches deep. Larger-size blades are used for deeper cuts.

CARBORUNDUM BLADES

Low-cost, abrasive blades are now widely used to cut green concrete made with soft aggregates, such as limestone, dolomite, coral, or slag. These blades are made from a mixture of silicon carbide grains and a resin bond. This mixture is pressed and baked. In many cases, some of the medium-hard aggregates can be cut if the step-cutting method is used. This method uses two or more saws to cut the same joint, each cutting only a part of the total depth. This principle is also used on the longitudinal saw, which has two individually adjustable cutting heads. When a total depth of 2 1/2 inches is to be cut, the leading blade cuts the first inch and the trailing blade, which is slightly narrower, cuts the remaining depth.

Abrasive blades come in 14- and 18-inch diameters. They are made in various thicknesses to cut joints from 1/4-inch to 1/2-inch wide.

When to Use

When is the best time to saw green concrete? In the case of abrasive blades, there is only one answer—as soon as the concrete can support the equipment and the joint can be cut with a minimum of chipping. In the case of diamond blades, two factors must be considered. In the interest of blade life, sawing should be delayed, but control of random cracking requires sawing at the transverse joints as early as possible. Where transverse joints are closely spaced, every second or third joint can be cut initially and the rest cut later. Sawing longitudinal joints can be delayed for 7 days or longer.

For proper operation and maintenance of the concrete saw, follow the manufacturer's manual.

PLACING CONCRETE

LEARNING OBJECTIVE: Upon completing this section, you should be able to describe the proper placing procedures for well-designed concrete.

You cannot obtain the full value of well-designed concrete without using proper placing procedures. Good concrete placing and compacting techniques produce a tight bond between the paste and aggregate and fill the forms completely. Both of these factors contribute to the full strength and best appearance of concrete. The following are some of the principles of concrete placement:

- **Segregation**—Avoid segregation during all operations, from the mixer to the point of placement, including final consolidation and finishing.
- **Consolidation**—Thoroughly consolidate the concrete, working solidly around all embedded reinforcement and filling all form angles and corners.
- **Bonding**—When placing fresh concrete against or upon hardened concrete, make sure that a good bond develops.
- **Temperature control**—Take appropriate steps to control the temperature of fresh concrete from mixing through final placement. Protect the concrete from temperature extremes after placement.
- **Maximum drop**—To save time and effort, you may be tempted to simply drop the concrete directly from the delivery chute regardless of form height. However, unless the free fall into the form is less than 4 feet, use vertical pipes, suitable drop chutes, or baffles. Figure 7-41 suggests several ways to control concrete fall. Good control prevents honeycombing and other undesirable results.

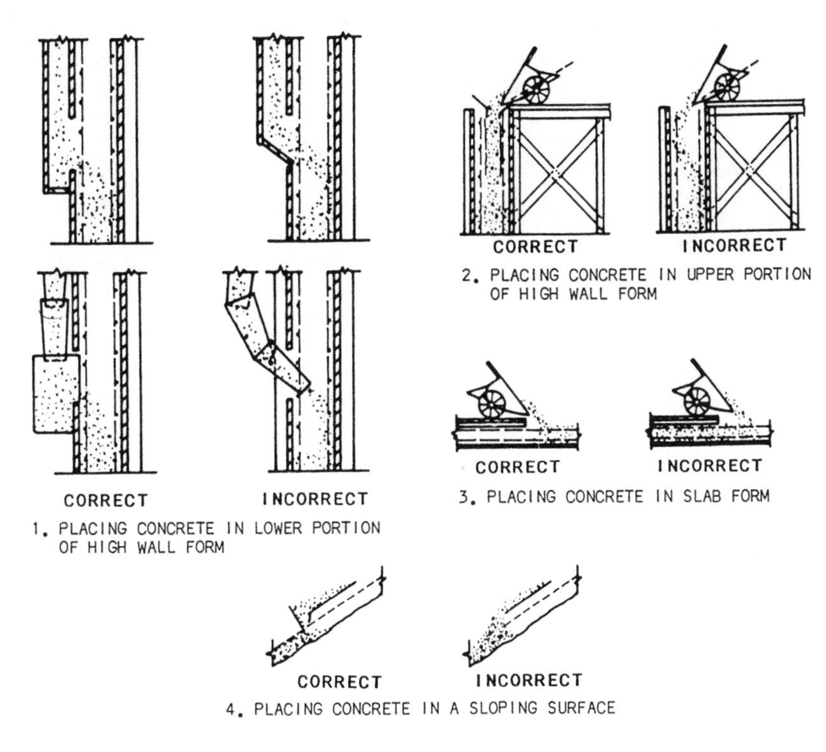

Figure 7-41.—Concrete placing techniques.

- **Layer thickness**—Try to place concrete in even horizontal layers. Do not attempt to puddle or vibrate it into the form. Place each layer in one operation and consolidate it before placing the next layer to prevent honeycombing and voids. This is particularly critical in wall forms containing considerable reinforcement. Use a mechanical vibrator or a hand spading tool for consolidation. Take care not to over vibrate. This can cause segregation and a weak surface. Do not allow the first layer to take its initial set before adding the next layer. Layer thickness depends on the type of construction, the width of the space between forms, and the amount of reinforcement.

- **Compacting**—(Note: This is different from soil compaction.) First, place concrete into its final position as nearly as possible. Then, work the concrete thoroughly around reinforcement and imbedded fixtures, into the corners, and against the sides of the forms. Because paste tends to flow ahead of aggregate, avoid horizontal movements that result in segregation.

- **Placing rate**—To avoid excessive pressure on large project forms, the filling rate should not exceed 4 vertical feet per hour, except for columns. Coordinate the placing and compacting so that the concrete is not deposited faster than it can be compacted properly. To avoid cracking during settlement, allow an interval of at least 4 hours, preferably 24 hours, between placing slabs, beams, or girders, and placing the columns and walls they support.

- **Wall construction**—When constructing walls, beams, or girders, place the first batches of each layer at the ends of the section, then proceed toward the center to prevent water from collecting at the form ends and corners. For walls, stop off the inside form at the construction level. Overfill the form for about 2 inches and remove the excess just before the concrete sets to ensure a rough, clean surface. Before placing the next lift of concrete, deposit a 1/2- to 1-inch-thick layer of sand-cement mortar. Make the mortar with the same water content ratio as the concrete and with a 6-inch slump to prevent stone pockets and help produce a watertight joint. View 1 of figure 7-41 shows the proper way to place concrete in the lower portion of high wall forms. Note the different types of drop chute that can be used to place concrete through port openings and into the lower portion of the wall. Space the port openings at about 10-foot intervals up the wall. The method used to place concrete in the upper portion of the wall is shown in view 2 of figure 7-41. When placing concrete for walls, be sure to remove the spreaders as you fill the forms.

- **Slab construction**—When constructing slabs, place the concrete at the far end of the slab first, and then place subsequent batches against previously placed concrete, as shown in view 3 of figure 7-41. Do not place the concrete in separate piles and then level the piles and work them together. Also, don't deposit the concrete in piles and then move them horizontally to their final position. These practices can result in segregation.

- **Placing concrete on slopes**—View 4 of figure 7-41 shows how to place concrete on slopes. Always deposit the concrete at the bottom of the slope first, then proceed up the slope placing each new batch against the previous one. When consolidated, the weight of the new concrete increases the compacting of the previously placed concrete.

CONSOLIDATING CONCRETE

LEARNING OBJECTIVE: Upon completing this section, you should be able to describe the methods available for consolidating concrete.

Except for concrete placed underwater, you must compact or consolidate all concrete after placement.

PURPOSE OF CONSOLIDATION

Consolidation eliminates rock pockets and air bubbles and brings enough fine material both to the surface and against the forms to produce the desired finish. You can use such hand tools as spades, puddling sticks, or tampers, but mechanical vibrators are best. Any compacting device must reach the bottom of the form and be small enough to pass between reinforcing bars. The process involves carefully working around all reinforcing steel with the compacting device to assure proper embedding of reinforcing steel in the concrete. Since the strength of the concrete member depends on proper reinforcement location, be careful not to displace the reinforcing steel.

VIBRATION

Vibrators consolidate concrete by pushing the coarse aggregate downward, away from the point of vibration. Vibrators allow placement of mixtures that are too stiff to place any other way, such as those having a 1- or 2-inch slump. Stiff mixtures are more economical because they require less cement and present fewer segregation or bleeding problems. However, do not use a mix so stiff that it requires too much labor to place it.

Mechanical Vibrators

The best compacting tool is a mechanical vibrator (figure 7-42). The best vibrators available in engineering construction battalions are called internal vibrators because the vibrating element is inserted into the concrete. When using an internal vibrator, insert it at approximately 18-inch intervals into air-entrained concrete for 5 to 10 seconds and into nonair-entrained concrete for 10 to 15 seconds. The exact period of time that you should leave a vibrator in the concrete depends on its slump. Overlap the vibrated areas somewhat at each insertion. Whenever possible, lower the vibrator into the concrete vertically and allow it to descend by gravity. The vibrator should not only pass through the layer just placed, but penetrate several inches into the layer underneath to ensure a good bond between the layers.

Vibration does not normally damage the lower layers, as long as the concrete disturbed in these lower layers becomes plastic under the vibrating action. You know that you have consolidated the concrete properly when a thin line of mortar appears along the form near the vibrator, the coarse aggregate disappears into the concrete, or the paste begins to appear near the vibrator head. Then, withdraw the vibrator vertically at about the same gravity rate that it descended.

Some hand spading or puddling should accompany all vibration. To avoid the possibility of segregation, do not vibrate mixes that you can consolidate easily by spading. Also, don't vibrate concrete that has a slump of 5 inches or more. Finally, do not use vibrators to move concrete in the form.

Hand Methods

Manual consolidation methods require spades, puddling sticks, or various types of tampers. To

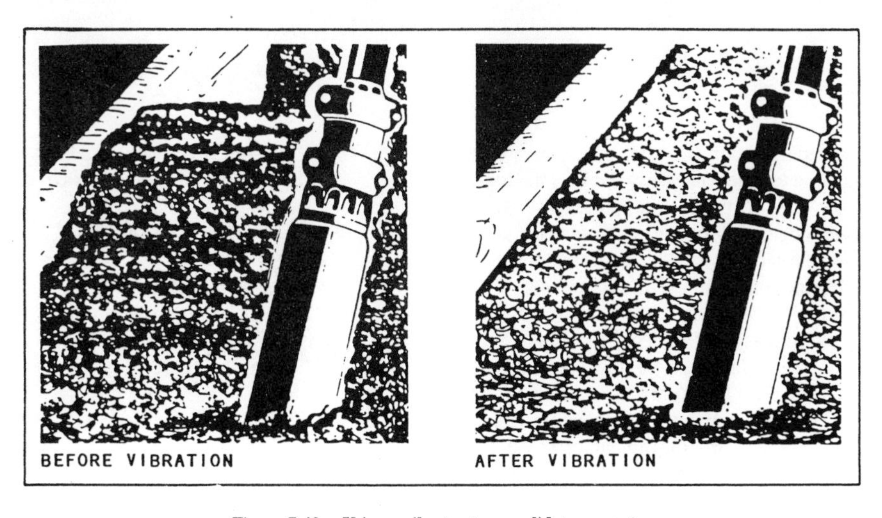

Figure 7-42.—Using a vibrator to consolidate concrete.

consolidate concrete by spading, insert the spade along the inside surface of the forms (figure 7-43), through the layer just placed, and several inches into the layer underneath. Continue spading or puddling until the coarse aggregate disappears into the concrete.

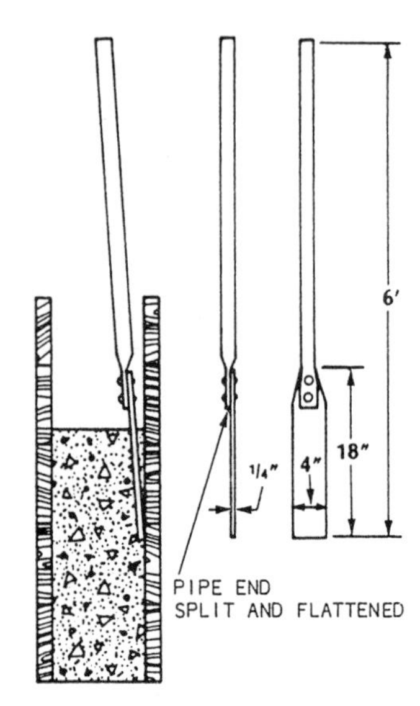

Figure 7-43.—Consolidation by spading and a spading tool.

FINISHING CONCRETE

LEARNING OBJECTIVE: Upon completing this section, you should be able to describe the finishing process for the final concrete surface.

The finishing process provides the final concrete surface. There are many ways to finish concrete surfaces, depending on the effect required. Sometimes you only need to correct surface defects, fill bolt holes, or clean the surface. Unformed surfaces may require only screeding to proper contour and elevation, or a broomed, floated, or trowelled finish may be specified.

SCREEDING

The top surface of a floor slab, sidewalk, or pavement is rarely placed at the exact specified elevation. Screeding brings the surface to the required elevation by striking off the excess concrete. Two types of screeds are used in concrete finishing operations: the hand screed and the mechanical screed.

Hand Screed

Hand screeding requires a tool called a screed. This is actually a templet (usually a 2-by-4) having a straight lower edge to produce a flat surface (or a curved lower edge to produce a curved surface). Move the screed back and forth across the concrete using a sawing motion, as shown in figure 7-44. With each sawing motion, move the screed forward an inch or so along the forms. This forces the concrete built up against the screed face into the low spots. If the screed tends to tear the surface, as it may on air-entrained concrete due to its sticky nature, either reduce the rate of forward movement or cover the lower edge of the screed with metal. This stops the tearing action in most cases.

You can hand-screed surfaces up to 30-feet wide, but the efficiency of this method diminishes on surfaces more than 10-feet wide. Three workers (excluding a vibrator operator) can screed approximately 200 square feet of concrete per hour. Two of the workers work the screed while the third pulls excess concrete from the front of the screed. You must screed the surface a second time to remove the surge of excess concrete caused by the first screeding.

Figure 7-44.—Screeding operation.

Mechanical Screed

The mechanical screed is being used more and more in construction for striking off concrete slabs on highways, bridge decks, and deck slabs. This screed incorporates the use of vibration and permits the use of stronger, and more economical, low-slump concrete. It can strike off this relatively dry material smoothly and quickly. The advantages of using a vibrating screed are greater density and stronger concrete. Vibrating screeds give a better finish, reduce maintenance, and save considerable time due to the speed at which they operate. Vibrating screeds are also much less fatiguing to operate than hand screeds.

A mechanical screed (figure 7-45) usually consists of a beam (or beams) and a gasoline engine, or an electric motor and a vibrating mechanism mounted in the center of the beam. Most mechanical screeds are quite heavy and usually equipped with wheels to help move them around. You may occasionally encounter lightweight screeds not equipped with wheels. These are easily lifted by two crewmembers and set back for the second pass if required.

The speed at which the screed is pulled is directly related to the slump of the concrete—the less the slump, the slower the speed; the more the slump, the faster the speed. On the finishing pass of the screed, there should be no transverse (crosswise) movement of the beam; the screed is merely drawn directly forward riding on the forms or rails. For a mechanical screed, a method is provided to quickly start or stop the vibration. This is important to prevent over vibration when the screed might be standing still.

Concrete is usually placed 15 to 20 feet ahead of the screed and shoveled as close as possible to its final resting place. The screed is then put into operation and pulled along by two crewmembers, one at each end of the screed. It is important that sufficient concrete is kept in front of the screed. Should the concrete be below the level of the screed beam, voids or bare spots will appear on the concrete surface as the screed passes over the slab. Should this occur, a shovelful or so of concrete is thrown on the bare spot, and the screed is lifted up and carried back past this spot for a second pass. In rare cases, the screed crew will work out the void or bare spot with a hand-operated bull float, rather than make a second pass with the screed.

Figure 7-45.—Mechanical screed.

The vibration speed will need to be adjusted for particular mixes and different beam lengths. Generally, the stiffer the mix and the longer the beam, the greater the vibration speed required. The speed at which the screed is moved also affects the resulting finish of the slab. After a few minutes of operation, a satisfactory vibration pulling speed can be established. After the vibrating screed has passed over the slab, the surface is then ready for broom or burlap finishing.

Where possible, it is advisable to lay out or engineer the concrete slab specifically for use of a vibrating screed. Forms should be laid out in lanes of equal widths, so that the same- length screed can be used on all lanes or slabs. It should also be planned, if possible, that any vertical columns will be next to the forms, so that the screed can easily be lifted or maneuvered around the column.

There are four important advantages of using a vibrating finishing screed. First, it allows the use of low-slump concrete, resulting in stronger slabs. Second, it reduces and sometimes eliminates the necessity of hand tamping and bull floating. Third, it increases the density of the concrete, resulting in a superior wearing surface. And fourth, in the case of floor slabs, troweling can begin sooner since drier mixes can be used, which set up more quickly.

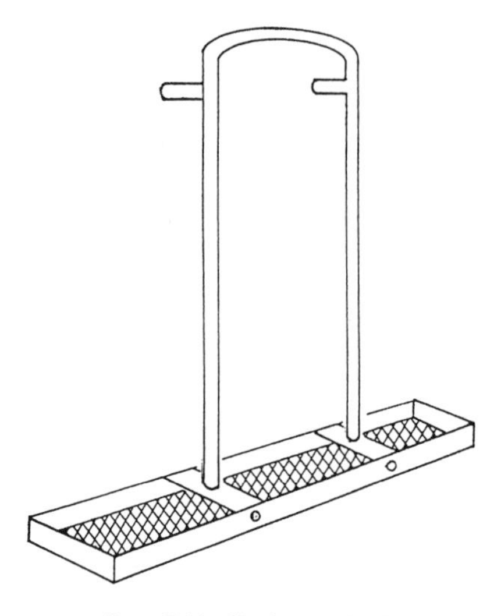

Figure 7-46.—Hand tamp (jitterbug).

HAND TAMPING

Hand tamping, or jitterbugging (figure 7-46), is done after the concrete has been screeded. Hand tamping is used to compact the concrete into a dense mass and to force the larger particles of coarse aggregate slightly below the surface. This enables you to put the desired finish on the surface. The tamping tool should be used only with a low-slump concrete, and bring only just enough mortar to the surface for proper finish. After using the jitterbug, you can go directly to using the bull float.

FLOATING

If a smoother surface is required than the one obtained by screeding, the surface should be worked sparingly with a wood or aluminum magnesium float (figure 7-47, view A) or with a finishing machine. In view B, the wood float is shown in use. A long-handled wood float is used for slab construction (view C). The aluminum float, which is used the same way as the wood float, gives the finished concrete a much smoother surface. To avoid cracking and dusting of the finished concrete, begin aluminum floating when the water sheen disappears from the freshly placed concrete surface. Do not use cement or water as an aid in finishing the surface.

Floating has three purposes: (1) to embed aggregate particles just beneath the surface; (2) to remove slight imperfections (high and low spots); and, (3) to compact the concrete at the surface in preparation for other finishing operations.

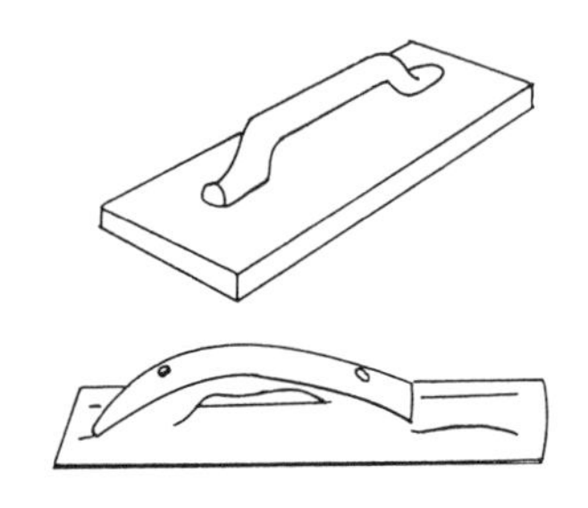

A. WOOD OR MAGNESIUM FLOAT

B. FLOATING OPERATION

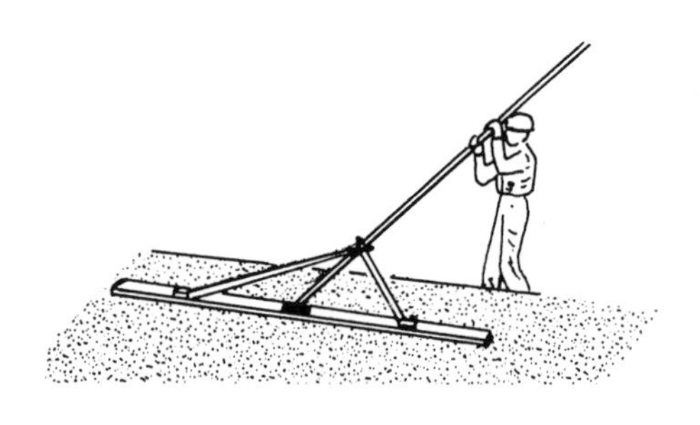

C. LONG-HANDLED WOOD FLOAT AND FLOATING OPERATION

Figure 7-47.—Wood floats and floating operations.

Begin floating immediately after screeding while the concrete is still plastic and workable. However, do not overwork the concrete while it is still plastic because you may bring an excess of water and paste to the surface. This fine material forms a thin, weak layer that will scale or quickly wear off under use. To remove a coarse texture as the final finish, you usually have to float the surface a second time after it partially hardens.

EDGING

As the sheen of water begins to leave the surface, edging should begin. All edges of a slab that do not abut another structure should be finished with an edger (figure 7-48). An edger dresses corners and rounds or bevels the concrete edges. Edging the slab helps prevent chipping at the corners and helps give the slab a finished appearance.

TROWELING

If a dense, smooth finish is desired, floating must be followed by steel troweling (figure 7-49). Troweling should begin after the moisture film or sheen disappears from the floated surface and when the concrete has hardened enough to prevent fine material and water from being worked to the surface. This step should be delayed as long as possible. Troweling too early tends to produce crazing and lack of durability. However, too long a delay in troweling results in a surface too hard to finish properly. The usual tendency is to start to trowel too soon. Troweling should leave the surface smooth, even, and free of marks and ripples. Spreading dry cement on a wet surface to take up excess water is not a good practice where a wear-resistant and durable surface is required. Wet spots must be avoided if possible. When they do occur, however, finishing operations should not be resumed until the water has been absorbed, has evaporated, or has been mopped up.

Figure 7-48.—Edger.

Figure 7-49.—Steel finishing tools and troweling operations.

Steel Trowel

An unslippery, fine-textured surface can be obtained by troweling lightly over the surface with a circular motion immediately after the first regular troweling. In this process, the trowel is kept flat on the surface of the concrete. Where a hard steel-troweled finish is required, follow the first regular troweling by a second troweling. The second troweling should begin after the concrete has become hard enough so that no mortar adheres to the trowel, and a ringing sound is produced as the trowel passes over the surface. During this final troweling, the trowel should be tilted slightly and heavy pressure exerted to thoroughly compact the surface. Hairline cracks are usually due to a concentration of water and extremely fine aggregates at the surface. This results from overworking the concrete during finishing operations. Such cracking is aggravated by drying and cooling too rapidly. Checks that develop before troweling can usually be closed by pounding the concrete with a hand float.

Mechanical Troweling Machine

The mechanical troweling machine (figure 7-50) is used to good advantage on flat slabs with a stiff consistency. Mechanical trowels come with a set of float blades that slip over the steel blades. With these blades, you can float a slab with the mechanical trowels. The concrete must be set enough to support the weight of the machine and the operator. Machine finishing is faster than hand finishing. However, it cannot be used with all types of construction. Refer to the manufacturer's manual for operation and maintenance of the machine you are using.

Figure 7-50.—Mechanical troweling machine.

BROOMING

A nonskid surface can be produced by brooming the concrete before it has thoroughly hardened. Brooming is carried out after the floating operation. For some floors and sidewalks where scoring is not desirable, a similar finish can be produced with a hairbrush after the surface has been troweled once. Where rough scoring is required, a stiff broom made of steel wire or coarse fiber should be used. Brooming should be done so that the direction of the scoring is at right angles to the direction of the traffic.

GRINDING

When grinding of a concrete floor is specified, it should be started after the surface has hardened sufficiently to prevent dislodgement of aggregate particles and should be continued until the coarse aggregate is exposed. The machines used should be of an approved type with stones that cut freely and rapidly. The floor is kept wet during the grinding process, and the cuttings are removed by squeegeeing and flushing with water.

After the surface is ground, air holes, pits, and other blemishes are filled with a thin grout composed of one part No. 80-grain carborundum grit and one part portland cement. This grout is spread over the floor and worked into the pits with a straightedge. Next, the grout is rubbed into the floor with the grinding machine. When the filings have hardened for 17 days, the floor receives a final grinding to remove the film and to give the finish a polish. All surplus material is then removed by washing thoroughly. When properly constructed of good-quality materials, ground floors are dustless, dense, easily cleaned, and attractive in appearance.

SACK-RUBBED FINISH

A sack-rubbed finish is sometimes necessary when the appearance of formed concrete falls considerably below expectations. This treatment is performed after all required patching and correction of major imperfections have been completed. The surfaces are thoroughly wetted, and sack rubbing is commenced immediately.

The mortar used consists of one part cement; two parts, by volume, of sand passing a No. 16 screen; and enough water so that the consistency of the mortar will be that of thick cream. It may be necessary to blend the cement with white cement to obtain a color matching that of the surrounding concrete surface. The mortar is rubbed thoroughly over the area with clean burlap or a sponge rubber float, so that it fills all pits. While the mortar in the pits is still plastic, the surface should be rubbed over with a dry mix of the same material. This removes all excess plastic material and places enough dry material in the pits to stiffen and solidify the mortar. The filings will then be flush with the surface. No material should remain on the surface above the pits. Curing of the surface is then continued.

RUBBED FINISH

A rubbed finish is required when a uniform and attractive surface must be obtained. A surface of satisfactory appearance can be obtained without rubbing if plywood or lined forms are used. The first rubbing should be done with coarse carborundum stones as soon as the concrete has hardened so that the aggregate is not pulled out. The concrete should then be cured until final rubbing. Finer carborundum stones are used for the final rubbing. The concrete should be kept damp while being rubbed. Any mortar used in this process and left on the surface should be kept damp for 1 to 2 days after it sets to cure properly. The mortar layer should be kept to a minimum thickness as it is likely to scale off and mar the appearance of the surface.

EXPOSED AGGREGATE FINISH

An exposed aggregate finish provides a nonskid surface. To obtain this, you must allow the concrete to harden sufficiently to support the finisher. The aggregate is exposed by applying a retarder over the surface and then brushing and flushing the concrete surface with water. Since timing is important, test panels should be used to determine the correct time to expose the aggregate.

CURING CONCRETE

Adding water to portland cement to form the water-cement paste that holds concrete together starts a chemical reaction that makes the paste into a bonding agent. This reaction, called hydration, produces a stone-like substance—the hardened cement paste. Both the rate and degree of hydration, and the resulting strength of the final concrete, depend on the curing process that follows placing and consolidating the plastic concrete. Hydration continues indefinitely at a decreasing rate as long as the mixture contains water and the temperature conditions are favorable. Once the water is removed, hydration ceases and cannot be restarted.

Curing is the period of time from consolidation to the point where the concrete reaches its design strength. During this period, you must take certain steps to keep the concrete moist and as near 73°F as practical. The properties of concrete, such as freeze and thaw resistance, strength, watertightness, wear resistance, and volume stability, cure or improve with age as long as you maintain the moisture and temperature conditions favorable to continued hydration.

The length of time that you must protect concrete against moisture loss depends on the type of cement used, mix proportions, required strength, size and shape of the concrete mass, weather, and future exposure conditions. The period can vary from a few days to a month or longer. For most structural use, the curing period for cast-in-place concrete is usually 3 days to 2 weeks. This period depends on such conditions as temperature, cement type, mix proportions, and so forth. Bridge decks and other slabs exposed to weather and chemical attack usually require longer curing periods. Figure 7-51 shows how moist curing affects the compressive strength of concrete.

Curing Methods

Several curing methods will keep concrete moist and, in some cases, at a favorable hydration temperature. They fall into two categories: those that

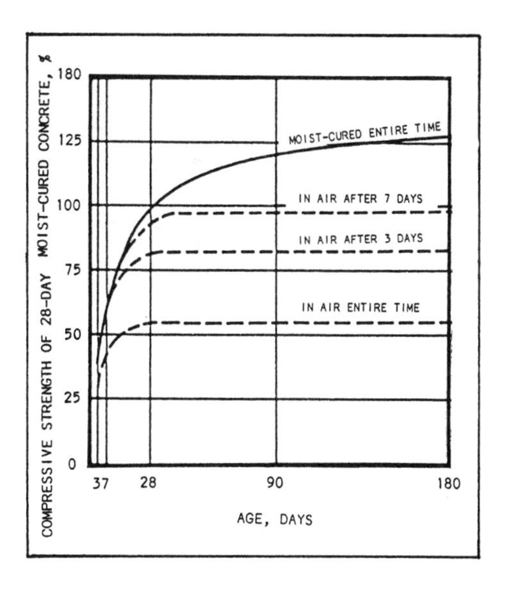

Figure 7-51.—Moist curing effect on compressive strength of concrete.

supply additional moisture and those that prevent moisture loss. Table 7-2 lists several of these methods and their advantages and disadvantages.

METHODS THAT SUPPLY ADDITIONAL MOISTURE.—Methods that supply additional moisture include sprinkling and wet covers. Both these methods add moisture to the concrete surface during the early hardening or curing period. They also provide some cooling through evaporation. This is especially important in hot weather.

Sprinkling continually with water is an excellent way to cure concrete. However, if you sprinkle at intervals, do not allow the concrete to dry out between applications. The disadvantages of this method are the expense involved and volume of water required.

Wet covers, such as straw, earth, burlap, cotton mats, and other moisture-retaining fabrics, are used extensively in curing concrete. Figure 7-52 shows a typical application of wet burlap. Lay the wet coverings as soon as the concrete hardens enough to prevent surface damage. Leave them in place and keep them moist during the entire curing period.

If practical, horizontal placements can be flooded by creating an earthen dam around the edges and submerging the entire concrete structure in water.

Table 7-2.—Curing Methods

METHOD	ADVANTAGE	DISADVANTAGES
Sprinkling with Water or Covering with Burlap	Excellent results if kept constantly wet	Likelihood of drying between sprinklings; difficult on vertical walls
Straw	Insulator in winter	Can dry out, blow away, or burn
Moist Earth	Cheap but messy	Stains concrete; can dry out; removal problem
Ponding on Flat Surfaces	Excellent results, maintains uniform temperature	Requires considerable labor; undesirable in freezing weather
Curing Compounds	Easy to apply and inexpensive	Sprayer needed; inadequate coverage allows drying out; film can be broken or tracked off before curing is completed; unless pigmented, can allow concrete to get too hot
Waterproof Paper	Excellent protection, prevents drying	Heavy cost can be excessive; must be kept in rolls; storage and handling problem
Plastic Film	Absolutely watertight, excellent protection. Light and easy to handle	Should be pigmented for heat protection; requires reasonable care and tears must be patched; must be weighed down to prevent blowing away

Figure 7-52.—Curing a wall with wet burlap sacks.

METHODS THAT PREVENT MOISTURE LOSS.—Methods that prevent moisture loss include laying waterproof paper, plastic film, or liquid-membrane-forming compounds, and simply leaving forms in place. All prevent moisture loss by sealing the surface.

Waterproof paper (figure 7-53) can be used to cure horizontal surfaces and structural concrete having relatively simple shapes. The paper should be large enough to cover both the surfaces and the edges of the concrete. Wet the surface with a fine water spray before covering. Lap adjacent sheets 12 inches or more and weigh their edges down to form a continuous cover with closed joints. Leave the coverings in place during the entire curing period.

Figure 7-53.—Waterproof paper used for curing.

Plastic film materials are sometimes used to cure concrete. They provide lightweight, effective moisture barriers that are easy to apply to either simple or complex shapes. However, some thin plastic sheets may discolor hardened concrete, especially if the surface was steel-troweled to a hard finish. The coverage, overlap, weighing down of edges, and surface wetting requirements of plastic film are similar to those of waterproof paper.

Curing compounds are suitable not only for curing fresh concrete, but to further cure concrete following form removal or initial moist curing. You can apply them with spray equipment, such as hand-operated pressure sprayers, to odd slab widths or shapes of fresh concrete, and to exposed concrete surfaces following form removal. If there is heavy rain within 3 hours of application, you must respray the surface. You can use brushes to apply curing compound to formed surfaces, but do not use brushes on unformed concrete because of the risk of marring the surface, opening the surface to too much compound penetration, and breaking the surface film continuity. These compounds permit curing to continue for long periods while the concrete is in use. Because curing compounds can prevent a bond from forming between hardened and fresh concrete, do not use them if a bond is necessary.

Forms provide adequate protection against moisture loss if you keep the exposed concrete surfaces wet. Keep wood forms moist by sprinkling, especially during hot, dry weather.

FORM REMOVAL

Forms should, whenever possible, be left in place for the entire curing period. Since early form removal is desirable for their reuse, a reliable basis for determining the earliest possible stripping time is necessary. Some of the early signs to look for during stripping are no excessive deflection or distortion and no evidence of cracking or other damage to the concrete due to the removal of the forms or the form supports. In any event, forms must not be stripped until the concrete has hardened enough to hold its own weight and any other weight it may be carrying. The surface must be hard enough to remain undamaged and unmarked when reasonable care is used in stripping the forms.

Curing Period

Haunch boards (side forms on girders and beams) and wall forms can usually be removed after 1 day. Column forms usually require 3 days before the forms can be removed. Removal of forms for soffits on girders and beams can usually be done after 7 days. Floor slab forms (over 20-foot clear span between supports) usually require 10 days before removing the forms.

Inspections

After removing the forms, the concrete should be inspected for surface defects. These defects may be rock pockets, inferior quality ridges at form joints, bulges, bolt holes, and form-stripping damage. Experience has proved that no steps can be omitted or carelessly performed without harming the serviceability of the work. If not properly performed, the repaired area may later become loose, crack at the edges, and not be watertight. Repairs are not always necessary, but when they are, they should be done immediately after stripping the forms (within 24 hours).

Defects can be repaired in various ways. Therefore, let's look at some common defects you may encounter when inspecting new concrete and how repairs can be made.

RIDGES AND BULGES.—Ridges and bulges can be repaired by careful chipping followed by rubbing with a grinding stone.

HONEYCOMB.—Defective areas, such as honeycomb, must be chipped out of the solid concrete. The edges must be cut as straight as possible at right angles to the surface or slightly undercut to provide a key at the edge of the patch. If a shallow layer of mortar is placed on top of the honeycomb concrete, moisture will form in the voids and subsequent weathering will cause the mortar to spall off. Shallow patches can be filled with mortar placed in layers not more than 1/2-inch thick. Each layer is given a scratch finish to match the surrounding concrete by floating, rubbing, or tooling or on formed surfaces by pressing the form material against the patch while the mortar is still in place.

Large or deep patches can be filled with concrete held in place by forms. These patches should be reinforced and doweled to the hardened concrete (figure 7-54). Patches usually appear darker than the surrounding concrete. Some white cement should be used in the mortar or concrete used for patching if appearance is important. A trial mix should be tried to determine the proportion of white and gray cements to use. Before mortar or concrete is placed in patches, the surrounding concrete should be kept wet for several hours. A grout of cement and water mixed to the consistency of paint should then be brushed into the surfaces to which the new material is to be bonded. Curing should be started as soon as possible to avoid early drying. Damp burlap, tarpaulins, and membrane-curing compounds are useful for this purpose.

BOLT HOLES.—Bolt holes should be filled with small amounts of grout carefully packed into place. The grout should be mixed as dry as possible, with just enough water so it compacts tightly when forced into place. Tie-rod holes extending through the concrete can be filled with grout with a pressure gun similar to an automatic grease gun.

ROCK POCKETS.—Rock pockets should be completely chipped out. The chipped out hole should have sharp edges and be so shaped that the grout patch will be keyed in place (figure 7-55). The surface of all holes that are to be patched should be kept moist for several hours before applying the grout. Grout should be placed in these holes in layers not over 1/4 inch thick and be well compacted. The grout should be allowed to set as long as possible before being used to reduce the amount of shrinkage and to make a better patch. Each layer should be scratched rough to improve the bond with the succeeding layer and the last layer smoothed to match the adjacent surface.

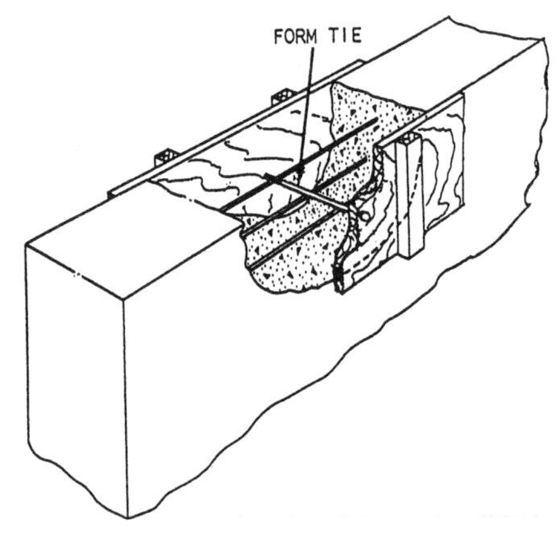

Figure 7-54.—Repair of large volumes of concrete.

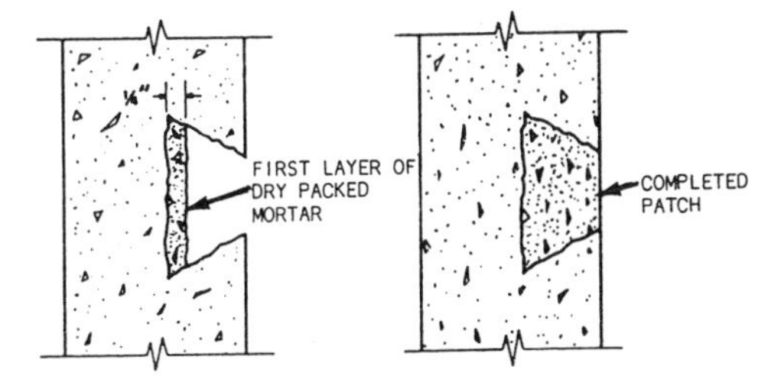

Figure 7-55.—Repairing concrete with dry-packed mortar.

Where absorptive form lining has been used, the patch can be made to match the rest of the surface by pressing a piece of form lining against the fresh patch.

View A of figure 7-56 shows an incorrectly installed patch. Feathered edges around a patch lack sufficient strength and will eventually break down. View B of the figure shows a correctly installed patch. The chipped area should be at least 1-inch deep with the edges at right angles to the surface. The correct method of screeding a patch is shown in view C. The new concrete should project slightly above the surface of the old concrete. It should be allowed to stiffen and then troweled and finished to match the adjoining surfaces.

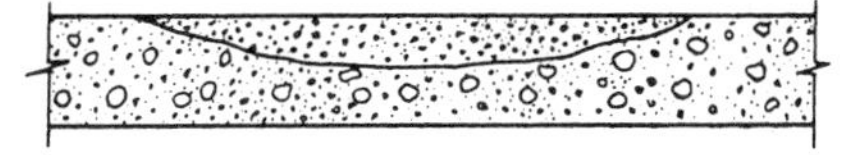

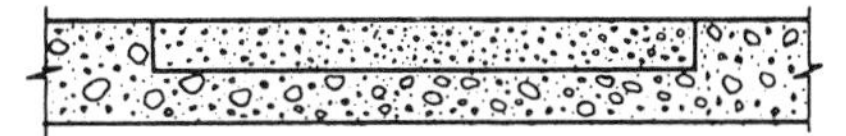

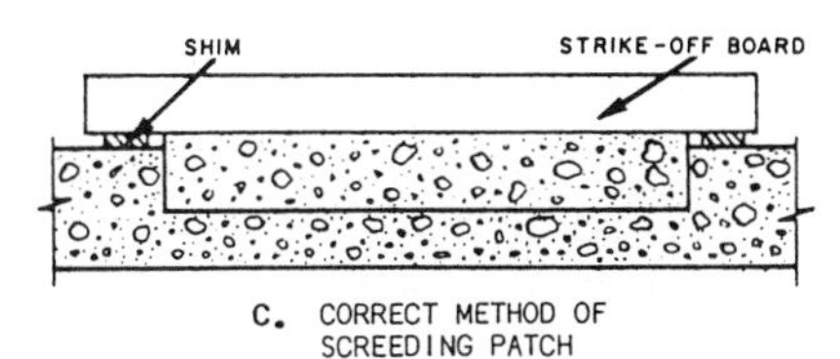

Figure 7-56.—Patching concrete.

RECOMMENDED READING LIST

NOTE

Although the following references were current when this TRAMAN was published, their continued currency cannot be assured. You therefore need to ensure that you are studying the latest version.

Concrete and Masonry, FM 5-742, Headquarters, Department of the Army, Washington, D.C., 1985.

Concrete Formwork, Koel, Leonard, American Technical Publishers, Homewood, Ill., 1988.

Steelworker 3 & 2, NAVEDTRA 10653-G, Naval Education and Training Program Management Support Activity, Pensacola, Fla., 1988.

CHAPTER 8

MASONRY

Originally, masonry was the art of building a structure from stone. Today, it refers to construction consisting of units held together with mortar, such as concrete block, stone, brick, clay tile products, and, sometimes, glass block. The characteristics of masonry work are determined by the properties of the masonry units and mortar and by the methods of bonding, reinforcing, anchoring, tying, and joining the units into a structure.

MASONRY TOOLS AND EQUIPMENT

LEARNING OBJECTIVE: Upon completing this section, you should be able to identify the basic masonry tools and equipment.

Masonry involves the use of a wide selection of tools and equipment. A set of basic mason's tools, including trowels, a chisel, hammer, and a jointer, is shown in figure 8-1.

TROWELS

A trowel (figure 8-1) is used to pick up mortar from the board, throw mortar on the unit, spread the mortar, and tap the unit down into the bed. A common trowel is usually triangular, ranging in size up to about 11 inches long and from 4 to 8 inches wide. Generally, short, wide trowels are best because they do not put too much strain on the wrist. Trowels used to point and strike joints are smaller, ranging from 3 to 6 inches long and 2 to 3 inches wide. We will talk more about pointing and striking joints later in the chapter.

CHISEL

A chisel (figure 8-1) is used to cut masonry units into parts. A typical chisel is 2 1/2 to 4 1/2 inches wide.

HAMMER

A mason's hammer (figure 8-1) has a square face on one end and a long chisel on the other. The hammer weighs from 1 1/2 to 3 1/2 pounds. You use it to split and rough-break masonry units.

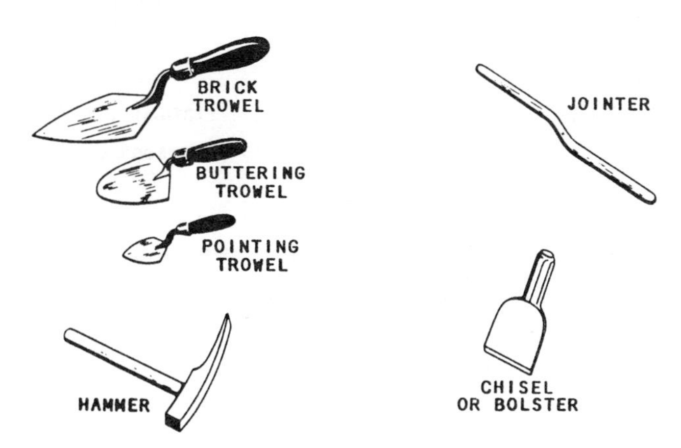

Figure 8-1.—Basic mason's tools.

JOINTER

As its name implies, you use a jointer (figure 8-1) to make various mortar joints. There are several different types of jointer—rounded, flat, or pointed—depending on the shape of the mortar joint you want.

SQUARE

You use the square (figure 8-2, view 1) to measure right angles and to lay out corners. Squares are usually made of metal and come in various sizes.

MASON'S LEVEL

The mason's level (figure 8-2, view 2) is used to establish "plumb" and "level" lines. A plumb line is absolutely **vertical**. A level line is absolutely **horizontal**. The level may be constructed of seasoned hardwood, various metals, or a combination of both. They are made as lightweight as possible without sacrificing strength to withstand fairly rough treatment. Levels may be equipped with single or double vials. Double-vial levels are preferred since they can be used either horizontally or vertically.

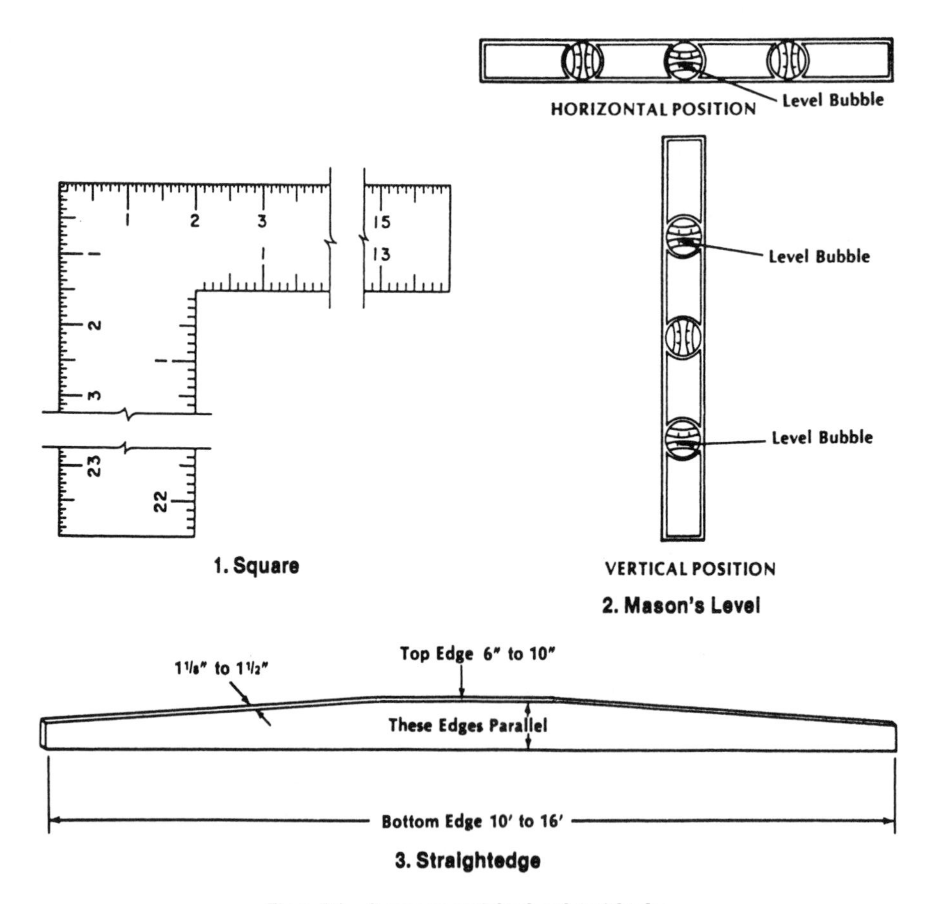

Figure 8-2.—Square, mason's level, and straightedge.

Levels are shaped similar to rulers and have vials enclosed in glass. Inside each vial is a bubble of air suspended in either alcohol or oil. When a bubble is located exactly between the two center marks on the vial, the object is either level or plumb, depending on the position in which the mason is using the level. In a level, alcohol is the more suitable since oil is more affected by heat and cold. The term "spirit level" indicates that alcohol is used in the vials. The vials are usually embedded in plaster or plastic so that they remain secure and true. Shorter levels are made for jobs where a longer level will not fit. The most popular of these are 24 and 18 inches long.

In a level constructed of wood, you should occasionally rub a small amount of linseed oil into the wood with a clean cloth. This treatment also stops mortar from sticking to the level. Do not use motor oil as this eventually rots the wood.

STRAIGHTEDGE

A straightedge (figure 8-2, view 3) can be any length up to 16 feet. Thickness can be from 1 1/8 inches to 1 1/2 inches, and the middle portion of the top edge from 6 to 10 inches wide. The middle portion of the top edge must be parallel to the bottom edge. You use a straightedge to extend a level to plumb or level distances longer than the level length.

MISCELLANEOUS ITEMS

Other mason's tools and equipment include shovels, mortar hoes, wheelbarrows, chalk lines, plumb bobs, and a 200-foot ball of good-quality mason's line. Be sure to keep wheelbarrows and mortar tools clean; hardened mortar is difficult to remove. Clean all tools and equipment thoroughly at the end of each day or when the job is finished.

A mortar mixing machine (figure 8-3) is used for mixing large quantities of mortar. The mixer consists primarily of a metal drum containing mixing blades mounted on a chassis equipped with wheels for towing the machine from one job site to another. The mixer is powered by either an electric motor or a gasoline engine. After mixing, the mortar is discharged into a mortar box or wheelbarrow, usually by tilting the mixer drum. As with any machine, refer to the manufacturer's operator and maintenance manuals for proper operation. Be sure to follow all safety requirements related to mixer operations.

CONCRETE MASONRY

LEARNING OBJECTIVE: Upon completing this section, you should be able to identify the components and requirements of concrete masonry construction.

One of the most common masonry units is the concrete block. It consists of hardened cement and may be completely solid or contain single or multiple hollows. It is made from conventional cement mixes and various types of aggregate. These include sand, gravel, crushed stone, air-cooled slag, coal cinders, expanded shale or clay, expanded slag, volcanic cinders (pozzolan), pumice, and "scotia" (refuse obtained from metal ore reduction and smelting). The term "concrete block" was formerly limited to only hollow masonry units made with such aggregates as sand, gravel, and crushed stone. Today, the term covers all types of concrete block—both hollow and solid—made with any kind of aggregate. Concrete blocks are also available with applied glazed surfaces, various pierced designs, and a wide variety of surface textures.

Figure 8-3.—Mortar mixing machine.

Although concrete block is made in many sizes and shapes (figure 8-4) and in both modular and nonmodular dimensions, its most common unit size is 7 5/8 by 7 5/8 by 15 5/8 inches. This size is known as 8-by-8-by-16-inch block nominal size. All concrete block must meet certain specifications covering size, type, weight, moisture content, compressive strength, and other characteristics. Properly designed and constructed, concrete masonry walls satisfy many building requirements, including fire prevention, safety, durability, economy, appearance, utility, comfort, and acoustics.

Concrete blocks are used in all types of masonry construction. The following are just a few of many examples:

- Exterior load-bearing walls (both below and above grade);
- Interior load-bearing walls;

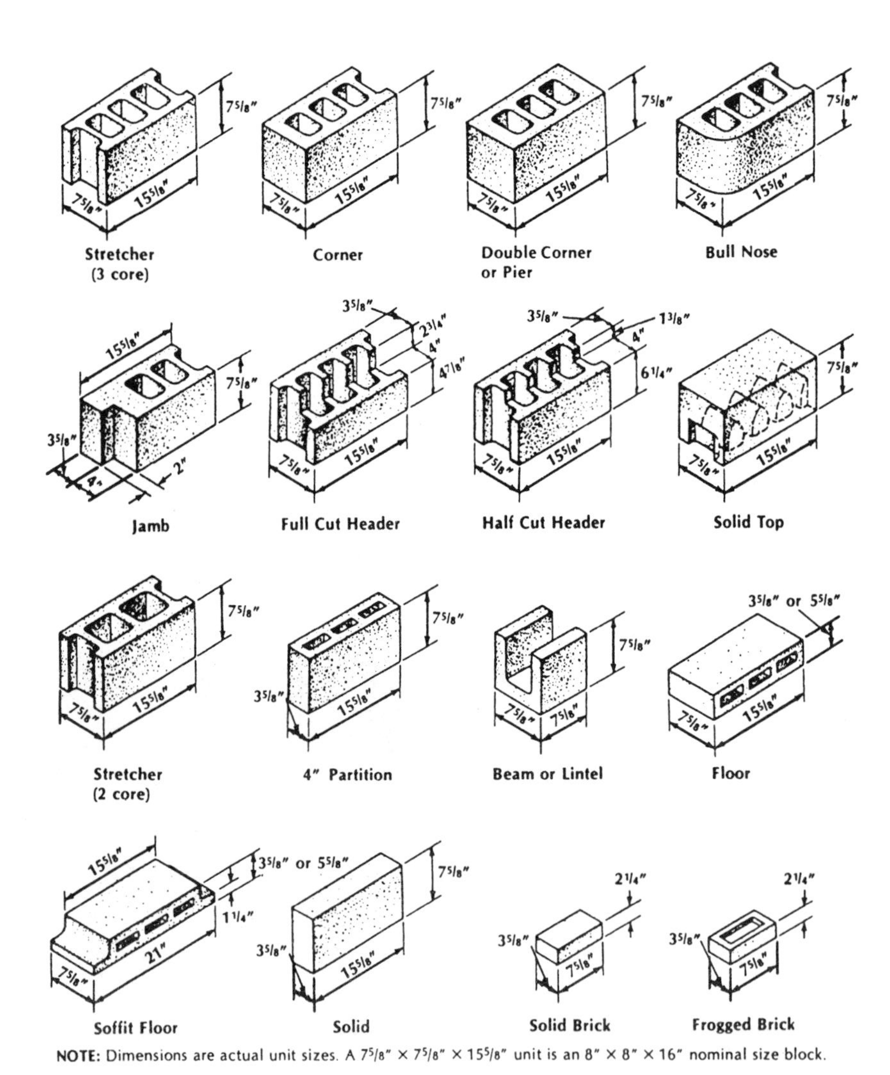

Figure 8-4.—Typical unit sizes and shapes of concrete masonry units.

- Fire walls and curtain walls;
- Partitions and panel walls;
- Backing for brick, stone, and other facings;
- Fireproofing over structural members;
- Fire safe walls around stairwells, elevators, and enclosures;
- Piers and columns;
- Retaining walls;
- Chimneys; and
- Concrete floor units.

There are five main types of concrete masonry units:

1. Hollow load-bearing concrete block;
2. Solid load-bearing concrete block;
3. Hollow nonload-bearing concrete block;
4. Concrete building tile; and
5. Concrete brick.

Load-bearing blocks are available in two grades: N and S. Grade N is for general use, such as exterior walls both above and below grade that may or may not be exposed to moisture penetration or weather. Both grades are also used for backup and interior walls. Grade S is for above-grade exterior walls with a weather-protective coating and for interior walls. The grades are further subdivided into two types. Type I consists of moisture-controlled units for use in arid climates. Type II consists of nonmoisture-controlled units.

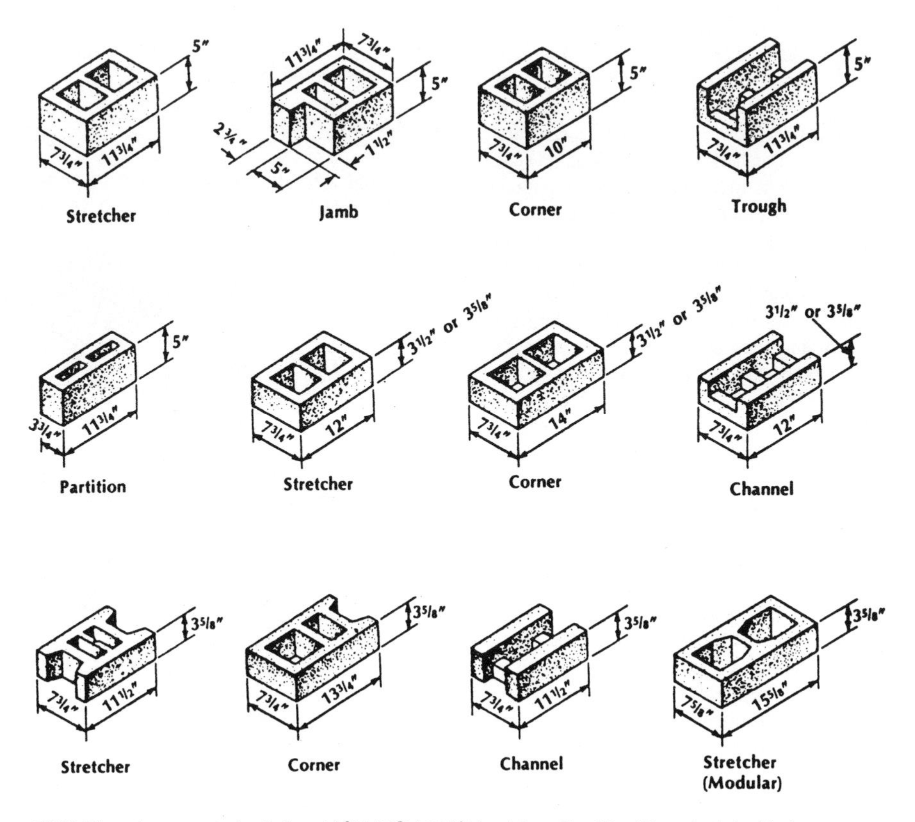

Figure 8-4.—Typical unit sizes and shapes of concrete masonry units—Continued.

BLOCK SIZES AND SHAPES

Concrete masonry units are available in many sizes and shapes to fit different construction needs. Both full- and half-length sizes are shown in figure 8-4. Because concrete block sizes usually refer to nominal dimensions, a unit actually measuring 7 5/8-by-7 5/8-by-15 5/8-inches is called an 8-by-8-by-16-inch block. When laid with 3/8-inch mortar joints, the unit should occupy a space exactly 8-by-8-by-16 inches.

ASTM (American Society for Testing and Materials) specifications define a solid concrete block as having a core area not more than 25 percent of the gross cross-sectional area. Most concrete bricks are solid and sometimes have a recessed surface like the frogged brick shown in figure 8-4. In contrast, a hollow concrete block has a core area greater than 25 percent of its gross cross-sectional area—generally 40 percent to 50 percent.

Blocks are considered heavyweight or lightweight, depending on the aggregate used in their production. A hollow load-bearing concrete block 8-by-8-by-16-inches nominal size weighs from 40 to 50 pounds when made with heavyweight aggregate, such as sand, gravel, crushed stone, or air-cooled slag. The same size block weighs only 25 to 35 pounds when made with coal cinders, expanded shale, clay, slag, volcanic cinders, or pumice. The choice of blocks depends on both the availability and requirements of the intended structure.

Blocks may be cut with a chisel. However, it is more convenient and accurate to use a power-driven masonry saw (figure 8-5). Be sure to follow the manufacturer's manual for operation and maintenance. As with all electrically powered equipment, follow all safety guidelines.

BLOCK MORTAR JOINTS

The sides and the recessed ends of a concrete block are called the shell. The material that forms the partitions between the cores is called the web. Each of the long sides of a block is called a face shell. Each of the recessed ends is called an end shell. The vertical ends of the face shells, on either side of the end shells, are called the edges.

Bed joints on first courses and bed joints in column construction are mortared by spreading a 1-inch layer of mortar. This procedure is referred to as "full mortar bedding." For most other bed joints, only the upper edges of the face shells need to be mortared. This is referred to as "face shell mortar bedding."

Head joints may be mortared by buttering both edges of the block being laid or by buttering one edge on the block being laid and the opposite edge on the block already in place.

MASONRY MORTAR

Properly mixed and applied mortar is necessary for good workmanship and good masonry service because it must bond the masonry units into a strong, well-knit structure. The mortar that bonds concrete block, brick, or clay tile will be the weakest part of the masonry unless you mix and apply it properly. When masonry leaks, it is usually through the joints. Both the strength of masonry and its resistance to rain penetration depend largely on the strength of the bond between the masonry unit and the mortar. Various factors affect bond strength, including the type and quantity of the mortar, its plasticity and workability, its water retentivity, the surface texture of the mortar bed, and the quality of workmanship in laying the units. You can correct irregular brick dimensions and shape with a good mortar joint.

Workability of Mortar

Mortar must be plastic enough to work with a trowel. You obtain good plasticity and workability by

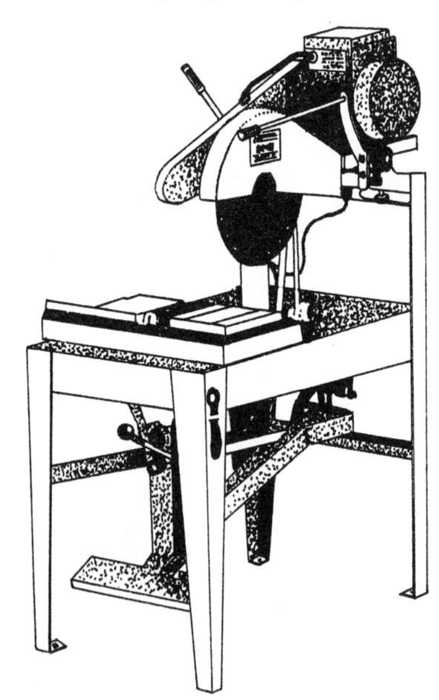

Figure 8-5.—Masonry saw.

using mortar having good water retentivity, using the proper grade of sand, and thorough mixing. You do not obtain good plasticity by using a lot of cementitious materials. Mortar properties depend largely upon the type of sand it contains. Clean, sharp sand produces excellent mortar, but too much sand causes mortar to segregate, drop off the trowel, and weather poorly.

Water Retentivity

Water retentivity is the mortar property that resists rapid loss of water to highly absorbent masonry units. Mortar must have water to develop the bond. If it does not contain enough water, the mortar will have poor plasticity and workability, and the bond will be weak and spotty. Sometimes, you must wet brick to control water absorption before applying mortar, but never wet concrete masonry units.

Mortar Strength and Durability

The type of service that the masonry must give determines the strength and durability requirements of mortar. For example, walls subject to severe stress or weathering must be laid with more durable, stronger mortar than walls for ordinary service. Table 8-1 gives mortar mix proportions that provide adequate mortar strength and durability for the conditions listed.

Types of Mortar

The following mortar types are proportioned on a volume basis:

- **Type M**—One part portland cement, one-fourth part hydrated lime or lime putty, and three parts sand; or, one part portland cement, one part type II masonry cement, and six parts sand. Type M mortar is suitable for general use, but is recommended specifically for below-grade masonry that contacts earth, such as foundations, retaining walls, and walks.

- **Type S**—One part portland cement, one-half part hydrated lime or lime putty, and four and one-half parts sand; or, one-half part portland cement, one part type II masonry cement, and four and one-half parts sand. Type S mortar is also suitable for general use, but is recommended where high resistance to lateral forces is required.

Table 8-1.—Recommended Mortar Mix Proportions by Unit Volume

TYPE OF SERVICE	CEMENT	HYDRATED LIME	MORTAR SAND IN DAMP, LOOSE CONDITION
ORDINARY	1 unit masonry cement[1]		2 1/4 to 3 units
	or		
	1 unit portland cement	1/2 to 1 1/4 units	4 1/2 to 6 units
ISOLATED PIERS SUBJECT TO EXTREMELY HEAVY LOADS, VIOLENT WINDS, EARTHQUAKES, OR SEVERE FROST ACTION	1 unit masonry cement[1] plus 1 unit portland cement		4 1/2 to 6 units
	or		
	1 unit portland cement	0 to 1/4 unit	2 1/4 to 3 units

[1]ASTM Specification C91 Type II

- **Type N**—One part portland cement, one part hydrated lime or lime putty, and six parts sand; or, one part type II masonry cement and three parts sand. Type N mortar is suitable for general use in above-grade exposed masonry where high compressive or lateral strength is not required.

- **Type O**—One part portland cement, two parts hydrated lime or lime putty, and nine parts sand; or, one part type I or type II masonry cement and three parts sand. Type O mortar is recommended for load-bearing, solid-unit walls when the compressive stresses do not exceed 100 pounds per square inch (psi) and the masonry is not subject to freezing and thawing in the presence of a lot of moisture.

MIXING MORTAR

The manner in which mortar is mixed has a lot to do with the quality of the final product. In addition to machine and hand mixing, you need to know the requirements for introducing various additives, including water, to the mix in order to achieve optimum results.

Machine Mixing

Machine mixing refers to mixing large quantities of mortar in a drum-type mixer. Place all dry ingredients in the mixer first and mix them for 1 minute before adding the water. When adding water, you should always add it slowly. Minimum mixing time is 3 minutes. The mortar should be mixed until a completely uniform mixture is obtained.

Hand Mixing

Hand mixing involves mixing small amounts of mortar by hand in a mortar box or wheelbarrow. Take care to mix all ingredients thoroughly to obtain a uniform mixture. As in machine mixing, mix all dry materials together first before adding water. Keep a steel drum of water close at hand to use as the water supply. You should also keep all your masonry tools free of hardened mortar mix and dirt by immersing them in water when not in use.

Requirements

You occasionally need to mix lime putty with mortar. When machine mixing, use a pail to measure the lime putty. Place the putty on top of the sand. When hand mixing, add the sand to the lime putty. Wet pails before filling them with mortar and clean them immediately after emptying.

Mixing water for mortar must meet the same quality requirements as mixing water for concrete. Do not use water containing large amounts of dissolved salts. Salts weaken the mortars.

You can restore the workability of any mortar that stiffens on the mortar board due to evaporation by remixing it thoroughly. Add water as necessary, but discard any mortar stiffened by initial setting. Because it is difficult to determine the cause of stiffening, a practical guide is to use mortar within 2 1/2 hours after the original mixing. Discard any mortar you do not use within this time.

Do not use an antifreeze admixture to lower the freezing point of mortars during winter construction. The quantity necessary to lower the freezing point to any appreciable degree is so large it will seriously impair the strength and other desirable properties of the mortar.

Do not add more than 2-percent calcium chloride (an accelerator) by weight of cement to mortar to accelerate its hardening rate and increase its early strength. Do not add more than 1-percent calcium chloride to masonry cements. Make a trial mix to find the percentage of calcium chloride that gives the desired hardening rate. Calcium chloride should not be used for steel-reinforced masonry. You can also obtain high early strength in mortars with high-early-strength portland cement.

MODULAR PLANNING

Concrete masonry walls should be laid out to make maximum use of full- and half-length units. This minimizes cutting and fitting of units on the job. Length and height of walls, width and height of openings, and wall areas between doors, windows, and corners should be planned to use full-size and half-size units, which are usually available (figure 8-6). This procedure assumes that window and door frames are of modular dimensions which fit modular full- and half-size units. Then, all horizontal dimensions should be in multiples of nominal full-length masonry units.

Both horizontal and vertical dimensions should be designed to be in multiples of 8 inches. Table 8-2 lists nominal length of concrete masonry walls by stretchers. Table 8-3 lists nominal height of concrete masonry walls by courses. When 8-by-4-by-16 units are used, the horizontal dimensions should be planned in multiples of 8 inches (half-length units) and the vertical dimensions in multiples of 4 inches. If the thickness of the wall is greater or less than the length of a half unit, a special-length unit is required at each corner in each course. Table 8-4 lists the average number of concrete masonry units by size and approximate number of cubic feet of mortar required for every 100 square feet of concrete masonry wall. Table 8-5 lists the number of 16-inch blocks per course for any wall.

You should always use **outside** measurements when calculating the number of blocks required per course. For example, a basement 22 feet by 32 feet should require 79 blocks for one complete course. Multiply 79 by the number of courses needed. Thus, a one-course basement requires a total of 790 blocks for a solid wall, from which deductions should be made for windows and doors. If any dimension is an odd number, use the nearest smaller size listed in the table. For example, for a 22-foot by 31-foot enclosure, use 22 feet by 30 feet and add one-half block per row.

As a Builder, you might find yourself in the field without the tables handy, so here is another method. Use 3/4 times the length and 3/2 times the height for figuring how many 8-by-8-by-16-inch blocks you need for a wall. Let's take an example:

Given: A wall 20 ft long × 8 ft high

$$\frac{3}{4} \times 20 = 60 \div 4 = 15 \text{ (8}'' \times 8'' \times 16'' \text{ block per course)}$$

$$\frac{3}{2} \times 8 = 24 \div 2 = 12 \text{ courses high}$$

$$15 \times 12 = 180 \text{ total blocks}$$

ESTIMATING MORTAR

You can use "rule 38" for calculating the raw material needed to mix 1 yard of mortar without a great deal of paperwork. This rule does not, however, accurately calculate the required raw materials for large masonry construction jobs. For larger jobs, use the absolute volume or weight formula. In most cases, though, and particularly in advanced base construction, you can use rule 38 to quickly estimate the quantities of the required raw materials.

Builders have found that it takes about 38 cubic feet of raw materials to make 1 cubic yard of mortar. In using rule 38 for calculating mortar, take the rule number and divide it by the sum of the quantity figures specified in the mix. For example, let's assume that the building specifications call for a 1:3 mix for mortar, *1 + 3 = 4*. Since *38 4 = 9 1/2*, you'll need 9 1/2 sacks, or 9 1/2 cubic feet, of cement. To calculate the amount of fine aggregate (sand), you multiply 9 1/2 by 3. The product (28 1/2 cubic feet) is the amount of sand you need to mix 1 cubic yard of mortar using a 1:3 mix. The sum of the two required quantities should always equal 38. This is how you can check whether you are using the correct amounts. In the above example, 9 1/2 sacks of cement plus 28 1/2 cubic feet of sand equal 38.

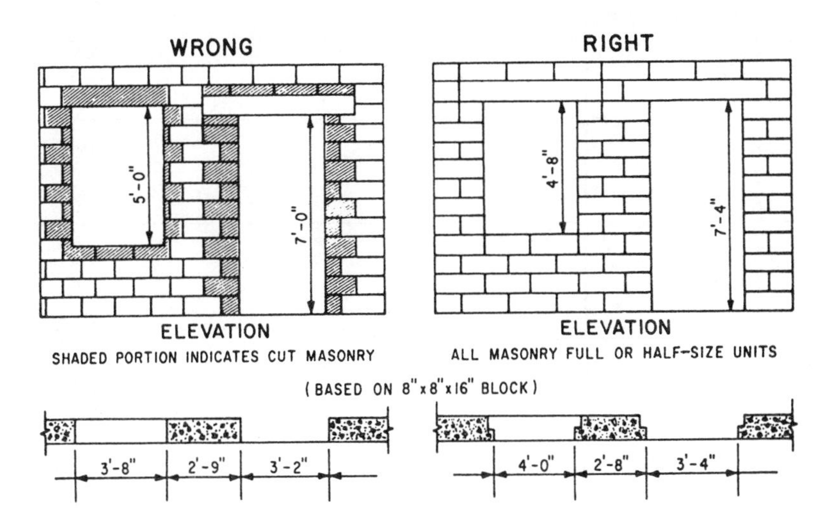

Figure 8-6.—Planning concrete masonry wall openings.

Table 8-2.—Nominal Lengths of Concrete Masonry Walls in Stretchers

NUMBER OF STRETCHERS	NOMINAL LENGTH OF CONCRETE MASONRY WALLS	
	Units 15 5/8" long and half units 7 5/8" long with 3/8" thick head joints	Units 11 5/8" long and half units 5 5/8" long with 3/8" thick head joints
1	1′4"	1′0"
1 1/2	2′0"	1′6"
2	2′	2′0"
2 1/2	3′4"	2′6"
3	4′0"	3′0"
3 1/2	4′8"	3′6"
4	5′4"	4′0"
4 1/2	6′0"	4′6"
5	6′8"	5′0"
5 1/2	7′4"	5′6"
6	8′0"	6′0"
6 1/2	8′8"	6′6"
7	9′4"	7′0"
7 1/2	10′0"	7′6"
8	10′8"	8′0"
8 1/2	11′4"	8′6"
9	12′0"	9′0"
9 1/2	12′8"	9′6"
10	13′4"	10′0"
10 1/2	14′0"	10′6"
11	14′8"	11′0"
11 1/2	15′4"	11′6"
12	16′0"	12′0"
12 1/2	16′8"	12′6"
13	17′4"	13′0"
13 1/2	18′0"	13′6"
14	19′4"	14′0"
14 1/2	20′0"	14′6"
15	20′0"	15′0"
20	26′8"	20′0"

NOTE: Actual wall length is measured from outside edge to outside edge of units, and equals the nominal length minus 3/8" (one mortar joint).

Table 8-3.—Nominal Heights of Modular Concrete Masonry Walls in Courses

NUMBER OF COURSES	NOMINAL HEIGHT OF CONCRETE MASONRY WALLS	
	Units 7 5/8" high and 3/8" thick bed joints	Units 3 5/8" high and 3/8" thick bed joints
1	0′8"	0′4"
2	1′4"	0′8"
3	2′0"	1′0"
4	2′8"	1′4"
5	3′4"	1′8"
6	4′0"	2′0"
7	4′8"	2′4"
8	5′4"	2′8"
9	6′0"	3′0"
10	6′8"	3′4"
15	10′0"	5′0"
20	13′4"	6′8"
25	16′8"	8′4"
30	20′0"	10′0"
35	23′4"	11′8"
40	26′8"	13′4"
45	30′0"	15′0"
50	33′4"	16′8"

NOTE: For concrete masonry units 7 5/8" and 3 5/8" in height laid with 3/8" mortar joints. Height is measured from center to center of mortar joints.

Table 8-4.—Average Concrete Masonry Units and Mortar per 100 sq. ft. of Wall

DESCRIPTION, SIZE OF BLOCK (IN.)	THICKNESS WALL (IN.)	NUMBER OF UNITS PER 100 SQ. FT. OF WALL AREA	MORTAR (CU. FT.)
8 × 8 × 16	8	110	3.25
8 × 8 × 12	8	146	3.5
8 × 12 × 16	12	110	3.25
8 × 3 × 16	3	110	2.75
9 × 3 × 18	3	87	2.5
12 × 3 × 12	3	100	2.5
8 × 3 × 12	3	146	3.5
8 × 4 × 16	4	110	3.25
9 × 4 × 18	4	87	3.25
12 × 4 × 12	4	100	3.25
8 × 4 × 12	4	146	4
8 × 6 × 16	6	110	3.25

Table 8-5.—Number of 16-Inch Blocks per Course

LENGTH IN FEET	WIDTH IN FEET																	
	6	8	10	12	14	16	18	20	22	24	26	28	30	32	34	36	38	40
8	19	22	25	28	31	34	37	40	43	46	49	52	55	58	61	64	67	70
10	22	25	28	31	34	37	40	43	46	49	52	55	58	61	64	67	70	73
12	25	28	31	34	37	40	43	46	49	52	55	58	61	64	67	70	73	76
14	28	31	34	37	40	43	46	49	52	55	58	61	64	67	70	73	76	79
16	31	34	37	40	43	46	49	52	55	58	61	64	67	70	73	76	79	82
18	34	37	40	43	46	49	52	55	58	61	64	67	70	73	76	79	82	85
20	37	40	43	46	49	52	55	58	61	64	67	70	73	76	79	82	85	88
22	40	43	46	49	52	55	58	61	64	67	70	73	76	79	82	85	88	91
24	43	46	49	52	55	58	61	64	67	70	73	76	79	82	85	88	91	94
26	46	49	52	55	58	61	64	67	70	73	76	79	82	85	88	91	94	97
28	49	52	55	58	61	64	67	70	73	76	79	82	85	88	91	94	97	100
30	52	55	58	61	64	67	70	73	76	79	82	85	88	91	94	97	100	103
32	55	58	61	64	67	70	73	76	79	82	85	88	91	94	97	100	103	106
34	58	61	64	67	70	73	76	79	82	85	88	91	94	97	100	103	106	109
36	61	64	67	70	73	76	79	82	85	88	91	94	97	100	103	106	109	112
38	64	67	70	73	76	79	82	85	88	91	94	97	100	103	106	109	112	115
40	67	70	73	76	79	82	85	88	91	94	97	100	103	106	109	112	115	118
42	70	73	76	79	82	85	88	91	94	97	100	103	106	109	112	115	118	121
44	73	76	79	82	85	88	91	94	97	100	103	106	109	112	115	118	121	124
46	76	79	82	85	88	91	94	97	100	103	106	109	112	115	118	121	124	127
48	79	82	85	88	91	94	97	100	103	106	109	112	115	118	121	124	127	130
50	82	85	88	91	94	97	100	103	106	109	112	115	118	121	124	127	130	133
52	85	88	91	94	97	100	103	106	109	112	115	118	121	124	127	130	133	136
54	88	91	94	97	100	103	106	109	112	115	118	121	124	127	130	133	136	139
56	91	94	97	100	103	106	109	112	115	118	121	124	127	130	133	136	139	142
58	94	97	100	103	106	109	112	115	118	121	124	127	130	133	136	139	142	145
60	97	100	103	106	109	112	115	118	121	124	127	130	133	136	139	142	145	148

SAFE HANDLING OF MATERIAL

When you handle cement or lime bags, wear goggles and snug-fitting neckbands and wristbands. Always practice good personal cleanliness and never wear clothing that has become stiff with cement. Cement-impregnated clothing irritates the skin and may cause serious infection. Any susceptibility of the skin to cement and lime burns should be reported. Personnel who are allergic to cement or lime should be transferred to other jobs.

Bags of cement or lime should not be piled more than 10 bags high on a pallet. The only exception is when storage is in bins or enclosures built for such storage. The bags around the outside of the pallet should be placed with the mouths of the bags facing the center. The first five tiers of bags each way from any corner must be cross piled. A setback starting with the sixth tier should be made to prevent piled bags from falling outward. If you have to pile bags above 10 tiers, another setback must be made. The back tier, when not resting against an interior wall of sufficient strength to withstand the pressure, should be set back one bag every five tiers, the same as the end tiers. During unpiling, the entire top of the pile should be kept level and the necessary setbacks maintained.

Lime and cement must be stored in a dry place. This helps prevent lime from crumbling and the cement from hydrating before it is used.

CONCRETE MASONRY CONSTRUCTION

LEARNING OBJECTIVE: Upon completing this section, you should be able to explain the elements of concrete masonry.

Good workmanship is a very important factor in building masonry walls. You should make every effort to lay each masonry unit plumb and true. In the following paragraphs, we will discuss the basic steps in laying up masonry walls.

STEPS IN CONSTRUCTION

The first step in building a concrete masonry wall is to locate the corners of the structure. In locating the corners, you should also make sure the footing or slab formation is level so that each Builder starts each section wall on a common plane. This also helps ensure that the bed joints are straight when the sections are connected. If the foundation is badly out of level, the entire first course should be laid before Builders begin working on other courses. If this is not possible, a level plane should be established with a transit or engineer's level.

The second step is to chase out bond, or lay out, by placing the first course of blocks without mortar (figure 8-7, view 1). Snap a chalk line to mark the

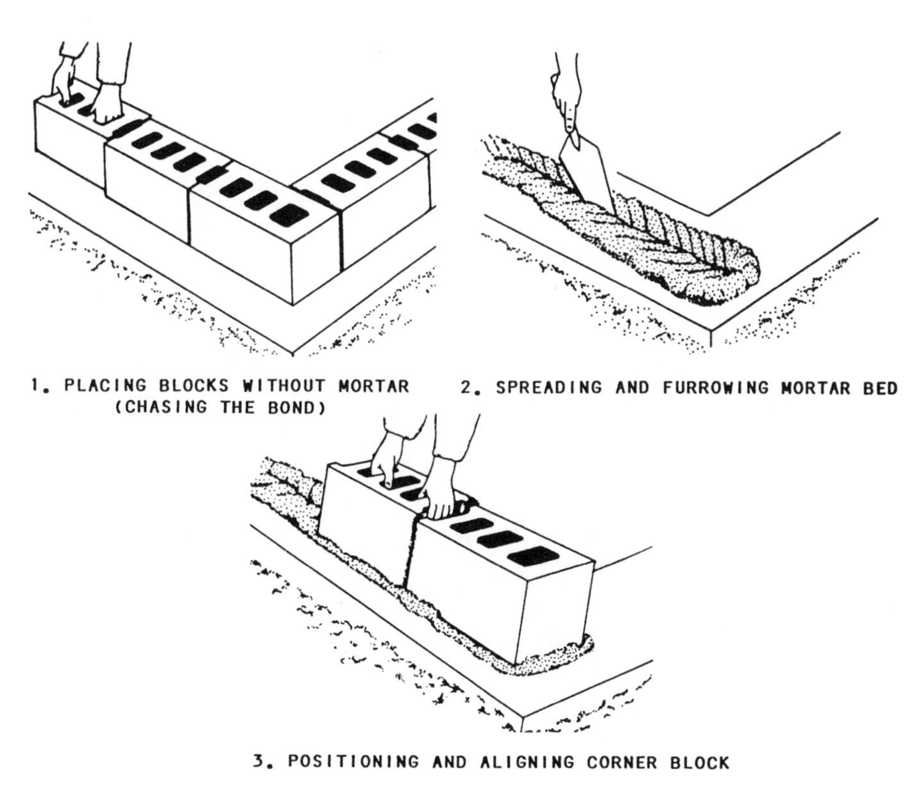

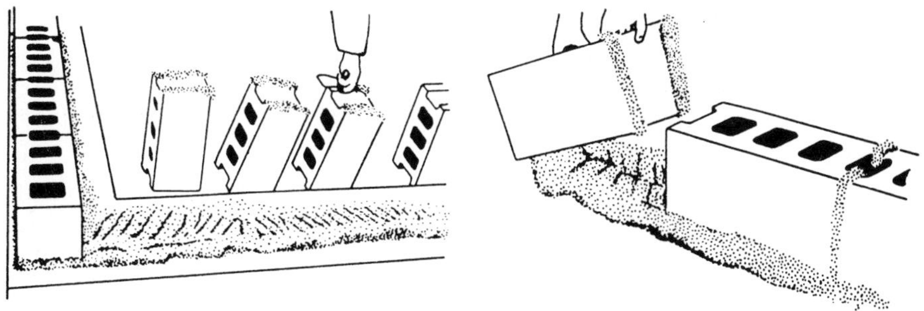

Figure 8-7.—Laying first course of blocks for a wall.

1. Leveling Block

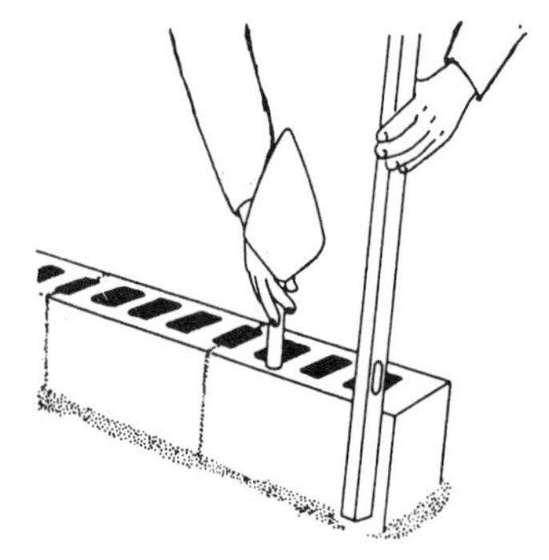

2. Plumbing Block

Figure 8-8.—Leveling and plumbing first course of blocks for a wall.

footing and align the blocks accurately. Then, use a piece of material 3/8 inch thick to properly space the blocks. This helps you get an accurate measurement.

The third step is to replace the loose blocks with a full mortar bed, spreading and furrowing it with a

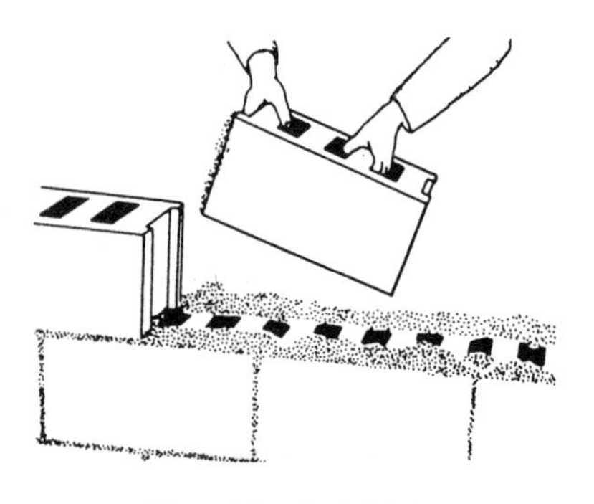

Figure 8-9.—Vertical joints.

1. Aligning

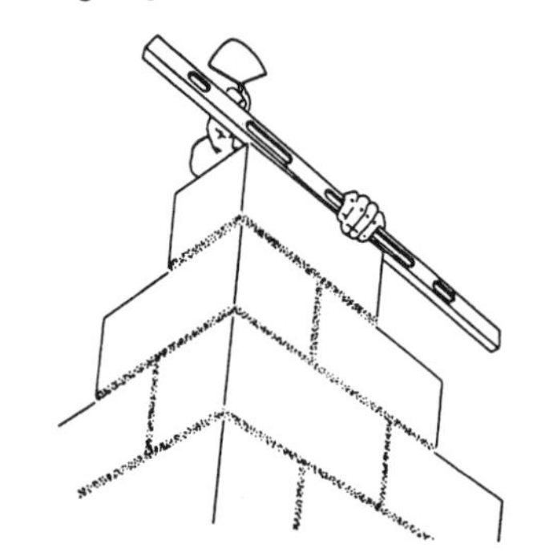

2. Leveling

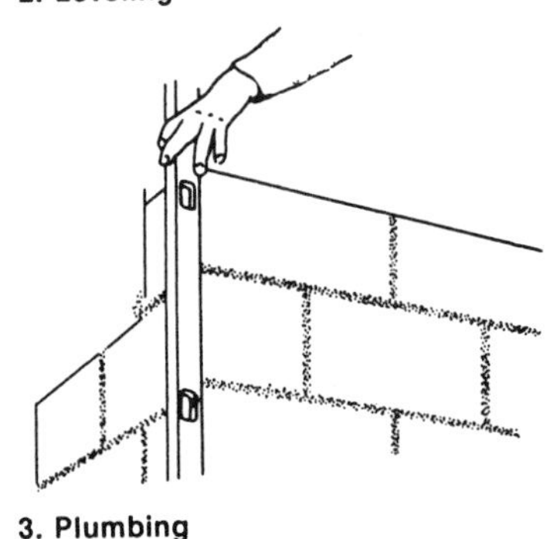

3. Plumbing

Figure 8-10.—Checking each course at the corner.

trowel to ensure plenty of mortar under the bottom edges of the first course (figure 8-7, view 2). Carefully position and align the corner block first (view 3 of figure 8-7). Lay the remaining first-course blocks with the thicker end up to provide a larger mortar-bedding area. For the vertical joints, apply mortar only to the block ends by placing several blocks on end and buttering them all in one operation (view 4). Make the joints 3/8 inch thick. Then, place each block in its final position, and push the block down vertically into the mortar bed and against the previously laid block. This ensures a well-filled vertical mortar joint (view 5). After laying three or four blocks, use a mason's level as a straightedge to check correct block alignment (figure 8-8, view 1). Then, use the level to bring the blocks to proper grade and plumb by tapping with a trowel handle as shown in view 2. Always lay out the first course of concrete masonry carefully and make sure that you properly align, level, and plumb it. This assures that succeeding courses and the final wall are both straight and true.

The fourth step is to build up the corners of the wall, usually four or five courses high. This is also called laying up a lead. Step back each course one-half block. For the horizontal joints, apply mortar only to the tops of the blocks already laid. For the vertical joints, you can apply mortar either to the ends of the new block or the end of the block previously laid, or both, to ensure well-filled joints (figure 8-9). As you lay each course at the corner, check the course with a level for alignment (figure 8-10, view 1), for level (view 2), and for plumb (view 3). Carefully check each block with a level or straightedge to make sure that all the block faces are in the same plane. This ensures true, straight walls. A story or course pole, which is a board with markings 8 inches apart (figure 8-11), helps accurately place each masonry course. Also check the horizontal block spacing by placing a level diagonally across the corners of the blocks (figure 8-12).

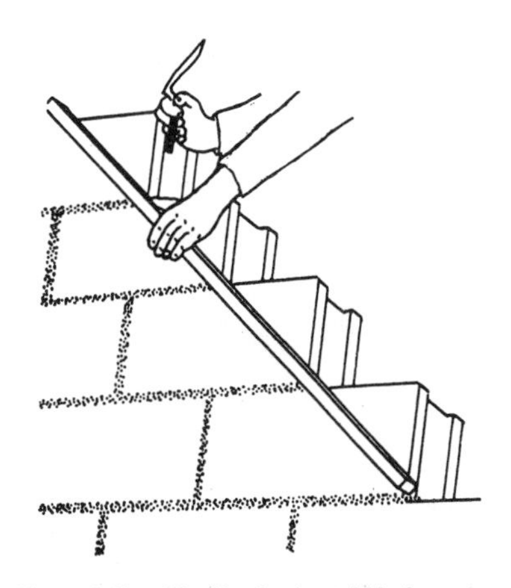

Figure 8-12.—Checking horizontal block spacing.

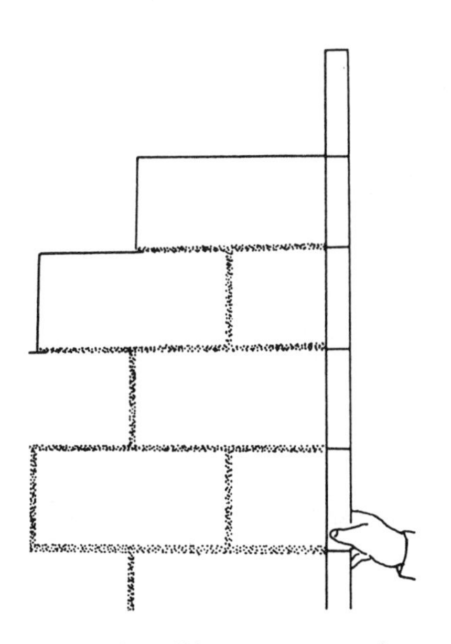

Figure 8-11.—Using a story or course pole.

When filling in the wall between the corners, first stretch a mason's line along the extensor block edges from corner to corner for each course. Then lay the top outside edge of each new block to this line (figure 8-13). How you grip a block before laying is

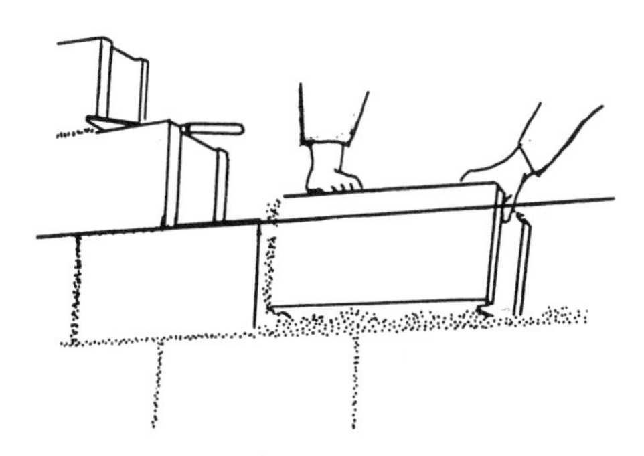

Figure 8-13.—Filling in the wall between corners.

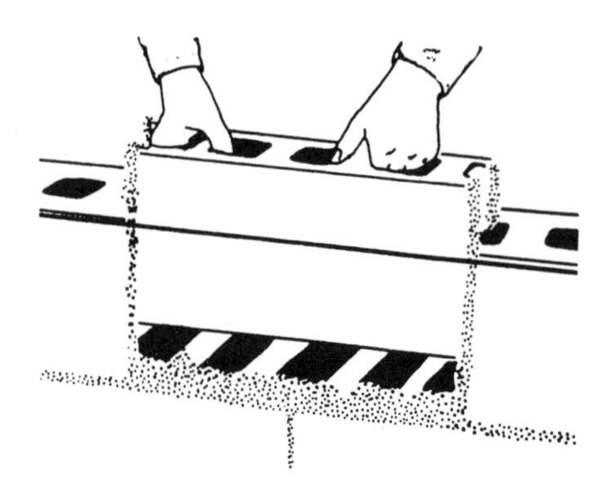

Figure 8-14.—Installing a closure block.

important. First, tip it slightly toward you so that you can see the edge of the course below. Then place the lower edge of the new block directly on the edges of the block below (figure 8-13). Make all position adjustments while the mortar is soft and plastic. Any adjustments you make after the mortar stiffens will break the mortar bond and allow water to penetrate. Level each block and align it to the mason's line by tapping it lightly with a trowel handle.

Fifth and last, before installing the closure block, butter both edges of the opening and all four vertical edges of the closure block with mortar. Then, lower the closure block carefully into place (figure 8-14). If any mortar falls out, leaving an open joint, remove the block and repeat the procedure.

To assure a good bond, do not spread mortar too far ahead when actually laying blocks. If you do, the mortar will stiffen and lose its plasticity. The recommended width of mortar joints for concrete masonry units is 3/8 inch. When properly made, these joints produce a weathertight, neat, and durable concrete masonry wall. As you lay each block, cut off excess mortar from the joints using a trowel (figure 8-15) and throw it back on the mortar board to rework into the fresh mortar. Do not, however, rework any mortar dropped on the scaffold or floor.

Weathertight joints and the neat appearance of concrete masonry walls depend on proper striking (tooling). After laying a section of the wall, tool the mortar joint when the mortar becomes "thumb print" hard. Tooling compacts the mortar and forces it tightly against the masonry on each side of the joint. Use either concave or V-shaped tooling on all joints (figure 8-16). Tool horizontal joints (figure 8-17, view 1) with a long jointer first, followed by tooling the vertical joints (view 2). Trim off mortar burrs from the tooling flush with the wall face using a trowel, soft bristle brush, or by rubbing with a burlap bag.

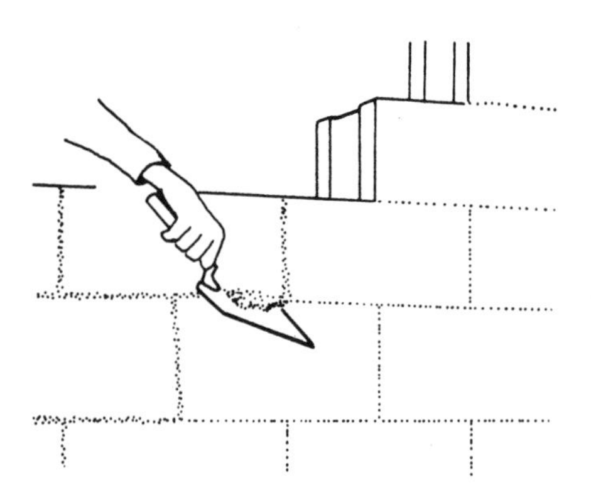

Figure 8-15.—Cutting off excess mortar from the joints.

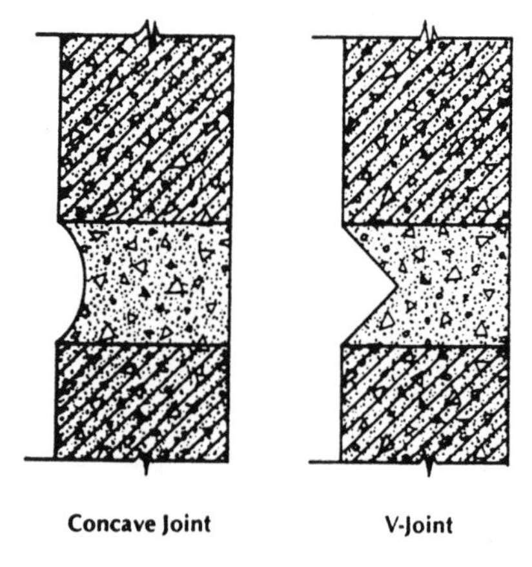

Figure 8-16.—Tooled mortar joints for weathertight exterior walls.

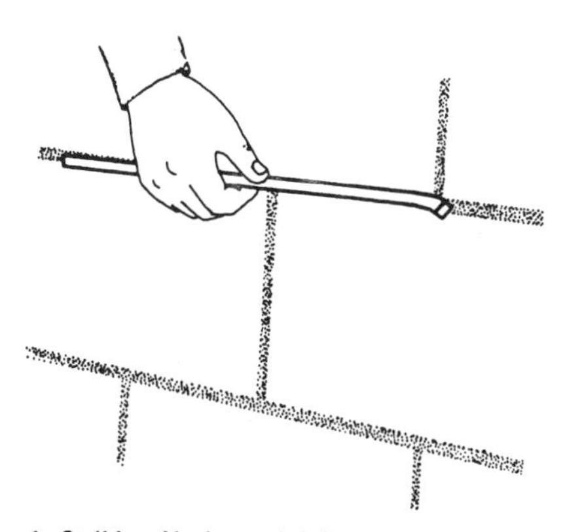

1. Striking Horizontal Joints

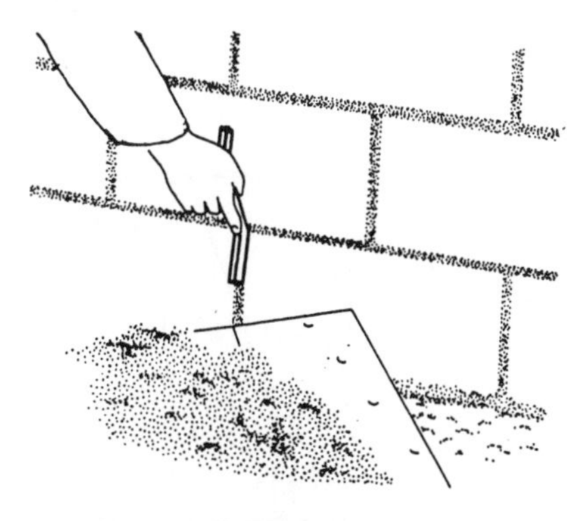

2. Striking Vertical Joints

Figure 8-17.—Tooling mortar joints.

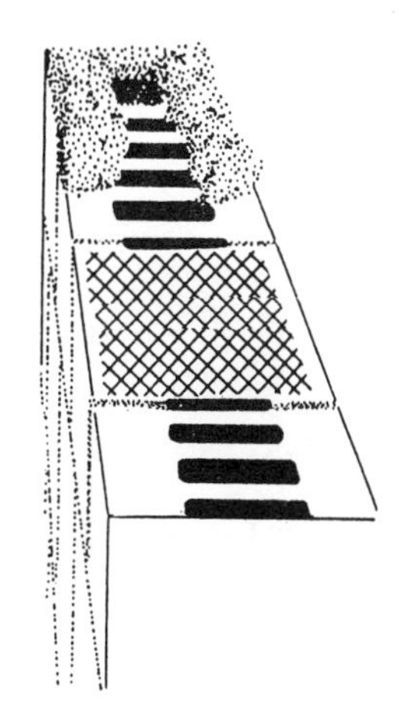

1. Placing Metal Lath under Cores

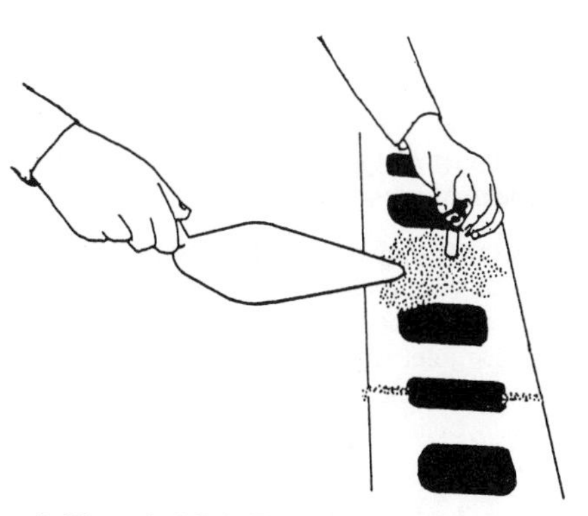

2. Threaded Bolt End Extends above Wall Top

Figure 8-18.—Installing anchor bolts for wood plates.

A procedure known as pointing may be required after jointing. Pointing is the process of inserting mortar into horizontal and vertical joints after the unit has been laid. Basically, pointing is done to restore or replace deteriorated surface mortar in old work. Pointing of this nature is called tuck pointing. However, even in freshly laid masonry, pointing may be necessary for filling holes or correcting defective joints.

You must prepare in advance for installing wood plates with anchor bolts on top of hollow concrete masonry walls. To do this, place pieces of metal lath in the second horizontal mortar joint from the top of the wall under the cores that will contain the bolts (figure 8-18, view 1). Use anchor bolts 1/2 inch in diameter and 18 inches long. Space them not more than 4 feet apart. Then, when you complete the top course, insert the bolts into the cores of the top two courses and fill the cores with concrete or mortar. The metal lath underneath holds the concrete or mortar filling in place. The threaded end of the bolt should extend above the top of the wall (view 2).

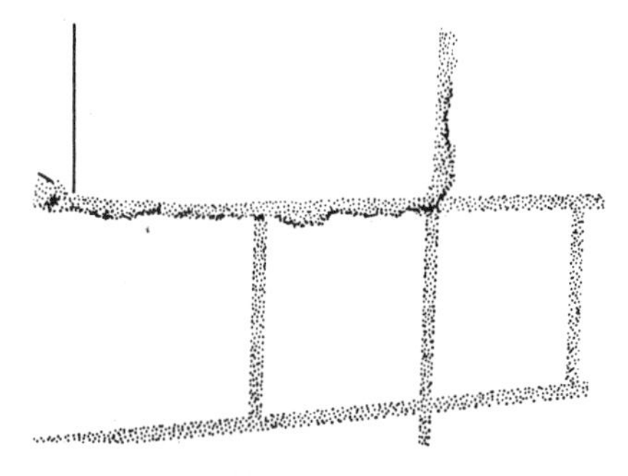

Figure 8-19.—Control joints.

CONTROL JOINTS

Control joints (figure 8-19) are continuous vertical joints that permit a masonry wall to move slightly under unusual stress without cracking. There are a number of types of control joints built into a concrete masonry wall.

The most preferred control joint is the Michigan type made with roofing felt. A strip of felt is curled into the end core, covering the end of the block on one side of the joint (figure 8-20, view 1). As the other side of the joint is laid, the core is filled with mortar. The filling bonds to one block, but the paper prevents bond to the block on the other side of the control joint.

View 2 of figure 8-20 shows the tongue-and-groove type of control joint. The special units are manufactured in sets consisting of full and half blocks. The tongue of one unit fits into the groove of another unit or into the open end of a regular flanged stretcher. The units are laid in mortar exactly the same as any other masonry units, including mortar in the head joint. Part of the mortar is allowed to remain in the vertical joint to form a backing against which the caulking can be packed.

View 3 shows a control joint that may be built with regular full- and half-length stretcher blocks with a Z-shaped bar across the joint or a 10- or 12-inch pencil rod (1/4-inch smooth bar) across each face shell. If a pencil rod is used, it must be greased on one side of the joint to prevent bond. These rods should be placed every other course. Lay up control joints in mortar just as any other joint. However, if

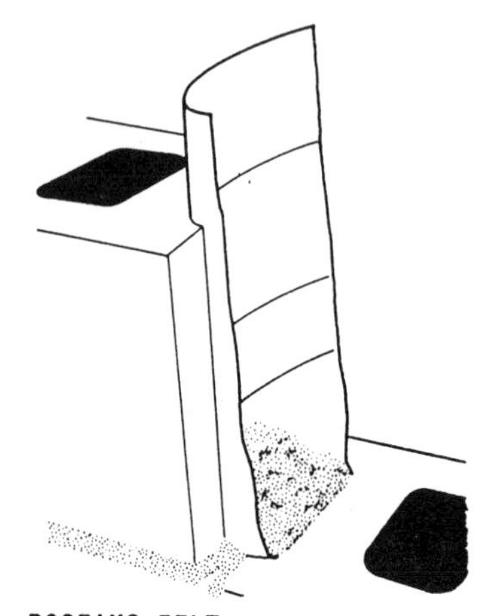

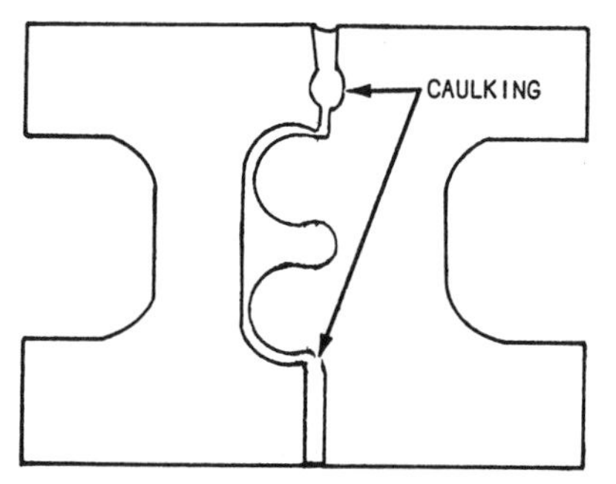

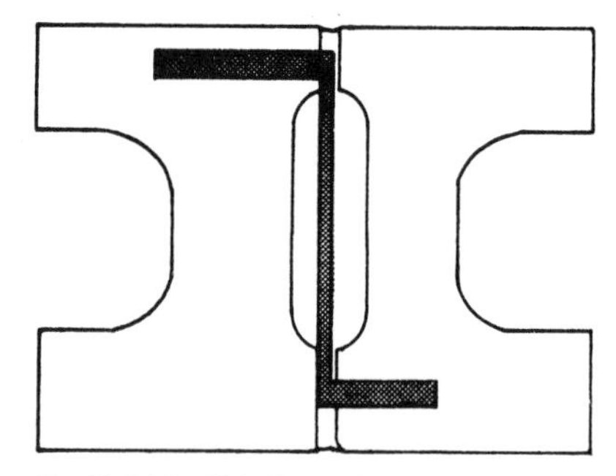

Figure 8-20.—Making control joints.

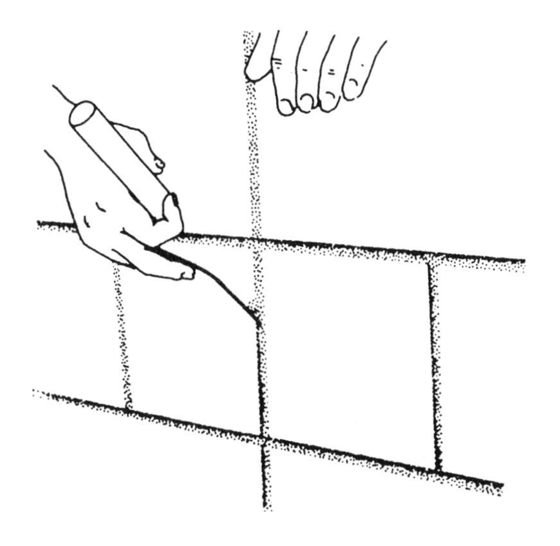

Raking Joint to Prepare for Caulking

Figure 8-21.—Making a control joint.

they are exposed to either the weather or to view, caulk them as well. After the mortar is stiff, rake it out to a depth of about 3/4-inch to make a recess for the caulking compound. Use a thin, flat caulking trowel to force the compound into the joint (figure 8-21).

The location of control joints is established by the architectural engineer and should be noted in the plans and specifications.

WALLS

Walls are differentiated into two types: load bearing and nonload bearing. Load-bearing walls not only separate spaces, but also provide structural support for whatever is above them. Nonload bearing walls function solely as partitions between spaces.

Load-bearing Walls

Do not join intersecting concrete block load-bearing walls with a masonry bond, except at the corners. Instead, terminate one wall at the face of the second wall with a control joint. Then, tie the intersecting walls together with Z-shaped metal tie bars 1/4-by-1/4-by-28 inches in size, having 2-inch right-angle bends on each end (figure 8-22, view 1).

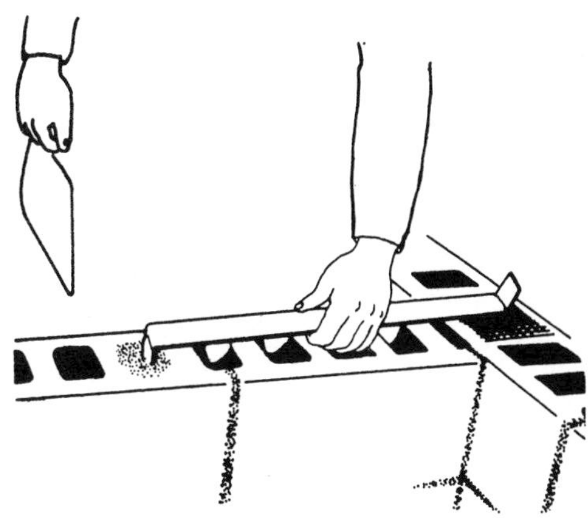

1. Z-Shaped Tie Bar Has Right Angle Bends at Each End

2. Filling Core With Mortar or Concrete

Figure 8-22.—Tying intersecting bearing walls.

Space the tie bars no more than 4 feet apart vertically and place pieces of metal lath under the block cores that will contain the tie bars ends (figure 8-18, view 1). Embed the right-angle bends in the cores by filling them with mortar or concrete (figure 8-22, view 2).

Nonload-bearing Walls

To join intersecting nonload-bearing block walls, terminate one wall at the face of the second with a control joint. Then, place strips of metal lath of

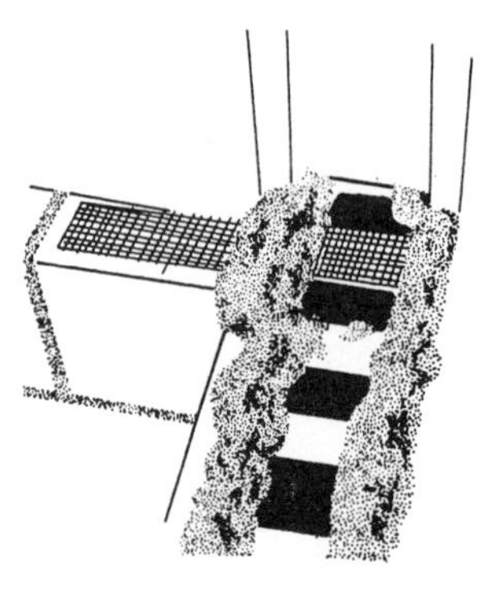

1. Metal Lath Spans the Joint between the Walls

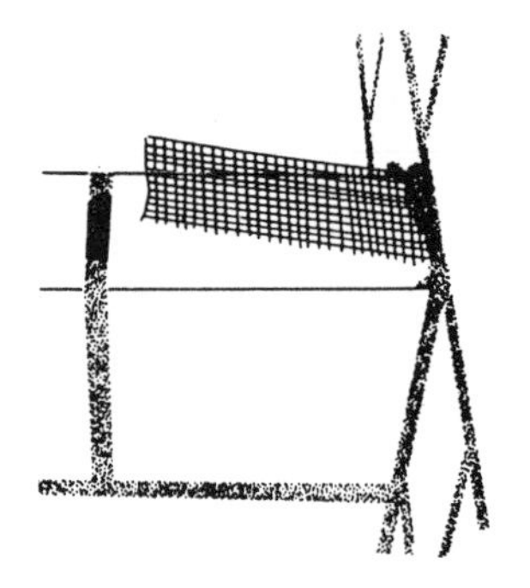

2. Set Lath in the Mortar Joint as You Construct the Second Wall

Figure 8-23.—Tying intersecting nonbearing walls.

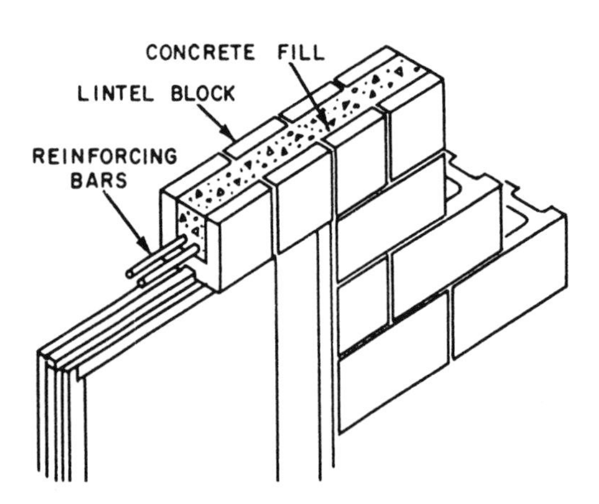

Figure 8-24.—Lintel made from blocks.

1/4-inch mesh galvanized hardware cloth across the joint between the two walls (figure 8-23, view 1) in alternate courses. Insert one-half of the metal stops into one wall as you build it, and then tie the other halves into the mortar joints as you lay the second wall (view 2).

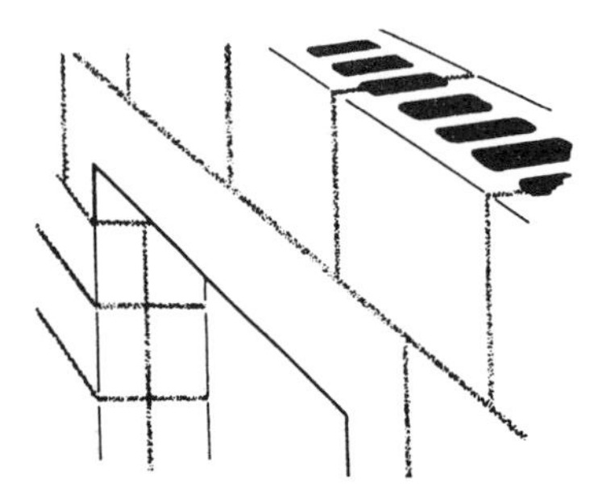

1. Precast Concrete Lintel

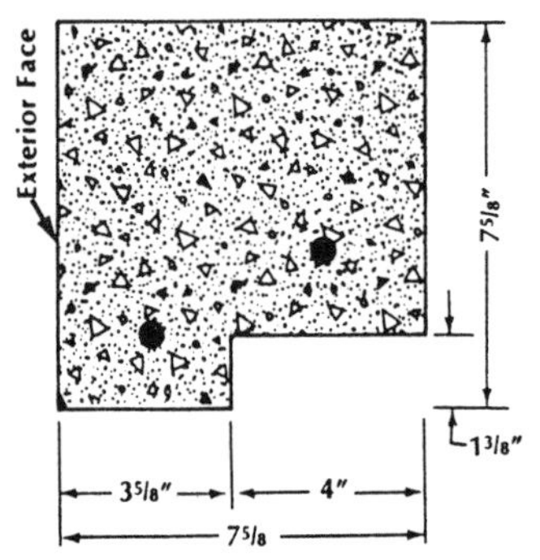

2. Precast Concrete Offset on Lintel Underside

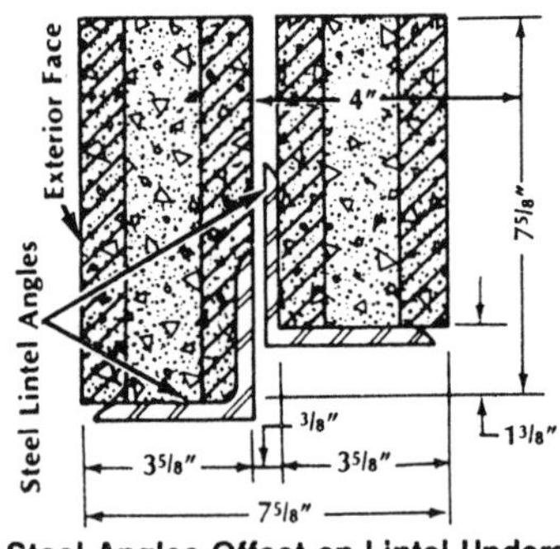

3. Steel Angles Offset on Lintel Underside

Figure 8-25.—Installing precast concrete lintels without and with steel angles.

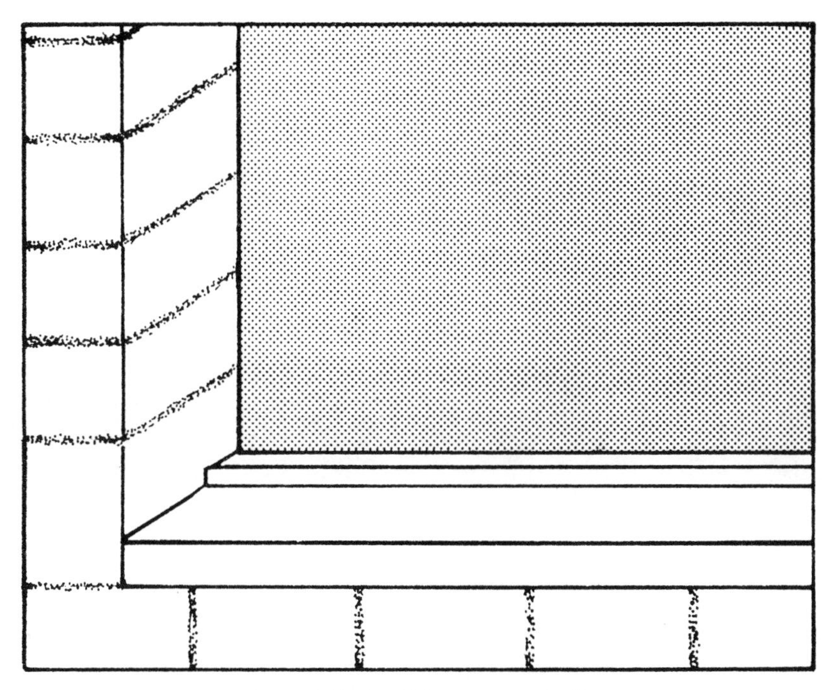

Figure 8-26.—Installed precast concrete sills.

BOND BEAMS, LINTELS, AND SILLS

Bond beams are reinforced courses of block that bond and integrate a concrete masonry wall into a stronger unit. They increase the bending strength of the wall and are particularly needed to resist the high winds of hurricanes and earthquake forces. In addition, they exert restraint against wall movement, reducing the formation of cracks.

Bond beams are constructed with special-shape masonry units (beam and lintel block) filled with concrete or grout and reinforced with embedded steel bars. These beams are usually located at the top of walls to stiffen them. Since bond beams have appreciable structural strength, they can be located to serve as lintels over doors and windows. Figure 8-24 shows the use of lintel blocks to place a lintel over a metal door, using the door case for support. Lintels should have a minimum bearing of 6 inches at each end. A rule of thumb is to provide 1 inch of bearing for every foot of clear space. When bond beams are located just above the floor, they act to distribute the wall weight (making the wall a deep beam) and thus help avoid wall cracks if the floor sags. Bond beams may also be located below a window sill.

Modular door and window openings usually require lintels to support the blocks over the openings. You can use precast concrete lintels (figure 8-25, view 1) that contain an offset on the underside (view 2) to fit the modular openings. You can also use steel lintel angles that you install with an offset on the underside (view 3) to fit modular openings. In either case, place a noncorroding metal plate under the lintel ends at the control joints to allow the lintel to slip and the control joints to function properly. Apply a full bed of mortar over the metal plate to uniformly distribute the lintel load.

You usually install precast concrete sills (figure 8-26) following wall construction. Fill the joints tightly at the ends of the sills with mortar or a caulking compound.

PIERS AND PILASTERS

Piers are isolated columns of masonry, whereas pilasters are columns or thickened wall sections built contiguous to and forming part of a masonry wall.

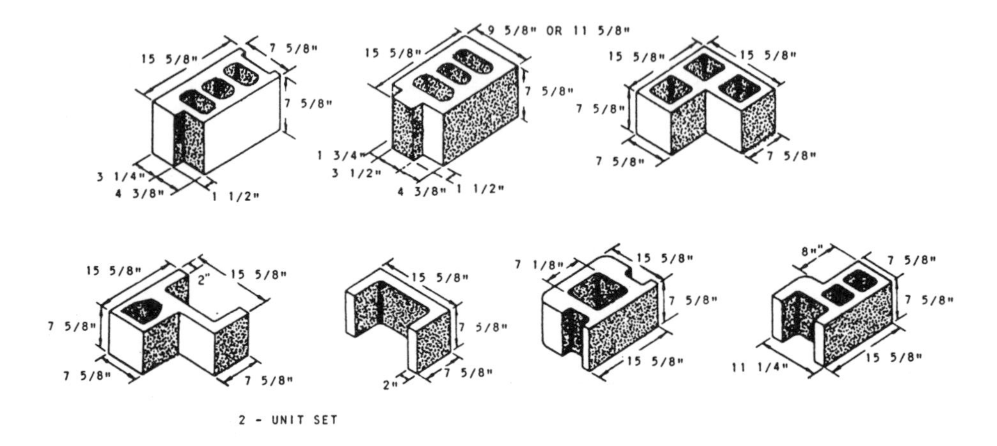

A. UNITS FOR SPECIAL CONDITIONS

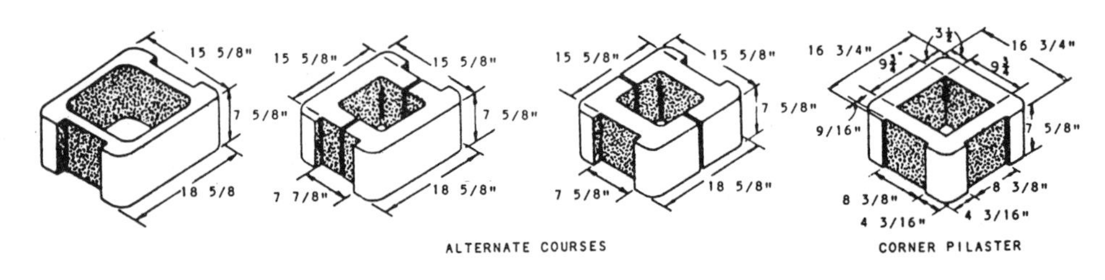

B. UNITS FOR 8-IN. WALLS

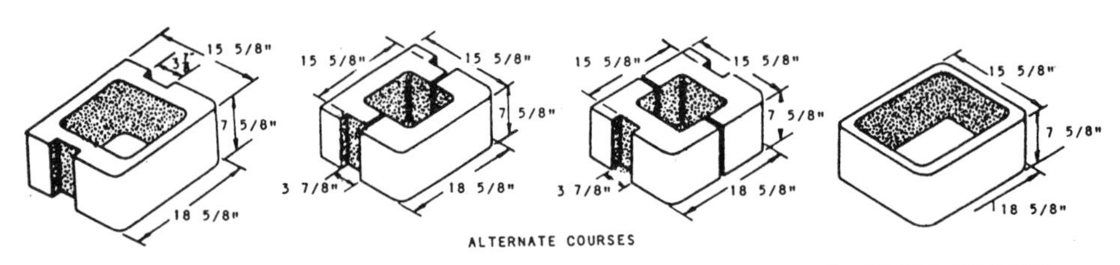

C. UNITS FOR KEY OR WOOD JAMB BLOCK

D. DOUBLE BULLNOSE PIER BLOCK

Figure 8-27.—Pilaster masonry units.

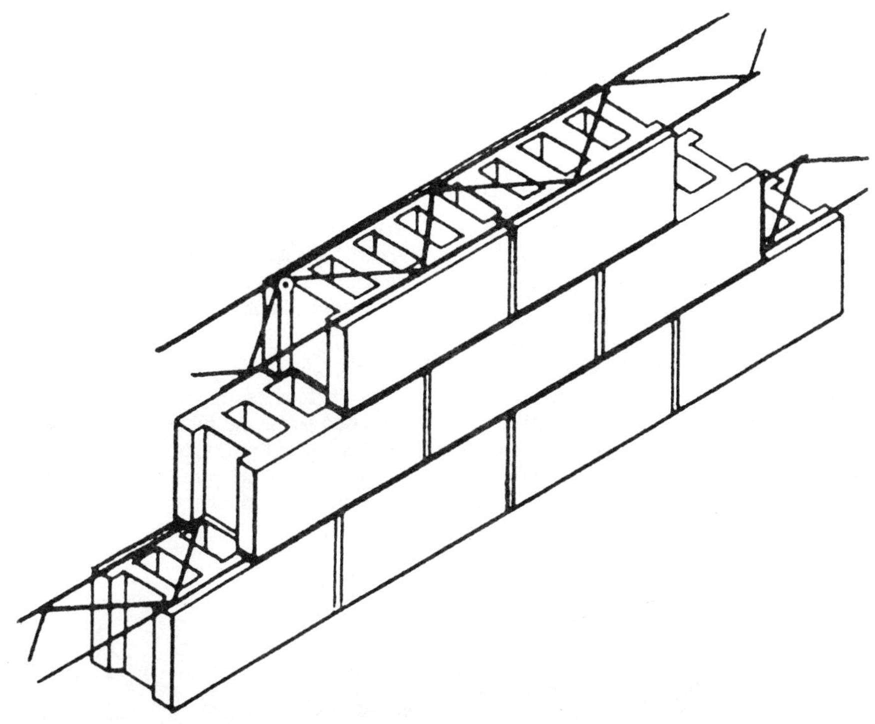

Figure 8-28.—Masonry wall horizontal joint reinforcement.

Both piers and pilasters are used to support heavy, concentrated vertical roof or floor loads. They also provide lateral support to the walls. Piers and pilasters offer an economic advantage by permitting construction of higher and thinner walls. They may be constructed of special concrete masonry units (figure 8-27) or standard units.

REINFORCED BLOCK WALLS

Block walls may be reinforced vertically or horizontally. To reinforce vertically, place reinforcing rods (called rebar) into the cores at the specified spacing and fill the cores with a relatively high-slump concrete. Rebar should be placed at each corner and at both sides of each opening. Vertical rebar should be spaced a maximum of 32 inches on center in walls. Where splices are required, the bars should be lapped 40 times the bar diameter. The concrete should be placed in one continuous pour from foundation to plate line. A cleanout block may be placed in the first course at every rebar stud for cleaning out excess mortar and to ensure proper alignment and laps of rebars.

Practical experience indicates that control of cracking and wall flexibility can be achieved with the use of horizontal joint reinforcing. The amount of joint reinforcement depends largely upon the type of construction. Horizontal joint reinforcing, where required, should consist of not less than two deformed longitudinal No. 9 or heavier cold-drawn steel wires. Truss-type cross wires should be l/8-inch diameter (or heavier) of the same quality. Figure 8-28 shows joint reinforcement on 16-inch vertical spacing. The location and details of bond beams, control joints, and joint reinforcing should all be shown on the drawings.

PATCHING AND CLEANING BLOCK WALLS

Always fill holes made by nails or line pins with fresh mortar and patch mortar joints. When laying concrete masonry walls, **be careful not to smear mortar on the block surfaces**. Once they harden, these smears cannot be removed, even with an acid wash, nor will paint cover them. Allow droppings to dry and harden. You can then chip off most of the mortar with a small piece of broken concrete block (figure 8-29, view 1) or with a trowel (view 2). A final brushing of the spot removes practically all the mortar (view 3).

RETAINING WALLS

The purpose of a retaining wall is to hold back a mass of soil or other material. As a result, concrete masonry retaining walls must have the structural strength to resist imposed vertical **and** lateral loads. The footing of a retaining wall should be large enough to support the wall and the load of the material that the wall is to retain. The reinforcing must be properly located as specified in the plans. Provisions to prevent the accumulation of water behind retaining walls should be made. This includes the installation of drain tiles or weep holes, or both.

PAINTING CONCRETE MASONRY

Several finishes are possible with concrete masonry construction. The finish to use in any specific situation should be governed by the type of structure in which the walls will be used and the climatic conditions to which they will be exposed.

Paints now commonly used on concrete masonry walls include portland cement paint, latex paint, oil-based paint, and rubber-based paint. For proper application and preparation of the different types of paint, refer to the plans, specifications, or manufacturer's instructions.

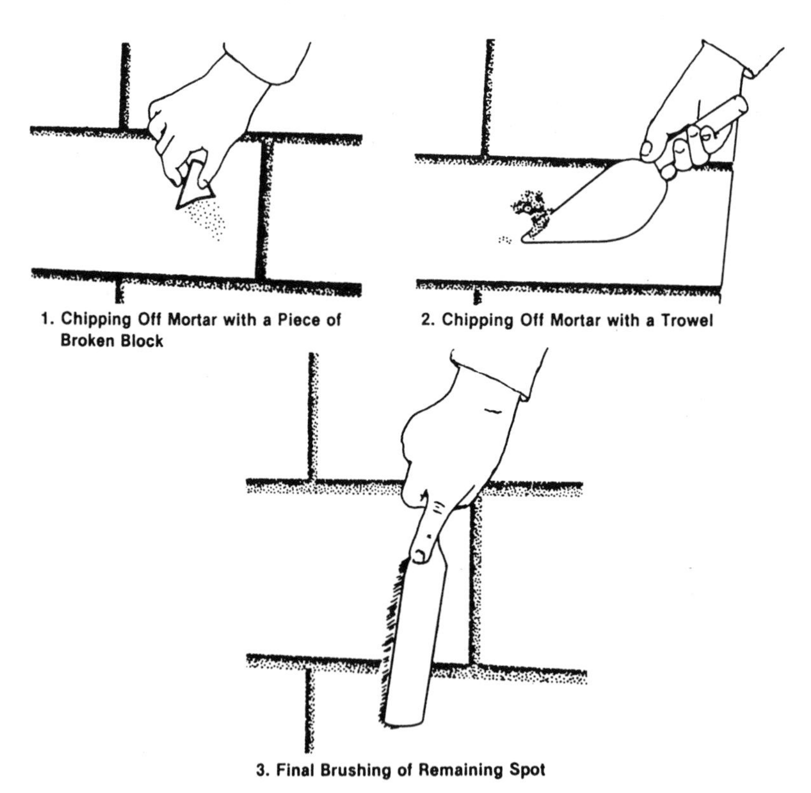

Figure 8-29.—Cleaning mortar droppings from a concrete block wall.

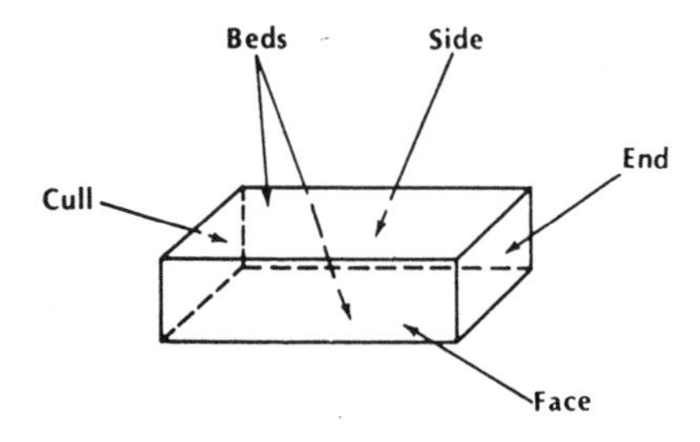

Figure 8-30.—Names of brick surfaces.

BRICK MASONRY

LEARNING OBJECTIVE: Upon completing this section, you should be able to explain the elements of brick masonry.

Brick masonry is construction in which uniform units ("bricks"), small enough to be placed with one hand, are laid in courses with mortar joints to form walls. Bricks are kiln baked from various clay and shale mixtures. The chemical and physical characteristics of the ingredients vary considerably. These characteristics and the kiln temperatures combine to produce brick in a variety of colors and harnesses. In some regions, individual pits yield clay or shale which, when ground and moistened, can be formed and baked into durable brick. In other regions, clay or shale from several pits must be mixed.

BRICK TERMINOLOGY

Standard U.S. bricks are 2 1/4-by-3 3/4-by-8 inches nominal size. They may have three core holes or ten core holes. Modular U.S. bricks are 2 1/4-by-3 5/8-by-7 5/8 inches nominal size. They usually have three core holes. English bricks are 3-by-4 1/2-by-9 inches; Roman bricks are 1 1/2-by-4-by-12 inches; and Norman bricks are 2 3/4-by-4-by-12 inches nominal size. Actual brick dimensions are smaller, usually by an amount equal to a mortar joint width. Bricks weigh from 100 to 150 pounds per cubic foot, depending on the ingredients and duration of firing. Fired brick is heavier than under-burned brick. The six surfaces of a brick are called cull, beds, side, end, and face, as shown in figure 8-30.

Occasionally you will have to cut brick into various shapes to fill in spaces at corners and other locations where a full brick does not fit. Figure 8-31 shows the more common cut shapes: half or bat, three-quarter closure, quarter closure, king closure, queen closure, and split.

TYPES OF BRICKS

Brick masonry units may be solid, hollow, or architectural terra cotta. All types can serve a structural function, a decorative function, or a combination of both. The various types differ in their formation and composition.

Building brick, also called common, hard, or kiln-run brick, is made from ordinary clay or shale and is fired in kilns. These bricks have no special shoring, markings, surface texture, or color. Because building bricks are generally used as the backing courses in either solid or cavity brick walls, the harder and more durable types are preferred.

Face brick is better quality and has better durability and appearance than building brick. Because of this, face bricks are used in exposed wall faces. The most common face brick colors are various shades of brown, red, gray, yellow, and white.

Clinker brick is over burned in the kiln. Clinker bricks are usually rough, hard, durable, and sometimes irregular in shape.

Pressed brick is made by a dry-press process rather than by kiln firing. Pressed bricks have regular smooth faces, sharp edges, and perfectly square corners. Ordinarily, they are used like face brick.

Glazed brick has one surface coated with a white or colored ceramic glazing. The glazing forms when mineral ingredients fuse together in a glass like coating during burning. Glazed bricks are particularly

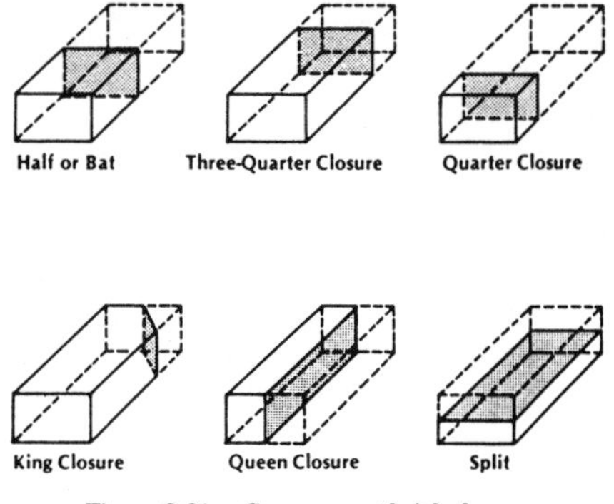

Figure 8-31.—Common cut brick shapes.

suited to walls or partitions in hospitals, dairies, laboratories, and other structures requiring sanitary conditions and ease of cleaning.

Fire brick is made from a special type of clay. This clay is very pure and uniform and is able to withstand the high temperatures of fireplaces, boilers, and similar constructions. Fire bricks are generally larger than other structural bricks and are often hand molded.

Cored bricks have ten holes—two rows of five holes each—extending through their beds to reduce weight. Walls built from cored brick are not much different in strength than walls built from solid brick. Also, both have about the same resistance to moisture penetration. Whether cored or solid, use the more available brick that meets building requirements.

European brick has strength and durability about equal to U.S. clay brick. This is particularly true of the English and Dutch types.

Sand-lime brick is made from a lean mixture of slaked lime and fine sand. Sand-lime bricks are molded under mechanical pressure and are hardened under steam pressure. These bricks are used extensively in Germany.

STRENGTH OF BRICK MASONRY

The main factors governing the strength of a brick structure include brick strength, mortar strength and elasticity, bricklayer workmanship, brick uniformity, and the method used to lay brick. In this section, we'll cover strength and elasticity. Workmanship is covered separately in the next section.

The strength of a single brick masonry unit varies widely, depending on its ingredients and manufacturing method. Brick can have an ultimate compressive strength as low as 1,600 psi. On the other hand, some well-burned brick has compressive strength exceeding 15,000 psi.

Because portland-cement-lime mortar is normally stronger than the brick, brick masonry laid with this mortar is stronger than an individual brick unit. The load-carrying capacity of a wall or column made with plain lime mortar is less than half that made with portland-cement-lime mortar. The compressive working strength of a brick wall or column laid with plain lime mortar normally ranges from 500 to 600 psi.

For mortar to bond to brick properly, sufficient water must be present to completely hydrate the portland cement in the mortar. Bricks sometimes have high absorption rates, and, if not properly treated, can "suck" the water out of the mortar, preventing complete hydration. Here is a quick field test to determine brick absorptive qualities. Using a medicine dropper, place 20 drops of water in a 1-inch circle (about the size of a quarter) on a brick. A brick that absorbs all the water in less than 1 1/2 minutes will suck the water out of the mortar when laid. To correct this condition, thoroughly wet the bricks and allow time for the surfaces to air-dry before placing.

BRICKLAYING METHODS

Good bricklaying procedure depends on good workmanship and efficiency. Efficiency involves doing the work with the fewest possible motions. Each motion should have a purpose and should accomplish a definite result. After learning the fundamentals, every Builder should develop methods for achieving maximum efficiency. The work must be arranged in such a way that the Builder is continually supplied with brick and mortar. The scaffolding required must be planned before the work begins. It must be built in such a way as to cause the least interference with other crewmembers.

Bricks should always be stacked on planks; they should never be piled directly on uneven or soft ground. Do not store bricks on scaffolds or runways. This does not, however, prohibit placing normal supplies on scaffolding during actual bricklaying operations. Except where stacked in sheds, brick piles should never be more than 7 feet high. When a pile of brick reaches a height of 4 feet, it must be tapered back 1 inch in every foot of height above the 4-foot level. The tops of brick piles must be kept level, and the taper must be maintained during unpiling operations.

MASONRY TERMS

To efficiently and effectively lay bricks, you must be familiar with the terms that identify the position of masonry units and mortar joints in a wall. The following list, which is referenced to figure 8-32, provides some of the basic terms you will encounter.

- **Course**—One of several continuous, horizontal layers (or rows) of masonry units bonded together.

- **Wythe**—Each continuous, vertical section of a wall, one masonry unit thick. Sometimes called a tier.

- **Stretcher**—A masonry unit laid flat on its bed along the length of a wall with its face parallel to the face of the wall.

- **Header**—A masonry unit laid flat on its bed across the width of a wall with its face perpendicular to the face of the wall. Generally used to bond two wythes.

- **Row lock**—A header laid on its face or edge across the width of a wall.

- **Bull header**—A rowlock brick laid with its bed perpendicular to the face of the wall.

- **Bull stretcher**—A rowlock brick laid with its bed parallel to the face of the wall.

- **Soldier**—A brick laid on its end with its face perpendicular to the face of the wall.

BONDS

The term "bond" as used in masonry has three different meanings: structural bond, mortar bond, or pattern bond.

Structural bond refers to how the individual masonry units interlock or tie together into a single structural unit. You can achieve structural bonding of brick and tile walls in one of three ways:

- Overlapping (interlocking) the masonry units;

- Embedding metal ties in connecting joints; and

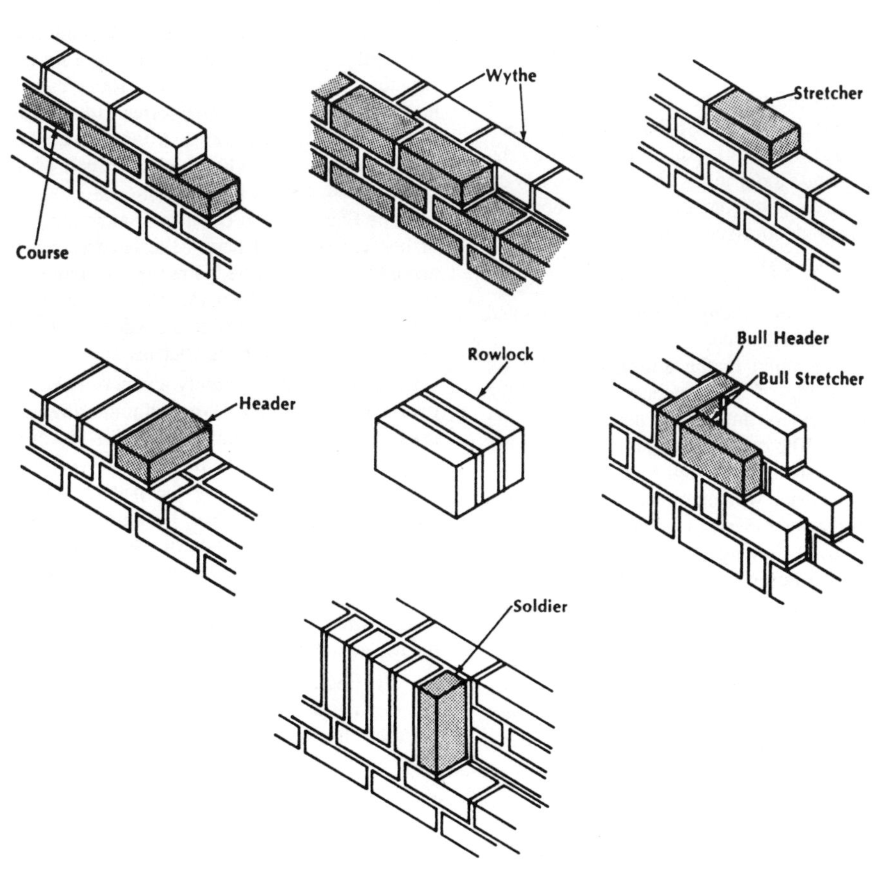

Figure 8-32.—Masonry units and mortar joints.

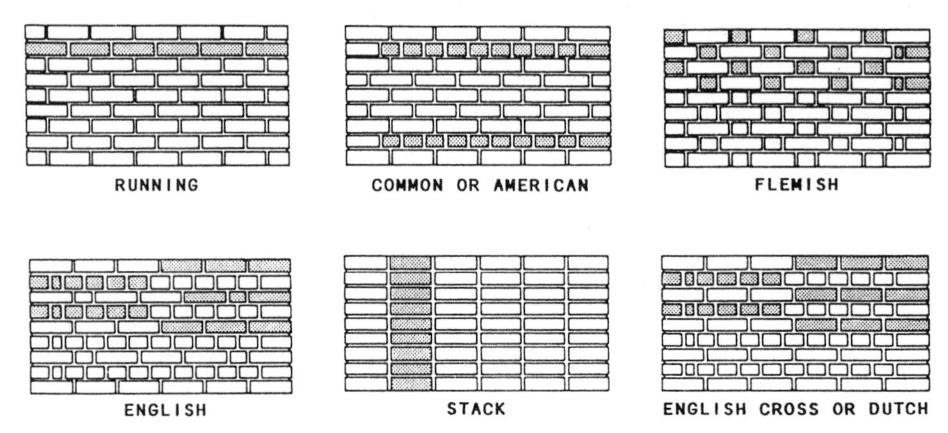

Figure 8-33.—Types of masonry bonds.

- Using grout to adhere adjacent wythes of masonry.

Mortar bond refers to the adhesion of the joint mortar to the masonry units or to the reinforcing steel.

Pattern bond refers to the pattern formed by the masonry units and mortar joints on the face of a wall. The pattern may result from the structural bond, or may be purely decorative and unrelated to the structural bond. Figure 8-33 shows the six basic pattern bonds in common use today: running, common or American, Flemish, English, stack, and English cross or Dutch bond.

The running bond is the simplest of the six patterns, consisting of all stretchers. Because the bond has no headers, metal ties usually form the structural bond. The running bond is used largely in cavity wall construction, brick veneer walls, and facing tile walls made with extra wide stretcher tile.

The common, or American, bond is a variation of the running bond, having a course of full-length headers at regular intervals that provide the structural bond as well as the pattern. Header courses usually appear at every fifth, sixth, or seventh course, depending on the structural bonding requirements. You can vary the common bond with a Flemish header course. In laying out any bond pattern, be sure to start the corners correctly. In a common bond, use a three-quarter closure at the corner of each header course.

In the Flemish bond, each course consists of alternating headers and stretchers. The headers in every other course center over and under the stretchers in the courses in between. The joints between stretchers in all stretcher courses align vertically. When headers are not required for structural bonding, you can use bricks called blind headers. You can start the corners in two different ways. In the Dutch corner, a three-quarter closure starts each course. In the English corner, a 2-inch or quarter closure starts the course.

The English bond consists of alternating courses of headers and stretchers. The headers center over and under the stretchers. However, the joints between stretchers in all stretcher courses do not align vertically. You can use blind headers in courses that are not structural bonding courses.

The stack bond is purely a pattern bond, with no overlapping units and all vertical joints aligning. You must use dimensionally accurate or carefully rematched units to achieve good vertical joint alignment. You can vary the pattern with combinations and modifications of the basic patterns shown in figure 8-33. This pattern usually bonds to the backing with rigid steel ties or 8-inch-thick stretcher units when available. In large wall areas or load-bearing construction, insert steel pencil rods into the horizontal mortar joints as reinforcement.

The English cross or Dutch bond is a variation of the English bond. It differs only in that the joints between the stretchers in the stretcher courses align vertically. These joints center on the headers in the courses above and below.

When a wall bond has no header courses, use metal ties to bond the exterior wall brick to the backing courses. Figure 8-34 shows three typical metal ties.

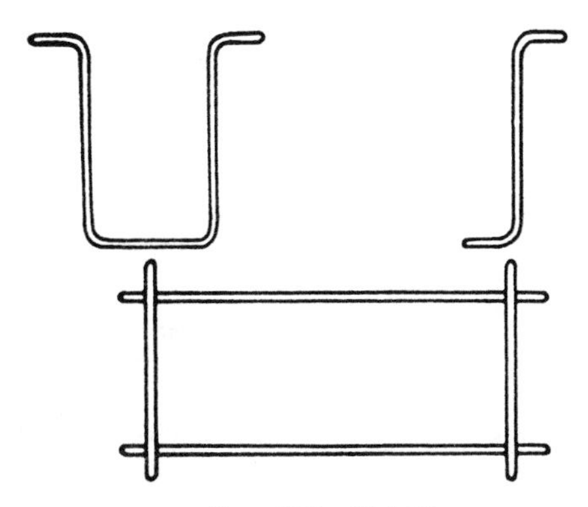

Figure 8-34.—Metal ties.

Install flashing at any spot where moisture is likely to enter a brick masonry structure. Flashing diverts the moisture back outside. Always install flashing under horizontal masonry surfaces, such as sills and copings; at intersections between masonry walls and horizontal surfaces, such as a roof and parapet or a roof and chimney; above openings (doors and windows, for example); and frequently at floor lines, depending on the type of construction. The flashing should extend through the exterior wall face and then turn downward against the wall face to form a drop.

You should provide weep holes at intervals of 18 to 24 inches to drain water to the outside that might accumulate on the flashing. Weep holes are even more important when appearance requires the flashing to stop behind the wall face instead of extending through the wall. This type of concealed flashing, when combined with tooled mortar joints, often retains water in the wall for long periods and, by concentrating the moisture at one spot, does more harm than good.

MORTAR JOINTS AND POINTING

There is no set rule governing the thickness of a brick masonry mortar joint. Irregularly shaped bricks may require mortar joints up to 1/2 inch thick to compensate for the irregularities. However, mortar joints 1/4 inch thick are the strongest. Use this thickness when the bricks are regular enough in shape to permit it.

A slushed joint is made simply by depositing the mortar on top of the head joints and allowing it to run down between the bricks to form a joint. You cannot make solid joints this way. Even if you fill the space between the bricks completely, there is no way you can compact the mortar against the brick faces; consequently a poor bond results. The only effective way to build a good joint is to trowel it.

The secret of mortar joint construction and pointing is in how you hold the trowel for spreading mortar. Figure 8-35 shows the correct way to hold a trowel. Hold it firmly in the grip shown, with your

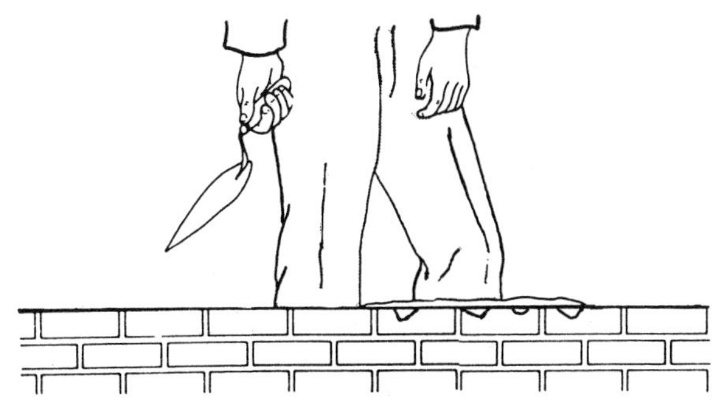

Figure 8-35.—Correct way to hold a trowel.

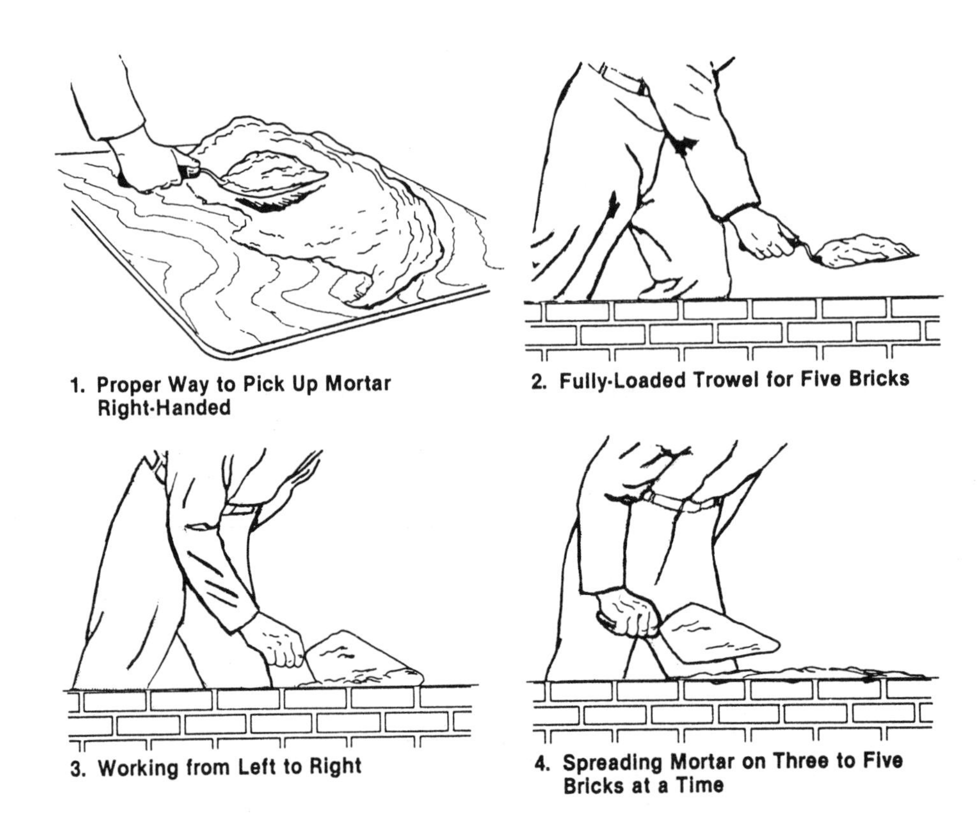

Figure 8-36.—Picking up and spreading mortar.

thumb resting on top of the handle, not encircling it. If you are right-handed, pick up mortar from the outside of the mortar board pile with the left edge of your trowel (figure 8-36, view 1). You can pick up enough to spread one to five bricks, depending on the wall space and your skill. A pickup for one brick forms only a small pile along the left edge of the trowel. A pickup for five bricks is a full load for a large trowel (view 2).

If you are right-handed, work from left to right along the wall. Holding the left edge of the trowel directly over the center line of the previous course, tilt the trowel slightly and move it to the right (view 3), spreading an equal amount of mortar on each brick until you either complete the course or the trowel is empty (view 4). Return any mortar left over to the mortar board.

Do not spread the mortar for a bed joint too far ahead of laying—four or five brick lengths is best. Mortar spread out too far ahead dries out before the bricks become bedded and causes a poor bond (figure 8-37). The mortar must be soft and plastic so that the brick will bed in it easily. Spread the mortar about 1 inch thick and then make a shallow furrow in

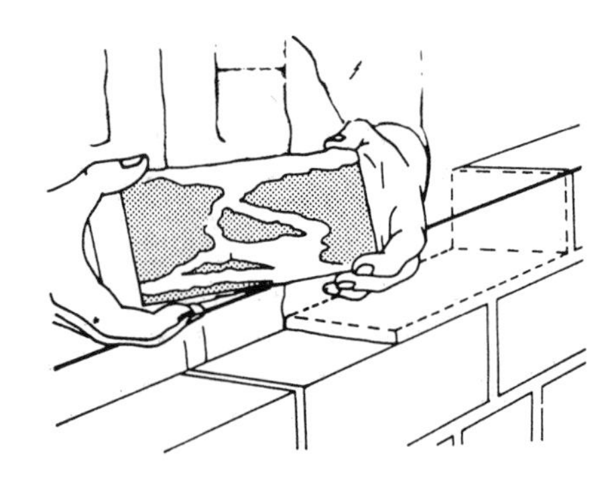

Figure 8-37.—A poorly bonded brick.

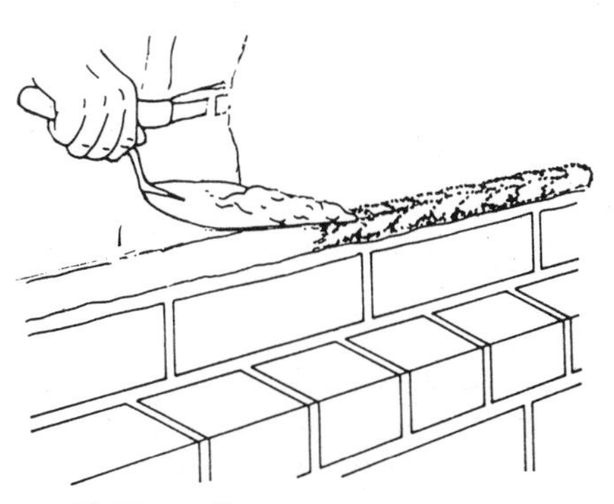

1. Making a Furrow

2. Cutting Off Excess Mortar

Figure 8-38.—Making a bed joint in a stretcher course.

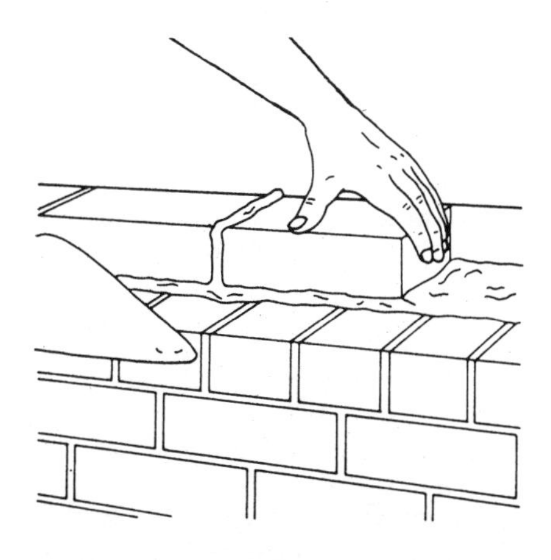

Figure 8-40.—Making a head joint in a stretcher course.

it (figure 8-38, view 1). A furrow that is too deep leaves a gap between the mortar and the bedded brick. This reduces the resistance of the wall to water penetration.

Using a smooth, even stroke, cut off any mortar projecting beyond the wall line with the edge of the trowel (figure 8-38, view 2). Retain enough mortar on the trowel to butter the left end of the first brick you will lay in the fresh mortar. Throw the rest back on the mortar board.

Pick up the first brick to be laid with your thumb on one side of the brick and your fingers on the other (figure 8-39). Apply as much mortar as will stick to the end of the brick and then push it into place. Squeeze out the excess mortar at the head joint and at the sides (figure 8-40). Make sure the mortar

Figure 8-39.—Proper way to hold a brick when buttering the end.

completely fills the head joint. After bedding the brick, cut off the excess mortar and use it to start the next end joint. Throw any surplus mortar back on the mortar board where it can be restored to workability.

Figure 8-41 shows how to insert a brick into a space left in a wall. First, spread a thick bed of mortar (view 1), and then shove the brick into the wall space (view 2) until mortar squeezes out of all four joints (view 3). This way, you know that the joints are full of mortar at every point.

To make a cross joint in a header course, spread the bed joint mortar several brick widths in advance. Then, spread mortar over the face of the header brick before placing it in the wall (figure 8-42, view 1). Next, shove the brick into place, squeezing out mortar at the top of the joint. Finally, cut off the excess mortar as shown in view 2.

Figure 8-43 shows how to lay a closure brick in a header course. First, spread about 1 inch of mortar on the sides of the brick already in place (view 1), as well

1. Spreading a Thick Bed of Mortar

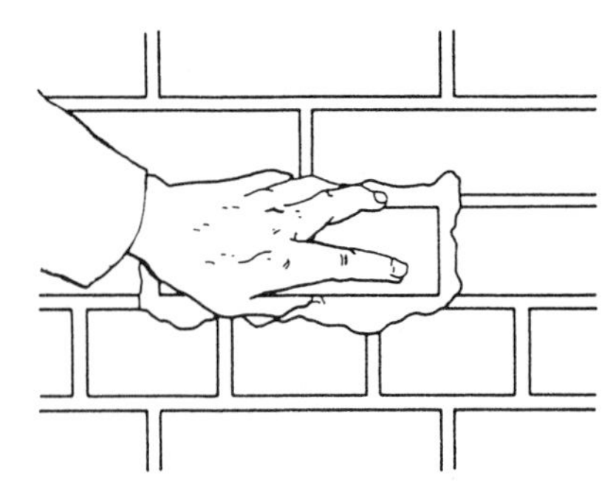

2. Shoving the Brick into Place

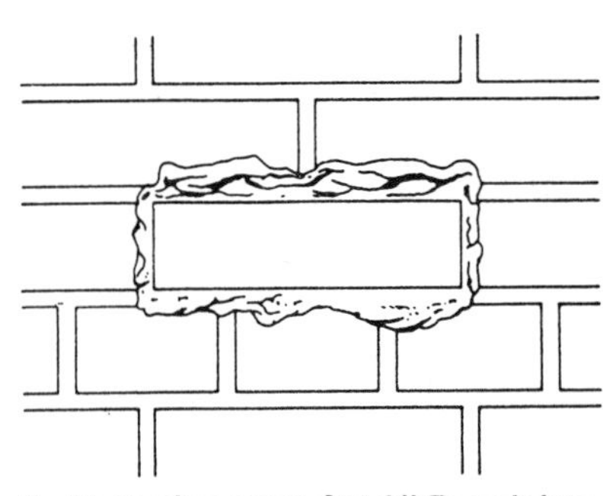

3. Mortar Squeezes Out All Four Joints

Figure 8-41.—Inserting a brick in a wall space.

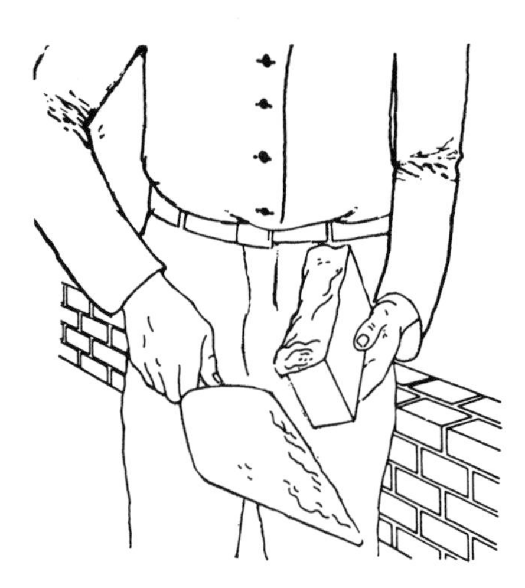

1. Spreading Mortar over Brick Face

2. Cutting Off Excess Mortar

Figure 8-42.—Making a cross joint in a header course.

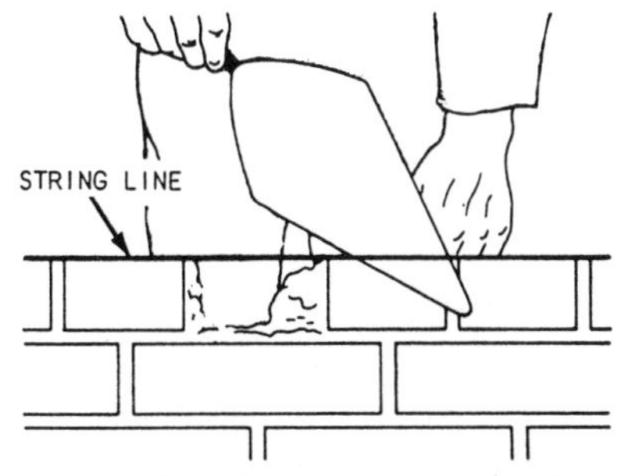

1. Spreading Mortar on Sides of Brick Already Laid

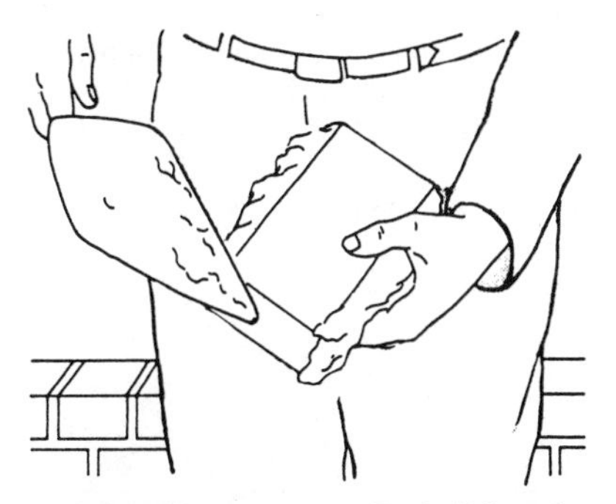

2. Spreading Mortar on Both Sides of Closure Brick

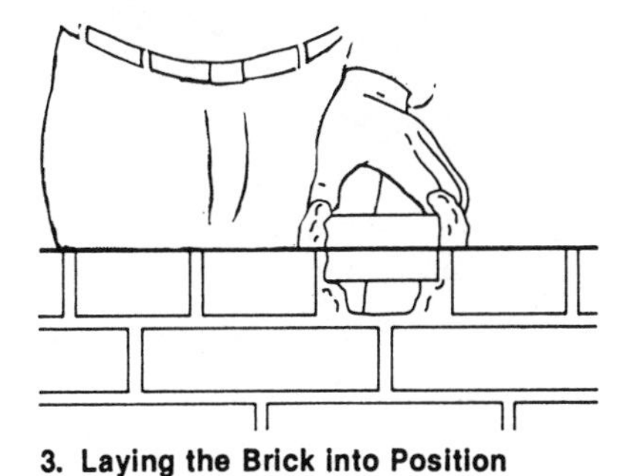

3. Laying the Brick into Position

Figure 8-43.—Making a closure joint in a header course.

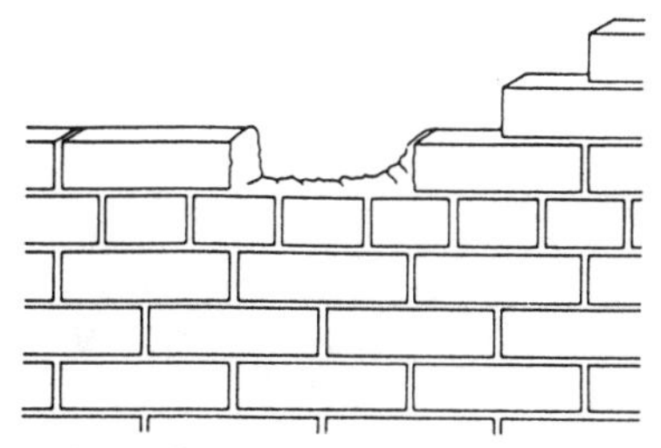

1. Spreading Mortar on Ends of Brick Already Laid

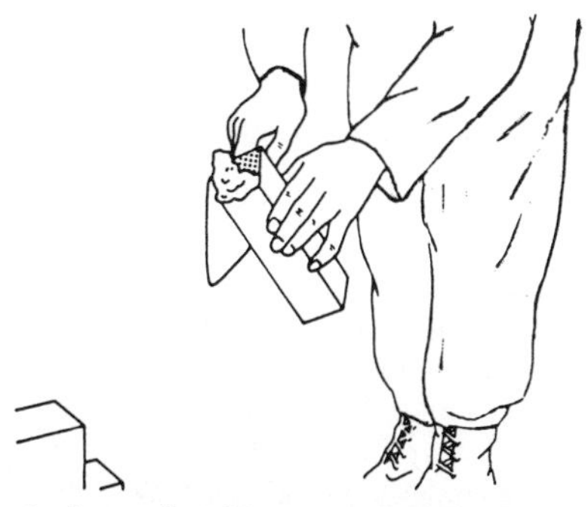

2. Spreading Mortar on Both Ends of Closure Brick

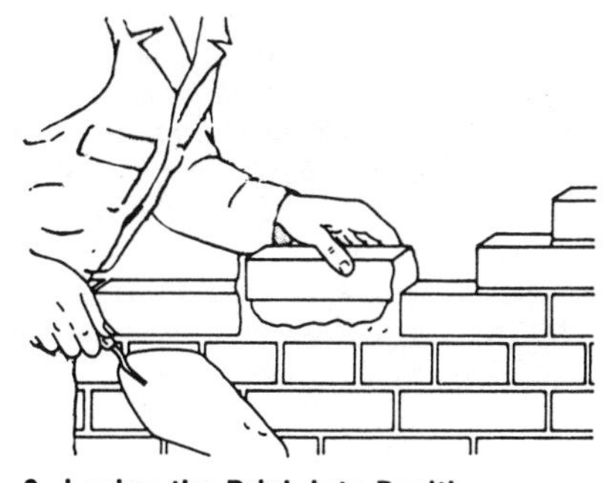

3. Laying the Brick into Position

Figure 8-44.—Making a closure joint in a stretcher course.

as on both sides of the closure brick (view 2). Then, lay the closure brick carefully into position without disturbing the brick already laid (view 3). If you do disturb any adjacent brick, cracks will form between the brick and mortar, allowing moisture to penetrate the wall. You should place a closure brick for a stretcher course (figure 8-44) using the same techniques as for a header course.

As we mentioned earlier, filling exposed joints with mortar immediately after laying a wall is called pointing. You can also fill holes and correct defective mortar joints by pointing, using a pointing trowel.

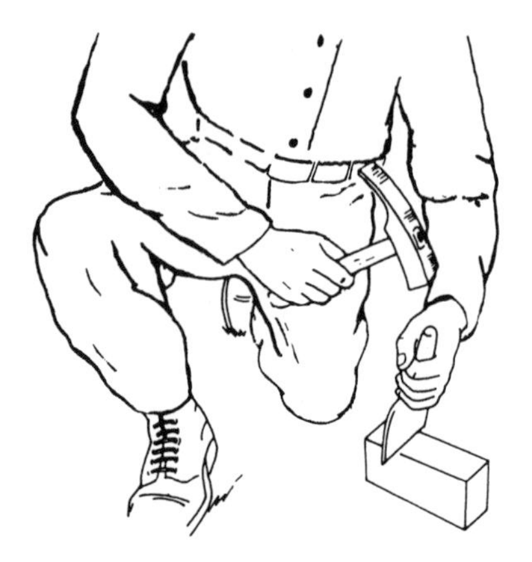

Figure 8-45.—Cutting brick with a chisel.

CUTTING BRICK

To cut a brick to an exact line, you should use a chisel (figure 8-45), or brick set. The straight side of the tool's cutting edge should face both the part of the brick to be saved and the bricklayer. One mason's hammer blow should break the brick. For extremely hard brick, first roughly cut it using the brick hammer head, but leave enough brick to cut accurately with the brick set.

Use a brick hammer for normal cutting work, such as making the closure bricks and bats around wall openings or completing corners. Hold the brick firmly while cutting it. First, cut a line all the way around the brick using light hammer head blows. Then, a sharp blow to one side of the cutting line should split the brick at the cutting line (figure 8-46, view 1). Trim rough spots using the hammer blade, as shown in view 2.

FINISHING JOINTS

The exterior surfaces of mortar joints are finished to make brick masonry waterproof and give it a better appearance. If joints are simply cut to the face of the brick and not finished, shallow cracks will develop immediately between the brick and the mortar. Always finish a mortar joint before the mortar hardens too much. Figure 8-47 shows several types of joint finishes, the more important of which are concave, flush, and weather.

Of all joints, the concave is the most weather tight. After removing the excess mortar with a trowel, make this joint using a jointer that is slightly larger than the joint. Use force against the tool to press the mortar tight against the brick on both sides of the mortar joint.

The flush joint is made by holding the trowel almost parallel to the face of the wall while drawing its point along the joint.

A weather joint sheds water from a wall surface more easily. To make it, simply push downward on the mortar with the top edge of the trowel.

ARCHES

A well-constructed brick arch can support a heavy load, mainly due to the way weight is distributed over its curved shape. Figure 8-48 shows two common arch shapes: elliptical and circular. Brick arches require full mortar joints. The joint width is narrower

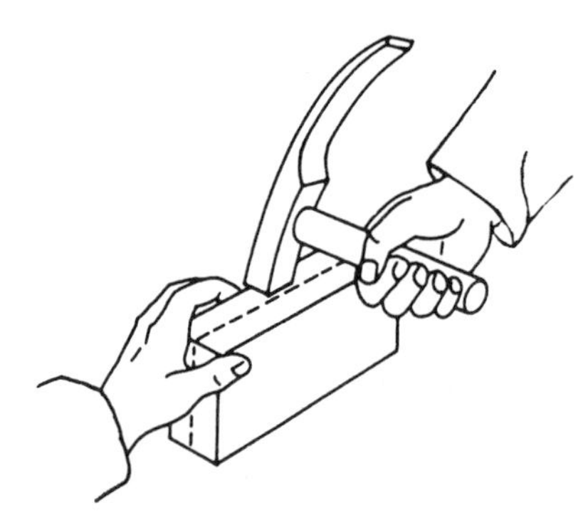

1. Striking Brick to One Side of Cutting Line

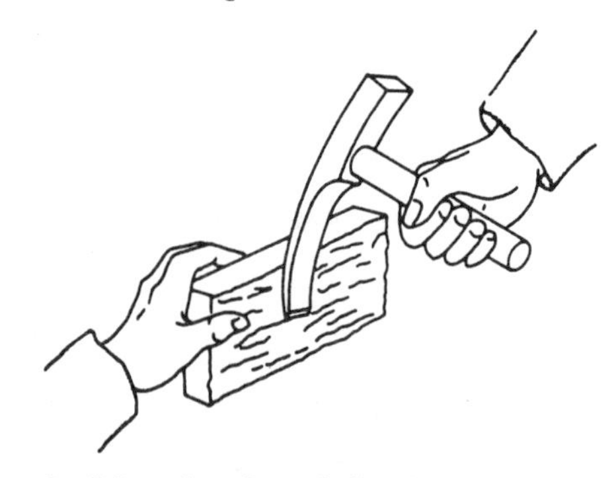

2. Trimming Rough Spots

Figure 8-46.—Cutting brick with a hammer.

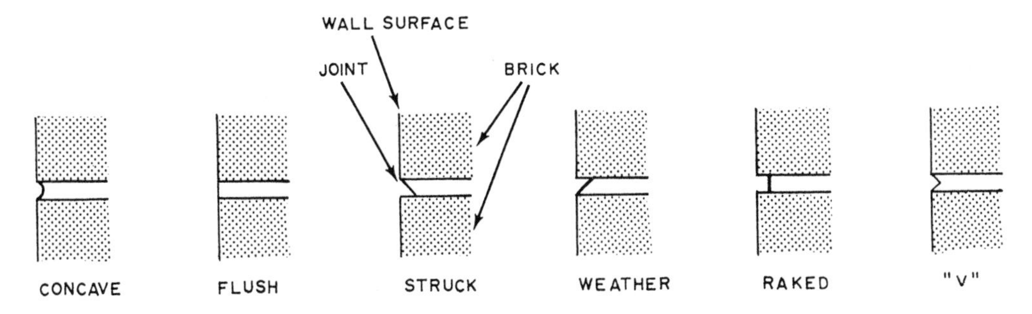

Figure 8-47.—Joint finishes.

at the bottom of the arch than at its top, but it should not narrow to less than 1/4 inch at any point. As laying progresses, make sure the arch does not bulge out of position.

Templet

It is obviously impossible to construct an arch without support from underneath. These temporary wooden supports must not only be able to support the masonry during construction but also provide the geometry necessary for the proper construction and appearance of the arch. Such supports are called templets.

DIMENSIONS.—Construct a brick arch over the templet (figure 8-49) that remains in place until the mortar sets. You can obtain the templet dimensions from the construction drawings. For arches spanning up to 6 feet, use 3/4-inch plywood to make the templet. Cut two pieces to the proper curvature, and nail them to 2-by-4 spacers that provide a surface wide enough to support the brick.

POSITIONING.—Use wedges to hold the templet in position until the mortar hardens enough to make the arch self-supporting. Then drive out the wedges.

Layout

Lay out the arch carefully so that you don't have to cut any bricks. Use an odd number of bricks so that the key, or middle, brick falls into place at the exact arch center, or crown. The key, or middle, brick is the last one laid. To determine how many bricks an arch requires, lay the templet on its side on level ground and set a trial number of bricks around the curve. Adjust the number of bricks and the joint spacing (not less than 1/4-inch) until the key brick is at the exact center of the curve. Then, mark the positions of the bricks on the templet and use them as a guide when laying the brick.

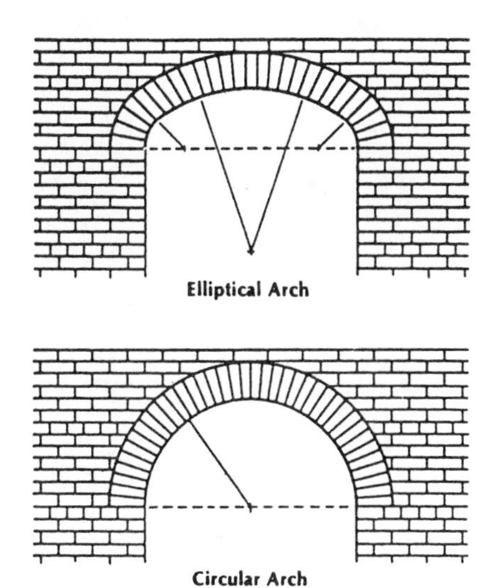

Figure 8-48.—Common arch shapes.

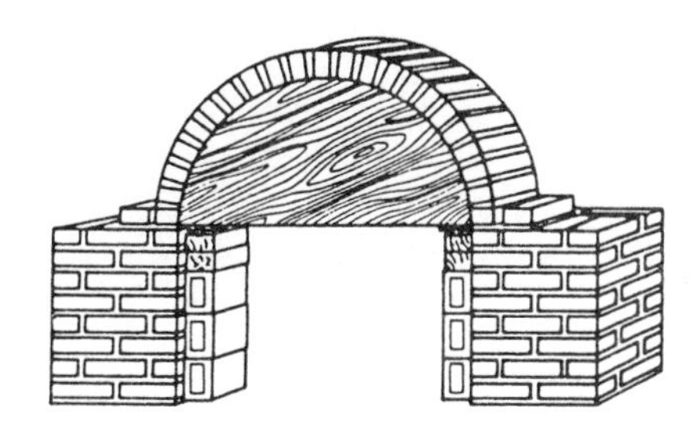

Figure 8-49.—Using a template to construct an arch.

RECOMMENDED READING LIST

Although the following reference was current when this TRAMAN was published, its continued currency cannot be assured. You therefore need to ensure that you are studying the latest revision.

Concrete and Masonry, FM 5-742, Headquarters, Department of the Army, Washington, D.C., 1985.

CHAPTER 9

PLANNING, ESTIMATING, AND SCHEDULING

Good construction planning and estimating procedures are essential for the Naval Construction Forces (NCFs) to provide quality construction response to the fleet's operational requirements. This chapter gives you helpful information for planning, estimating, and scheduling construction projects normally undertaken by Seabees. The material is designed to help you understand the concepts and principles involved; it is not intended to be a reference or to establish procedures. The techniques described are suggested methods that have been proved with use and can result in effective planning and estimating. It is your responsibility to decide how and when to apply these techniques.

Later in the chapter, you will encounter helpful tables to aid you in effective planning and estimating. Keep in mind that these tables are not intended to establish production standards. They should be used with sound judgment and in accordance with established regulations and project specifications. Man-hour tables are based upon direct labor and do not include allowances for indirect or overhead labor.

We provide helpful references at the end of the chapter. You are encouraged to study these, as required, for additional information on the topics discussed.

DEFINITIONS

LEARNING OBJECTIVE: Upon completing this section, you should be able to identify basic planning, estimating, and scheduling terms.

In planning any project, you must be familiar with the vocabulary commonly associated with planning, estimating, and scheduling. Here, we'll define a number of terms you need to know as a Builder.

PLANNING

Planning is the process of determining requirements, and devising and developing methods and action for constructing a project. Good construction planning is a combination of many elements: the activity, material, equipment, and manpower estimates; project layout; project location; material delivery and storage; work schedules; quality control; special tools required; environmental protection; safety; and progress control. All of these elements depend upon each other. They must all be considered in any well-planned project.

ESTIMATING

Estimating is the process of determining the amount and type of work to be performed and the quantities of material, equipment, and labor required. Lists of these quantities and types of work are called estimates.

PRELIMINARY ESTIMATES

Preliminary estimates are made from limited information, such as the general description of projects or preliminary plans and specifications having little or no detail. Preliminary estimates are prepared to establish costs for the budget and to program general manpower requirements.

DETAILED ESTIMATES

Detailed estimates are precise statements of quantities of material, equipment, and manpower required to construct a given project. Underestimating quantities can cause serious delays in construction and even result in unfinished projects. A detailed estimate must be accurate to the smallest detail to correctly quantify requirements.

ACTIVITY ESTIMATES

An activity estimate is a listing of all the steps required to construct a given project, including specific descriptions as to the limits of each clearly definable quantity of work (activity). Activity quantities provide the basis for preparing the material, equipment, and manpower estimates. They are used to provide the basis for scheduling material deliveries, equipment, and manpower. Because activity estimates are used to prepare other estimates and schedules, errors in these estimates can multiply many times. Be careful in their preparation!

MATERIAL ESTIMATES

A material estimate consists of a listing and description of the various materials and the quantities required to construct a given project. Information for preparing material estimates is obtained from the activity estimates, drawings, and specifications. A material estimate is sometimes referred to as "a Bill of Material (BM)" or "a Material Takeoff (MTO) Sheet." (We will discuss the BM and the MTO a little later in the chapter.)

EQUIPMENT ESTIMATES

Equipment estimates are listings of the various types of equipment, the amount of time, and the number of pieces of equipment required to construct a given project. Information, such as that obtained from activity estimates, drawings, specifications, and an inspection of the site, provides the basis for preparing the equipment estimates.

MANPOWER ESTIMATES

The manpower estimate consists of a listing of the number of direct labor man-days required to complete the various activities of a specific project. These estimates may show only the man-days for each activity, or they may be in sufficient detail to list the number of man-days for each rating in each activity—Builder (BU), Construction Electrician (CE), Equipment Operator (EO), Steelworker (SW), and Utilitiesman (UT). Man-day estimates are used in determining the number of personnel and the ratings required on a deployment. They also provide the basis for scheduling manpower in relation to construction progress.

When the *Seabee Planner's and Estimator's Handbook*, NAVFAC P-405, is used, a man-day is a unit of work performed by one person in one 8-hour day or its equivalent. One man-day is equivalent to a 10-hour day when the *Facilities Planning Guide*, NAVFAC P-437, is used.

Battalions set their own schedules, as needed, to complete their assigned tasks. In general, the work schedule of the battalion is based on an average of 55 hours per man per week. The duration of the workday is 10 hours per day, which starts and ends at the jobsite. This includes 9 hours for direct labor and 1 hour for lunch.

Direct labor includes all labor expended directly on assigned construction tasks, either in the field or in the shop, that contributes directly to the completion of the end product. Direct labor must be reported separately for each assigned construction item. In addition to direct labor, the estimator must also consider overhead labor and indirect labor. Overhead labor is considered productive labor that does not contribute directly or indirectly to the product. It includes all labor that must be performed regardless of the assigned mission. Indirect labor includes labor required to support construction operations but does not, in itself, produce an end product.

ESTIMATOR

An estimator is a person who evaluates the requirements of a task. A construction estimator must be able to mentally picture the separate operations of the job as the work progresses through the various stages of construction and be able to read and obtain accurate measurements from drawings. The estimator must have an understanding of math, previous construction experience, and a working knowledge of all branches of construction. The estimator must use good judgment when determining what effect numerous factors and conditions have on construction of the project and what allowances should be made for each of them. The estimator must be able to do careful and accurate work. A Seabee estimator must have ready access to information about the material, equipment, and labor required to perform various types of work under conditions encountered in Seabee deployments. The collection of such information on construction performance is part of estimating. Since this kind of reference information may change from time to time, information should be frequently reviewed.

SCHEDULING

Scheduling is the process of determining when an action must be taken and when material, equipment, and manpower are required. There are four basic types of schedules: progress, material, equipment, and manpower.

Progress schedules coordinate all the projects of a Seabee deployment or all the activities of a single project. They show the sequence, the starting time, the performance time required, and the time required for completion. Material schedules show when the material is needed on the job. They may also show the sequence in which materials should be delivered. Equipment schedules coordinate all the equipment to be used on a project. They also show when it is to be used and the amount of time each piece of equipment is required to perform the work. Manpower schedules coordinate the manpower requirements of a project and show the number of personnel required for each activity. In addition, the number of personnel of each rating (Builder, Construction Electrician, Equipment Operator, Steelworker, and Utilitiesman) required for each activity for each period of time may be shown. The time unit shown in a schedule should be some convenient interval, such as a day, a week, or a month.

NETWORK ANALYSIS

Network analysis is a method of planning and controlling projects by recording their interdependence in diagram form. This enables you to undertake each problem separately. The diagram form, known as a network diagram, is drawn so that each job is represented by an activity on the diagram, as shown in figure 9-1. The direction in which the activities are linked indicates the dependencies of the jobs on each other.

PROGRESS CONTROL

Progress control is the comparing of actual progress with scheduled progress and the steps necessary to correct deficiencies or to balance activities to meet overall objectives.

PLANNING DOCUMENTATION

LEARNING OBJECTIVE: Upon completing this section, you should be able to give the documentation requirements necessary in planning a construction project.

There are two basic ground rules in analyzing a project. **First**, planning and scheduling are separate operations. **Second**, planning must always precede scheduling. If you don't plan sequentially, you will end up with steps out of sequence and may substantially delay the project. Everyone concerned should know precisely the following aspects of a project:

- What it is;
- Its start and finish points;
- Its external factors, such as the schedule dates and requirements of other trade groups;

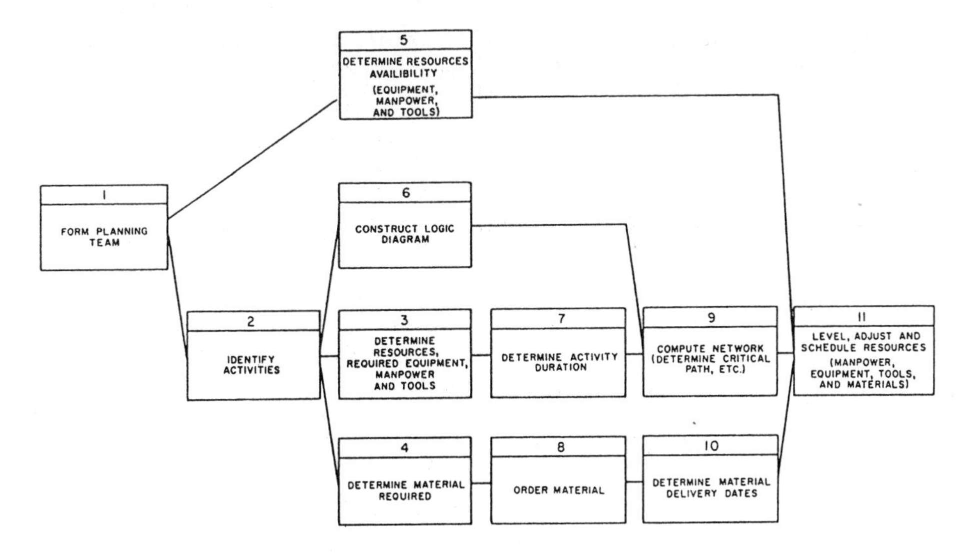

Figure 9-1.—Planning and estimating a precedence diagram.

- The availability of resources, such as manpower and equipment; and
- What you need to make up the project planning files.

PROJECT FOLDER

The project folder, or package, consists of nine individual project files. These files not only represent the project in a paper format, but also give you, as the project crew leader, supervisor, or crewmember, exposure to the fundamentals of construction management.

File No. 1—General Information File

File No. 1 is the General Information File and contains the following information:

LEFT SIDE—The left side of the General Information File basically contains information authorizing the project. The file should have the following items:

- Project scope sheet;
- Tasking letter;
- Project planning check list; and
- Project package sign-off sheet.

RIGHT SIDE—The right side of the General Information File contains basic information relating to coordinating the project. The file should have the following items:

- Project organization;
- Deployment calendar;
- Preconstruction conference notes; and
- Predeployment visit summary.

File No. 2—Correspondence File

File No. 2 is the Correspondence File and consists of the following items:

LEFT SIDE—The left side contains outgoing messages and correspondence.

RIGHT SIDE—The right side of the file contains incoming messages and correspondence.

File No. 3—Activity File

File No. 3, the Activity File, contains the following information:

LEFT SIDE—The left side contains the Construction Activity Summary Sheets of completed activities.

RIGHT SIDE—The right side of the file contains the following form sheets:

- Master activity sheets;
- Level II;
- Level II precedence diagram;
- Master activity summary sheets; and
- Construction activity summary sheets.

File No. 4—Network File

File No. 4 is the Network File. It contains the following information:

LEFT SIDE—The left side contains the following documents:

- Computer printouts;
- Level III; and
- Level III precedence diagram.

RIGHT SIDE—The right side of the Network File contains the following items:

- Resource leveled plan for manpower and equipment; and
- Equipment requirement summary.

File No. 5—Material File

File No. 5 is the Material File. It contains the following information:

LEFT SIDE—The left side contains the worksheets that you, as a project planner, must assemble. The list includes the following items:

- List of long lead items;
- 45-day material list;
- Material transfer list;
- Add-on/reorder justification forms;

- Bill of materials/material take-off comparison worksheets; and
- Material take-off worksheets.

RIGHT SIDE—The right side of the Material File contains the Bill of Materials (including all add-on/reorder BMs) supplied by the Naval Construction Regiment.

File No. 6—Quality Control File

File No. 6, the Quality Control File, contains the following information:

LEFT SIDE—The left side of this file contains various quality control forms and the field adjustment request.

RIGHT SIDE—The right side of the Quality Control File contains daily quality control inspection reports and your quality control plan.

File No. 7—Safety/Environmental File

File No. 7 is the Safety/Environmental File and consists of the following information:

LEFT SIDE—The left side of the Safety/Environmental File contains the following items:

- Required safety equipment;
- Stand-up safety lectures;
- Safety reports; and
- Accident reports.

RIGHT SIDE—The right side of the Safety/Environmental File contains the following:

- Safety plan, which you must develop;
- Highlighted EM 385; and
- Environmental plan (if applicable).

File No. 8—Plans File

File No. 8 is the Plans File and contains the following information:

LEFT SIDE—The left side contains the following planning documents:

- Site layout;
- Shop drawings;
- Detailed slab layout drawings (if applicable); and
- Rebar bending schedule.

RIGHT SIDE—The right side of the Plans File contains the actual project plans. Depending on thickness, plans should be either rolled or folded.

File No. 9—Specifications File

File No. 9 is the Specifications File; it contains the following information:

LEFT SIDE—The left side of this file is reserved for technical data.

RIGHT SIDE—The right side of the Specifications File has highlighted project specifications.

ESTIMATING

LEARNING OBJECTIVE: Upon completing this section, you should be able to explain the estimating requirements for a construction project.

As project estimator, you will need to assemble information about various conditions affecting the construction of the project. This enables you to prepare a detailed and accurate estimate. Drawings should be detailed and complete. Specifications should be exact and leave no doubt as to their intent. Information should be available about local material, such as quarries, gravel pits, spoil areas, types of soil, haul roads and distances, foundation conditions, the weather expected during construction, and the time allotted for completion. You should know the number and types of construction equipment available for use. Consider all other items and conditions that might affect the production or the progress of construction.

USING BLUEPRINTS

The construction drawings are your main basis for defining the required activities for measuring the quantities of material. Accurate estimating requires a thorough examination of the drawings. You should carefully read all notes and references and examine all details and reference drawings. The orientation of sectional views should be carefully checked. Dimensions shown on drawings or computed figures shown from those drawings should be used in preference to those obtained by scaling distances.

You should check the "Revision" section near the title section to ensure that the indicated changes were made in the drawing itself. You must ensure that the construction plan, the specifications, and the drawings are discussing the same project. When there are inconsistencies between general drawings and details, details should be followed unless they are obviously wrong. When there are inconsistencies between drawings and specifications, you should follow the specifications.

As an estimator, you must first study the specifications and then use them with the drawings when preparing quantity estimates. You should become thoroughly familiar with all the requirements stated in the specifications. Some estimators may have to read the specifications more than once to fix these requirements in their mind. You are encouraged to make notes as you read the specifications. These notes will be helpful to you later as you examine the drawings. In the notes, list any unusual or unfamiliar items of work or materials and reminders for use during examination of the drawings. A list of activities and materials that are described or mentioned in the specifications is helpful in checking quantity estimates.

The tables and diagrams in the *Seabee Planner's and Estimator's Handbook*, NAVFAC P-405, should save you time in preparing estimates and, when understood and used properly, provide accurate results. Whenever possible, the tables and the diagrams used were based on Seabee experience. Where suitable information was not available, construction experience was adjusted to represent production under the range of conditions encountered in Seabee construction. A thorough knowledge of the project drawings and specifications makes you alert to the various areas where errors may occur.

Accuracy as a Basis for Ordering and Scheduling

Quantity estimates are used as a basis for purchasing materials, determining equipment, and determining manpower requirements. They are also used in scheduling progress, which provides the basis for scheduling material deliveries, equipment, and manpower. Accuracy in preparing quantity estimates is extremely important; these estimates have widespread uses and errors can be multiplied many times. Say, for example, a concrete slab is to measure 100 feet by 800 feet. If you misread the dimension for the 800-foot side as 300 feet, the computed area of the slab will be 30,000 square feet, when it should actually be 80,000 square feet. Since area is the basis for ordering materials, there will be shortages. For example, concrete ingredients, lumber, reinforcing materials, and everything else involved in mixing and placing the concrete, including equipment time, manpower, and man-hours, will be seriously underestimated and ordered.

Checking Estimates

The need for accuracy is vital, and quantity estimates should be checked to eliminate as many errors as possible. One of the best ways to check your quantity estimate is to have another person make an independent estimate and then to compare the two. Any differences should be checked to determine which is right. A less effective way of checking is for another person to take your quantity estimate and check all measurements, recordings, computations, extensions, and copy work, keeping in mind the most common error sources (listed in the next section).

Error Sources

Failure to read **all** the notes on a drawing or failure to examine reference drawings results in many omissions. For example, you may overlook a note that states "symmetrical about the center line" and thus compute only half the required quantity.

Errors in scaling obviously mean erroneous quantities. Great care should be taken in scaling drawings so correct measurements are recorded. Common scaling errors include using the wrong scale, reading the wrong side of a scale, and failing to note that a detail being scaled is drawn to a scale different from that of the rest of the drawing. Remember: Some drawings are **not** drawn to scale. Since these cannot be scaled for dimensions, you must obtain dimensions from other sources.

Sometimes wrongly interpreting a section of the specifications causes errors in the estimate. If there is any doubt concerning the meaning of any part of the specification, you should request an explanation of that particular part.

Omissions are usually the result of careless examination of the drawings. Thoroughness in examining drawings and specifications usually eliminates errors of omission. Checklists should be used to assure that all activities or materials have been included in the estimate. If drawings are revised after material takeoff, new issues must be compared with the copy used for takeoff and appropriate revisions made in the estimate.

Construction materials are subject to waste and loss through handling, cutting to fit, theft, normal breakage, and storage loss. Failure to make proper allowance for waste and loss results in erroneous estimates.

Other error sources are inadvertent figure transpositions, copying errors, and math errors.

ACTIVITY ESTIMATES

The activity estimate provides a basis for preparing the estimates of material, equipment, and manpower requirements. An activity estimate, for example, might call for rough-in piping in a floor slab. In an activity estimate, your immediate concern is to identify the material necessary to do the task—pipe, fittings, joining materials, and so forth. The equipment estimate for this activity should consider vehicles for movement of material and special tools, such as portable power tools, a threader, and a power vise. From the scope of the activity and the time restraints, you can estimate the manpower required. The information shown in the activity estimate is also useful in scheduling progress and in providing the basis for scheduling deliveries of material, equipment, and manpower to the jobsite.

The techniques discussed in the next paragraphs will help you produce satisfactory activity estimates. But, before doing anything, you should become knowledgeable about the project by studying the drawings. Read the specifications and examine all available information concerning the site and local conditions. Only after becoming familiar with the project are you ready to identify individual activities. Now, here are two ideas that will help you make good estimates.

First, define activities. They may vary depending on the scope of the project. An activity is a clearly definable quantity of work. For estimating and scheduling, an activity for a single building or job should be a specific task or work element done by a single trade. For scheduling of large-scale projects, however, a complete building may be defined as an activity. But, for estimating it should remain at the single-task, single-trade level.

Second, after becoming familiar with the project and defining its scope, proceed with identifying the individual activities required to construct the project. To identify activities, be sure each activity description shows a specific quantity of work with clear, definite limitations or cutoff points that can be readily understood by everyone concerned with the project. Prepare a list of these activities in a logical sequence to check for completeness.

Material

Material estimates are used to procure construction material and to determine whether sufficient material is available to construct or complete a project. The sample forms shown in figures 9-2, 9-3, and 9-4 may be used in preparing material estimates. The forms show one method of recording the various steps taken in preparing a material estimate. Each step can readily be understood when the work sheets are reviewed. A work sheet must have the following headings: Project Title, Project Location, Drawing Number, Sheet Number, Project Section, Prepared By, Checked By, and Date Prepared.

ESTIMATING WORK SHEET.—The Estimating Work Sheet (figure 9-2), when completed, shows the various individual activities for a project with a listing of the required material. Material scheduled for several activities or uses is normally shown in the "Remarks" section. The work sheet should also contain an activity description, the item number, a material description, the cost, the unit of issue, the waste factors, the total quantities, and the remarks. The Estimating Work Sheets should be kept by the field supervisor during construction to ensure the use of the material as planned.

MATERIAL TAKEOFF SHEET.—The Material Takeoff Sheet (MTO) is shown in figure 9-3. In addition to containing some of the information on the Estimating Work Sheet, the MTO also contains the suggested vendors or sources, supply status, and the required delivery date.

BILL OF MATERIAL.—The Bill of Material (BM) sheet (figure 9-4) is similar in content to the Material Takeoff Sheet. Here, though, the information is presented in a format suitable for data processing. Use this form for requests of supply status, issue, or location of material, and for preparing purchase documents. When funding data is added, use these sheets for drawing against existing supply stocks.

Between procurement and final installation, construction material is subject to loss and waste.

ESTIMATING WORK SHEET								
PREPARED BY:	PROJ. LOCATION. DIEGO GARCIA	SHEET 1 OF 5	DRAWING NO 1,337,494/7,604,988		PROJ. TITLE CANTONMENT AREA INTERIM WATER SYSTEM BLDG.			
CHECKED BY:	PROJ SECTION. ARCHITECTURAL	ACTIVITY NO NODE 71 TO NODE 64	BM NO. DIW-112		MTO NO.		DATE PREPARED 19 FEB '92	

ITEM NO	DESCRIPTION	PREFAB FORMS / REFER TO DIW QC SECT. V PAR B&C PP 7+8	BM NO.	BM LINE ITEM NO	UNIT OF ISSUE		TOTAL QTY	REMARKS USE, LOC, PROCEDURES, ETC
	BUILDING FOOTING	L W T						SLAB/FOOTING
		26'-8" 20'-0" 12"						EDGE FORMS-TO BE
1.	3/4" PLYWOOD	2(26'-8")+(20')2 = 53'-4"+40' = 93'-4"		1	SH	O	3	USED AT TRANSMITTER
	BB EXTERIOR TYPE							SITE BLDG.
	4'x8'	8'x4' PLYWOOD RIPPED 12" = 32'						
		93.33/32' = 3 SHEETS						
2.	LUMBER 1x6xRL	6' LENGTH x 2 EA. CORNER		2	BF	15	30	BLDG LAYOUT
	GR 2 OR BETTER	x 4 CORNERS = 8 PCS / 6 LONG						BATTER BOARDS
3.	LUMBER 2x4xRL	16' -48 PCS -16'x2x4x48 PCS		3	BF	15	590	USE REFUSABLE
								2x4 AT TRANSMITTER
								SITE BUILDING
	RAMP AND DOOR STOOP FORMS							
4.	3/4" PLYWOOD 4'x8'	(13'-8")+2(6')+3(4') = 37'-8"		1	SH	O	1	EDGE FORMS REUSE
		RIP PLYWOOD INTO 8" STRIPS						AT TRANSMITTER
		= 6x8 = 48						SITE BUILDING
	BEAMS							
	B-1	2 EACH 26'-8" BOND BEAMS						
5.	3/4" PLYWOOD GR BB							B-1 SIDE FORMS
	EXT TYPE 4'x8'	26'-8" x 4 SIDES = 106'-8"		1	SH	O	10	REUSE AT
								TRANSMITTER SITE
								BUILDING

Figure 9-2.—Typical Estimating Work Sheet.

MATERIAL TAKEOFF SHEET

MATERIAL COST SECTION $ 1938.97	PROJECT TITLE: CANTONMENT AREA, INTERIM WATER SYS. BLDG	ARCHITECTURAL		1,337,494/7,604,988
	DIEGO GARCIA	19 FEB '92	31ST NCR	SHEET 1 - 5

P96	2	N	62583	4081	N	Y	1W112	A	TN	W	ZM9	05	4K6404

LINE ITEM	DOC IDENT	FSC / FSCM	PART NUMBER	U/I	QUANTITY	SERIAL	DIST COG	RDD	ADV	UNIT COST	TOTAL COST	REMARKS DESCRIPTION / VENDOR	SUPPLY STATUS
1	AOE			SH	69	7903		094		1200	82800	PLYWOOD, 3/4"x4'x8' BB EXTERIOR TYPE, SUGGESTED VENDOR: THOMPSON LBR CO	
2				BF	30	7904				18	540	LUMBER, SOFTWOOD, 1"x6"x12' STANDARD CONSTRUCTION GRADE 2 OR BETTER SUGGESTED VENDOR: THOMPSON LBR CO	
3				BF	2422	7905				28	67816	LUMBER, SOFTWOOD, 2"x4"x16' STANDARD CONSTRUCTION GRADE 2 OR BETTER SUGGESTED VENDOR: THOMPSON LBR CO	
4				BF	144	7906				28	4032	LUMBER, SOFTWOOD, 2"x6"x16' STANDARD CONSTRUCTION GRADE 2 OR BETTER, SUGGESTED VENDOR: THOMPSON LBR CO	
5				BF	1173	7907				33	38709	LUMBER, SOFTWOOD, 4"x4"x16' STANDARD CONSTRUCTION GRADE 2 OR BETTER SUGGESTED VENDOR: THOMPSON LUMBER CO	

Figure 9-3.—Typical Material Takeoff (MTO) Sheet.

SHIP APRIL '92

BILL OF MATERIAL

PROJECT: INTERIM WATER SYSTEM REINDEER STATION (CANTONMENT AREA) — L.W 112

ROUT IDENT	M&S	SERV & REQNR	SIG	SERV & SUPP ADDRESS	SIG	FUND CODE	DIST	PROJECT	PRI	RDD	JOB ORDER
P96	2	N62583	N	YIW112	A	TN	W	ZM9	05	094	4K6404

DOC IDENT	FSC	STOCK NUMBER NIIN	UNIT OF ISSUE	QUANTITY	DOCUMENT NUMBER DATE SERIAL	COG	ADV	L.I.	CRIT FLAG	DESCRIPTION/SPECIFICATION/VENDORS	UNIT COST	TOTAL COST	QTY/DATE RECD	QTY/DATE ISSUED
APE			SH	69	40817903			1		PLYWOOD, 3/4" X 4' X 8' BB EXTERIOR	12.00	828.00		
										TYPE SUGGESTED VENDOR:				
										THOMPSON LUMBER CO.				
			BF	30	7904			2		LUMBER, SOFTWOOD, 1" X 6" X 12'	.18	5.40		
										STANDARD CONSTRUCTION GRADE 2				
										OR BETTER. SUGGESTED VENDOR:				
										THOMPSON LUMBER CO.				
			BF	2242	7905			3		LUMBER, SOFTWOOD, 2" X 4" X 16'	.28	627.76		
										STANDARD CONSTRUCTION GRADE 2				
										OR BETTER. SUGGESTED VENDOR:				
										THOMPSON LUMBER CO.				
			BF	144	7906			4		LUMBER, SOFTWOOD, 2" X 6" X 16'	.28	40.32		
										STANDARD CONSTRUCTION GRADE 2				
										OR BETTER. SUGGESTED VENDOR				
										THOMPSON LUMBER CO.				

E. B. JASON | 20 MAR 92 | T.J. ABERNATHY | 5-29-92 | TOTAL 1501.48

Figure 9-4.—Sample Bill of Material (BM) sheet.

This loss may occur during shipping, handling, storage, or from the weather. Waste is inevitable where material is subject to cutting or final fitting before installation. Frequently, material, such as lumber, conduit, or pipe, has a standard issue length longer than required. More often than not, however, the excess is too short for use and ends up as waste. Waste and loss factors vary depending on the individual item and should be checked against the conversion and waste factors found in NAVFAC P-405, appendix C.

CHECKLISTS.—Use checklists to eliminate any omissions from the material estimates. Prepare a list for each individual project when you examine the drawings, specifications, and activity estimates. This is the practical way to prepare a listing for the variety of material used in a project. The listing applies only to the project for which it has been prepared. If no mistakes or omissions have been made in either the checklist or estimate, the material estimate will contain a quantity for each item on the list.

LONG LEAD TIMES.—Long lead items are not readily available through the normal supply system. They require your special attention to ensure timely delivery. Items requiring a long lead time are nonshelf items, such as steam boilers, special door and window frames, items larger than the standard issue, and electrical transformers for power distribution systems. Identify and order these items early. Make periodic status checks of the orders to avoid delays in completing the project.

PREPARING MATERIAL ESTIMATES.—There are several steps for preparing a material estimate. First, determine the activity by using the activity description with the detailed information furnished by the drawings and plans to provide a quantity of work. Convert this quantity to the material required. Next, enter the conversion on a work sheet to show how each quantity was computed, as shown in figure 9-2. Include sufficient detail; work sheets need to be self-explanatory. Anyone examining them should be able to determine how the quantities were computed without having to consult the estimator. Allowances for waste and loss are added after determining the total requirement. All computations should appear on the estimate work sheet, as must all notes relative to the reuse of the material. Material quantities for similar items of a project are entered on the Material Takeoff Sheet or

Bill of Material. Figures 9-3 and 9-4 become the material estimate for the project.

Equipment

Equipment estimates are used with production schedules to determine the construction equipment requirements and constraints for Seabee deployment. Of these constraints, the movement of material over roadways is frequently miscalculated. In the past, estimators used the posted speed limit as an average rate for moving material. This was wrong. Equipment speed usually averages between 40 to 56 percent of the posted speed limit. Factors, such as the road conditions, the number of intersections, the amount of traffic, and the hauling distances, vary the percentage of the posted speed limit. You should consider the types of material hauled; damp sand or loam, for example, is much easier to handle than clay. Safety (machine limitations), operator experience, condition of the equipment, work hours, and the local climate are other factors.

Equipment production must be determined so that the amount and type of equipment can be selected. Equipment production rates are available in the

SHEET 1 OF 2

ESTIMATED BY Brown DATE 6/13/92

CHECKED BY Green DATE 6/23/92

EQUIPMENT ESTIMATE

NMCB ______ LOCATION GUAM YEAR 1992

PROJECT No. 013 DESCRIPTION Site Preparation

Earth Fill — 36,000 CY loose measurement required.
Haul one way 2-1/2 miles.
Use 2-1/2 CY endloader and 10 CY dump trucks.

Endloader capacity 100 CY/hours.

$\frac{36,000}{100}$ =360 hours or 45 eight-hour days.

$\frac{100}{10}$ = 10 trucks loaded per hour.

Average hauling speed estimated at 15 MPH.

2 x 2.5 = 5 miles round trip.

5/15 x 60 = 20 minutes hauling time.

60/10 = 6 minutes loading time.

Estimated 4 minutes dumping time.

30 minutes total time per truckload.

60/30 = 2 loads per hour per truck.

10/2 = 5 trucks required to keep endloader working at capacity.

100 x 8 = 800 CY hauled per 8-hour day.

Need one bulldozer (can spread 1400 CY daily).

Need one grader to keep haul road in shape.

1 bulldozer (can spread 1400 CY daily).

1 tractor & tandem sheepfoot roller (can compact 1200 CY daily).

1 water truck with sprinkler for moisture control.

1 rubber-tired wobbly wheel roller on standby for compaction and sealing fill when rain is expected. (Can be towed by above bulldozer or tractor.)

Figure 9-5.—Sample equipment estimate (sheet 1 of 2).

Seabee Planner's and Estimator's Handbook. The tables in this handbook provide information about the type of equipment required. Estimate the production rate per day for each piece of equipment. You should consider the factors discussed above, along with information obtained from NAVFAC P-405 and your experience. The quantity of work divided by the production rate per day produces the number of days required to perform the project. After determining the number of days of required equipment operation, consult the project schedule to find the time allotted to complete the activities. Prepare the schedule for the total deployment. Use the project schedule to determine when the work will be performed. The schedule should also indicate peak usage. It may have to be revised for more even distribution of equipment loading, thereby reducing the amount of equipment required during the deployment.

ESTIMATE SHEETS.—After the reviews and revisions, prepare a list of equipment required. The list must include anticipated downtime. Sufficient reserve pieces must be added to cover any downtime.

To aid you in preparing the equipment estimate schedule, use such forms as those shown in figures 9-5 and 9-6. The important information on the forms

SHEET 2 OF 2

ESTIMATED BY Brown DATE 6/13/92

CHECKED BY Green DATE 6/23/92

EQUIPMENT ESTIMATE

NMCB ______ LOCATION GUAM YEAR 1992

PROJECT No. 013 DESCRIPTION Site Preparation

NOTE: Preceding is not very efficient, as spreading equipment is not used to full capacity. Suppose that when the work schedule is prepared, completion of fill will be required in 18 days. Assume that climate is such that 3 days in every 17 working days will be lost due to rain. Therefore, 15 working days would be available in an 18 day schedule.

3,600/15 = 2,400 cu. yd. must be hauled daily to complete the work on schedule.

2,400/800 = 3 times the output of loading and hauling spread shown previously.

Equipment required for loading and hauling:

3 — 2-1/2 cu.yd. endloaders.
1 — bulldozer to keep pit in shape.
1 — grader to keep haul road in shape.
15 — 10-ton trucks hauling (1 or 2 extra trucks should be added to assure that a truck will always be waiting to be loaded so that endloader will work at full capacity).

2,400 cu. yd. will be hauled each day.

2,400/1,200 = 2 tractors and tandem sheepfoot roller for compaction.
2 bulldozers to spread earth.

2,400/1,400 = 1 water truck with sprinkler.

1 wobbly-wheel roller (standby for sealing of fill before rains).

NOTE: This is a more efficient operation, as production has been tripled but equipment has not, and total equipment working at or as close to capacity as can be expected.

Figure 9-6.—Sample equipment estimate (sheet 2 of 2).

includes the sheet number, the name of the estimator, the name of the checker, date checked, battalion and detachment number, location of deployment, year of deployment, project number, and a brief description of the project.

TOA AND EQUIPMENT CHARACTERISTICS.—The table of allowance (TOA) of the Naval Mobile Construction Battalion (NMCB) contains specific information on the quantities and characteristics of construction equipment available to the NMCBs. Table 9-1 contains an abbreviated listing of such equipment.

Labor

There are two types of labor estimates: preliminary manpower estimates and detailed manpower estimates.

Table 9-1.—NMCB Construction Equipment Characteristics

QUANTITY	EQUIPMENT DESCRIPTION
	TRUCKS
12	Dump, 6 × 6, 5 ton, 5 cu. yd. capacity
8	Dump, 6 × 4, 15 ton, 10 cu. yd. capacity
	GRADERS
6	Motor, road, 12 ft. blade, 6 × 4, with scarifier
	LOADERS
4	Scoop, full tracked, 2 1/2 cu. yd. multipurpose bucket
2	Scoop, wheeled, 4 × 4, 2 1/2 cu. yd. std. bucket with forks
2	Scoop, wheeled, 4 × 2, 2 1/2 cu. yd. std. bucket with forks backhoe, crane, dozer blade
	ROLLERS
2	Oscillating, self-propelled, 9 wheel, pneumatic tired
3	Vibrating, self-propelled, pneumatic tired, single drum
	SCRAPERS
6	Tractor, wheeled, 14 to 20 cu. yd., hydraulic
	TRACTORS
5	Crawler, hydraulic semi-U-tilt dozer
2	Crawler, hydraulic angle dozer, winch
1	Crawler, hydraulic semi-U dozer, hydraulic ripper
	CRANES
2	Truck, mounted, 8 × 4, 35 ton, 60-ft. boom with extension
1	Truck, mounted, 8 × 4, 25 ton, hydraulic
1	Tractor, wheel mounted, 4 × 4, 12 1/2 ton, telescoping boom, hydraulic
	SPECIALIZED EQUIPMENT
1	Distributor, bituminous material, truck mounted, 6 × 4, 2,000 gal. capacity
2	Distributor, water, truck mounted, 6 × 6, 2,000 gal. capacity
2	Distributor, water, wagon mounted, 8,000 gal. capacity
2	Ditching machine, ladder type, 8- to 24-in. width by 7-ft. depth, crawler mounted
2	Excavator, multipurpose, hydraulic, 6 × 6, 11 ft. 1 in digging depth, truck mounted
2	Auger, earth, truck mounted
8	Truck, forklift, rough terrain, 6,000-lb. capacity, pneumatic tired

PRELIMINARY.—Use preliminary manpower estimates to establish budget costs and to project manpower requirements for succeeding projects and deployments. The estimates are prepared from limited information, such as general descriptions or preliminary plans and specifications that contain little or no detailed information. In some cases, you can make a comparison with similar facilities of the same basic design, size, and type of construction. A good preliminary estimate varies less than 15 percent from the detailed estimate.

DETAILED.—Use detailed manpower estimates to determine the manpower requirements for constructing a given project and the total direct labor requirements of a deployment. Take the individual activity quantities from the activity work sheet to prepare detailed estimates. Then, select the man-hours per unit figure from the appropriate table in NAVFAC P-405 and multiply it by the quantity to obtain the total man-hours required. When preparing the activity estimates in the format discussed earlier, you may use a copy of the activity estimates as a manpower estimate work sheet by adding four columns to it with the headings of Activity, Quantity, Man-Hours Per Unit, and Total Man-Days Required. Work sheets, whether on the activity work sheet or on another format, should be prepared in sufficient detail to provide the degree of progress control desired. For example, the work sheets should show the following information:

DESCRIPTION	QUANTITY	MAN-HOURS* PER UNIT	TOTAL MAN-DAYS
Install 12-inch-diameter concrete pipe	2,500 feet	20/100	62.5
Install 30-inch-diameter concrete pipe	2,500 feet	80/100	250.0
TOTALS	5,000 feet		312.5

* 8 man-hours equals 1 man-day.

If the control is to be exercised only on concrete pipe installation without regard to detail, the manpower estimate should show the following information on the summary sheet:

DESCRIPTION	QUANTITY	MAN-HOURS PER UNIT	TOTAL MAN-DAYS
Install concrete pipe	5,000 feet	50/100	312.5

Table 9-2.—Production Efficiency Guide Chart

ELEMENTS	LOW PRODUCTION	AVERAGE PRODUCTION	HIGH PRODUCTION
	Production Elements in Percent 25 35 45	55 65 75 85	90 95 100
1. Work Load	Construction requirement high, miscellaneous overhead high	Construction requirement normal, miscellaneous overhead normal	Construction requirement low, miscellaneous overhead low
2. Site Area	Cramped working area, no area for material storage, work restricted to design, poor job layout	Work area limited slightly, partial material storage, some variation from design, average job layout	Large work area, adequate material storage, wide latitude from design, good job layout
3. Labor	Poorly trained, low strength, low morale, high sick call	Average trained, normal strength, fair morale, normal sick call	Highly trained, over strength, high morale, low sick call
4. Supervision	Poor management, poorly trained personnel, low strength	Average management, average trained personnel, normal strength	Efficient management, highly trained personnel, over strength
5. Job Conditions	High quality work required, unfavorable site materials, short time operations, insect annoyance high	Average work required, average site materials, reasonable operation time, insect annoyance normal	Passable work required, good site materials, long time operation, no insect annoyance
6. Weather	Abnormal rain, abnormal heat, abnormal cold	Moderate rain, moderate heat, moderate cold	Some rain, occasional heat, occasional cold
7. Equipment	Improper job application, equipment in poor condition, repair and maintenance inadequate	Fair job application, equipment in average condition, repair and maintenance average	Efficient job application, equipment in good condition, efficient repair and maintenance
8. Tactical and Logistical	Slow supply delivery, frequent tactical delays	Normal supply delivery, occasional tactical delays	Prompt supply delivery, no tactical delays

The man-hours per unit on the work sheet is obtained by dividing the total man-days shown in the detail estimate by the total feet of concrete pipe times the unit to obtain the average man-hours. The man-hours per unit should be used for checking actual progress. You should check manpower estimates against the activity estimate to ensure that no activities have been omitted. NAVFAC P-405 provides labor estimates for the various projects undertaken by the Engineering Aids.

The *Facilities Planning Guide*, NAVFAC P-437, volumes 1 and 2, is an excellent source for preliminary estimates. Use it to find estimates for a wide range of facilities and assemblies commonly constructed. The P-437 not only gives the man-hours required, but it also gives a breakdown of the construction effort by ratings (BU, CE, UT, and so forth) as well as lapsed day estimates.

You must bear in mind that the lapse time from the P-437 is calculated using the contingency norm of a 10-hour man-day instead of the 8-hour man-day used in the P-405. For example, a specific task from the P-437 requires 100 man-hours (MH) of effort by the Utilitiesman. The optimum crew size is four UTs. This yields the following lapse time:

$$100\frac{MH}{4\ UTs \times 10\ hr} = 2.5 \text{ days (lapse time)}$$

Using the P-405 and an 8-hour man-day, you will find that the same task yields the following:

$$100\frac{MH}{4\ UTs \times 8\ hr} = 3.1 \text{ days (lapse time)}$$

In preparing manpower estimates, weigh the various factors affecting the amount of labor required to construct a project. These include weather conditions during the construction period, skill and experience of personnel who will perform the work, time allotted for completing the job, size of the crew to be used, accessibility of the site, and types of material and equipment to be used.

The production efficiency guide chart (table 9-2) lists eight elements that directly affect production. Each production element is matched with three areas for evaluation. Each element contains two or more foreseen conditions from which to select for the job in question. Evaluate each production element at some percentage between 25 and 100, according to your analysis of the foreseen conditions. The average of the eight evaluations is the overall production efficiency percentage. Now, convert the percentage

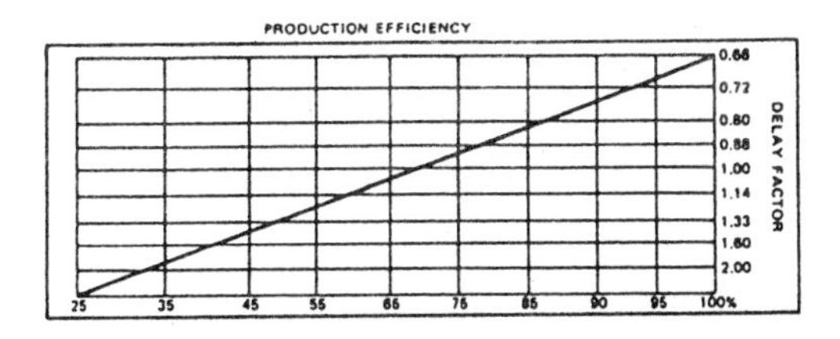

Figure 9-7.—Production efficiency graph.

to a delay factor, using the production efficiency graph (figure 9-7). It is strongly recommended that the field or project supervisors reevaluate the various production elements and make the necessary adjustments to man-day figures based on actual conditions at the jobsite.

NOTE

> The estimate of average Seabee production used in the NAVFAC P-405 tables falls at 67-percent production efficiency on the graph shown in figure 9-7. As you see, this represents a delay factor of 1.00. A delay factor of 0.66 represents peak production efficiency, equivalent to 100 percent.

In reading the graph, note that the production elements have been computed into percentages of production efficiency, which are indicated at the bottom of the graph. First, place a straightedge so that it extends up vertically from the desired percentage, and then place it horizontally from the point at which it intersected the diagonal line. You can now read the delay factor from the values given on the right-hand side of the chart. Let's look at an example of the process of adjusting man-hour estimates.

Assume that from the work estimate taken from the tables in P-405, you find that 6 man-hours are needed for a given unit of work. To adjust this figure to the conditions evaluated on your job, assume that the average of foreseen conditions rated by you is 87 percent. The corresponding delay factor read from the production efficiency graph is 0.80. You find the adjusted man-hour estimate by multiplying this delay factor by the man-hours from the estimating tables ($6\,MH \times 0.8 = 4.8$ as the adjusted man-hour estimate).

The man-hour labor estimating tables are arranged and grouped together into the 16 major divisions of work. This is the same system used to prepare government construction specifications. The 16 major divisions of work are as follows:

1. General;
2. Site work;
3. Concrete;
4. Masonry;
5. Metal;
6. Carpentry;
7. Moisture protection;
8. Doors, windows, glass;
9. Finishes;
10. Specialties;
11. Architectural equipment;
12. Furnishings;
13. Special construction;
14. Conveying systems;
15. Mechanical; and
16. Electrical.

The activities in the various labor estimating tables are divided into units of measurement commonly associated with each craft and material takeoff quantities. There is only one amount of man-hour effort per unit of work. This number represents normal Seabee production under average conditions. As used herein, 1 man-day equals 8 man-hours of direct labor. Man-day figures do not include overhead items, such as dental or personnel visits, transportation to and from the jobsite, or inclement weather.

No two jobs are exactly alike, nor do they have exactly the same conditions. Therefore, you, as the estimator, must exercise some judgment about the project that is being planned. The production efficiency guide chart and graph (table 9-2 and figure 9-7) are provided to assist you in weighing the many factors that contribute to varying production conditions and the eventual completion of a project. You can then translate what is known about a particular project and produce a more accurate quantity from the average figures given on the labor estimating tables.

SCHEDULING

LEARNING OBJECTIVE: Upon completing this section, you should be able to explain the scheduling requirements for a construction project.

After World War II, the construction industry experienced the same critical examination that the manufacturing industry had experienced 50 years before. Large construction projects came under the same pressures of time, resources, and cost that prompted studies in scientific management in the factories.

The emphasis, however, was not on actual building methods, but upon the management techniques of programming and scheduling. The only planning methods being used at that time were those developed for use in factories. Management tried to use these methods to control large construction projects. These techniques suffered from serious limitations in project work. The need to overcome these limitations led to the development of network analysis techniques.

BASIC CONCEPTS

In the late 1950s, this new system of project planning, scheduling, and control came into widespread use in the construction industry. The critical path analysis (CPA), critical path method (CPM), and project evaluation and review technique (PERT) are samples of about 50 different approaches. The basis of each of these approaches is the analysis of a network of events and activities. For this reason, the generic title covering the various networks is "network analysis."

Network analysis techniques are now the accepted method of construction planning in many organizations. They form the core of project planning and control systems.

Advantages and Disadvantages

There are many advantages of network analysis. As a management tool, it readily separates planning from scheduling of time. The analysis diagram, a pictorial representation of the project, enables you to see the interdependencies between events and the overall project to prevent unrealistic or superficial planning. Resource and time restraints are easily detachable, to permit adjustments in the plan before its evaluation.

Because the system splits the project into individual events, estimates and lead times are more accurate. Deviations from the schedule are quickly noticed. Manpower, material, and equipment resources are easily identifiable. Since the network remains constant throughout its duration, it is also a statement of logic and policy. Modifications of the policy are allowed, and the impact on events is assessed quickly.

Identification of the critical path is useful when you have to advance the completion date. Attention can then be concentrated toward speeding up those relatively few critical events. The network allows you to accurately analyze critical events and provides an effective basis for the preparation of charts. This results in better control of the entire project.

The main disadvantage of network analysis as a planning tool is that it is a tedious and exacting task when attempted manually. Depending upon what the project manager wants as output, the number of activities that can be handled without a computer varies but is never high.

Calculations are in terms of the sequence of activities. Now, a project involving several hundred activities **may** be attempted manually. However, the chance for error is high. Suppose the jobs are to be sorted by rating, so jobs undertaken by Utilitiesmen are together as are those for Equipment Operators or Construction Electricians. The time required for manual operation would become costly.

On the other hand, standard computer programs for network analysis can handle project plans of 5,000 activities or more and can produce output in various forms. However, a computer assists only with the calculations and print plans of operations sorted into various orders. The project manager, **not** the computer, is responsible for planning and must make decisions based on information supplied by the computer. Also, computer output is only as accurate as its input, supplied by people. The phrase "garbage in, garbage out" applies.

Elements

A network represents any sequencing of priorities among the activities that form a project. This sequencing is determined by hard or soft dependencies. Hard dependencies are based upon the physical characteristics of the job, such as the necessity for placing a foundation before building the walls. A hard dependency is normally inflexible. Soft dependencies are based upon practical considerations of policy and may be changed if circumstances demand. The decision to start at the north end of a building rather than at the south end is an example.

PRECEDENCE DIAGRAMING

Network procedures are based upon a system that identifies and schedules key events into precedence-related patterns. Since the events are interdependent, proper arrangement helps in monitoring the independent activities and in evaluating project progress. The basic concept is known as the critical path method (CPM). Because the CPM places great emphasis upon task accomplishment, a means of activity identification must be established to track the progress of an activity. The method currently in use is the activity-on-node precedence diagramming method (PDM), where a node is simply the graphic representation of an activity. An example of this is shown in figure 9-8.

Precedence diagramming does not require the use of dummy activities. It is also easier to draw, and has greater applications and advantages when networks are put in the computer. In precedence diagrams, the activity is "on the node."

Activities and Events

To build a flexible CPM network, the manager needs a reliable means of obtaining project data to be represented by a node. An activity in a precedence diagram is represented by a rectangular box and identified by an activity number.

The left side of the activity box represents the start of the activity. The right side represents the

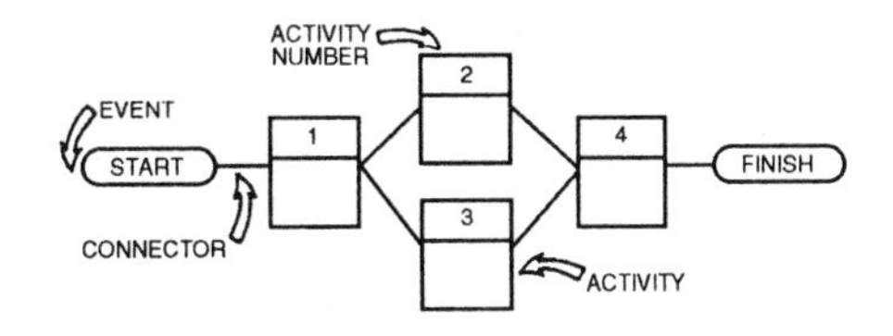

Figure 9-8.—Precedence diagram.

completion. Lines linking the boxes are called connectors. The general direction of flow is evident in the connectors themselves.

Activities may be divided into three distinct groups:

1. **Working activities**—Activities that relate to particular tasks;
2. **Milestone events**—Intermediate goals with no time duration, but that require completion of prior events before the project can proceed; and
3. **Critical activities**—Activities that, together, comprise the longest path through the network. This is represented by a heavy- or hash-marked line.

The activities are logically sequenced to show the activity flow for the project. The activity flow can be determined by answering the following questions: What activities must precede the activity being examined? What activities can be concurrent with this activity? What activities must follow this activity?

WORKING ACTIVITIES.—With respect to a given activity, these representations indicate points in time for the associated activities. Although the boxes in the precedence diagram represent activities, they do not represent time and, therefore, are not normally drawn to scale. They only reflect the logical sequence of events.

MILESTONE EVENTS.—The network may also contain certain precise, definable points in time, called events. Examples of events are the start and finish of the project as a whole. Events have no duration and are represented by oval boxes in a network, as shown in figure 9-8.

Milestones are intermediate goals within a network. For instance, "ready for print" is an important event that represents a point in time but has no time duration of its own. To reach this particular activity, all activities leading up to it must be completed.

CRITICAL ACTIVITIES.—A critical activity is an activity within the network that has zero float time. The critical activities of a network make up the longest path through the network (critical path) that controls the project finish date. Slashes drawn through an activity connector, as shown in figure 9-9, denote a critical path.

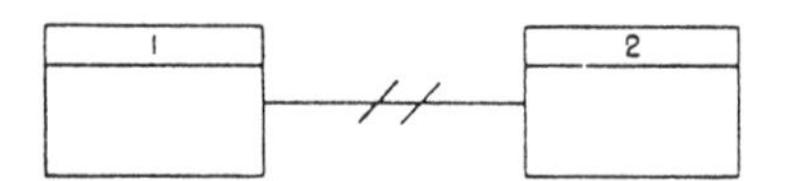

Figure 9-9.—Designation of a critical path.

The rule governing the drawing of a network is that the start of an activity must be linked to the ends of all completed activities before that start may take place. Activities taking place at the same time are not linked in any way. In figure 9-8, both Activity 2 and Activity 3 start as soon as Activity 1 is complete. Activity 4 requires the completion of both Activities 2 and 3 before it may start.

Use of Diagram Connectors

Within a precedence diagram, connectors are lines drawn between two or more activities to establish logic sequence. In the next paragraphs, we will look at the diagram connectors commonly used in the NCFs.

REPRESENTING A DELAY.—In certain cases, there may be a delay between the start of one activity and the start of another. In this case, the delay may be indicated on the connector itself, preceded by the letter *d* as in figure 9-10. Here, Activity 2 may start as soon as Activity 1 is complete, but Activity 3 must wait 2 days. The delay is stated in the basic time units of the project, so the word "days" can be omitted.

REPRESENTING A PARALLEL ACTIVITY.— Some activities may parallel others. This can be achieved in precedence diagrams without increasing the number of activities. For instance, it is possible to start laying a long pipeline before the excavations are completed. This type of overlap is known as a lead. It is also possible to start a job independently, but to not complete it before another is

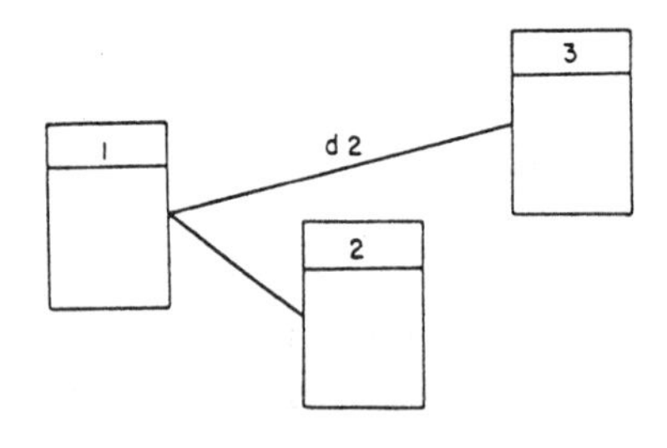

Figure 9-10.—Representation of delay.

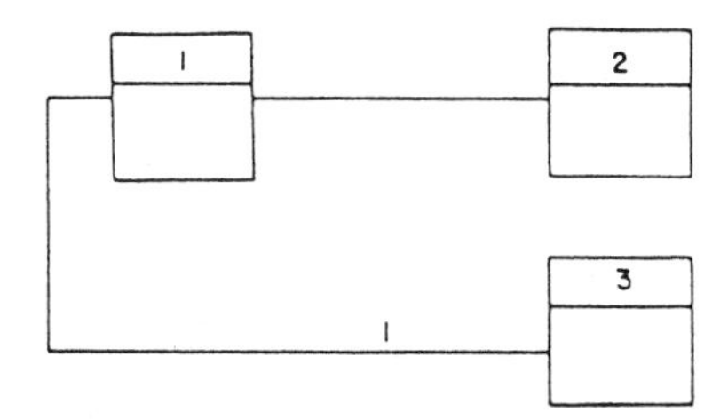

Figure 9-11.—Lead on start of a preceding activity.

completed. This type of overlap is known as a lag. It is also a common occurrence that both the start and the finish of two activities may be linked, but, in this case, they are accommodated by a combination of lead and lag.

As seen in figure 9-11, a lead (or partial start) is indicated by drawing the connector from the start of the preceding activity (1). In figure 9-12, a lag (or partial finish) is indicated by drawing the connector from the end of the following activity (3). The values may be given in the basic time units of the project, as with a delay, or as a percentage of overlap. In certain circumstances, they can be stated as quantities if the performance of the activity can be measured on a quantitative basis. The indication of the type and amount of delay, lead, or lag is generally referred to as a "lag factor."

In figure 9-11, Activity 3 may start when Activity 1 is 1-day completed, although Activity 2 must wait for the final completion of Activity 1. In figure 9-12, Activity 3 may start when Activity 2 is completed but will still have 1 day to go when Activity 1 is completed. The last phase of Activity 3 may not begin until Activity 1 has been completed. In figure 9-13, Activity 2 may start when Activity 1 is

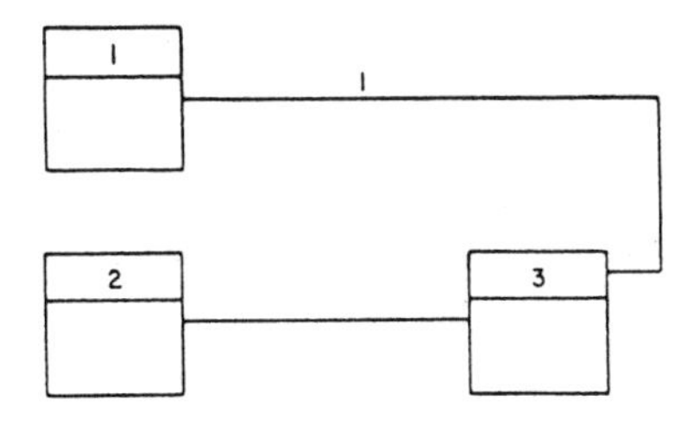

Figure 9-12.—Lag on finish a of following activity.

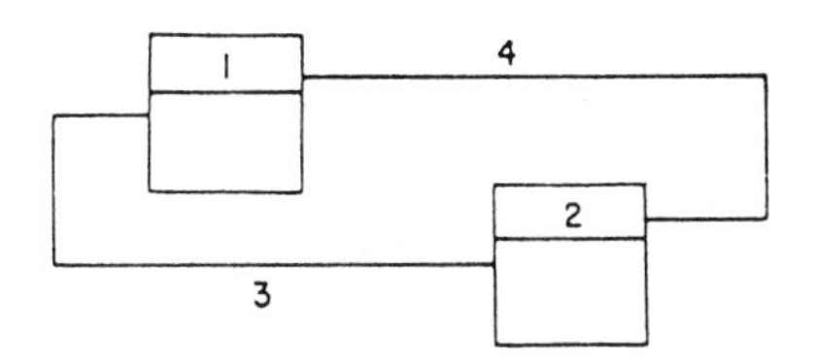

Figure 9-13.—Start and finish lags on same activity.

advanced 3 days but will still have 4 days of work left when Activity 1 is completed.

SPLITTING CONNECTORS.—The number of sequencing connectors becomes large when a network is of a great size. When two activities are remote from each other and have to be connected, the lines tend to become lost or difficult to follow. In such cases, it is not necessary to draw a continuous line between the two activities. Their relationship is shown by circles with the following-activity number in one and the preceding-activity number in the other. In figure 9-14, both Activities 2 and 6 are dependent upon Activity 1.

DIRECT LINKING USING AN EVENT.—When the number of common preceding and succeeding activities in a particular complex is large, as in figure 9-15, a dummy event or focal activity of zero duration may be introduced to simplify the network. The use of such a dummy event is shown in figure 9-16, which is a simplification of figure 9-15. Although the effect in terms of scheduling is the same, the introduction of the dummy improves the clarity of the diagram.

JOINING CONNECTORS.—In many instances, there are opportunities to join several

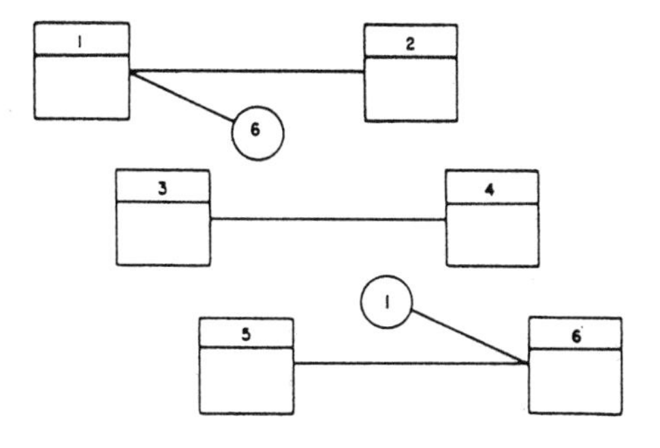

Figure 9-14.—Splitting connectors.

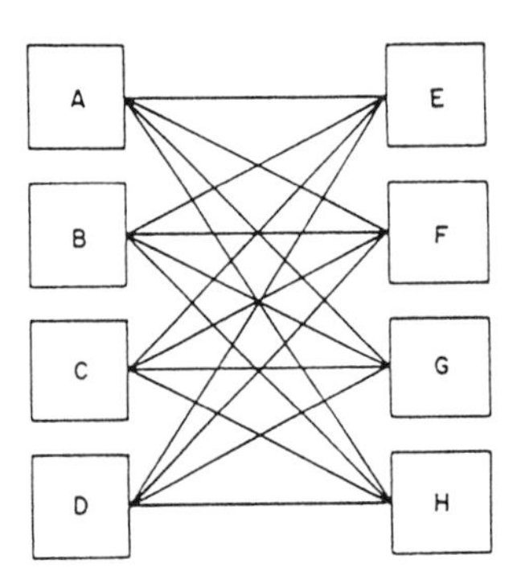

Figure 9-15.—Multiple predecessors and successors (direct linking).

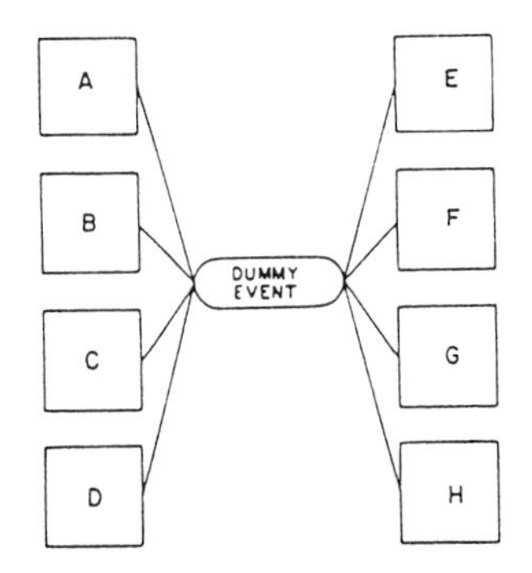

Figure 9-16.—Multiple predecessors and successors (using dummy collector).

connectors going to a common point to reduce congestion in the drawing. This practice is, however, discouraged.

The diagrams in figures 9-17 and 9-18 have precisely the same interpretation. The danger with

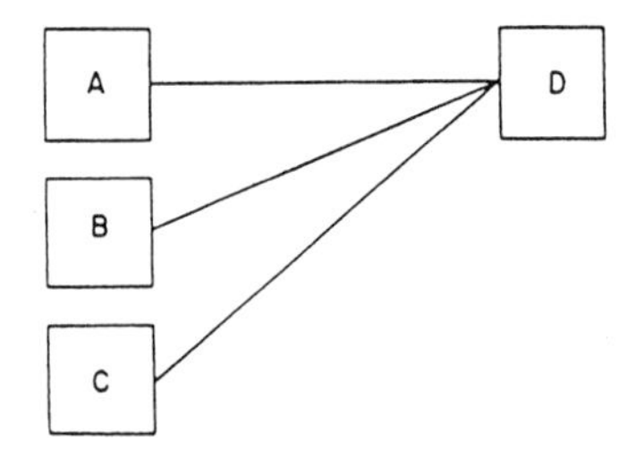

Figure 9-17.—Direct representation of dependencies.

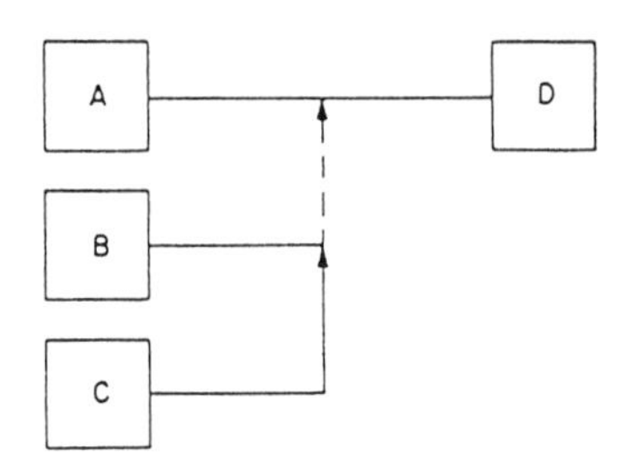

Figure 9-18.—Indirect linking of dependencies.

the form of representation is evident in figure 9-18, where several connectors have been joined. When the network is coded for the computer, you may lose sight of the fact that Activity D has **three** preceding activities since only **one** line actually enters Activity D.

PRECEDENCE DIAGRAMS

Scheduling involves putting the network on a working timetable. Information relating to each activity is contained within an activity box, as shown in figure 9-19.

Forward and Backward Pass Calculations

To place the network on a timetable, you must make time and duration computations for the entire project. These computations establish the critical path and provide the start and finish dates for each activity.

Each activity in the network can be associated with four time values:

- **Early start (ES)**—Earliest time an activity may be started;

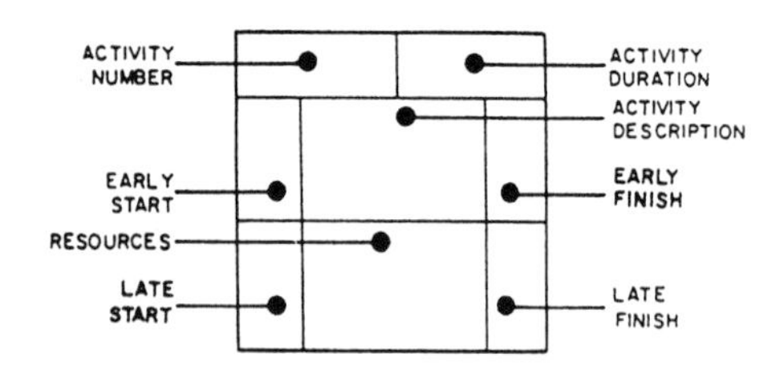

Figure 9-19.—Information for a precedence activity.

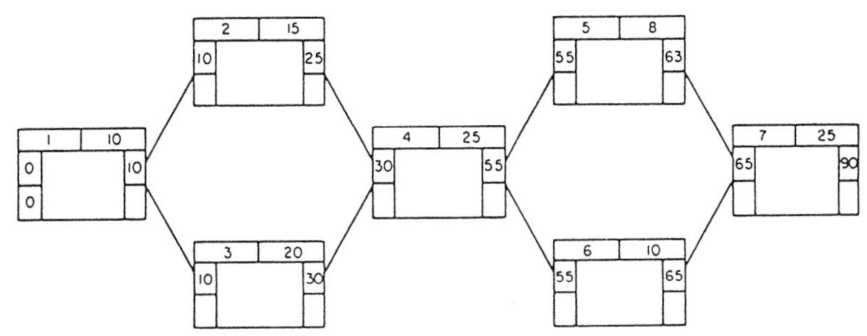

Figure 9-20.—Example of forward-pass calculations.

- **Early finish (EF)**—Earliest time an activity may be finished;
- **Late start (LS)**—Latest time an activity may be started and still remain on schedule; and
- **Late finish (LF)**—Latest time an activity may be finished and still remain on schedule.

The main objective of forward-pass computations is to determine the duration of the network. The forward pass establishes the early start and finish of each activity and determines the longest path through the network (critical path).

The common procedure for calculating the project duration is to add activity durations successively, as shown in figure 9-20, along chains of activities until a merge is found. At the merge, the largest sum entering the activity is taken at the start of succeeding activities. The addition continues to the next point of merger, and the step is repeated. The formula for forward-pass calculations is as follows:

ES = EF of preceding activity

EF = ES + activity duration

The backward-pass computations provide the latest possible start and finish times that may take place without altering the network relationships. These values are obtained by starting the calculations at the last activity in the network and working backward, subtracting the succeeding duration of an activity from the early finish of the activity being calculated. When a "burst" of activities emanating from the same activity is encountered, each path is calculated. The smallest or multiple value is recorded as the late finish.

The backward pass is the opposite of the forward pass. During the forward pass, the early start is added to the activity duration to become the early finish of that activity. During the backward pass, the activity duration is subtracted from the late finish to provide the late start time of that activity. This late start time then becomes the late finish of the next activity within the backward flow of the diagram.

LS = LF – activity duration

Figure 9-21 shows a network with forward- and backward-pass calculations entered.

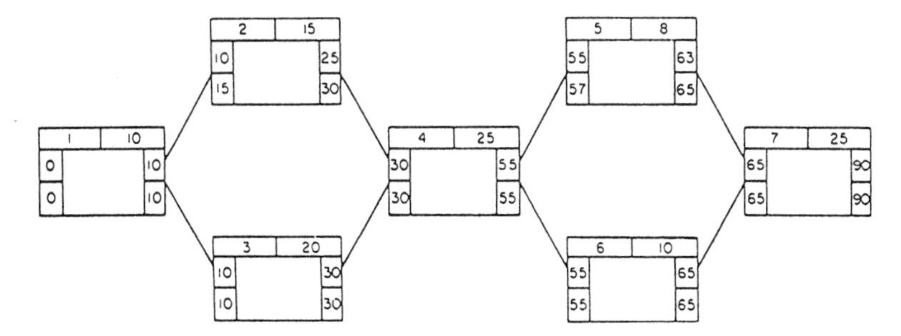

Figure 9-21.—Example of forword- and backward-pass calculations.

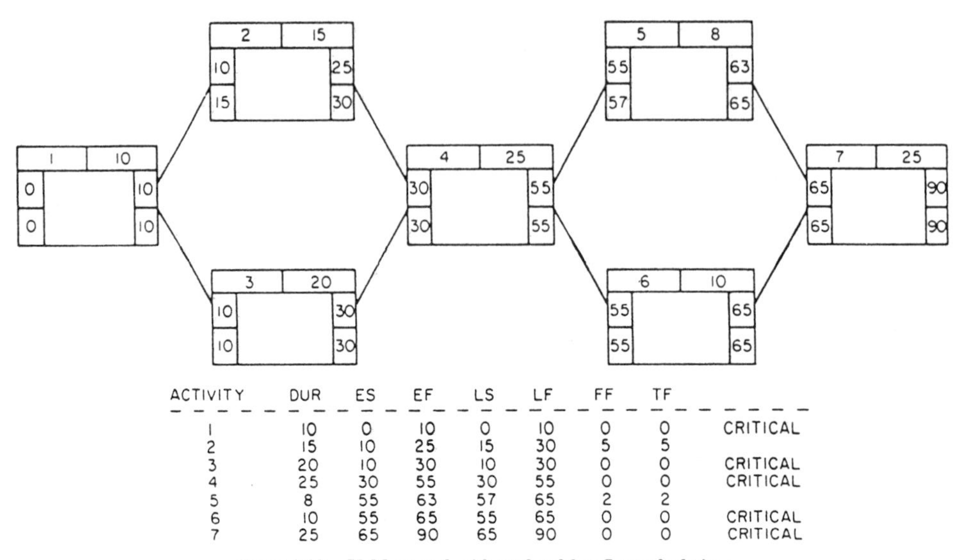

ACTIVITY	DUR	ES	EF	LS	LF	FF	TF	
1	10	0	10	0	10	0	0	CRITICAL
2	15	10	25	15	30	5	5	
3	20	10	30	10	30	0	0	CRITICAL
4	25	30	55	30	55	0	0	CRITICAL
5	8	55	63	57	65	2	2	
6	10	55	65	55	65	0	0	CRITICAL
7	25	65	90	65	90	0	0	CRITICAL

Figure 9-22.—PDM network with total and free float calculations.

The free and total float times are the amount of scheduled leeway allowed for a network activity, and are referred to as float or slack. For each activity, it is possible to calculate two float values from the results of the forward and backward passes.

TOTAL FLOAT.—The accumulative time span in which the completion of all activities may occur and not delay the termination date of the project is the total float. If the amount of total float is exceeded for any activity, the project end date extends to equal the exceeded amount of the total float.

Calculating the total float consists of subtracting the earliest finish (EF) date from the latest finish (LF) date, that is:

$$Total\ float = LF - EF$$

FREE FLOAT.—The time span in which the completion of an activity may occur and not delay the finish of the project or the start of a successor activity is the free float. If this value is exceeded, it may not affect the project end date but will affect the start of succeeding, dependent activities.

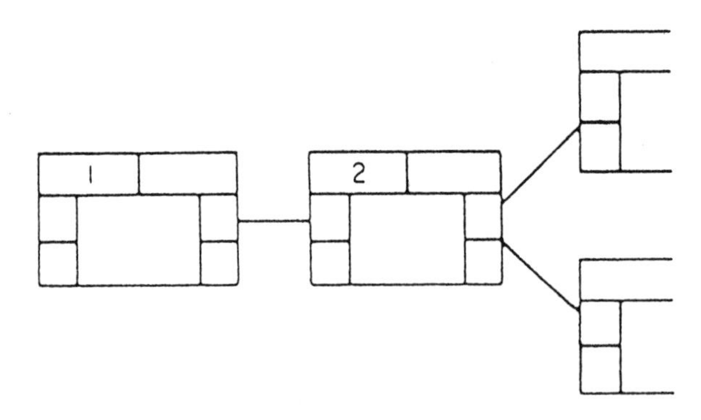

Figure 9-23.—Independent activity.

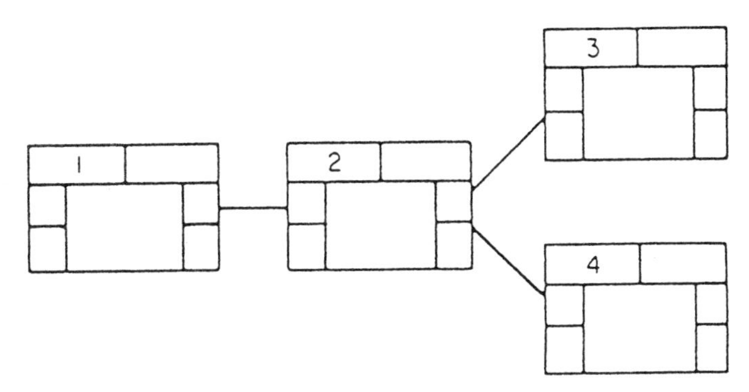

Figure 9-24.—Dependent activity.

Calculating the free float consists of subtracting the earliest start (ES) date from the latest start (LS) date, or:

Free float = *LS* – *ES*

Figure 9-22 is an example of an activity-on-node precedence diagramming method (PDM) network with total and free float calculations completed.

INDEPENDENT ACTIVITY.—An independent activity is an activity that is not dependent upon another activity to start. Activity 1, diagrammed in figure 9-23, is an example of an independent activity.

DEPENDENT ACTIVITY.—A dependent activity is an activity that is dependent upon one or more preceding activities being completed before it can start. The relationship in figure 9-24 states that the start of Activity 2 is dependent upon the finish of Activity 1.

Frequently, an activity cannot start until two or more activities have been completed. This appears in the diagram as a merge or junction. In figure 9-25, Activities 3 and 4 must be completed before the start of Activity 5.

Earlier we mentioned a "burst" of activities. A burst is similar to a merge. A burst exists when two or more activities cannot be started until a third activity is completed. In figure 9-24, when Activity 2 is finished, Activities 3 and 4 may start.

Advantages of Diagraming

Precedence networks are easy to draw because all the activities can be placed on small cards, laid out on a flat surface, and easily manipulated until a realistic logic is achieved. It is also easy to show the interrelationships and forward progress of the activities. Just

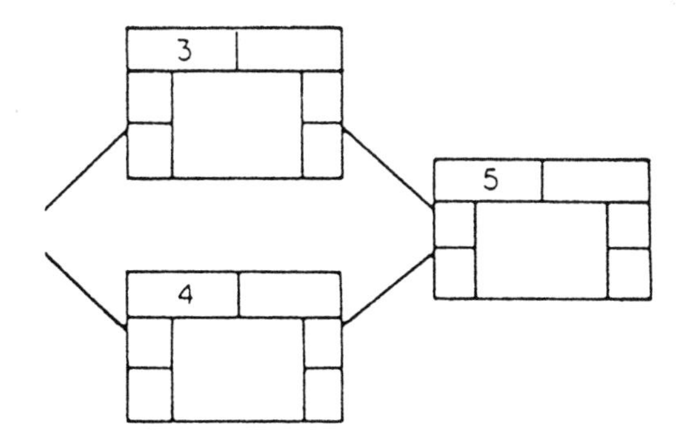

Figure 9-25.—Merge.

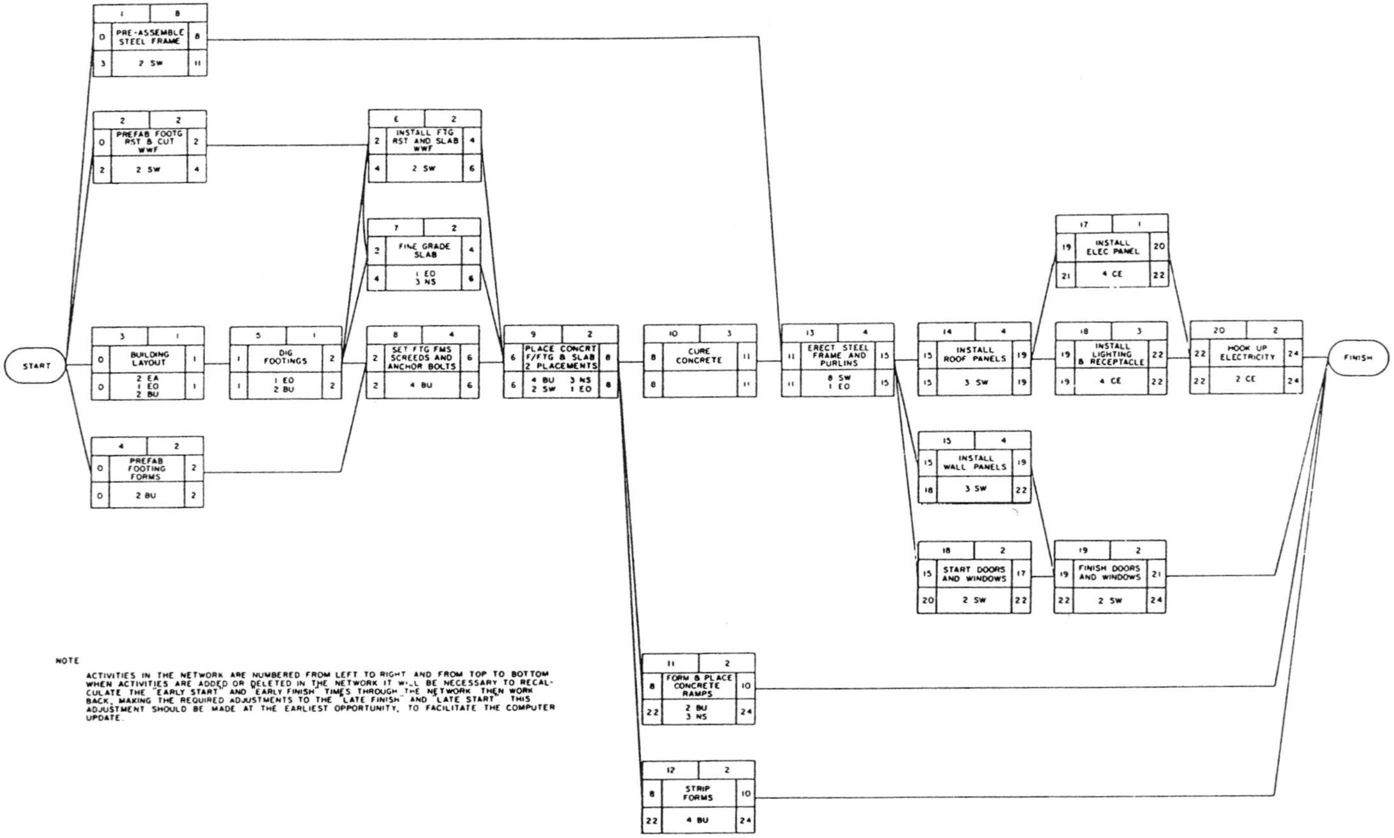

NOTE

ACTIVITIES IN THE NETWORK ARE NUMBERED FROM LEFT TO RIGHT AND FROM TOP TO BOTTOM. WHEN ACTIVITIES ARE ADDED OR DELETED IN THE NETWORK IT WILL BE NECESSARY TO RECALCULATE THE "EARLY START" AND "EARLY FINISH" TIMES THROUGH THE NETWORK, THEN WORK BACK, MAKING THE REQUIRED ADJUSTMENTS TO THE "LATE FINISH" AND "LATE START". THIS ADJUSTMENT SHOULD BE MADE AT THE EARLIEST OPPORTUNITY, TO FACILITATE THE COMPUTER UPDATE.

Figure 9-26.—Typical precedence diagram for a 40-by-100- foot rigid-frame building.

draw connector lines. Figure 9-26 shows a typical precedence diagram for a 40-by- 100-foot rigid-frame building.

RECOMMENDED READING LIST

NOTE

Although the following references were current when this TRAMAN was published, their continued currency cannot be assured. You therefore need to ensure that you are studying the latest revision.

Facilities Planning Guide, NAVFAC P-437, Naval Facilities Engineering Command, Alexandria, Va., 1982.

Operations Officer's Handbook, COMCBPAC/ COMCBLANTINST 5200.2A, Commander, Naval Construction Battalions, U.S. Pacific Fleet, Pearl Harbor, Hawaii, and Commander, Naval Construction Battalions, U.S. Atlantic Fleet, Norfolk, Va., 1988.

Seabee Planner's and Estimator's Handbook, NAVFAC P-405, Chapter 5, Naval Facilities Engineering Command, Alexandria, Va., 1983.

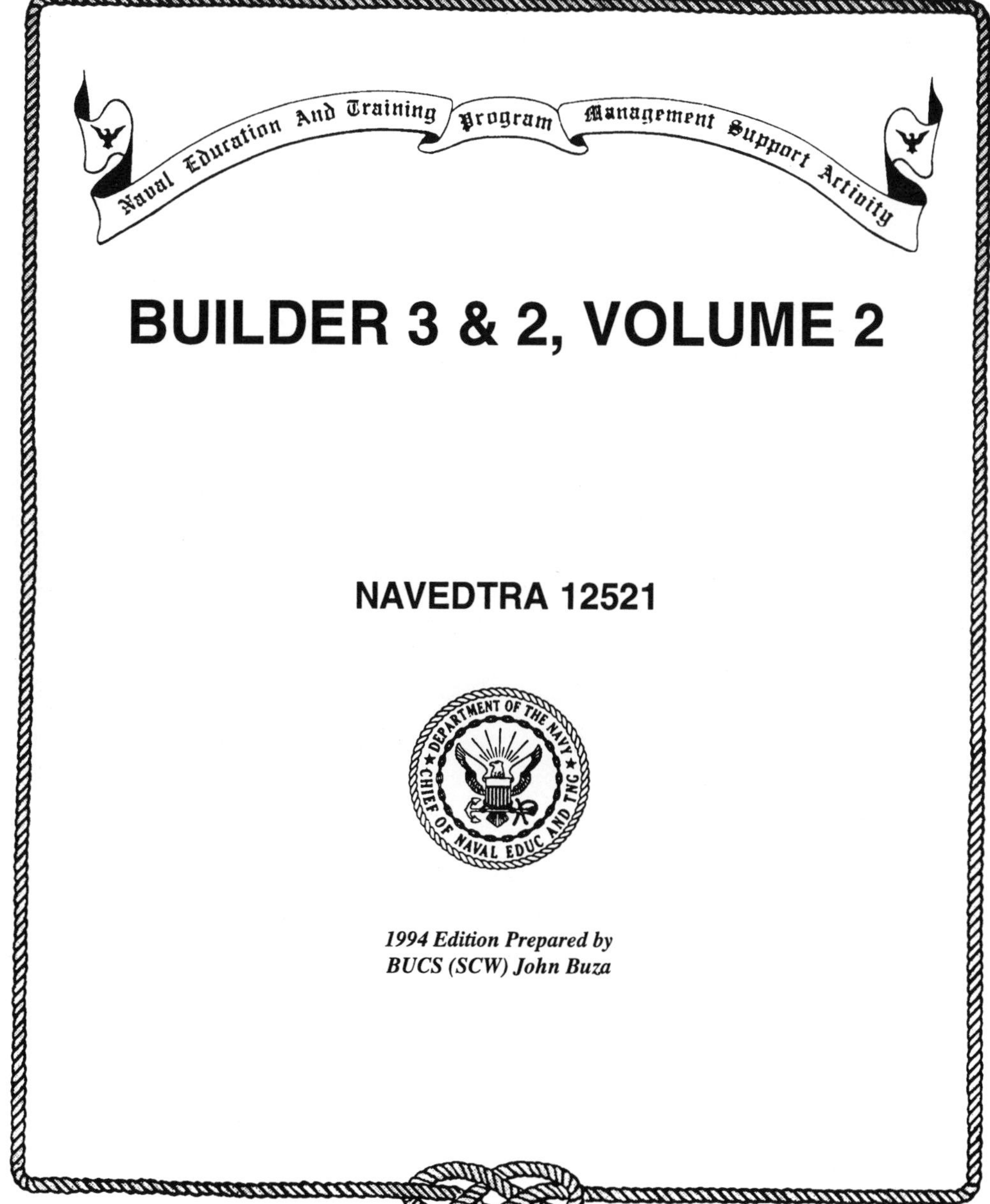

BUILDER 3 & 2, VOLUME 2

NAVEDTRA 12521

1994 Edition Prepared by
BUCS (SCW) John Buza

CHAPTER 1

LIGHT FLOOR AND WALL FRAMING

In the normal sequence of construction events, the floor and wall activities follow the completed foundation work. In this chapter, we'll examine established methods of frame construction and discuss in general how floor and wall framing members are assembled. An explanation of subflooring installation, exterior sheathing, interior partitions, and rough openings for doors and windows is also given.

WOOD SILL FRAMING

LEARNING OBJECTIVE: Upon completing this section, you should be able to describe sill layout and installation.

Framing of the structure begins after completion of the foundation. The lowest member of the frame structure resting on the foundation is the sill plate, often called the mud sill. This sill provides a nailing base for joists or studs resting directly over the foundation. Work in this area is critical as it is the real point of departure for actual building activities.

LAYOUT

The box sill is usually used in platform construction. It consists of a sill plate and header joist anchored to the foundation wall. Floor joists are supported and held in position by the box sill (fig. 1-1). Insulation material and metal termite shields are placed under the sill if desired or when specified. Sills are usually single, but double sills are sometimes used.

Following construction of the foundation wall, the sill is normally the first member laid out. The edge of the sill is set back from the outside face of the foundation a distance equal to the thickness of the exterior sheathing. When laying out sills, remember the corners should be halved together, but are often butted or mitered. If splicing is necessary to obtain required

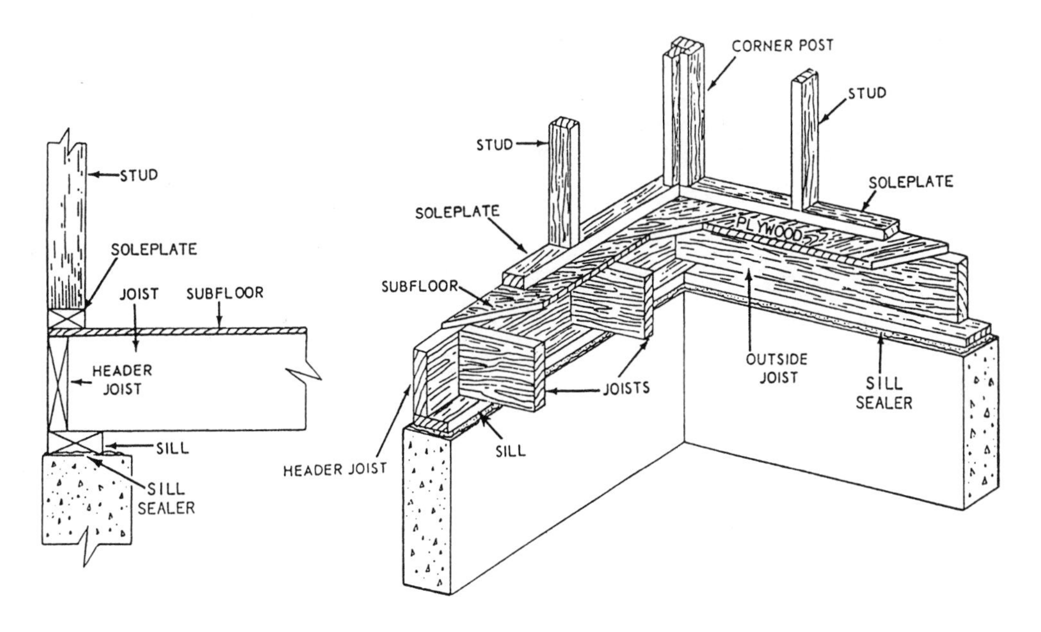

Figure 1-1.—Box-sill assembly.

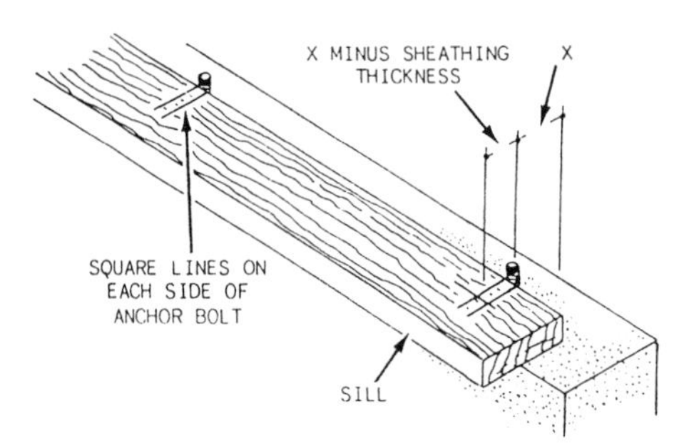

Figure 1-2.—Anchor bolt layout.

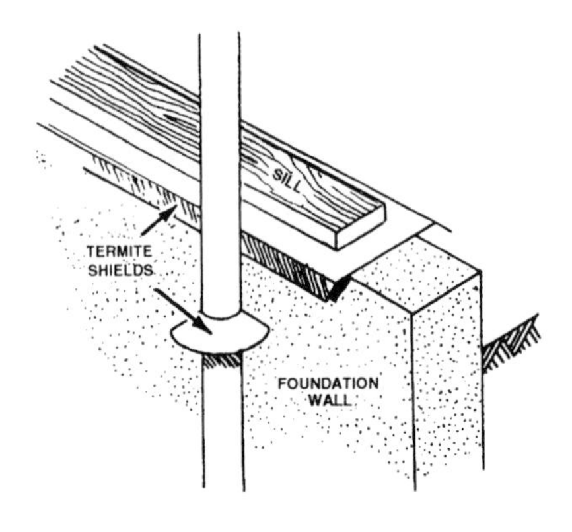

Figure 1-4.—Installing termite shields.

length, you should halve the splice joint at least 2 feet and bolt together.

Once the required length has been determined, the next step is to lay out the locations of the anchor bolt holes. Use the following steps:

1. Establish the building line points at each of the corners of the foundation.
2. Pull a chalk line at these established points and snap a line for the location of the sill.
3. Square the ends of the sill stock. (Stock received at jobsites is not necessarily squared at both ends.)
4. Place the sill on edge and mark the locations of the anchor bolts.
5. Extend these marks with a square across the width of the sill. The distance **X** in figure 1-2 shows how far from the edge of the sill to bore the holes; that is, **X** equals the thickness of the exterior sheathing.

After all the holes are marked, bore the holes. Each should be about 1/4 inch larger than the diameter of the bolts to allow some adjustment for slight inaccuracies in the layout. As each section is bored, position that section over the bolts.

When all sill sections are fitted, remove them from the anchor bolts. Install sill sealer (insulation) as shown in figure 1-3. The insulation compresses, filling the irregularities in the foundation. It also stops drafts and reduces heat loss. Also install a termite shield (fig. 1-4) if specified. A termite shield should be at least 26-gauge aluminum, copper, or galvanized sheet metal. The outer edges should be slightly bent down. Replace the sills and

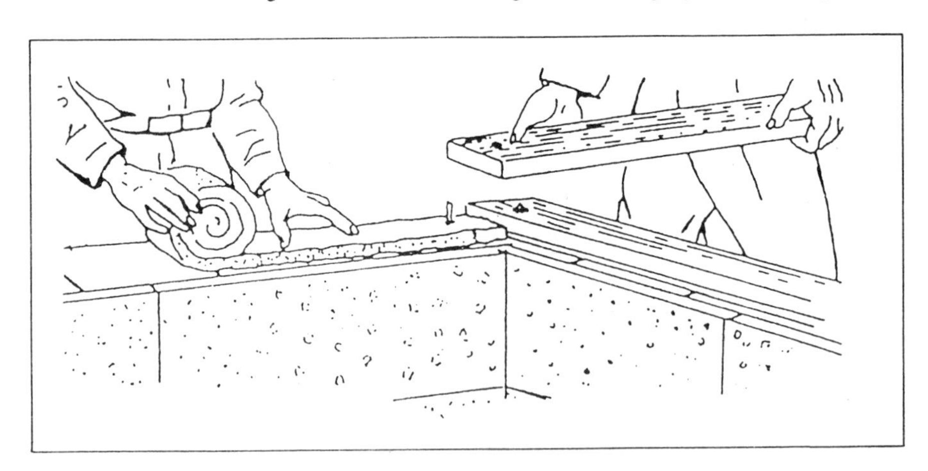

Figure 1-3.—Installing sill sealer.

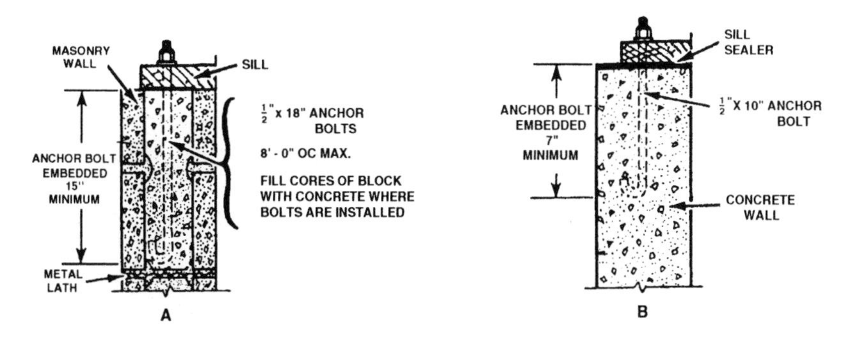

Figure 1-5.—Methods of sill fastening to foundations.

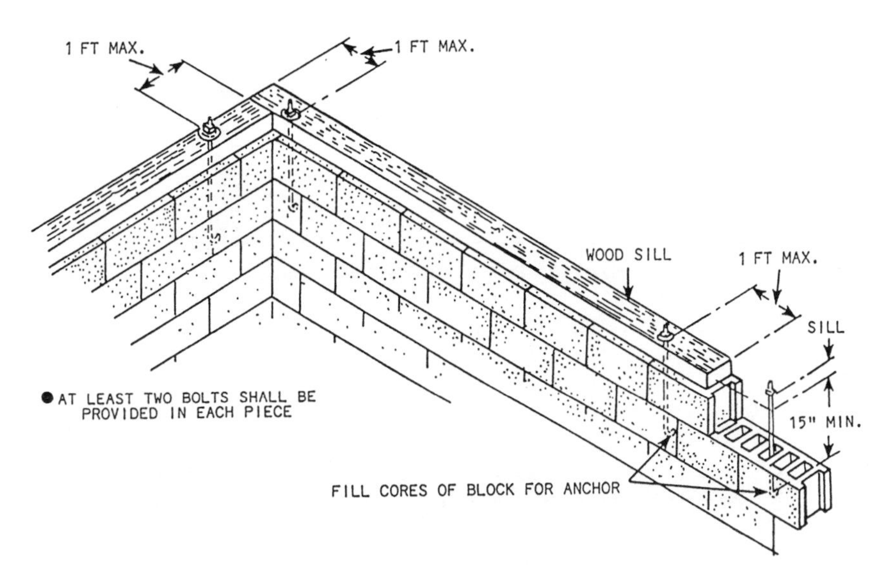

Figure 1-6.—Spacing of anchor bolts.

install the washers and nuts. As the nuts are tightened, make sure the sills are properly aligned. Also, check the distance from the edge of the foundation wall. The sill must be level and straight. Low spots can be shimmied with wooden wedges, but it is better to use grout or mortar.

FASTENING TO FOUNDATION WALLS

Wood sills are fastened to masonry walls by 1/2-inch anchor bolts. These bolts, also known as j-bolts because of their shape, should be embedded 15 inches or more into the wall in unreinforced concrete (fig. 1-5, view A) and a minimum of 7 inches into reinforced concrete (view B). The length of the anchor bolt is found in the specifications; the spacing and location of the bolts are shown on the drawings. If this information is not available, anchor bolt spacing should not exceed 6 feet on center (OC). Also, a bolt must be placed within 1 foot of the ends of each piece (as shown in fig. 1-6).

There are alternative ways to fasten sill plates to foundations. Location and building codes will dictate which to use. Always consult the job specifications before proceeding with construction.

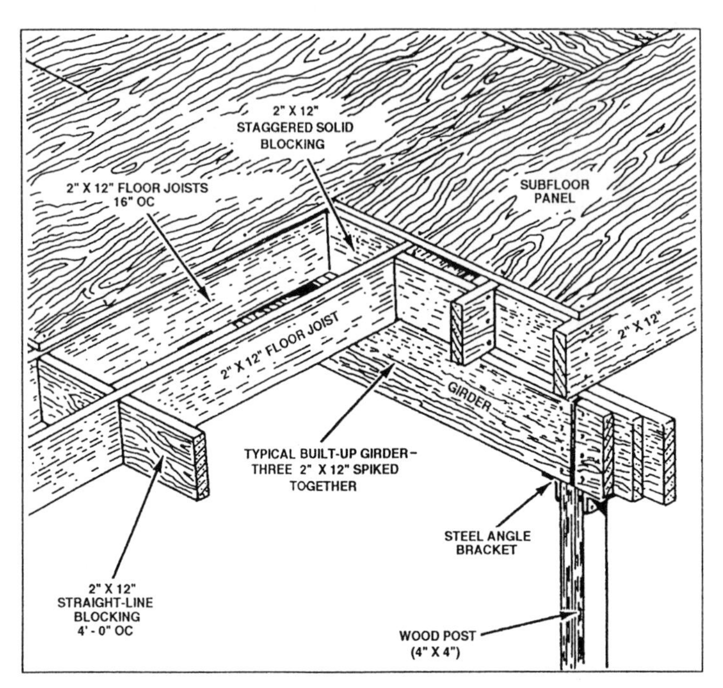

Figure 1-7.—Basic components of floor framing.

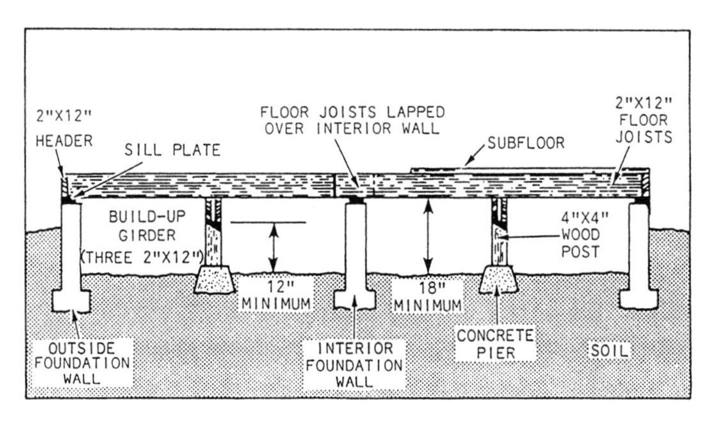

Figure 1-8.—Floor framing on sill plates with intermediate posts and built-up girders.

FLOOR FRAMING

LEARNING OBJECTIVE: Upon completing this section, you should be able to identify members used in floor construction, and the construction methods used with subfloor and bridging.

Floor framing consists specifically of the posts, girders, joists, and subfloor. When these are assembled, as in figure 1-7, they form a level anchored platform for the rest of the construction.

POSTS

Wood or steel posts and girders support floor joists and the subfloor. Sizes depend on the loads carried. The

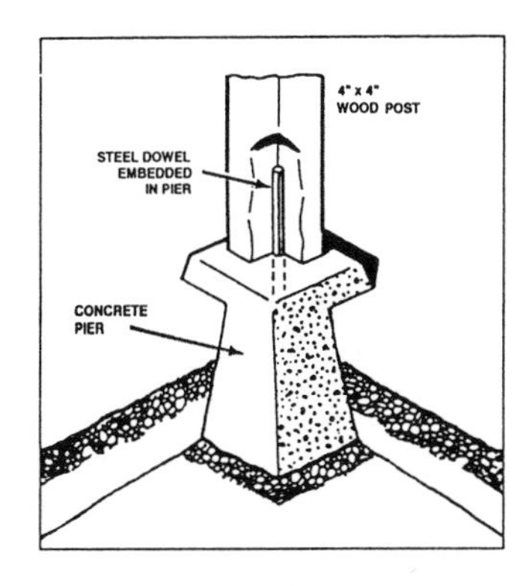

Figure 1-9.—Post fastened using dowel method.

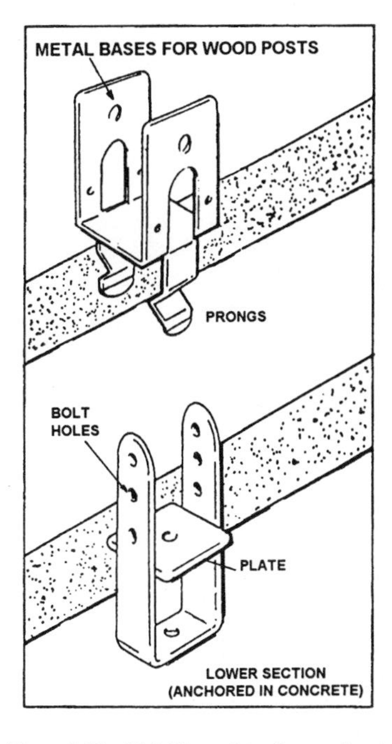

Figure 1-10.—Metal base plates for wood posts.

dimensions and locations are shown on the foundation plan. When required, posts give central support to the long span of girders. Also, girders can be used to support other girders. There should be at least 18 inches clearance between the bottoms of the floor joists and the ground and at least 12 inches between the bottom of the girder and the ground (fig. 1-8).

Wood

Wood posts are placed directly below wood girders. As a general rule, the width of the wood post should be equal to the width of the girder it supports. For example, a 4-inch-wide girder requires a 4- by 4- or 4- by 6-inch post.

A wood post can be secured to a concrete pillar in several ways. The post can be nailed to a pier block secured to the top of a concrete pier; it can be placed over a previously inserted 1/2-inch steel dowel in the concrete; or, it can be placed into a metal base set into the concrete pier at the time of the pour. When using the dowel method, make sure the dowel extends at least 3 inches into the concrete and the post, as shown in figure 1-9. A metal base embedded in the concrete (fig. 1-10) is the preferred method since nothing else is needed to secure the base.

As with the bottom of the post, the top must also be secured to the girder. This can be done using angle iron brackets or metal plates. Figure 1-11 shows two metal post caps used with posts and girders, either nailed or bolted to the girders.

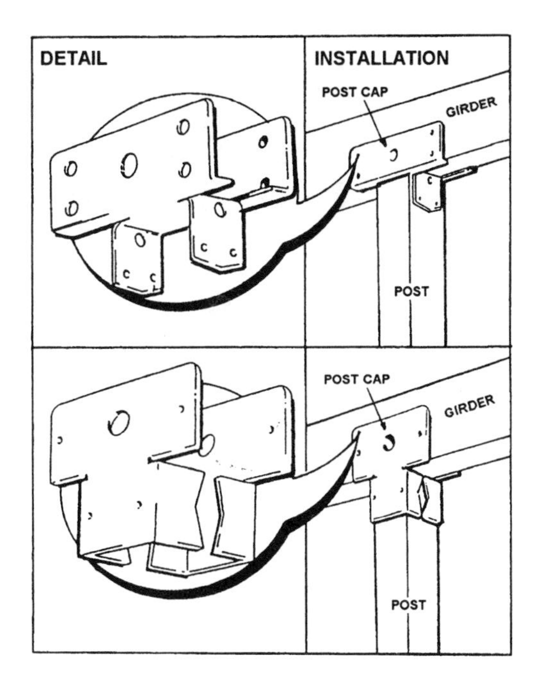

Figure 1-11.—Metal post caps.

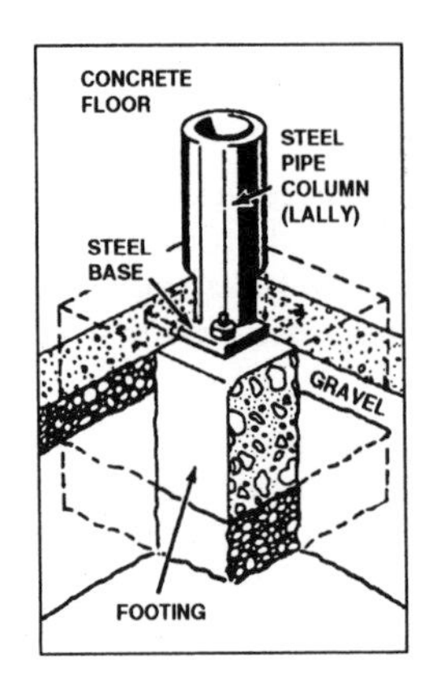

Figure 1-12.—Bolting of steel column.

Steel

Steel pipe columns are often used in wood-frame construction, with both wood and steel girders. When using wood girders, secure the post to the girder with lag bolts. For steel girders, machine bolts are required. The base of the steel post is bolted to the top of the pier, as shown in figure 1-12. The post can also be bolted to anchor bolts inserted in the slab prior to pouring.

GIRDERS

Girders are classified as bearing and nonbearing according to the amount and type of load supported. Bearing girders must support a wall framed directly above, as well as the live load and dead load of the floor. Nonbearing girders support just the dead and live loads of the floor system directly above. The dead load is the weight of the material used for the floor unit itself. The live load is the weight created by people, furniture, appliances, and so forth.

Wood

Wood girders may be a single piece of timber, or they may be laminated (that is, built up) of more than one plank. The built-up girder in figure 1-13, for example, consists of three 2- by 12-inch planks. The

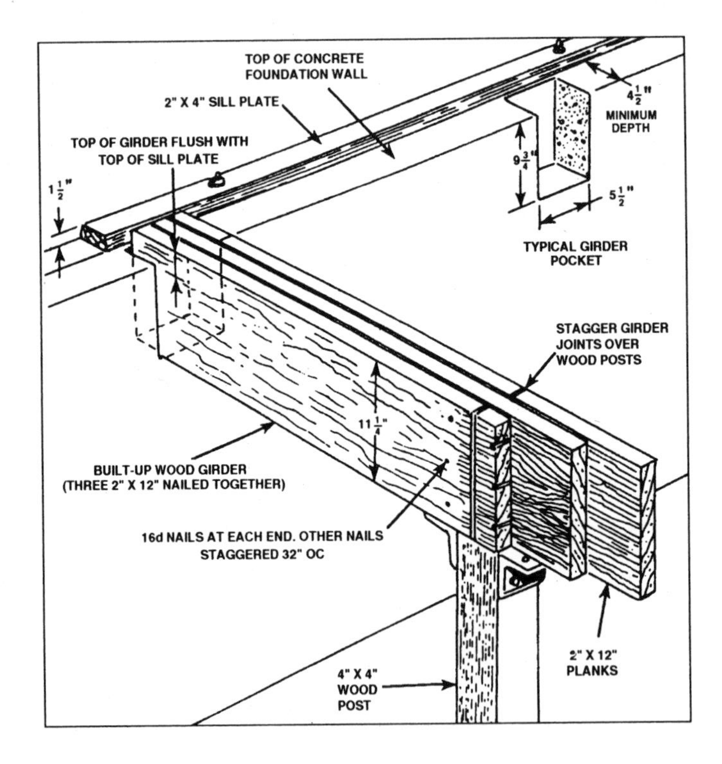

Figure 1-13.—Built-up girder.

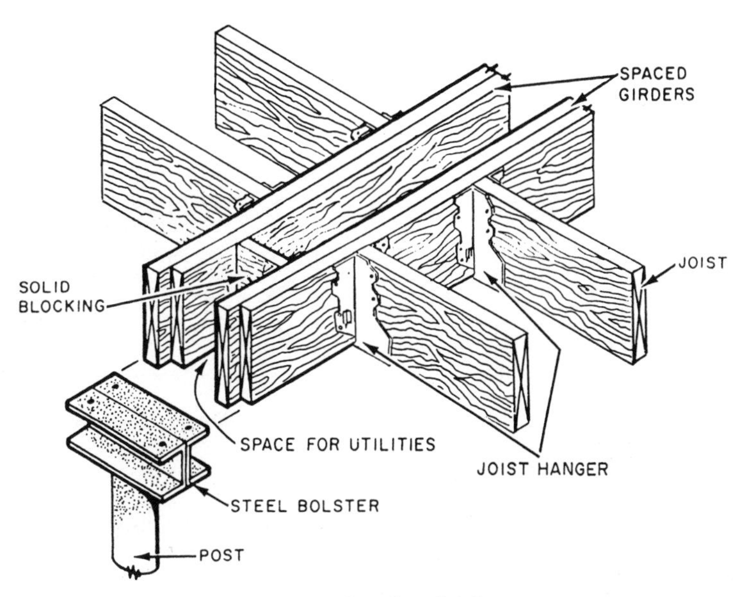

Figure 1-14.—Spaced wood girders.

joints between the planks are staggered. In framing, a built-up girder is placed so that the joints on the outside of the girder fall directly over a post. Three 16-penny (16d) nails are driven at the ends of the planks, and other nails are staggered 32 inches OC. As shown in figure 1-13, the top of the girder is flush with the top sill plate.

When space is required for heat ducts in a partition supported on a girder, a spaced wood girder, such as that shown in figure 1-14, is sometimes necessary. Solid blocking is used at intervals between the two members. A single-post support for a spaced girder usually requires a bolster, preferably metal, with a sufficient span to support the two members.

The ends of a girder often rest in pockets prepared in a concrete wall (fig. 1-13). Here, the girder ends must bear at least 4 inches on the wall, and the pocket should be large enough to provide a 1/2-inch air space around the sides and end of the girder. To protect against termites, treat the ends of the girder with a preservative. As a further precaution, line the pockets with metal.

Steel

S-beams (standard) or W-beams (wide flange), both shown in figure 1-15, are most often used as girders in wood-framed construction. Whether the beam is wood or steel, make sure it aligns from end to end and side to side. Also make sure the length of the bearing post under the girder is correct to ensure the girder is properly supported.

PLACING POSTS AND GIRDERS

Posts must be cut to length and set up before the girders can be installed. The upper surface of the girder may be in line with the foundation plate sill, or the girder ends may rest on top of the walls. Long girders must be

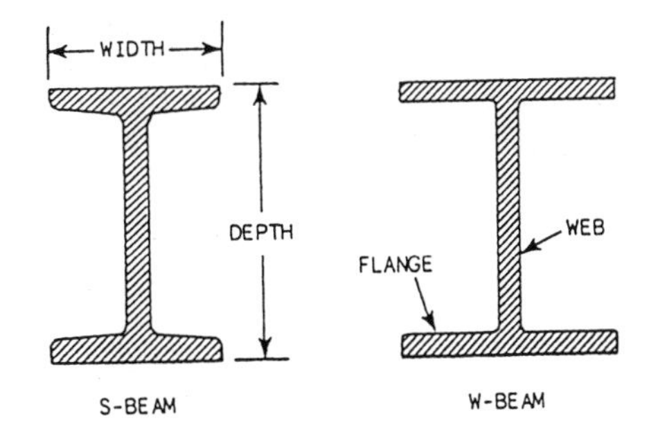

Figure 1-15.—Types of steel beams.

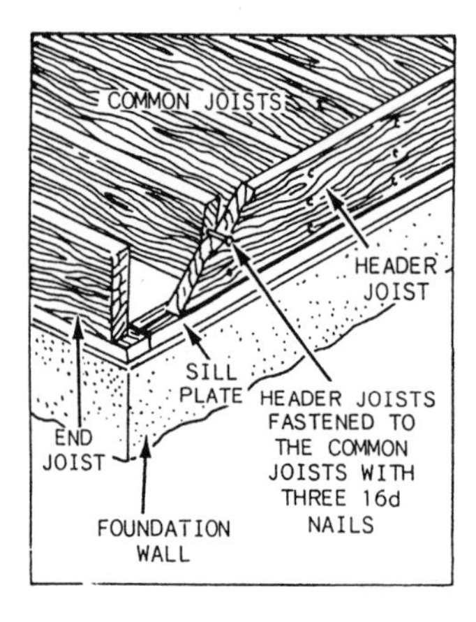

Figure 1-16.—Header joist.

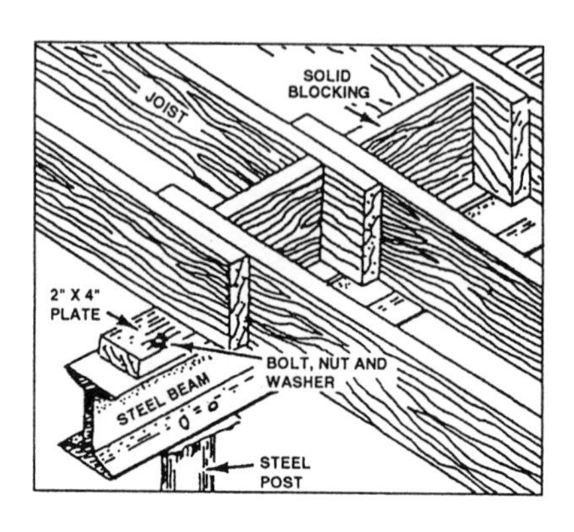

Figure 1-17.—Lapped joists.

placed in sections. Solid girders must be measured and cut so that the ends fall over the center of a post. Built-up girders should be placed so their outside joints fall over the posts (fig. 1-13).

FLOOR JOISTS

In platform framing, one end of the floor joist rests directly on the sill plate of the exterior foundation wall or on the top plate of a framed outside wall. The bearing should be at least 1 1/2 inches. The opposite end of the joist laps over or butts into an interior girder or wall. The size of joist material (2 by 6, 2 by 10, 2 by 12, and so forth) must be chosen with consideration for the span and the amount of load to be carried. The foundation plan usually specifies the joist size, the spacing between joists, and what direction the joists should travel.

The usual spacing of floor joists is 16 inches OC. Floor joists are supported and held in position over exterior walls by header joists or by solid blocking between the joists. The header-joist system is used most often.

Header

Header joists run along the outside walls. Three 16d nails are driven through the header joists into the ends of the common joists, as shown in figure 1-16. The header and joists are toenailed to the sill with 16d nails. The header joists prevent the common joists from rolling or tipping. They also help support the wall above and fill in the spaces between the common joists.

Lapped

Joists are often lapped over a girder running down the center of a building. The lapped ends of the joists may also be supported by an interior foundation or framed wall. It is standard procedure to lap joists the full width of the girder or wall. The minimum lap should be 4 inches. Figure 1-17 shows lapped joists resting on a steel girder. A 2- by 4-inch plate has been bolted to the top of a steel beam. The joists are toenailed into the plate. Solid blocking may be installed between the lapped ends after all the joists have been nailed down. Another system is to put in the blocks at the time the joists are placed.

Double

Joists should be doubled under partitions running in the same direction as the joists. Some walls have water pipes, vent stacks, or heating ducts coming up from the basement or the floor below. Place bridging between double joists to allow space for these purposes (fig. 1-18).

Cantilevered

Cantilevered joists are used when a floor or balcony of a building projects past the wall below, as shown in figure 1-19. A header piece is nailed to the ends of the

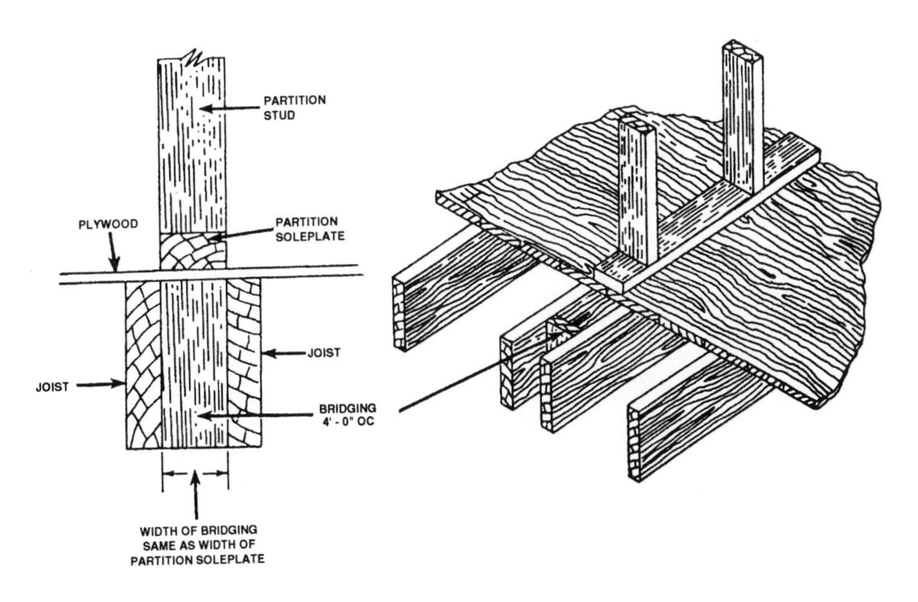

Figure 1-18.—Double joists.

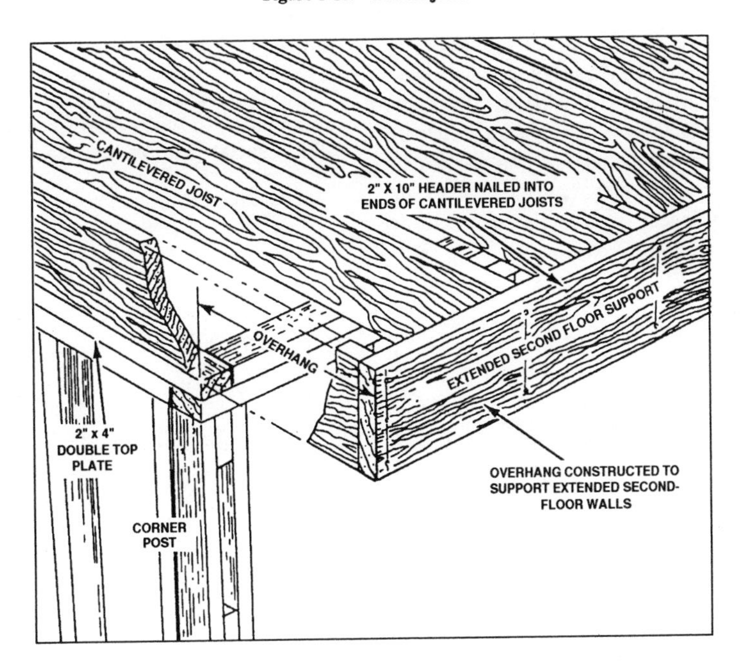

Figure 1-19.—Cantilevered joists.

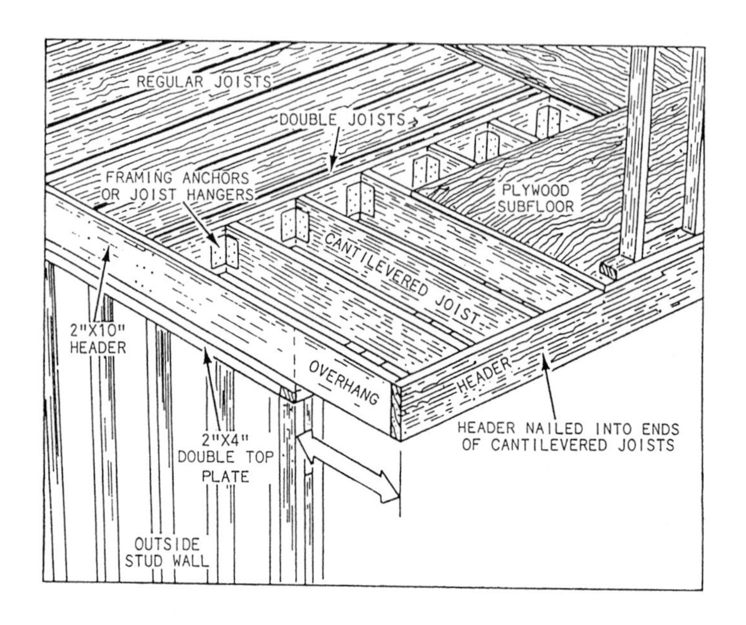

Figure 1-20.—Framing for cantilevered joists.

joists. When regular floor joists run parallel to the intended overhang, the inside ends of the cantilevered joists are fastened to a pair of double joists (fig. 1-20). Nailing should be through the first regular joist into the ends of the cantilevered joists. Framing anchors are strongly recommended and often required by the specifications. A header piece is also nailed to the outside ends of the cantilevered joists.

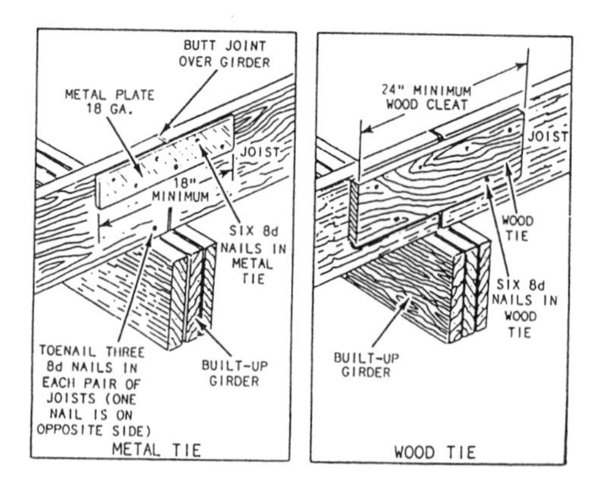

Figure 1-21.—Butting joists over a girder.

Butted over a Girder

Joist ends can also be butted (rather than lapped) over a girder. The joists should then be cleated together with a metal plate or wooden cleat, as shown in

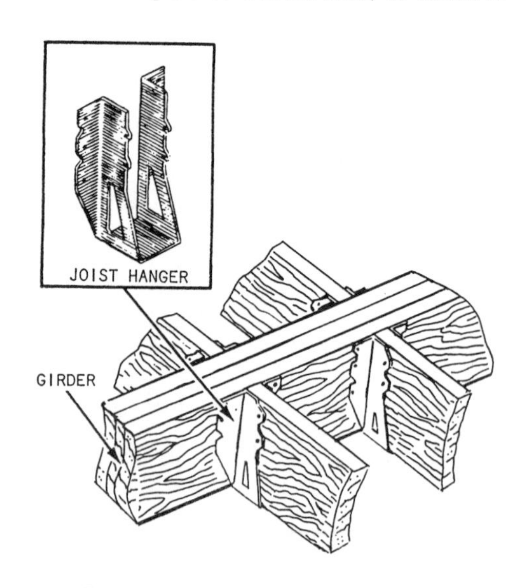

Figure 1-22.—Butting joists against a girder.

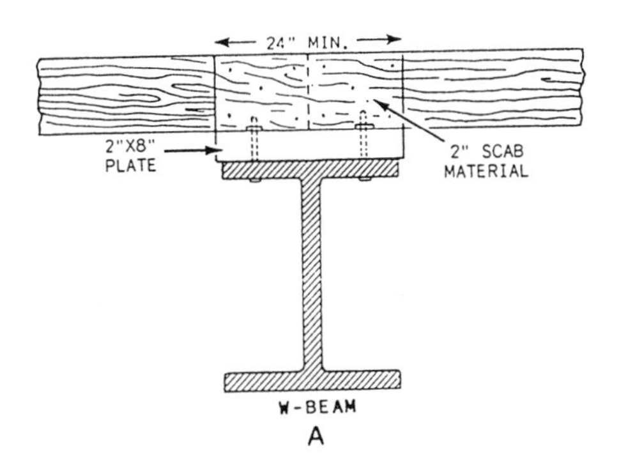

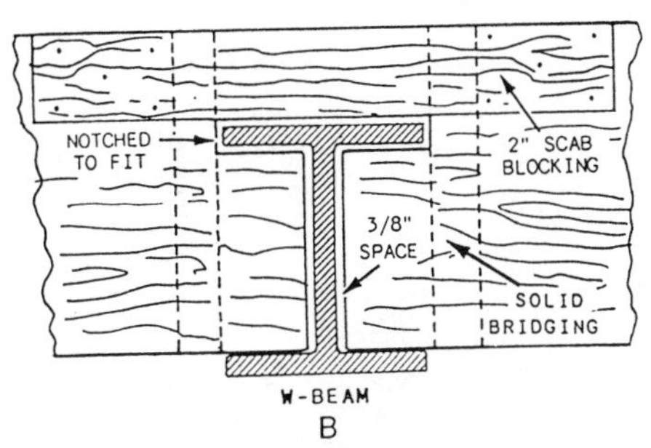

Figure 1-23.—Joists supported by steel beams.

figure 1-21. These can be left out if the line of panels from the plywood subfloor straddles the butt joints.

Butted against a Girder

Butting joists against (rather than over) a girder allows more headroom below the girder. When it is necessary for the underside of the girder to be flush with the joists to provide an unbroken ceiling surface, the joists should be supported with joist hangers (fig. 1-22).

Blocking between Joists

Another system of providing exterior support to joists is to place solid blocking between the outside ends of the joists. In this way, the ends of the joists have more bearing on the outside walls.

Interior Support

Floor joists usually run across the full width of the building. However, extremely long joists are expensive and difficult to handle. Therefore, two or more shorter joists are usually used. The ends of these joists are supported by lapping or butting them over a girder, butting them against a girder, or lapping them over a wall.

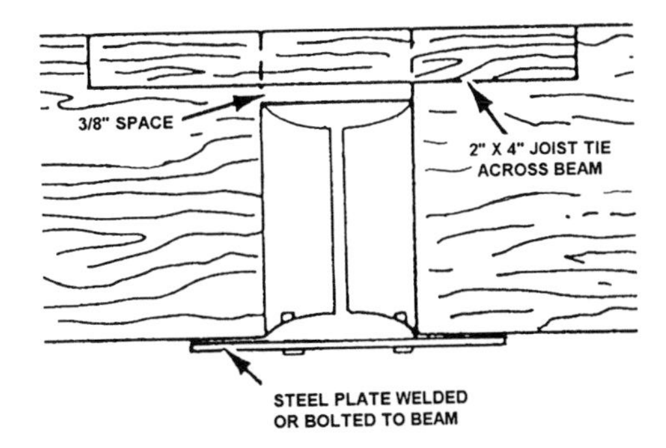

Figure 1-24.—Joists supported on steel plates.

Supported by a Steel Beam

Wood joists are often supported by a steel beam rather than a wood girder. The joists may rest on top of the steel beam (fig. 1-23, view A), or they may be butted (and notched to fit) against the sides of the beam (view B). If the joists rest on top of a steel beam, a plate is fastened to the beam and the joists are toenailed into the plate. When joists are notched to fit against the sides of the beam, allowance must be made for joist shrinkage while the steel beams remain the same size. For average work with a 2- by 10-inch joist, an allowance of 3/8 inch above the top flange of the steel girder or beam is usually sufficient.

Another method of attaching butted joists to a steel girder is shown in figure 1-24. A 3/8-inch space is shown above the beam to allow for shrinkage. Notching the joists so they rest on the lower flange of an S-beam is not recommended; the flange surface does not provide sufficient bearing surface. A wide plate may be bolted or welded to the bottom of the S-beams to provide better support. Wooden blocks may be placed at the bottoms of the joists to help keep them in position. Wide-flanged beams, however, do provide sufficient support surface for this method of

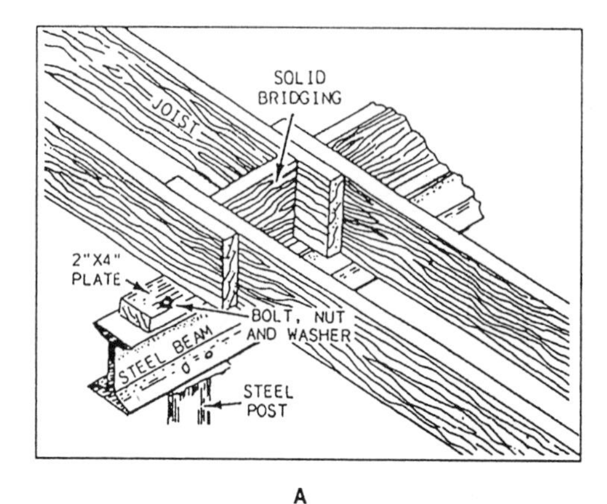

A

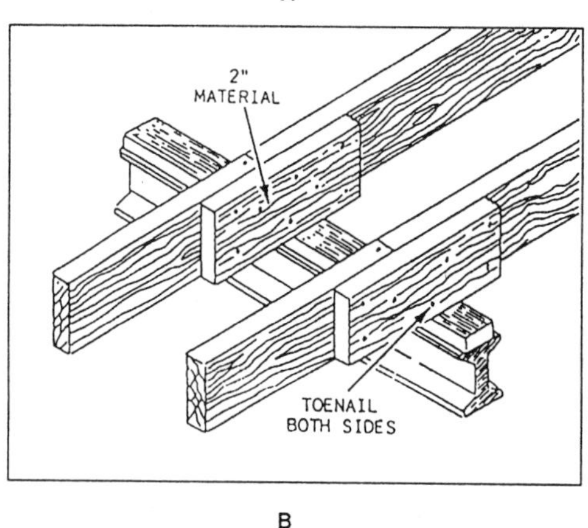

B

Figure 1-25.—Joists supported by S-beam using wooden blocks.

construction. Figure 1-25 shows the lapped (view A) and butt (view B) methods of framing over girders.

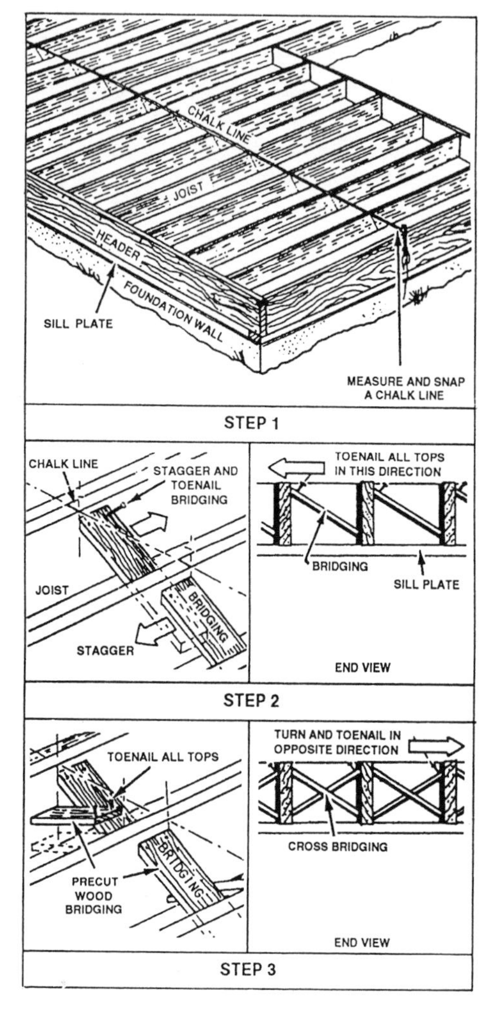

Figure 1-26.—Wood cross bridging.

Bridging between Joists

Floor plans or specifications usually call for bridging between joists. Bridging holds the joists in line and helps distribute the load carried by the floor unit. It is usually required when the joist spans are more than 8 feet. Joists spanning between 8 and 15 feet need one row of bridging at the center of the span. For longer spans, two rows of bridging spaced 6 feet apart are required.

CROSS BRIDGING.—Also known as herringbone bridging, cross bridging usually consists of 1- by 3-inch or 2- by 3-inch wood. It is installed as shown in figure 1-26. Cross bridging is toenailed at each end with 6d or 8d nails. Pieces are usually precut on a radial-arm saw. Nails are started at each end before the cross bridging is placed between the joists. The usual procedure is to fasten only the top end of the cross bridging. The nails at the bottom end are not driven in until the subfloor has been placed. Otherwise the joist could be pushed out of line when the bridging is nailed in.

An efficient method for initial placement of cross bridging is shown in figure 1-26. In step 1, snap a chalk line where the bridging is to be nailed between the joists. In step 2, moving in one direction, stagger and nail the

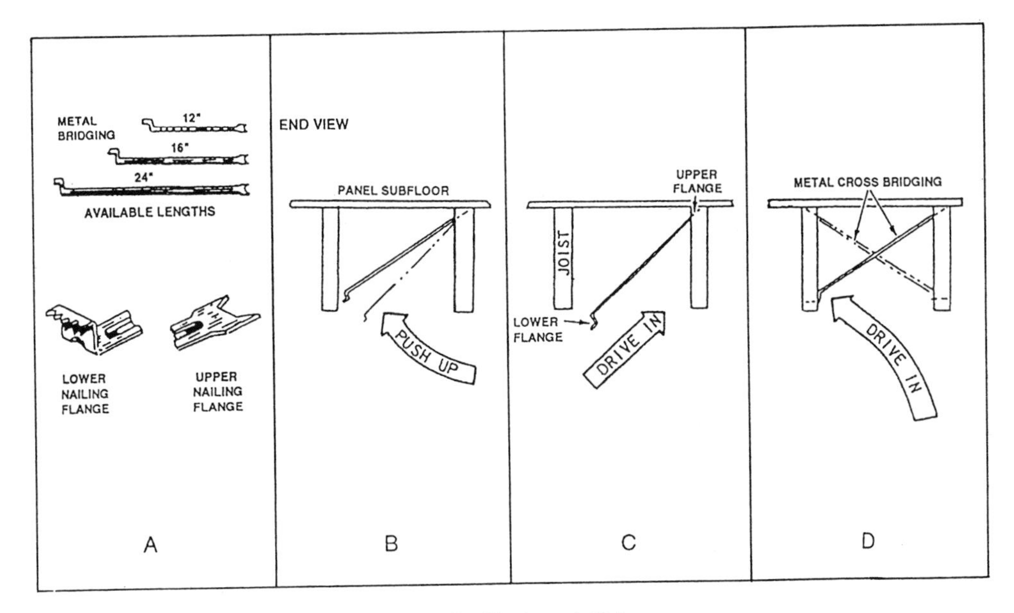

Figure 1-27.—Metal cross bridging.

tops of the bridging. In step 3, reverse direction and nail tops of the opposite pieces into place.

Another approved system of cross bridging uses metal pieces instead of wood and requires no nails. The pieces are available for 12-, 16-, and 24-inch joist spacing (fig. 1-27, view A). You can see how to install this type of cross bridging in views B, C, and D. In view B, strike the flat end of the lower flange, driving the flange close to the top of the joist. In view C, push the lower end of the bridging against the opposite joist. In view D, drive the lower flange into the joist.

SOLID BRIDGING.—Also known as solid blocking, solid bridging (fig. 1-28) serves the same purpose as cross bridging. This method is preferred by many Builders to cross bridging. The pieces are cut from lumber the same width as the joist material. They can be installed in a straight line by toenailing or staggering. If staggered, the blocks can be nailed from both ends, resulting in a faster nailing operation. Straight lines of blocking may be required every 4 feet OC to provide a nailing base for a plywood subfloor.

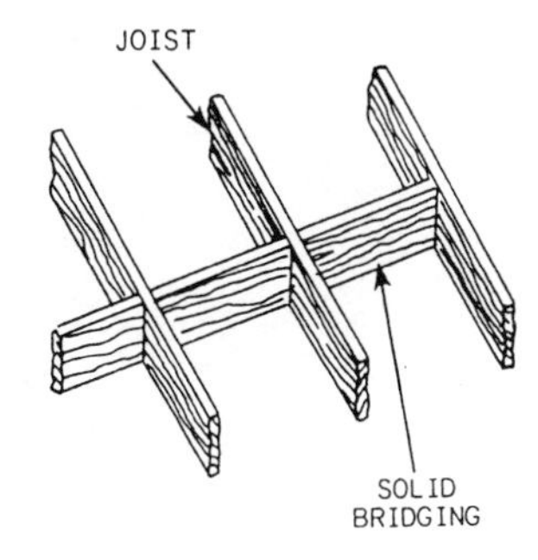

Figure 1-28.—Solid bridging.

Placing Floor Joists

Before floor joists are placed, the sill plates and girders must be marked to show where the joists are to be nailed. As we mentioned earlier, floor joists are usually placed 16 inches OC.

For joists resting directly on foundation walls, layout marks may be placed on the sill plates or the header joists. Lines must also be marked on top of the girders or walls over which the joists lap. If framed walls are below the floor unit, the joists are laid out on top of the double plate. The floor layout should also show where any joists are to be doubled. Double joists are required where partitions resting on the floor run in the same direction as the floor joists. Floor openings for stairwells must also be marked.

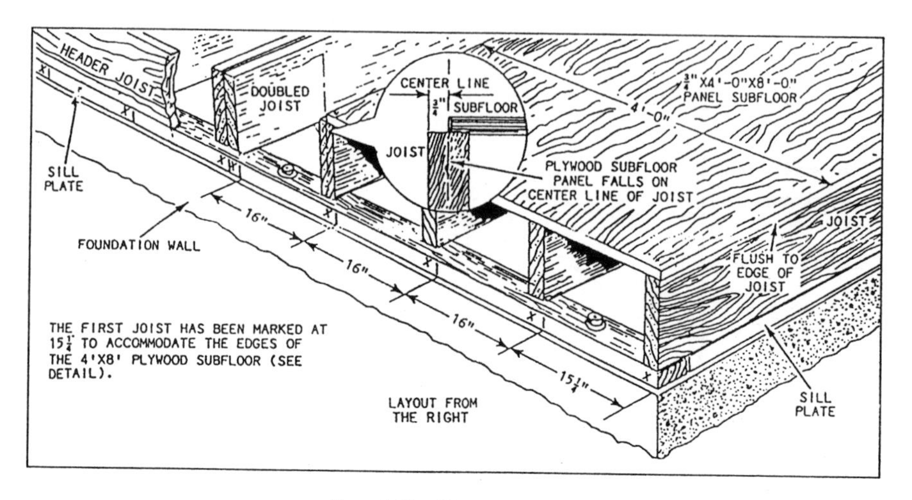

Figure 1-29.—Floor joists layout.

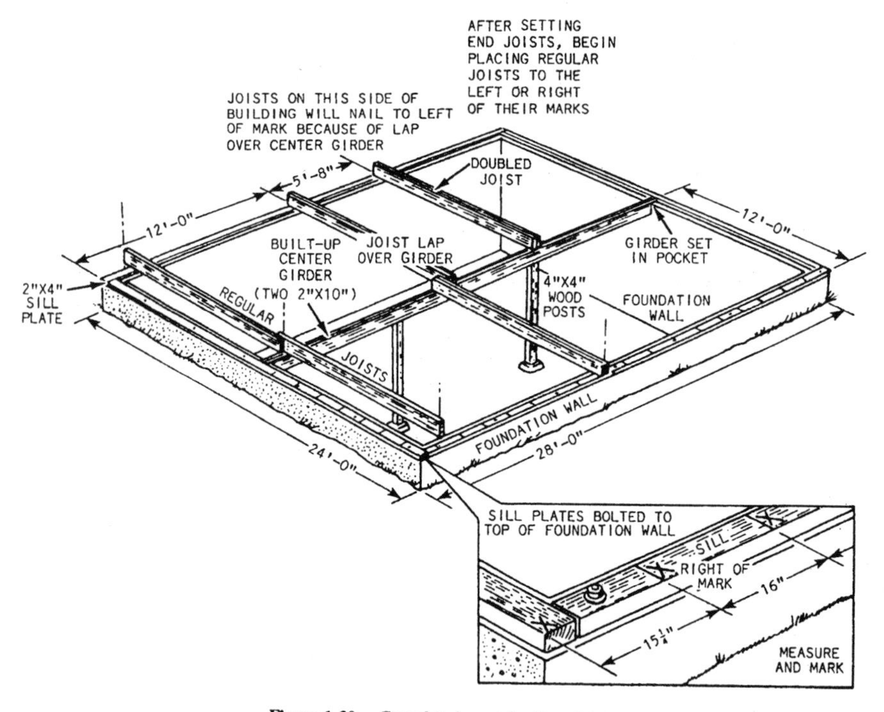

Figure 1-30.—Complete layout for floor joists.

Joists should be laid out so that the edges of standard-size subfloor panels break over the centers of the joists (see insert, fig. 1-29). This layout eliminates additional cutting of panels when they are being fitted and nailed into place. One method of laying out joists this way is to mark the first joists 15 1/4 inches from the edge of the building. From then on, the layout is 16 inches OC. A layout for the entire floor is shown in figure 1-30.

Most of the framing members should be precut before construction begins. The joists should all be trimmed to their proper lengths. Cross bridging and

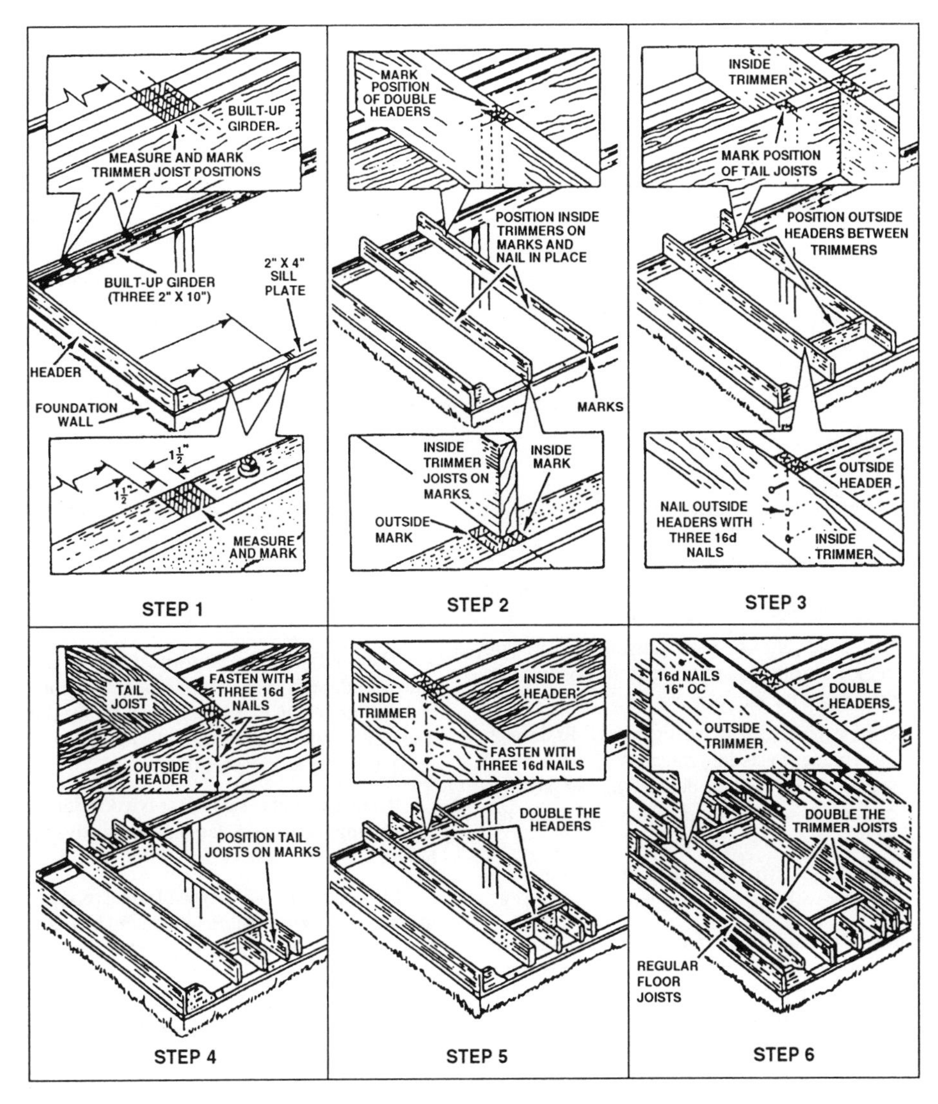

Figure 1-31.—Steps in framing a floor opening.

solid blocks should be cut to fit between the joists having a common spacing. The distance between joists is usually 14 1/2 inches for joists spaced 16 inches OC. Blocking for the odd spaces is cut afterwards.

Framing Floor Openings

Floor openings, where stairs rise to the floor or large duct work passes through, require special framing. When the joists are cut for such openings, there is a loss of strength in the area of the opening. You need to frame the opening in a way that restores this strength. The procedure is shown in figure 1-31. Refer to the figure as you study the following steps:

1. Measure and mark the positions of the trimmers on the outside wall and interior wall or girder.
2. Position and fasten the inside trimmers and mark the position of the double headers.

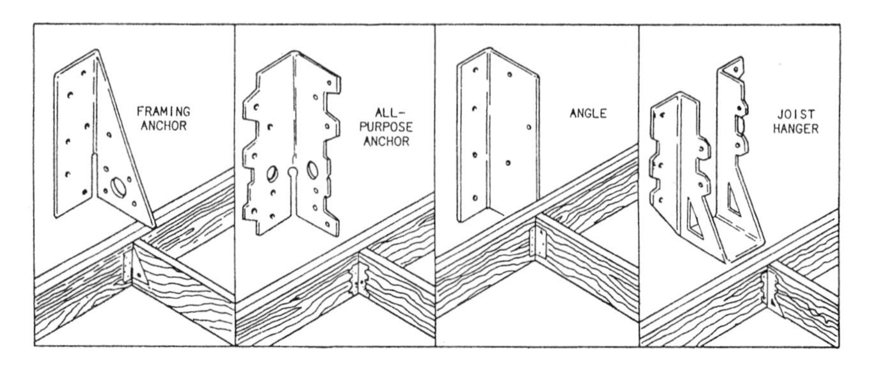

Figure 1-32.—Types of framing anchors.

3. Place the outside pieces between the inside trimmers. Drive three 16d nails through the trimmers into the headers. Mark the position of the tail joists on the headers (the tail joists should follow the regular joist layout).
4. Fasten the tail joists to the outside headers with three 16d nails driven through the headers into the ends of the tail joists.
5. Double the header. Drive three 16d nails through the trimmer joists into the ends of the doubled header pieces. Nail the doubled header pieces to each other with 16d nails staggered 16 inches OC.
6. Double the trimmer joists and fasten them together with 16d nails staggered 16 inches OC.

A pair of joists, called trimmers, is placed at each side of the opening. These trimmers support the headers. The headers should be doubled if the span is more than 4 feet. Nails supporting the ends of the headers are driven through the trimmer joists into the ends of the header pieces. Tail joists (cripple joists) run from the header to a supporting wall or girder. Nails are driven through the header into the ends of the tail joist. Various metal anchors, such as those shown in figure 1-32, are also used to strengthen framed floor openings.

Crowns

Most joists have a crown (a bow shape) on one side. Each joist should be sighted before being nailed in place to make certain the crown is turned up. The joist will later settle from the weight of the floor and straighten out. Caution should be exercised when sighting the board for the crown. Some crowns are too large and cannot be turned up for use as a joist.

SUBFLOOR

The subfloor, also known as rough flooring, is nailed to the top of the floor frame. It strengthens the entire floor unit and serves as a base for the finish floor. The walls of the building are laid out, framed, and raised into place on top of the subfloor.

Panel products, such as plywood, are used for subflooring. Plywood is less labor intensive than board lumber.

Plywood is the oldest type of panel product. It is still the most widely used subfloor material in residential and other light-framed construction. Other types of material available for use as subflooring include nonveneered (reconstituted wood) panels, such as structural particleboard, waferboard, oriented strandboard, and compositeboard.

Plywood is available in many grades to meet a broad range of end uses. All interior grades are also available with fully waterproof adhesive identical with that used in exterior plywood. This type is useful where prolonged moisture is a hazard. Examples are underlayments, subfloors adjacent to plumbing fixtures, and roof sheathing that may be exposed for long periods during construction. Under normal conditions and for sheathing used on walls, standard sheathing grades are satisfactory.

Plywood suitable for the subfloor, such as standard sheathing, structural I and II, and C-C exterior grades, has a panel identification index marking on each sheet.

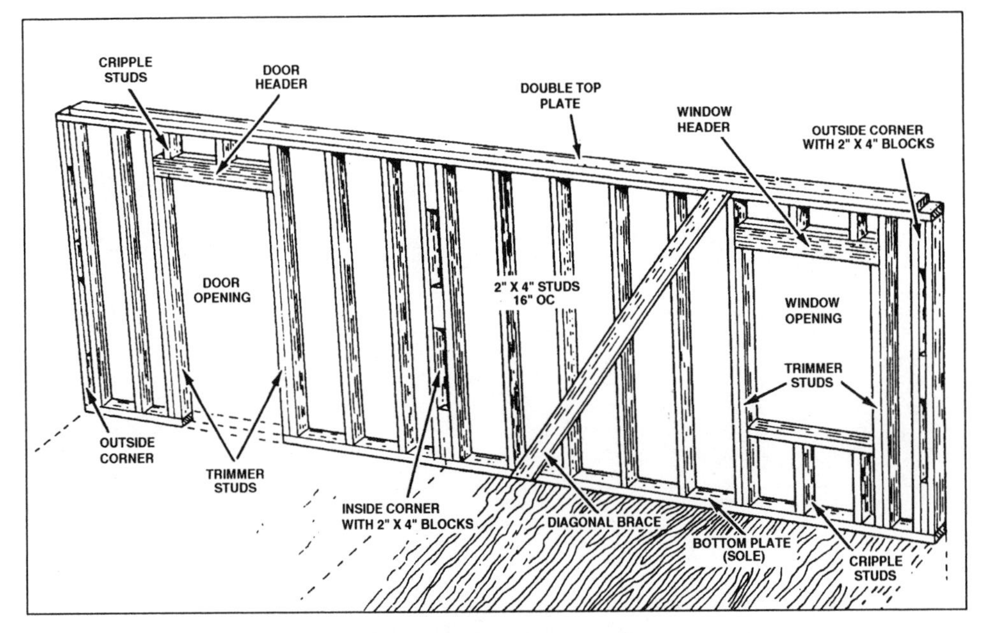

Figure 1-33.—Typical exterior wall.

These markings indicate the allowance spacing of rafters and floor joists for the various thicknesses when the plywood is used as roof sheathing or subfloor. For example, an index mark of 32/16 indicates the plywood panel is suitable for a maximum spacing of 32 inches for rafters and 16 inches for floor joists. Thus, no problem of strength differences between species is involved, as the correct identification is shown for each panel.

Plywood should be installed with the grain of the outer plies at right angles to the joists. Panels should be staggered so that end joints in adjacent panels break over different joists. The nailing schedule for most types of subfloor panels calls for 6d common nails for materials up to 7/8 inch thick and for 8d nails for heavier panels up to 1 1/8 inches thick. Deformed-shank nails are strongly recommended. They are usually spaced 6 inches OC along the edges of the panel and 10 inches OC over intermediate joists.

For the best performance, **do not** lay up plywood with tight joints, whether interior or exterior. Allow for expansion if moisture should enter the joints.

WALL FRAMING

LEARNING OBJECTIVE: Upon completing this section, you should be able to identify wall framing members and explain layout and installation procedures for these members in building construction.

Wall construction begins after the subfloor has been nailed in place. The wall system of a wood-framed building consists of exterior (outside) and interior (inside) walls. The typical exterior wall has door and window openings, as shown in figure 1-33. Interior walls, usually referred to as "partitions," divide the inside area into separate rooms. Some interior walls have door openings or archways.

Partitions are either bearing or nonbearing. Bearing partitions support the ends of the floor joists or ceiling joists. Nonbearing partitions run in the same direction as the joists and therefore carry little weight from the floor or ceiling above.

Traditionally, 2- by 4-inch structural lumber is used for the framed walls of one-story buildings, although the use of heavier structural lumber is specified at certain locations for particular projects. Multistory buildings,

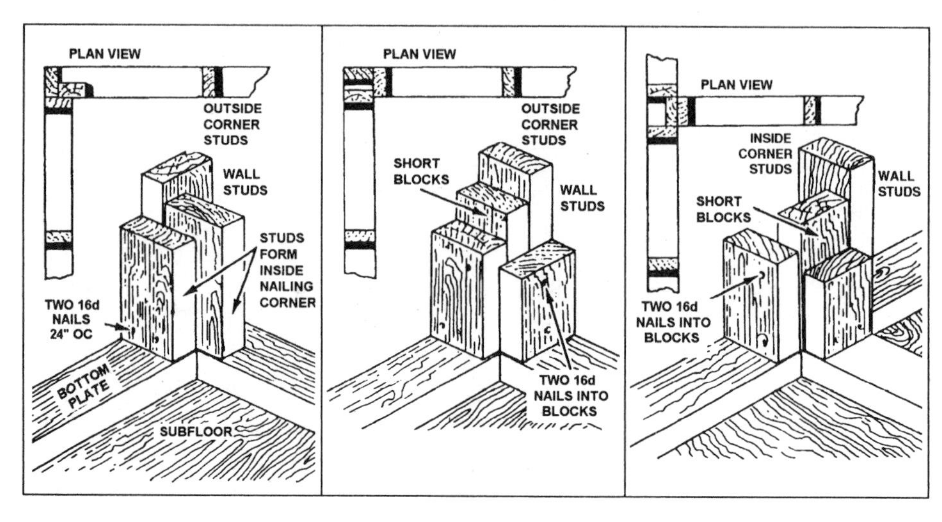

Figure 1-34.—Corner posts.

for example, require heavier structural lumber. This requirement is specific to the lower levels in order to support the weight of the floors above.

STRUCTURAL PARTS

A wood-framed wall consists of structural parts referred to as "wall components" or "framing members." The components (shown in fig. 1-33) typically include studs, plates, headers, trimmers, cripples, sills, corner posts, and diagonal braces. Each component is essential to the integrity of the total wall structure.

Studs

Studs are upright (vertical) framing members running between the top and bottom plates. Studs are usually spaced 16 inches OC, but job specifications sometimes call for 12-inch and 24-inch OC stud spacing.

Plates

The plate at the bottom of a wall is the soleplate, or bottom plate. The plate at the top of the wall is the top plate. A double top plate is normally used. It strengthens the upper section of the wall and helps carry the weight of the joists and roof rafters. Since top and bottom plates are nailed into all the vertical wall members, they serve to tie the entire wall together.

Corner Posts

Corner posts are constructed wherever a wall ties into another wall. Outside corners are at the ends of a wall. Inside corners occur where a partition ties into a wall at some point between the ends of the wall.

Three typical designs for corner assemblies are shown in figure 1-34. View A shows outside corner construction using only three studs. View B shows outside corner construction using two studs with short blocks between them at the center and ends. A third full-length stud can be used instead of blocks. View C shows inside corner construction using a block laid flat. A full-length stud can be used instead of a block. Note that all corner assemblies should be constructed from straight stud material and should be well nailed. When framing corners, you can use full-length studs or short blocks.

Rough Door and Window Openings

A rough opening must be framed into a wall wherever a door or window is planned. The dimensions of the rough opening must allow for the final frame and for the required clearance around the frame.

Figure 1-35 shows details of rough openings for doors and windows in wood-frame construction. The rough opening for a typical door is framed with a header,

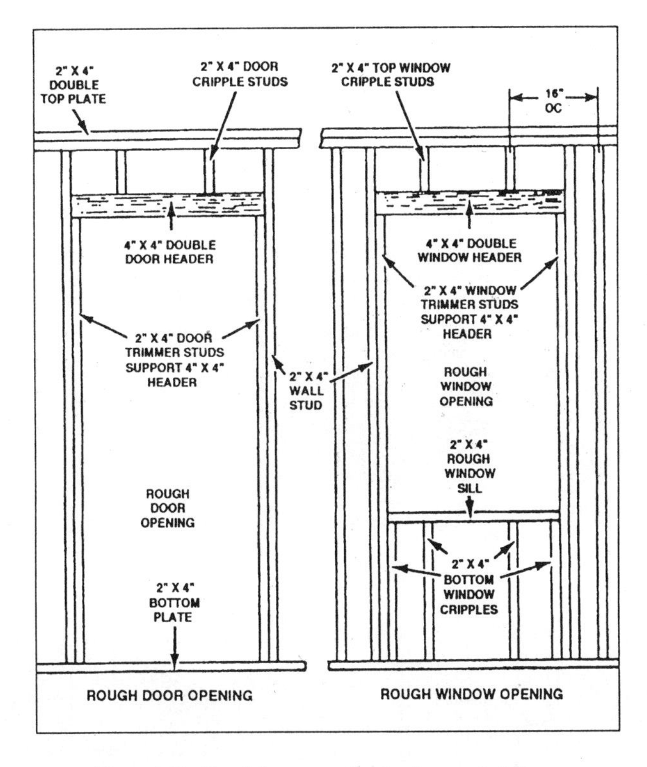

Figure 1-35.—Rough frame openings for doors and windows.

trimmer studs, and, in some cases, top cripple studs. The rough opening for a typical window includes the same members as for a door, plus a rough window sill and bottom cripples.

A header is placed at the top of a rough opening. It must be strong enough to carry the weight bearing down on that section of the wall. The header is supported by trimmer studs fitting between the soleplate and the bottom of the header. The trimmer studs are nailed into the regular studs at each side of the header. Nails are also driven through the regular studs into the ends of the header.

The header may be either solid or built up of two 2 by 4 pieces with a 1/2-inch spacer. The spacer is needed to bring the width of the header to 3 1/2 inches. This is the actual width of a nominal 2 by 4 stud wall. A built-up header is as strong as or stronger than a solid piece.

The type and size of header is shown in the blueprints. Header size is determined by the width of the opening and by how much weight is bearing down from the floor above.

The tops of all door and window openings in all walls are usually in line with each other. Therefore, all headers are usually the same height from the floor. The standard height of walls in most wood-framed buildings is either 8 feet 3/4 inch or 8 feet 1 inch from the subfloor to the ceiling joists. The standard height of the doors is 6 feet 8 inches.

Cripple studs are nailed between the header and the double top plate of a door opening. These help carry the weight from the top plate to the header. The cripple studs are generally spaced 16 inches OC.

A rough window sill is added to the bottom of a rough window opening. The sill provides support for the finished window and frame to be placed in the wall. The distance between the sill and the header is determined by the dimensions of the window, the window frame, and the necessary clearances at the top and bottom of the frame. Cripple studs, spaced 16 inches OC, are

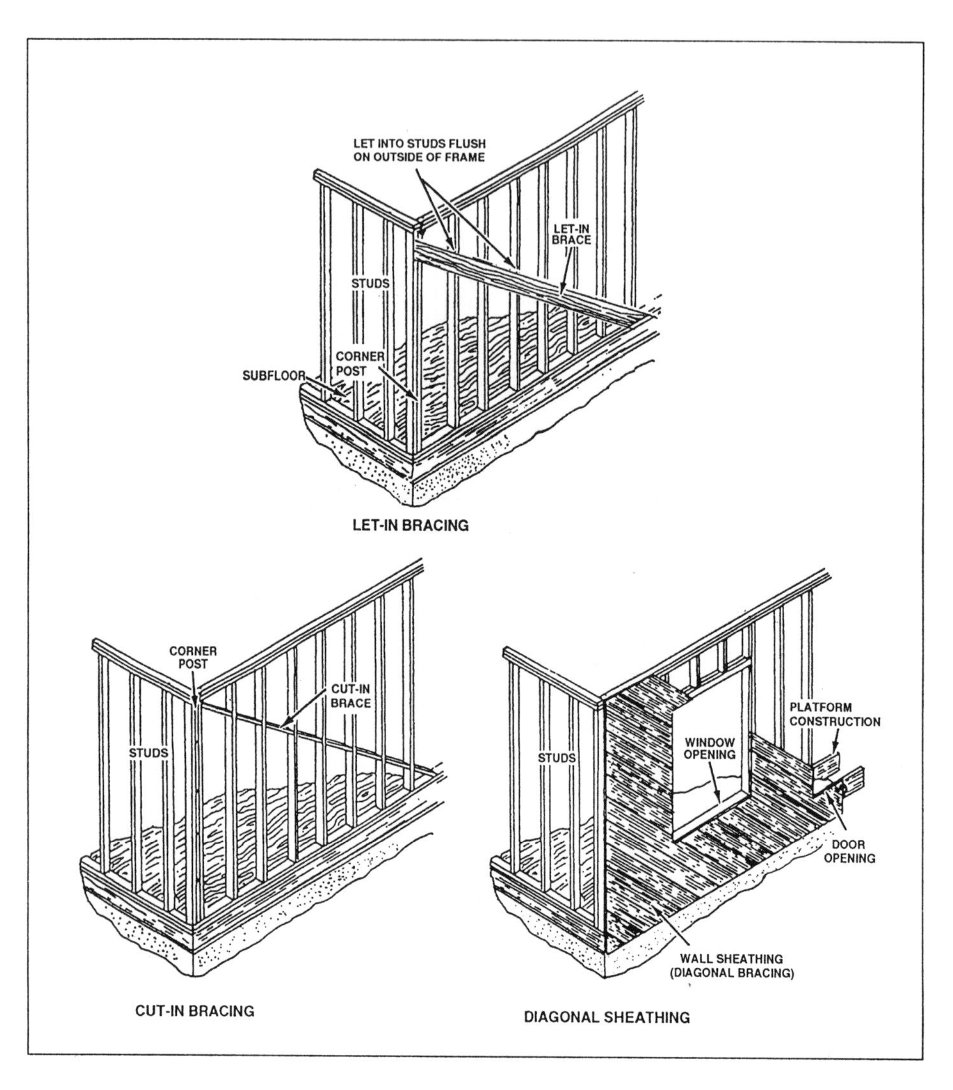

Figure 1-36.—Types of bracing.

nailed between the sill and soleplate. Additional cripple studs may be placed under each end of the sill.

Bracing

Diagonal bracing is necessary for the lateral strength of a wall. In all exterior walls and main interior partitions, bracing should be placed at both ends (where possible) and at 25-foot intervals. An exception to this requirement is an outside wall covered with structural sheathing nailed according to building specifications. This type of wall does not require bracing.

Diagonal bracing is most effective when installed at a 45° to 60° angle. You can do this after the wall has been squared and still lying on the subfloor. The most widely used bracing system is the 1 by 4 let-in type, as shown in figure 1-36. The studs are notched so that the 1 by 4 piece is flush with the surface of the studs.

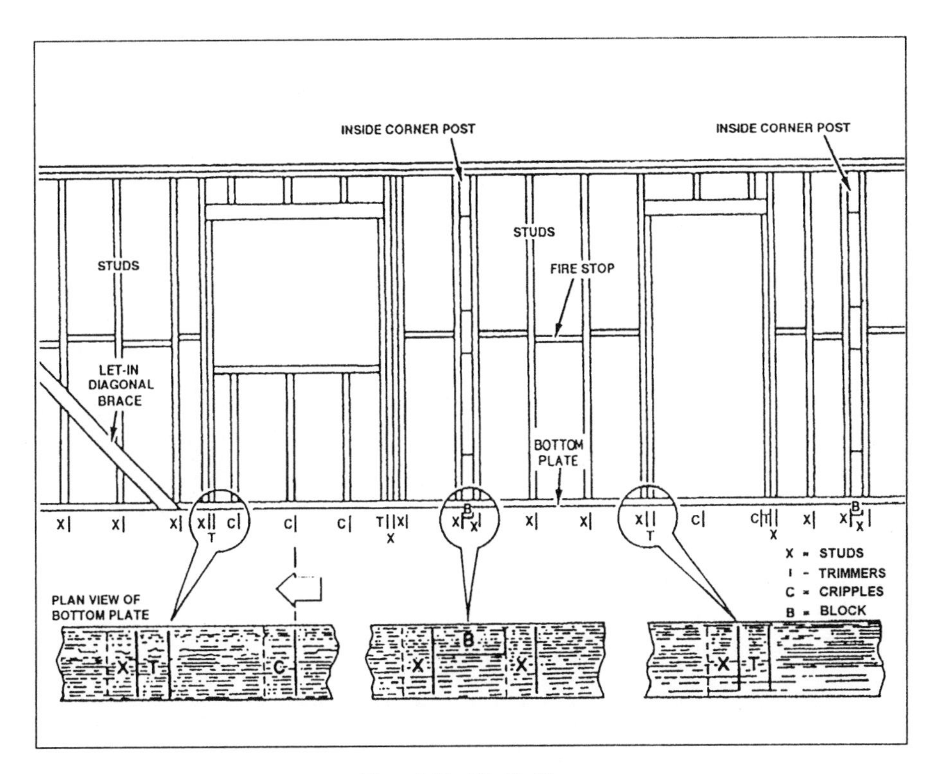

Figure 1-37.—Fire blocking.

Cut-in bracing (fig. 1-36) is another type of diagonal bracing. It usually consists of 2 by 4s cut at an angle and toenailed between studs at a diagonal from the top of a corner post down to the soleplate.

Diagonal sheathing (fig. 1-36) is the strongest type of diagonal bracing. Each board acts as a brace for the wall. When plywood or other panel sheathing is used, other methods of bracing may be omitted.

Fire Stops

Most local building codes require fire stops (also known as fire blocks) in walls over 8 foot 1 inch high. Fire stops slow down fire travel inside walls. They can be nailed between the studs before or after the wall is raised. Fire stops can be nailed in a straight line or staggered for easier nailing. Figure 1-37 shows a section of a framed wall with fire stops.

It is not necessary to nail fire stops at the midpoint of the wall. They can be positioned to provide additional backing for nailing the edges of drywall or plywood.

CONSTRUCTION

All major components of a wall should be cut before assembly. By reading the blueprints, you can determine the number of pieces and lengths of all components. The different parts of the wall are then assembled. Any hard, level surface can be used for assembly. After completing nailing, raise the walls in place for securing.

Two layout procedures are used in wall layout: horizontal plate and vertical layout. In horizontal plate layout, the location of the wall is determined from the dimensions found in the floor plan of the blueprints. For vertical layout, the dimension can be found in the sectional views of the building's blueprints.

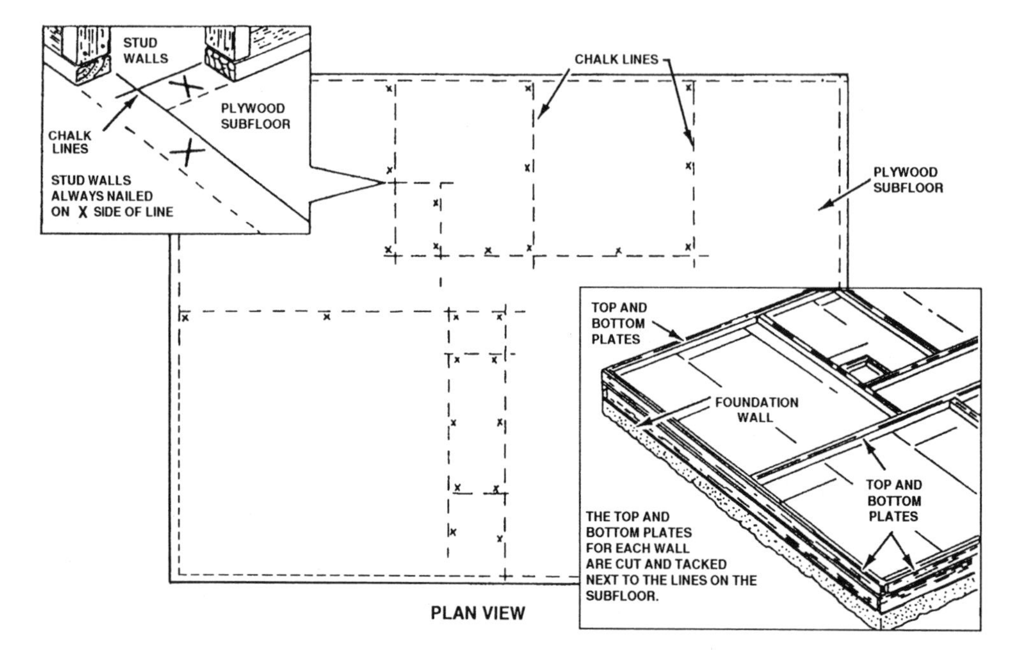

Figure 1-38.—Layout and cutting of plates.

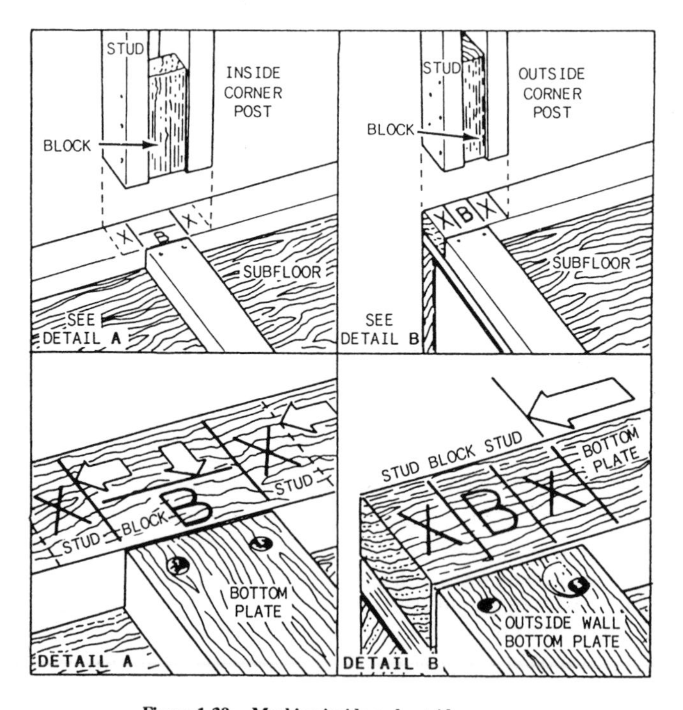

Figure 1-39.—Marking inside and outside corners.

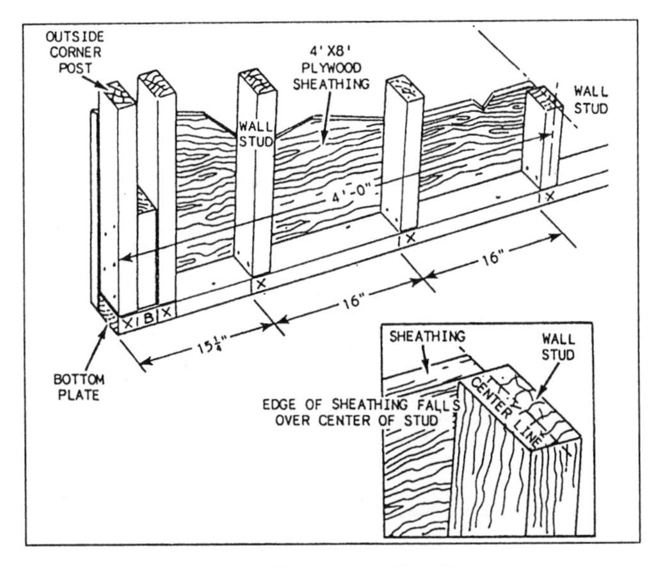

Figure 1-40.—First exterior wall stud layout.

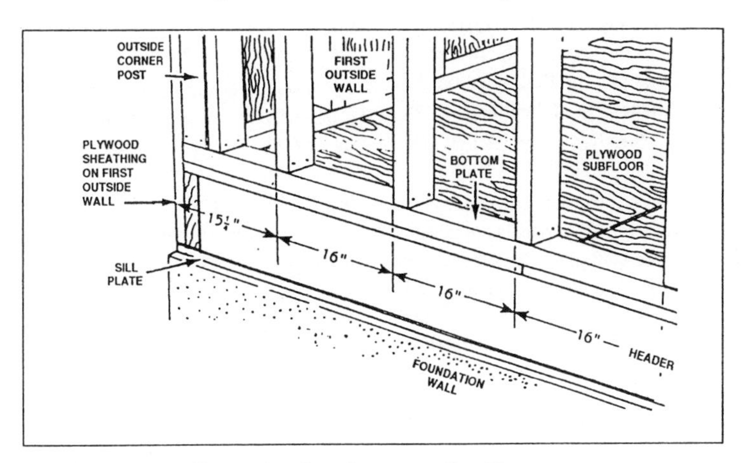

Figure 1-41.—Second exterior wall stud layout.

Horizontal Plate Layout

After all the lines are snapped, the wall plates are cut and tacked next to the lines (fig. 1-38). The plates are then marked off for corner posts and regular studs, as well as for the studs, trimmers, and cripples for the rough openings. All framing members must be clearly marked on the plates. This allows for efficient and error-free framing. Figure 1-37 shows a wall with framing members nailed in place according to layout markings.

A procedure for marking outside and inside corners for stud-and-block corner post construction is shown in figure 1-39. For laying out studs for the first exterior wall, see figure 1-40. In figure 1-40, the plates are marked for the first stud from a corner to be placed 15 1/4 inches from the end of the corner. Studs after the first stud follow 16 inches OC layout. This ensures the edges of standard-size panels used for sheathing or wallboard fall on the centers of the studs. Cripples are laid out to follow the layout of the studs.

A procedure for laying out studs for the second exterior wall is shown in figure 1-41. The plates are

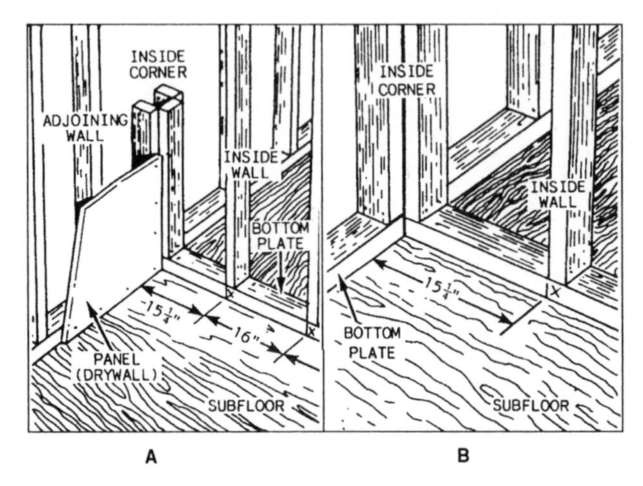

Figure 1-42.—Starting measurement for interior wall.

marked for the first stud to be placed 15 1/4 inches from the outside edge of the panel thickness on the first wall. This layout allows the corner of the first panel on the second wall to line up with the edge of the first panel on the second wall. Also, the opposite edge of the panel on the second wall will break on the center of a stud.

A procedure for laying out studs for interior walls (partitions) is shown in figure 1-42. If panels are placed on the exterior wall first, the wall plates for the interior wall are marked for the first stud to be placed 15 1/4 inches from the edge of the panel thickness on the exterior wall. If panels are to be placed on the interior wall, the wall plates of the interior wall are marked for the first stud to be placed 15 1/4 inches from the unpaneled exterior wall.

If drywall or other interior finish panels are to be nailed to an adjoining wall (fig. 1-42, view A), you must measure 15 1/4 inches plus the thickness of the material. When panels are to be nailed on a wall first (view B), measure and mark the 15 1/4 inches from the front surface of the bottom plate. These procedures ensure stud alignment remains accurate throughout the nailing process.

Rough openings for doors and windows must also be marked on the wall plates. The rough opening dimensions for a window (fig. 1-43, view A) or wood door (view B) are calculated based on the window or door width, the thickness of the finish frame, and 1/2-inch clearance for shim materials at the sides of the frame. Some blueprint door and window schedules give the rough opening dimensions, simplifying the layout.

A rough opening for a metal window often requires a 1/2-inch clearance around the entire frame. When the measurements are not given in the window schedule, take them from the manufacturer's installation instructions supplied with the windows.

A completely laid out bottom plate includes markings for corner posts, rough openings, studs, and cripples. The corner posts are laid out first. Next, the 16-inch marks for the studs and cripples are marked, and then the marks for the rough openings are made.

Some Builders prefer to lay out the rough openings before the studs and cripples are marked. There is, however, an advantage to laying out the 16-inch OC marks first. Studs and trimmers framing a door and window often fall very close to a 16-inch OC stud mark. Slightly shifting the position of the rough opening may eliminate an unnecessary stud from the wall frame.

Vertical Layout

Vertical layout is the procedure for calculating the lengths of the different vertical members of a wood-framed wall. This makes it possible to precut all studs, trimmers, and cripples required for a building.

Some blueprints contain section views giving the exact rough heights of walls. The rough height is the distance from the subfloor to the bottom of the ceiling

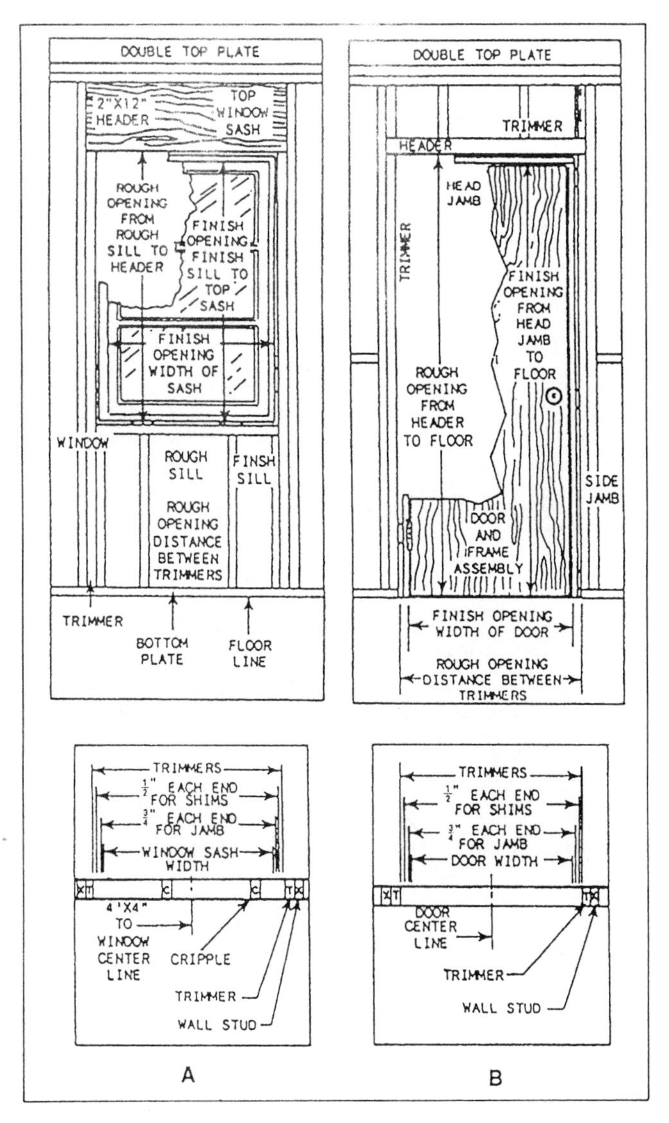

Figure 1-43.—Measurements for windows and doors.

joists. The rough height to the top of the door (the distance from the subfloor to the bottom of the door header) may also be noted on the section drawing. In addition, it may be given in the column for rough opening measurements on the door schedule. The rough height to the top of the door establishes the measurement for the rough height to the top of the window, as window headers are usually in line with door headers.

The distance from the bottom to the top of a rough window opening can be found by measuring down from the bottom of the window header using dimensions provided in the rough opening column of the window schedule.

Many Builders prefer to frame the door and window openings before assembling the wall. View A of

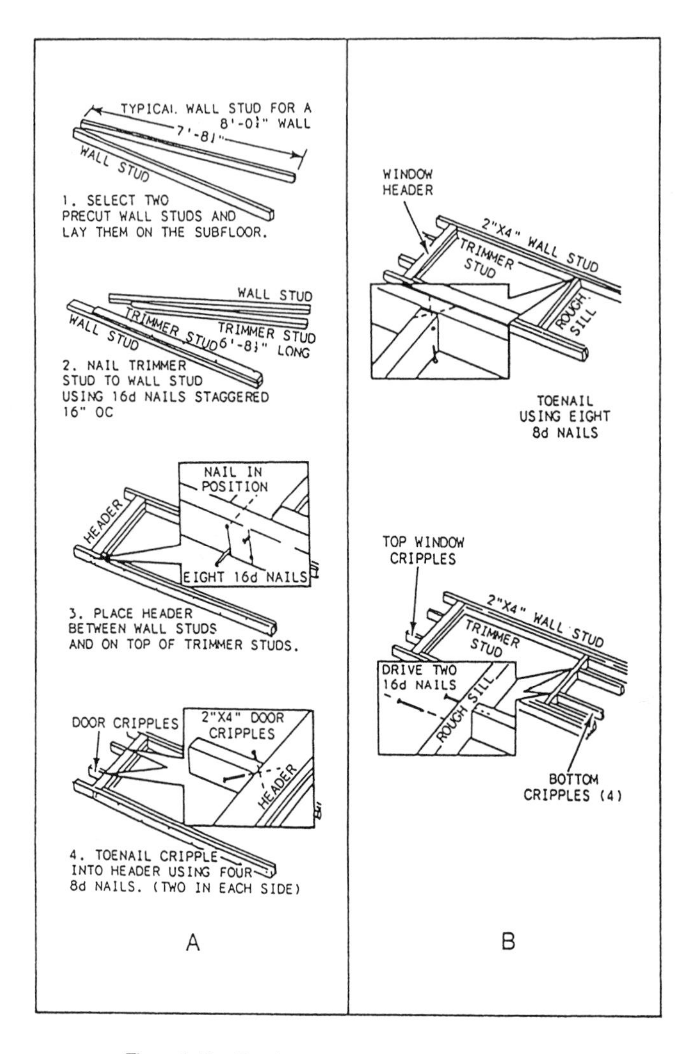

Figure 1-44.—Framing typical door and window openings.

figure 1-44 shows typical door framing; view B shows typical window framing. After stud layout, cripple studs are laid out (usually 16 inches OC) and nailed between the header and top plate and rough window sill and soleplate. It is a good practice to place a cripple stud under each end of a sill.

ASSEMBLY

After the corners and openings for doors and windows have been made up, the entire wall can be nailed together on the subfloor (fig. 1-45). Place top and bottom plates at a distance slightly greater than the length of the studs. Position the corners and openings between the plates according to the plate layout. Place studs in position with the crown side up. Nail the plates into the studs, cripples, and trimmers. On long walls, the breaks in the plates should occur over a stud or cripple.

Placing the Double Top Plate

The double top plate (fig. 1-46) can be placed while the wall is still on the subfloor or after all the walls have

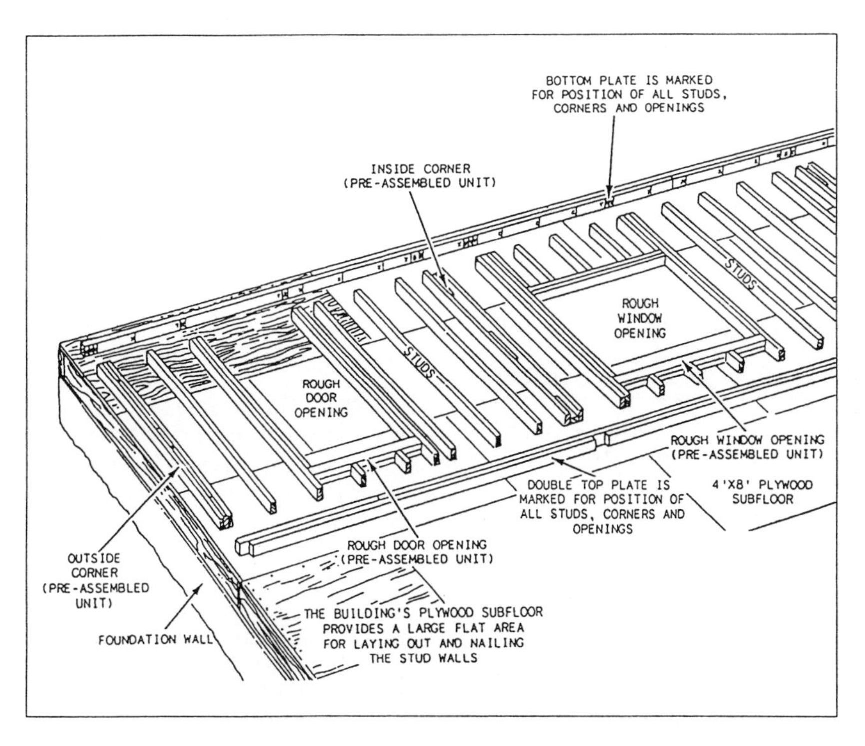

Figure 1-45.—Assembly of wall components.

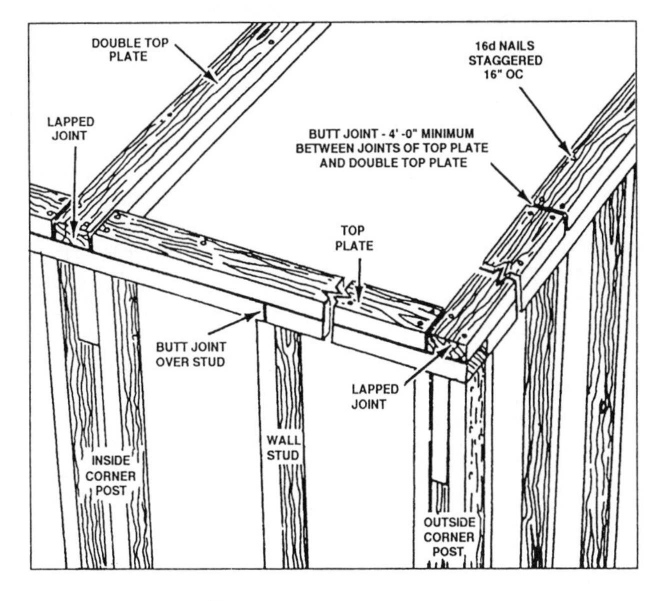

Figure 1-46.—Double top plate.

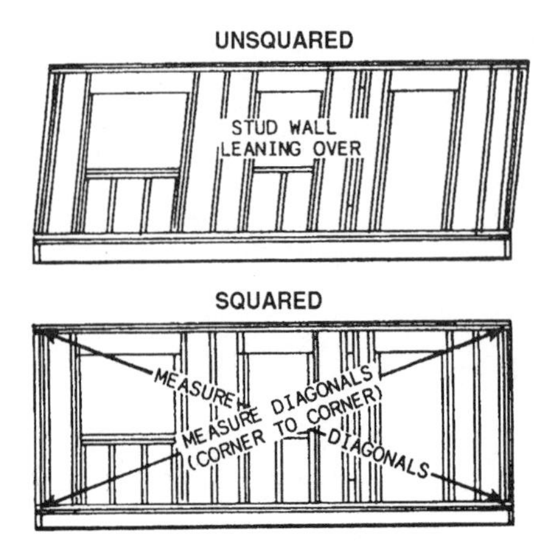

Figure 1-47.—Squaring a wall.

been raised. The topmost plates are nailed so that they overlap the plates below at all corners. This helps to tie the walls together. All ends are fastened with two 16d nails. Between the ends, 16d nails are staggered 16 inches OC. The butt joints between the topmost plates should be at least 4 feet from any butt joint between the plates below them.

Squaring Walls and Placing Braces

A completely framed wall is often squared while it is still lying on the subfloor. In this way, bracing, plywood, or other exterior wall covering can be nailed before the wall is raised. When diagonal measurements are equal, the wall is square. Figure 1-47 shows examples of unsquared and squared walls.

A let-in diagonal brace may be placed while the wall is still on the subfloor. Lay out and snap a line on the studs to show the location of the brace (fig. 1-48). The studs are then notched for the brace. Tack the brace to the studs while the wall is still lying on the subfloor. Tacking instead of nailing allows for some adjustment after the wall is raised. After any necessary adjustment is made, the nails can be securely driven in.

Raising

Most walls can be raised by hand if enough help is available. It is advisable to have one person for every 10 feet of wall for the lifting operation.

The order in which walls are framed and raised may vary from job to job. Generally, the longer exterior walls are raised first. The shorter exterior walls are then raised, and the corners are nailed together. The order of framing interior partitions depends on the floor layout.

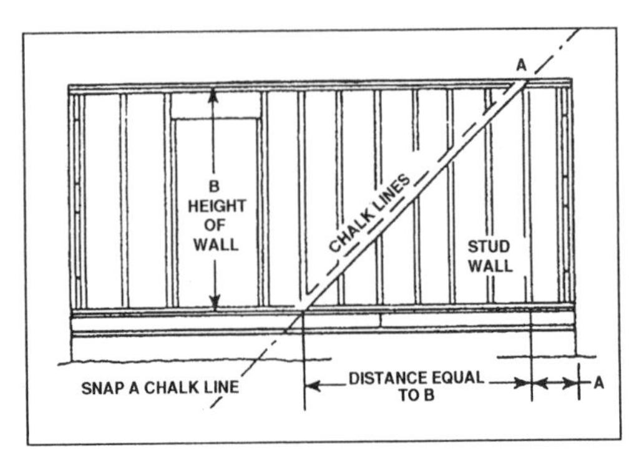

Figure 1-48.—Let-in diagonal brace.

After a wall has been raised, its bottom plates must be nailed securely to the floor. Where the wall rests on a wood subfloor and joists, 16d nails should be driven through the bottom plate and into the floor joists below the wall.

Plumbing and Aligning

Accurate plumbing of the corners is possible only after all the walls are up. Most framing materials are not perfectly straight; walls should never be plumbed by applying a hand level directly to an end stud. Always use a straightedge along with the level, as shown in figure 1-49, view A. The straightedge can be a piece ripped out of plywood or a straight piece of 2 by 4 lumber. Blocks 3/4 inch thick are nailed to each end. The blocks make it possible to accurately plumb the wall from the bottom plate to the top plate.

Plumbing corners requires two persons working together—one working the bottom area of the brace and the other watching the level. The bottom end of the brace is renailed when the level shows a plumb wall.

The tops of the walls (fig. 1-49, view B) are straightened (aligned or lined up) after all the corners have been plumbed. Prior to nailing the floor or ceiling joists to the tops of the walls, make sure the walls are aligned. Here's how: Fasten a string from the top plate at one corner of the wall to the top plate at another corner of the wall. You then cut three small blocks from 1 by 2 lumber. Place one block under each end of the string so that the line is clear of the wall.

The third block is used as a gauge to check the wall at 6- or 8-foot intervals. At each checkpoint, a temporary brace is fastened to a wall stud.

When fastening the temporary brace to the wall stud, adjust the wall so that the string is barely touching the gauge block. Nail the other end of the brace to a short 2 by 4 block fastened to the subfloor. These temporary

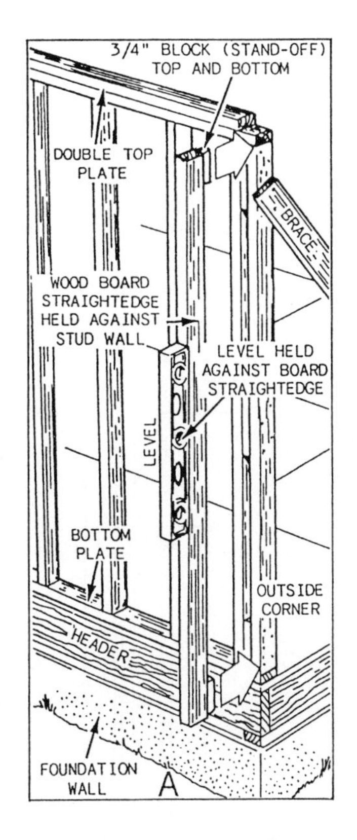

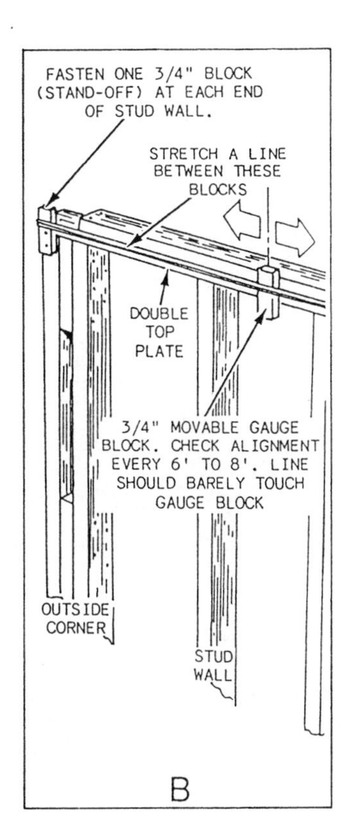

Figure 1-49.—Plumbing and aligning corners and walls.

braces are not removed until the framing and sheathing for the entire building have been completed.

Framing over Concrete Slabs

Often, the ground floor of a wood-framed building is a concrete slab. In this case, the bottom plates of the walls must be either bolted to the slab or nailed to the slab with a powder-actuated driver. If bolts are used, they must be accurately set into the slab at the time of the concrete pour. Holes for the bolts are laid out and drilled in the bottom plate when the wall is framed. When the wall is raised, it is slipped over the bolts and secured with washers and nuts.

Occasionally, on small projects, the soleplate is bolted or fastened down first. The top plate is nailed to the studs, and the wall is lifted into position. The bottom ends of the studs are toenailed into the plate. The rest of the framing procedure is the same as for walls nailed on top of a subfloor.

SHEATHING THE WALLS

Wall sheathing is the material used for the exterior covering of the outside walls. In the past, nominal 1-inch-thick boards were nailed to the wall horizontally or at a 45° angle for sheathing. Today, plywood and other types of panel products (waferboard, oriented strandboard, compositeboard) are usually used for sheathing. Plywood and nonveneered panels can be applied much quicker than boards. They add considerable strength to a building and often eliminate the need for diagonal bracing.

Generally, wall sheathing does not include the finished surface of a wall. Siding, shingles, stucco, or brick veneer are placed over the sheathing to finish the wall. Exterior finish materials are discussed later in this TRAMAN.

Plywood

Plywood is the most widely used sheathing material. Plywood panels usually applied to exterior

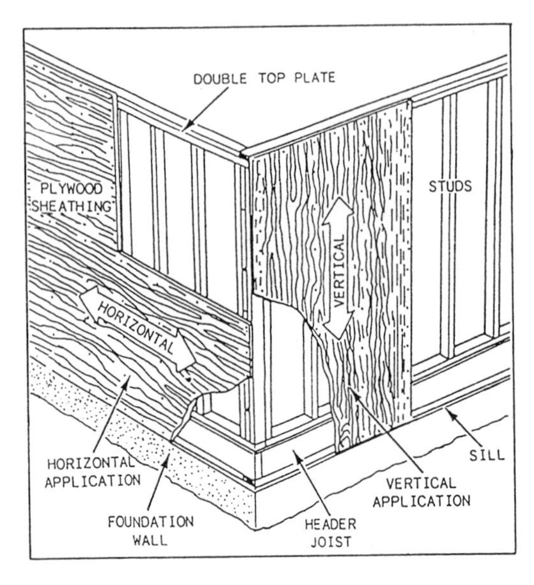

Figure 1-50.—Plywood sheathing.

walls range in size from 4 by 8 feet to 4 by 12 feet with thicknesses from 5/16 inch to 3/4 inch. The panels may be placed with the grain running vertically or horizontally (fig. 1-50). Specifications may require blocking along the long edges of horizontally placed panels.

Typical nailing specifications require 6d nails with panels 1/2 inch or less in thickness and 8d nails for panels more than 1/2 inch thick. The nails should be spaced 6 inches apart along the edges of the panels and 12 inches apart at the intermediate studs.

When nailing the panels, leave a 1/8-inch gap between the horizontal edges of the panels and a 1/16-inch gap between the vertical edges. These gaps allow for expansion caused by moisture and prevent panels from buckling.

In larger wood-framed buildings, plywood is often nailed to some of the main interior partitions. The result is called a shear wall and adds considerable strength to the entire building.

Plywood sheathing can be applied when the squared wall is still lying on the subfloor. However, problems can occur after the wall is raised if the floor is not perfectly straight and level. For this reason, some Builders prefer to place the plywood after the entire building has been framed.

Nonveneered Panels

Although plywood is the most commonly used material for wall sheathing, specs sometimes call for

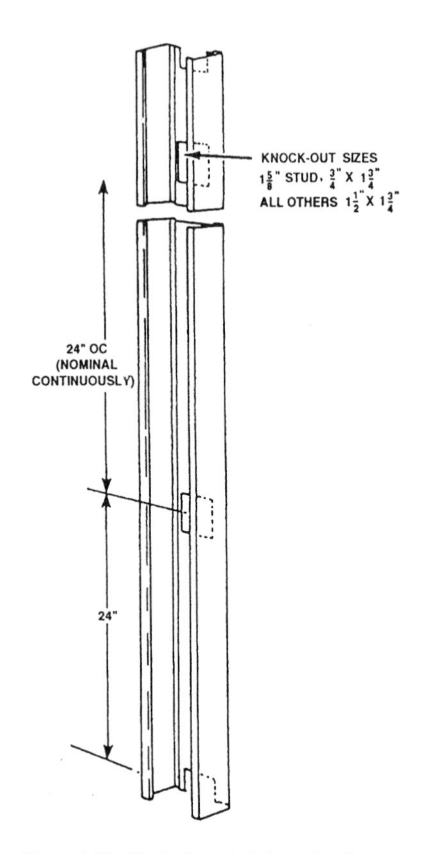

Figure 1-51.—Typical metal stud construction.

nonveneered (reconstituted wood) panels. Panels made of waferboard, oriented strandboard, and composite-board have been approved by most local building codes for use as wall sheathing. Like plywood, these panels resist racking, so no corner bracing is necessary in normal construction. However, where maximum shear strength is required, conventional veneered plywood panels are still recommended.

The application of nonveneered wall sheathing is similar to that for plywood. Nailing schedules usually call for 6d common nails spaced 6 inches OC above the panel edges, and 12 inches OC when nailed into the intermediate studs. Nonveneered panels are usually applied with the long edge of the panel in a vertical position.

METAL FRAMING

Metal is an alternative to wood framing. Many buildings are framed entirely of metal, whereas some

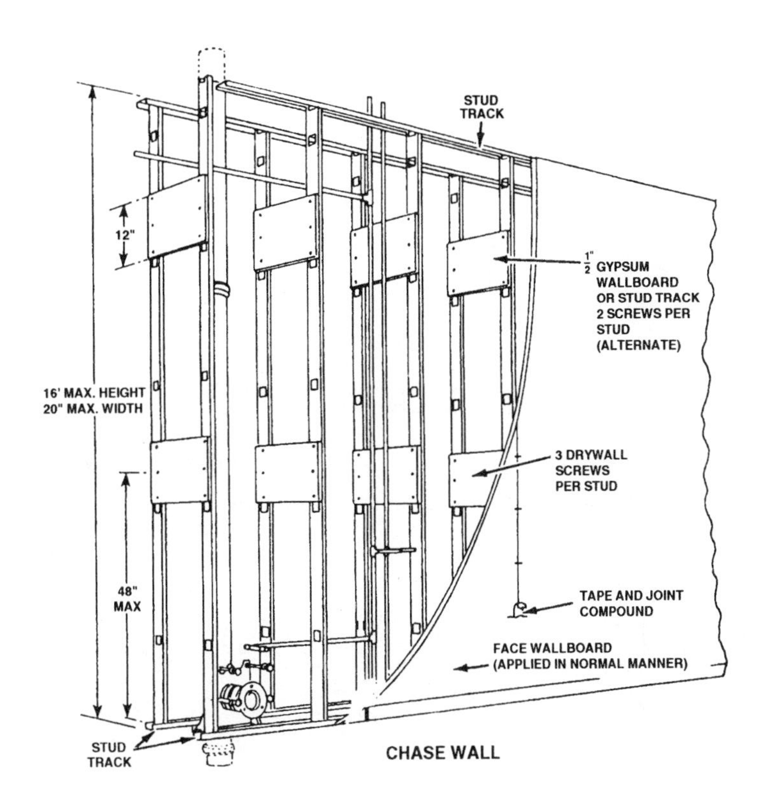

Figure 1-52.—Chase wall construction.

buildings are framed in a combination of metal and wood.

The metal framing members generally used are cold-formed steel, electrogalvanized to resist corrosion. Thicknesses range from 18 gauge to 25 gauge, the latter being most common. Most metal studs have notches at each end and knockouts located about 24 inches OC (fig. 1-51) to facilitate pipe and conduit installation. The size of the knockout, not the size of the stud, determines the maximum size of pipe or other material that can be passed through horizontally.

Chase (or double stud) walls (fig. 1-52) are often used when large pipes, ducts, or other items must pass vertically or horizontally in the walls. Studs are generally available in thicknesses of 1 5/8, 2 1/2, 3 5/8, 4, and 6 inches. The metal runners used are also 25-gauge (or specified gauge) steel or aluminum, sized to complement the studs. Both products have features advantageous to light-frame construction. The metal studs and runners do not shrink, swell, twist, or warp. Termites cannot affect them, nor are they susceptible to dry rot. Also, when combined with proper covering material, they have a high fire-resistance rating.

A variety of systems have been developed by manufacturers to meet various requirements of attachment, sound control, and fire resistance. Many of the systems are designed for ease in erection, yet they are still demountable for revising room arrangements.

The framing members are assembled with power screwdrivers and using self-drilling, self-tapping screws. The floor assembly is fastened to the foundation or concrete slab with studs (special nails) driven through the stud track (runner) by a powder-actuated stud driver. The plywood subfloor is installed over the metal floor framing system with self-drilling, self-tapping screws and structural adhesive. Wall sections are assembled at the jobsite or delivered as preassembled panels from an off-site prefabrication shop. Conventional sheathing is attached to the framework with self-tapping screws.

Door frames for both the interior partitions and exterior walls are integral with the system. They are prepainted and may come complete with necessary

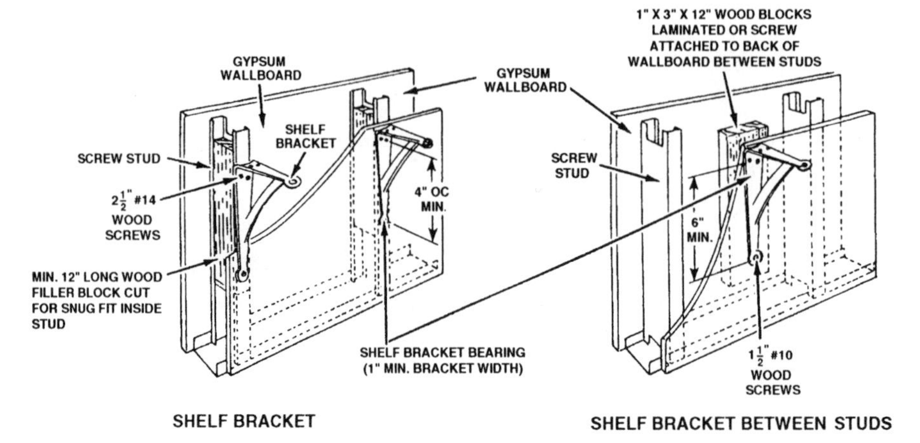

Figure 1-53.—Wood blocking for ceiling or wall-mounted fixtures.

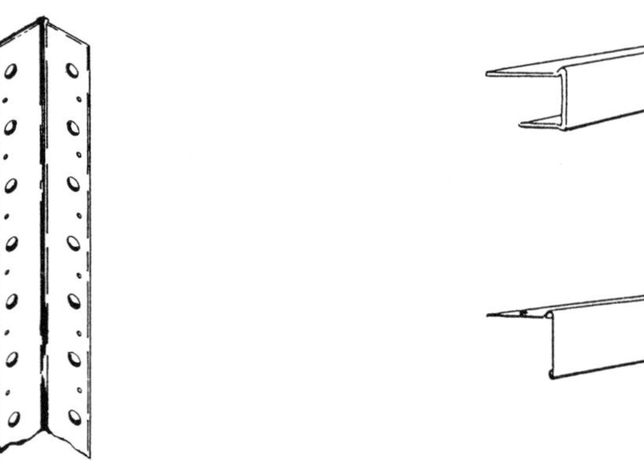

Figure 1-54.—Standard corner bead.

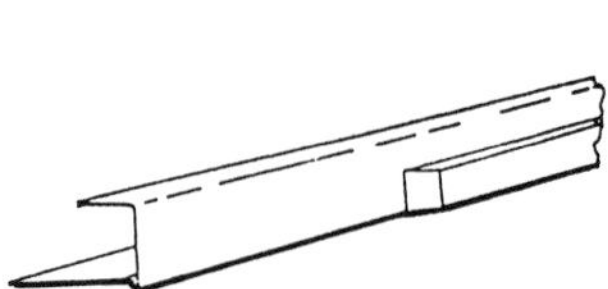

Figure 1-56.—Casing and trim beads.

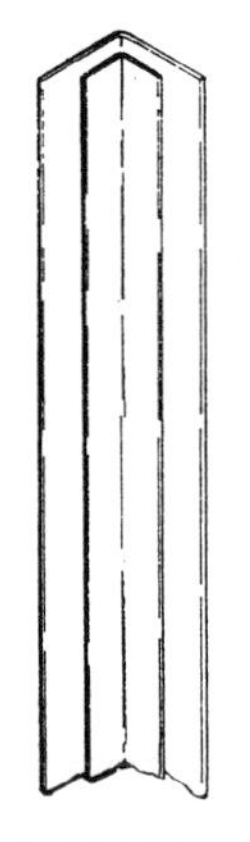

Figure 1-55.—Multiflex tape bead.

hinges, locks, rubber stops, and weather stripping. The windows are also integral to the system, prefabricated and painted. These units may include interior and exterior trim designed to accept 1/2-inch wallboard and 1/2-inch sheathing plus siding on the outside.

Plumbing is installed in prepunched stud webs. Wiring is passed through insulated grommets inserted in the prepunched webs of the studs and plates. Wall and ceiling fixtures are mounted by attaching wood blocking spaced between the flanges of the wall studs or trusses

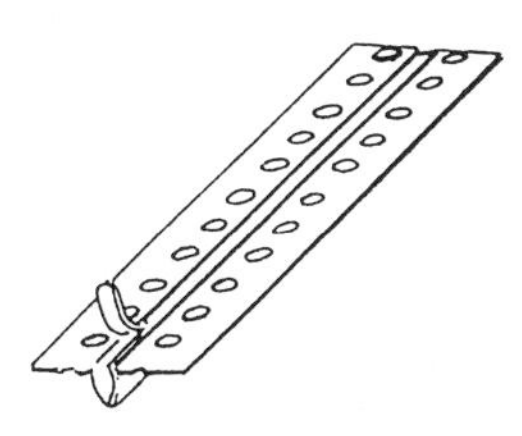

Figure 1-57.—Expansion joint.

(fig. 1-53). Friction-tight insulation is installed by placing the batts (bundles of insulating material) between the studs on the exterior walls. Studs are spaced 12, 16, or 24 inches OC as specified in the blueprints.

Corner and Casing Beads

Standard wallboard corner bead is manufactured from galvanized steel with perforated flanges, as shown in figure 1-54. It provides a protective reinforcement of straight corners. The corner bead is made with 1-inch by 1-inch flanges for 3/8- or 1/2-inch single-layer wallboard; 1 inch by 1 1/4 inches for 1/2-inch or 5/8-inch single-layer wallboard; 1 1/4 inches by 1 1/4 inches for two-layer wallboard application. It is available in 10-foot lengths.

Multiflex tape bead consists of two continuous metal strips on the undersurface of 2 1/8-inch-wide reinforcing tape (fig. 1-55). This protects corners formed at any angle. Multiflex tape bead comes in 100-foot rolls.

Casing and trim beads (examples are shown in fig. 1-56) are used as edge protection and trim around window and door openings and as moldings at ceiling angles. They are made from galvanized steel in three styles to fit 3/8-inch, 1/2-inch, and 5/8-inch wallboard and come in 10-foot lengths.

Expansion Joints

Expansion joints are vinyl extrusions used as control joints in drywall partitions and ceilings. A typical form is shown in figure 1-57.

Figure 1-58 shows a typical metal frame layout and use of corner and casing beads for corners, partition intersections, and partition ends. It also shows a typical

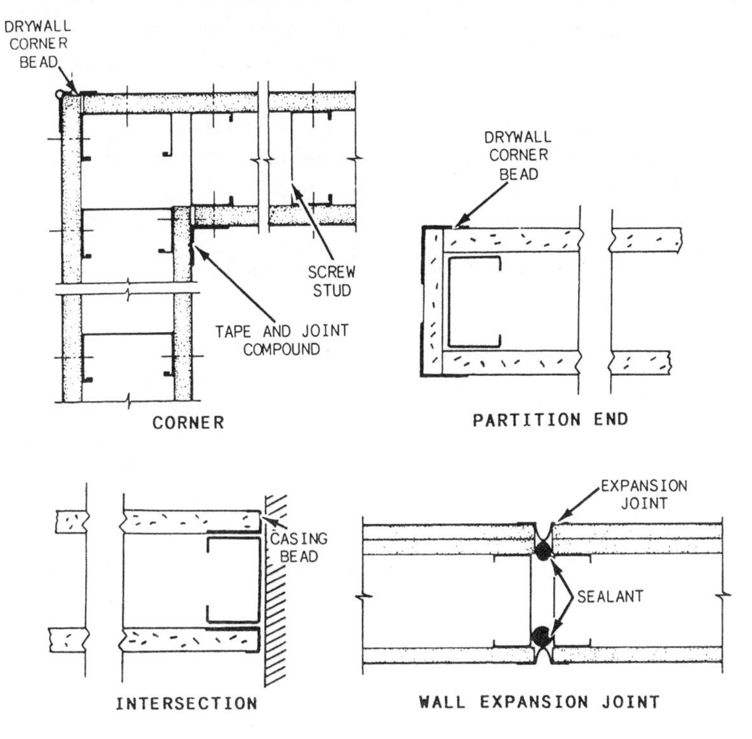

Figure 1-58.—Metal frame layout with various beads and joints.

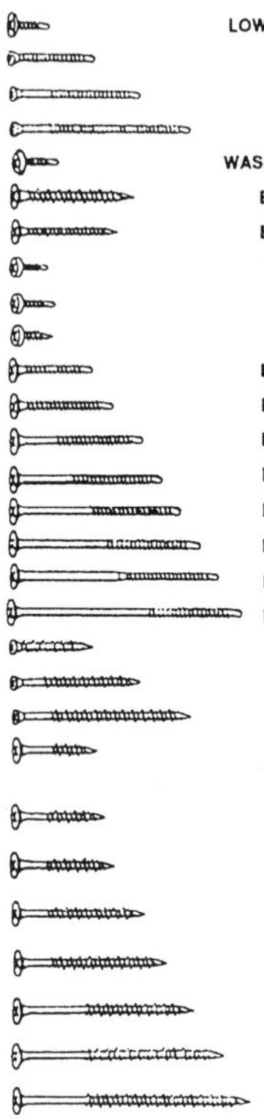

HEAD	LENGTH	FASTENING APPLICATION
LOW PROFILE	$\frac{1}{2}$"	METAL-TO-METAL, REDUCES DRYWALL BULGE OVER SCREW.*
TRIM	1"	BATTEN TO METAL STUD.*
TRIM	$1\frac{5}{8}$"	WOOD TRIM ON SINGLE LAYER DRYWALL TO METAL FRAMING.*
TRIM	$2\frac{1}{4}$"	WOOD TRIM ON DOUBLE LAYER DRYWALL TO METAL FRAMING.*
WASHER HEAD	$\frac{1}{2}$"	METAL LATH TO STEEL FRAMING.*
BUGLE	$1\frac{1}{2}$"	BOARD-TO-BOARD DURING LAMINATION.
BUGLE	$1\frac{1}{4}$"	DRYWALL OR CHANNELS TO WOOD FRAMING.
PAN	$\frac{3}{8}$"	METAL-TO-METAL.*
PAN	$\frac{1}{2}$"	METAL STUDS TO RUNNERS. DOOR FRAME CLIPS TO METAL STUDS.*
PAN	$\frac{3}{8}$"	METAL STUDS TO METAL RUNNERS SPLICING STUDS.
BUGLE	1"	$\frac{1}{2}$" OR $\frac{5}{8}$" DRYWALL TO METAL FRAMING.*
BUGLE	$1\frac{1}{4}$"	METAL LATH TO METAL STUDS.*
BUGLE	$1\frac{5}{8}$"	DOUBLE LAYER $\frac{1}{2}$" OR $\frac{5}{8}$" DRYWALL TO METAL FRAMING.*
BUGLE	$1\frac{7}{8}$"	MULTI-LAYER DRYWALL TO METAL FRAMING.*
BUGLE	2"	LAMINATED DRYWALL UP TO $1\frac{5}{8}$" THICK TO METAL FRAMING.*
BUGLE	$2\frac{3}{8}$"	DOUBLE LAYER 1" SHAFTLINER TO METAL FRAMING.*
BUGLE	$2\frac{5}{8}$"	LAMINATED DRYWALL UP TO $2\frac{1}{4}$" THICK TO FRAMING.*
BUGLE	3"	MULTI-LAYER DRYWALL UP TO $2\frac{1}{2}$" THICK TO FRAMING.*
TRIM	1"	BATTEN OVER DRYWALL TO METAL STUD.
TRIM	$1\frac{5}{8}$"	WOOD TRIM OVER DRYWALL TO METAL STUD.
TRIM	$2\frac{1}{4}$"	WOOD TRIM OVER DOUBLE LAYER DRYWALL TO METAL STUD.
BUGLE	1"	$\frac{1}{2}$" OR $\frac{5}{8}$" DRYWALL TO METAL STUD UP TO 20 GAUGE. STEEL CHANNELS TO WOOD FRAMING.
BUGLE	$1\frac{1}{8}$"	$\frac{5}{8}$" DRYWALL CEILING ATTACHMENTS.
BUGLE	$1\frac{1}{4}$"	$\frac{5}{8}$" DRYWALL CEILING ON RESILIENT FURRING CHANNELS.
BUGLE	$1\frac{5}{8}$"	DOUBLE LAYER $\frac{5}{8}$" DRYWALL TO METAL STUD.
BUGLE	$1\frac{7}{8}$"	$\frac{1}{2}$" DRYWALL THROUGH SHAFTLINER TO METAL RUNNERS.
BUGLE	$2\frac{1}{4}$"	$\frac{5}{8}$" DRYWALL THROUGH SHAFTLINER TO METAL RUNNERS.
BUGLE	$2\frac{5}{8}$"	DOUBLE LAYER 1" SHAFTLINER TO METAL STUD.
BUGLE	3"	$\frac{1}{2}$" DRYWALL OVER 2 LAYERS OF 1" SHAFTLINER TO METAL FRAMING.

*FRAMING NOT EXCEEDING .1084 THICKNESS (12 GAUGE).

Figure 1-59.—Drywall screws and fastening application.

cross section of a metal frame stud wall control joint. Figure 1-59 lists the different types of fasteners used in metal frame construction and explains the application of each type.

CEILING FRAMING

LEARNING OBJECTIVE: Upon completing this section, you should be able to state the purpose of ceiling frame members and describe layout and installation procedures.

Ceiling construction begins after all walls have been plumbed, aligned, and secured. One type of ceiling supports an attic area beneath a sloping (pitched) roof. Another type serves as the framework of a flat roof. When a building has two or more floors, the ceiling of a lower story is the floor of the story above.

One of the main structural functions of a ceiling frame is to tie together the outside walls of the building. When located under a pitched roof, the ceiling frame also resists the outward pressure placed on the walls by the roof rafters (fig. 1-60). The tops of interior partitions are fastened to the ceiling frame. In addition to supporting the attic area beneath the roof, the ceiling frame supports the weight of the finish ceiling materials, such as gypsum board or lath and plaster.

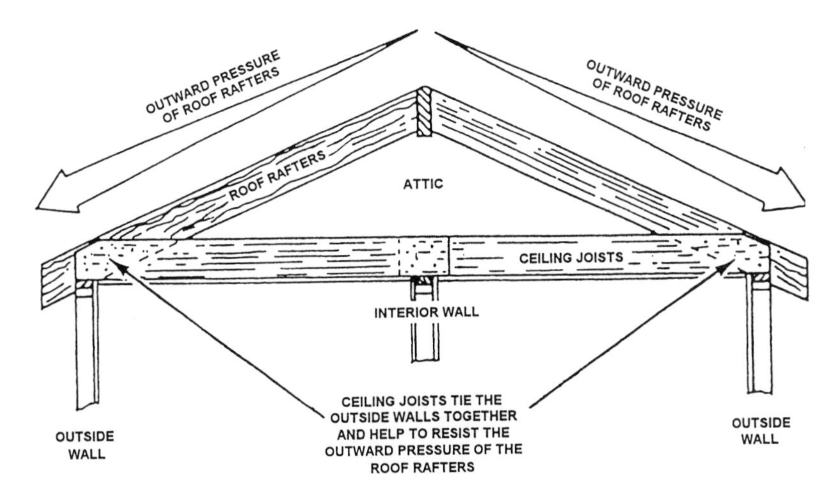

Figure 1-60.—Ceiling frame tying exterior walls together.

JOISTS

Joists are the most important framing members of the ceiling. Their size, spacing, and direction of travel are given on the floor plan. As mentioned earlier, the spacing between ceiling joists is usually 16 inches OC, although 24-inch spacing is also used. The size of a ceiling joist is determined by the weight it carries and the span it covers from wall to wall. Refer to the blueprints and specifications for size and OC spacing. Although it is more convenient to have all the joists running in the same direction, plans sometimes call for different sets of joists running at right angles to each other.

Interior Support

One end of a ceiling joist rests on an outside wall. The other end often overlaps an interior bearing partition or girder. The overlap should be at least 4 inches. Ceiling joists are sometimes butted over the partition or girder. In this case, the joists must be cleated with a 3/4-inch-thick plywood board, 24 inches long, or an 18-gauge metal strap, 18 inches long.

Ceiling joists may also butt against the girder, supported by joist hangers in the same manner as floor joists.

Roof Rafters

Whenever possible, the ceiling joists should run in the same direction as the roof rafters. Nailing the outside end of each ceiling joist to the heel of the rafter as well as to the wall plates (fig. 1-61) strengthens the tie between the outside walls of the building.

A building may be designed so that the ceiling joists do not run parallel to the roof rafters. The rafters are therefore pushing out on walls not tied together by ceiling joists. In this case, 2 by 4 pieces are added to run

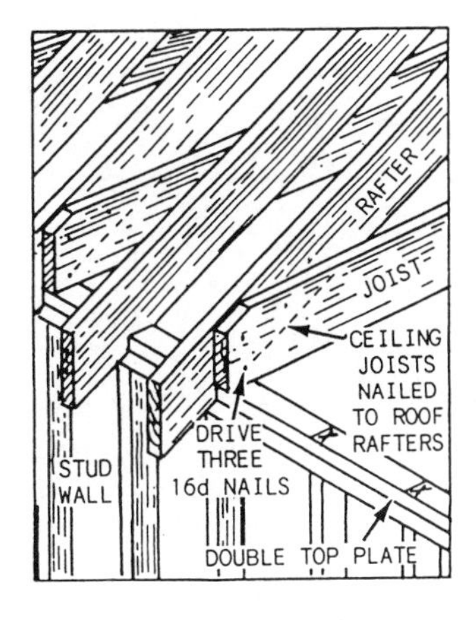

Figure 1-61.—Nailing of ceiling joists.

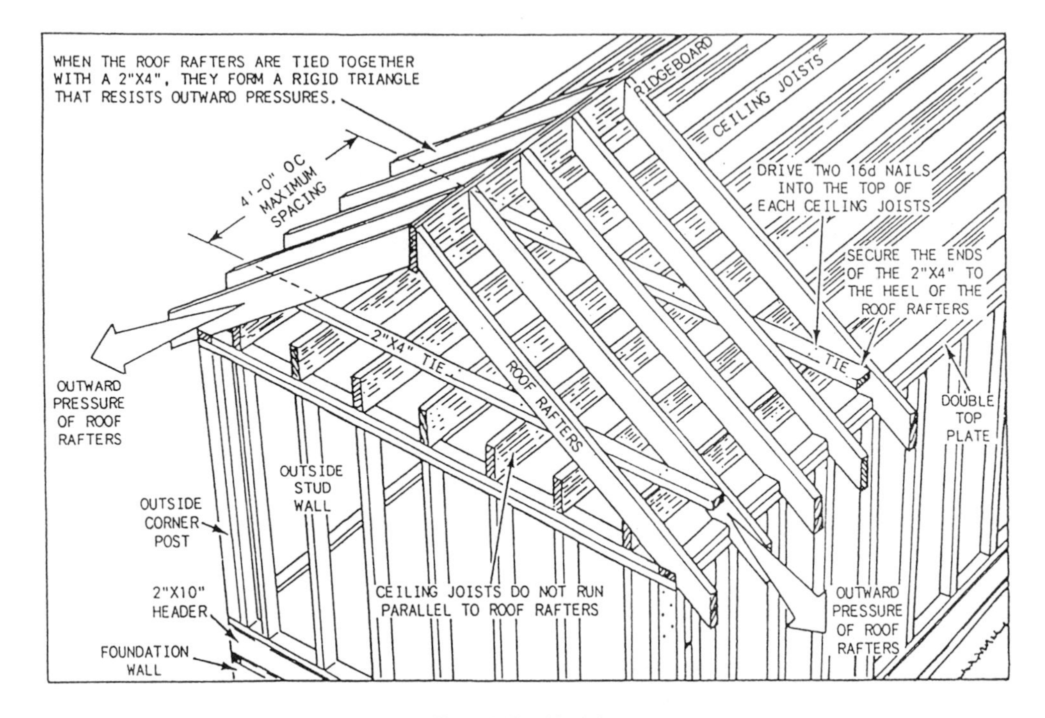

Figure 1-62.—2 by 4 ties.

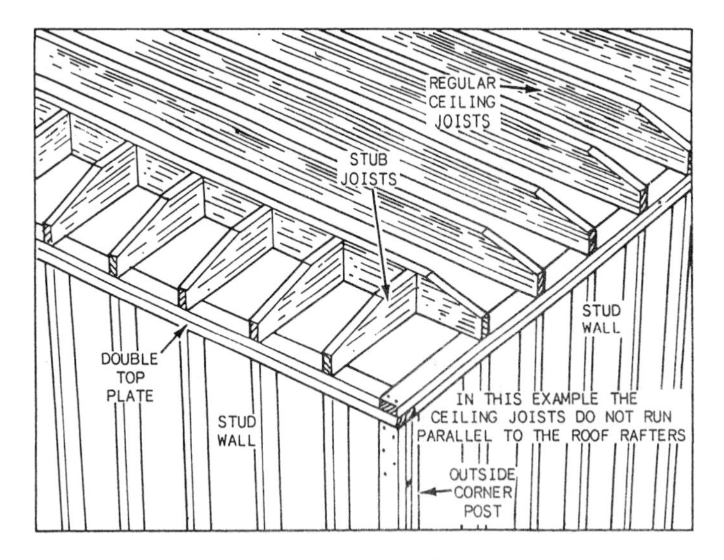

Figure 1-63.—Stub joists.

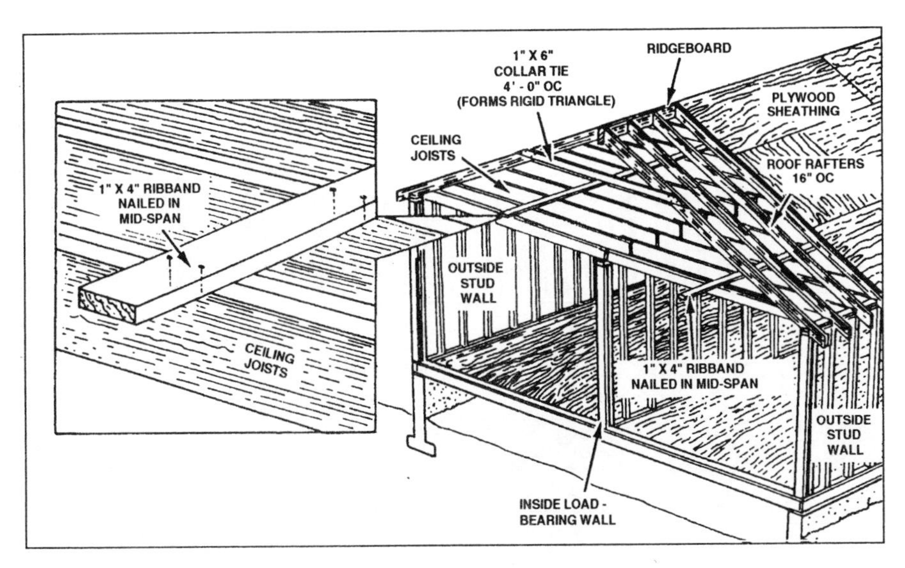

Figure 1-64.—Ribband installation.

in the same direction as the rafters, as shown in figure 1-62. The 2 by 4s should be nailed to the top of each ceiling joist with two 16d nails. The 2 by 4 pieces should be spaced no more than 4 feet apart, and the ends secured to the heels of the rafters or to blocking over the outside walls.

Roof Slope

When ceiling joists run in the same direction as the roof rafters, the outside ends must be cut to the slope of the roof. Ceiling frames are sometimes constructed with stub joists (fig. 1-63). Stub joists are necessary when, in certain sections of the roof, rafters and ceiling joists do not run in the same direction. For example, a low-pitched hip roof requires stub joists in the hip section of the roof.

Ribbands and Strongbacks

Ceiling joists not supporting a floor above require no header joists or blocking. Without the additional header joists, however, ceiling joists may twist or bow at the centers of their span. To help prevent this, nail a 1 by 4 piece called a ribband at the center of the spans (fig. 1-64). The ribband is laid flat and fastened to the top of each joist with two 8d nails. The end of each ribband is secured to the outside walls of the building.

A more effective method of preventing twisting or bowing of the ceiling joists is to use a strongback. A strongback is made of 2 by 6 or 2 by 8 material nailed to the side of a 2 by 4 piece. The 2 by 4 piece is fastened with two 16d nails to the top of each ceiling joist, as shown in figure 1-65. The strongbacks are blocked up and supported over the outside walls and interior partitions. Each strongback holds a ceiling joist in line and also helps support the joist at the center of its span.

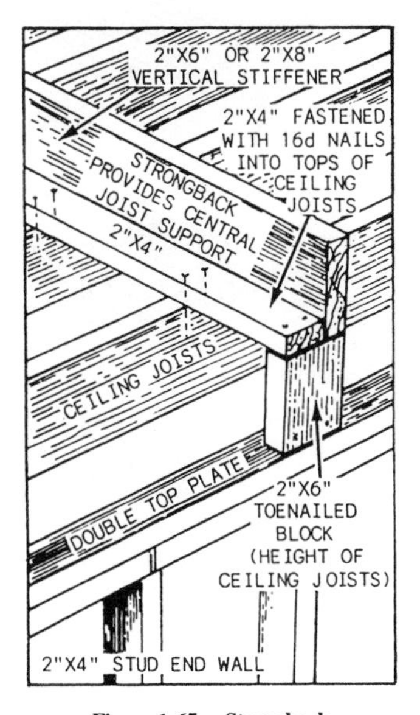

Figure 1-65.—Strongback.

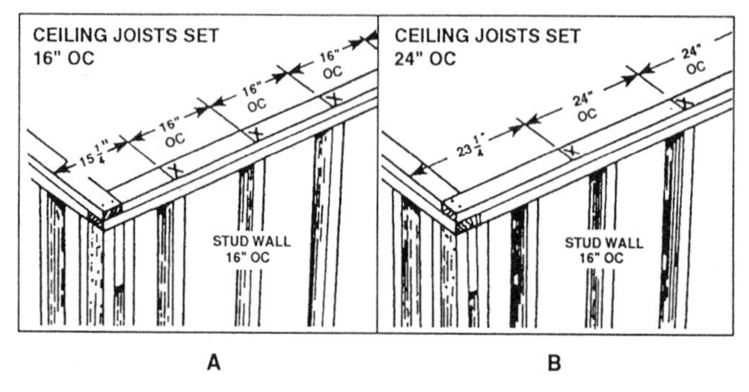

Figure 1-66.—Ceiling joist spacing.

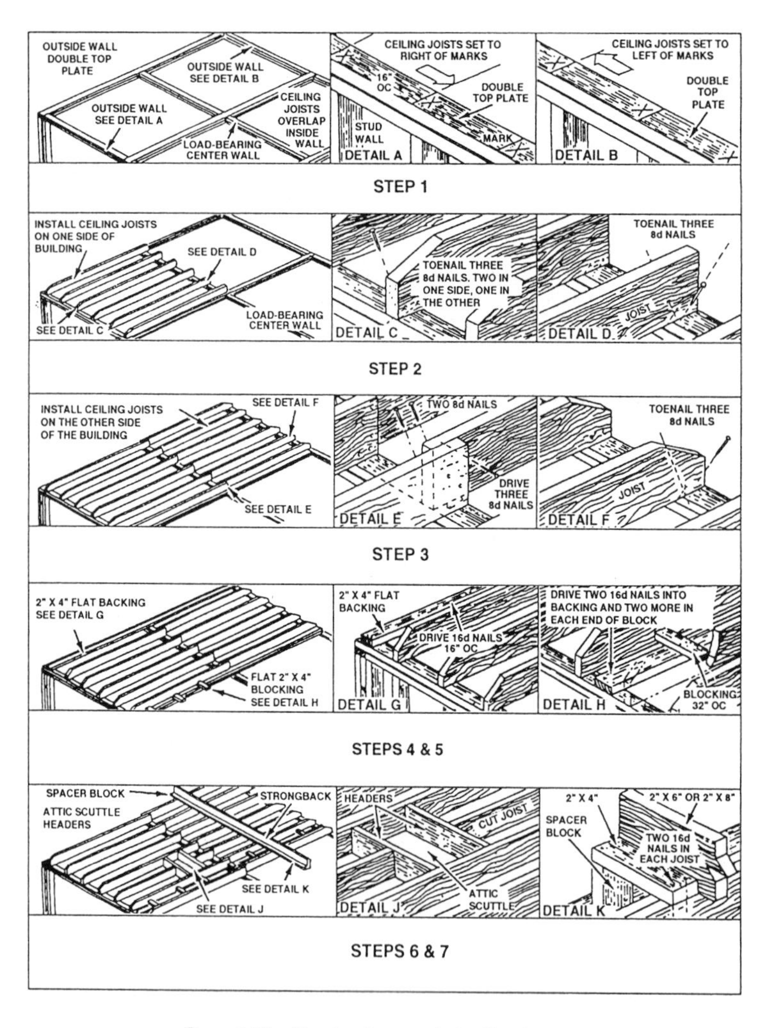

Figure 1-67.—Constructing a typical ceiling frame.

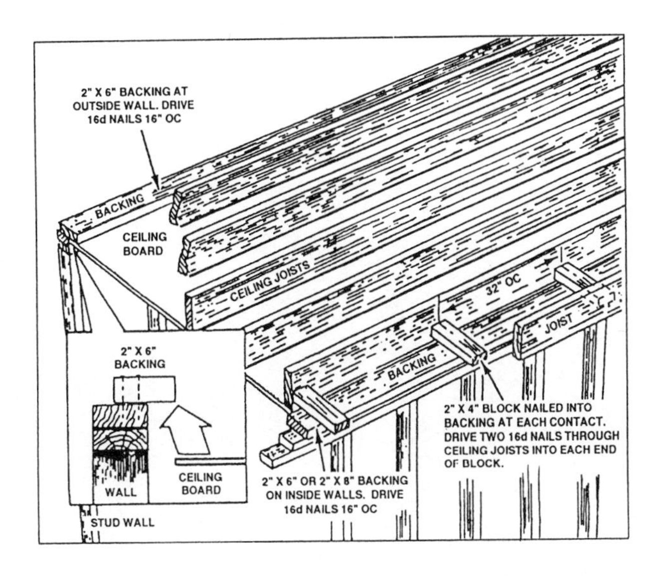

Figure 1-68.—Backing for nailing joists to ceiling frame.

Layout

Ceiling joists should be placed directly above the studs when the spacing between the joists is the same as between the studs. This arrangement makes it easier to install pipes, flues, or ducts running up the wall and through the roof. However, for buildings with walls having double top plates, most building codes do not require ceiling joists to line up with the studs below.

If the joists are being placed directly above the studs, they follow the same layout as the studs below (fig. 1-66, view A). If the joist layout is different from that of the studs below (for example, if joists are laid out 24 inches OC over a 16 inch OC stud layout), mark the first joist at 23 1/4 inches and then at every 24 inches OC (fig. 1-66, view B).

It is a good practice to mark the positions of the roof rafters at the time the ceiling joists are being laid out. If the spacing between the ceiling joists is the same as between the roof rafters, there will be a rafter next to every joist. Often, the joists are laid out 16 inches OC and the roof rafters 24 inches OC. Therefore, every other rafter can be placed next to a ceiling joist.

FRAME

All the joists for the ceiling frame should be cut to length before they are placed on top of the walls. On structures with pitched roofs, the outside ends of the joists should also be trimmed for the roof slope. This angle must be cut on the crown (top) side of the joist. The prepared joists can then be handed up to the Builders working on top of the walls. The joists are spread in a flat position along the walls, close to where they will be nailed. Figure 1-67 shows one procedure for constructing the ceiling frame. In this example, the joists lap over an interior partition. Refer to the figure as you study the following steps:

1. Measure and mark for the ceiling joists.
2. Install the ceiling joists on one side of the building.
3. Install the ceiling joists on the opposite side of the building.
4. Place backing on walls running parallel to the joists.
5. Install 2 by 4 blocks flat between joists where needed to fasten the tops of inside walls running parallel to the joists.
6. Cut and frame the attic scuttle.
7. Place strongbacks at the center of the spans.

Fastening Walls

The tops of walls running in the same direction as the ceiling joists must be securely fastened to the ceiling frame. The method most often used is shown in figure 1-68. Blocks, 2 inches by 4 inches, spaced 32 inches OC, are laid flat over the top of the partition. The ends of

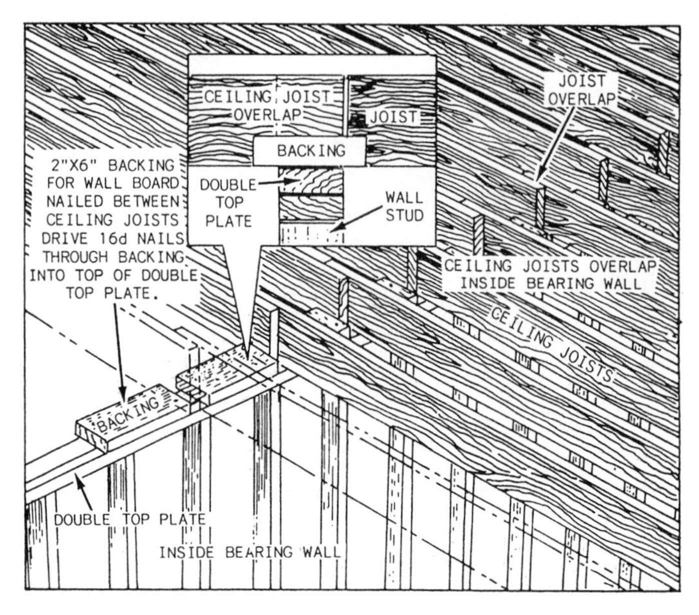

Figure 1-69.—Backing for interior wall plates.

each block are fastened to the joists with two 16d nails. Two 16d nails are also driven through each block into the top of the wall.

Applying Backing

Walls running in the same direction as the ceiling joists require backing. Figure 1-68 (insert) shows how backing is nailed to the top plates to provide a nailing surface for the edges of the finish ceiling material. Lumber used for backing usually has 2-inch nominal thickness, although 1-inch boards are sometimes used.

Figure 1-68 shows backing placed on top of walls. The 2 by 4 pieces nailed to the exterior wall projects from one side of the wall. The interior wall requires a 2 by 6 or 2 by 8 piece extending from both sides of the wall. Backing is fastened to the top plates with 16d nails spaced 16 inches OC. Backing is also used where joists run at right angles to the partition (fig. 1-69).

Attic Scuttle

The scuttle is an opening framed in the ceiling to provide an entrance into the attic area. The size of the opening is decided by specification requirements and should be indicated in the blueprints. It must be large enough for a person to climb through easily.

The scuttle is framed in the same way as a floor opening. If the opening is no more than 3 feet square, it is not necessary to double the joists and headers. Scuttles must be placed away from the lower areas of a sloping roof. The opening may be covered by a piece of plywood resting on stops. The scuttle opening can be cut out after all the regular ceiling joists have been nailed in place.

RECOMMENDED READING LIST

NOTE

Although the following references were current when this TRAMAN was published, their continued currency cannot be assured. You therefore need to ensure that you are studying the latest revisions.

Carpentry, Leonard Koel, American Technical Publishers, Alsip, Ill., 1985.

Design of Wood Frame Structures for Permanence, National Forest Products Association, Washington, D.C., 1988.

Exterior and Interior Trim, John E. Ball, Delmar Publishers, Inc., Albany, N.Y., 1975.

CHAPTER 2

ROOF FRAMING

In this chapter, we will introduce you to the fundamentals of roof design and construction. But, before discussing roof framing, we will first review some basic terms and definitions used in roof construction; we will then discuss the framing square and learn how it's used to solve some basic construction problems. Next, we'll examine various types of roofs and rafters, and techniques for laying out, cutting, and erecting rafters. We conclude the chapter with a discussion of the types and parts of roof trusses.

TERMINOLOGY

LEARNING OBJECTIVE: Upon completing this section, you should be able to identify the types of roofs and define common roof framing terms.

The primary object of a roof in any climate is protection from the elements. Roof slope and rigidness are for shedding water and bearing any extra additional weight. Roofs must also be strong enough to withstand high winds. In this section, we'll cover the most common types of roofs and basic framing terms.

TYPES OF ROOFS

The most commonly used types of pitched roof construction are the gable, the hip, the intersecting, and the shed (or lean-to). An example of each is shown in figure 2-1.

Gable

A gable roof has a ridge at the center and slopes in two directions. It is the form most commonly used by the Navy. It is simple in design, economical to construct, and can be used on any type of structure.

Hip

The hip roof has four sloping sides. It is the strongest type of roof because it is braced by four hip rafters. These hip rafters run at a 45° angle from each corner of the building to the ridge. A disadvantage of the hip roof is that it is more difficult to construct than a gable roof.

Intersecting

The intersecting roof consists of a gable and valley, or hip and valley. The valley is formed where the two different sections of the roof meet, generally at a 90° angle. This type of roof is more complicated than the

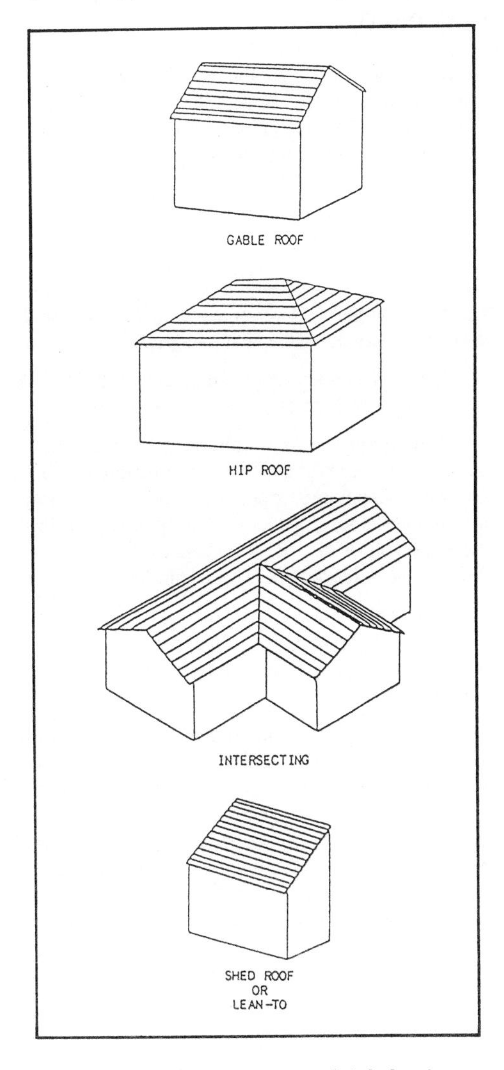

Figure 2-1.—Most common types of pitched roofs.

other types and requires more time and labor to construct.

Shed

The shed roof, or lean-to, is a roof having only one slope, or pitch. It is used where large buildings are framed under one roof, where hasty or temporary construction is needed, and where sheds or additions are erected. The roof is held up by walls or posts where one wall or the posts on one side are at a higher level than those on the opposite side.

FRAMING TERMS

Knowing the basic vocabulary is a necessary part of your work as a Builder. In the following section, we'll cover some of the more common roof and rafter terms you'll need. Roof framing terms are related to the parts of a triangle.

Roof

Features associated with basic roof framing terms are shown in figure 2-2. Refer to the figure as you study the terms discussed in the next paragraphs.

Span is the horizontal distance between the outside top plates, or the base of two abutting right triangles.

Unit of run is a fixed unit of measure, always 12 inches for the common rafter. Any measurement in a horizontal direction is expressed as run and is always measured on a level plane. Unit of span is also fixed, twice the unit of run, or 24 inches. Unit of rise is the distance the rafter rises per foot of run (unit of run).

Total run is equal to half the span, or the base of one of the right triangles. Total rise is the vertical distance from the top plate to the top of the ridge, or the altitude of the triangle.

Pitch is the ratio of unit of rise to the unit of span. It describes the slope of a roof. Pitch is expressed as a fraction, such as 1/4 or 1/2 pitch. The term "pitch" is gradually being replaced by the term "cut." Cut is the angle that the roof surface makes with a horizontal plane. This angle is usually expressed as a fraction in which the numerator equals the unit of rise and the denominator equals the unit of run (12 inches), such as 6/12 or 8/12. This can also be expressed in inches per foot; for example, a 6- or 8-inch cut per foot. Here, the unit of run (12 inches) is understood. Pitch can be converted to cut by using the following formula: *unit of span (24 in.)* $\times$ *pitch* = unit of rise. For example,

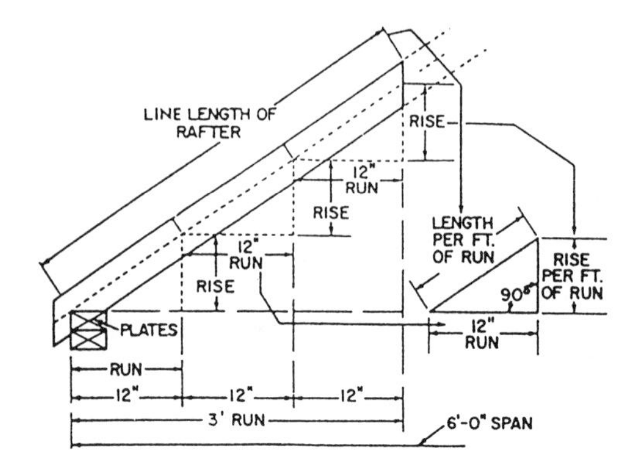

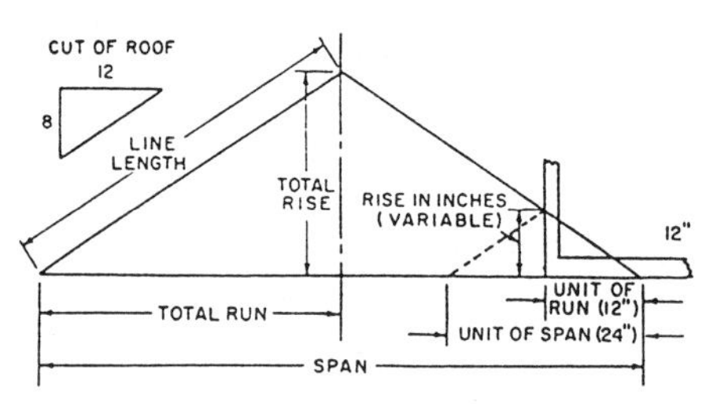

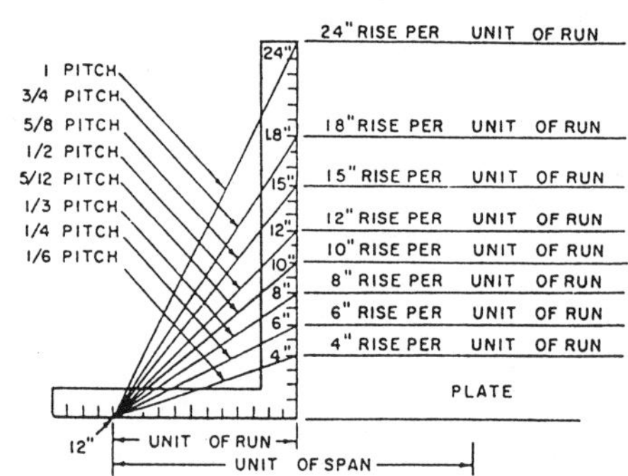

Figure 2-2.—Roof framing terms.

1/8 pitch is given, so 24 $\times$ 1/8 equals 3, or unit of rise in inches. If the unit of rise in inches is 3, then the cut is the unit of rise and the unit of run (12 inches), or 3/12.

Line length is the hypotenuse of the triangle whose base equals the total run and whose altitude equals the total rise. The distance is measured along the rafter from the outside edge of the top plate to the center line of the ridge. Bridge measure is the hypotenuse of the triangle with the unit of run for the base and unit of rise for the altitude.

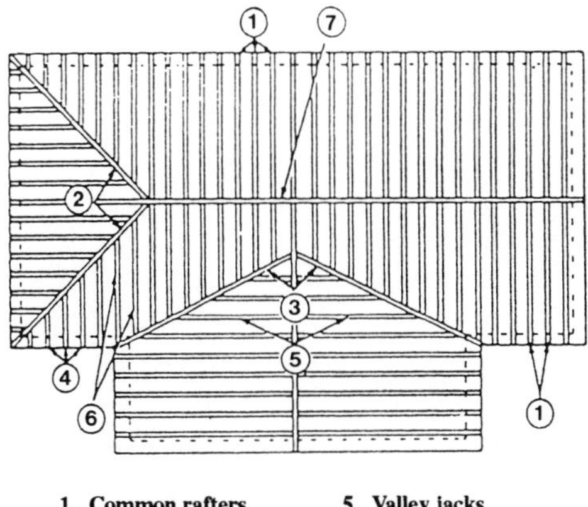

1. **Common rafters**
2. **Hip rafters**
3. **Valley rafters**
4. **Hip jack**
5. **Valley jacks**
6. **Cripple jacks**
7. **Ridgeboard**

Figure 2-3.—Rafter terms.

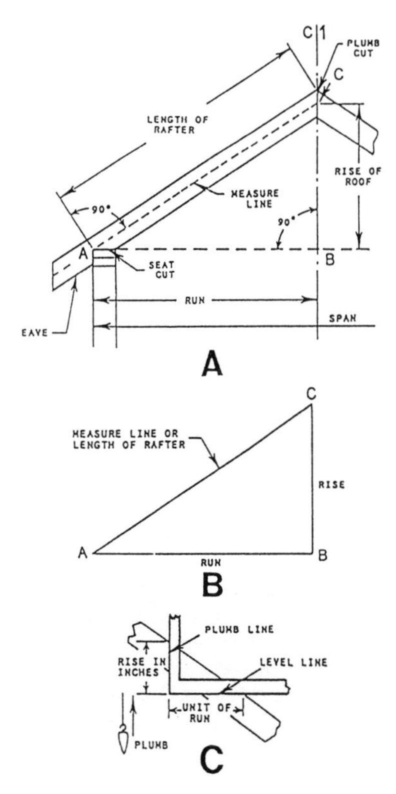

Figure 2-4.—Rafter layout.

Rafter

The members making up the main body of the framework of all roofs are called rafters. They do for the roof what the joists do for the floor and what the studs do for the wall. Rafters are inclined members spaced from 16 to 48 inches apart. They vary in size, depending on their length and spacing. The tops of the inclined rafters are fastened in one of several ways determined by the type of roof. The bottoms of the rafters rest on the plate member, providing a connecting link between the wall and the roof. The rafters are really functional parts of both the walls and the roof.

The structural relationship between the rafters and the wall is the same in all types of roofs. The rafters are not framed into the plate, but are simply nailed to it. Some are cut to fit the plate, whereas others, in hasty construction, are merely laid on top of the plate and nailed in place. Rafters usually extend a short distance beyond the wall to form the eaves (overhang) and protect the sides of the building. Features associated with various rafter types and terminology are shown in figure 2-3.

Common rafters extend from the plate to the ridgeboard at right angles to both. Hip rafters extend diagonally from the outside corner formed by perpendicular plates to the ridgeboard. Valley rafters extend from the plates to the ridgeboard along the lines where two roofs intersect. Jack rafters never extend the full distance from plate to ridgeboard. Jack rafters are subdivided into the hip, valley, and cripple jacks.

In a hip jack, the lower ends rest on the plate and the upper ends against the hip rafter. In a valley jack, the lower ends rest against the valley rafters and the upper ends against the ridgeboard. A cripple jack is nailed between hip and valley rafters.

Rafters are cut in three basic ways (shown in fig. 2-4, view A). The top cut, also called the plumb cut, is made at the end of the rafter to be placed against the ridgeboard or, if the ridgeboard is omitted, against the opposite rafters. A seat, bottom, or heel cut is made at the end of the rafter that is to rest on the plate. A side cut (not shown in fig. 2-4), also called a cheek cut, is a bevel cut on the side of a rafter to make it fit against another frame member.

Rafter length is the shortest distance between the outer edge of the top plate and the center of the ridge line. The eave, tail, or overhang is the portion of the rafter extending beyond the outer edge of the plate. A measure line (fig. 2-4, view B) is an imaginary reference line laid out down the middle of the face of a rafter. If a portion of a roof is represented by a right triangle, the measure line corresponds to the hypotenuse; the rise to the altitude; and, the run to the base.

A plumb line (fig. 2-4, view C) is any line that is vertical (plumb) when the rafter is in its proper position. A level line (fig. 2-4, view C) is any line that is horizontal (level) when the rafter is in its proper position.

FRAMING SQUARE

LEARNING OBJECTIVE: Upon completing this section, you should be able to describe and solve roof framing problems using the framing square.

The framing square is one of the most frequently used Builder tools. The problems it can solve are so many and varied that books have been written on the square alone. Only a few of the more common uses of the square can be presented here. For a more detailed discussion of the various uses of the framing square in solving construction problems, you are encouraged to obtain and study one of the many excellent books on the square.

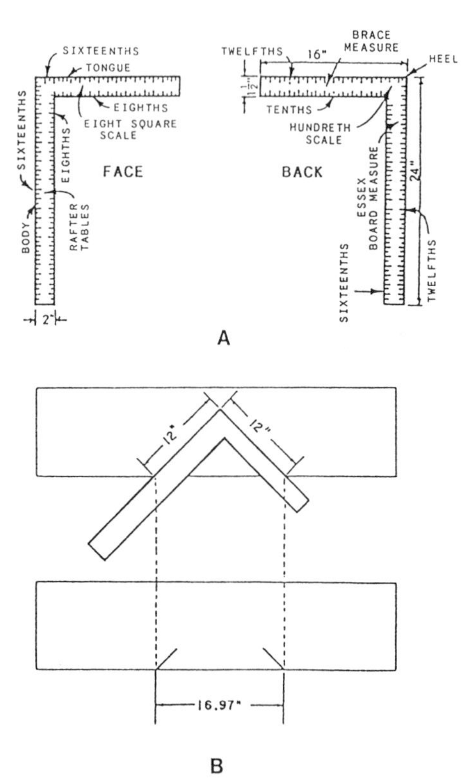

Figure 2-5.—Framing square: A. Nomenclature; B. Problem solving.

DESCRIPTION

The framing square (fig. 2-5, view A) consists of a wide, long member called the blade and a narrow, short member called the tongue. The blade and tongue form a right angle. The face of the square is the side one sees when the square is held with the blade in the left hand, the tongue in the right hand, and the heel pointed away from the body. The manufacturer's name is usually stamped on the face. The blade is 24 inches long and 2 inches wide. The tongue varies from 14 to 18 inches long and is 1 1/2 inches wide, measured from the outer corner, where the blade and the tongue meet. This corner is called the heel of the square.

The outer and inner edges of the tongue and the blade, on both face and back, are graduated in inches. Note how inches are subdivided in the scale on the back of the square. In the scales on the face, the inch is subdivided in the regular units of carpenter's measure (1/8 or 1/16 inch). On the back of the square, the outer edge of the blade and tongue is graduated in inches and twelfths of inches. The inner edge of the tongue is graduated in inches and tenths of inches. The inner edge of the blade is graduated in inches and thirty-seconds of inches on most squares. Common uses of the twelfths scale on the back of the framing square will be described later. The tenths scale is not normally used in roof framing.

SOLVING BASIC PROBLEMS WITH THE FRAMING SQUARE

The framing square is used most frequently to find the length of the hypotenuse (longest side) of a right triangle when the lengths of the other two sides are known. This is the basic problem involved in determining the length of a roof rafter, a brace, or any other member that forms the hypotenuse of an actual or imaginary right triangle.

Figure 2-5, view B, shows you how the framing square is used to determine the length of the hypotenuse of a right triangle with the other sides each 12 inches long. Place a true straightedge on a board and set the square on the board so as to bring the 12-inch mark on

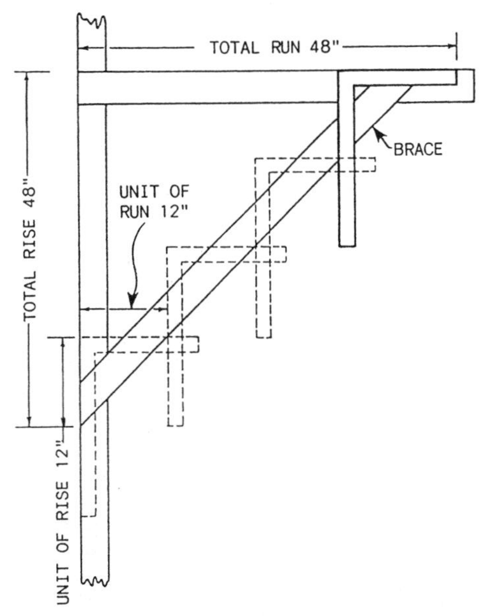

Figure 2-6.—"Stepping off" with a framing square.

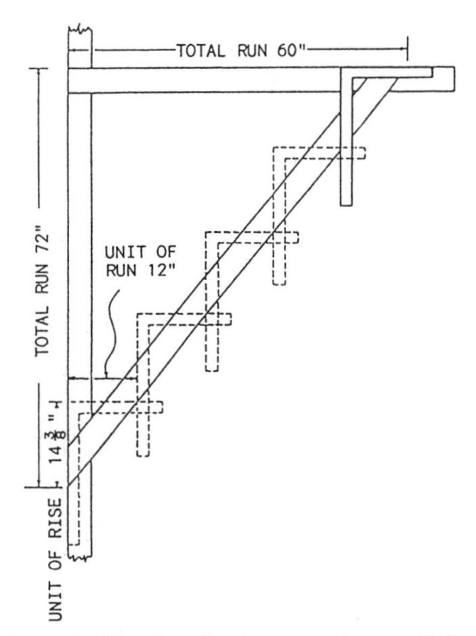

Figure 2-7.—"Stepping off" with a square when the unit of run and unit of rise are different.

the tongue and the blade even with the edge of the board. Draw the pencil marks as shown. The distance between these marks, measured along the edge of the board, is the length of the hypotenuse of a right triangle with the other sides each 12 inches long. You will find that the distance, called the bridge measure, measures just under 17 inches—16.97 inches, as shown in the figure. For most practical Builder purposes, though, round 16.97 inches to 17 inches.

Solving for Unit and Total Run and Rise

In figure 2-5, the problem could be solved by a single set (called a cut) of the framing square. This was due to the dimensions of the triangle in question lying within the dimensions of the square. Suppose, though, you are trying to find the length of the hypotenuse of a right triangle with the two known sides each being 48 inches long. Assume the member whose length you are trying to determine is the brace shown in figure 2-6. The total run of this brace is 48 inches, and the total rise is also 48 inches.

To figure the length of the brace, you first reduce the triangle in question to a similar triangle within the dimensions of the framing square. The length of the vertical side of this triangle is called unit of rise, and the length of the horizontal side is called the unit of run. By a general custom of the trade, unit of run is always taken as 12 inches and measured on the tongue of the framing square.

Now, if the total run is 48 inches, the total rise is 48 inches, and the unit of run is 12 inches, what is the unit of rise? Well, since the sides of similar triangles are proportional, the unit of rise must be the value of x in the proportional equation *48:48::12:x.* In this case, the unit of rise is obviously 12 inches.

To get the length of the brace, set the framing square to the unit of run (12 inches) on the tongue and to the unit of rise (also 12 inches) on the blade, as shown in figure 2-6. Then, "step off" this cut as many times as the unit of run goes into the total run. In this case, 48/12, or 4 times, as shown in the figure.

In this problem, the total run and total rise were the same, from which it followed that the unit of run and unit of rise were also the same. Suppose now that you want to know the length of a brace with a total run of 60 inches and a total rise of 72 inches, as in figure 2-7. Since the unit of run is 12 inches, the unit of rise must be the value of x in the proportional equation *60:72::12:x.* That is, the proportion 60:72 is the same as the proportion 12:x. Working this out, you find the unit of rise is

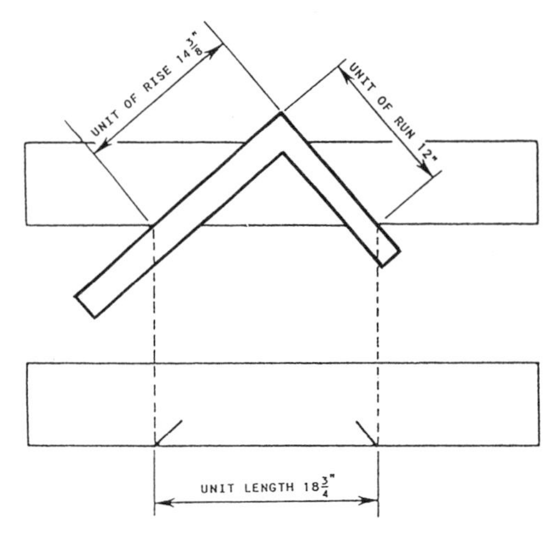

Figure 2-8.—Unit length.

14.4 inches. For practical purposes, you can round this to 14 3/8.

To lay out the full length of the brace, set the square to the unit of rise (14 3/8 inches) and the unit of run (12 inches), as shown in figure 2-7. Then, step off this cut as many times as the unit of run goes into the total run (60/12, or 5 times).

Determining Line Length

If you do not go through the stepping-off procedure, you can figure the total length of the member in question by first determining the bridge measure. The bridge measure is the length of the hypotenuse of a right triangle with the other sides equal to the unit of run and unit of rise. Take the situation shown above in figure 2-7. The unit of run here is 12 inches and the unit of rise is 14 3/8 inches. Set the square to this cut, as shown in figure 2-8, and mark the edges of the board as shown. If you measure the distance between the marks, you will find it is 18 3/4 inches. Bridge measure can also be found by using the Pythagorean theorem: ($a^2 + b^2 = c^2$). Here, the unit of rise is the altitude (a), the unit or run is the base (b), and the hypotenuse (c) is the bridge measure.

To get the total length of the member, you simply multiply the bridge measure in inches by the total run in feet. Since that is 5, the total length of the member is 18 3/4 × 5, or 93 3/4 inches. Actually, the length of the hypotenuse of a right triangle with the other sides 60 and 72 inches long is slightly more than 93.72 inches, but 93 3/4 inches is close enough for practical purposes.

Once you know the total length of the member, just measure it off and make the end cuts. To make these cuts at the proper angles, set the square to the unit of run on the tongue and the unit of rise on the blade and draw a line for the cut along the blade (lower end cut) or the tongue (upper end cut).

SCALES

A framing square contains four scales: tenths, twelfths, hundredths, and octagon. All are found on the face or along the edges of the square. As we mentioned earlier, the tenths scale is not used in roof framing.

Twelfths Scale

The graduations in inches, located on the back of the square along the outer edges of the blade and tongue, are called the twelfths scale. The chief purpose of the twelfths scale is to provide various shortcuts in problem solving graduated in inches and twelfths of inches. Dimensions in feet and inches can be reduced to 1/12th by simply allowing each graduation on the twelfths scale to represent 1 inch; for example, 2 6/12 inches on the twelfths scale may be taken to represent 2 feet 6 inches. A few examples will show you how the twelfths scale is used.

Suppose you want to know the total length of a rafter with a total run of 10 feet and a total rise of 6 feet 5 inches. Set the square on a board with the twelfths scale on the blade at 10 inches and the twelfths scale on the tongue at 6 5/12 inches and make the usual marks. If you measure the distance between the marks, you will find it is 11 11/12 inches. The total length of the rafter is 11 feet 11 inches.

Suppose now that you know the unit of run, unit of rise, and total run of a rafter, and you want to find the total rise and the total length. Use the unit of run (12 inches) and unit of rise (8 inches), and total run of 8 feet 9 inches. Set the square to the unit of rise on the tongue and unit of run on the blade (fig. 2-9, top view). Then, slide the square to the right until the 8 9/12-inch mark on the blade (representing the total run of 8 feet 9 inches) comes even with the edge of the board, as shown in the second view. The figure of 5 10/12 inches, now indicated on the tongue, is one-twelfth of the total rise. The total rise is, therefore, 5 feet 10 inches. The distance between pencil marks (10 7/12 inches) drawn along the tongue and the blade is one-twelfth of the total length. The total length is, therefore, 10 feet 7 inches.

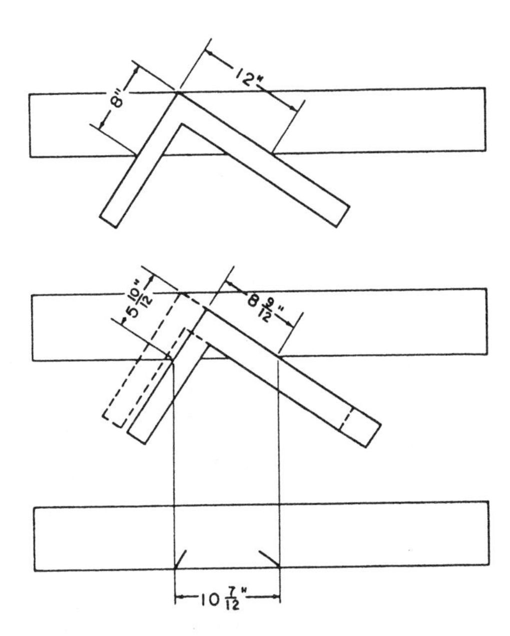

Figure 2-9.—Finding total rise and length when unit of run, unit of rise, and total run are known.

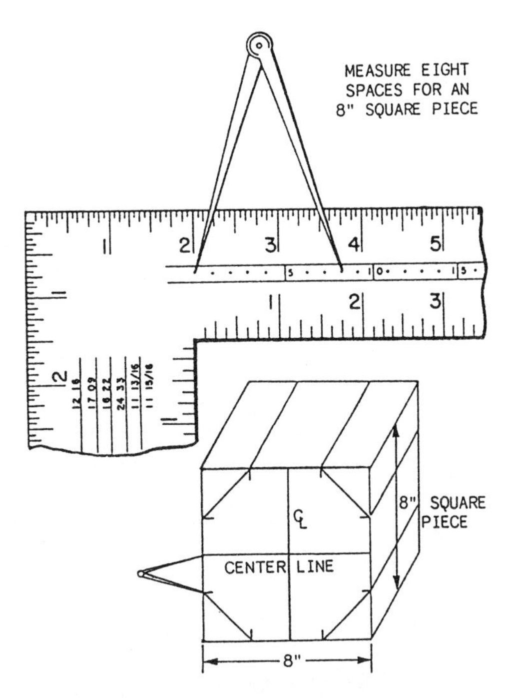

Figure 2-10.—Using the octagon square.

The twelfths scale may also be used to determine dimensions by inspection for proportional reductions or enlargements. Suppose you have a panel 10 feet 9 inches long by 7 feet wide. You want to cut a panel 7 feet long with the same proportions. Set the square, as shown in figure 2-9, but with the blade at 10 9/12 inches and the tongue at 7 inches. Then slide the blade to 7 inches and read the figure indicted on the tongue, which will be 4 7/12 inches if done correctly. The smaller panel should then be 4 feet 7 inches wide.

Hundredths Scale

The hundredths scale is on the back of the tongue, in the corner of the square, near the brace table. This scale is called the hundredths scale because 1 inch is divided into 100 parts. The longer lines indicate 25 hundredths, whereas the next shorter lines indicate 5 hundredths, and so forth. By using dividers, you can easily obtain a fraction of an inch.

The inch is graduated in sixteenths and located below the hundredths scale. Therefore, the conversion from hundredths to sixteenths can be made at a glance without the use of dividers. This can be a great help when determining rafter lengths, using the figures of the rafter tables where hundredths are given.

Octagon Scale

The octagon scale (sometimes called the eight-square scale) is located in the middle of the face of the tongue. The octagon scale is used to lay out an octagon (eight-sided figure) in a square of given even-inch dimensions.

Let's say you want to cut an 8-inch octagonal piece for a stair newel. First, square the stock to 8 by 8 inches and smooth the end section. Then, draw crossed center lines on the end section, as shown in figure 2-10. Next, set a pair of dividers to the distance from the first to the eighth dot on the octagon scale, and lay off this distance on either side of the center line on the four slanting sides of the octagon. This distance equals one-half the length of a side of the octagon.

When you use the octagon scale, set one leg of the dividers on the first dot and the other leg on the dot whose number corresponds to the width in inches of the square from which you are cutting the piece.

FRAMING SQUARE TABLES

There are three tables on the framing square: the unit length rafter table, located on the face of the blade; the

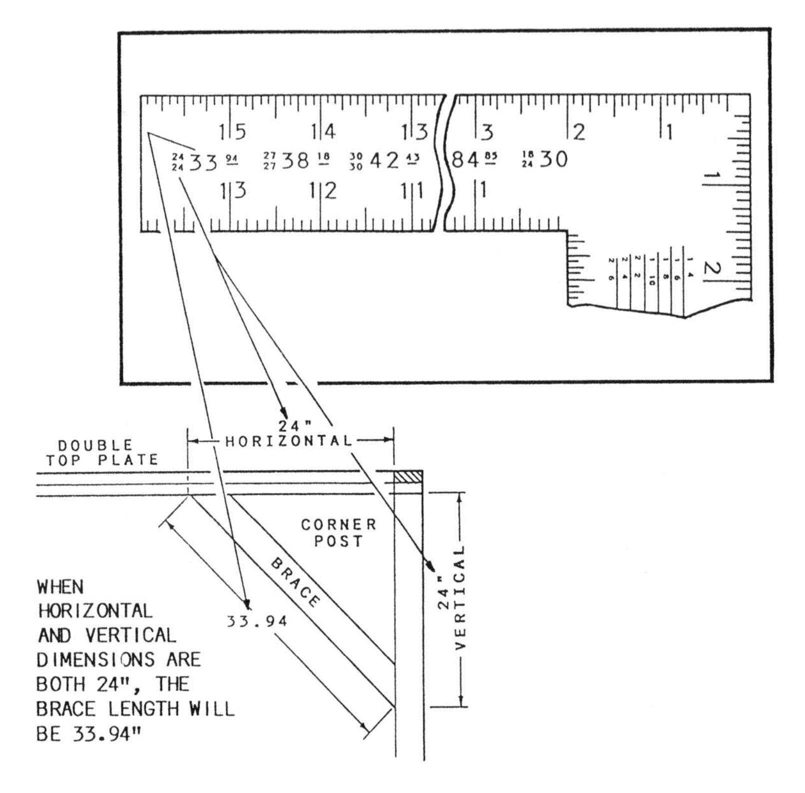

Figure 2-11.—Brace table.

brace table, located on the back of the tongue; and the Essex board measure table, located on the back of the blade. Before you can use the unit length rafter table, you must be familiar with the different types of rafters and with the methods of framing them. The use of the unit length rafter table is described later in this chapter. The other two tables are discussed below.

Brace

The brace table sets forth a series of equal runs and rises for every three-units interval from 24/24 to 60/60, together with the brace length, or length of the hypotenuse, for each given run and rise. The table can be used to determine, by inspection, the length of the hypotenuse of a right triangle with the equal shorter sides of any length given in the table. For example, in the segment of the brace table shown in figure 2-11, you can see that the length of the hypotenuse of a right triangle with two sides 24 units long is 33.94 units; with two sides 27 units long, 38.18 units; two sides 30 units long, 42.43 units; and so on.

By applying simple arithmetic, you can use the brace table to determine the hypotenuse of a right triangle with equal sides of practically any even-unit length. Suppose you want to know the length of the hypotenuse of a right triangle with two sides 8 inches long. The brace table shows that a right triangle with two sides 24 inches long has a hypotenuse of 33.94 inches. Since 8 amounts to 24/3, a right triangle with two shorter sides each 8 inches long must have a hypotenuse of 33.94 ÷ 3, or approximately 11.31 inches.

Suppose you want to find the length of the hypotenuse of a right triangle with two sides 40 inches each. The sides of similar triangles are proportional, and any right triangle with two equal sides is similar to any other right triangle with two equal sides. The brace table shows that a right triangle with the two shorter sides

7	8	9	10	11	12	13
4 8	5 4	6	6 8	7 4	8	8 8
5 3	6	6 9	7 6	8 3	9	9 9
5 10	6 8	7 6	8 4	9 2	10	10 10
6 5	7 4	8 3	9 2	10 1	11	11 11
7 7	8 8	9 9	10 10	11 11	13	14 1
8 2	9 4	10 6	11 8	12 10	14	15 2
8 9	10	11 3	12 6	13 9	15	16 3

5 6 7 8 9 10 11

A

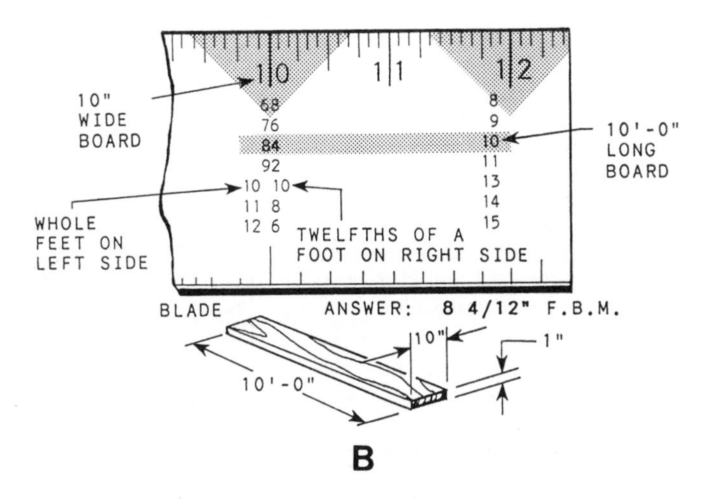

Figure 2-12.—Segment of Essex board measure table.

being 30 inches long has a hypotenuse of 42.43 inches. The length of the hypotenuse of a right triangle with the two shorter sides being 40 inches long must be the value of x in the proportional equation *30:42.43::40:x*, or about 56.57 inches.

Notice that the last item in the brace table (the one farthest to the right in fig. 2-11) gives you the hypotenuse of a right triangle with the other proportions 18:24:30. These proportions are those of the most common type of unequal-sided right triangle, with sides in the proportions of 3:4:5.

Essex Board

The primary use of the Essex board measure table is for estimating the board feet in lumber of known dimensions. The inch graduations (fig. 2-12, view A) above the table (1, 2, 3, 4, and so on) represent the width in inches of the piece to be measured. The figures under the 12-inch graduation (8, 9, 10, 11, 13, 14, and 15, arranged in columns) represent lengths in feet. The figure 12 itself represents a 12-foot length. The column headed by the figure 12 is the starting point for all calculations.

To use the table, scan down the figure 12 column to the figure that represents the length of the piece of lumber in feet. Then go horizontally to the figure directly below the inch mark that corresponds to the width of the stock in inches. The figure you find will be the number of board feet and twelfths of board feet in a 1-inch-thick board.

Let's take an example. Suppose you want to figure the board measure of a piece of lumber 10 feet long by 10 inches wide by 1 inch thick. Scan down the column (fig. 2-12, view B) headed by the 12-inch graduation to 10, and then go horizontally to the left to the figure directly below the 10-inch graduation. You will find the figure to be 84, or 8 4/12 board feet. For easier calculating purposes, you can convert 8 4/12 to a decimal (8.33).

To calculate the cost of this piece of lumber, multiply the cost per board foot by the total number of board feet. For example, a 1 by 10 costs $1.15 per board foot. Multiply the cost per board foot ($1.15) by the number of board feet (8.33). This calculation is as follows:

$$\begin{array}{r} \$1.15 \\ \times\ 8.33 \\ \hline \end{array}$$

$9.5795 (rounded off to $9.58).

What do you do if the piece is more than 1 inch thick? All you have to do is multiply the result obtained for a 1-inch-thick piece by the actual thickness of the piece in inches. For example, if the board described in the preceding paragraph were 5 inches thick instead of 1 inch thick, you would follow the procedure described and then multiply the result by 5.

The board measure scale can be read only for lumber from 8 to 15 feet in length. If your piece is longer than 15 feet, you can proceed in one of two ways. If the length of the piece is evenly divisible by one of the lengths in the table, you can read for that length and multiply the result by the number required to equal the piece you are figuring. Suppose you want to find the number of board feet in a piece 33 feet long by 7 inches wide by 1 inch thick. Since 33 is evenly divisible by 11, scan down the 12-inch column to 11 and then go left to the 7-inch column. The figure given there (which is 6 5/12, or 6.42 bd. ft.) is one-third of the total board feet. The total number of board feet is 6 5/12 (or 6.42) × 3, or 19 3/12 (or 19.26) board feet.

If the length of the piece is not evenly divisible by one of the tabulated lengths, you can divide it into two tabulated lengths, read the table for these two, and add the results together. For example, suppose you want to find the board measure of a piece 25 feet long by 10 inches wide by 1 inch thick. This length can be divided into 10 feet and 15 feet. The table shows that the 10-foot length contains 8 4/12 (8.33) board feet and the 15-foot length contains 12 6/12 (12.5) board feet. The total length then contains 8 4/12 (8.33) plus 12 6/12 (12.5), or 20 10/12 (20.83) board feet.

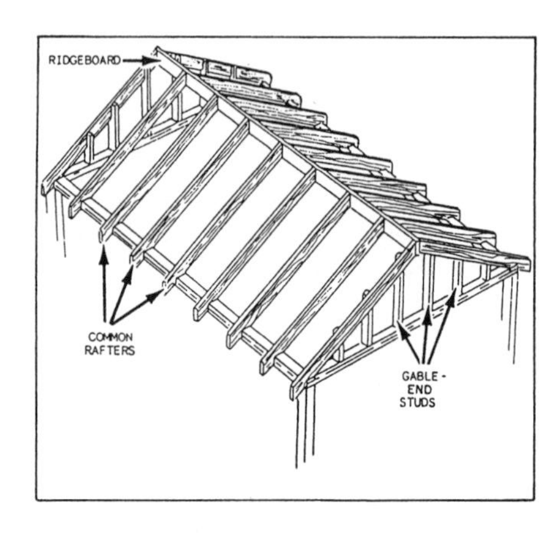

Figure 2-13.—Framework of a gable roof.

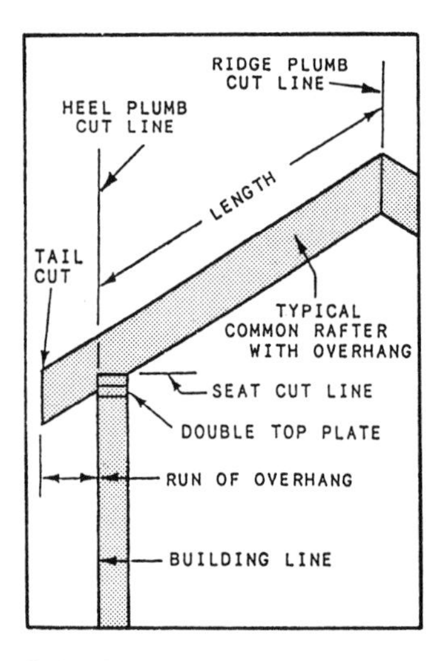

Figure 2-14.—Typical common rafter with an overhang.

DESIGNS

LEARNING OBJECTIVE: Upon completing this section, you should be able to describe procedures for the layout and installation of members of gable, hip, intersecting, and shed roof designs.

As we noted earlier, the four most common roof designs you will encounter as a Builder are gable, hip, intersecting, and shed. In this section, we will examine

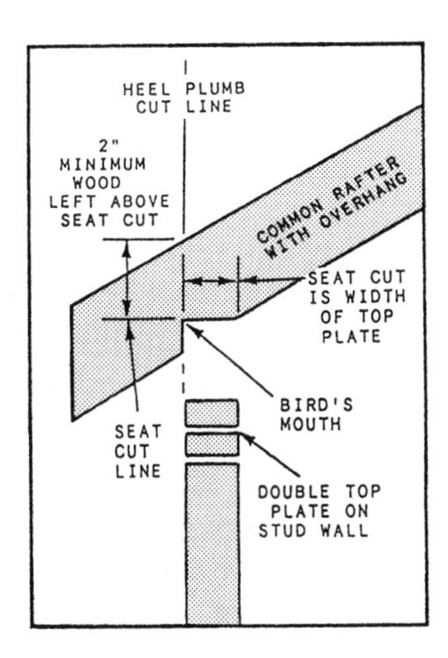

Figure 2-15.—A "bird's-mouth" is formed by the heel plumb line and seat line.

various calculations, layouts, cutting procedures, and assembly requirements required for efficient construction.

GABLE

Next to the shed roof, which has only one slope, the gable roof is the simplest type of sloping roof to build because it slopes in only two directions. The basic structural members of the gable roof are the ridgeboard, the common rafters, and the gable-end studs. The framework is shown in figure 2-13.

The ridgeboard is placed at the peak of the roof. It provides a nailing surface for the top ends of the common rafters. The common rafters extend from the top wall plates to the ridge. The gable-end studs are upright framing members that provide a nailing surface for siding and sheathing at the gable ends of the roof.

Common Rafters

All common rafters for a gable roof are the same length. They can be precut before the roof is assembled. Today, most common rafters include an overhang. The overhang (an example is shown in fig. 2-14) is the part of the rafter that extends past the building line. The run of the overhang, called the projection, is the horizontal distance from the building line to the tail cut on the rafter. In figure 2-14, note the plumb cuts at the ridge, heel, and tail of the rafter. A level seat cut is placed where the rafter rests on the top plate. The notch formed by the seat and heel cut line (fig. 2-15) is often called the bird's-mouth.

The width of the seat cut is determined by the slope of the roof: the lower the slope, the wider the cut. At least 2 inches of stock should remain above the seat cut. The procedure for marking these cuts is explained later in this chapter. Layout is usually done after the length of the rafter is calculated.

CALCULATING LENGTHS OF COMMON RAFTERS.—The length of a common rafter is based on the unit of rise and total run of the roof. The unit of rise and total run are obtained from the blueprints. Three different procedures can be used to calculate common rafter length: use a framing square printed with a rafter table; use a book of rafter tables; or, use the step-off method where rafter layout is combined with calculating length.

Framing squares are available with a rafter table printed on the face side (fig. 2-16). The rafter table makes it possible to find the lengths of all types of rafters for pitched roofs, with unit of rises ranging from 2 inches to 18 inches. Let's look at two examples:

Example 1. The roof has a 7-inch unit of rise and a 16-foot span.

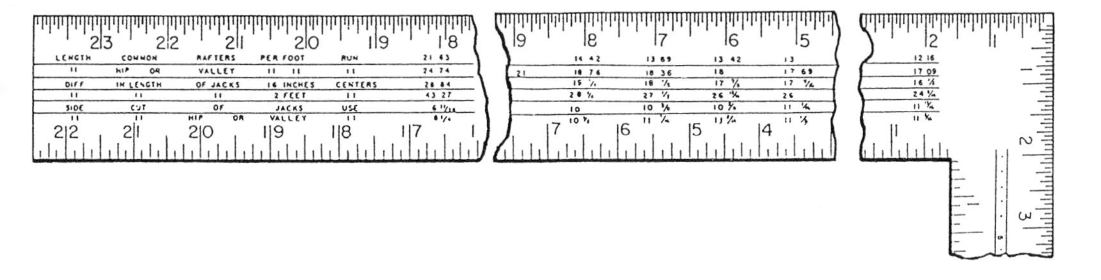

Figure 2-16.—Rafter table on face of a steel square.

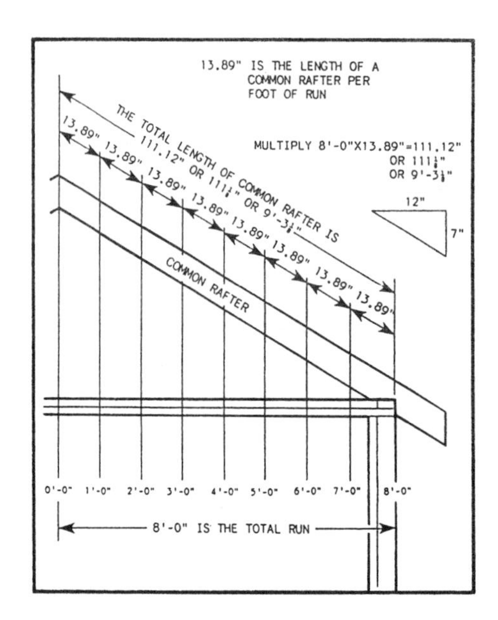

Figure 2-17.—Rafter length.

Look at the first line of the rafter table on a framing square to find LENGTH COMMON RAFTERS PER FOOT RUN (also known as the bridge measure). Since the roof in this example has a 7-inch unit of rise, locate the number 7 at the top of the square. Directly beneath the number 7 is the number 13.89. This means that a common rafter with a 7-inch unit of rise will be 13.89 inches long for every unit of run. To find the length of the rafter, multiply 13.89 inches by the number of feet in the total run. (The total run is always one-half the span.) The total run for a roof with a 16-foot span is 8 feet; therefore, multiply 13.89 inches by 8 to find the rafter length. Figure 2-17 is a schematic of this procedure.

If a framing square is not available, the bridge measure can be found by using the Pythagorean theorum using the same cut of 7/12: $7^2 + 12^2 = 193^2$; the square root of 193 is 13.89.

Two steps remain to complete the procedure.

Step 1. Multiply the number of feet in the total run (8) by the length of the common rafter per foot of run (13.89 inches):

$$\begin{array}{r} 13.89 \text{ inches} \\ \times \quad 8 \\ \hline 111.12 \text{ inches} \end{array}$$

Step 2. To change .12 of an inch to a fraction of an inch, multiply by 16:

$$\begin{array}{r} .12 \\ \times \ 16 \\ \hline 1.92 \end{array}$$

The number 1 to the left of the decimal point represents 1/16 inch. The number .92 to the right of the decimal represents ninety-two hundredths of 1/16 inch. For practical purposes, 1.92 is calculated as being equal to 2 × 1/16 inch, or 1/8 inch. As a general rule in this kind of calculation, if the number to the right of the decimal is 5 or more, add 1/16 inch to the figure on the left side of the decimal. The result of steps 1 and 2 is a total common rafter length of 111 1/8 inches, or 9 feet 3 1/8 inches.

Example 2. A roof has a 6-inch unit of rise and a 25-foot span. The total run of the roof is 12 feet 6 inches. You can find the rafter length in four steps.

Step 1. Change 6 inches to a fraction of a foot by placing the number 6 over the number 12:

$$\frac{6}{12} = \frac{1}{2} \quad \text{(1/2 foot = 6 inches).}$$

Step 2. Change the fraction to a decimal by dividing the bottom number (denominator) into the top number (numerator):

$$\frac{1}{2} = .5 \quad \text{(.5 foot = 6 inches).}$$

Step 3. Multiply the total run (12.5) by the length of the common rafter per foot of run (13.42 inches) (fig. 2-16):

$$\begin{array}{r} 12.5 \\ \times 13.42 \\ \hline 167.75 \text{ inches.} \end{array}$$

Step 4. To change .75 inch to a fraction of an inch, multiply by 16 (for an answer expressed in sixteenths of an inch).

$$.75 \times 16 = 12$$

$$12 = \frac{12}{16} \text{ inch, or } \frac{3}{4} \text{ inch.}$$

The result of these steps is a total common rafter length of 167 3/4 inches, or 13 feet 11 3/4 inches.

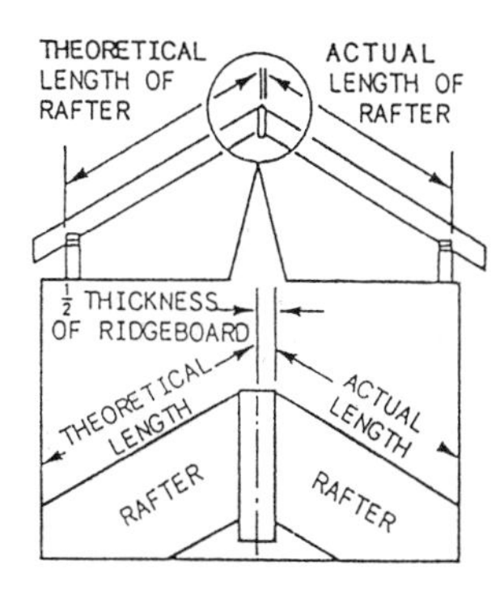

Figure 2-18.—The actual (versus theoretical) length of a common rafter.

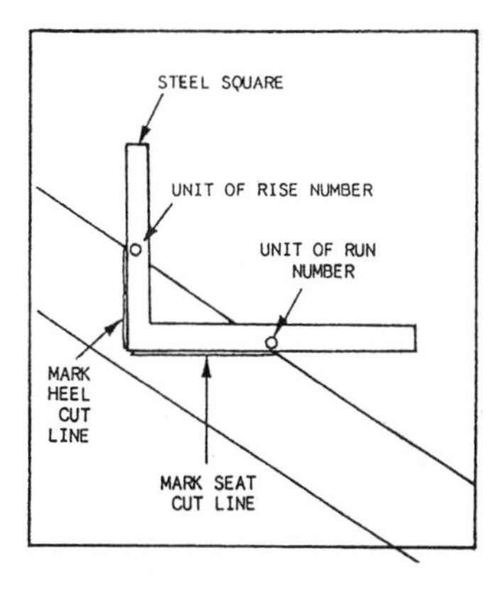

Figure 2-19.—Steel square used to lay out plumb and seat cuts.

SHORTENING.—Rafter length found by any of the methods discussed here is the measurement from the heel plumb line to the center of the ridge. This is known as the theoretical length of the rafter. Since a ridgeboard, usually 1 1/2 inches thick, is placed between the rafters, one-half of the ridgeboard (3/4 inch) must be deducted from each rafter. This calculation is known as shortening the rafter. It is done at the time the rafters are laid out. The actual length (as opposed to the theoretical length) of a rafter is the distance from the heel plumb line to the shortened ridge plumb line (fig. 2-18).

LAYING OUT.—Before the rafters can be cut, the angles of the cuts must be marked. Layout consists of marking the plumb cuts at the ridge, heel, and tail of the rafter, and the seat cut where the rafter will rest on the wall. The angles are laid out with a framing square, as shown in figure 2-19. A pair of square gauges is useful in the procedure. One square gauge is secured to the tongue of the square next to the number that is the same as the unit of rise. The other gauge is secured to the blade of the square next to the number that is the same as the unit of run (always 12 inches). When the square is placed on the rafter stock, the plumb cut can be marked along the tongue (unit of rise) side of the square. The seat cut can be marked along the blade (unit of run) side of the square.

Rafter layout also includes marking off the required overhang, or tail line length, and making the shortening calculation explained earlier. Overhang, or tail line length, is rarely given and must be calculated before laying out rafters. Projection, the horizontal distance from the building line to the rafter tail, must be located from drawings or specifications. To determine tail line length, use the following formula: bridge measure (in inches) times projection (in feet) equals tail line length (in inches). Determine the bridge measure by using the rafter table on the framing square or calculate it by using the Pythagorean theorem. Using figure 2-20 as a guide, you can see there are four basic steps remaining.

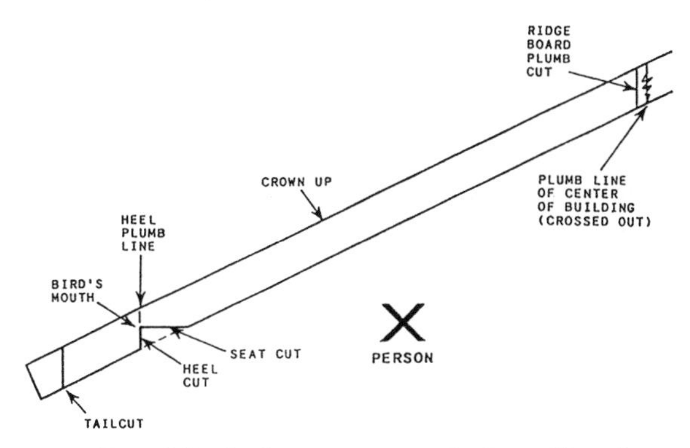

Figure 2-20.—Laying out a common rafter for a gable roof.

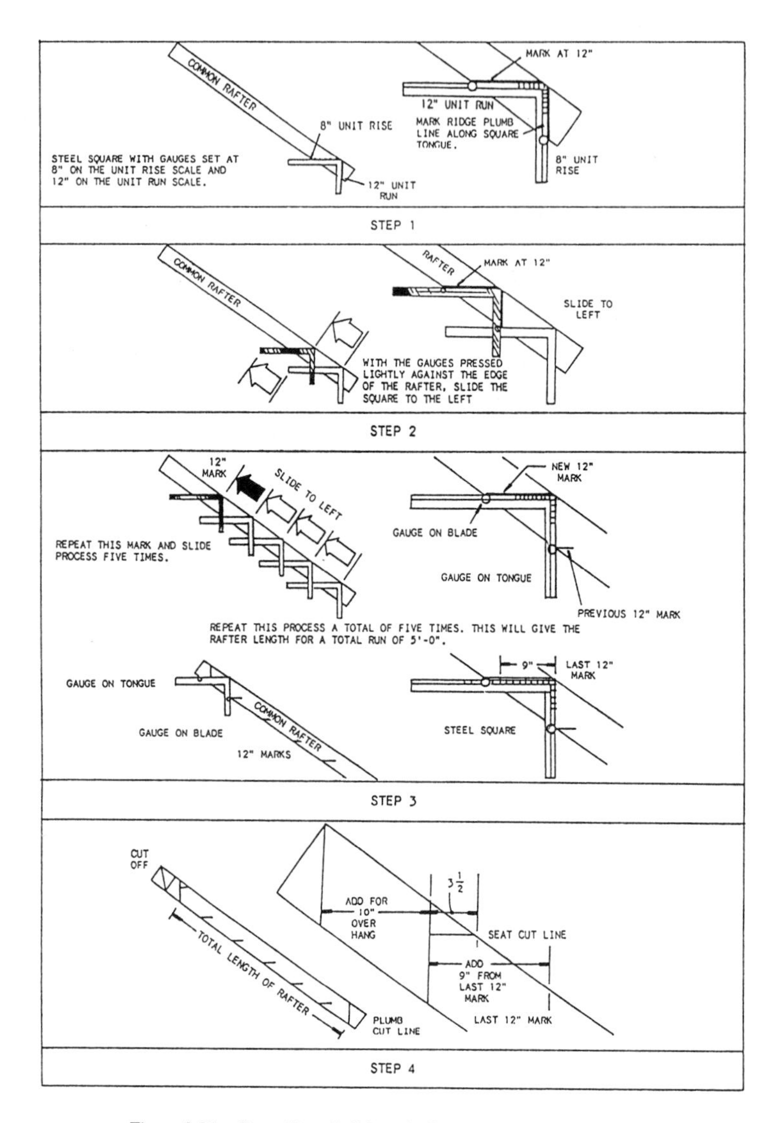

Figure 2-21.—Step-off method for calculating common rafter length.

Step 1. Lay out the rafter line length. Hold the framing square with the tongue in your right hand, the blade in the left, and the heel away from your body. Place the square as near the right end of the rafter as possible with the unit of rise on the tongue and the unit of run on the blade along the edge of the rafter stock. Strike a plumb mark along the tongue on the wide part of the material. This mark represents the center line of the roof. From either end of this mark, measure the line length of the rafter and mark the edge of the rafter stock. Hold the framing square in the same manner with the 6 on the tongue on the mark just made and the 12 on the blade along the edge. Strike a line along the tongue. This mark represents the plumb cut of the heel.

Step 2. Lay out the bird's-mouth. Measure 1 1/2 inches along the heel plumb line up from the bottom of the rafter. Set the blade of the square along the plumb line with the heel at the mark just made and strike a line along the tongue. This line represents the seat of the bird's-mouth.

Step 3. Lay out the tail line length. Measure the tail line length from the bird's-mouth heel plumb line. Strike a plumb line at this point in the same manner as the heel plumb line of the common rafter.

Step 4. Lay out the plumb cut at the ridgeboard. Measure and mark the point along the line length half the thickness of the ridgeboard. (This is the ridgeboard shortening allowance.) Strike a plumb line at this point. This line represents the plumb cut of the ridgeboard.

Step-Off Calculations and Layout

The step-off method for rafter layout is old but still practiced. It combines procedures for laying out the rafters with a procedure of stepping off the length of the rafter (see fig. 2-21). In this example, the roof has an 8-inch unit of rise, a total run of 5 feet 9 inches, and a 10-inch projection.

First, set gauges at 8 inches on the tongue and 12 inches on the blade. With the tongue in the right hand, the blade in the left hand, and the heel away from the body, place the square on the right end of the rafter stock. Mark the ridge plumb line along the tongue. Put a pencil line at the 12-inch point of the blade.

Second, with the gauges pressed lightly against the rafter, slide the square to the left. Line the tongue up with the last 12-inch mark and make a second 12-inch mark along the bottom of the blade.

Third, to add the 9-inch remainder of the total run, place the tongue on the last 12-inch mark. Draw another mark at 9 inches on the blade. This will be the total length of the rafter.

Last, lay out and cut the plumb cut line and the seat cut line.

Roof Assembly

The major part of gable-roof construction is setting the common rafters in place. The most efficient method is to precut all common rafters, then fasten them to the ridgeboard and the wall plates in one continuous operation.

The rafter locations should be marked on the top wall plates when the positions of the ceiling joists are laid out. Proper roof layout ensures the rafters and joists tie into each other wherever possible.

The ridgeboard, like the common rafters, should be precut. The rafter locations are then copied on the ridgeboard from the markings on the wall plates (fig. 2-22). The ridgeboard should be the length of the building plus the overhang at the gable ends.

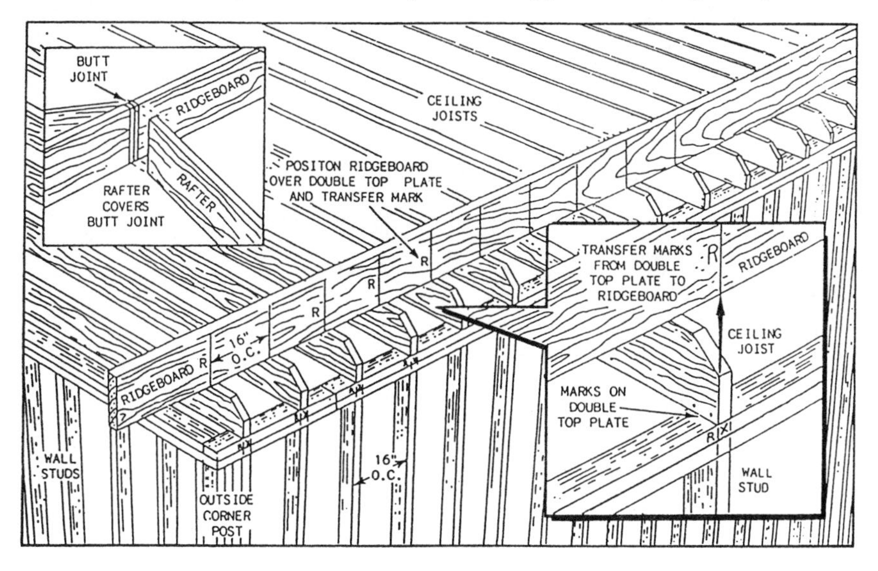

Figure 2-22.—Ridgeboard layout.

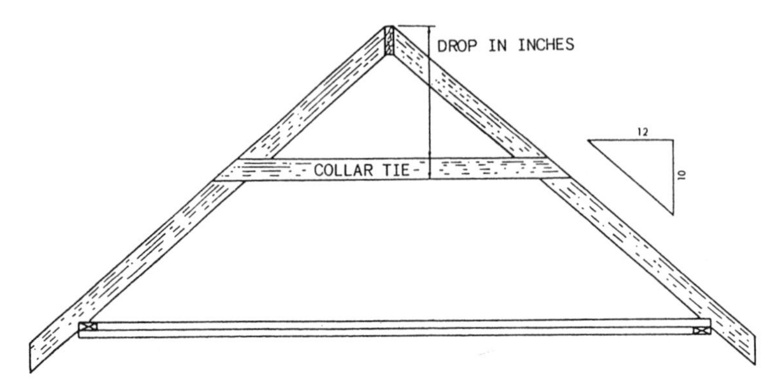

Figure 2-23.—Calculation for a collar tie.

The material used for the ridgeboard is usually wider than the rafter stock. For example, a ridgeboard of 2- by 8-inch stock would be used with rafters of 2- by 6-inch stock. Some buildings are long enough to require more than one piece of ridge material. The breaks between these ridge pieces should occur at the center of a rafter.

One pair of rafters should be cut and checked for accuracy before the other rafters are cut. To check the first pair for accuracy, set them in position with a 1 1/2-inch piece of wood fitted between them. If the rafters are the correct length, they should fit the building. If, however, the building walls are out of line, adjustments will have to be made on the rafters.

After the first pair of rafters is checked for accuracy (and adjusted if necessary), one of the pair can be used as a pattern for marking all the other rafters. Cutting is usually done with a circular or radial-arm saw.

COLLAR TIE.—Gable or double-pitch roof rafters are often reinforced by horizontal members

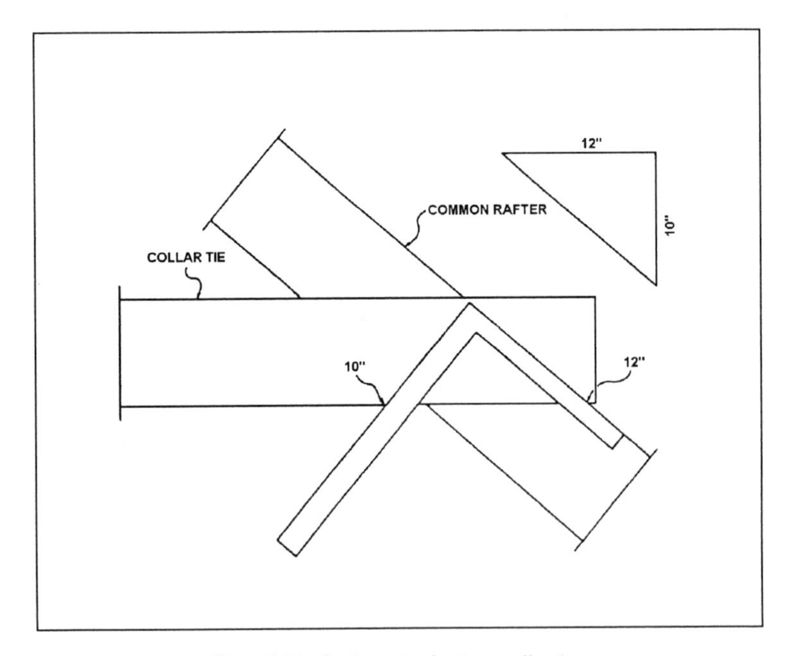

Figure 2-24.—Laying out end cut on a collar tie.

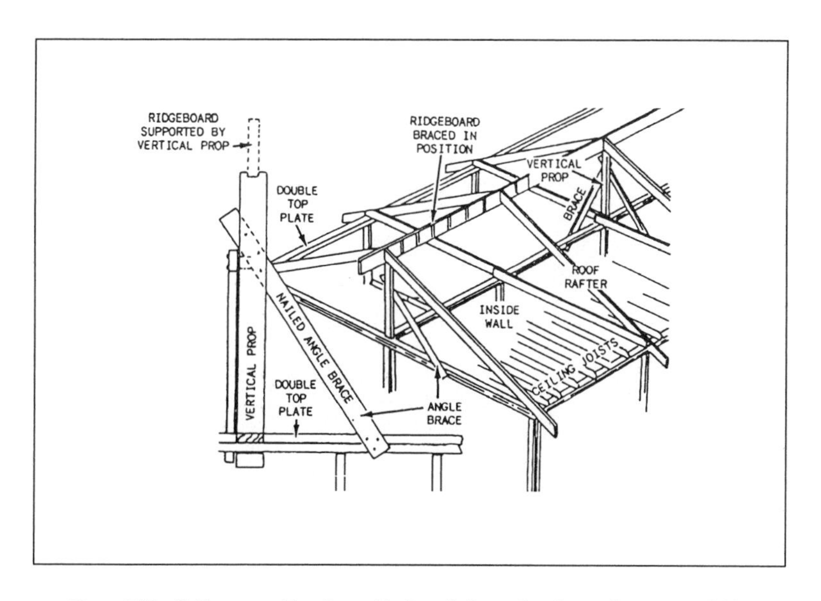

Figure 2-25.—Setting up and bracing a ridgeboard when only a few workers are available.

called collar ties (fig. 2-23). In a finished attic, the ties may also function as ceiling joists.

To find the line length of a collar tie, divide the amount of drop of the tie in inches by the unit of rise of the common rafter. This will equal one-half the length of the tie in feet. Double the result for the actual length. The formula is as follows: Drop in inches times 2, divided by unit or rise, equals the length in feet.

The length of the collar tie depends on whether the drop is measured to the top or bottom edge of the collar tie (fig. 2-23). The tie must fit the slope of the roof. To obtain this angle, use the framing square. Hold the unit of run and the unit of rise of the common rafter. Mark and cut on the unit of run side (fig. 2-24).

METHODS OF RIDGE BOARD ASSEMBLY.—Several different methods exist for setting up the ridgeboard and attaching the rafters to it. When only a few Builders are present, the most convenient procedure is to set the ridgeboard to its required height (total rise) and hold it in place with temporary vertical props (fig. 2-25). The rafters can then be nailed to the ridgeboard and the top wall plates.

Plywood panels should be laid on top of the ceiling joists where the framing will take place. The panels provide safe and comfortable footing. They also provide a place to put tools and materials.

Common rafter overhang can be laid out and cut before the rafters are set in place. However, many Builders prefer to cut the overhang after the rafters are fastened to the ridgeboard and wall plates. A line is snapped from one end of the building to the other, and the tail plumb line is marked with a sliding T-bevel, also called a bevel square. These procedures are shown in figure 2-26. The rafters are then cut with a circular saw.

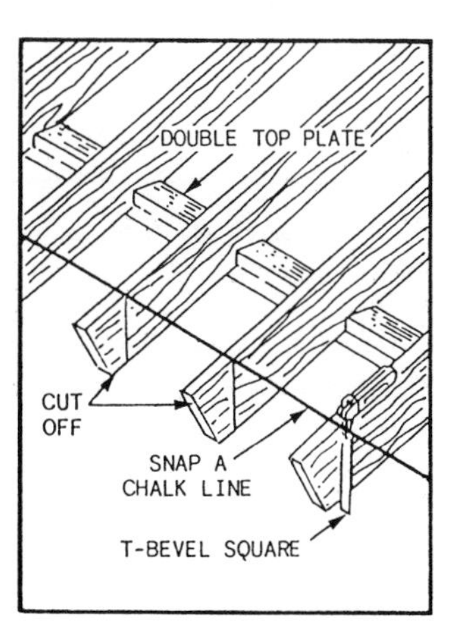

Figure 2-26.—Snapping a line and marking plumb cuts for a gable-end overhang.

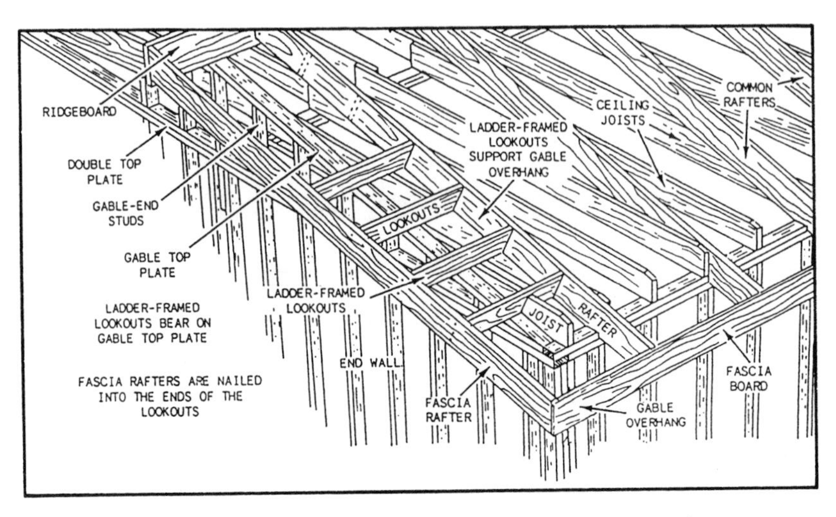

Figure 2-27.—Gable-end overhang with the end wall framed under the overhang.

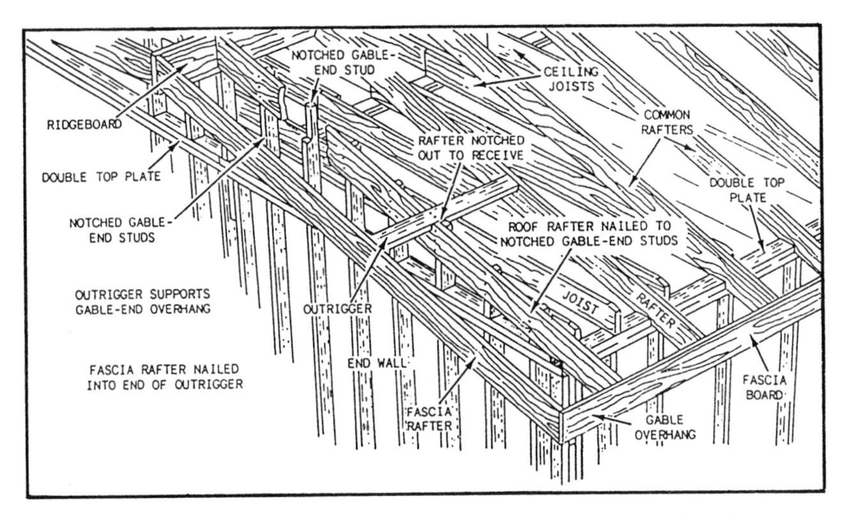

Figure 2-28.—Gable-end overhang with the end wall framed directly beneath the rafters.

This method guarantees that the line of the overhang will be perfectly straight, even if the building is not.

Over each gable end of the building, another overhang can be framed. The main framing members of the gable-end overhang are the fascia, also referred to as "fly" (or "barge") rafters. They are tied to the ridgeboard at the upper end and to the fascia board at the lower end. Fascia boards are often nailed to the tail ends of the common rafters to serve as a finish piece at the edge of the roof. By extending past the gable ends of the house, common rafters also help to support the basic rafters.

Figures 2-27 and 2-28 show different methods used to frame the gable-end overhang. In figure 2-27, a fascia rafter is nailed to the ridgeboard and to the fascia board. Blocking (not shown in the figures) rests on the end wall and is nailed between the fascia rafter and the rafter next to it. This section of the roof is further strengthened when the roof sheathing is nailed to it. In figure 2-28, two common rafters are placed directly over the gable

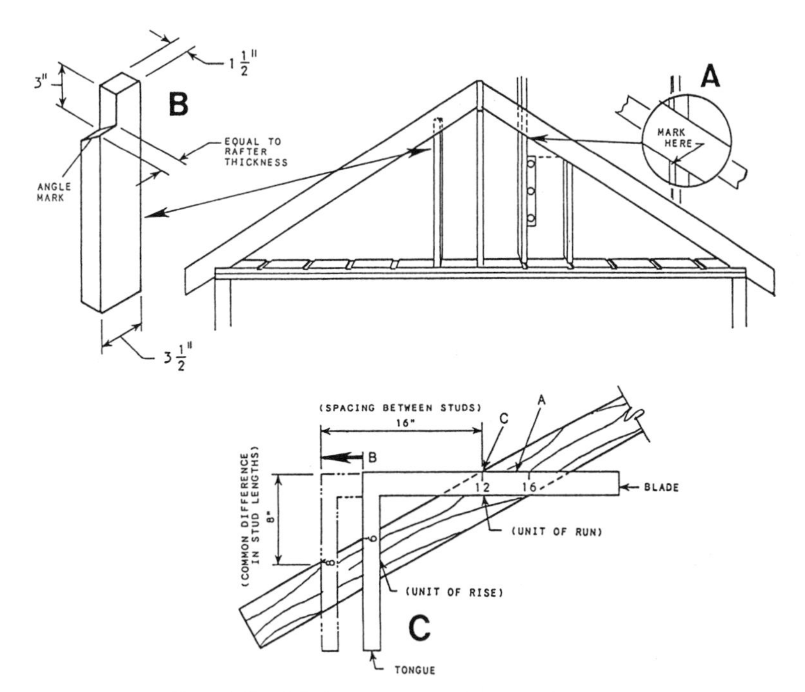

Figure 2-29.—Calculating common difference of gable-end studs.

ends of the building. The fascia rafters (fly rafters) are placed between the ridgeboard and the fascia boards. The gable studs should be cut to fit against the rafter above.

End Framing

Gable-end studs rest on the top plate and extend to the rafter line in the ends of a gable roof. They may be placed with the edge of the stud even with the outside wall and the top notched to fit the rafter (as shown in fig. 2-28), or they may be installed flatwise with a cut on the top of the stud to fit the slope of the rafter.

The position of the gable-end stud is located by squaring a line across the plate directly below the center of the gable. If a window or vent is to be installed in the gable, measure one-half of the opening size on each side of the center line and make a mark for the first stud. Starting at this mark, lay out the stud spacing (that is, 16 or 24 inches on center [OC]) to the outside of the building. Plumb the gable-end stud on the first mark and mark it where it contacts the bottom of the rafter, as shown in figure 2-29, view A. Measure and mark 3 inches above this mark and notch the stud to the depth equal to the thickness of the rafter, as shown in view B. The lengths of the other gable studs depend on the spacing.

The common difference in the length of the gable studs may be figured by the following method:

$$\frac{\text{24 inches (OC spacing)}}{\text{12 inches (unit of run)}} = 2$$

and, 2 × 6 inches (unit of rise) or 12 inches (common difference).

The common difference in the length of the gable studs may also be laid out directly with the framing square (fig. 2-29, view C). Place the framing square on the stud to the cut of the roof (6 and 12 inches for this example). Draw a line along the blade at A. Slide the square along this line in the direction of the arrow at B until the desired spacing between the studs (16 inches for this example) is at the intersection of the line drawn at A and the edge of the stud. Read the dimension on the tongue aligned with the same edge of the stud (indicated by C). This is the common difference (8 inches for this example) between the gable studs.

Toenail the studs to the plate with two 8d nails in each side. As the studs are nailed in place, care must be taken not to force a crown into the top of the rafter.

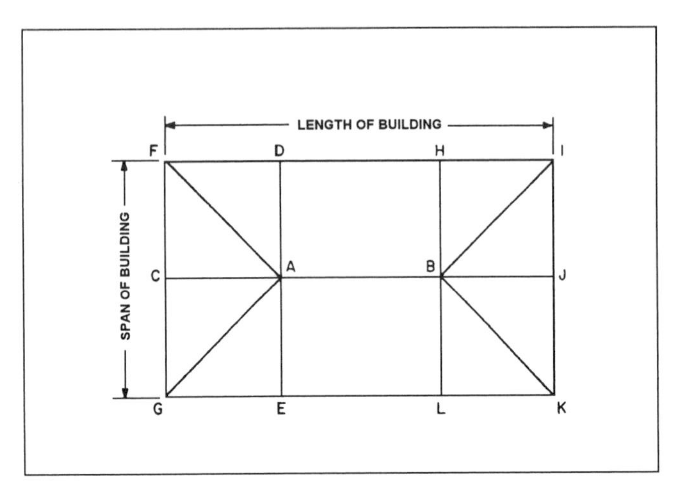

Figure 2-30.—Equal-pitch hip roof framing diagram.

HIP

Most hip roofs are equal pitch. This means the angle of slope on the roof end or ends is the same as the angle of slope on the sides. Unequal-pitch hip roofs do exist, but they are quite rare. They also require special layout methods. The unit length rafter table on the framing square applies only to equal-pitch hip roofs. The next paragraphs discuss an equal-pitch hip roof.

The length of a hip rafter, like the length of a common rafter, is calculated on the basis of bridge measure multiplied by the total run (half span). Any of the methods previously described for a common rafter may be used, although some of the dimensions for a hip rafter are different.

Figure 2-30 shows part of a roof framing diagram for an equal-pitch hip roof. A roof framing diagram may be included among the working drawings; if not, you should lay one out for yourself. Determine what scale will be used, and lay out all framing members to scale. Lay the building lines out first. You can find the span and the length of the building on the working drawings. Then, draw a horizontal line along the center of the span.

In an equal-pitch hip roof framing diagram, the lines indicating the hip rafters (AF, AG, BI, and BK in figure 2-30) form 45° angles with the building lines. Draw these lines at 45°, as shown. The points where they meet the center line are the theoretical ends of the ridge piece. The ridge-end common rafters AC, AD, AE, BH, BJ, and BL join the ridge at the same points.

A line indicating a rafter in the roof framing diagram is equal in length to the total run of the rafter it represents. You can see from the diagram that the total run of a hip rafter (represented by lines AF-AG-BI-BK) is the hypotenuse of a right triangle with the altitude and base equal to the total run of a common rafter. You know the total run of a common rafter: It is one-half the span, or one-half the width of the building. Knowing this, you can find the total run of a hip rafter by applying the Pythagorean theorem.

Let's suppose, for example, that the span of the building is 30 feet. Then, one-half the span, which is the same as the total run of a common rafter, is 15 feet. Applying the Pythagorean theorem, the total run of a hip rafter is:

$$\sqrt{(15^2 + 15^2)} = 21.21 \text{ feet.}$$

What is the total rise? Since a hip rafter joins the ridge at the same height as a common rafter, the total rise for a hip rafter is the same as the total rise for a common rafter. You know how to figure the total rise of a common rafter. Assume that this roof has a unit of run of 12 and a unit of rise of 8. Since the total run of a common rafter in the roof is 15 feet, the total rise of common rafter is the value of x in the proportional equation *12:8::15:x*, or 10 feet.

Knowing the total run of the hip rafter (21.21 feet) and the total rise (10 feet), you can figure the line length by applying the Pythagorean theorem. The line length is:

$\sqrt{(21.21^2 + 10^2)}$ = 23.45 feet, or about 23 feet 5 3/8 inches.

To find the length of a hip rafter on the basis of bridge measure, you must first determine the bridge measure. As with a common rafter, the bridge measure of a hip rafter is the length of the hypotenuse of a triangle with its altitude and base equal to the unit of run and unit of rise of the rafter. The unit of rise of a hip rafter is always the same as that of a common rafter, but the unit of run of a hip rafter is a fixed unit of measure, always 16.97.

The unit of run of a hip rafter in an equal-pitch roof is the hypotenuse of a right triangle with its altitude and base equal to the unit of run of a common rafter, 12. Therefore, the unit of run of a hip rafter is:

$$\sqrt{(12^2 + 12^2)} = 16.97$$

If the unit of run of a hip rafter is 16.97 and the unit of rise (in this particular case) is 8, the bridge measure of the hip rafter must be:

$$\sqrt{(16.97^2 + 8^2)} = 18.76$$

This means that for every unit of run (16.97) the rafter has a line length of 18.76 inches. Since the total run of the rafter is 21.21 feet, the length of the rafter must be the value of x in the proportional equation *16.97:18.76::21.21:x*, or 23.45 feet.

Like the unit length of a common rafter, the bridge measure of a hip rafter can be obtained from the unit length rafter table on the framing square. If you turn back to figure 2-16, you will see that the second line in the table is headed LENGTH HIP OR VALLEY PER FT RUN. This means "per foot run of a common rafter in the same roof." Actually, the unit length given in the tables is the unit length for every 16.97 units of run of the hip rafter itself. If you go across to the unit length given under 8, you will find the same figure, 18.76 units, that you calculated above.

An easy way to calculate the length of an equal-pitch hip roof is to multiply the bridge measure by the number of feet in the total run of a common rafter, which is the same as the number of feet in one-half of the building span. One-half of the building span, in this case, is 15 feet. The length of the hip rafter is therefore 18.76×15, or 281.40 inches—23.45 feet once converted.

You step off the length of an equal-pitch hip roof just as you do the length of a common rafter, except that you set the square to a unit of run of 16.97 inches instead of to a unit of run of 12 inches. Since 16.97 inches is the same as 16 and 15.52 sixteenths of an inch, setting the square to a unit of run of 17 inches is close enough for most practical purposes. Bear in mind that for any plumb cut line on an equal-pitch hip roof rafter, you set the square to the unit of rise of a common rafter and to a unit of run of 17.

You step off the same number of times as there are feet in the total run of a common rafter in the same roof; only the size of each step is different. For every 12-inch step in a common rafter, a hip rafter has a 17-inch step. For the roof on which you are working, the total run of common rafter is exactly 15 feet; this means that you would step off the hip-rafter cut (17 inches and 8 inches) exactly 15 times.

Suppose, however, that there was an odd unit in the common rafter total run. Assume, for example, that the total run of a common rafter is 15 feet 10 1/2 inches. How would you make the odd fraction of a step on the hip rafter?

You remember that the unit of run of a hip rafter is the hypotenuse of a right triangle with the other side each equal to the unit of run of a common rafter. In this case, the run of the odd unit on the hip rafter must be the hypotenuse of a right triangle with the altitude and base equal to the odd unit of run of the common rafter (in this case, 10 1/2 inches). You can figure this using the Pythagorean theorem

$$\sqrt{(10.5^2 + 10.5^2)}$$

or you can set the square on a true edge to 10 1/2 inches on the blade and measure the distance between the marks. It comes to 14.84 inches. Rounded off to the nearest 1/16 inch, this equals 14 13/16 inches.

To lay off the odd unit, set the tongue of the framing square to the plumb line for the last full step made and measure off 14 13/16 inches along the blade. Place the tongue of the square at the mark, set the square to the hip rafter plumb cut of 8 inches on the tongue and 17 inches on the blade, and draw the line length cut.

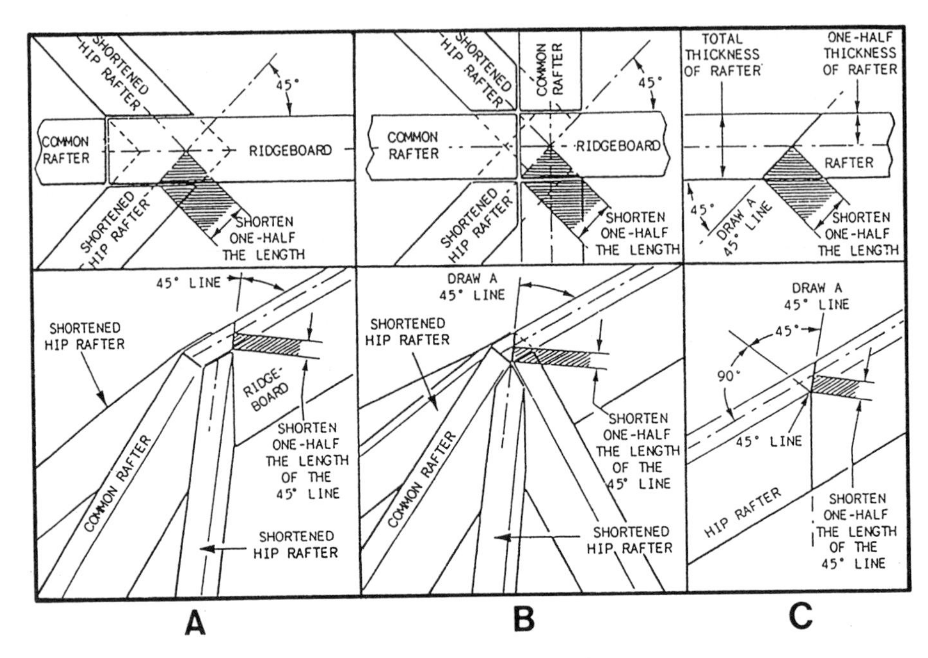

Figure 2-31.—Shortening a hip rafter.

Rafter Shortening Allowance

As in the case with a common rafter, the line length of a hip rafter does not take into account the thickness of the ridge piece. The size of the ridge-end shortening allowance for a hip rafter depends upon the way the ridge end of the hip rafter is joined to the other structural members. As shown in figure 2-31, the ridge end of the hip rafter can be framed against the ridgeboard (view A) or against the ridge-end common rafters (view B). To calculate the actual length, deduct one-half the 45° thickness of the ridge piece that fits between the rafters from the theoretical length.

When no common rafters are placed at the ends of the ridgeboard, the hip rafters are placed directly against the ridgeboard. They must be shortened one-half the length of the 45° line (that is, one-half the thickness of the ridgeboard). When common rafters are placed at the ends of the ridgeboard (view B), the hip rafter will fit between the common rafters. The hip rafter must be shortened one-half the length of the 45° line (that is, one-half the thickness of the common rafter).

If the hip rafter is framed against the ridge piece, the shortening allowance is one-half of the 45° thickness of the ridge piece (fig. 2-31, view C). The 45° thickness of stock is the length of a line laid at 45° across the thickness dimension of the stock. If the hip rafter is framed against the common rafter, the shortening allowance is one-half of the 45° thickness of a common rafter.

To lay off the shortening allowance, first set the tongue of the framing square to the line length ridge cut line. Then, measure off the shortening allowance along the blade, set the square at the mark to the cut of the rafter (8 inches and 17 inches), draw the actual ridge plumb cut line. (To find the 45° thickness of a piece of lumber, draw a 45° line across the edge, and measure the length of the line and divide by 2.)

Rafter Projection

A hip or valley rafter overhang, like a common rafter overhang, is figured as a separate rafter. The projection, however, is not the same as the projection of a common rafter overhang in the same roof. The projection of the hip or valley rafter overhang is the hypotenuse of a right triangle whose shorter sides are each equal to the run of a common rafter overhang (fig. 2-32). If the run of the common rafter overhang is

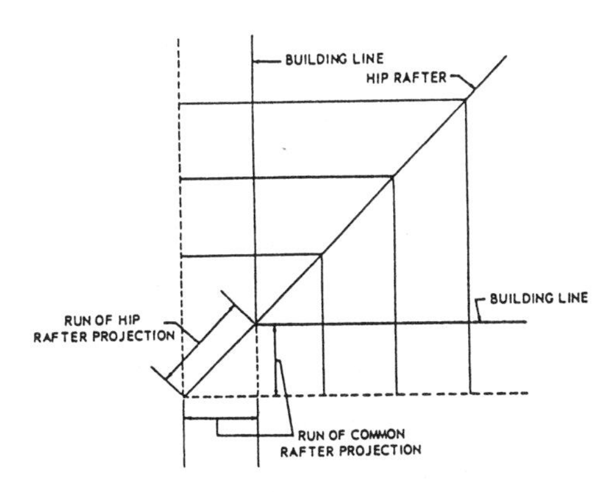

Figure 2-32.—Run of hip rafter projection.

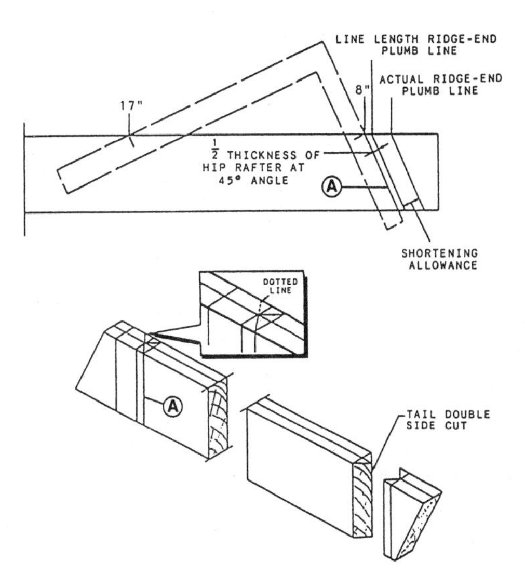

Figure 2-33.—Laying out hip rafter side cut.

18 inches for a roof with an 8-inch unit of rise, the length of the hip or valley rafter tail is figured as follows:

1. Find the bridge measure of the hip or valley rafter on the framing square (refer to figure 2-16). For this roof, it is 18.76 inches.

2. Multiply the bridge measure (in inches) of the hip or valley rafter by the projection (in feet) of the common rafter overhang:

   18.76 inches (bridge measure)
   × 1.5 feet (projection of the common rafter)

   28.14, or 28 1/8 inches.

3. Add this product to the theoretical rafter length.

The overhang may also be stepped off as described earlier for a common rafter. When stepping off the length of the overhang, set the 17-inch mark on the blade of the square even with the edge of the rafter. Set the unit of rise, whatever it might be, on the tongue even with the same rafter edge.

Rafter Side Cuts

Since a common rafter runs at 90° to the ridge, the ridge end of a common rafter is cut square, or at 90° to the lengthwise line of the rafter. A hip rafter, however, joins the ridge, or the ridge ends of the common rafter, at other than a 90° angle, and the ridge end of a hip rafter must therefore be cut to a corresponding angle, called a side cut. The angle of the side cut is more acute for a high rise than it is for a low one.

The angle of the side cut is laid out as shown in figure 2-33. Place the tongue of the framing square along the ridge cut line, as shown, and measure off one-half the thickness of the hip rafter along the blade. Shift the tongue to the mark, set the square to the cut of the rafter (17 inches and 8 inches), and draw the plumb line marked "A" in the figure. Then, turn the rafter edge-up, draw an edge center line, and draw in the angle of the side cut, as indicated in the lower view of figure 2-33. For a hip rafter to be framed against the ridge, there will be only a single side cut, as indicated by the dotted line in the figure. For one to be framed against the ridge ends of the common rafters, there will be a double side cut, as shown in the figure. The tail of the rafter must have a double side cut at the same angle, but in the reverse direction.

The angle of the side cut on a hip rafter may also be laid out by referring to the unit length rafter table on the framing square. (Look ahead to figure 2-41.) You will see that the bottom line in the table is headed SIDE CUT HIP OR VALLEY USE. If you follow this line over to the column headed by the figure 8 (for a unit of rise of 8), you will find the figure 10 7/8. If you place the framing square faceup on the rafter edge with the tongue on the ridge-end cut line, and set the square to a cut of 10 7/8 inches on the blade and 12 inches on the tongue, you can draw the correct side-cut angle along the tongue.

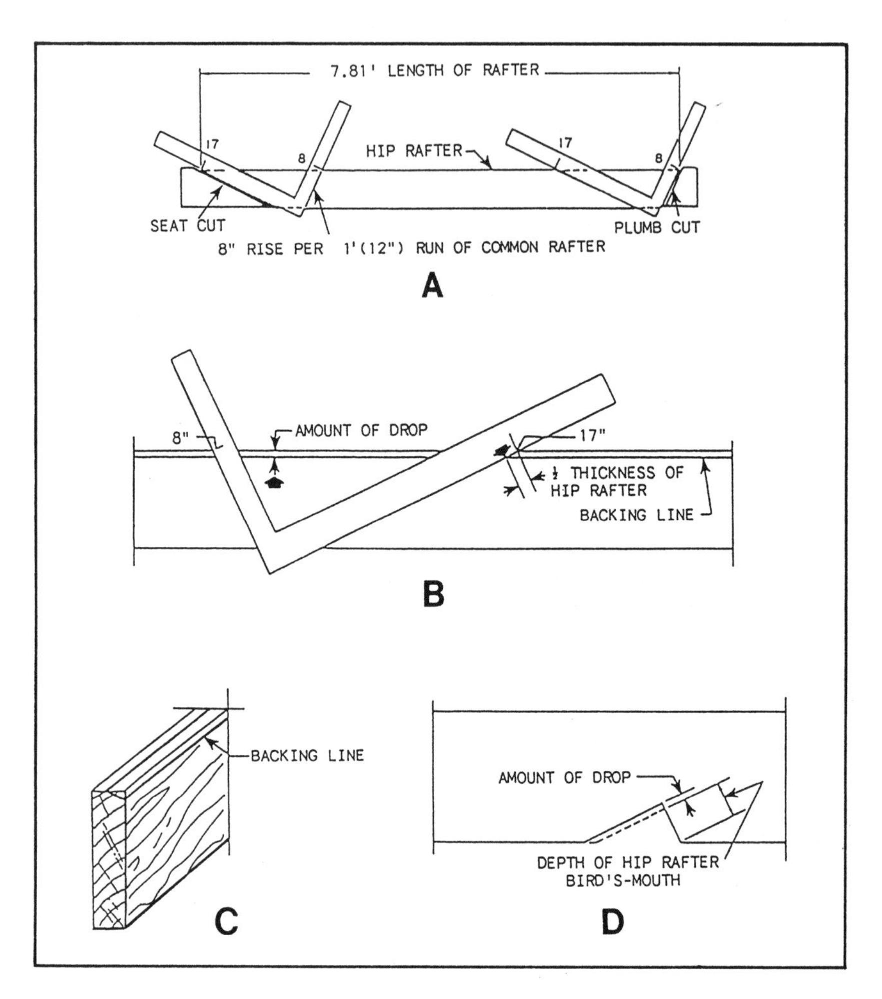

Figure 2-34.—Backing or dropping a hip rafter: A. Marking the top (plumb) cut and the seat (level) cut of a hip rafter; B. Determining amount of backing or drop; C. Bevel line for backing the rafter; D. Deepening the bird's-mouth for dropping the rafter.

Bird's-Mouth

Laying out the bird's-mouth for a hip rafter is much the same as for a common rafter. However, there are a couple of things to remember. When the plumb (heel) cut and level (seat) cut lines are laid out for a bird's-mouth on a hip rafter, set the body of the square at 17 inches and the tongue to the unit of rise (for example, 8 inches—depending on the roof pitch) (fig. 2-34, view A). When laying out the depth of the heel for the bird's-mouth, measure along the heel plumb line down from the top edge of the rafter a distance equal to the same dimension on the common rafter. This must be done so that the hip rafter, which is usually wider than a common rafter, will be level with the common rafters.

If the bird's-mouth on a hip rafter has the same depth as the bird's-mouth on a common rafter, the edge of the hip rafter will extend above the upper ends of the jack rafters. You can correct this by either backing or dropping the hip rafter. Backing means to bevel the top edges of the hip rafter (see fig. 2-35). The amount of backing is taken at a right angle to the roof surface on

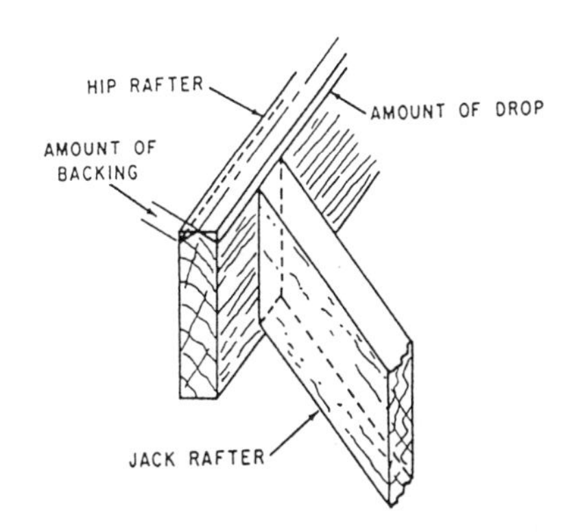

Figure 2-35.—Backing or dropping a hip rafter.

the top edge of the hip rafters. Dropping means to deepen the bird's-mouth so as to bring the top edge of the hip rafter down to the upper ends of the jacks. The amount of drop is taken on the heel plumb line (fig. 2-34, view D).

The backing or drop required is calculated, as shown in figure 2-34, view B. Set the framing square to the cut of the rafter (8 inches and 17 inches) on the upper edge, and measure off one-half the thickness of the rafter from the edge along the blade. A line drawn through this mark and parallel to the edge (view C) indicates the bevel angle if the rafter is to be backed. The perpendicular distance between the line and the edge of the rafter is the amount of the drop. This represents the amount the depth of the hip rafter bird's-mouth should exceed the depth of the common rafter bird's-mouth (view D).

INTERSECTING

An intersecting roof, also known as a combination roof, consists of two or more sections sloping in different directions. A valley is formed where the different sections come together.

The two sections of an intersecting roof may or may not be the same width. If they are the same width, the roof is said to have equal spans. If they are not the same width, the roof is said to have unequal spans.

Spans

In a roof with equal spans, the height (total rise) is the same for both ridges (fig. 2-36). That is, both sections are the same width, and the ridgeboards are the same height. A pair of valley rafters is placed where the slopes of the roof meet to form a valley between the two sections. These rafters go from the inside corners formed by the two sections of the building to the corners

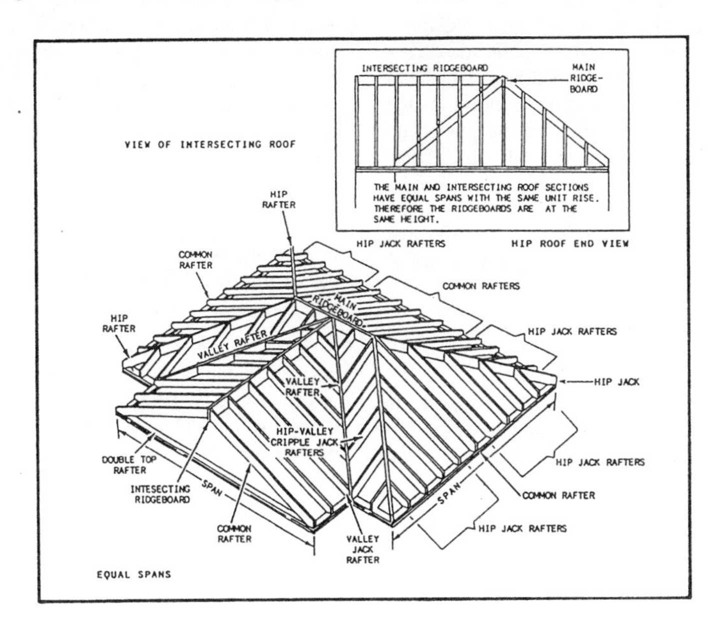

Figure 2-36.—Intersecting roof with equal spans.

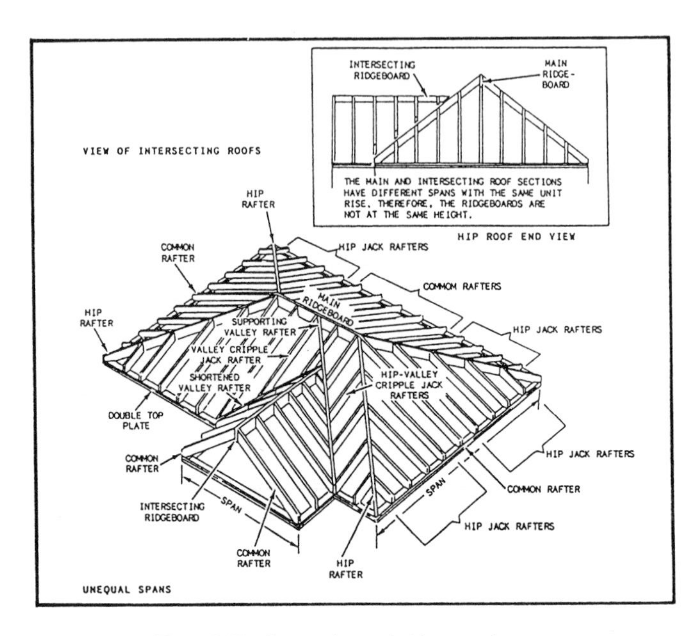

Figure 2-37.—Intersecting roof with unequal spans.

formed by the intersecting ridges. Valley jack rafters run from the valley rafters to both ridges. Hip-valley cripple jack rafters are placed between the valley and hip rafters.

An intersecting roof with unequal spans requires a supporting valley rafter to run from the inside corner formed by the two sections of the building to the main ridge (fig. 2-37). A shortened valley rafter runs from the other inside corner of the building to the supporting valley rafter. Like an intersecting roof with equal spans, one with unequal spans also requires valley jack rafters and hip-valley cripple jack rafters. In addition, a valley cripple jack rafter is placed between the supporting and shortened valley rafters. Note that the ridgeboard is lower on the section with the shorter span.

Valley Rafters

Valley rafters run at a 45° angle to the outside walls of the building. This places them parallel to the hip rafters. Consequently, they are the same length as the hip rafters.

A valley rafter follows the line of intersection between a main-roof surface and a gable-roof addition or a gable-roof dormer surface. Most roofs having valley rafters are equal-pitch roofs, in which the pitch of the addition or dormer roof is the same as the pitch of the main roof. There are unequal-pitch valley-rafter roofs, but they are quite rare and require special framing methods.

In the discussion of valley rafter layout, it is assumed that the roof is equal pitch. Also, the unit of run and unit of rise of an addition or dormer common rafter are assumed to be the same as the unit of run and rise of a main-roof common rafter. In an equal-pitch roof, the valley rafters always run at 45° to the building lines and the ridge pieces.

Figure 2-38 shows an equal-span framing situation, in which the span of the addition is the same as the span of the main roof. Since the pitch of the addition roof is the same as the pitch of the main roof, equal spans bring the ridge pieces to equal heights.

Looking at the roof framing diagram in the figure, you can see the total run of a valley rafter (indicated by AB and AC in the diagram) is the hypotenuse of a right triangle with the altitude and base equal to the total run of a common rafter in the main roof. The unit of run of a valley rafter is therefore 16.97, the same as the unit of run for a hip rafter. It follows that figuring the length of an equal-span valley rafter is the same as figuring the length of an equal-pitch hip roof hip rafter.

A valley rafter, however, does not require backing or dropping. The projection, if any, is figured just as it is for a hip rafter. Side cuts are laid out as they are for a

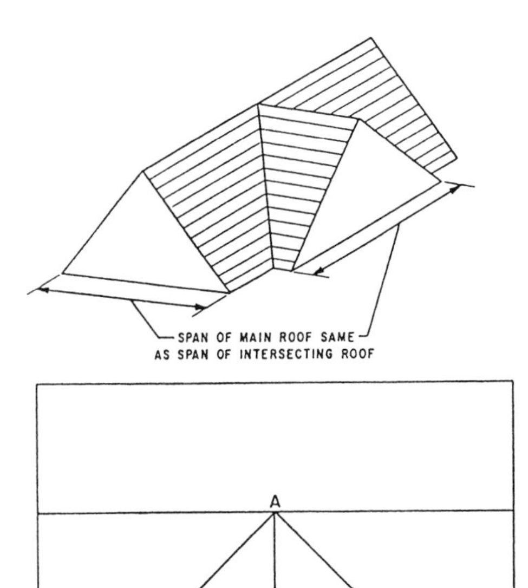

Figure 2-38.—Equal-span intersecting roof.

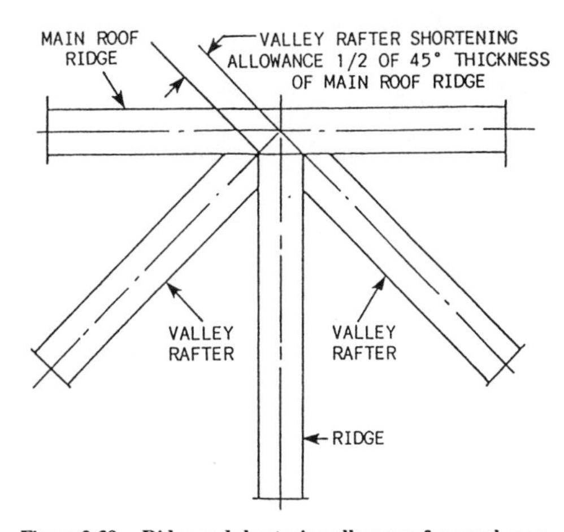

Figure 2-39.—Ridge-end shortening allowance for equal-span intersecting valley rafter.

hip rafter. The valley-rafter tail has a double side cut (like the hip-rafter tail) but in the reverse direction. This is because the tail cut on a valley rafter must form an inside, rather than an outside, corner. As indicated in figure 2-39, the ridge-end shortening allowance in this framing situation amounts to one-half of the 45° thickness of the ridge.

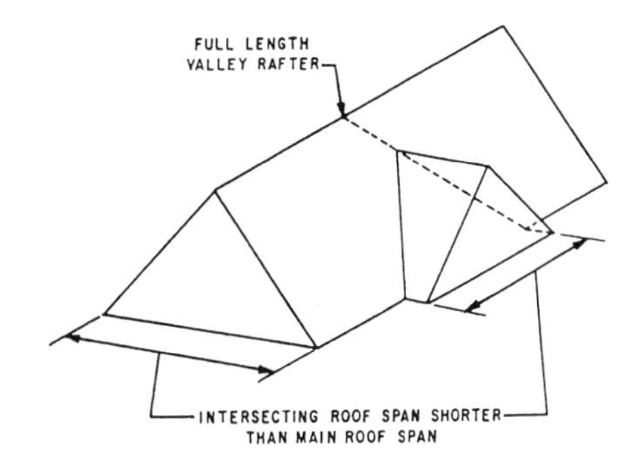

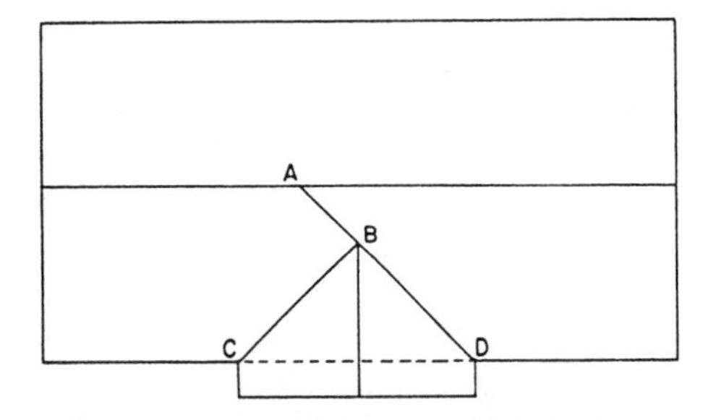

Figure 2-40.—Equal pitch but unequal span framing.

Figure 2-40 shows a framing situation in which the span of the addition is shorter than the span of the main roof. Since the pitch of the addition roof is the same as the pitch of the main roof, the shorter span of the addition brings the addition ridge down to a lower level than that of the main-roof ridge.

There are two ways of framing an intersection of this type. In the method shown in figure 2-40, a full-length valley rafter (AD in the figure) is framed between the top plate and the main-roof ridgeboard. A shorter valley rafter (BC in the figure) is then framed to the longer one. If you study the framing diagram, you can see that the total run of the longer valley rafter is the hypotenuse of a right triangle with the altitude and base equal to the total run of a common rafter in the main roof. The total run of the shorter valley rafter, on the other hand, is the hypotenuse of a right triangle with the altitude and base equal to the total run of a common rafter in the addition. The total run of a common rafter in the main roof is equal to one-half the span of the main roof. The total run of a common rafter in the addition is equal to one-half the span of the addition.

Knowing the total run of a valley rafter, or of any rafter for that matter, you can always find the line length by applying the bridge measure times the total run.

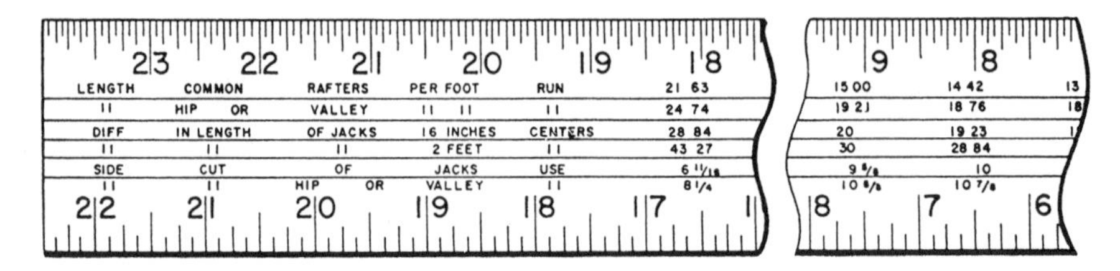

Figure 2-41.—Rafter table method.

Suppose, for example, that the span of the addition in figure 2-40 is 30 feet and that the unit of rise of a common rafter in the addition is 9. The total run of the shorter valley rafter is:

$$\sqrt{(15^2 + 15^2)} = 21.21 \text{ feet.}$$

Referring to the unit length rafter table in figure 2-41, you can see the bridge measure for a valley rafter in a roof with a common rafter unit of rise of 9 is 19.21. Since the unit of run of a valley rafter is 16.97, and the total run of this rafter is 21.21 feet, the line length must be the value of x in the proportional equation *16.97:19.21::21.21:x*, or 24.01 feet.

An easier way to find the length of a valley rafter is to multiply the bridge measure by the number of feet in one-half the span of the roof. The length of the longer valley rafter in figure 2-40, for example, would be 19.21 times one-half the span of the main roof. The length of the shorter valley rafter is 19.21 times one-half the span of the addition. Since one-half the span of the addition is 15 feet, the length of the shorter valley rafter is $15 \times 9.21 = 288.15$ inches, or approximately 24.01 feet.

Figure 2-42 shows the long and short valley rafter shortening allowances. Note that the long valley rafter has a single side cut for framing to the main-roof ridge piece, whereas the short valley rafter is cut square for framing to the long valley rafter.

Figure 2-43 shows another method of framing an equal-pitch unequal-span addition. In this method, the inboard end of the addition ridge is nailed to a piece that hangs from the main-roof ridge. As shown in the framing diagram, this method calls for two short valley rafters (AB and AC), each of which extends from the top plate to the addition ridge.

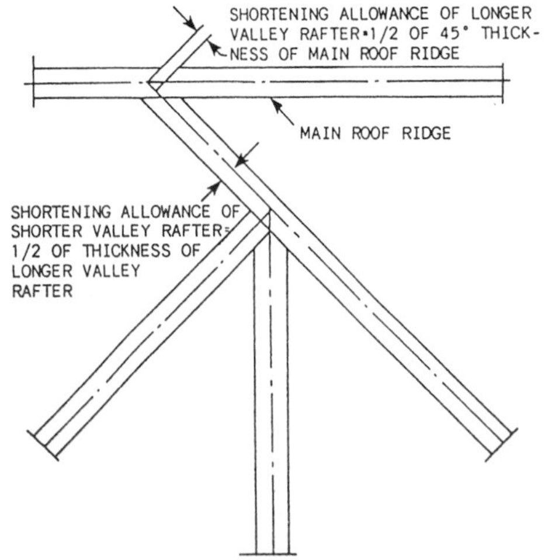

SHORTENING ALLOWANCE OF LONGER VALLEY RAFTER•1/2 OF 45° THICKNESS OF MAIN ROOF RIDGE

MAIN ROOF RIDGE

SHORTENING ALLOWANCE OF SHORTER VALLEY RAFTER=1/2 OF THICKNESS OF LONGER VALLEY RAFTER

Figure 2-42.—Long and short valley rafter shortening allowance.

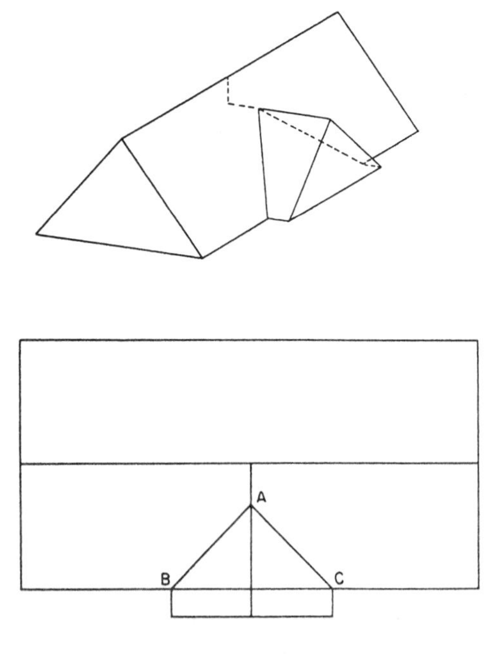

Figure 2-43.—Another method of framing equal-pitch unequal-span intersection.

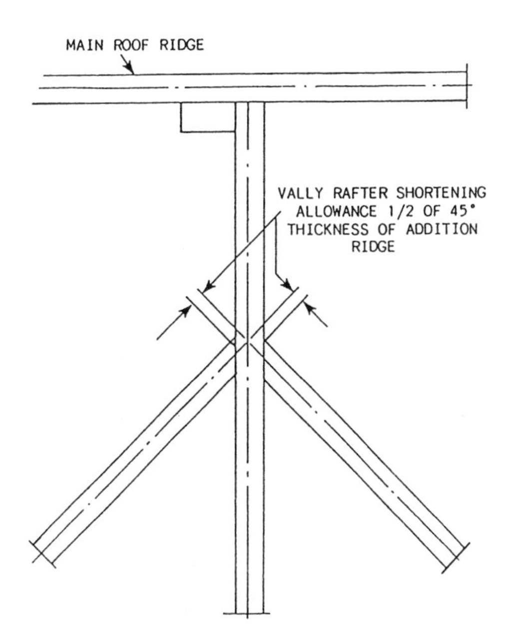

Figure 2-44.—Shortening allowance of valley rafters in suspended ridge method of intersecting roof framing.

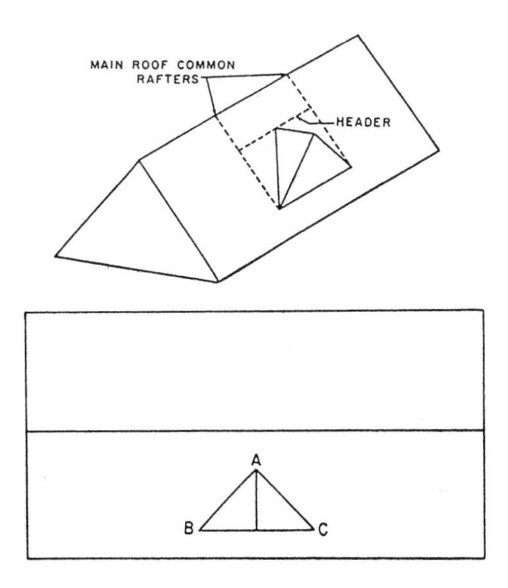

Figure 2-45.—Method of framing dormer without sidewalls.

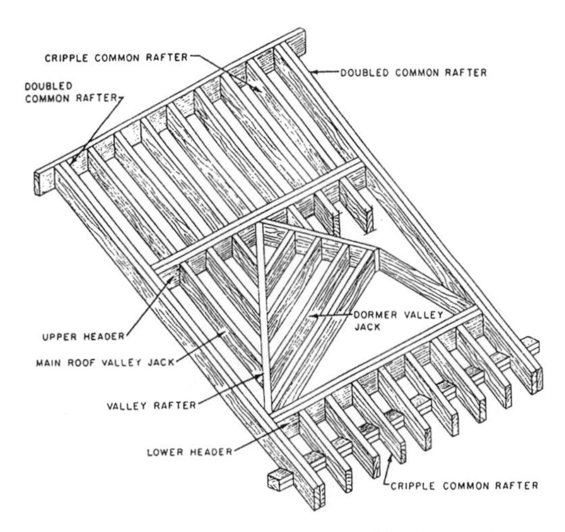

Figure 2-46.—Arrangement and names of framing members for dormer without sidewalls.

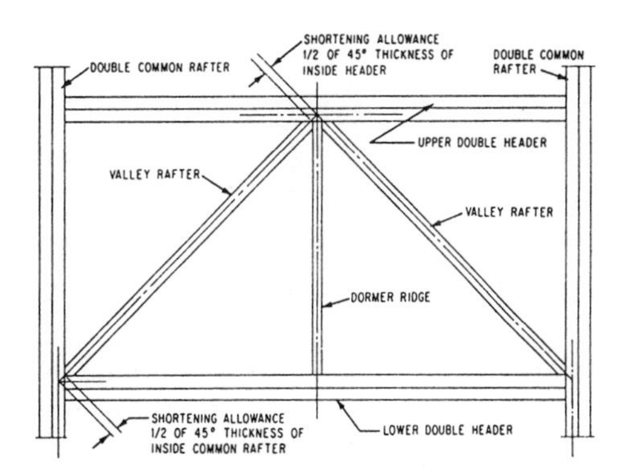

Figure 2-47.—Valley rafter shortening allowance for dormer without sidewalls.

As indicated in figure 2-44, the shortening allowance of each of the short valley rafters is one-half the 45° thickness of the addition ridge. Each rafter is framed to the addition ridge with a single side cut.

Figure 2-45 shows a method of framing a gable dormer without sidewalls. The dormer ridge is framed to a header set between a pair of doubled main-roof common rafters. The valley rafters (AB and AC) are framed between this header and a lower header. As indicated in the framing diagram, the total run of a valley rafter is the hypotenuse of a right triangle with the shorter sides equal to the total run of a common rafter in the dormer. Figure 2-46 shows the arrangement and names of framing members in this type of dormer framing.

The upper edges of the header must be beveled to the cut of the main roof. Figure 2-47 shows that in this

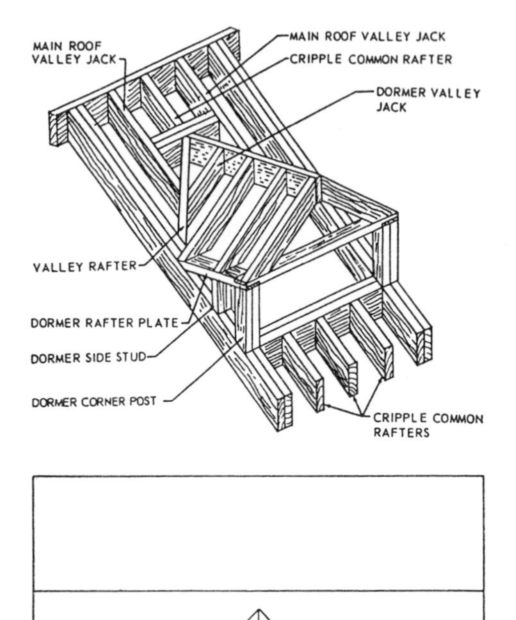

Figure 2-48.—Method of framing gable dormer with sidewalls.

method of framing, the shortening allowance for the upper end of a valley rafter is one-half the 45° thickness of the inside member in the upper doubled header. There is also a shortening allowance for the lower end, consisting of one-half the 45° thickness of the inside member of the doubled common rafter. The figure also shows that each valley rafter has a double side cut at the upper and lower ends.

Figure 2-48 shows a method of framing a gable dormer with sidewalls. As indicated in the framing diagram, the total run of a valley rafter is again the hypotenuse of a right triangle with the shorter sides each equal to the run of a common rafter in the dormer. You figure the lengths of the dormer corner posts and side studs just as you do the lengths of gable-end studs, and you lay off the lower end cutoff angle by setting the square to the cut of the main roof.

Figure 2-49 shows the valley rafter shortening allowance for this method of framing a dormer with sidewalls.

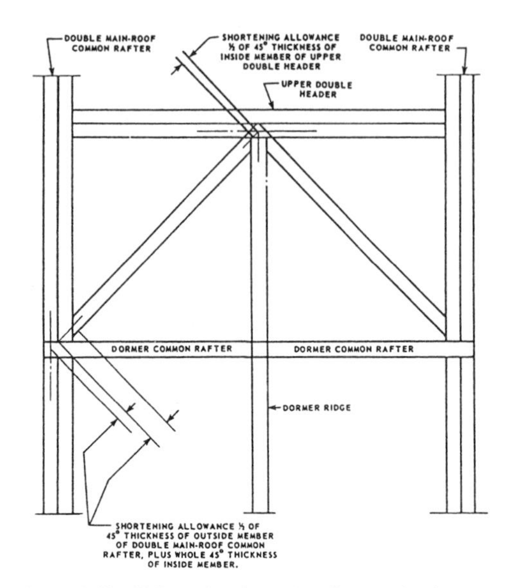

Figure 2-49.—Valley rafter shortening allowance for dormers with sidewalls.

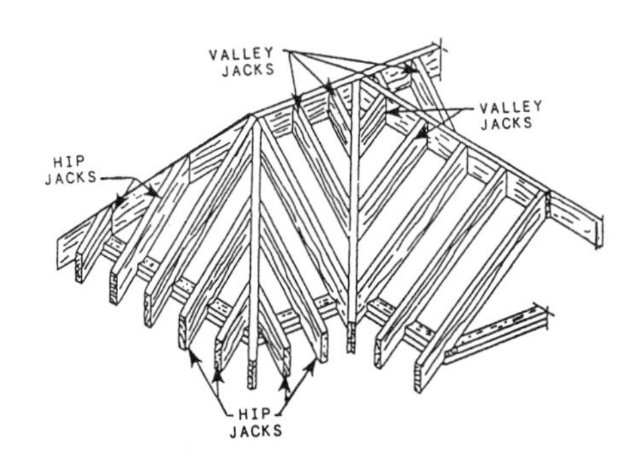

Figure 2-50.—Types of jack rafters.

Jack Rafters

A jack rafter is a part of a common rafter, shortened for framing a hip rafter, a valley rafter, or both. This means that, in an equal-pitch framing situation, the unit of rise of a jack rafter is always the same as the unit of rise of a common rafter. Figure 2-50 shows various types of jack rafters.

A hip jack rafter extends from the top plate to a hip rafter. A valley jack rafter extends from a valley rafter to a ridge. (Both are shown in fig. 2-51.) A cripple jack rafter does not contact either a top plate or a ridge. A

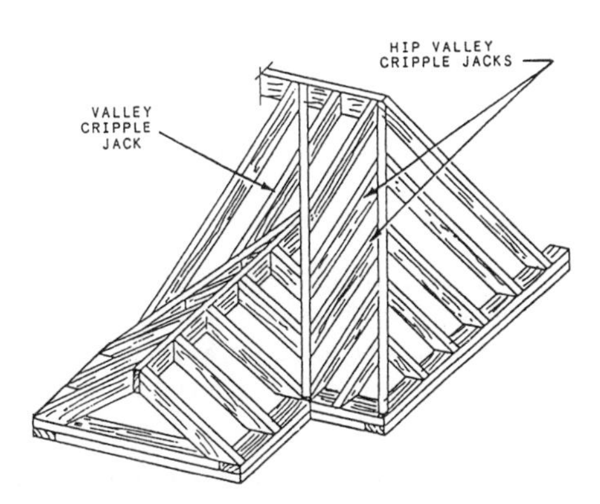

Figure 2-51.—Valley cripple jack and hip-valley cripple jack.

valley cripple jack extends between two valley rafters in the long and short valley rafter method of framing. A hip-valley cripple jack extends from a hip rafter to a valley rafter.

LENGTHS.—Figure 2-52 shows a roof framing diagram for a series of hip jack rafters. The jacks are always on the same OC spacing as the common rafters.

Now, suppose the spacing, in this instance, is 16 inches OC. You can see that the total run of the shortest jack is the hypotenuse of a right triangle with the shorter sides each 16 inches long. The total run of the shortest jack is therefore:

$$\sqrt{(16^2 + 16^2)} = 22.62 \text{ inches.}$$

Suppose that a common rafter in this roof has a unit of rise of 8. The jacks have the same unit of rise as a common rafter. The unit length of a jack in this roof is:

$$\sqrt{(12^2 + 8^2)} = 14.42 \text{ inches.}$$

This means that a jack is 14.42 units long for every 12 units of run. The length of the shortest hip jack in this roof is therefore the value of x in the proportional equation $12{:}14.42{::}16{:}x$, or 19.23 inches.

This is always the length of the shortest hip jack when the jacks are spaced 16 inches OC and the common rafter in the roof has a unit of rise of 8. It is also the common difference of jacks, meaning that the next hip jack will be 2 times 19.23 inches.

The common difference for hip jacks spaced 16 inches OC, or 24 inches OC, is given in the unit length rafter table on the framing square for unit of rise ranging from 2 to 18, inclusive. Turn back to figure 2-41, which shows a segment of the unit length rafter table. Note the third line in the table, which reads DIFF IN LENGTH OF JACKS 16 INCHES CENTERS. If you follow this line over to the figure under 8 (for a unit of rise of 8), you'll find the same unit length (19.23) that you worked out above.

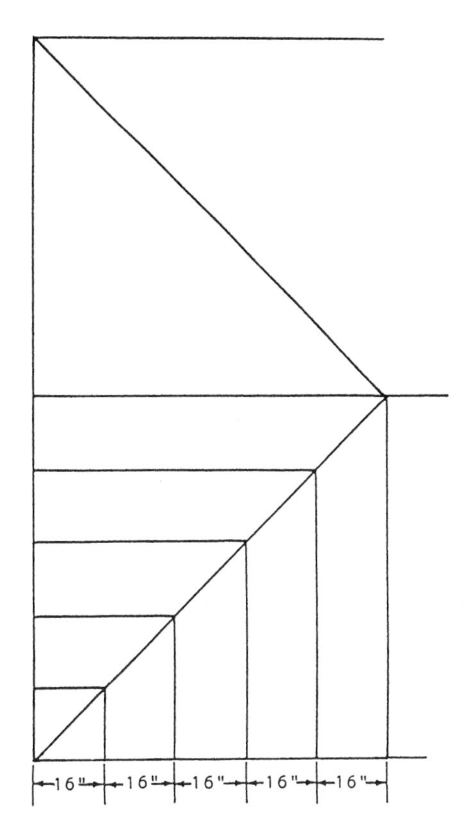

Figure 2-52.—Hip jack framing diagram.

The best way to determine the length of a valley jack or a cripple jack is to apply the bridge measure to the total run. The bridge measure of any jack is the same as the bridge measure of a common rafter having the same unit of rise as the jack. Suppose the jack has a unit of rise of 8. In figure 2-41, look along the line on the unit length rafter tables headed LENGTH COMMON RAFTER PER FOOT RUN for the figure in the column under 8; you'll find a unit length of 14.42. You should know by this time how to apply this to the total run of a jack to get the line length.

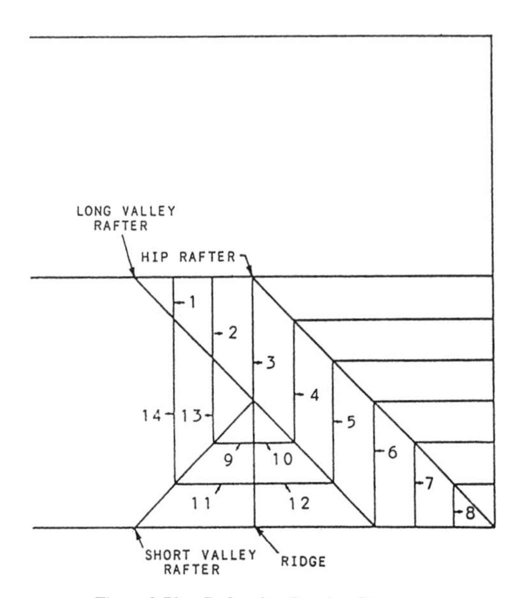

Figure 2-53.—Jack rafter framing diagram.

The best way to figure the total runs of valley jacks and cripple jacks is to lay out a framing diagram and study it to determine what these runs must be. Figure 2-53 shows part of a framing diagram for a main hip roof with a long and short valley rafter gable addition. By studying the diagram, you can figure the total runs of the valley jacks and cripple jacks as follows:

- The run of valley jack No. 1 is obviously the same as the run of hip jack No. 8, which is the run of the shortest hip jack. The length of valley jack No. 1 is therefore equal to the common difference of jacks.
- The run of valley jack No. 2 is the same as the run of hip jack No. 7, and the length is therefore twice the common difference of jacks.
- The run of valley jack No. 3 is the same as the run of hip jack No. 6, and the length is therefore three times the common difference of jacks.
- The run of hip-valley cripple Nos. 4 and 5 is the same as the run of valley jack No. 3.

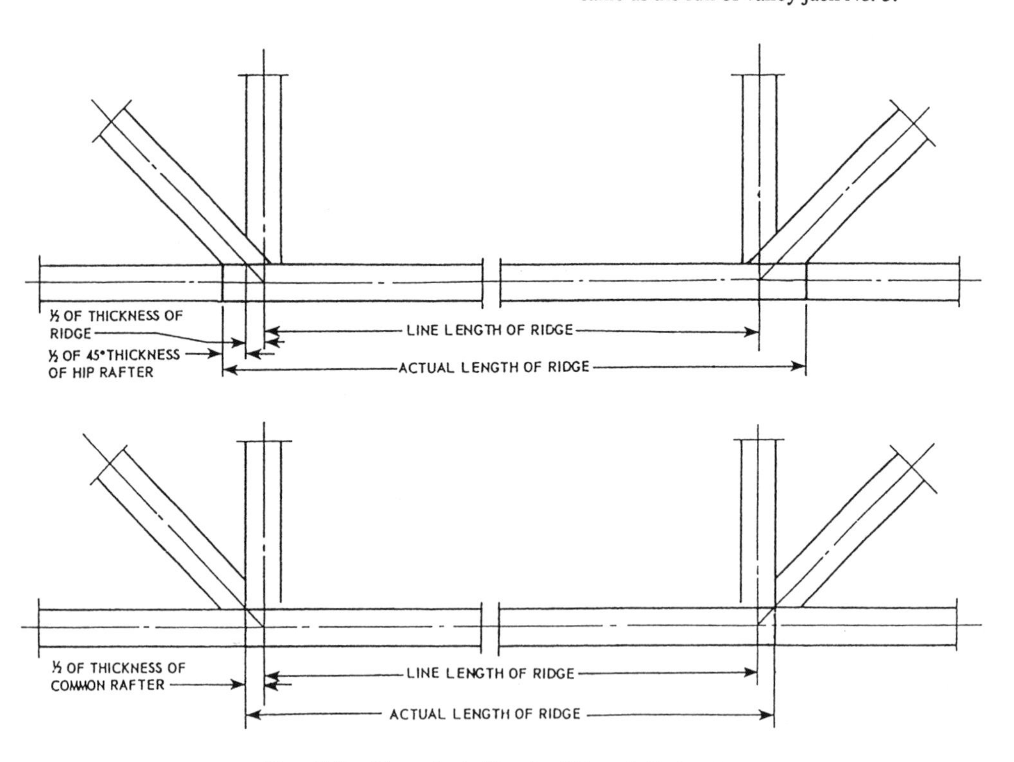

Figure 2-54.—Line and actual lengths of hip roof ridgeboard.

- The run of valley jack Nos. 9 and 10 is equal to the spacing of jacks OC. Therefore, the length of one of these jacks is equal to the common difference of jacks.
- The run of valley jacks Nos. 11 and 12 is twice the run of valley jacks Nos. 9 and 10, and the length of one of these jacks is therefore twice the common difference of jacks.
- The run of valley cripple No. 13 is twice the spacing of jacks OC, and the length is therefore twice the common difference of jacks.
- The run of valley cripple No. 14 is twice the run of valley cripple No. 13, and the length is therefore four times the common difference of jacks.

SHORTENING ALLOWANCES.—A hip jack has a shortening allowance at the upper end, consisting of one-half the 45° thickness of the hip rafter. A valley jack rafter has a shortening allowance at the upper end, consisting of one-half the 45° thickness of the ridge, and another at the lower end, consisting of one-half the 45° thickness of the valley rafter. A hip-valley cripple has a shortening allowance at the upper end, consisting of one-half the 45° thickness of the hip rafter, and another at the lower end, consisting of one-half the 45° thickness of the valley rafter. A valley cripple has a shortening allowance at the upper end, consisting of one-half the 45° thickness of the long valley rafter, and another at the lower end, consisting of one-half the 45° thickness of the short valley rafter.

SIDE CUTS.—The side cut on a jack rafter can be laid out using the same method as for laying out the side cut on a hip rafter. Another method is to use the fifth line of the unit length rafter table, which is headed SIDE CUT OF JACKS USE (fig. 2-41). If you follow that line over to the figure under 8 (for a unit of rise of 8), you will see that the figure given is 10. To lay out the side cut on a jack, set the square faceup on the edge of the rafter to 12 inches on the tongue and 10 inches on the blade, and draw the side-cut line along the tongue.

BIRD'S-MOUTH AND PROJECTION.—A jack rafter is a shortened common rafter; consequently, the bird's-mouth and projection on a jack rafter are laid out just as they are on a common rafter.

Ridge Layout

Laying out the ridge for a gable roof presents no particular problem since the line length of the ridge is equal to the length of the building. The actual length includes any overhang. For a hip main roof, however, the ridge layout requires a certain amount of calculation.

As previously mentioned, in an equal-pitch hip roof, the line length of the ridge amounts to the length of the building minus the span. The actual length depends upon the way the hip rafters are framed to the ridge.

As indicated in figure 2-54, the line length ends of the ridge are at the points where the ridge center line and the hip rafter center line cross. In the figure, the hip rafter is framed against the ridge. In this method of framing, the actual length of the ridge exceeds the line length, at each end, by one-half the thickness of the ridge, plus one-half the 45° thickness of the hip rafter. In the figure, the hip rafter is also framed between the common rafters. In this method of framing, the actual length of the ridge exceeds the line length at each end by one-half the thickness of a common rafter.

Figure 2-55, view A, shows that the length of the ridge for an equal-span addition is equal to the length of the addition top plate, plus one-half the span of the building, minus the shortening allowance at the

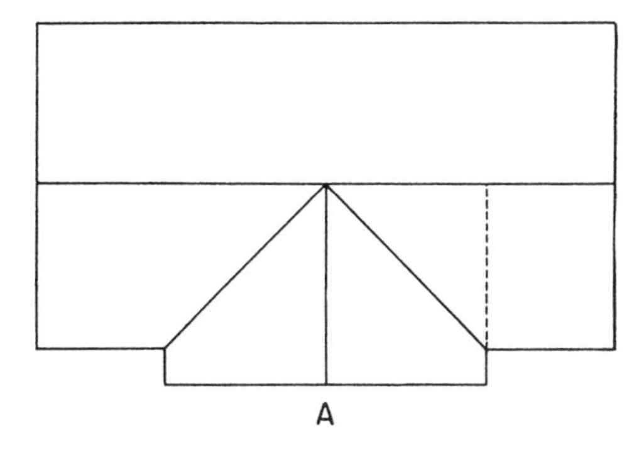

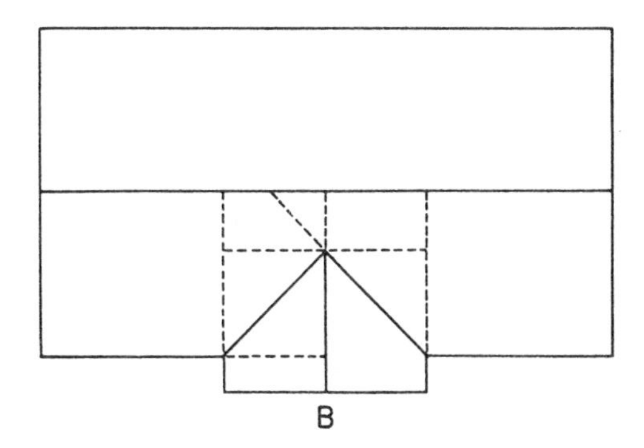

Figure 2-55.—Lengths of addition ridge.

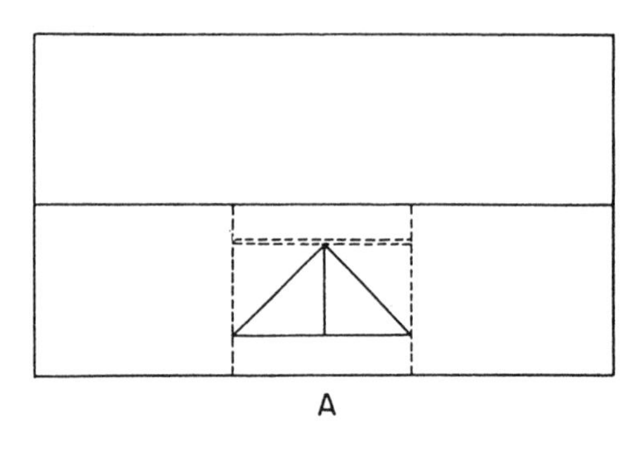

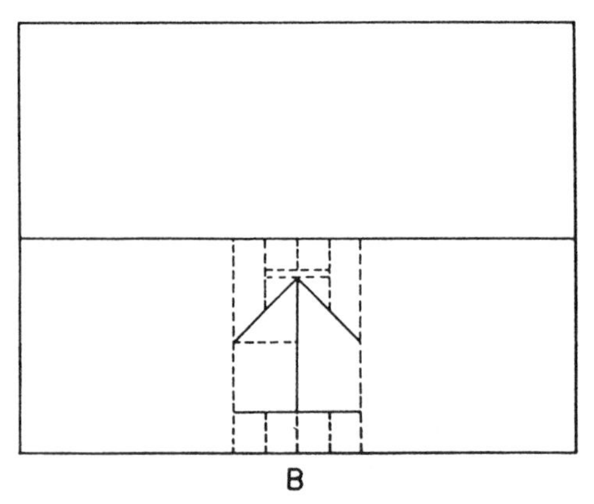

Figure 2-56.—Lengths of dormer ridge.

main-roof ridge. The shortening allowance amounts to one-half the thickness of the main-roof ridge.

View B shows that the length of the ridge for an unequal-span addition varies with the method of framing the ridge. If the addition ridge is suspended from the main-roof ridge, the length is equal to the length of the addition top plate, plus one-half the span of the building. If the addition ridge is framed by the long and short valley rafter method, the length is equal to the length of the addition top plate, plus one-half the span of the addition, minus a shortening allowance one-half the 45° thickness of the long valley rafter. If the addition ridge is framed to a double header set between a couple of double main-roof common rafters, the length of the ridge is equal to the length of the addition sidewall rafter plate, plus one-half the span of the addition, minus a shortening allowance one-half the thickness of the inside member of the double header.

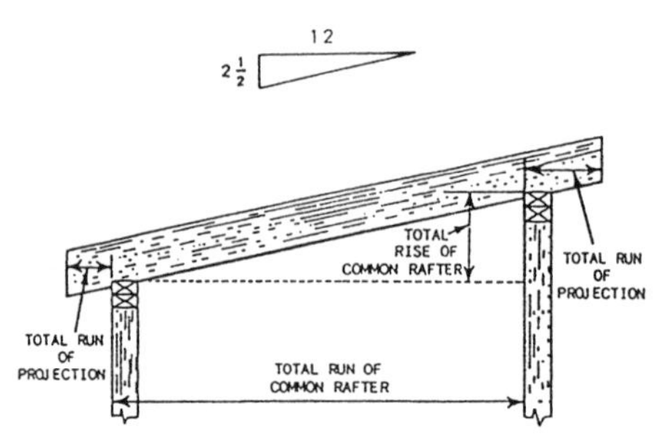

Figure 2-57.—Shed roof framing.

Figure 2-56, view A, shows that the length of the ridge on a dormer without sidewalls is equal to one-half the span of the dormer, less a shortening allowance one-half the thickness of the inside member of the upper double header. View B shows that the length of the ridge on a dormer with sidewalls is the length of the dormer rafter plate, plus one-half the span of the dormer, minus a shortening allowance one-half the thickness of the inside member of the upper double header.

SHED

A shed roof is essentially one-half of a gable roof. Like the full-length rafters in a gable roof, the full-length rafters in a shed roof are common rafters. However, the total run of a shed roof common rafter is equal to the span of the building minus the width of the top plate on the higher rafter-end wall (fig. 2-57). Also, the run of the overhang on the higher wall is measured from the inner edge of the top plate. With these exceptions, shed roof common rafters are laid out like gable roof common rafters. A shed roof common rafter has two bird's-mouths, but they are laid out just like the bird's-mouth on a gable roof common rafter.

For a shed roof, the height of the higher rafter-end wall must exceed the height of the lower by an amount equal to the total rise of a common rafter.

Figure 2-58 shows a method of framing a shed dormer. This type of dormer can be installed on almost any type of roof. There are three layout problems to be solved here: determining the total run of a dormer rafter; determining the angle of cut on the inboard ends of the dormer rafters; and determining the lengths of the dormer sidewall studs.

To determine the total run of a dormer rafter, divide the height of the dormer end wall, in inches, by the

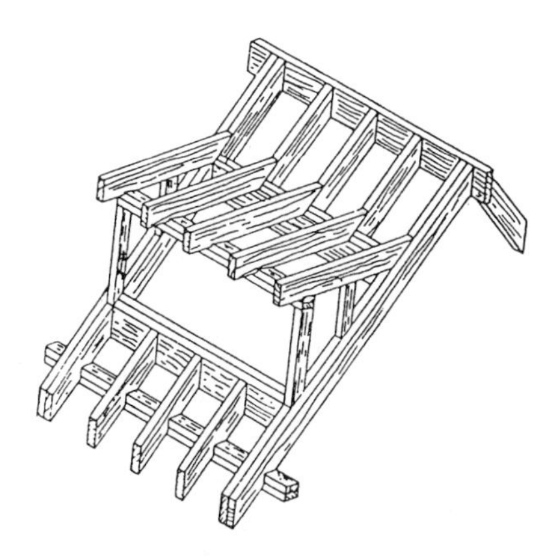

Figure 2-58.—Method of framing a shed dormer.

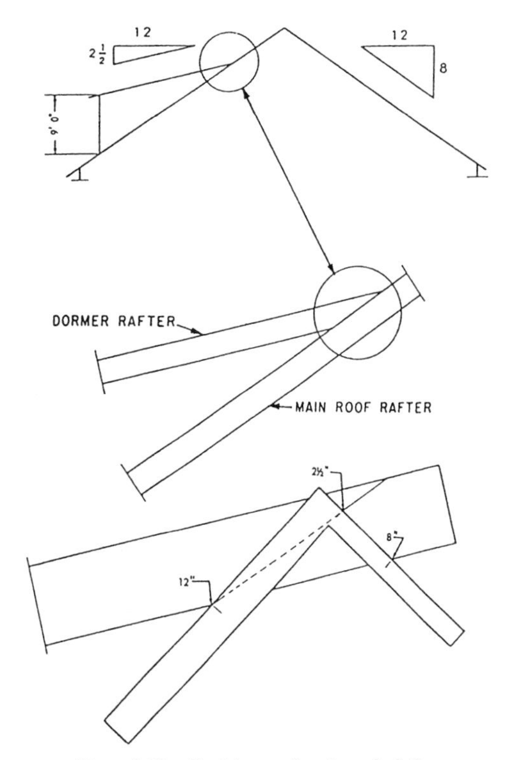

Figure 2-59.—Shed dormer framing calculation.

difference between the unit of rise of the dormer roof and the unit of rise of the main roof. Take the dormer shown in figure 2-59, for example. The height of the dormer end wall is 9 feet, or 108 inches. The unit of rise of the main roof is 8; the unit of rise of the dormer roof is 2 1/2; the difference is 5 1/2. The total run of a dormer rafter is therefore 108 divided by 5 1/2, or 19.63 feet. Knowing the total run and the unit of rise, you can figure the length of a dormer rafter by any of the methods already described.

As indicated in figure 2-59, the inboard ends of the dormer rafters must be cut to fit the slope of the main roof. To get the angle of this cut, set the square on the rafter to the cut of the main roof, as shown in the bottom view of figure 2-59. Measure off the unit of rise of the dormer roof from the heel of the square along the tongue as indicated and make a mark at this point. Draw the cutoff line through this mark from the 12-inch mark.

You figure the lengths of the sidewall studs on a shed dormer as follows: In the roof shown in figure 2-59, a dormer rafter raises 2 1/2 units for every 12 units of run. A main-roof common rafter rises 8 units for every 12 units of run. If the studs were spaced 12 inches OC, the length of the shortest stud (which is also the common difference of studs) would be the difference between 8 and 2 1/2 inches, or 5 1/2 inches. If the stud spacing is 16 inches, the length of the shortest stud is the value of x in the proportional equation $12{:}5\ 1/2{::}16{:}x$, or 7 5/16 inches. The shortest stud, then, will be 7 5/16 inches long. To get the lower end cutoff angle for studs, set the square on the stud to the cut of the main roof. To get the upper end cutoff angle, set the square to the cut of the dormer roof.

INSTALLATION

Rafter locations are laid out on wall plates and ridgeboards with matching lines and marked with X's, as used to lay out stud and joist locations. For a gable roof, the rafter locations are laid out on the rafter plates first. The locations are then transferred to the ridge by matching the ridge against a rafter plate.

Rafter Locations

The rafter plate locations of the ridge-end common rafters in an equal-pitch hip roof measure one-half of the span (or the run of a main-roof common rafter) away from the building corners. These locations, plus the rafter plate locations of the rafters lying between the ridge-end common rafters, can be transferred to the ridge by matching the ridgeboards against the rafter plates.

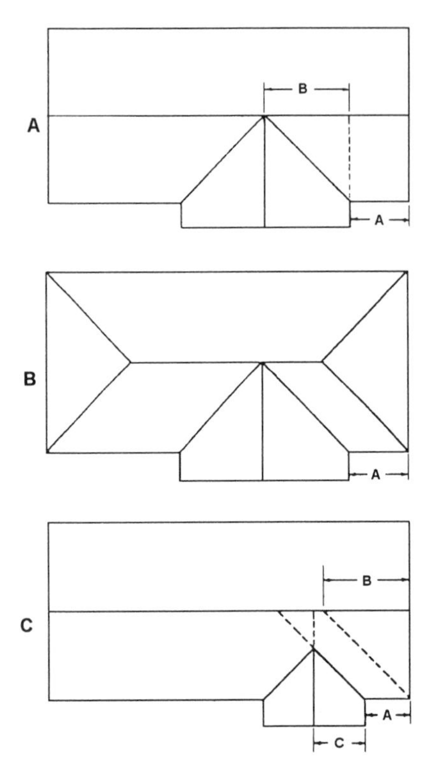

Figure 2-60.—Intersection ridge and valley rafter location layout.

The locations of additional ridge and valley rafters can be determined as indicated in figure 2-60. In an equal-span situation (views A and B), the valley rafter locations on the main-roof ridge lie alongside the addition ridge location. In view A, the distance between the end of the main-roof ridge and the addition ridge location is equal to A plus distance B, distance B being one-half the span of the addition. In view B, the distance between the line length end of the main-roof ridge and the addition ridge location is the same as distance A. In both cases, the line length of the addition ridge is equal to one-half the span of the addition, plus the length of the addition sidewall rafter plate.

Figure 2-60, view C, shows an unequal-span situation. If framing is by the long and short valley rafter method, the distance from the end of the main-roof ridge to the upper end of the longer valley rafter is equal to distance A plus distance B, distance B being one-half the span of the main roof. To determine the location of the inboard valley rafter, first calculate the unit length of the longer valley rafter, or obtain it from the unit length rafter tables. Let's suppose that the common rafter unit of rise is 8. In that case, the unit length of a valley rafter is 18.76.

The total run of the longer valley rafter between the shorter rafter tie-in and the rafter plate is the hypotenuse of a right triangle with the altitude and base equal to one-half of the span of the addition. Suppose the addition is 20 feet wide. Then, the total run is:

$$\sqrt{(10^2 = 10^2)} = 14.14 \text{ feet.}$$

You know that the valley rafter is 18.76 units long for every 16.97 units of run. The length of rafter for 14.14 feet of run must therefore be the value of x in the proportional equation *16.97:18.76::14.14:x*, or 15.63 feet. The location mark for the inboard end of the shorter valley rafter on the longer valley rafter, then, will be 15.63 feet, or 15 feet 7 9/16 inches, from the heel plumb cut line on the longer valley rafter. The length of the additional ridge will be equal to one-half the span of the addition, plus the length of the additional sidewall top plate, minus a shortening allowance one-half the 45° thickness of the longer valley rafter.

If framing is by the suspended ridge method, the distance between the suspension point on the main-roof and the end of the main-roof ridge is equal to distance A plus distance C. Distance C is one-half the span of the addition. The distance between the point where the inboard ends of the valley rafters (both short in this method of framing) tie into the addition ridge and the outboard end of the ridge is equal to one-half the span of the addition, plus the length of the additional ridge (which is equal to one-half of the span of the main roof), plus the length of the addition sidewall rafter plate.

Roof Frame Erection

Roof framing should be done from a scaffold with planking not less than 4 feet below the level of the main-roof ridge. The usual type of roof scaffold consists of diagonally braced two-legged horses, spaced about 10 feet apart and extending the full length of the ridge.

If the building has an addition, as much as possible of the main roof is framed before the addition framing is started. Cripples and jack rafters are usually left out until after the headers, hip rafters, valley rafters, and ridges to which they will be framed have been installed. For a gable roof, the two pairs of gable-end rafters and the ridge are usually erected first.

Two crewmembers, one at each end of the scaffold, hold the ridge in position. Another crewmember sets the gable-end rafters in place and toenails them at the rafter plate with 8d nails, one on each side of a rafter. Before we proceed any further, see table 2-1 as to the type and

Table 2-1.—Recommended Schedule for Nailing the Framing and Sheathing of a Wood-Frame Structure

JOINING	NAILING METHOD	NAILS		
		Nr.	Size	Placement
Header to joist	End-nail	3	16d	
Joist to sill or girder	Toenail	2 3	10d or 8d	
Header and stringer joist to sill	Toenail		10d	16 in. OC
Bridging to joist	Toenail each end	2	8d	
Ledger strip to beam, 2 in. thick		3	16d	At each joist
Subfloor, boards: 1 by 6 in. and smaller 1 by 8 in.		 2 3	 8d 8d	 To each joist To each joist
Subfloor, plywood: At edges At intermediate joists			 8d 8d	 6 in. OC 8 in. OC
Subfloor (2 by 6 in., T&G) to joist or girder	Blind nail (casing) and face-nail	2	16d	
Soleplate to stud, horizontal assembly	End-nail	2	16d	At each stud
Top plate to stud	End-nail	2	16d	
Stud to soleplate	Toenail	4	8d	
Soleplate to joist or blocking	Face-nail		16d	16 in. OC
Doubled studs	Face-nail, stagger		10d	16 in. OC
End stud of intersecting wall to exterior wall stud	Face-nail		16d	16 in. OC
Upper top plate to lower top plate	Face-nail		16d	16 in. OC
Upper top plate, laps and intersections	Face-nail	2	16d	
Continuous header, two pieces, each edge			12d	12 in. OC
Ceiling joist to top wall plates	Toenail	3	8d	
Ceiling joist laps at partition	Face-nail	4	16d	
Rafter to top plate	Toenail	2	8d	
Rafter to ceiling joist	Face-nail	5	10d	
Rafter to valley or hip rafter	Toenail	3	10d	
Ridgeboard to rafter	End-nail	3	10d	
Rafter to rafter through ridgeboard	Toenail Edge-nail	2 4	10d 8d	
Collar tie to rafter: 2-in. member 1-in. member	 Face-nail Face-nail	 2 3	 12d 8d	
1-in. diagonal let-in brace to each stud and plate (four nails at top)		2	8d	
Built-up corner studs: Studs to blocking Intersecting stud to corner studs	 Face-nail Face-nail	 2	 10d 6d	 Each side 12 in. OC
Built-up girders and beams, three or more members	Face-nail		20d	32 in. OC, each side
Wall sheathing: 1 by 8 in. or less, horizontal 1 by 6 in. or greater, diagonal	 Face-nail Face-nail	 2 3	 8d 8d	 At each stud At each stud

Table 2-1.—Recommended Schedule for Nailing the Framing and Sheathing of a Wood-Frame Structure—Continued

JOINING	NAILING METHOD	NAILS		
		Nr.	Size	Placement
Wall sheathing, vertically applied plywood: 3/8 in. and less thick 1/2 in. and over thick	 Face-nail Face-nail		 6d 8d	 6 in. edge 12 in. intermediate
Wall sheathing, vertically applied fiberboard: 1/2 in. thick 25/32 in. thick	 Face-nail Face-nail		1 1/2-in. roofing nail 3 in. from edge and 1 3/4-in. roofing nail 6 in. intermediately spaced	
Roof sheathing, boards, 4-, 6-, 8-in. width	Face-nail	2	8d	At each rafter
Roof sheathing, plywood: 3/8 in. and less thick 1/2 in. and over thick	 Face-nail Face-nail		 6d 8d	 6 in. edge and 12 in. intermediate

size nails used in roof framing erection. Each crew-member on the scaffold then end-nails the ridge to the end of the rafter. They then toenail the other rafter to the ridge and to the first rafter with two 10d nails, one on each side of the rafter.

Temporary braces, like those for a wall, should be set up at the ridge ends to hold the rafter approximately plumb, after which the rafters between the end rafters should be erected. The braces should then be released, and the pair of rafters at one end should be plumbed with a plumb line, fastened to a stick extended from the end of the ridge. The braces should then be reset, and they should be left in place until enough sheathing has been installed to hold the rafters plumb. Collar ties, if any, are nailed to common rafters with 8d nails, three to each end of a tie. Ceiling-joist ends are nailed to adjacent rafters with 10d nails.

On a hip roof, the ridge-end common rafters and ridges are erected first, in about the same manner as for a gable roof. The intermediate common rafters are then filled in. After that, the ridge-end common rafters extending from the ridge ends to the midpoints on the end walls are erected. The hip rafters and hip jacks are installed next. The common rafters in a hip roof do not require plumbing. When correctly cut and installed, hip rafters will bring the common rafters to plumb. Hip rafters are toenailed to plate corners with 10d nails. Hip jacks are toenailed to hip rafters with 10d nails.

For an addition or dormer, the valley rafters are usually erected first. Valley rafters are toenailed with 10d nails. Ridges and ridge-end common rafters are erected next, other addition common rafters next, and valley and cripple jacks last. A valley jack should be held in position for nailing, as shown in figure 2-61. When properly nailed, the end of a straightedge laid along the top edge of the jack should contact the center line of the valley rafter, as shown.

TRUSSES

LEARNING OBJECTIVE: Upon completing this section, you should be able to describe the types and parts of roof trusses, and explain procedures for fabricating, handling, and erecting them.

Roof truss members are usually connected at the joints by gussets. Gussets are made of boards, plywood, or metal. They are fastened to the truss by nails, screws, bolts, or adhesives. A roof truss is capable of supporting loads over a long span without intermediate supports.

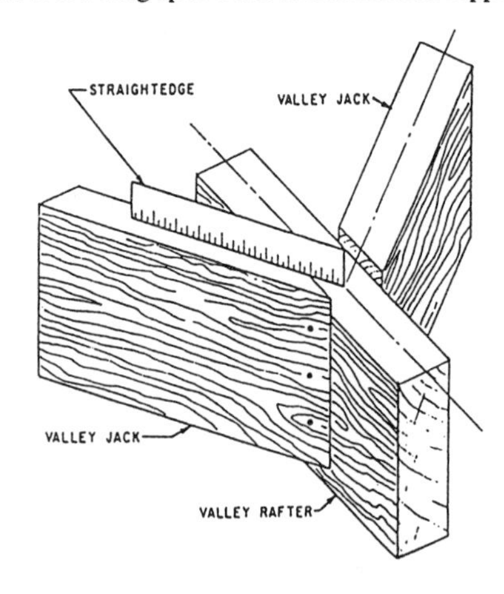

Figure 2-61.—Correct position for nailing a valley jack rafter.

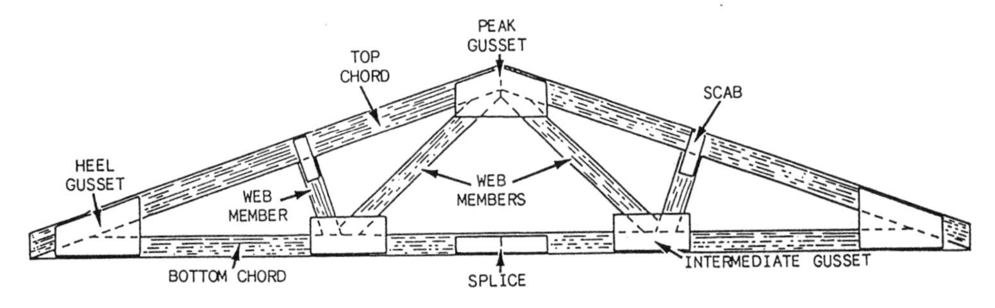

Figure 2-62.—Truss construction.

Roof trusses save material and on-site labor costs. It is estimated that a material savings of about 30 percent is made on roof members and ceiling joists. When you are building with trusses, the double top plates on interior partition walls and the double floor joists under interior bearing partitions are not necessary. Roof trusses also eliminate interior bearing partitions because trusses are self-supporting.

The basic components of a roof truss are the top and bottom chords and the web members (fig. 2-62). The top chords serve as roof rafters. The bottom chords act as ceiling joists. The web members run between the top and bottom chords. The truss parts are usually made of 2- by 4-inch or 2- by 6-inch material and are tied together with metal or plywood gusset plates. Gussets shown in this figure are made of plywood.

TYPES

Roof trusses come in a variety of shapes. The ones most commonly used in light framing are the king post, the W-type (or fink), and the scissors. An example of each is shown in figure 2-63.

King Post

The simplest type of truss used in frame construction is the king-post truss. It consists of top and bottom chords and a vertical post at the center.

W-Type (Fink)

The most widely used truss in light-frame construction is the W-type (fink) truss. It consists of top and bottom chords tied together with web members. The W-type truss provides a uniform load-carrying capacity.

Scissors

The scissor truss is used for building with sloping ceilings. Many residential, church, and commercial buildings require this type of truss. Generally, the slope of the bottom chord of a scissor truss equals one-half the slope of the top chord.

DESIGN PRINCIPLES

A roof truss is an engineered structural frame resting on two outside walls of a building. The load carried by the truss is transferred to these outside walls.

Weight and Stress

The design of a truss includes consideration of snow and wind loads and the weight of the roof itself. Design also takes into account the slope of the roof. Generally, the flatter the slope, the greater the stresses. Flatter slopes, therefore, require larger members and stronger connections in roof trusses.

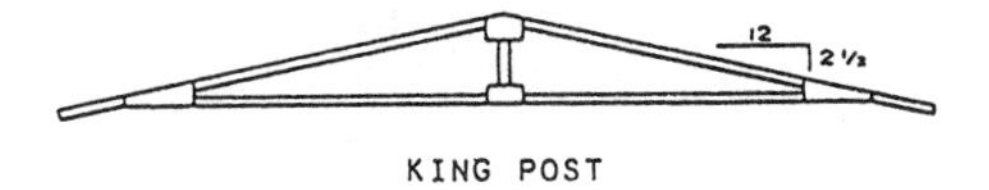

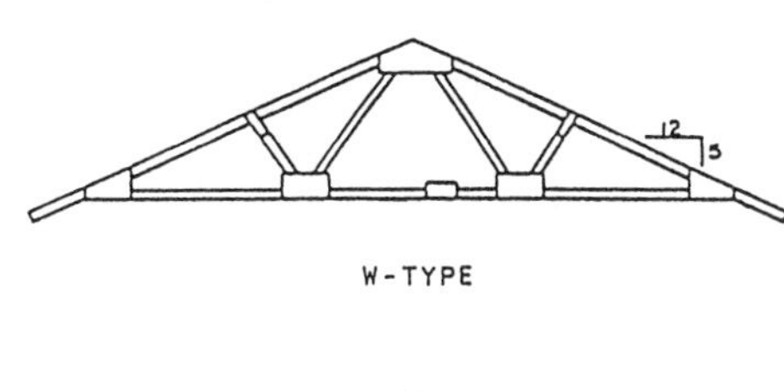

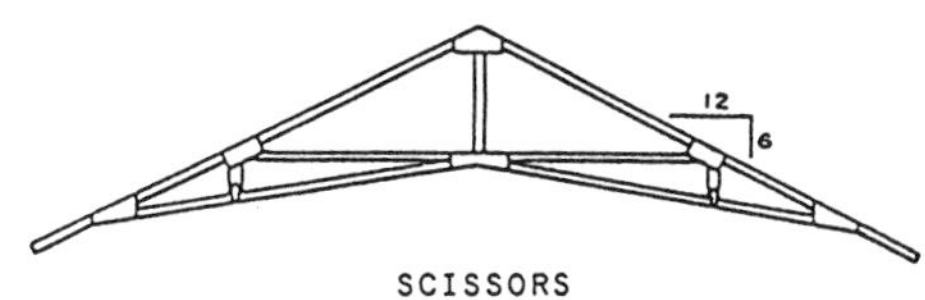

Figure 2-63.—Truss types.

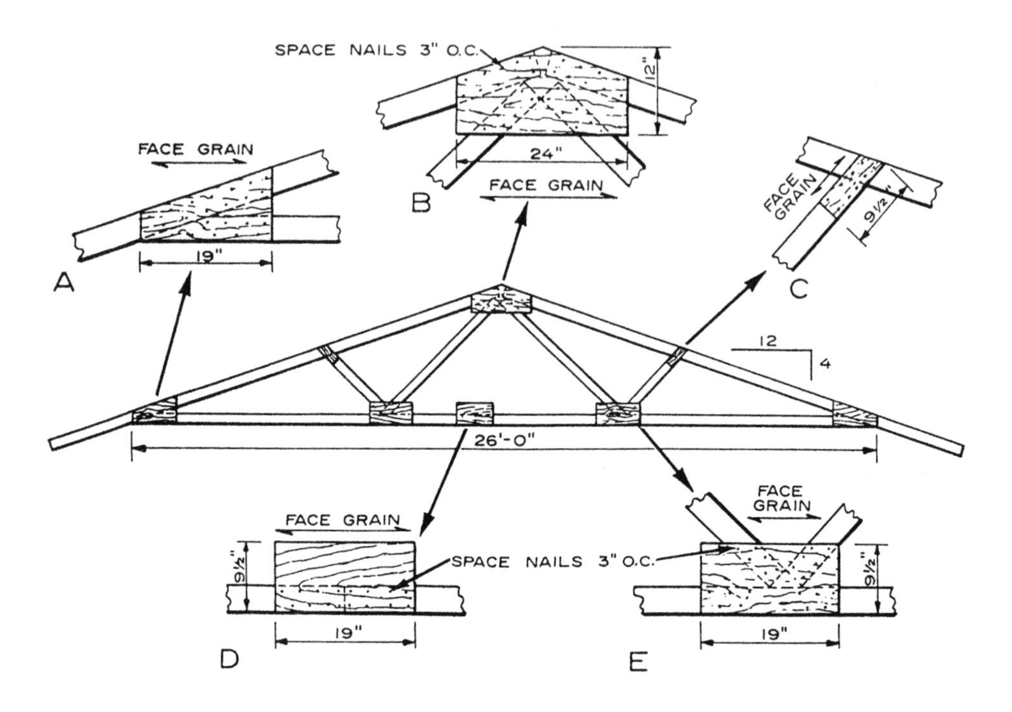

Figure 2-64.—Plywood gussets.

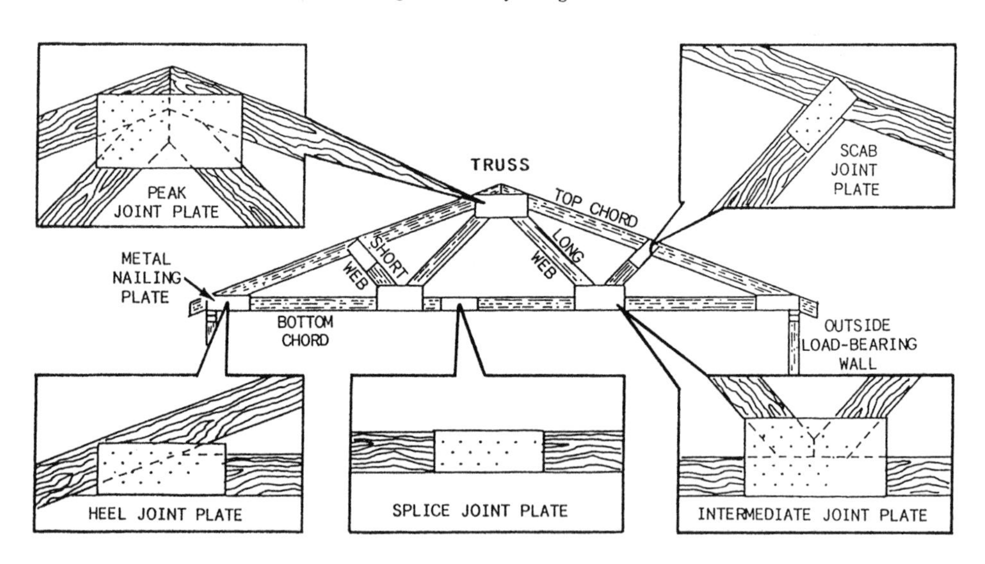

Figure 2-65.—Metal gusset plates.

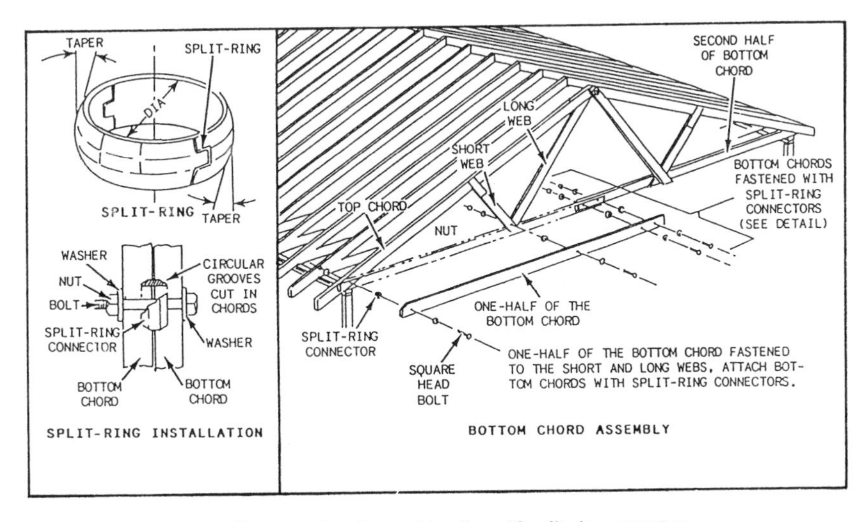

Figure 2-66.—Truss members fastened together with split-ring connectors.

A great majority of the trusses used are fabricated with plywood gussets (fig. 2-64, views A through E), nailed, glued, or bolted in place. Metal gusset plates (fig. 2-65) are also used. These are flat pieces usually manufactured from 20-gauge zinc-coated or galvanized steel. The holes for the nails are prepunched. Others are assembled with split-ring connectors (fig. 2-66) that prevent any movement of the members. Some trusses are designed with a 2- by 4-inch soffit return at the end of each upper chord to provide nailing for the soffit of a wide box cornice.

Tension and Compression

Each part of a truss is in a state of either tension or compression (see fig. 2-67). The parts in a state of tension are subjected to a pulling-apart force. Those under compression are subjected to a pushing-together force. The balance of tension and compression gives the truss its ability to carry heavy loads and cover wide spans.

In view A of figure 2-67, the ends of the two top chords (A-B and A-C) are being pushed together (compressed). The bottom chord prevents the lower ends (B and C) of the top chords from pushing out; therefore, the bottom chord is in a pulling-apart state (tension). Because the lower ends of the top chords cannot pull apart, the peak of the truss (A) cannot drop down.

In view B, the long webs are secured to the peak of the truss (A) and also fastened to the bottom chord at points D and E. This gives the bottom chord support along the outside wall span. The weight of the bottom chord has a pulling-apart effect (tension) on the long webs.

In view C, the short webs run from the intermediate points F and G of the top chord to points D and E of the bottom chord. Their purpose is to provide support to the top chord. This exerts a downward, pushing-together force (compression) on the short web.

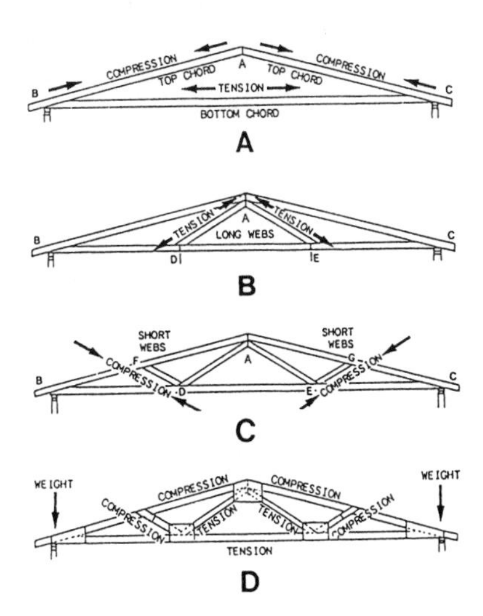

Figure 2-67.—Tension and compression in a truss.

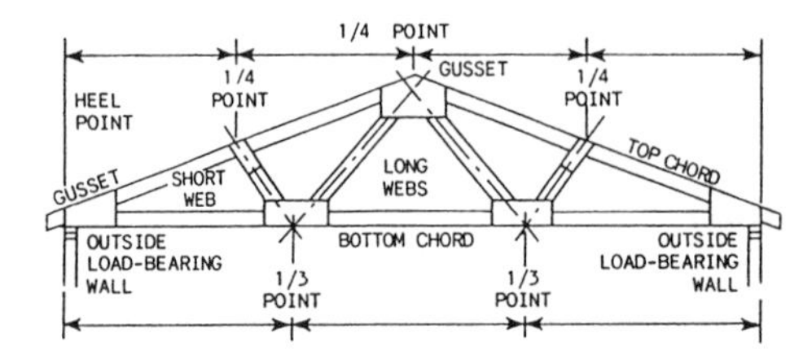

Figure 2-68.—Layout for a W-type (fink) truss.

In view D, you can see that the overall design of the truss roof transfers the entire load (roof weight, snow load, wind load, and so forth) down through the outside walls to the foundation.

Web members must be fastened at certain points along the top and bottom chords in order to handle the stress and weight placed upon the truss. A typical layout for a W-type (fink) truss is shown in figure 2-68. The points at which the lower ends of the web members fasten to the bottom chord divide the bottom chord into

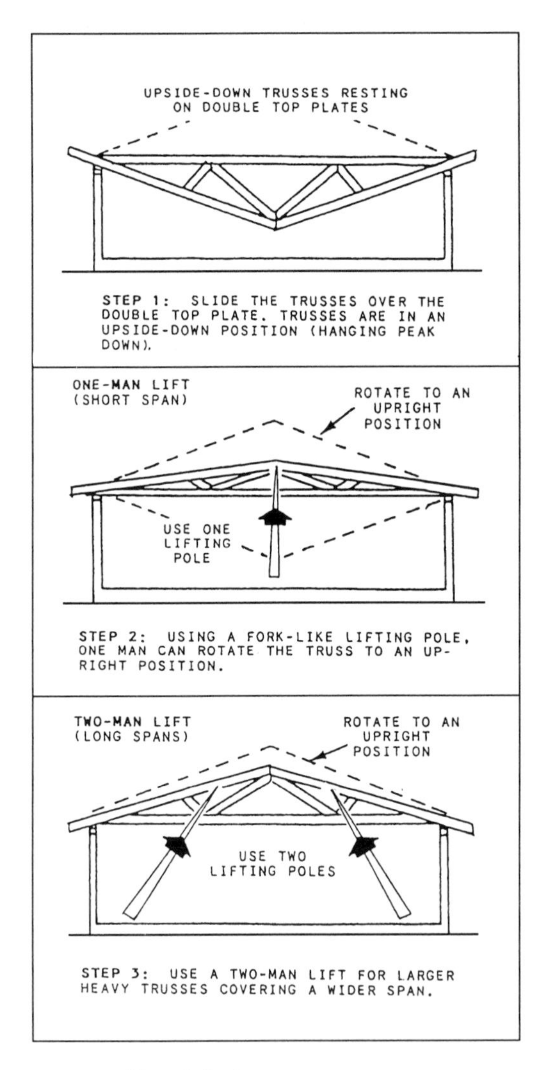

Figure 2-69.—Placing trusses by hand.

three equal parts. Each short web meets the top chord at a point that is one-fourth the horizontal distance of the bottom chord.

FABRICATION

The construction features of a typical W-truss are shown in figure 2-64. Also shown are gusset cutout sizes and nailing patterns for nail-gluing. The span of this truss is 26 feet and roof cut is 4/12. When spaced 24 inches apart and made of good- quality 2- by 4-inch members, the trusses should be able to support a total roof load of 40 pounds per square foot.

Gussets for light wood trusses are cut from 3/8- or 1/2-inch standard plywood with an exterior glue line, or from sheathing-grade exterior plywood. Glue is spread on the clean surfaces of the gussets and truss members. Staples are used to supply pressure until the glue is set. Under normal conditions and where the relative humidity of air in attic spaces tends to be high, a resorcinol glue is applied. In areas of low humidity, a casein or similar glue is used. Two rows of 4d nails are used for either the 3/8- or 1/2-inch-thick gusset. The nails are spaced so that they are 3 inches apart and 3/4 inches from the edges of the truss members. Gussets are nail-glued to both sides of the truss.

Plywood-gusset, king-post trusses are limited to spans of 26 feet or less if spaced 24 inches apart and fabricated with 2- by 4-inch members and a 4/12 roof cut. The spans are somewhat less than those allowed for W-trusses having the same-sized members. The shorter span for the king-post truss is due, in part, to the unsupported upper chord. On the other hand, because it has more members than the king-post truss and distances between connections are shorter, the W-truss can span up to 32 feet without intermediate support, and its members can be made of lower grade lumber.

INSTALLATION

Trusses are usually spaced 24 inches OC. They must be lifted into place, fastened to the walls, and braced. Small trusses can be placed by hand, using the procedure shown in figure 2-69. Builders are required on the two opposite walls to fasten the ends of the trusses. One or two workers on the floor below can push the truss to an upright position. If appropriate equipment is available, use it to lift trusses into place.

In handling and storing completed trusses, avoid placing unusual stresses on them. They were designed to carry roof loads in a vertical position; thus it is important that they be lifted and stored upright. If they must be handled in a flat position, enough support should be used along their length to minimize bending deflections. Never support the trusses only at the center or only at each end when they are in a flat position.

Bracing

After the truss bundles have been set on the walls, they are moved individually into position, nailed down, and temporarily braced. Without temporary bracing, a truss may topple over, cause damage to the truss, and possibly injure workers. A recommended procedure for bracing trusses as they are being set in place is shown in figure 2-70. Refer to the figure as you study the following steps:

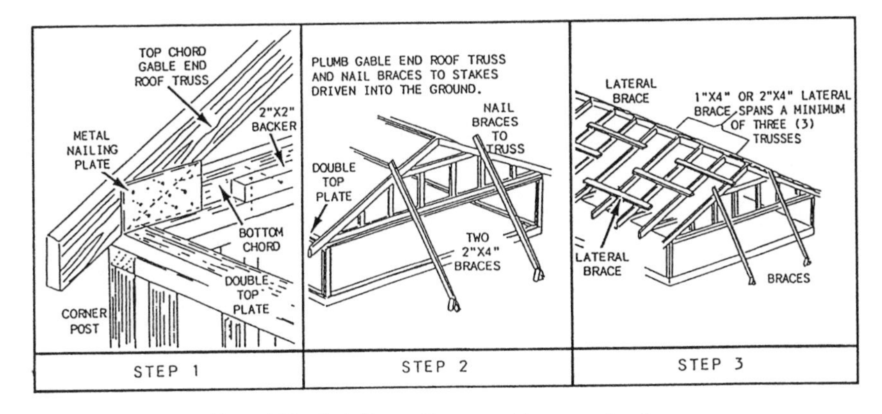

Figure 2-70.—Installing roof trusses and temporary bracing.

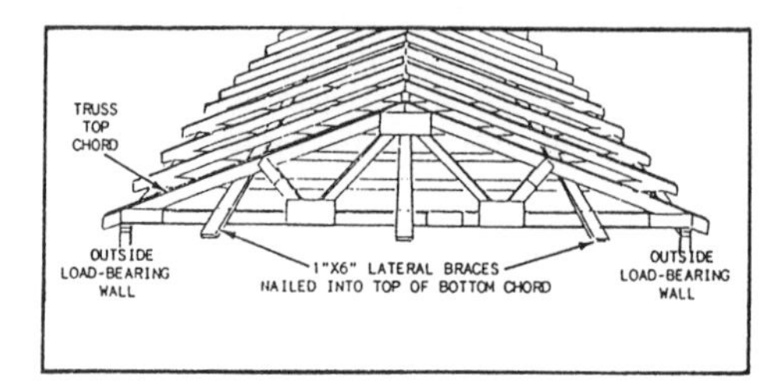

Figure 2-71.—Permanent lateral bracing in a truss.

Step 1. Position the first roof truss. Fasten it to the double top plate with toenails or metal anchor brackets. A 2- by 2-inch backer piece is sometimes used for additional support.

Step 2. Fasten two 2 by 4 braces to the roof truss. Drive stakes at the lower ends of the two braces. Plumb the truss and fasten the lower ends of the braces to the stakes driven into the ground.

Step 3. Position the remaining roof trusses. As each truss is set in place, fasten a lateral brace to tie it to the preceding trusses. Use 1 by 4 or 2 by 4 material for lateral braces. They should overlap a minimum of three trusses. On larger roofs, diagonal bracing should be placed at 20-foot intervals.

The temporary bracing is removed as the roof sheathing is nailed. Properly nailed plywood sheathing is sufficient to tie together the top chords of the trusses. Permanent lateral bracing of 1- by 4-inch material is recommended at the bottom chords (fig. 2-71). The braces are tied to the end walls and spaced 10 feet OC.

Anchoring Trusses

When fastening trusses, you must consider resistance to uplift stresses as well as thrust. Trusses are fastened to the outside walls with nails or framing anchors. The ring-shank nail provides a simple connection that resists wind uplift forces. Toenailing is sometimes done, but this is not always the most satisfactory method. The heel gusset and a plywood gusset or metal gusset plate are located at the wall plate and make toenailing difficult. However, two 10d nails on each side of the truss (fig. 2-72, view A) can be used in nailing the lower chord to the plate. Predrilling may be necessary to prevent splitting. Because of the single-member thickness of the truss and the presence of gussets at the wall plates, it is usually a good idea to use some type of metal connector to supplement the toenailings.

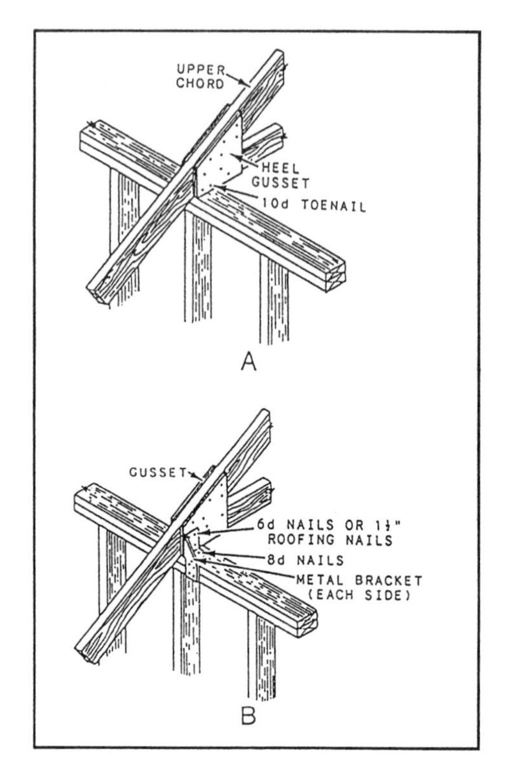

Figure 2-72.—Fastening trusses to the wall plate: A. Toenailing; B. Metal bracket.

The same types of metal anchors (fig. 2-72, view B) used to tie regular rafters to the outside walls are equally effective for fastening the ends of the truss. The brackets are nailed to the wall plates at the side and top with 8d nails and to the lower chords of the truss with 6d or 1 1/2-inch roofing nails.

INTERIOR PARTITION INSTALLATION

Where partitions run parallel to, but between, the bottom truss chords, and the partitions are erected before the ceiling finish is applied, install 2- by 4-inch blocking

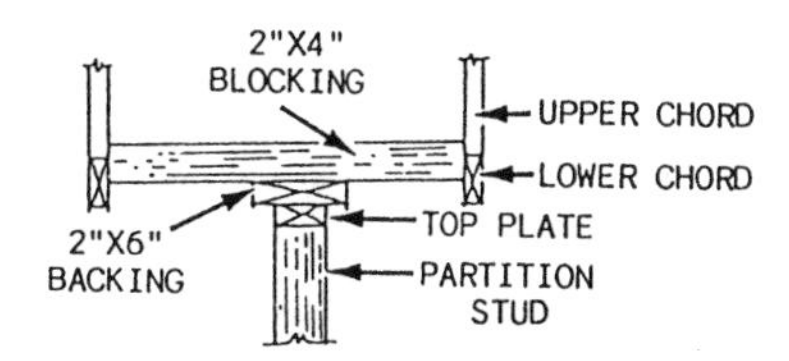

Figure 2-73.—Construction details for partitions that run parallel to the bottom truss chords.

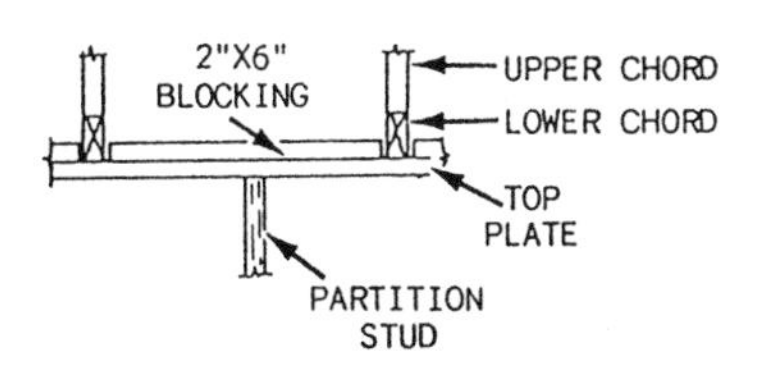

Figure 2-74.—Construction details for partitions that run at right angles to the bottom of the truss chords.

between the lower chords (fig. 2-73). This blocking should be spaced not over 4 feet OC. Nail the blocking to the chords with two 16d nails in each end. To provide nailing for lath or wallboard, nail a 1- by 6-inch or 2- by 6-inch continuous backer to the blocking. Set the bottom face level with the bottom of the lower truss chords.

When partitions are erected after the ceiling finish is applied, 2- by 4-inch blocking is set with the bottom edge level with the bottom of the truss chords. Nail the blocking with two 16d nails in each end.

If the partitions run at right angles to the bottom of the truss chords, the partitions are nailed directly to lower chord members. For applying ceiling finish, nail 2- by 6-inch blocking on top of the partition plates between the trusses (fig. 2-74).

RECOMMENDED READING LIST

NOTE

Although the following reference was current when this TRAMAN was published, its continued currency cannot be assured. You therefore need to ensure that you are studying the latest revision.

Basic Roof Framing, Benjamin Barnow, Tab Books, Inc., Blue Ridge Summit, Pa., 1986.

CHAPTER 3

ROOF CONSTRUCTION AND TRIM CARPENTRY

The previous chapters have dealt with framing wood structures, including joists, studs, rafters, and other structural members. These constitute "rough carpentry" and are the main supports of a wood-frame structure. (Subflooring and wall and roof sheathing strengthen and brace the frame.)

The remaining work on the structure involves installing the nonstructural members. This work, referred to as "finish carpentry," includes installing the roof covering, door and window frames, and the doors and windows themselves. Some nonstructural members are purely ornamental, such as casings on doors and windows, and the moldings on cornices and inside walls. Installation of purely ornamental members is known as trim carpentry.

Finish carpentry is divided into exterior and interior finish. Exterior finish materials consist of roof sheathing, exterior trim, roof coverings, outside wall covering, and exterior doors and windows. Exterior finish materials are installed after the rough carpentry has been completed. Examples of interior finish materials include all coverings applied to the rough walls, ceilings, and floors. We will cover these topics in a later chapter.

In this chapter, we'll cover the exterior finishing of roofs. In the next chapter, we'll examine the exterior finishing of walls.

ROOF SHEATHING

LEARNING OBJECTIVE: Upon completing this section, you should be able to identify various types of roof sheathing and describe their installation requirements.

Roof sheathing covers the rafters or roof joists. The roof sheathing is a structural element and, therefore, part of the framing. Sheathing provides a nailing base for the finish roof covering and gives rigidity and strength to the roof framing. Lumber and plywood roof sheathing are the most commonly used materials for pitched roofs. Plank or laminated roof decking is sometimes used in structures with exposed ceilings. Manufactured wood fiber roof decking is also adaptable to exposed ceiling applications.

LUMBER

Roof sheathing boards are generally No. 3 common or better. These are typically softwoods, such as Douglas fir, redwood, hemlock, western larch, fir, and spruce. If you're covering the roof with asphalt shingles, you should use only thoroughly seasoned wood for the sheating. Unseasoned wood will dry and shrink, which may cause the shingles to buckle or lift along the full length of the sheathing board.

Nominal 1-inch boards are used for both flat and pitched roofs. Where flat roofs are to be used for a deck or a balcony, thicker sheathing boards are required. Board roof sheathing, like board wall sheathing and subflooring, can be laid either horizontally or diagonally. Horizontal board sheathing may be closed (laid with no space between the courses) or open (laid with space between the courses). In areas subject to wind-driven snow, a solid roof deck is recommended.

Installation

Roof boards used for sheathing under materials requiring solid, continuous support must be laid closed. This includes such applications as asphalt shingles, composition roofing, and sheet-metal roofing. Closed roof sheathing can also be used for wood shingles. The boards are nominal 1 inch by 8 inches and may be square-edged, dressed and matched, shiplapped, or tongue and groove. Figure 3-1 shows the installation of both closed and open lumber roof sheathing.

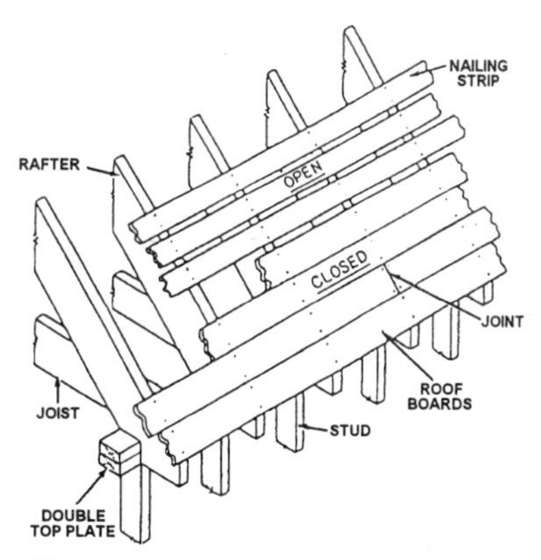

Figure 3-1.—Closed and open roof sheathing.

Open sheathing can be used under wood shingles or shakes in blizzard-free areas or damp climates. Open sheathing usually consists of 1- by 4-inch strips with the on-center (OC) spacing equal to the shingle weather exposure, but not over 10 inches. (A 10-inch shingle lapped 4 inches by the shingle above it is said to be laid 6 inches to the weather.) When applying open sheathing, you should lay the boards without spacing to a point on the roof above the overhang.

Nailing

Nail lumber roof sheathing to each rafter with two 8-penny (8d) nails. Joints must be made on the rafters just as wall sheathing joints must be made over the studs. When tongue-and-groove boards are used, joints may be made between rafters. In no case, however, should the joints of adjoining boards be made over the same rafter space. Also, each board should bear on at least two rafters.

PLYWOOD

Plywood offers design flexibility, construction ease, economy, and durability. It can be installed quickly over large areas and provides a smooth, solid base with a minimum number of joints. A plywood deck is equally effective under any type of shingle or built-up roof. Waste is minimal, contributing to the low in-place cost.

Plywood is one of the most common roof sheathing materials in use today. It comes in 4- by 8-foot sheets in a variety of thicknesses, grades, and qualities. For sheathing work, a lower grade called CDX is usually used. A large area (32 square feet) can be applied at one time. This, plus its great strength relative to other sheathing materials, makes plywood a highly desirable choice.

The thickness of plywood used for roof sheathing is determined by several factors. The distance between rafters (spacing) is one of the most important. The larger the spacing, the greater the thickness of sheathing that should be used. When 16-inch OC rafter spacing is used, the minimum recommended thickness is 3/8 inch. The type of roofing material to be applied over the sheathing also plays a role. The heavier the roof covering, the thicker the sheathing required. Another factor determining sheathing thickness is the prevailing weather. In areas where there are heavy ice and snow loads, thicker sheathing is required. Finally, you have to consider allowable dead and live roof loads established by calculations and tests.

These are the controlling factors in the choice of roof sheathing materials. Recommended spans and plywood grades are shown in table 3-1.

Installation

Plywood sheathing is applied after rafters, collar ties, gable studs, and extra bracing (if necessary) are in place. Make sure there are no problems with the roof frame. Check rafters for plumb, make sure there are no badly deformed rafters, and check the tail cuts of all the rafters for alignment. The crowns on all the rafters should be in one direction—**up**.

Figure 3-2 shows two common methods of starting the application of sheathing at the roof eaves. In view A, the sheathing is started flush with the tail cut of the rafters. Notice that when the fascia is placed, the top edge of the fascia is even with the top of the sheathing. In view B, the sheathing overlaps the tail end of the rafter by the thickness of the fascia material. You can see that the edge of the sheathing is flush with the fascia.

If you choose to use the first method (view A) to start the sheathing, measure the two end rafters the width of the plywood panel (48 inches). From the rafter tail ends, and using the chalk box, strike a line on the top edge of all the rafters. If you use the second method,

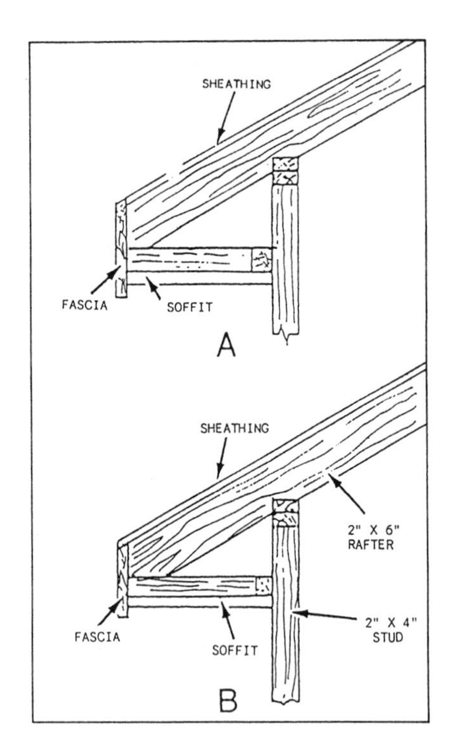

Figure 3-2.—Two methods of starting the first sheet of roof sheathing at the eaves of a roof: A. Flush with rafter; B. Overlapping rafter.

Table 3-1.—Plywood Roof Sheathing Application Specifications

Plywood roof sheathing [1, 2, 3] (Plywood continuous over two or more spans; grain of face plies across supports)

PANEL IDENTIFICATION INDEX	PLYWOOD THICKNESS (inch)	MAX. SPAN (inches)[4]	UNSUPPORTED EDGE-MAX. LENGTH (inches)[5]	ALLOWABLE ROOF LOADS (psf)[6, 7] Spacing of Supports (inches center to center)										
				12	16	20	24	30	32	36	42	48	60	72
12/0	5/16	12	12	100 (130)										
16/0	5/16, 3/8	16	16	130 (170)	55 (75)									
20/0	5/16, 3/8	20	20		85 (110)	45 (55)								
24/0	3/8, 1/2	24	24		150 (160)	75 (100)	45 (60)							
30/12	5/8	30	26			145 (165)	85 (110)	40 (55)						
32/16	1/2, 5/8	32	28				90 (105)	45 (60)	40 (50)					
36/16	3/4	36	30				125 (145)	65 (85)	55 (70)	35 (50)				
42/20	5/8, 3/4 7/8	42	32					80 (105)	65 (90)	45 (60)	(35) (40)			
48/24	3/4, 7/8	48	36						105 (115)	75 (90)	55 (55)	40 (40)		
2-4-1	1 1/8	72	48							175 (175)	105 (105)	80 (80)	50 (50)	30 (35)
1 1/8 G1&2	1 1/8	72	48							145 (145)	85 (85)	65 (65)	40 (40)	30 (30)
1 1/4 G3&4	1 1/4	72	48							160 (165)	95 (95)	75 (75)	45 (45)	25 (35)

Notes:
1. Applies to Standard, Structural I and II and C-C grades only
2. For applications where the roofing is to be guaranteed by a performance bond, recommendations may differ somewhat from these values. Contact American Plywood Association for bonded roof recommendations
3. Use 6d common smooth, ring-shank or spiral thread nails for plywood 1/2" thick or less, and 8d common smooth, ring-shank or spiral thread for panels over 1/2" but not exceeding 1" thick (if ring-shank or spiral thread nails same diameter as common). Use 8d ring-shank or spiral thread or 10d common smooth shank nails for 2-4-1, 1 1/8" and 1 1/4" panels. Space nails 6" at panel edges and 12" at intermediate supports except those where spans are 48" or more, nails must be 6" at all supports
4. These spans must not be exceeded for any load conditions
5. Provide adequate blocking, tongue-and-groove edges or other suitable edge support, such as ply clips when spans exceed indicated value. Use two ply clips for 48" or greater spans and one for lesser spans
6. Uniform load deflection limitation: 1/180th of the span under live load plus dead load. 1/240th under live load only. In the table, allowable live load is shown above with allowable total load shown below in parentheses
7. Allowable roof loads were established by laboratory tests and calculations assuming evenly distributed loads

measure the width of the panel minus the actual thickness of the fascia material. Use this chalk line to position the upper edge of the sheathing panels. If the roof rafters are at right angles to the ridge and plates, this line will place the sheathing panels parallel to the outer ends of the rafters.

WARNING

Be particularly careful when handling sheet material on a roof during windy conditions. You may be thrown off balance and possibly off the roof entirely. Also, the sheet may be blown off the roof and strike someone.

Placing

Notice in figure 3-2 that sheathing is placed before the trim is applied. Sheathing is always placed from the lower (eaves) edge of the roof up toward the ridge. It can be started from the left side and worked toward the right, or you can start from the right and work toward the left. Usually, it is started at the same end of the house from which the rafters were laid out.

The first sheet of plywood is a full 4- by 8-foot panel. The top edge is placed on the chalk line. If the sheathing is started from the left side of the roof, make sure the right end falls in the middle of a rafter. This must be done so that the left end of the next sheet has a surface upon which it can bear weight and be nailed.

The plywood is placed so that the grain of the top ply is at right angles (perpendicular) to the rafters. Placing the sheathing in this fashion spans a greater number of rafters, spreads the load, and increases the strength of the roof. Figure 3-3 shows plywood panels laid perpendicular to the rafters with staggered joints. Note that a small space is left between sheets to allow for expansion.

The sheets that follow are butted against spacers until the opposite end is reached. If there is any panel hanging over the edge, it is trimmed after the panel is fastened in place. A chalk line is snapped on the sheathing flush with the end of the house, and the panel is then cut with a circular saw. Read the manufacturer's specification stamp and allow proper spacing at the ends and edges of the sheathing. This will compensate for any swelling that might take place with changes in moisture content.

The cutoff piece of sheathing can be used to start the second course (row of sheathing), provided it spans two or more rafters. If it doesn't span two rafters, start the second course with a half sheet (4 by 4) of plywood.

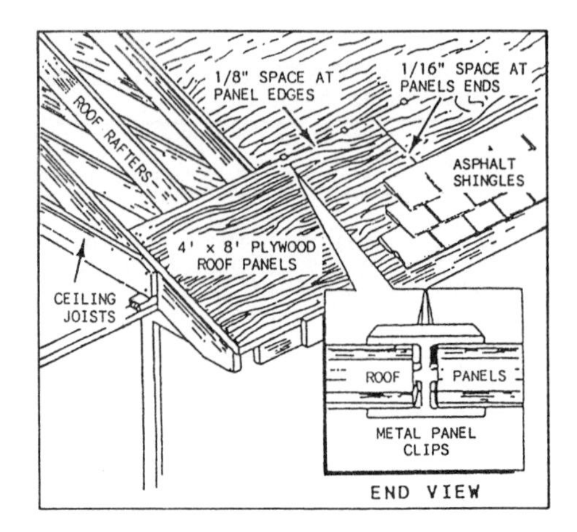

Figure 3-3.—Plywood roofing panel installation.

It is important to stagger all vertical joints. All horizontal joints need blocking placed underneath or a metal clip (ply clip). Ply clips (H clips or panel clips) are designed to strengthen the edges of sheathing panels between supports or rafters. The use of clips is determined by the rafter spacing and specifications (see figure 3-3).

The pattern is carried to the ridge. The final course is fastened in place, a chalk line is snapped at the top edge of the rafters, and the extra material cut off. The opposite side of the roof is then sheeted using the same pattern.

Nailing

When nailing plywood sheathing, follow the project specifications for nailing procedures. Use 6d common smooth, ring-shank or spiral thread nails for plywood 1/2 inch thick or less. For plywood more than 1/2 inch but not exceeding 1 inch thick, use 8d common smooth, ring-shank or spiral thread nails. When using a nail gun for roof sheathing, follow all applicable safety regulations.

ROOF DECKING

In this section, we'll discuss the two most common types of roof decking you will encounter as a Builder: plank and wood fiber.

Plank

Plank roof decking, consisting of 2-inch (and thicker) tongue-and-groove planking, is commonly used for flat or low-pitched roofs in post-and-beam construction. Single tongue-and-groove decking in nominal 2 by 6 and 2 by 8 sizes is available with the V-joint pattern only.

Decking comes in nominal widths of 4 to 12 inches and in nominal thicknesses of 2 to 4 inches. Three- and 4-inch roof decking is available in random lengths of 6 to 20 feet or longer (odd and even).

Laminated decking is also available in several different species of softwood lumber: Idaho white pine, inland red cedar, Idaho white fir, ponderosa pine, Douglas fir, larch, and yellow pine. Because of the laminating feature, this material may have a facing of one wood species and back and interior laminations of different woods. It is also available with all laminations of the same species. For all types of decking, make sure the material is the correct thickness for the span by checking the manufacturer's recommendations. Special load requirements may reduce the allowable spans. Roof decking can serve both as an interior ceiling finish and as a base for roofing. Heat loss is greatly reduced by adding fiberboard or other rigid insulation over the wood decking.

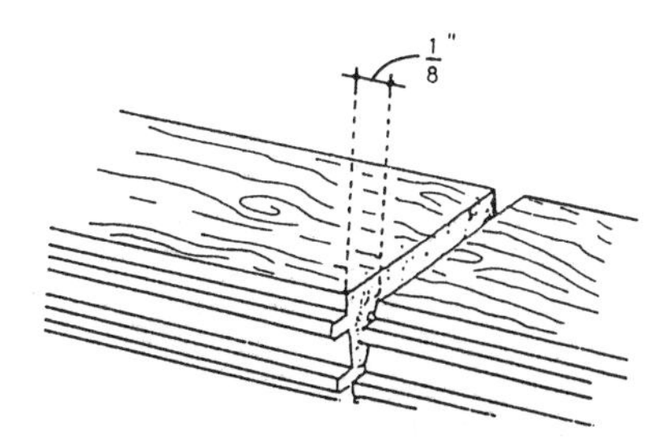

Figure 3-4.—Ends of roof decking cut at a 2° angle.

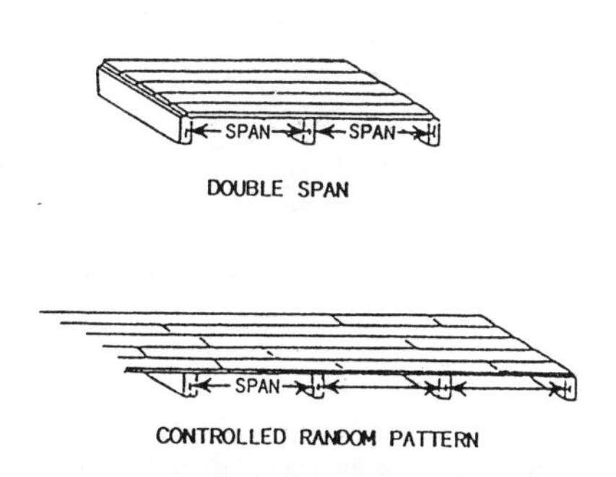

Figure 3-5.—Plank decking span arrangements.

INSTALLATION.—Roof decking applied to a flat roof should be installed with the tongue away from the worker. Roof decking applied to a sloping roof should be installed with the tongue up. The butt ends of the pieces are bevel cut at approximately a 2° angle (fig. 3-4). This provides a bevel cut from the face to the back to ensure a tight face butt joint when the decking is laid in a random-length pattern. If there are three or more supports for the decking, a controlled random laying pattern (shown in figure 3-5) can be used. This is an economical pattern because it makes use of random-plank lengths, but the following rules must be observed:

- Stagger the end joints in adjacent planks as widely as possible and not less than 2 feet.
- Separate the joints in the same general line by at least two courses.
- Minimize joints in the middle one-third of all spans.
- Make each plank bear on at least one support.
- Minimize the joints in the end span.

The ability of the decking to support specific loads depends on the support spacing, plank thickness, and span arrangement. Although two-span continuous layout offers structural efficiency, use of random-length planks is the most economical. Random-length double tongue-and-groove decking is used when there are three or more spans. It is not intended for use over single spans, and it is not recommended for use over double spans (see figure 3-5).

NAILING.—Fasten decking with common nails twice as long as the nominal plank thickness. For widths 6 inches or less, toenail once and face-nail once at each support. For widths over 6 inches, toenail once and face-nail twice. Decking 3 and 4 inches thick must be predrilled and toenailed with 8-inch spikes. Bright common nails may be used, but dipped galvanized common nails have better holding power and reduce the possibility of rust streaks. End joints not over a support should be side-nailed within 10 inches of each plank end. Splines are recommended on end joints of 3- and 4-inch material for better alignment, appearance, and strength.

Wood Fiber

All-wood fiber roof decking combines strength and insulation advantages that make possible quality construction with economy. This type of decking is weather resistant and protected against termites and rot.

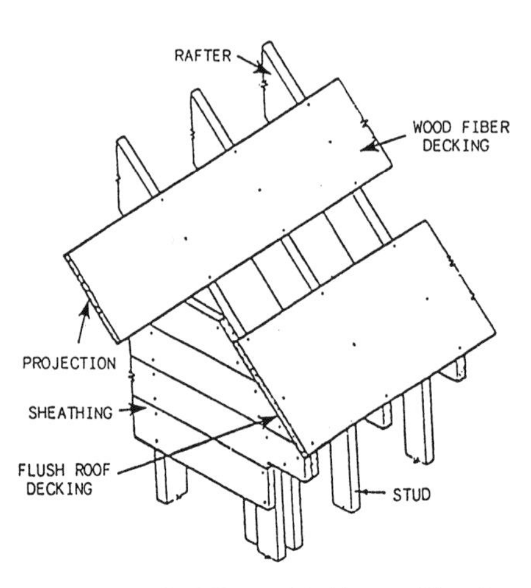

Figure 3-6.—Wood fiber roof decking at gable ends.

It is ideally suited for built-up roofing, as well as for asphalt and wood shingles on all types of buildings. Wood fiber decking is available in four thicknesses: 2 3/8 inches, 1 7/8 inches, 1 3/8 inches, and 15/16 inch. The standard panels are 2 inches by 8 feet with tongue-and-groove edges and square ends. The surfaces are coated on one or both sides at the factory in a variety of colors.

INSTALLATION.—Wood fiber roof decking is laid with the tongue-and-groove joint at right angles to the support members. The decking is started at the eave line with the groove edge opposite the applicator. Staple wax paper in position over the rafter before installing the roof deck. The wax paper protects the exposed interior finish of the decking if the beams are to be stained. Caulk the end joints with a nonstaining caulking compound. Butt the adjacent piece up against the caulked joint. Drive the tongue-and-groove edges of each unit firmly together with a wood block cut to fit the grooved edge of the decking. End joints must be made over a support member.

NAILING.—Although the wood fiber roof panels have tongue-and-groove edges, they are nailed through the face into the wood, rafters, or trusses. Face-nail 6 inches OC with 6d nails for 15/16-inch, 8d for 1 3/8-inch, 10d for 1 7/8-inch, and 16d for 2 3/8-inch thicknesses.

If you aren't going to apply the finish roofing material immediately after the roof is sheeted, cover the deck with building felt paper. The paper will protect the sheathing in case of rain. Wet panels tend to separate.

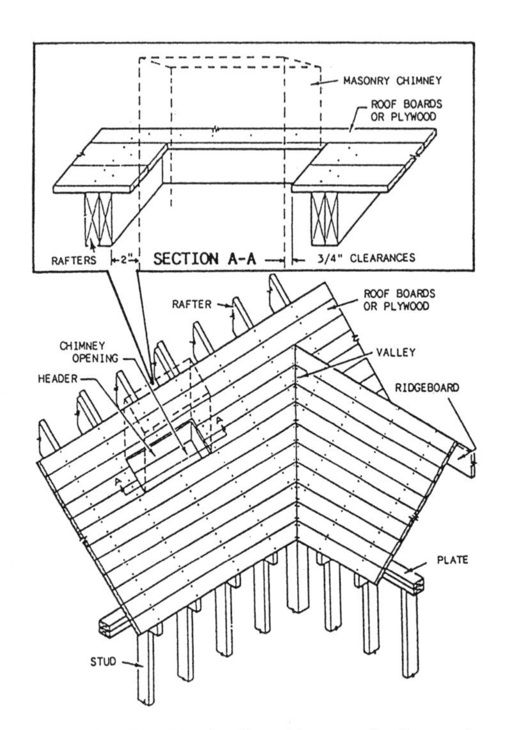

Figure 3-7.—Sheathing details at chimney and valley openings.

Roof decking that extends beyond gable-end walls for the overhang should span not less than three rafter spaces. This is to ensure anchorage to the rafters and to prevent sagging (see figure 3-6). When the projection is greater than 16 to 20 inches, special ladder framing is used to support the sheathing.

Table 3-2.—Determining Roof Area from a Plan

RISE (Inches)	FACTOR	RISE	FACTOR
3"	1.031	8"	1.202
3 1/2"	1.042	8 1/2"	1.225
4"	1.054	9"	1.250
4 1/2"	1.068	9 1/2"	1.275
5"	1.083	10"	1.302
5 1/2"	1.100	10 1/2"	1.329
6"	1.118	11"	1.357
6 1/2"	1.137	11 1/2"	1.385
7"	1.158	12"	1.414
7 1/2"	1.179		

Table 3-3.—Lumber Sheathing Specifications and Estimating Factor

	NOMINAL SIZE	WIDTH		AREA FACTOR
		Dress	Face	
Shiplap	1 × 6	5 7/16	4 15/16	1.22
	1 × 8	7 1/8	6 5/8	1.21
	1 × 10	9 1/8	8 5/8	1.16
	1 × 12	11 1/8	10 5/8	1.13
Tongue and Groove	1 × 4	3 7/16	3 3/16	1.26
	1 × 6	5 7/16	5 3/16	1.16
	1 × 8	7 1/8	6 7/8	1.16
	1 × 10	9 1/8	8 7/8	1.13
	1 × 12	11 1/8	10 7/8	1.10
S4S	1 × 4	3 1/2	3 1/2	1.14
	1 × 6	5 1/2	5 1/2	1.09
	1 × 8	7 1/4	7 1/4	1.10
	1 × 10	9 1/4	9 1/4	1.08
	1 × 12	11 1/4	11 1/4	1.07

Plywood extension beyond the end wall is usually governed by the rafter spacing to minimize waste. Thus, a 16-inch rake (gable) projection is commonly used when rafters are spaced 16 inches OC. Butt joints of the plywood sheets should be alternated so they do not occur on the same rafter.

DETAILS AT CHIMNEY AND VALLEY OPENINGS

Where chimney openings occur in the roof structure, the roof sheathing should have a 3/4-inch clearance on all sides from the finished masonry. Figure 3-7 shows sheathing details at the valley and chimney opening. The detail at the top shows the clearances between masonry and wood-framing members. Framing members should have a 2-inch clearance for fire protection. The sheathing should be securely nailed to the rafters and to the headers around the opening.

Wood or plywood sheathing at the valleys and hips should be installed to provide a tight joint and should be securely nailed to hip and valley rafters. This provides a smooth solid base for metal flashing.

ESTIMATING SHEATHING MATERIAL

To figure the roof area without actually getting on the roof and measuring, find the dimensions of the roof on the plans. Multiply the length times the width of the roof, including the overhang. Then multiply by the factor shown opposite the rise of the roof in table 3-2. The result will be the roof area.

For example, assume a building is 70 feet long and 30 feet wide (including the overhang), and the roof has a rise of 5 1/2 inches: 70 feet × 30 feet = 2,100 square feet. For a rise of 5 1/2 inches, the factor on the chart is 1.100: 2,100 square feet × 1.100 = 2,310 square feet. So, the total area to be covered is 2,310 square feet. Use this total area for figuring roofing needs, such as sheathing, felt underlayment, or shingles.

Lumber Sheathing

To decide how much lumber will be needed, first calculate the total area to be covered. Determine the size boards to be used, then refer to table 3-3. Multiply the total area to be covered by the factor from the chart. For example, if 1- by 8-inch tongue-and-groove sheathing

Table 3-4.—Plank Decking Estimating Factor

SIZE	AREA FACTOR
2" × 6"	2.40
2" × 8"	2.29
3" × 6"	3.43
4" × 6"	4.57

boards are to be used, multiply the total roof area by 1.16. To determine the total number of board feet needed, add 5 percent for trim and waste.

Plywood Sheathing

To determine how much plywood will be needed, find the total roof area to be covered and divide by 32 (the number of square feet in one 4- by 8-foot sheet of plywood). This gives you the number of sheets required to cover the area. Be sure to add 5 percent for a trim and waste allowance.

Decking or Planking

To estimate plank decking, first determine the area to be covered, then refer to the chart in table 3-4. In the left column, find the size planking to be applied. For example, if 2- by 6-inch material is selected, the factor is 2.40. Multiply the area to be covered by this factor and add a 5 percent trim and waste allowance.

Wood Fiber Roof Decking

To estimate the amount of wood fiber decking required, first find the total roof area to be covered. For every 100 square feet of area, you will need 6.25 panels, 2 by 8 feet in size. So, divide the roof area by 100 and multiply by 6.25. Using our previous example with a roof area of 2,310 square feet, you will need 145 panels.

EXTERIOR TRIM

LEARNING OBJECTIVE: Upon completing this section, you should be able to identify the types of cornices and material used in their construction.

Exterior trim includes door and window trim, cornice trim, facia boards and soffits, and rake or gable-end trim. Contemporary designs with simple cornices and moldings contain little of this material; traditional designs have considerably more. Much of the exterior trim, in the form of finish lumber and moldings, is cut and fitted on the job. Other materials or assemblies, such as shutters, louvers, railings, and posts, are shop fabricated and arrive on the job ready to be fastened in place.

The properties desired in materials used for exterior trim are good painting and weathering characteristics, easy working qualities, and maximum freedom from warp. Decay resistance is desirable where materials may absorb moisture. Heartwood from cedar, cypress, and redwood has high decay resistance. Less durable species can be treated to make them decay resistant. Many manufacturers pre-dip materials, such as siding, window sash, door and window frames, and trim, with a water-repellent preservative. On-the-job dipping of end joints or miters cut at the building site is recommended when resistance to water entry and increased protection are desired.

Rust-resistant trim fastenings, whether nails or screws, are preferred wherever they may be in contact with weather. These include galvanized, stainless steel, or aluminum fastenings. When a natural finish is used, nails should be stainless steel or aluminum to prevent staining and discoloration. Cement-coated nails are not rust-resistant.

Siding and trim are normally fastened in place with a standard siding nail, which has a small flathead. However, finish or casing nails might also be used for some purposes. Most of the trim along the shingle line, such as at gable ends and cornices, is installed before the roof shingles are applied.

The roof overhangs (eaves) are the portions of the roof that project past the sidewalls of the building. The cornice is the area beneath the overhangs. The upward slopes of the gable ends are called rakes. Several basic designs are used for finishing off the roof overhangs and cornices. Most of these designs come under the category of open cornice or closed cornice. They not only add to the attractiveness of a building but also help protect the sidewalls of the building from rain and snow. Wide overhangs also shade windows from the hot summer sun.

Cornice work includes the installation of the lookout ledger, lookouts, plancier (soffit), ventilation screens, fascia, frieze, and the moldings at and below the eaves, and along the sloping sides of the gable end (rake). The ornamental parts of a cornice are called cornice trim and consist mainly of molding; the molding running up the side of the rakes of a gable roof is called gable cornice trim. Besides the main roof, the additions and dormers may have cornices and cornice trim.

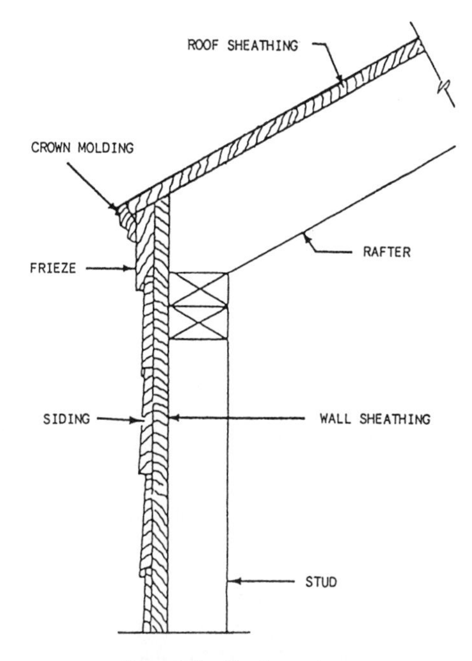

Figure 3-8.—Simple cornice.

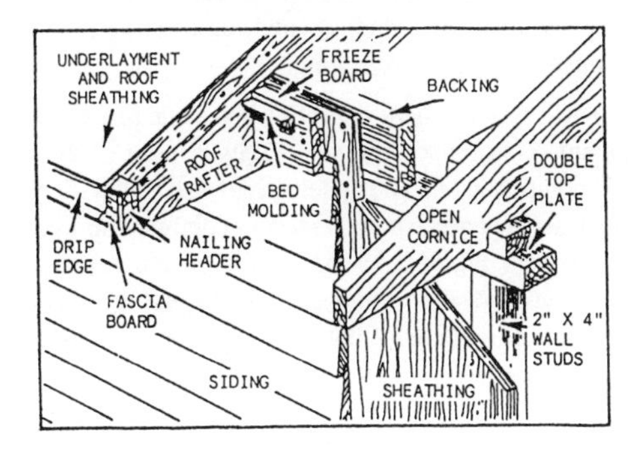

Figure 3-9.—Open cornice.

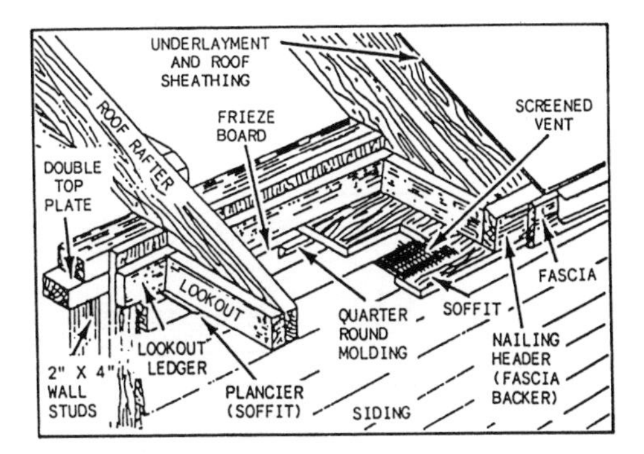

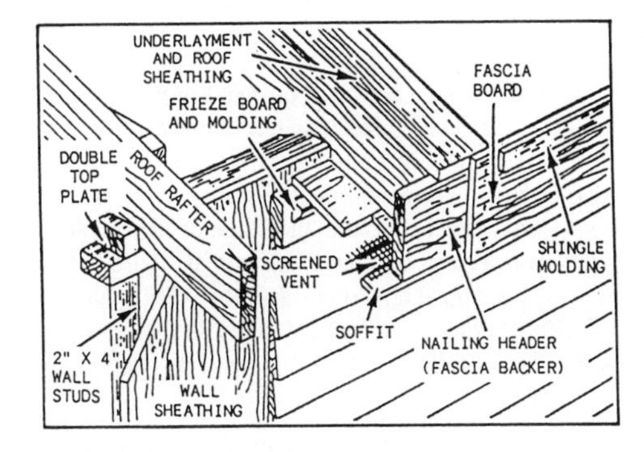

Figure 3-10.—Closed cornices: A. Flat boxed cornice; B. Sloped boxed cornice.

CORNICES

The type of cornice required for a particular structure is indicated on the wall sections of the drawings, and there are usually cornice detail drawings as well. A roof with no rafter overhang or eave usually has the simple cornice shown in figure 3-8. This cornice consists of a single strip or board called a frieze. It is beveled on the upper edge to fit under the overhang or eave and rabbeted on the lower edge to overlap the upper edge of the top course of siding. If trim is used, it usually consists of molding placed as shown in figure 3-8. Molding trim in this position is called crown molding.

A roof with a rafter overhang may have an open cornice or a closed (also called a box) cornice. In open-cornice construction (fig. 3-9), the undersides of the rafters and roof sheathing are exposed. A nailing header (fascia backer) is nailed to the tail ends of the rafters to provide a straight and solid nailing base for the fascia board. Most spaces between the rafters are blocked off. Some spaces are left open (and screened) to allow attic ventilation. Usually, a frieze board is nailed to the wall below the rafters. Sometimes the frieze board is notched between the rafters and molding is nailed over it. Molding trim in this position is called bed molding. In closed-cornice construction, the bottom of the roof overhang is closed off. The two most common types of closed cornices are the flat boxed cornice and the sloped boxed cornice (shown in figure 3-10, views A and B, respectively).

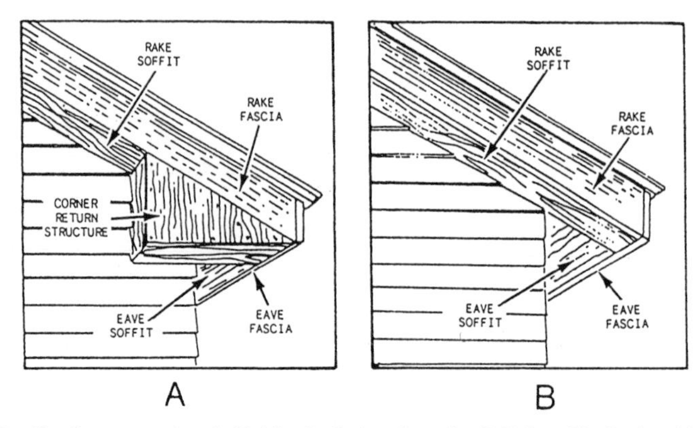

Figure 3-11.—Cornice construction: A. Finish rake for boxed cornice; B. Rake soffit of a sloped box cornice.

The flat boxed cornice requires framing pieces called lookouts. These are toenailed to the wall or to a lookout ledger and face-nailed to the ends of the rafters. The lookouts provide a nailing base for the soffit, which is the material fastened to the underside of the cornice. A typical flat boxed cornice is shown in figure 3-10, view A. For a sloped boxed cornice, the soffit material is nailed directly to the underside of the rafters (fig. 3-10, view B). This design is often used on buildings with wide overhangs.

The basic rake trim pieces are the frieze board, trim molding, and the fascia and soffit material. Figure 3-11, view A, shows the finish rake for a flat boxed cornice. It requires a cornice return where the eave and rake soffits join. View B shows the rake of a sloped boxed cornice. Always use rust-resistant nails for exterior finish work. They may be aluminum, galvanized, or cadmium-plated steel.

PREFABRICATED WOOD AND METAL TRIM

Because cornice construction is time-consuming, various prefabricated systems are available that provide a neat, trim appearance. Cornice soffit panel materials include plywood, hardboard, fiberboard, and metal. Many of these are factory-primed and available in a variety of standard widths (12 to 48 inches) and in lengths up to 12 feet. They also may be equipped with factory-installed screen vents.

When installing large sections of wood fiber panels, you should fit each panel with clearance for expansion. Nail 4d rust-resistant nails 6 inches apart along the edges and intermediate supports (lookouts). Start nailing at the end butted against a previously placed panel. First, nail the panel to the main supports and then along the edges. Drive nails carefully so the underside of the head is just flush with the panel surface. Remember, this is finish work; no hammer head marks please. Always read and follow manufacturer's directions and recommended installation procedures. Cornice trim and soffit systems are also available in aluminum and come in a variety of prefinished colors and designs.

Soffit systems made of prefinished metal panels and attachment strips are common. They consist of three basic components: wall hanger strips (also called frieze strips); soffit panels (solid, vented, or combination); and fascia covers. Figure 3-12 shows the typical installation configuration of the components. Soffit panels include a vented area and are available in a variety of lengths.

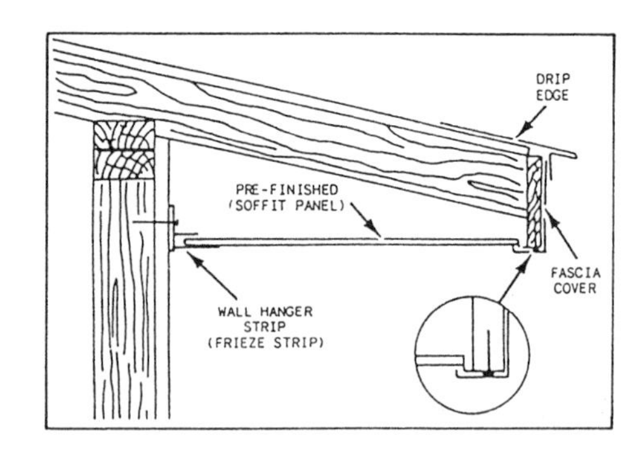

Figure 3-12.—Basic components of prefinished metal soffit system.

To install a metal panel system, first snap a chalk line on the sidewall level with the bottom edge of the fascia board. Use this line as a guide for nailing the wall hanger strip in place. Insert the panels, one at a time, into the wall strip. Nail the outer end to the bottom edge of the fascia board.

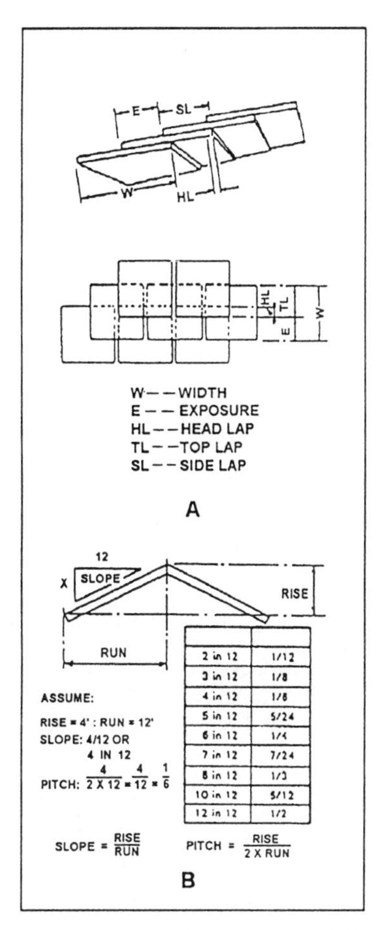

Figure 3-13.—Roofing terminology: A. Surfaces; B. Slope and pitch.

After all soffit panels are in place, cut the fascia cover to length and install it. The bottom edge of the cover is hooked over the end of the soffit panels. It is then nailed in place through prepunched slots located along the top edge. Remember to use nails compatible with the type of material being used to avoid electrolysis between dissimilar metals. Again, always study and follow the manufacturer's directions when making an installation of this type.

ROOFING TERMS AND MATERIALS

LEARNING OBJECTIVE: Upon completing this section, you should be able to define roofing terms and identify roofing materials.

The roof covering, or roofing, is a part of the exterior finish. It should provide long-lived waterproof protection for the building and its contents from rain, snow, wind, and, to some extent, heat and cold.

Before we begin our discussion of roof coverings, let's first look at some of the most common terms used in roof construction.

TERMINOLOGY

Correct use of roofing terms is not only the mark of a good worker, but also a necessity for good construction. This section covers some of the more common roofing terms you need to know.

Square

Roofing is estimated and sold by the square. A square of roofing is the amount required to cover 100 square feet of the roof surface.

Coverage

Coverage is the amount of weather protection provided by the overlapping of shingles. Depending on the kind of shingle and method of application, shingles may furnish one (single coverage), two (double coverage), or three (triple coverage) thicknesses of material over the roof surface. Shingles providing single coverage are suitable for re-roofing over existing roofs. Shingles providing double and triple coverage are used for new construction. Multiple coverage increases weather resistance and provides a longer service life.

Shingle Surfaces

The various surfaces of a shingle are shown in view A of figure 3-13. "Shingle width" refers to the total measurement across the top of either a strip type or individual type of shingle. The area that one shingle overlaps a shingle in the course (row) below it is referred to as "top lap." "Side lap" is the area that one shingle overlaps a shingle next to it in the same course. The area that one shingle overlaps a shingle two courses below it is known as head lap. Head lap is measured from the bottom edge of an overlapping shingle to the nearest top edge of an overlapped shingle. "Exposure" is the area that is exposed (not overlapped) in a shingle. For the best protection against leakage, shingles (or shakes) should be applied only on roofs with a unit rise of 4 inches or more. A lesser slope creates slower water runoff, which increases the possibility of leakage as a result of windblown rain or snow being driven underneath the butt ends of the shingles.

Slope

"Slope" and "pitch" are often incorrectly used synonymously when referring to the incline of a sloped roof. View B of figure 3-13 shows some common roof slopes with their corresponding roof pitches.

"Slope" refers to the incline of a roof as a ratio of vertical rise to horizontal run. It is expressed sometimes as a fraction but typically as X-in-12; for example, a 4-in-12 slope for a roof that rises at the rate of 4 inches for each foot (12 inches) of run. The triangular symbol above the roof in figure 3-13, view B, conveys this information.

Pitch

"Pitch" is the incline of a roof as a ratio of the vertical rise to twice the horizontal run. It is expressed as a fraction. For example, if the rise of a roof is 4 feet and the run 12 feet, the roof is designated as having a pitch of 1/6 (4/24 = 1/6).

MATERIALS

In completing roofing projects, you will be working with a number of different materials. In the following section, we will discuss the most common types of underlayments, flashing, roofing cements, and exterior materials you will encounter. We will also talk about built-up roofing.

Materials used for pitched roofs include shingles of asphalt, fiberglass, and wood. Shingles add color, texture, and pattern to the roof surface. To shed water, all shingles are applied to roof surfaces in some overlapping fashion. They are suitable for any roof with enough slope to ensure good drainage. Tile and slate are also popular. Sheet materials, such as roll roofing, galvanized steel, aluminum, copper, and tin, are sometimes used. For flat or low-pitched roofs, composition or built-up roofing with a gravel topping or cap sheet are frequent combinations. Built-up roofing consists of a number of layers of asphalt-saturated felt mopped down with hot asphalt or tar. Metal roofs are sometimes used on flat decks of dormers, porches, or entryways.

The choice of materials and the method of application are influenced by cost, roof slope, expected service life of the roofing, wind resistance, fire resistance, and local climate. Because of the large amount of exposed surface of pitched roofs, appearance is also important.

Underlayments

There are basically four types of underlayments you will be working with as a Builder: asphalt felt, organic, glass fiber, and tarred.

Once the roof sheathing is in place, it is covered with an asphalt felt underlayment commonly called roofing felt. Roofing felt is asphalt-saturated and serves three basic purposes. First, it keeps the roof sheathing dry until the shingles can be applied. Second, after the shingles have been laid, it acts as a secondary barrier against wind-driven rain and snow. Finally, it also protects the shingles from any resinous materials, which could be released from the sheathing.

Roofing felt is designated by the weight per square. As we mentioned earlier, a square is equal to 100 square feet and is the common unit to describe the amount of roofing material. Roofing felt is commonly available in rolls of 15 and 30 pounds per square. The rolls are usually 36 inches wide. A roll of 15-pound felt is 144 feet long, whereas a roll of 30-pound felt is 72 feet long. After you allow for a 2-inch top lap, a roll of 15-pound felt will cover 4 squares; a roll of 30-pound felt will cover 2 squares.

Underlayment should be a material with low vapor resistance, such as asphalt-saturated felt. Do not use materials, such as coated felts or laminated waterproof papers, which act as a vapor barrier. These allow moisture or frost to accumulate between the underlayment and the roof sheathing. Underlayment requirements for different kinds of shingles and various roof slopes are shown in table 3-5.

Apply the underlayment as soon as the roof sheathing has been completed. For single underlayment, start at the eave line with the 15-pound felt. Roll across

Table 3-5.—Underlayment Recommendations for Shingle Roofs

TYPE OF ROOFING	SHEATHING	TYPE OF UNDERLAYMENT	NORMAL SLOPE		LOW SLOPE	
Asbestos-Cement Shingles	Solid	No. 15 asphalt saturated asbestos (inorganic) felt, OR No. 30 asphalt saturated felt	5/12 and up	Single layer over entire roof	3/12 to 5/12	Double layer over entire roof[1]
Asphalt/Fiberglass Shingles	Solid	No. 15 asphalt saturated felt	4/12 and up	Single layer over entire roof	2/12 to 4/12	Double layer over entire roof[2]
Wood Shakes	Spaced	No. 30 asphalt saturated felt (interlayment)	4/12 and up	Underlayment starter course; interlayment over entire roof	Shakes not recommended on slopes less than 4/12 with spaced sheathing	
	Solid[3,5]	No. 30 asphalt saturated felt (interlayment)	4/12 and up	Underlayment starter course; interlayment over entire roof	3/12 to 4/12[4]	Single layer underlayment over entire roof; interlayment over entire roof
Wood Shingles	Spaced	None required	5/12 and up	None required	3/12 to 5/12[4]	None required
	Solid[5]	No. 15 asphalt saturated felt	5/12 and up	None required[6]	3/12 to 5/12[4]	None required[6]

Notes:
1. May be single layer on 4/12 slope in areas where outside design temperature is warmer than 0°F
2. Square-butt strip shingles only; requires wind-resistant shingles or cemented tabs
3. Recommended in areas subject to wind-driven snow
4. Requires reduced weather exposure
5. May be desirable for added insulation and to minimize air infiltration
6. May be desirable for protection of sheathing

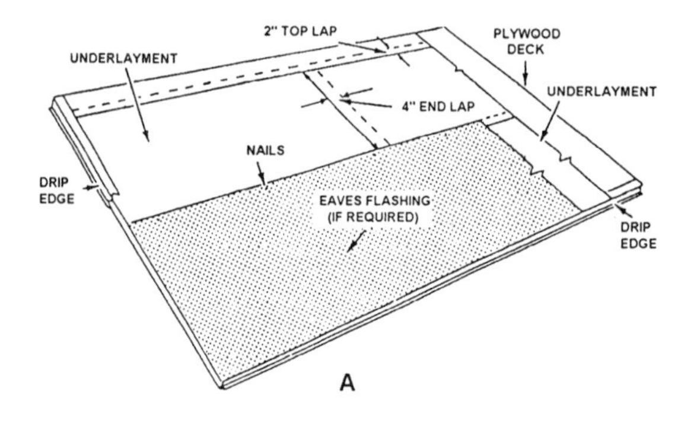

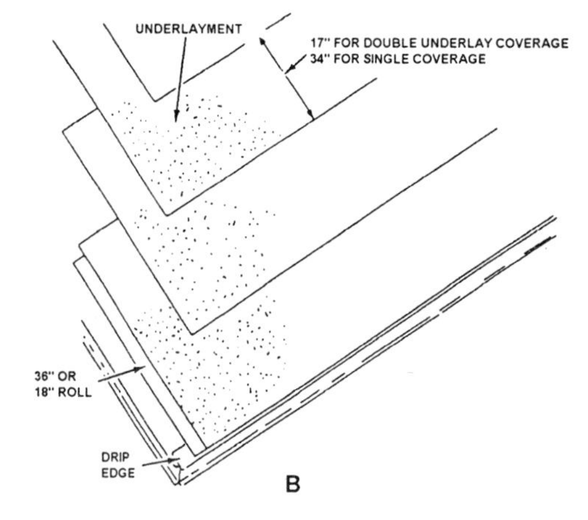

Figure 3-14.—Roofing underlayment: A. Single coverage; B. Double coverage.

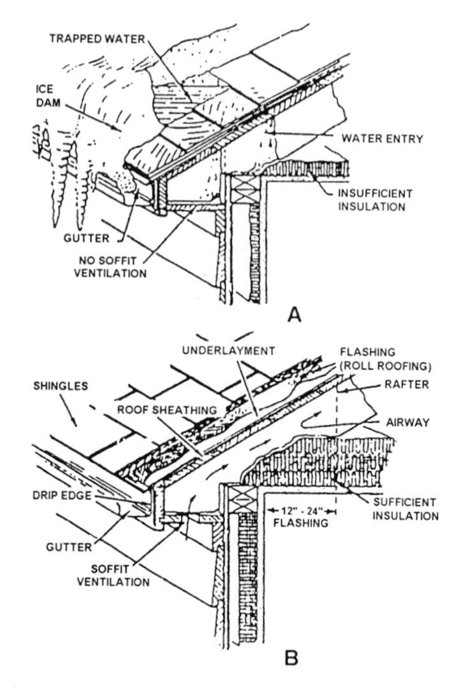

Figure 3-15.—Protection from ice dams: A. Refreezing snow and ice; B. Cornice ventilation.

the roof with a top lap of at least 2 inches at all horizontal points and a 4-inch side lap at all end joints (fig. 3-14, view A). Lap the underlayment over all hips and ridges 6 inches on each side. A double underlayment can be started with two layers at the eave line, flush with the fascia board or molding. The second and remaining strips have 19-inch head laps with 17-inch exposures (fig. 3-14, view B). Cover the entire roof in this manner. Make sure that all surfaces have double coverage. Use only enough fasteners to hold the underlayment in place until the shingles are applied. Do not apply shingles over wet underlayment.

In areas where moderate-to-severe snowfall is common and ice dams occur, melting snow refreezes at the eave line (fig. 3-15, view A). It is a good practice to apply one course of 55-pound smooth-surface roll roofing as a flashing at the eaves. It should be wide enough to extend from the roof edge to between 12 and 24 inches inside the wall line. The roll roofing should be installed over the underlayment and metal drip edge. This will lessen the chance of melting snow to back up under the shingles and fascia board of closed cornices. Damage to interior ceilings and walls results from this water seepage. Protection from ice dams is provided by eave flashing. Cornice ventilation by means of soffit vents and sufficient insulation will minimize the melting (fig. 3-15, view B).

ASPHALT FELT.—Roofing felts are used as underlayment for shingles, for sheathing paper, and for reinforcements in the construction of built-up roofs. They are made from a combination of shredded wood fibers, mineral fibers, or glass fibers saturated with asphalt or coal-tar pitch. Sheets are usually 36 inches wide and available in various weights from 10 to 50 pounds. These weights refer to weight per square (100 feet).

ORGANIC FELTS.—Asphalt-saturated felts composed of a combination of felted papers and organic shredded wood fibers are considered felts. They are among the least expensive of roofing felts and are widely used not only as roofing, but also as water and vapor retarders. Fifteen-pound felt is used under wood siding and exterior plaster to protect sheathing or wood studs. It is generally used in roofing for layers or plies in gravel-surfaced assemblies and is available perforated. Perforated felts used in built-up roofs allow entrapped moisture to escape during application. Thirty-pound felt requires fewer layers in a built-up roof. It is usually used as underlayment for heavier cap sheets or tile on steeper roofs.

GLASS-FIBER FELTS.—Sheets of glass fiber, when coated with asphalt, retain a high degree of porosity, assuring a maximum escape of entrapped moisture or vapor during application and maximum bond between felts. Melted asphalt is applied so that the finished built-up roof becomes a monolithic slab reinforced with properly placed layers of glass fibers. The glass fibers, which are inorganic and do not curl, help create a solid mass of reinforced waterproof roofing material.

TARRED FELTS.—Coal-tar pitch saturated organic felts are available for use with bitumens of the same composition. Since coal-tar and asphalt are not compatible, the components in any construction must be limited to one bitumen or the other unless approved by the felt manufacturer.

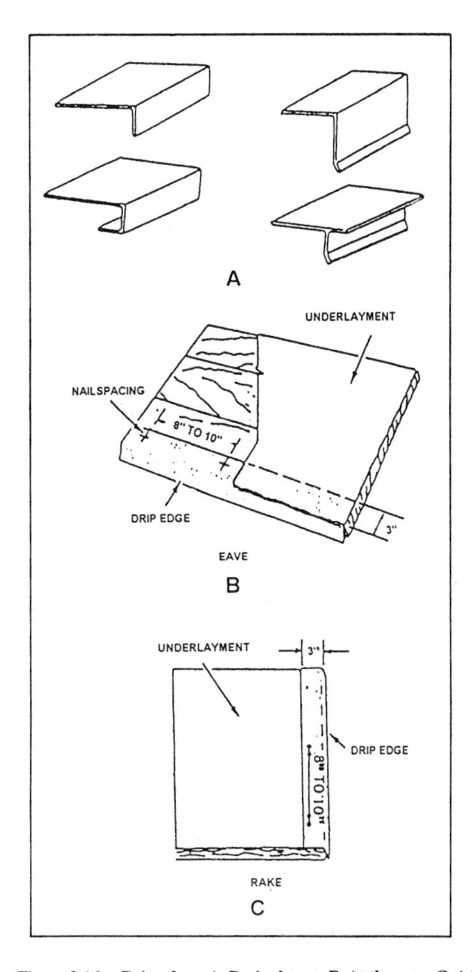

Figure 3-16.—Drip edges: A. Basic shapes: B. At the eave; C. At the rake.

Flashing

The roof edges along the eaves and rake should have a metal drip edge, or flashing. Flashing is specially constructed pieces of sheet metal or other materials used to protect the building from water seepage. Flashing must be made watertight and be water shedding. Flashing materials used on roofs may be asphalt-saturated felt, metal, or plastic. Felt flashing is generally used at the ridges, hips, and valleys. However, metal flashing, made of aluminum, galvanized steel, or copper, is considered superior to felt. Metal used for flashing must be corrosion resistant. It should be galvanized steel (at least 26 gauge), 0.019-inch-thick aluminum, or 16-ounce copper.

Flashing is available in various shapes (fig. 3-16, view A), formed from 26-gauge galvanized steel. It should extend back approximately 3 inches from the roof edge and bend downward over the edge. This causes the water to drip free of underlying cornice construction. At the eaves, the underlayment should be laid over the drip edge (view B). At the rake (view C), place the underlayment under the drip edge. Galvanized nails, spaced 8 to 10 inches apart, are recommended for fastening the drip edge to the sheathing.

The shape and construction of different types of roofs can create different types of water leakage problems. Water leakage can be prevented by placing flashing materials in and around the vulnerable areas of the roof. These areas include the point of intersection between roof and soil stack or ventilator, the valley of a roof, around chimneys, and at the point where a wall intersects a roof.

Figure 3-17.—Flashing around a roof projection.

As you approach a soil stack, apply the roofing up to the stack and cut it to fit (fig. 3-17). You then install a corrosion-resistant metal sleeve, which slips over the stack and has an adjustable flange to fit the slope of the roof. Continue shingling over the flange. Cut the shingles to fit around the stack and press them firmly into the cement.

The open or closed method can be used to construct valley flashing. A valley underlayment strip of 15-pound asphalt- saturated felt, 36 inches wide, is applied first. The strip is centered in the valley and secured with enough nails to hold it in place. The horizontal courses of underlayment are cut to overlap this valley strip a minimum of 6 inches.

Open valleys can be flashed with metal or with 90-pound mineral-surfaced asphalt roll roofing. The color can match or contrast with the roof shingles. An 18-inch-wide strip of mineral-surfaced roll roofing is placed over the valley underlayment. It is centered in the valley with the surfaced side down and the lower edge cut to conform to and be flush with the eave flashing. When it is necessary to splice the material, the ends of the upper segments are laid to overlap the lower segments 12 inches and are secured with asphalt plastic cement. This method is shown in figure 3-18. Only enough nails are used 1 inch in from each edge to hold the strip smoothly in place.

Another 36-inch-wide strip is placed over the first strip. It is centered in the valley with the surfaced side up and secured with nails. It is lapped the same way as the underlying 18-inch strip.

Before shingles are applied, a chalk line is snapped on each side of the valley. These lines should start 6 inches apart at the ridge and spread wider apart (at the rate of 1/8 inch per foot) to the eave (fig. 3-18). The chalk lines serve as a guide in trimming the shingle units to fit the valley and ensure a clean, sharp edge. The upper corner of each end shingle is clipped to direct water into the valley and prevent water penetration between courses. Each shingle is cemented to the valley lining with asphalt cement to ensure a tight seal. No exposed nails should appear along the valley flashing.

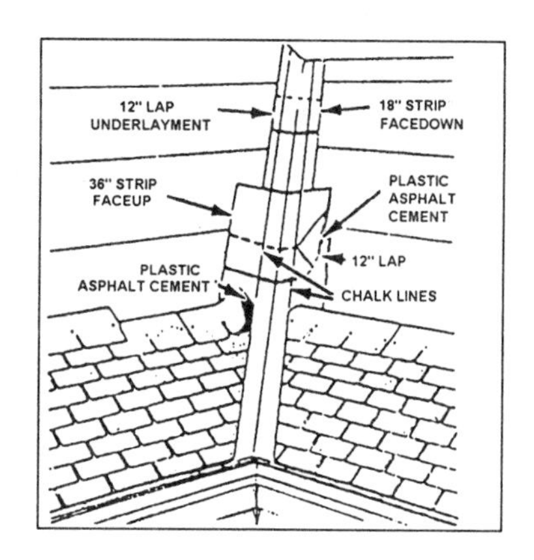

Figure 3-18.—Open valley flashing using roll roofing.

Closed (woven) valleys can be used only with strip shingles. This method has the advantage of doubling the coverage of the shingles throughout the length of the valley. This increases the weather resistance at this vulnerable point. A valley lining made from a 36-inch-wide strip of 55-pound (or heavier) roll roofing is placed over the valley underlayment and centered in the valley (fig. 3-19).

Valley shingles are laid over the lining by either of two methods:

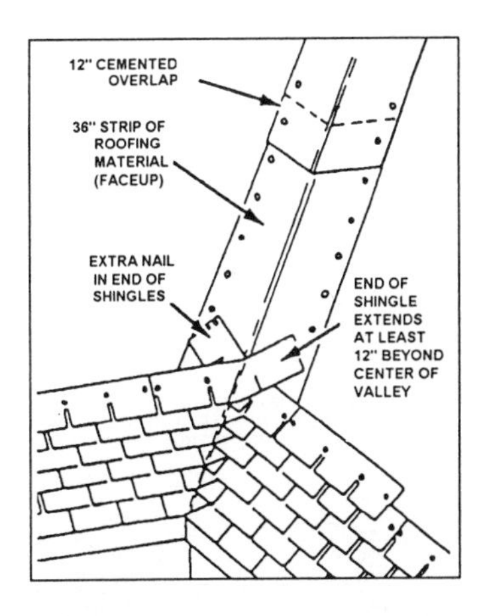

Figure 3-19.—Closed valley flashing.

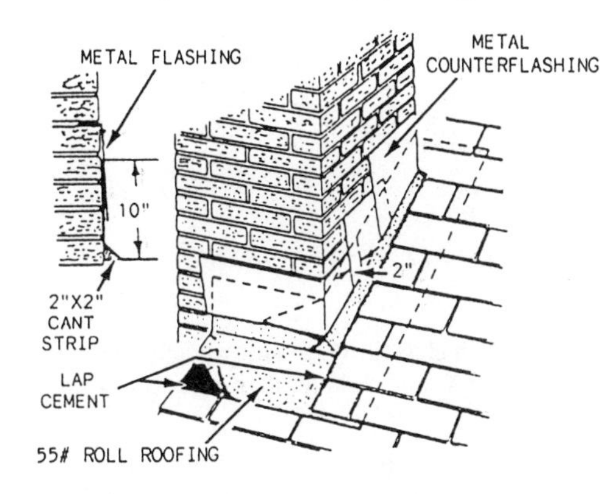

Figure 3-20.—Flashing around a chimney.

- They can be applied on both roof surfaces at the same time with each course, in turn, woven over the valley.
- Each surface can be covered to the point approximately 36 inches from the center of the valley and the valley shingles woven in place later.

In either case, the first course at the valley is laid along the eaves of one surface over the valley lining and extended along the adjoining roof surface for a distance of at least 12 inches. The first course of the adjoining roof surface is then carried over the valley on top of the previously applied shingle. Succeeding courses are then laid alternately, weaving the valley shingles over each other.

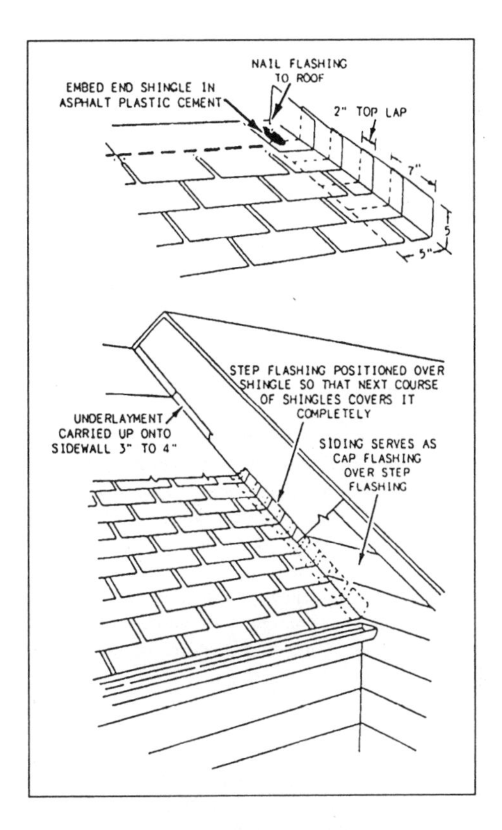

Figure 3-21.—Step flashing.

The shingles are pressed tightly into the valley and nailed in the usual manner. No nail should be located closer than 6 inches to the valley center line, and two nails should be used at the end of each terminal strip.

As you approach a chimney, apply the shingles over the felt up to the chimney face. If 90-pound roll roofing is to be used for flashing, cut wood cant strips and install them above and at the sides of the chimney (fig. 3-20). The roll roofing flashing should be cut to run 10 inches up the chimney. Working from the bottom up, fit metal counterflashing over the base flashing and insert it 1 1/2 inches into the mortar joints. Refill the joints with mortar or roofing cement. The counterflashing can also be installed when the chimney masonry work is done.

Where the roof intersects a vertical wall, it is best to install metal flashing shingles. They should be 10 inches long and 2 inches wider than the exposed face of the regular shingles. The 10-inch length is bent so that it will extend 5 inches over the roof and 5 inches up the wall (see figure 3-21). Apply metal flashing with each course. This waterproofs the joint between a sloping roof and vertical wall. This is generally called step flashing.

As each course of shingles is laid, a metal flashing shingle is installed and nailed at the top edge as shown. Do not nail flashing to the wall; settling of the roof frame could damage the seal.

Wall siding is installed after the roof is completed. It also serves as a cap flashing. Position the siding just above the roof surface. Allow enough clearance to paint the lower edges.

Roof Cements

Roofing cements are used for installing eave flashing, for flashing assemblies, for cementing tabs of asphalt shingles and laps in sheet material, and for repairing roofs. There are several types of cement, including plastic asphalt cements, lap cements, quick-setting asphalt adhesives, roof coatings, and primers. The type and quality of materials and methods of application on a shingle roof should follow the recommendation of the manufacturer of the shingle roofing.

Exterior

Basically, exterior roof treatment consists of applying various products, including shingles, roll roofing, tiles, slate, and bituminous coverings. Treatment also includes specific construction considerations for ridges, hips, and valleys.

SHINGLES.—The two most common shingle types are asphalt and fiberglass, both of which come in various strip shapes.

Asphalt.—Asphalt (composition) shingles are available in several patterns. They come in strip form or as individual shingles. The shingles are manufactured on a base of organic felt (cellulose) or an inorganic glass mat. The felt or mat is covered with a mineral-stabilized coating of asphalt on the top and bottom. The top side is coated with mineral granules of specified color. The bottom side is covered with sand, talc, or mica.

Fiberglass.—Improved technologies have made the fiberglass mat competitive with organic felt. The weight and thickness of a fiberglass mat is usually less than that of organic felt. A glass fiber mat may be 0.030 inch thick versus 0.055 inch thick for felt. The popularity of fiberglass-based shingles is their low cost. The mat does not have to be saturated in asphalt. ASTM standards specify 3 pounds per 100 feet. The combination of glass fiber mats with recently developed resins has significantly lowered the price of composition shingles.

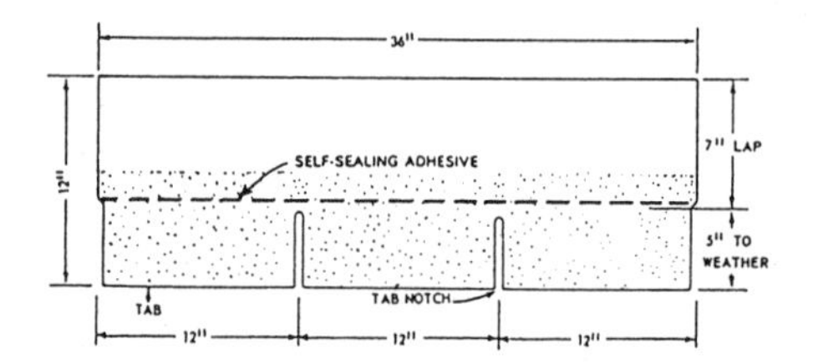

Figure 3-22.—A typical 12- by 36-inch shingle.

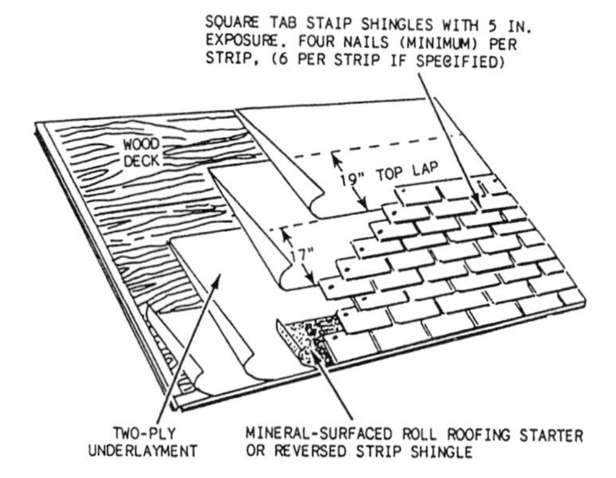

Figure 3-23.—Special shingle application.

Strip.—One of the most common shapes of asphalt or fiberglass shingles is a 12- by 36-inch strip (fig. 3-22) with the exposed surface cut or scored to resemble three 9- by 12-2- inch shingles. These are called strip shingles. They are usually laid with 5 inches exposed to the weather. A lap of 2 to 3 inches is usually provided over the upper edge of the shingle in the course directly below. This is called the head lap.

The thickness of asphalt shingles may be uniform throughout, or, as with laminated shingles, slotted at the butts to give the illusion of individual units. Strip shingles are produced with either straight-tab or random-tab design to give the illusion of individual units or to simulate the appearance of wood shakes. Most strip shingles have factory-applied adhesive spaced at intervals along the concealed portion of the strip. These strips of adhesive are activated by the warmth of the sun and hold the shingles firm through wind, rain, and snow.

Strip shingles are usually laid over a single thickness of asphalt-saturated felt if the slope of the roof

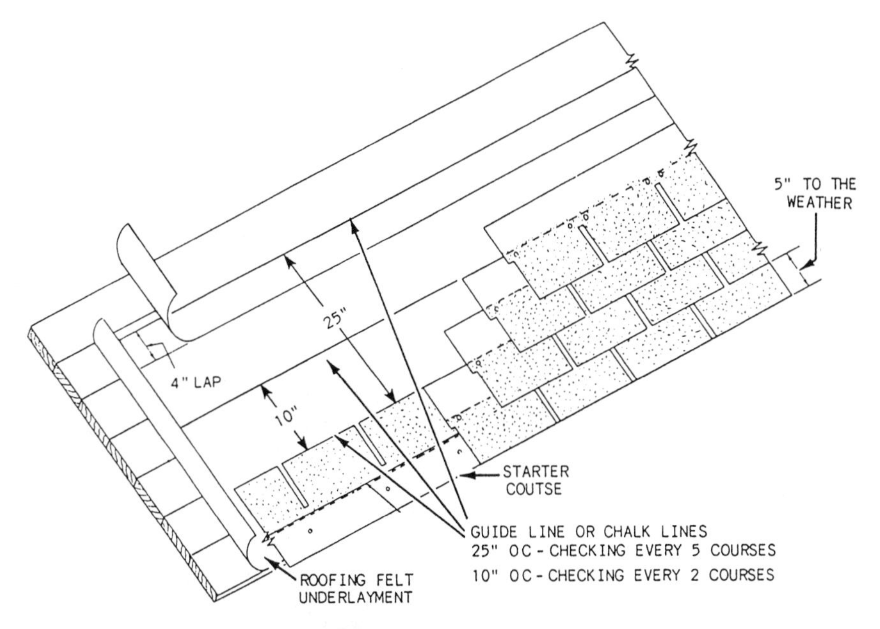

Figure 3-24.—Laying out a shingle roof.

is 4:12 or greater. When special application methods are used, organic- or inorganic-base-saturated or coated-strip shingles can be applied to decks having a slope of 4:12, but not less than 2:12. Figure 3-23 shows the application of shingles over a double layer of underlayment. Double underlayment is recommended under square-tab strip shingles for slopes less than 4:12.

When roofing materials are delivered to the building site, they should be handled with care and protected from damage. Try to avoid handling asphalt shingles in extreme heat or cold. They are available in one-third-square bundles, 27 strip shingles per bundle. Bundles should be stored flat so the strips will not curl after the bundles are open. To get the best performance from any roofing material, always study the manufacturer's directions and install as directed.

On small roofs (up to 30 feet long), strip shingles can be laid starting at either end. When the roof surface is over 30 feet long, it is usually best to start at the center and work both ways. Start from a chalk line perpendicular to the eaves and ridge.

Asphalt shingles will vary slightly in length (plus or minus 1/4 inch in a 36-inch strip). There may also be some variations in width. Thus, chalk lines are required to achieve the proper horizontal and vertical placement of the shingles (fig. 3-24).

The first chalk line from the eave should allow for the starter strip and/or the first course of shingles to overhang the drip edge 1/4 to 3/8 inch.

When laying shingles from the center of the roof toward the ends, snap a number of chalk lines between the eaves and ridge. These lines will serve as reference marks for starting each course. Space them according to the shingle type and laying pattern.

Chalk lines, parallel to the eaves and ridge, will help maintain straight horizontal lines along the butt edge of the shingle. Usually, only about every fifth course needs to be checked if the shingles are skillfully applied. Inexperienced workers may need to set up chalk lines for every second course.

The purpose of a starter strip is to back up the first course of shingles and fill in the space between the tabs. Use a strip of mineral-surfaced roofing 9 inches or wider of a weight and color to match the shingles. Apply the strip so it overhangs the drip edge 1/4 to 3/8 inch above the edge. Space the nails so they will not be exposed at the cutouts between the tabs of the first course of shingles. Sometimes an inverted (tabs to ridge) row of shingles is used instead of the starter strip. When you

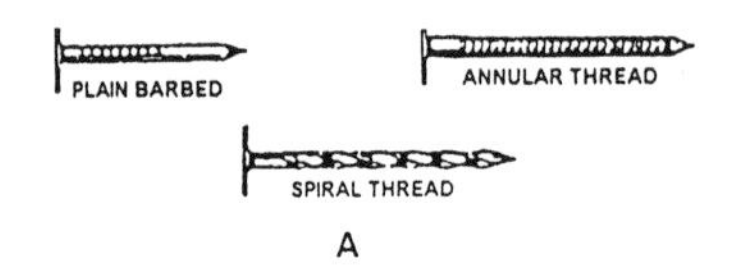

APPLICATION	1" SHEATHING	3/8" PLYWOOD
STRIP OR INDIVIDUAL SHINGLE (NEW CONSTRUCTION)	1 1/4"	7/8"
OVER ASPHALT ROOFING (REROOFING)	1 1/2"	1"
OVER WOOD SHINGLES (REROOFING)	1 3/4"	–

B

SELF-SEALING ADHESIVE

SELF-SEALING ADHESIVE

C

Figure 3-25.—Nails suitable for installing strip shingles, recommended nail lengths, and nail placement.

are laying self-sealing strip shingles in windy areas, the starter strip is often formed by cutting off the tabs of the shingles being used. These units are then nailed in place, right side up, and provide adhesive under the tabs of the first course.

Nails used to apply asphalt roofing must have a large head (3/8- to 7/16-inch diameter) and a sharp point. Figure 3-25 shows standard nail designs (view A) and recommended lengths (view B) for nominal 1-inch sheathing. Most manufacturers recommend 12-gauge galvanized steel nails with barbed shanks. Aluminum nails are also used. The length should be sufficient to penetrate the full thickness of the sheathing or 3/4 inch into the wood.

The number of nails and correct placement are both vital factors in proper application of roofing material. For three-tab square-butt shingles, use a minimum of four nails per strip (fig. 3-25, view C). Specifications may require six nails per shingle (view C). Align each shingle carefully and start the nailing from the end next to the one previously laid. Proceed across the shingle. This will prevent buckling. Drive nails straight so that the edge of the head will not cut into the shingle. The nail head should be driven flush, not sunk into the surface. If, for some reason, the nail fails to hit solid sheathing, drive another nail in a slightly different location.

WOOD SHINGLES AND SHAKES.—Wood shingles are available in three standard lengths: 16, 18, and 24 inches. The 16-inch length is the most popular. It has five-butt thicknesses per 2 inches of width when it is green (designated a 5/2). These shingles are packed in bundles. Four bundles will cover 100 square feet of wall or roof with 5-inch exposure. The 18- or 24-inch-long shingles have thicker butts—five in 2 1/4 inches for the 18-inch shingles and four in 2 inches for 24-inch shingles. The recommended exposures for the standard wood-shingle size are shown in table 3-6.

Figure 3-26 shows the proper method of applying a wood-shingle roof. Underlayment or roofing felt is not required for wood shingles except for protection in ice jam areas. Although spaced or solid sheathing is optional, spaced roof sheathing under wood shingles is most common. Observe the following steps when applying wood shingles:

1. Extend the shingles 1 1/2 inches beyond the eave line and 3/4 inch beyond the rake (gable) edge.

Table 3-6.—Recommended Exposure for Wood Shingles

SHINGLE LENGTH	SHINGLE THICKNESS (Green)	MAXIMUM EXPOSURE	
		Slope less than 4 in 12	Slope 5 in 12 and over
Inches		Inches	Inches
16	5 butts in 2"	3 3/4	5
18	5 butts in 2 1/4"	4 1/4	5 1/2
24	4 butts in 2"	5 3/4	7 1/2

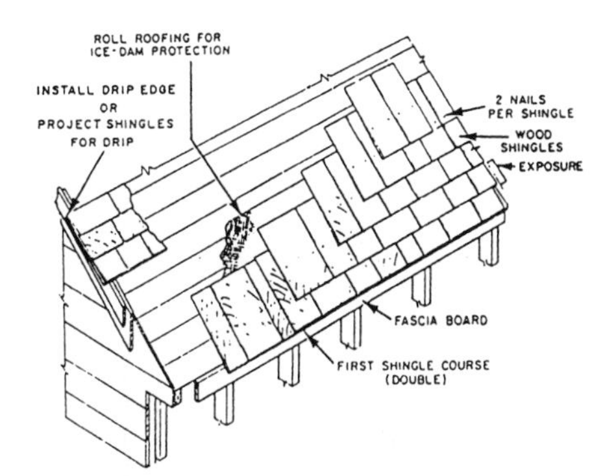

Figure 3-26.—Installation of wood shingles.

2. Use two rust-resistant nails in each shingle. Space them 3/4 inch from the edge and 1 1/2 inches above the butt line of the next course.
3. Double the first course of shingles. In all courses, allow 1/8- to 1/4-inch space between each shingle for expansion when they are wet. Offset the joints between the shingles at least 1 1/2 inches from the joints in the course below. In addition, space the joints in succeeding courses so that they do not directly line up with joints in the second course below.
4. Where valleys are present, shingle away from them. Select and precut wide valley shingles.
5. Use metal edging along the gable end to aid in guiding the water away from the sidewalls.
6. Use care when nailing wood shingles. Drive the nails just flush with the surface. The wood in shingles is soft and can be easily crushed and damaged under the nail heads.

Wood shakes are usually available in several types, but the split-and-resawed type is the most popular. The sawed face is used as the back face and is laid flat on the roof. The butt thickness of each shake ranges between 3/4 inch and 1 1/2 inches. They are usually packed in bundles of 20 square feet with five bundles to the square.

Wood shakes are applied in much the same way as wood shingles. Because shakes are much thicker (longer shakes have the thicker butts), use long galvanized nails. To create a rustic appearance, lay the butts unevenly. Because shakes are longer than shingles, they have greater exposure. Exposure distance is usually 7 1/2 inches for 18-inch shakes, 10 inches for 24-inch shakes, and 13 inches for 32-inch shakes. Shakes are not smooth on both faces, and because wind-driven rain or snow might enter, it is essential to use an underlayment between each course. A layer of felt should be used between each course with the bottom edge positioned above the butt edge of the shakes a distance equal to double the weather exposure. A 36-inch-wide strip of the asphalt felt is used at the eave line. Solid sheathing should be used when wood shakes are used for roofs in areas where wind-driven snow is common.

ROLL ROOFING.—Roll roofing is made of an organic or inorganic felt saturated with an asphalt coating and has a viscous bituminous coating. Finely ground talc or mica can be applied to both sides of the saturated felt to produce a smooth roofing. Mineral granules in a variety of colors are rolled into the upper surface while the final coating is still soft. These mineral granules protect the underlying bitumen from the deteriorating effects of sun rays. The mineral aggregates are nonflammable and increase the fire resistance and improve the appearance of the underlying bitumen. Mineral-surfaced roll roofing comes in weights of 75 to 90 pounds per square. Roll roofing may have one surface completely covered with granules or have a 2-inch plain-surface salvage along one side to allow for laps.

Roll roofing can be installed by either exposed or concealed nailing. Exposed nailing is the cheapest but doesn't last as long. This method uses a 2-inch lap at the side and ends. It is cemented with special cement and nailed with large-headed nails. In concealed-nailing installations, the roll roofing is nailed along the top of the strip and cemented with lap cement on the bottom edge. Vertical joints in the roofing are cemented into place after the upper edge is nailed. This method is used when maximum service life is required.

Double-coverage roll roofing is produced with slightly more than half its surface covered with granules. This roofing is also known as 19-inch salvage edge. It is applied by nailing and cementing with special adhesives or hot asphalt. Each sheet is lapped 19 inches, blind-nailed in the lapped salvage portion, and then cemented to the sheet below. End laps are cemented into place.

TILES.—Roofing tile was originally a thin, solid unit made by shaping moist clay in molds and drying it in the sun or in a kiln. Gradually, the term has come to include a variety of tile-shaped units made of clay, portland cement, and other materials. Tile designs have come down to us relatively unchanged from the Greeks and Romans. Roofing tiles are durable, attractive, and

Table 3-7.—Weight of Roofing Materials

MATERIAL	LB WEIGHT/SQUARE (100 sq ft)	KG WEIGHT/ 9.29 m²
Tin	100	45
Roll roofing	100	45
Asphalt shingles	130-320	59-145
Copper	150	68
Corrugated iron	200	91
Wood shingles	300	136
Asbestos-cement shingles	500	227
Portland cement shingles	500-900	227-408
Built-up roof	600	272
Sheet lead	600-800	272-363
Slate		
1/4"	700	318
3/8"	1,000	454
3/4"	1,500	680
Flat clay tile	1200	544
Clay shingles	1,100-1,400	499-635
Spanish clay tile	1,900	862
laid in mortar	2,900	1,315

resistant to fire; however, because of their weight (table 3-7), they usually require additional structural framing members and heavier roof decks.

Clay.—The clays used in the manufacture of roofing tile are similar to those used for brick. Unglazed tile comes in a variety of shades, from a yellow-orange to a deep red, and in blends of grays and greens. Highly glazed tiles are often used on prominent buildings and for landmark purposes.

Clay roofing tiles are produced as either flat or roll tile. Flat tile may be English (interlocking shingle) or French. Roll tiles are produced in Greek or Roman pan-and-cover, Spanish or Mission style (fig. 3-27).

Roll Tile.—Roll tile is usually installed over two layers of hot-mopped 15-pound felt. Double-coverage felts, laid shingle fashion, lapped 19 inches, and mopped with hot asphalt, may be required as an underlayment. The individual tiles are nailed to the sheathing through prepunched holes. Special shapes are available for starter courses, rakes, hips, and ridges. Some manufacturers produce tiles in special tile-and-a-half units for exposed locations, such as gables and hips.

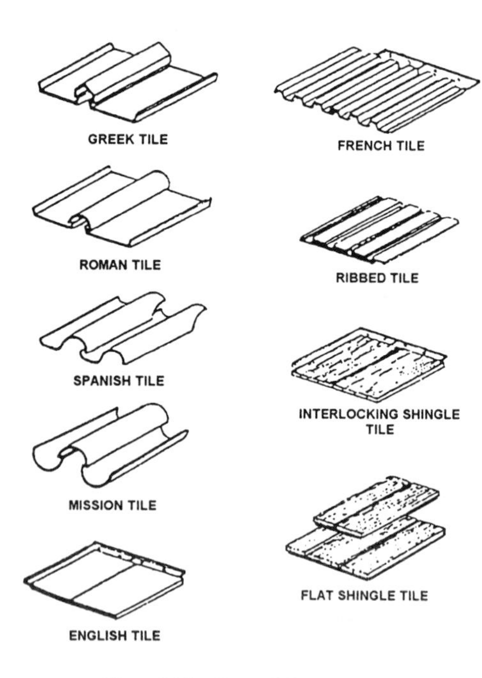

Figure 3-27.—Types of clay roof tiles.

Mission Tile.—Mission tiles are slightly tapered half-round units and are set in horizontal courses. The convex and concave sides are alternated to form pans and covers. The bottom edges of the covers can be laid with a random exposure of 6 to 14 inches to weather. Mission tile can be fastened to the prepared roof deck with copper nails, copper wire, or specially designed brass strips. The covers can be set in portland cement mortar. This gives the roof a rustic appearance, but it adds approximately 10 pounds per square to the weight of the finished roof.

Flat Tile.—Flat tile can be obtained as either flat shingle or interlocking. Single tiles are butted at the sides and lapped shingle fashion. They are produced in various widths from 5 to 8 inches with a textured surface to resemble wood shingles, with smooth colored surfaces, or with highly glazed surfaces. Interlocking shingle tiles have side and top locks, which permit the use of fewer pieces per square. The back of this type of tile is ribbed. This reduces the weight without sacrificing strength. Interlocking flat tile can be used in combination with lines of Greek pan-and-cover tile as accents.

Concrete.—The acceptance of concrete tile as a roofing material has been slow in the United States. However, European manufacturers have invested heavily in research and development to produce a uniformly high-quality product at a reasonable cost. Concrete tile is now used on more than 80 percent of all new residences in Great Britain. Modern high-speed machinery and techniques have revolutionized the industry in the United States, and American-made concrete tiles are now finding a wide market, particularly in the West.

Concrete roof tile, made of portland cement, sand, and water, is incombustible. It is also a poor conductor of heat. These characteristics make it an ideal roofing material in forested or brushy areas subject to periodic threats of fire. In addition, concrete actually gains strength with age and is unaffected by repeated freezing and thawing cycles.

Color pigments may be mixed with the basic ingredients during manufacture. To provide a glazed surface, cementitious mineral-oxide pigments are sprayed on the tile immediately after it is extruded. This glaze becomes an integral part of the tile. The surface of these tiles may be scored to give the appearance of rustic wood shakes.

Most concrete tiles are formed with side laps consisting of a series of interlocking ribs and grooves. These are designed to restrict lateral movement and provide weather checks between the tiles. The underside of the tile usually contains weather checks to halt wind-blown water. Head locks, in the form of lugs, overlap wood battens nailed to solid sheathing or strips of spaced sheathing. Nail holes are prepunched. The most common size of concrete tile is 12 3/8 by 17 inches. This provides for maximum coverage with minimum lapping.

Concrete tiles are designed for minimum roof slopes of 2 1/2:12. For slopes up to 3 1/2:12, roof decks are solidly sheeted and covered with roofing felt. For slopes greater than 3 1/2:12, the roof sheathing can be spaced. Roofing felt is placed between each row to carry any drainage to the surface of the next lower course of tile. The lugs at the top of the tiles lock over the sheathing or stripping. Generally, only every fourth tile in every fourth row is nailed to the sheathing, except where roofs are exposed to extreme winds or earthquake conditions. The weight of the tile holds it in place.

Lightweight concrete tile is now being produced using fiberglass reinforcing and a lightweight perlite aggregate. These tiles come in several colors and have the appearance of heavy cedar shakes. The weight of these shingles is similar to that of natural cedar shakes, so roof reinforcing is usually unnecessary.

SLATE.—Slate roofing is hand split from natural rock. It varies in color from black through blue-gray, gray, purple, red, and green. The individual slates may have one or more darker streaks running across them. These are usually covered during the laying of the slate. Most slate roofing is available in sizes from 10 by 6 to 26 by 14 inches. The standard thickness is 3/16 inch, but thicknesses of 1/4, 3/8, 1/2, and up to 2 inches can be obtained. Slate may be furnished in a uniform size or in random widths. The surface may be left with the rough hand-split texture or ground to a smoother texture.

The weight of a slate roof ranges from 700 to 1,500 pounds per square, depending upon thickness. The size of framing members supporting a slate roof must be checked against the weight of the slate and method of laying. The type of underlayment used for a slate roof varies, depending on local codes. The requirement ranges from one layer of 15-pound asphalt-saturated felt to 65-pound rolled asphalt roofing for slate over 3/4 inch thick.

Slate is usually laid like shingles with each course lapping the second course below at least 3 inches. The slates can be laid in even rows or at random. Each slate is predrilled with two nail holes and is held in place with two large-headed slaters' nails. These are made of hard copper wire, cut copper, or cut brass. On hips, ridges, and in other locations where nailing is not possible, the slates are held in place with waterproof elastic slaters' cement colored to match the slate. Exposed nail heads are covered with the same cement.

BITUMENS.—Hot bituminous compounds (bitumens) are used with several types of roofing systems. Both asphalt and coal-tar pitch are bitumens. Although these two materials are similar in appearance, they have different characteristics. Asphalt is usually a product of the distillation of petroleum, whereas coal-tar pitch is a byproduct of the coking process in the manufacture of steel.

Some asphalts are naturally occurring or are found in combination with porous rock. However, most roofing asphalts are manufactured from petroleum crudes from which the lighter fractions have been removed. Roofing asphalts are available in a number of different grades for different roof slopes, climatic conditions, or installation methods.

Roofing asphalts are graded on the basis of their softening points, which range from a low of 135°F (57.2°C) to a high of 225°F (107.2°C). The softening point is not the point at which the asphalt begins to flow, but is determined by test procedures established by the ASTM. Asphalts begin to flow at somewhat lower temperatures than their softening points, depending on the slope involved and the weight of the asphalt and surfacing material.

Generally, the lower the softening point of an asphalt, the better its self-healing properties and the less tendency it has to crack. Dead-flat roofs, where water may stand, or nearly flat roofs, require an asphalt that has the greatest waterproofing qualities and the self-healing properties of low-softening asphalts. A special asphalt known as dead-flat asphalt is used in such cases. As the slope of the roof increases, the need for waterproofing is lessened, and an asphalt that will not flow at expected normal temperatures must be used. For steeper roofing surfaces, asphalt with a softening point of 185°F to 205°F (85°C to 96.1°C) is used. This material is classed as steep asphalt. In hot, dry climates only the high-temperature asphalts can be used.

The softening point of coal-tar pitch generally ranges from 140°F to 155°F (60.0°C to 68.3°C). The softening point of coal-tar pitch limits its usefulness; however, it has been used successfully for years in the eastern and middle western parts of the United States on dead-level or nearly level roofs. In the southwest, where

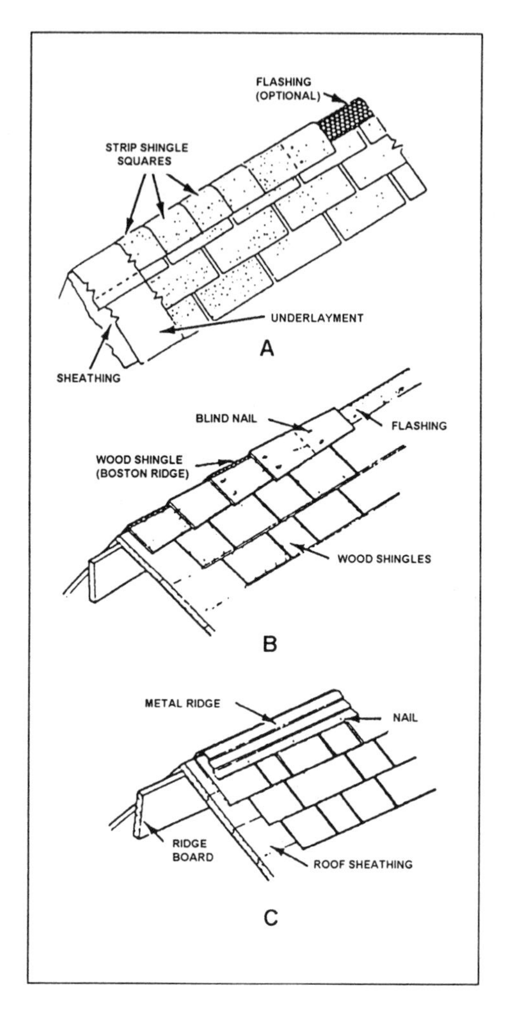

Figure 3-28.—Finish at the ridge: A. Boston ridge with strip shingles; B. Boston ridge with wood shingles; C. Metal ridge.

roof surfaces often reach temperatures of 126°F to 147°F (52.2°C to 63.9°C) in the hot desert sun, the low-softening point of coal-tar pitch makes it unsuitable as a roof surfacing material.

When used within its limitations on flat and low-pitched roofs in suitable climates, coal-tar pitch provides one of the most durable roofing membranes. Coal-tar pitch is also reputed to have cold-flow, or self-healing, qualities. This is because the molecular structure of pitch is such that individual molecules have a physical attraction for each other, so self-sealing is not

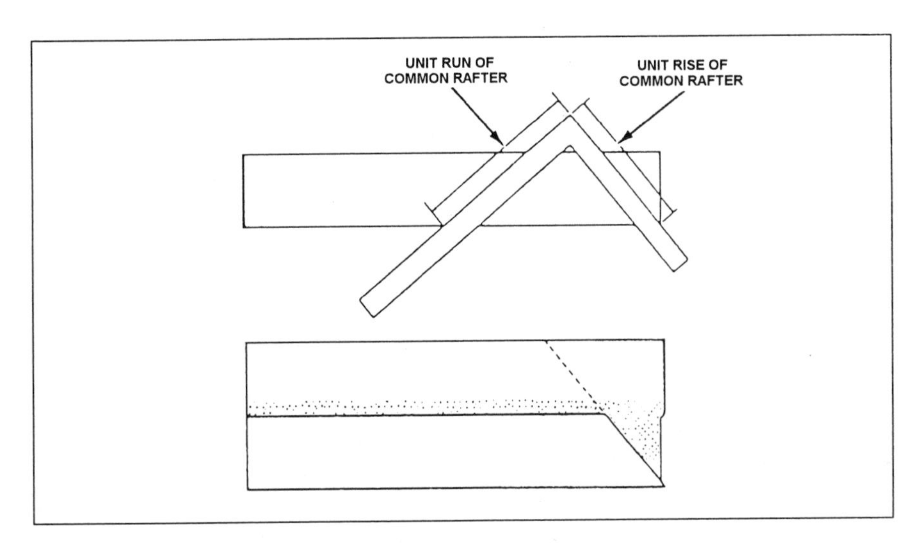

Figure 3-29.—Layout pattern for hip and valley shingles.

dependent on heat. Coal-tar pitch roofs are entirely unaffected by water. When covered by mineral aggregate, standing water may actually protect the volatile oils.

CONSTRUCTION CONSIDERATIONS.— Laying roofing on a flat surface is a relatively easy procedure. Correctly applying materials to irregular surfaces, such as ridges, hips, and valleys, is somewhat more complex.

Ridge.—The most common type of ridge and hip finish for wood and asphalt shingles is the Boston ridge. Asphalt-shingle squares (one-third of a 12- by 36-inch strip) are used over the ridge and blind-nailed (fig. 3-28, view A). Each shingle is lapped 5 to 6 inches to give double coverage. In areas where driving rains occur, use metal flashing under the shingle ridge to help prevent seepage. The use of a ribbon of asphalt roofing cement under each lap will also help.

A wood-shingle roof should be finished with a Boston ridge (fig. 3-28, view B). Shingles, 6 inches wide, are alternately lapped, fitted, and blind-nailed. As shown, the shingles are nailed in place so that the exposed trimmed edges are alternately lapped. Preassembled hip and ridge units for wood-shingle roofs are available and save both time and money.

A metal ridge can also be used on asphalt-shingle or wood-shingle roofs (fig. 3-28, view C). This ridge is formed to the roof slope and should be copper, galvanized iron, or aluminum. Some metal ridges are formed so that they provide an outlet ventilating area. However, the design should be such that it prevents rain or snow from blowing in.

Hips and Valleys.—One side of a hip or valley shingle must be cut at an angle to obtain an edge that will match the line of the hip or valley rafter. One way to cut these shingles is to use a pattern. First, select a 3 foot long 1 by 6. Determine the unit length of a common rafter in the roof (if you do not already know it). Set the framing square on the piece to get the unit run of the common rafter on the blade and the unit rise of the common rafter on the tongue (fig. 3-29). Draw a line along the tongue; then saw the pattern along this line. Note: The line cannot be used as a pattern to cut a hip or valley.

Built-up Roofing

A built-up roof, as the name indicates, is built up in alternate plies of roofing felt and bitumen. The bitumen forms a seamless, waterproof, flexible membrane that conforms to the surface of the roof deck and protects all angles formed by the roof deck and projecting surfaces. Without the reinforcement of the felts, the bitumens would crack and alligator and thus lose their volatile oils under solar radiation.

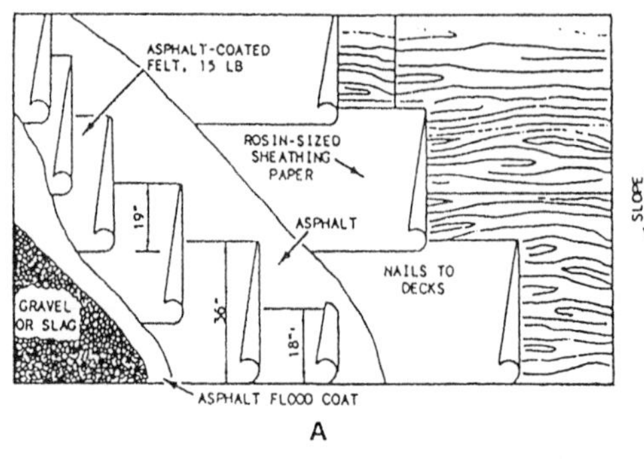

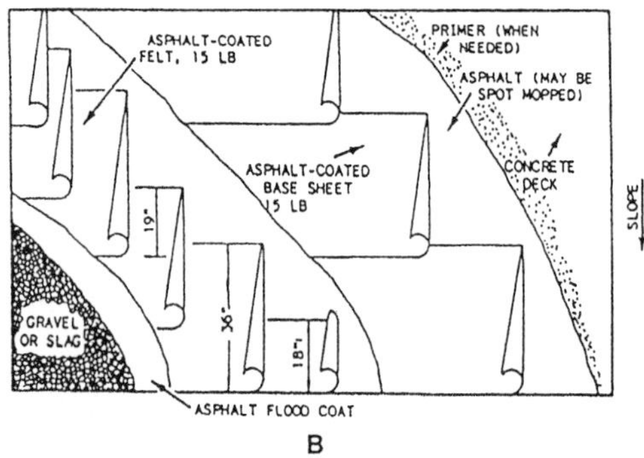

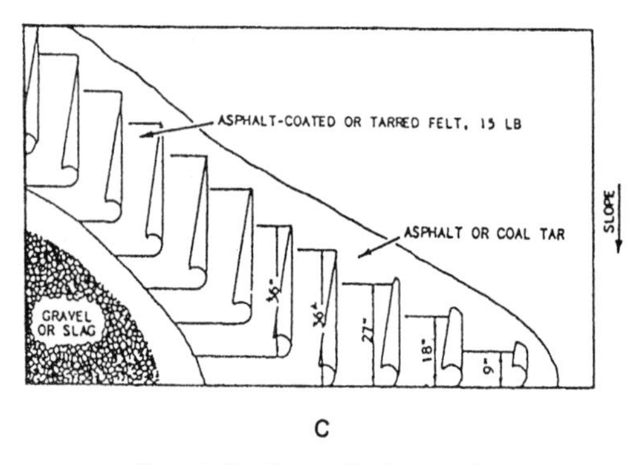

Figure 3-30.—Types of built-up roofing.

APPLICATION OF BITUMENS.—The method of applying roofing depends on the type of roof deck. Some roof decks are nailable and others are not. Figure 3-30 shows examples of wood deck (nailable), concrete deck (not nailable), and built-up roof over insulation. Nailable decks include such materials as wood or fiberboard, poured or precast units of gypsum, and nailable lightweight concrete. Non-nailable decks of concrete or steel require different techniques of roofing. View A of figure 3-30 shows a three-ply built-up roof over a nailable deck, with a gravel or slag surface. View B shows a three-ply built-up roof over a non-nailable deck, with a gravel or slag surface. View C shows a four-ply built-up roof over insulation, with a gravel or slag surface.

The temperatures at which bitumens are applied are very critical. At high temperatures, asphalt is seriously damaged and its life considerably shortened. Heating asphalt to over 500°F (260°C) for a prolonged period may decrease the weather life by as much as 50 percent. Coal-tar pitch should not be heated above 400°F (204°C). Asphalt should be applied to the roof at an approximate temperature of 375°F to 425°F (190.6°C to 218.3°C), and coal-tar pitch should be applied at 275°F to 375°F (135°C to 190°C).

Bitumens are spread between felts at rates of 25 to 35 pounds per square, depending on the type of ply or roofing felt. An asphalt primer must be used over concrete before the hot asphalt is applied. It usually is unnecessary to apply a primer under coal-tar pitch. With wood and other types of nailable decks, the ply is nailed to the deck to seal the joints between the units and prevent dripping of the bitumens through the deck.

Built-up roofs are classed by the number of plies of felt that is used in their construction. The roof may be three-ply, four-ply, or five-ply, depending on whether the roofing material can be nailed to the deck, whether insulation is to be applied underneath it, the type of surfacing desired, the slope of the deck, the climatic conditions, and the life expectancy of the roofing.

The ply-and-bitumen membrane of a built-up roof must form a flexible covering that has sufficient strength to withstand normal structure expansion. Most built-up roofs have a surfacing over the last felt ply. This protective surfacing can be applied in several ways.

SURFACING.—Glaze-coat and gravel surfaces are the most commonly seen bituminous roofs.

Glaze Coat.—A coat of asphalt can be flooded over the top layer of felt. This glaze coat protects the top layer of felt from the rays of the sun. The glaze coat is black, but it may be coated with white or aluminum surfacing to provide a reflective surface.

Gravel.—A flood coat of bitumen (60 pounds of asphalt or 70 pounds of coal-tar pitch per square) is applied over the top ply. Then a layer of aggregate, such as rock, gravel, slag, or ceramic granules, is applied while the flood coat is still hot. The gravel weighs

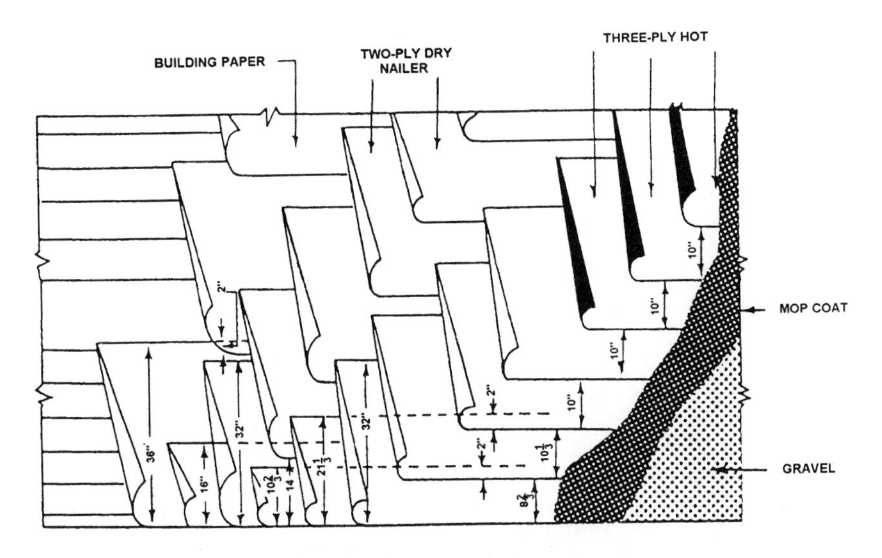

Figure 3-31.—Laying a five-ply built-up roof.

approximately 400 pounds per square and the slag 325 pounds per square. Other aggregates would be applied at a rate consistent with their weight and opacity. The surface aggregate protects the bitumen from the sun and provides a fire-resistant coating.

CAP SHEETS.—A cap sheet surface is similar to gravel-surfaced roofings, except that a mineral-surface is used in place of the flood coat and job-applied gravel. Cap-sheet roofing consists of heavy roofing felts (75 to 105 pounds per square) of organic or glass fibers. Mineral-surfaced cap sheets are coated on both sides with asphalt and surfaced on the exposed side with mineral granules, mica, or similar materials. The cap sheets are applied with a 2-inch lap for single-ply construction or a 19-inch lap if two-ply construction is desired. The mineral surfacing is omitted on the portion that is lapped. The cap sheets are laid in hot asphalt along with the base sheet. Cap sheets are used on slopes between 1/2:12 and 6:12 where weather is moderate.

COLD-PROCESS ROOFING.—Cold-applied emulsions, cutback asphalts, or patented products can be applied over the top ply of a hot-mopped roof or as an adhesive between plies. If emulsified asphalt is to be used as an adhesive between plies, special plies (such as glass fiber) must be used that are sufficiently porous to allow vapors to escape. Decorative and reflective coatings with asphalt-emulsion bases have been developed to protect and decorate roofing.

DRAINAGE.—When required, positive drainage should be established before the installation of built-up roofing. This can be achieved by the use of lightweight concrete or roofing insulation placed as specified with slopes toward roof drains, gutters, or scuppers.

APPLICATION PROCEDURES.—Built-up roofing consists of several layers of tar-rag-felt, asphalt-rag-felt, or asphalt-asbestos-felt set in a hot binder of melted pitch or asphalt.

Each layer of built-up roofing is called a **ply**. In a five-ply roof, the first two layers are laid without a binder; these are called the dry nailers. Before the nailers are nailed in place, a layer of building paper is tacked down to the roof sheating.

A built-up roof, like a shingled roof, is started at the eaves so the strips will overlap in the direction of the watershed. Figure 3-31 shows how 32-inch building paper is laid over a wood-sheathing roof to get five-ply coverage at all points in the roof. There are basically seven steps to the process.

1. Lay the building paper with a 2-inch overlap. Spot-nail it down just enough to keep it from blowing away.
2. Cut a 16-inch strip of saturated felt and lay it along the eaves. Nail it down with nails placed 1 inch from the back edge and spaced 12 inches OC.

3. Nail a full-width (32-inch) strip over the first strip, using the same nailing schedule.
4. Nail the next full-width strip with the outer edge 14 inches from the outer edges of the first two strips to obtain a 2-inch overlap over the edge of the first strip laid. Continue laying full-width strips with the same exposure (14 inches) until the opposite edge of the roof is reached. Finish off with a half-strip along this edge. This completes the two-ply dry nailer.
5. Start the three-ply hot with one-third of a strip, covered by two-thirds of a strip, and then by a full strip, as shown. To obtain a 2-inch overlap of the outer edge of the second full strip over the inner edge of the first strip laid, you must position the outer edge of the second full strip 8 2/3 inches from the outer edges of the first three strips. To maintain the same overlap, lay the outer edge of the third full strip 10 1/3 inches from the outer edge of the second full strip. Subsequent strips can be laid with an exposure of 10 inches. Finish off at the opposite edge of the roof with a full strip, two-thirds of a strip, and one-third of a strip to maintain three plies throughout.
6. Spread a layer of hot asphalt (the flood coat) over the entire roof.
7. Sprinkle a layer of gravel, crushed stone, or slag over the entire roof.

Melt the binder and maintain it at the proper temperature in a pressure fuel kettle. Make sure the kettle is suitably located. Position it broadside to the wind, if possible. The kettle must be set up and kept level. If it is not level, it will heat unevenly, creating a hazard. The first duty of the kettle operator is to inspect the kettle, especially to ensure that it is perfectly dry. Any accumulation of water inside will turn to steam when the kettle gets hot. This can cause the hot binder to bubble over, which creates a serious fire hazard. Detailed procedure for lighting off, operating, servicing, and maintaining the kettle is given in the manufacturer's manual. Never operate the kettle unattended, while the trailer is in transit, or in a confined area.

The kettle operator must maintain the binder at a steady temperature, as indicated by the temperature gauge on the kettle. Correct temperature is designated in binder manufacturer's specifications. For asphalt, it is about 400°F. The best way to keep an even temperature is to add material at the same rate as melted material is tapped off. Pieces must not be thrown into the melted mass, but placed on the surface, pushed under slowly, and then released. If the material is not being steadily tapped off, it may eventually overheat, even with the burner flame at the lowest possible level. In that case, the burner should be withdrawn from the kettle and placed on the ground to be reinserted when the temperature falls. Prolonged overheating causes flashing and impairs the quality of the binder.

Asphalt or pitch must not be allowed to accumulate on the exterior of the kettle because it creates a fire hazard. If the kettle catches fire, close the lid immediately, shut off the pressure and burner valves, and, if possible, remove the burner from the kettle. Never attempt to extinguish a kettle fire with water. Use sand, dirt, or a chemical fire extinguisher.

A hot roofing crew consists of a mopper and as many felt layers, broomers, nailers, and carriers as the size of the roof requires. The mopper is in charge of the roofing crew. It is the mopper's personal responsibility to mop on only binder that is at the proper temperature. Binder that is too hot will burn the felt, and the layer it makes will be too thin. A layer that is too thin will eventually crack, and the felt may separate from the binder. Binder that is too cold goes on too thick, so more material is used than is required.

The felt layer must get the felt down as soon as possible after the binder has been placed. If the interval between mopping and felt laying is too long, the binder will cool to the point where it will not bond well with the felt. The felt layer should follow the mopper at an interval of not more than 3 feet. The broomer should follow immediately behind the felt layer, brooming out all air bubbles and embedding the felt solidly in the binder.

Buckets of hot binder should never be filled more than three-fourths full, and they should never be carried any faster than a walk. Whenever possible, the mopper should work downwind from the felt layer and broomer to reduce the danger of spattering. The mopper must take every precaution against spattering at all times. The mopper should lift the mop out of the bucket, not drag it across the rim. Dragging the mop over the rim may upset the bucket, and the hot binder may quickly spread to the feet, or worse still to the knees, of nearby members of the roofing crew.

RECOMMENDED READING LIST

NOTE

Although the following references were current when this TRAMAN was published, their continued currency cannot be assured. You therefore need to ensure that you are studying the latest revisions.

Basic Roof Framing, Benjamin Barnow, Tab Books, Inc., Blue Ridge Summit, Pa., 1986.

Design of Wood Frame Structures for Permanence, National Forest Products Association, Washington, D.C., 1988.

Exterior and Interior Trim, John E. Ball, Delmar Pub., Albany, N.Y., 1975.

Manual of Built-up Roof Systems, C. W. Griffin, McGraw-Hill Book Co., New York, 1982.

Modern Carpentry, Willis H. Wagner, Goodheart-Wilcox Co., South Holland, Ill., 1983.

CHAPTER 4

EXTERIOR FINISH OF WALLS

In this chapter, we'll continue our discussion of exterior finishing. In chapter 3, we covered roof finishing; here, we'll examine the exterior finishing of walls, including exterior doors, windows, and glass.

EXTERIOR WALL COVERINGS

LEARNING OBJECTIVE: Upon completing this section, you should be able to identify the types of exterior wall coverings and describe procedures for installing siding.

Because siding and other types of exterior wall covering affect the appearance and the maintenance of a structure, the material and pattern should be selected carefully. Wood siding can be obtained in many different patterns and can be finished naturally, stained, or painted. Wood shingles, plywood, wood siding (paneling), fiberboard, and hardboard are some of the types of material used as exterior coverings. Masonry, veneers, metal or plastic siding, and other nonwood materials are additional choices. Many prefinished sidings are available, and the coatings and films applied to several types of base materials may eliminate the need of refinishing for many years.

WOOD SIDING

One of the materials most used for structure exteriors is wood siding. The essential properties required for siding are good painting characteristics, easy working qualities, and freedom from warp. Such properties are present to a high degree in cedar, eastern white pine, sugar pine, western white pine, cypress, and redwood; to a good degree in western hemlock, spruce, and yellow popular; and to a fair degree in Douglas fir and yellow pine.

Material

The material used for exterior siding that is to be painted should be of a high grade and free from knots, pitch pockets, and uneven edges. Vertical grain and mixed grain (both vertical and flat) are available in some species, such as redwood and western red cedar.

The moisture content at the time of application should be the same as what it will attain in service. To minimize seasonal movement due to changes in moisture content, choose vertical-grain (edge-grain) siding. While this is not as important for a stained finish, the use of edge-grain siding for a paint finish will result in longer paint life. A 3-minute dip in a water-repellent preservative before siding is installed will result in longer paint life and resist moisture entry and decay. Some manufacturers supply siding with this treatment. Freshly cut ends should be brush-treated on the job.

Patterns

Some wood siding patterns are used only horizontally and others only vertically. Some may be used in either manner if adequate nailing areas are provided. A description of each of the general types of horizontal siding follows.

PLAIN BEVEL.—Plain bevel siding (fig. 4-1) can be obtained in sizes from 1/2 by 4 inches to 1/2 by

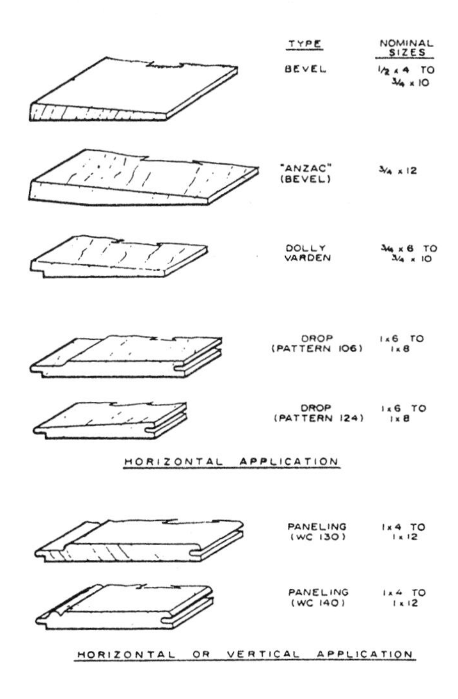

Figure 4-1.—Types of wood siding.

8 inches and also in sizes of 3/4 by 8 inches and 3/4 by 10 inches. "Anzac" siding is 3/4 by 12 inches in size. Usually, the finished width of bevel siding is about one-half inch less than the size listed. One side of beveled siding has a smooth planed surface, whereas the other has a rough resawn surface. For a stained finish, the rough or sawn side is exposed because wood stain works best and lasts longer on rough wood surfaces.

DOLLY VARDEN.—Dolly Varden siding is similar to true bevel siding except that it has shiplap edges. The shiplap edges have a constant exposure distance (fig. 4-1). Because it lays flat against the studs, it is sometimes used for garages and similar buildings without sheathing. Diagonal bracing is therefore needed to stiffen the building and help the structure withstand strong winds and other twist and strain forces.

DROP SIDING.—Regular drop siding can be obtained in several patterns, two of which are shown in figure 4-1. This siding, with matched or shiplap edges, is available in 1- by 6-inch and 1- by 8-inch sizes. It is commonly used for low-cost dwellings and for garages, usually without sheathing. Tests have shown that the tongue-and-grooved (matched) patterns have greater resistance to the penetration of wind-driven rain than the shiplap patterns, when both are treated with a water-repellent preservative.

Fiberboard and Hardboard

Fiberboard and hardboard sidings are also available in various forms. Some have a backing to provide rigidity and strength, whereas others are used directly over sheathing. Plywood horizontal lap siding, with a medium-density overlaid surface, is also available as an exterior covering material. It is usually 3/8 inch thick and 12 or 16 inches wide. It is applied in much the same manner as wood siding, except that a shingle wedge is used behind each vertical joint.

A number of siding or paneling patterns can be used horizontally or vertically (fig. 4-1). These are manufactured in nominal 1-inch thicknesses and in widths from 4 to 12 inches. Both dressed and matched and shiplapped edges are available. The narrow and medium-width patterns are usually more satisfactory under moderate moisture content changes. Wide patterns are more successful if they are vertical-grained (to keep shrinkage to a minimum). The correct moisture content is necessary in tongue-and-groove material to prevent shrinkage and tongue exposure.

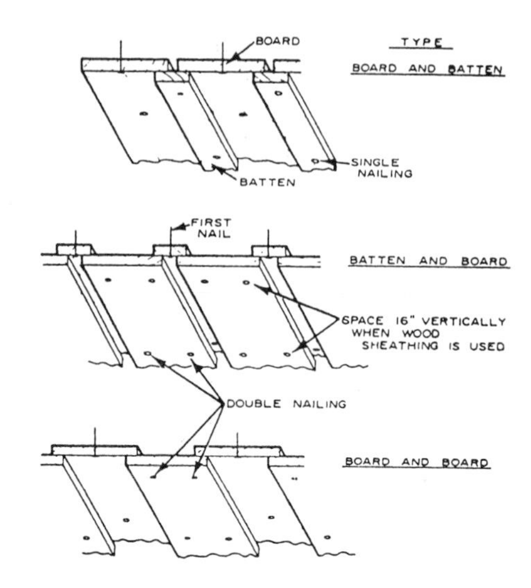

Figure 4-2.—Vertical board siding.

Treatment

Treating the edges of drop, matched, and shiplapped sidings with water-repellent preservative helps prevent wind-driven rain from penetrating the joints exposed to the weather. In areas under wide overhangs or in porches or other protected sections, the treatment is not as important. Some manufacturers provide siding with this treatment already applied.

Applications

A method of siding application, popular for some architectural styles, uses rough-sawn boards and battens applied vertically. These can be arranged in three ways: board and batten, batten and board, and board and board (fig. 4-2).

Sheet Materials

A number of sheet materials are now available for use as siding. These include plywood in a variety of face treatments and species, and hardboard. Plywood or paper-overlaid plywood, also known as panel siding, is sometimes used without sheathing. Paper-overlaid plywood has many of the advantages of plywood besides providing a satisfactory base for paint. A medium-density overlaid plywood is not common. Stud spacing of 16 inches requires a minimum thickness of panel siding of three-eighths inch. However, 1/2- or

5/8-inch-thick sheets perform better because of their greater thickness and strength.

Standard siding sheets are 4 by 8 feet; larger sizes are available. They must be applied vertically with intermediate and perimeter nailing to provide the desired rigidity. Most other methods of applying sheet materials require some type of sheathing beneath. Where horizontal joints are necessary, they should be protected by simple flashing.

An exterior-grade plywood should always be used for siding and can be obtained in grooved, brushed, and saw-textured surfaces. These surfaces are usually finished with stain. If shiplap or matched edges are not provided, the joints should be waterproofed. Waterproofing often consists of caulking and a batten at each joint and a batten at each stud if closer spacing is desired for appearance. An edge treatment of water-repellent preservative will also aid in reducing moisture penetration. When plywood is being installed in sheet form, allow a 1/16-inch edge and end spacing.

Exterior-grade particle board might also be considered for panel siding. Normally, a 5/8-inch thickness is required for 16-inch stud spacing and 3/4-inch thickness for 24-inch stud spacing. The finish must be an approved paint, and the stud wall behind must have corner bracing.

Medium-density fiberboards might also be used in some areas as exterior coverings over certain types of sheathing. Many of these sheet materials resist the passage of water vapor. Hence, when they are used, it is important that a good vapor barrier, well insulated, be used on the warm side of the insulated walls.

NONWOOD SIDING

Nonwood materials are used in some types of architectural design. Stucco or a cement-plaster finish, preferably over a wire mesh base, is common in the Southwest and the West Coast areas. Masonry veneers can be used effectively with wood siding in various finishes to enhance the beauty of both materials.

Some structures require an exterior covering with minimum maintenance. Although nonwood materials are often chosen for this reason, the paint industry is providing comparable long-life coatings for wood-base materials. Plastic films on wood siding and plywood are also promising because little or no refinishing is necessary for the life of the building.

Installation

Siding can be installed only after the window and doorframes are installed. In order to present a uniform appearance, the siding must line up properly with the drip caps and the bottom of the window and door sills. At the same time, it must line up at the corners. Siding must be properly lapped to increase wind resistance and watertightness. In addition, it must be installed with the proper nails and in the correct nailing sequence.

Fasteners

One of the most important factors in the successful performance of various siding materials is the type of fasteners used. Nails are the most common, and it is poor economy to use them sparingly. Galvanized, aluminum, and stainless steel corrosive-resistant nails may cost more, but their use will ensure spot-free siding under adverse conditions. Ordinary steel-wire nails should not be used to attach siding since they tend to rust in a short time and stain the face of the siding. In some cases, the small-head nails will show rust spots through the putty and paint. Noncorrosive nails that will not cause rust are readily available.

Two types of nails are commonly used with siding: the small-head finishing nail and the moderate-size flathead siding nail.

The small-head finishing nail is set (driven with a nail set) about 1/16 inch below the face of the siding. The hole is filled with putty after the prime coat of paint has been applied. The more commonly used flathead siding nail is nailed flush with the face of the siding and the head later covered with paint.

If the siding is to be natural finished with a water-repellent preservative or stain, it should be fastened with stainless steel or aluminum nails. In some types of prefinished sidings, nails with color-matched heads are supplied.

Nails with modified shanks are available. These include the annularly (ring) threaded shank nail and the spirally threaded shank nail. Both have greater withdrawal resistance than the smooth-shank nail, and, for this reason, a shorter nail is often used.

In siding, exposed nails should be driven flush with the surface of the wood. Overdriving may not only show the hammer mark, but may also cause objectionable splitting and crushing of the wood. In sidings with prefinished surfaces or overlays, the nails should be driven so as not to damage the finished surface.

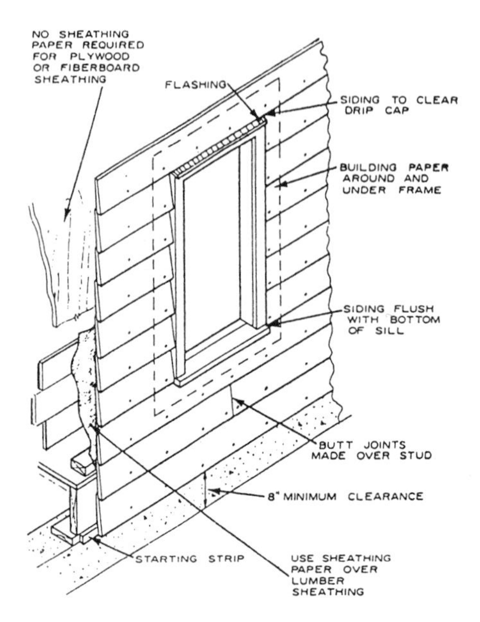

Figure 4-3.—Installation of bevel siding.

Exposure

The minimum lap for bevel siding is 1 inch. The average exposure distance is usually determined by the distance from the underside of the window sill to the top of the drip cap (fig. 4-3). From the standpoint of weather resistance and appearance, the butt edge of the first course of siding above the window should coincide with the top of the window drip cap. In many one-story structures with an overhang, this course of siding is often replaced with a frieze board. It is also desirable that the bottom of a siding course be flush with the underside of the window sill. However, this may not always be possible because of varying window heights and types that might be used in a structure.

One system used to determine the siding exposure width so that it is approximately equal above and below the window sill is as follows:

1. Divide the overall height of the window frame by the approximate recommended exposure distance for the siding used (4 inches for 6-inch-wide siding, 6 inches for 8-inch-wide siding, 8 inches for 10-inch-wide siding, and 10 inches for 12-inch-wide siding). This result will be the number of courses between the top and the bottom of the window. For example, the overall height of our sample window from the top of the drip cap to the bottom of the sill is 61 inches. If 12-inch-wide siding is used, the number of courses would be 61/10 = 6.1, or six courses. To obtain the exact exposure distance, divide 61 by 6 and the result would be 10 1/6 inches.
2. Determine the exposure distance from the bottom of the sill to just below the top of the foundation wall. If this distance is 31 inches, use three courses of 10 1/3 inches each. Thus, the exposure distance above and below the window would be almost the same (fig. 4-3).

When this system is not satisfactory because of big differences in the two areas, it is preferable to use an equal exposure distance for the entire wall height and notch the siding at the window sill. The fit should be tight to prevent moisture from entering.

Installation

Siding may be installed starting with a bottom course. It is normally blocked out with a starting strip the same thickness as the top of the siding board (fig. 4-3). Each succeeding course overlaps the upper edge of the course below it. Siding should be nailed to each stud or on 16-inch centers. When plywood, wood sheathing, or spaced wood nailing strips are used over nonwood sheathing, 7d or 8d nails may be used for 3/4-inch-thick siding. However, if gypsum or fiberboard sheathing is used, 10d nails are recommended to properly penetrate the stud. For 1/2-inch-thick siding, nails may be 1/4 inch shorter than those used for 3/4-inch siding.

The nails should be located far enough up from the butt to miss the top of the lower siding course (fig. 4-4). The clearance distance is usually 1/8 inch. This allows for slight movement of the siding because of moisture changes without causing splitting. Such an allowance is especially required for the wider (8 to 12 inch) siding.

Joints

It is good construction practice to avoid butt joints whenever possible. Use the longer sections of siding under windows and other long stretches, and use the shorter lengths for areas between windows and doors. When a butt joint is necessary, it should be made over a stud and staggered between courses.

Siding should be square cut to provide good joints. Open joints permit moisture to enter and often lead to paint deterioration. It is a good practice to brush or dip

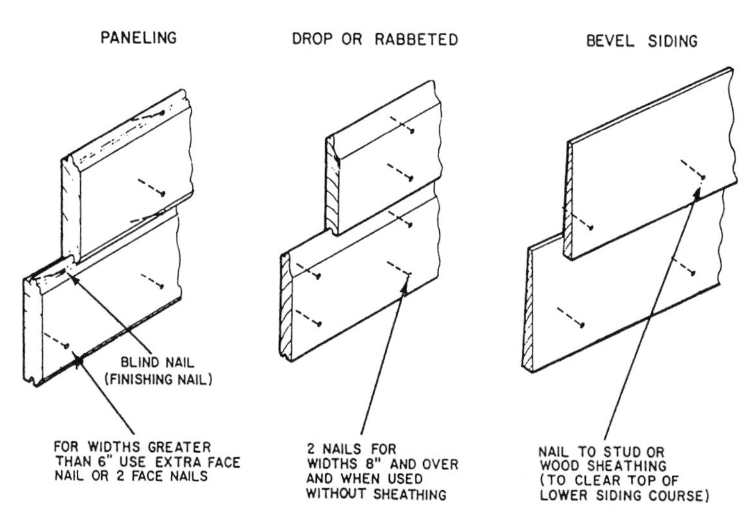

Figure 4-4.—Nailing the siding.

the fresh cut ends of the siding in a water-repellent preservative before boards are nailed in place. After the siding is in place, it is helpful to use a small finger-actuated oil can to apply the water-repellent preservative to the ends and butt joints.

Drop siding is installed in much the same way as lap siding except for spacing and nailing. Drop, Dolly Varden, and similar sidings have a constant exposure distance. The face width is normally 5 1/4 inches for 1- by 6-inch siding and 7 1/4 inches for 1- by 8-inch siding. Normally, one or two nails should be used at each stud, depending on the width (fig. 4-4). The length of the nail depends on the type of sheathing used, but penetration into the stud or through the wood backing should be at least 1 1/2 inches.

Application

There are two ways to apply nonwood siding: horizontally and vertically. Note that these are manufactured items. Make sure you follow the recommended installation procedures.

HORIZONTALLY.—Horizontally applied matched paneling in narrow widths should be blind-nailed at the tongue with a corrosion-resistant finishing nail (fig. 4-4). For widths greater than 6 inches, an additional nail should be used as shown.

Other materials, such as plywood, hardboard, or medium-density fiberboard, are used horizontally in widths up to 12 inches. They should be applied in the same manner as lap or drop siding, depending on the pattern. Prepackaged siding should be applied according to the manufacturer's directions.

VERTICALLY.—Vertically applied siding and sidings with interlapping joints should be nailed in the same manner as those applied horizontally. However, they should be nailed to blocking used between studs or to wood or plywood sheathing. Blocking should be spaced from 16 to 24 inches OC. With plywood or nominal 1-inch board sheathing, nails should be spaced on 16-inch centers only.

When the various combinations of boards and battens are used, they should also be nailed to blocking spaced from 16 to 24 inches OC between studs, or closer for wood sheathing. The first boards or battens should be fastened with nails at each blocking to provide at least 1 1/2 inches of penetration. For wide underboards, two nails spaced about 2 inches apart may be used rather than the single row along the center (fig. 4-2). Nails of the top board or batten should always miss the underboards and should not be nailed through them (fig. 4-2). In such applications, double nails should be spaced closely to prevent splitting if the board shrinks. It is also a good practice to use sheathing paper, such as 15-pound asphalt felt, under vertical siding.

Exterior-grade plywood, paper-overlaid plywood, and similar sheet materials used for siding are usually applied vertically. The nails should be driven over the studs, and the total effective penetration into the wood should be at least 1 1/2 inches. For example, 3/8-inch plywood siding over 3/4-inch wood sheathing would require a 7d nail (which is 2 1/4 inches long). This would result in a 1 1/8-inch penetration into the stud, but a total effective penetration of 1 7/8 inches into the wood sheathing.

The joints of all types of sheet material should be caulked with mastic unless the joints are of the interlapping or matched type of battens. It is a good practice to place a strip of 15-pound asphalt felt under joints.

CORNER COVERINGS

The outside corners of a wood-framed structure can be finished in several ways. Siding boards can be miter-joined at the corners. Shingles can be edge-lapped alternately. The ends of siding boards can be butted at the corners and then covered with a metal cap.

Corner Boards

A type of corner finish that can be used with almost any kind of outside-wall covering is called a corner board. This corner board can be applied to the corner with the siding or shingles end-or-edge-butted against the board.

A corner board usually consists of two pieces of stock: one piece 3 inches wide and the other 4 inches wide if an edge-butt joint between the corner boards is used. The boards are cut to a length that will extend from the top of the water table to the bottom of the frieze. They are edge-butted and nailed together before they are nailed to the corner. This procedure ensures a good tight joint (fig. 4-5). A strip of building paper should be tacked over the corner before the corner board is nailed in position (always allow an overlap of paper to cover the subsequent crack formed where the ends of the siding butts against the corner board).

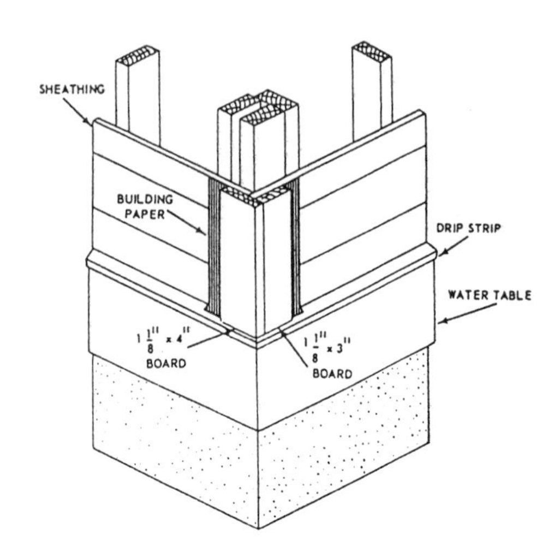

Figure 4-5.—Corner board.

Interior Corners

Interior corners (fig. 4-6, view A) are butted against a square corner board of nominal 1 1/4- or 1 3/8-inch size, depending on the thickness of the siding.

Mitered Corners

Mitering the corners (fig. 4-6, view B) of bevel and similar sidings is often not satisfactory, unless it is carefully done to prevent openings. A good joint must fit tightly the **full depth of the miter.** You should also treat the ends with a water-repellent preservative before nailing.

Metal Corners

Metal corners (fig. 4-6, view C) are perhaps more commonly used than the mitered corner and give a mitered effect. They are easily placed over each corner as the siding is installed. The metal corners should fit tightly and should be nailed on each side to the sheathing or corner stud beneath. When made of galvanized iron, they should be cleaned with a mild acid wash and primed with a metal primer before the structure is painted to prevent early peeling of the paint. Weathering of the metal will also prepare it for the prime paint coat.

Corner boards (fig. 4-5) of various types and sizes may be used for horizontal sidings of all types. They also provide a satisfactory termination for plywood and similar sheet materials. Vertical applications of matched paneling or of boards and battens are terminated by lapping one side and nailing into the edge of this member, as well as to the nailing members beneath. Corner boards are usually 1 1/8 or 1 3/8 inches wide. To give a distinctive appearance, they should be quite narrow. Plain outside casing, commonly used for window and doorframes, can be adapted for corner boards.

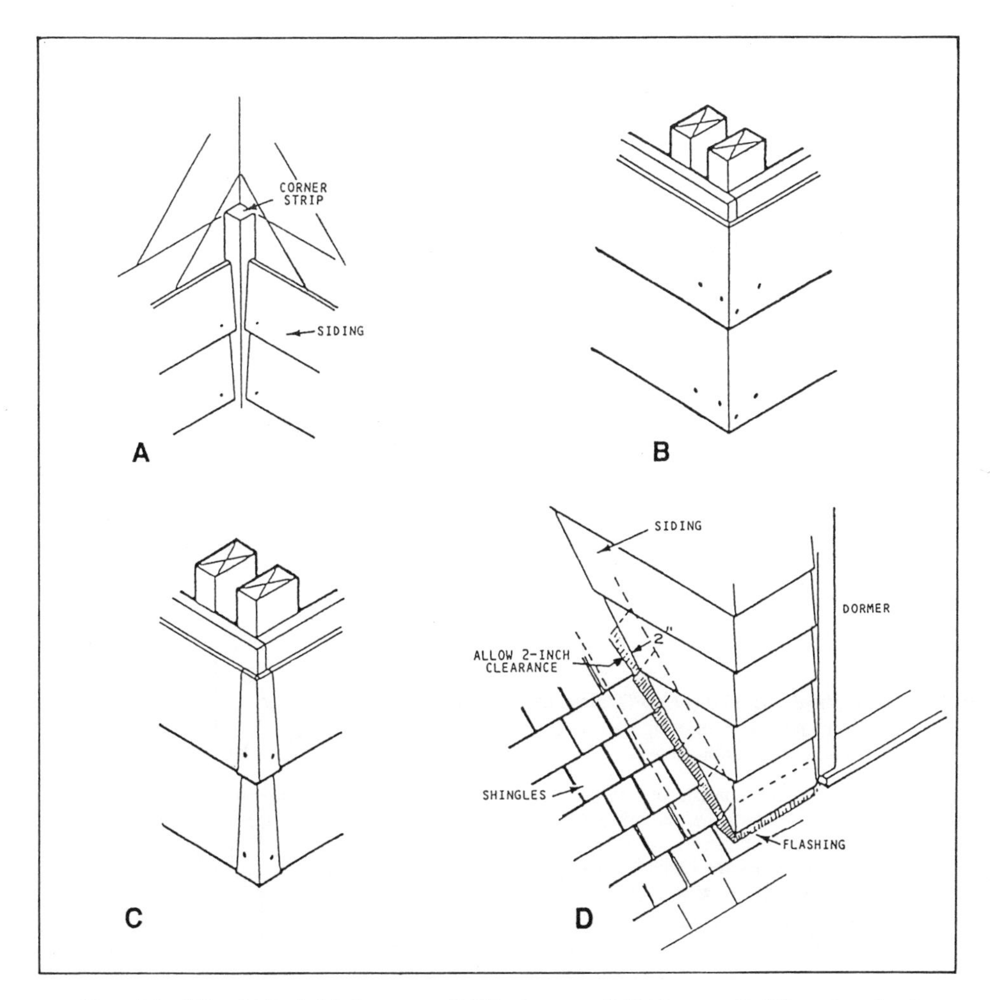

Figure 4-6.—Siding details: A. Interior corners; B. Mitered corners; C. Metal corners; D. Siding return of roof.

Shingles and Shakes

Prefinished shingle or shake exteriors are sometimes used with color-matched metal corners. They can also be lapped over the adjacent corner shingle, alternating each course. This kind of corner treatment, called lacing, usually requires that flashing be used beneath.

When siding returns against a roof surface, such as at the bottom of a dormer wall, there should be a 2-inch clearance (fig. 4-6, view D). Siding that is cut for a tight fit against the shingles retains moisture after rains and usually results in peeiing paint. Shingle flashing extending well up on the dormer wall will provide the necessary resistance to entry of wind-driven rain. Here again, a water-repellent preservative should be used on the ends of the siding at the roof line.

GABLE ENDS

At times, the materials used in the gable ends and in the walls below differ in form and application. The details of construction used at the juncture of the two materials should be such that good drainage is assured. For example, when vertical boards and battens are used at the gable end and horizontal siding below, a drip cap

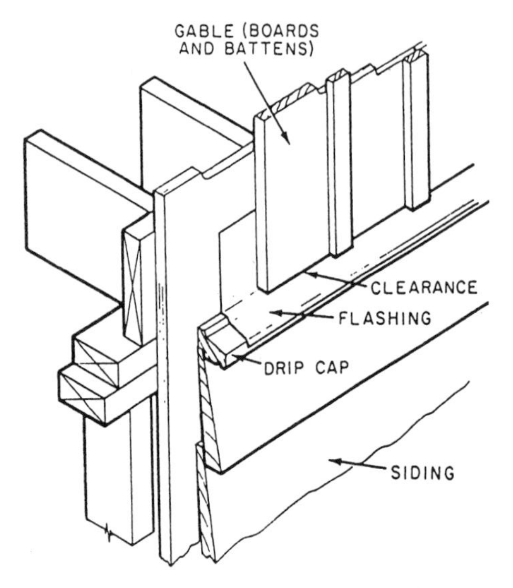

Figure 4-7.—Gable-end finish (material transition).

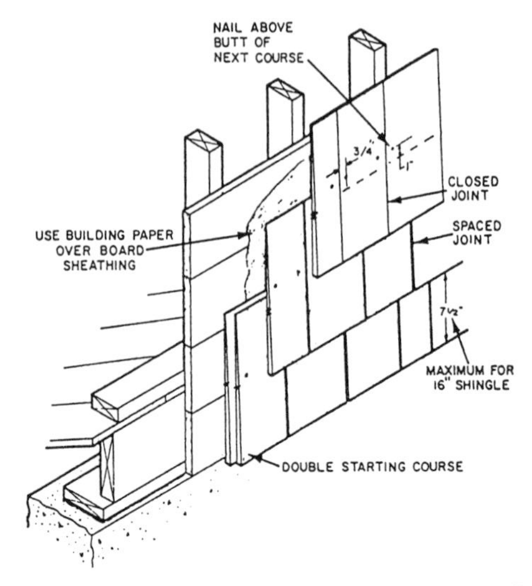

Figure 4-8.—Single coursing of sidewalls (wood shingles and shakes).

or similar molding should be used (fig. 4-7). Flashing should be used over and above the drip cap so that moisture cannot enter this transition area.

PATTERNS

Wood shingles and shakes are applied in a single- or double-course pattern. They may be used over wood or plywood sheathing. When sheathing with 3/8-inch plywood, use threaded nails. For nonwood sheathing, 1- by 3-inch or 1- by 4-inch wood nailing strips are used as a base.

In the single-course method, one course is simply laid over the other as lap siding is applied. The shingles can be second grade because only one-half or less of the butt portion is exposed (fig. 4-8). Shingles should not be soaked before application but should usually be laid with about 1/8- to 1/4-inch space between adjacent shingles to allow for expansion during rainy weather. When a siding effect is desired, shingles should be laid so that they are in contact, but only lightly. Pre-stained or treated shingles provide the best results.

In a double-course system, the undercourse is applied over the wall, and the top course is nailed directly over a 1/4- to 1/2-inch projection of the butt (fig. 4-9). The first course should be nailed only enough to hold it in place while the outer course is being applied.

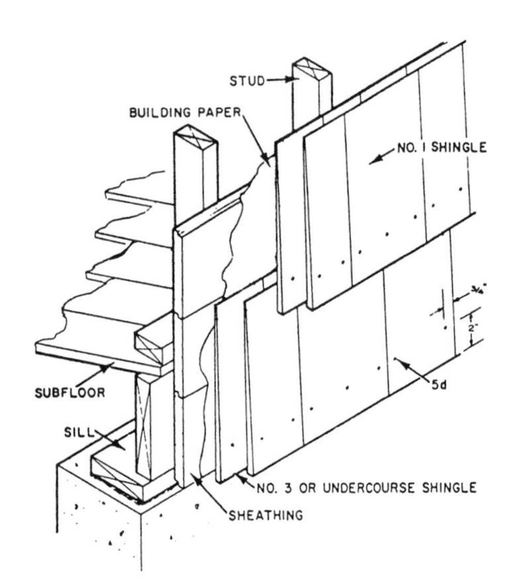

Figure 4-9.—Double coursing of sidewalls (wood shingles and shakes).

The first shingles can be a lower quality. Because much of the shingle length is exposed, the top course should be first-grade shingles.

Shingles and shakes should be applied with rust-resistant nails long enough to penetrate into the

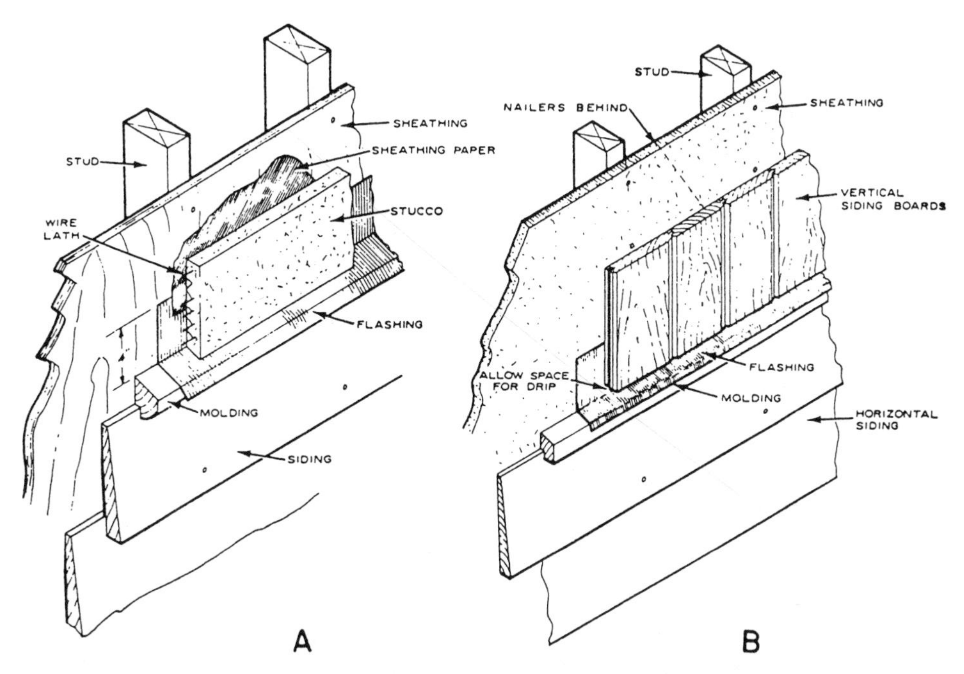

Figure 4-10.—Flashing of material changes: A. Stucco above, siding below; B. Vertical siding above, horizontal below.

wood backing strips or sheathing. In a single course, a 3d or 4d zinc-coated shingle nail is commonly used. In a double course, where nails are exposed, a 5d zinc-coated nail with a small flat head is used for the top course, and a 3d or 4d size for the undercourse. Use building paper over lumber sheathing.

FLASHING

Flashing should be installed at the junction of material changes, chimneys, and roof-wall intersections. It should also be used over exposed doors and windows, roof ridges and valleys, along the edge of a pitched roof, and any other place where rain and melted snow may penetrate.

To prevent corrosion or deterioration where unlike metals come together, use fasteners made of the same kind of metal as the flashing. For aluminum flashing, use only aluminum or stainless steel nails, screws, hangers, and clips. For copper flashing, use copper nails and fittings. Galvanized sheet metal or terneplate should be fastened with galvanized or stainless steel fasteners. (Terneplate is a steel plate coated with an alloy of lead and a small amount of tin.)

One wall area that requires flashing is at the intersection of two types of siding materials. For example, a stucco-finish gable end and a wood-siding lower wall should be flashed (fig. 4-10, view A). A wood molding, such as a drip cap, separates the two materials and is covered by the flashing, which extends at least 4 inches above the intersection. When sheathing paper is used, it should lap the flashing (fig. 4-10, view A).

When a wood-siding pattern change occurs on the same wall, the intersection should also be flashed. A vertical board-sided upper wall with horizontal siding below usually requires some type of flashing (fig. 4-10, view B). A small space above the molding provides a drip for rain. This will prevent paint peeling, which could occur if the boards were in tight contact with the molding. A drip cap (fig. 4-7) is sometimes used as a terminating molding.

DOOR AND WINDOW FLASHING

The same type of flashing as shown in figure 4-10, view A, should be used over door and window openings exposed to driving rain. However, window and door heads protected by wide overhangs in a single-story structure with a hip roof do not ordinarily require the

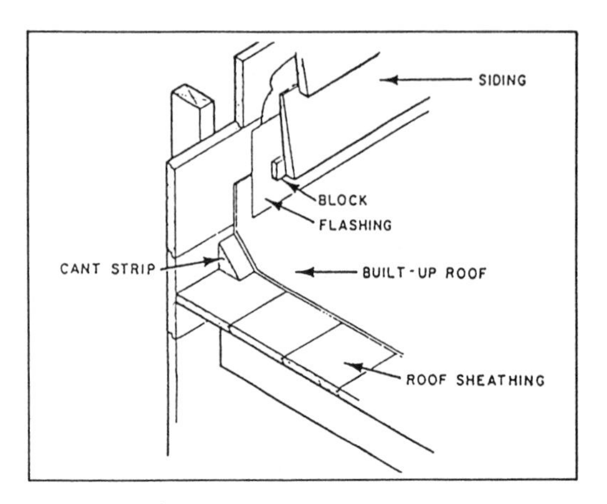

Figure 4-11.—Flashing at the intersection of an exterior wall and a flat or low-pitched roof.

flashing. When building paper is used on the sidewalls, it should lap the top edge of the flashing. To protect the walls behind the window sill in a brick veneer exterior, extend the flashing under the masonry sill up the underside of the wood sill.

Flashing is also required at the junctions of an exterior wall and a flat or low-pitched built-up roof (fig. 4-11). Where a metal roof is used, the metal is turned up on the wall and covered by the siding. A clearance should be allowed at the bottom of the siding to protect against melted snow and rain.

GUTTERS AND DOWNSPOUTS

Several types of gutters are available to carry the rainwater to the downspouts and away from the

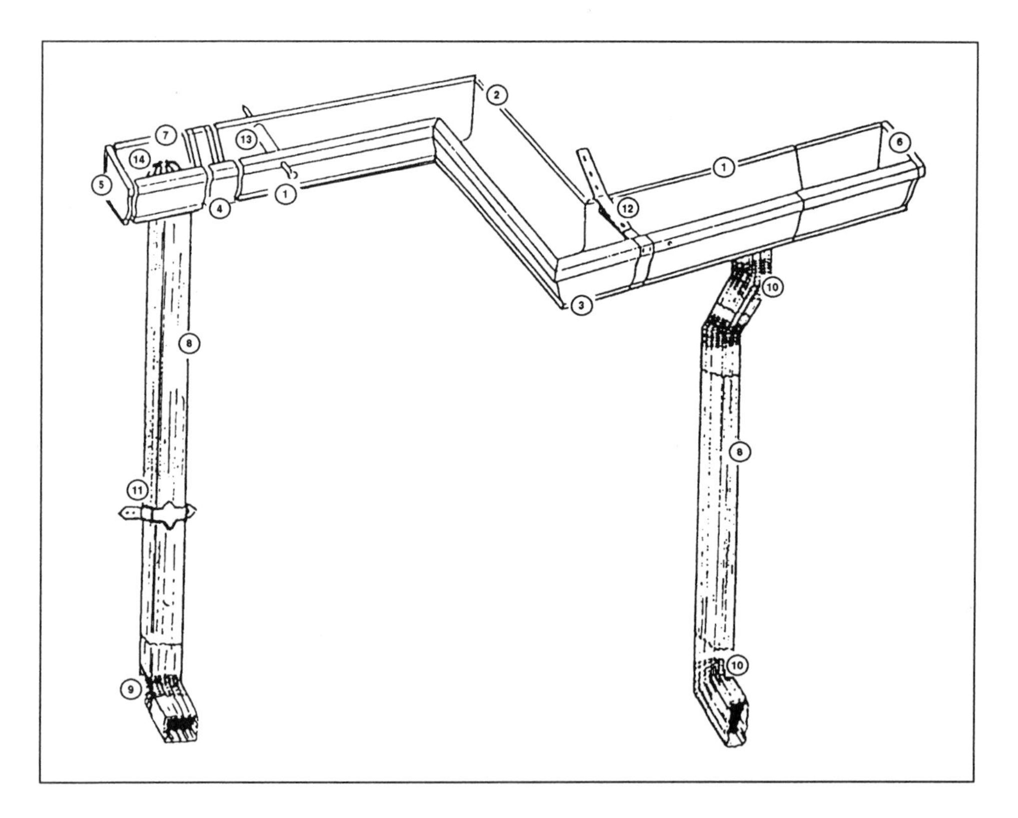

1. Gutter	4. Slip joint connectors	7. "D" end pieces	10. "B" elbows	13. 7" spike, 5" ferrule
2. Inside miter	5. End cap "L"	8. Downspout	11. Leader straps	14. Strainer basket
3. Outside miter	6. End cap "R"	9. "A" elbows	12. Hangers	

Figure 4-12.—Parts of a metal gutter system.

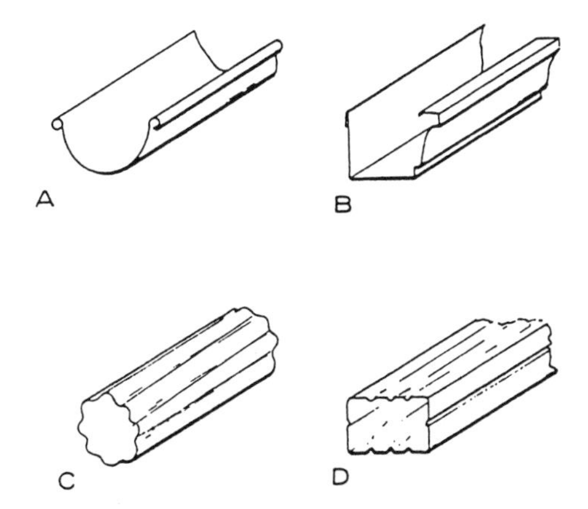

Figure 4-13.—Gutters and downspouts: A. Half-round gutter; B. "K" style gutter; C. Round downspout; D. Rectangular downspout.

foundation. On flat roofs, water is often drained from one or more locations and carried through an inside wall to an underground drain. All downspouts connected to an underground drain should be fitted with basket strainers (fig. 4-12) at the junctions of the gutter.

Perhaps the most commonly used gutter is the type hung from the edge of the roof or fastened to the edge of the cornice fascia. Metal gutters may be the half-round (fig. 4-13, view A) or "K" style (view B) and may be made of galvanized metal, copper, or aluminum. Some have a factory-applied enamel finish.

Downspouts are round or rectangular (fig. 4-13, views C and D). The round type is used for the half-round gutters. They are usually corrugated to provide extra stiffness and strength. Corrugated patterns are less likely to burst when plugged with ice.

On long runs of gutters, such as required around a hip-roof structure, at least four downspouts are desirable. Gutters should be installed with a pitch of 1 inch per 16 feet toward the downspouts. Formed or half-round gutters are suspended with flat metal hangers (fig. 4-14, views A and B). Spike and ferrule hangers are also used with formed gutters (view C). Gutter hangers should be spaced 3 feet OC.

Gutter splices, corner joints, and downspout connections should be watertight. Downspouts should be fastened to the wall by leaderstraps (fig. 4-12) or hooks. One strap should be installed at the top, one at the bottom, and one at each intermediate joint. An elbow is used at the bottom to guide the water to a splash block

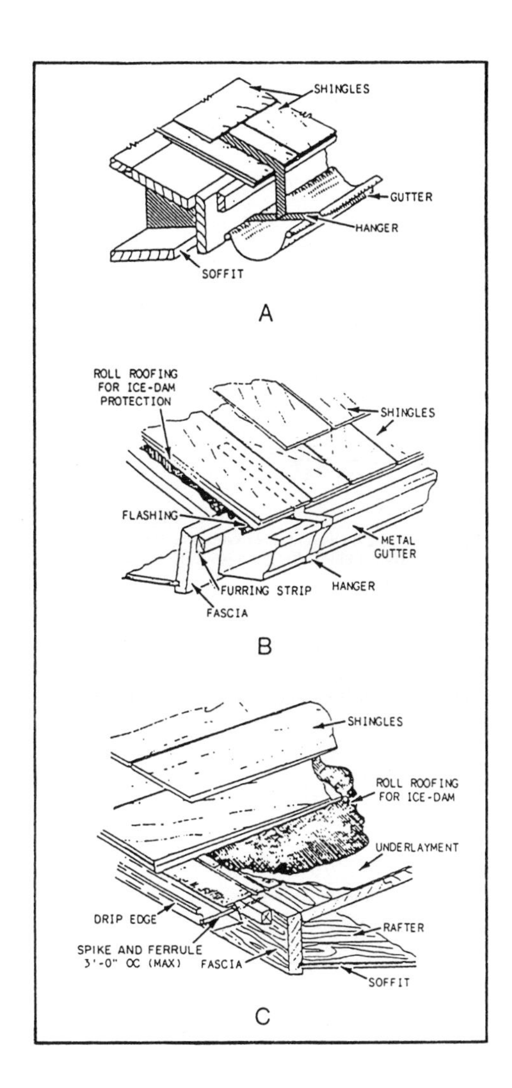

Figure 4-14.—Gutter hangers: A. Flat metal hanger with half-round gutter; B. Flat metal hanger with "K" style metal gutter; C. Spike and ferrule with formed gutter.

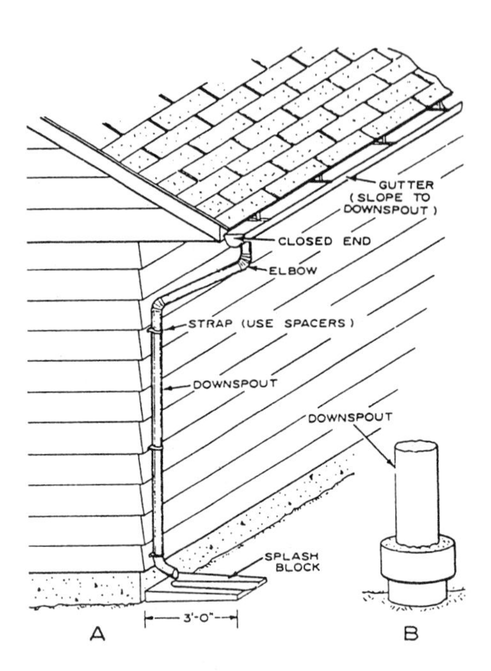

Figure 4-15.—Downspout installation: A. Downspout with splash block; B. Drain to storm sewer.

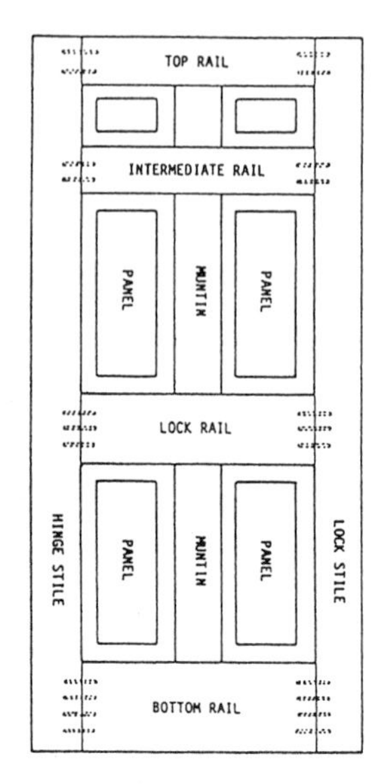

Figure 4-16.—Parts of a six-panel door.

(fig. 4-15, view A), which carries the water away from the foundation. The minimum length of a splash block should be 3 feet. In some areas, the downspout drains directly into a tile line, which carries the water to a storm sewer (view B).

EXTERIOR DOORS

LEARNING OBJECTIVE: Upon completing this section, you should be able to identify the types of exterior doors and describe basic exterior doorjamb installation procedures.

Many types of exterior doors are available to provide access, protection, safety, and privacy. Wood, metal, plastic, glass, or a combination of these materials are used in the manufacture of doors. The selection of door type and material depends on the degree of protection or privacy desired, architectural compatibility, psychological effect, fire resistance, and cost.

DOOR TYPES

Better quality exterior doors are of solid-core construction. The core is usually fiberglass, or the door is metal-faced with an insulated foam core. Solid-core doors are used as exterior doors because of the heavy service and the additional fireproofing. Hollow-core doors are normally used for interior applications. Wood doors are classified by design and method of construction as panel or flush doors.

Panel Doors

A panel door, or stile-and-rail door, consists of vertical members called stiles and horizontal members called rails. The stiles and rails enclose panels of solid wood, plywood, louvers, or glass (fig. 4-16). The stiles extend the full height at each side of the door. The vertical member at the hinged side of the door is called the hinge, or hanging, stile, and the one to which the latch, lock, or push is attached is called the closing, or lock, stile. Three rails run across the full width of the door between the stiles: the top rail, the intermediate or lock rail, and the bottom rail. Additional vertical or horizontal members, called muntins, may divide the door into any number of panels. The rails, stiles, and muntins may be assembled with either glued dowels or mortise-and-tenon joints.

Sash Doors

Panel doors in which one or more panels are glass are classed as sash (glazed) doors. Fully glazed panel

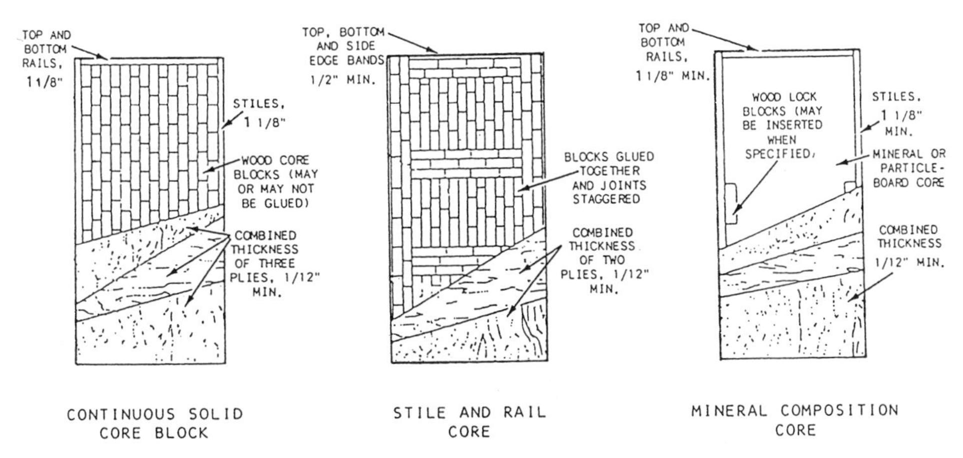

Figure 4-17.—Three types of solid-core doors.

doors with only a top and bottom rail, without horizontal or vertical muntins, are referred to as "casement" or "French doors." Storm doors are lightly constructed glazed doors. They are used in conjunction with exterior doors to improve weather resistance. Combination doors consist of interchangeable or hinged glass and screen panels.

Flush Doors

Flush doors are usually made up of thin sheets of veneer over a core of wood, particle board, or fiberboard. The veneer faces act as stressed-skin panels and tend to stabilize the door against warping. The face veneer may be of ungraded hardwood suitable for a plain finish or selected hardwood suitable for a natural finish. The appearance of flush doors may be enhanced by the application of plant-on decorative panels. Both hollow-core and solid-core doors usually have solid internal rails and stiles so that hinges and other hardware may be set in solid wood.

Two types of solid wood cores are widely used in flush-door construction (fig. 4-17). The first type, called a continuous-block, strip- or wood-stave core, consists of low-density wood blocks or strips that are glued together in adjacent vertical rows, with the end joints staggered. This is the most economical type of solid core. However, it is subject to excessive expansion and contraction unless it is sealed with an impervious skin, such as a plastic laminate.

The second type is the stile-and-rail core, in which blocks are glued up as panels inside the stiles and rails. This type of core is highly resistant to warpage and is more dimensionally stable than the continuous-block core.

In addition to the solid lumber cores, there are two types of composition solid cores. Mineral cores (see fig. 4-17) consist of inert mineral fibers bonded into rigid panels. The panels are framed within the wood rails and stiles, resulting in a core that is light in weight and little affected by moisture. Because of its low density, this type of door should not be used where sound control is important.

The other type (not shown) has particleboard, flakeboard, or waferboard cores, consisting of wood chips or vegetable fibers mixed with resins or other binders, formed under heat and pressure into solid panels. This type of core requires a solid-perimeter frame. Since particleboard has no grain direction, it provides exceptional dimensional stability and freedom from warpage. Because of its low screw-holding ability, it is usually desirable to install wood blocks in the core at locations where hardware will be attached.

DOORJAMBS

The doorjamb is the part of the frame that fits inside the masonry opening or rough frame opening. Jambs may be wood or metal. The jamb has three parts: the two side jambs and the head jamb across the top. Exterior doorjambs have a stop as part of the jamb. The stop is the portion of the jamb that the face of the door closes against. The jamb is 1 1/4 inches thick with a 1/2-inch rabbet serving as a stop.

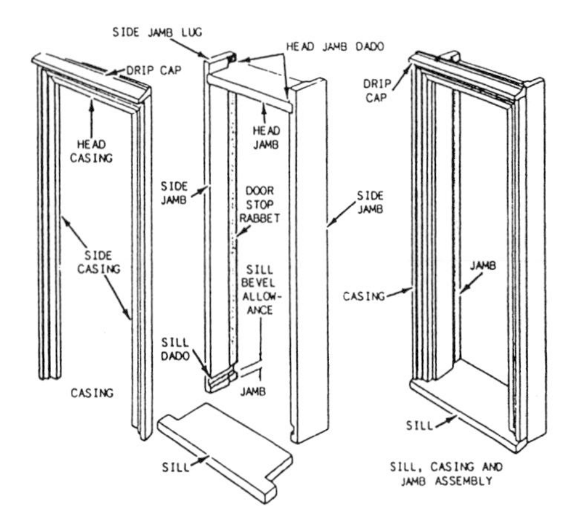

Figure 4-18.—Parts of an exterior doorframe.

Wood

Wood jambs are manufactured in two standard widths: 5 1/4 inches for lath and plaster and 4 1/2 inches for drywall. Jambs may be easily cut to fit walls of any thickness. If the jamb is not wide enough, strips of wood are nailed on the edges to form an extension. Jambs may also be custom made to accommodate various wall thicknesses.

Metal

Standard metal jambs are available for lath and plaster, concrete block, and brick veneer in 4 3/4-, 5 3/4-, 6 3/4-, and 8 3/4-inch widths. For drywall construction, the common widths available are 5 1/2 and 5 5/8 inches.

The sill is the bottom member in the doorframe. It is usually made of oak for wear resistance. When softer wood is used for the sill, a metal nosing and wear strips are generally included.

The brick mold or outside casings are designed and installed to serve as stops for the screen or combination door. The stops are provided for by the edge of the jamb and the exterior casing thickness (fig. 4-18).

Doorframes can be purchased knocked down (K.D.) or preassembled with just the exterior casing or brick mold applied. In some cases, they come preassembled with the door hung in the opening. When the doorframe is assembled on the job, nail the side

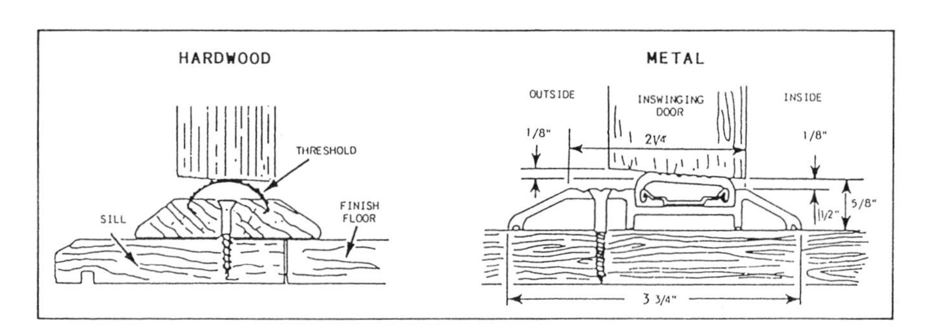

Figure 4-19.—Thresholds.

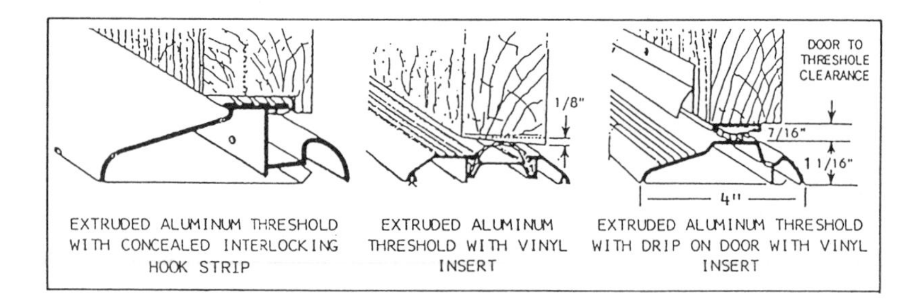

Figure 4-20.—Thresholds providing weatherproof seals.

jambs to the head jamb and sill with 10d casing nails. Then nail the casings to the front edges of the jambs with 10d casing nails spaced 16 inches OC.

Exterior doors are 1 3/4 inches thick and not less than 6 feet 8 inches high. The main entrance door is 3 feet wide, and the side or rear service door is 2 feet 8 inches wide. A hardwood or metal threshold (fig. 4-19) covers the joint between the sill and the finished floor.

The bottom of an exterior door may be equipped with a length of hooked metal that engages with a specially shaped threshold to provide a weatherproof seal. Wood and metal thresholds are available with flexible synthetic rubber tubes that press tightly against the bottom of the door to seal out water and cold or hot air. These applications are shown in figure 4-20. Manufacturers furnish detailed instruction for installation.

DOOR SWINGS

Of the various types of doors, the swinging door is the most common (fig. 4-21). The doors are classed as either right hand or left hand, depending on which side is hinged. Stand outside the door. If the hinges are on your left-hand side, it is a left-hand door. If the hinges are on your right, it is a right-hand door. For a door to swing freely in an opening, the vertical edge opposite the hinges must be beveled slightly. On a left-hand door that swings away from the viewer, a left-hand regular bevel is used; if the door opens toward the viewer, it has a left-hand reverse bevel. Similarly, if the hinges are on the right and the door swings toward the viewer, it has a right-hand reverse bevel.

A door that swings both ways through an opening is called a double-acting door. Two doors that are hinged on opposite sides of a doorway and open from the center are referred to as "double doors"; such doors are frequently double acting. One leaf of a double door may be equipped with an astragal—an extended lip that fits over the crack between the two doors. A Dutch door is one that is cut and hinged so that top and bottom portions open and close independently.

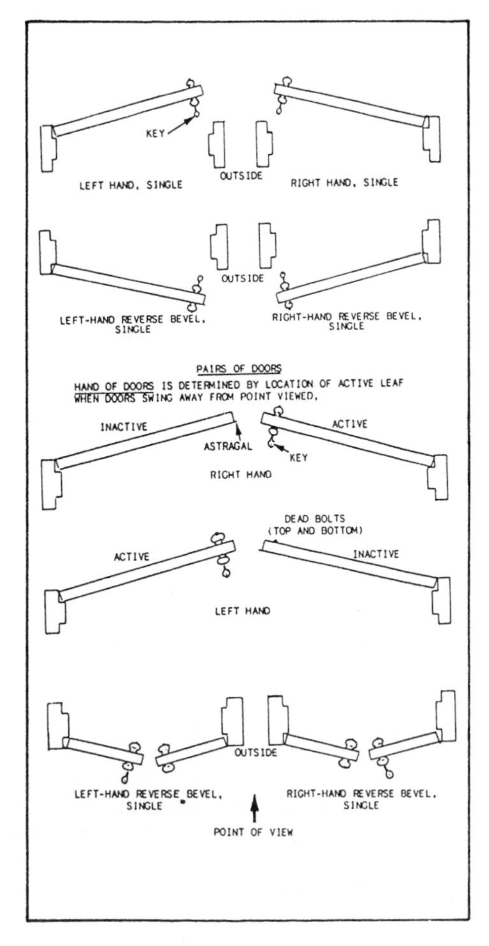

Figure 4-21.—Determining door swings.

INSTALLING THE EXTERIOR DOORFRAME

Before installing the exterior doorframe, prepare the rough opening to receive the frame. The opening should be approximately 3 inches wider and 2 inches higher than the size of the door. The sill should rest firmly on the floor framing, which normally must be notched to accommodate the sill. The subfloor, floor joists, and stringer or header joist must be cut to a depth that places the top of the sill flush with the finished floor surface.

Line the rough opening with a strip of 15-pound asphalt felt paper, 10 or 12 inches wide. In some structures, it may be necessary to install flashing over the bottom of the opening. The assembled frame is then set into the opening. Set the sill of the assembled doorframe on the trimmed-out area in the floor framing, tip the frame into place, center it horizontally, and then secure it with temporary braces.

Using blocking and wedges, you should level the sill and bring it to the correct height (even with the finished floor). Be sure the sill is level and well supported. For masonry wall and slab floors, the sill is usually placed on a bed of mortar.

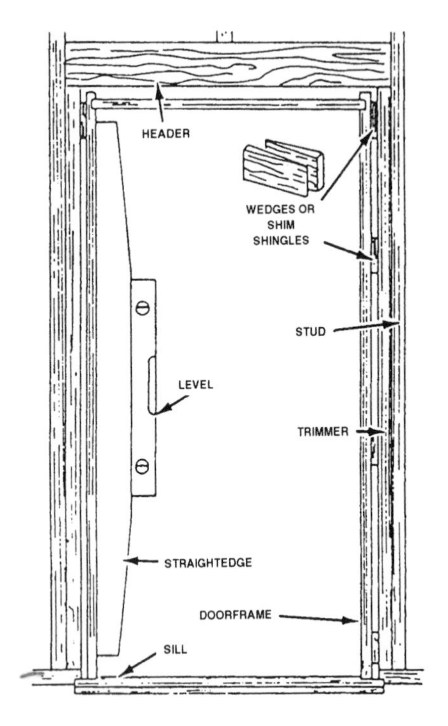

Figure 4-22.—Plumbing an exterior doorjamb.

With the sill level, drive a 16d casing nail through the side casing into the wall frame at the bottom of each side. Insert blocking or wedges between the trimmer studs and the top of the jambs. Adjust the wedges until the frame is plumb. Use a level and straightedge for this procedure (fig. 4-22).

NOTE

When setting doorframes, never drive any of the nails completely into the wood until all nails are in place and a final check has been made to make sure that no adjustments are necessary.

Place additional wedges between the jambs and stud frame in the approximate location of the lock strike plate and hinges. Adjust the wedges until the side jambs are well supported and straight. Then secure the wedges by driving a 16d casing nail through the jamb, wedge, and into the trimmer stud. Finally, nail the casing in place with 16d casing nails. These nails should be placed 3/4 inch from the outer edges of the casing and spaced 16 inches OC.

After the installation is complete, a piece of 1/4- or 3/8- inch plywood should be lightly tacked over the sill to protect it during further construction work. At this time, many Builders prefer to hang a temporary door so the interior of the structure can be secured and provide a place to store tools and materials.

Hanging the door and installing door hardware are a part of the interior finishing operation and will be described later in this TRAMAN.

PREHUNG EXTERIOR DOOR UNITS

A variety of prehung exterior door units are available. They include single doors, double doors, and doors with sidelights. Millwork plants provide detailed instructions for installing their products.

First, check the rough opening. Make sure the size is correct and that it is plumb, square, and level. Apply a double bead of caulking compound to the bottom of the opening, and set the unit in place. Spacer shims, located between the frame and door, should not be removed until the frame is firmly attached to the rough opening.

Insert shims between the side jambs and trimmer studs. They should be located at the top, bottom, and midpoint of the door. Drive 16d finishing nails through the jambs, shims, and into the structural frame members. Manufacturers usually recommend that at least two of the screws in the top hinge be replaced with 2 1/4-inch screws. Finally, adjust the threshold so that it makes smooth contact with the bottom edge of the door. After a prehung exterior door unit is installed, the door should be removed from the hinges and carefully stored. A temporary door can be used until final completion of the project.

WINDOWS

LEARNING OBJECTIVE: Upon completing this section, you should be able to identify the types of windows used in frame structures, and describe installation procedures.

The primary purpose of windows is to allow the entry of light and air, but they may also be an important part of the architectural design of a building. Windows and their frames are millwork units that are usually fully assembled at the factory, ready for use in buildings. These units often have the sash fitted and weather stripped, frame assembled, and exterior casing in place. Standard combination storms and screens or separate units can also be included. Wood components are treated with a water-repellent preservative at the factory to

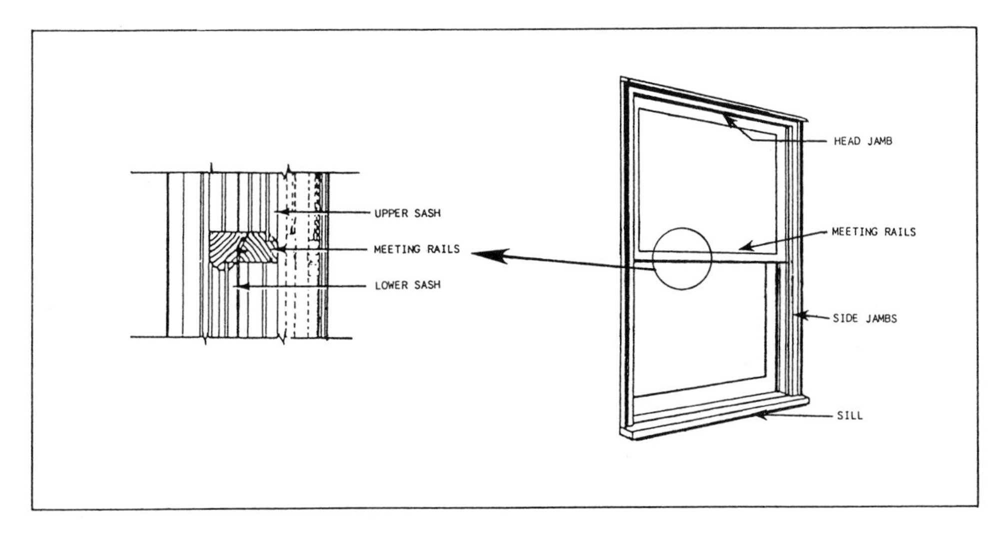

Figure 4-23.—Typical double-hung window.

provide protection before and after they are placed in the walls.

Insulated glass, used both for stationary and moveable sash, consists of two or more sheets of spaced glass with hermetically sealed edges. It resists heat loss more than a single thickness of glass and is often used without a storm sash.

Window frames and sashes should be made from a clear grade of decay-resistant heartwood stock, or from wood that has been given a preservative treatment. Examples include pine, cedar, cypress, redwood, and spruce.

Frames and sashes are also available in metal. Heat loss through metal frames and sash is much greater than through similar wood units. Glass blocks are sometimes used for admitting light in places where transparency or ventilation is not required.

Windows are available in many types. Each type has its own advantage. The principal types are double-hung, casement, stationary, awning, and horizontal sliding. In this chapter, we'll cover just the first three.

DOUBLE-HUNG WINDOWS

The double-hung window is perhaps the most familiar type of window. It consists of upper and lower sashes (fig. 4-23 detail) that slide vertically in separate grooves in the side jambs or in full-width metal weather stripping. This type of window provides a maximum face opening for ventilation of one-half the total window area. Each sash is provided with springs, balances, or compression weather stripping to hold it in place in any location. Compression weather stripping, for example, prevents air infiltration, provides tension, and acts as a counterbalance. Several types allow the sash to be removed for easy painting or repair.

The jambs (sides and top of the frames) are made of nominal 1-inch lumber; the width provides for use with drywall or plastered interior finish. Sills are made from nominal 2-inch lumber and sloped at about 3 inches in 12 inches for good drainage. Wooden sash is normally 1 3/8 inches thick. Figure 4-24 shows an assembled window stool and apron.

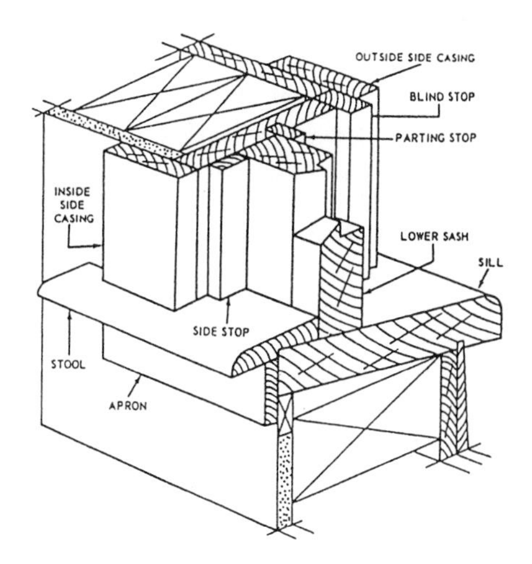

Figure 4-24.—Window stool with apron.

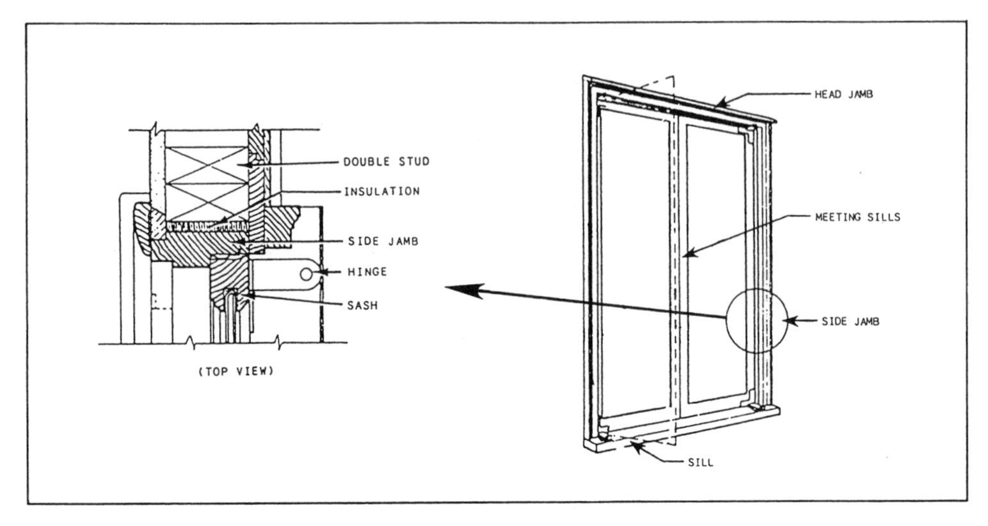

Figure 4-25.—Out-swinging casement sash.

Sash may be divided into a number of lights (glass panes or panels) by small wood members called muntins. Some manufacturers provide preassembled dividers, which snap in place over a single light, dividing it into six or eight lights. This simplifies painting and other maintenance.

Assembled frames are placed in the rough opening over strips of building paper put around the perimeter to minimize air infiltration. The frame is plumbed and nailed to side studs and header through the casings or the blind stops at the sides. Where nails are exposed, such as on the casing, use the corrosion-resistant type.

Hardware for double-hung windows includes the sash lifts that are fastened to the bottom rail. These are sometimes eliminated by providing a finger groove in the rail. Other hardware consists of sash locks or fasteners located at the meeting rail. They lock the window and draw the sash together to provide a wind-tight fit.

Double-hung windows can be arranged in a number of ways—as a single unit, doubled (or mullion), or in groups or three or more. One or two double-hung windows on each side of a large stationary insulated window are often used to create a window wall. Such large openings must be framed with headers large enough to carry roof loads.

CASEMENT WINDOWS

Casement windows consist of side-hinged sash, usually designed to swing outward (fig. 4-25). This type can be made more weathertight than the in-swinging style. Screens are located inside these out-swinging windows, and winter protection is obtained with a storm sash or by using insulated glass in the sash. One advantage of the casement window over the double-hung type is that the entire window area can be opened for ventilation.

Weather stripping is also provided for this type of window, and units are usually received from the factory entirely assembled with hardware in place. Closing hardware consists of a rotary operator and sash lock. As in the double-hung units, casement sash can be used in a number of ways—as a pair or in combinations of two or more pairs. Style variations are achieved by divided lights. Snap-in muntins provide a small, multiple-pane appearance for traditional styling.

Metal sash is sometimes used but, because of low-insulating value, should be installed carefully to prevent condensation and frosting on the interior surfaces during cold weather. A full storm-window unit is sometimes necessary to eliminate this problem in cold climates.

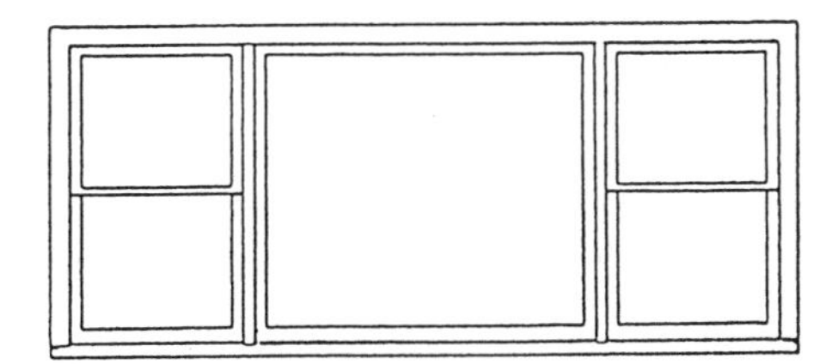

Figure 4-26.—Typical use of stationary window in combination with other types.

STATIONARY WINDOWS

Stationary windows, used alone or in combination with double-hung or casement windows (fig. 4-26), usually consist of a wood sash with a large single pane of insulated glass. They are designed to provide light, as well as be attractive, and are fastened permanently into the frame. Because of their size (sometimes 6 to 8 feet wide), stationary windows require a 1 3/4-inch-thick sash to provide strength. This thickness is required because of the thickness of the insulating glass.

Other types of stationary windows may be used without a sash. The glass is set directly into rabbeted frame members and held in place with stops. As with all window-sash units, back puttying and face puttying of the glass (with or without a stop) will assure moisture-resistance windows (fig. 4-27).

GLASS

LEARNING OBJECTIVE: Upon completing this section, you should be able to identify the different types of glass, glazing materials, and describe procedures for cutting, glazing, and installing glass.

It is surprising how many types of glass and glass-like materials are used in construction. Each has its own characteristics, advantages, and best uses. In this section, we'll cover the various types of glass and materials, and the methods used in assembling glass features ("glazing").

TYPES

The "Glass and Glazing" section of construction specifications contains a wide range of materials. These may include sheet glass, plate glass, heat- and glare-reducing glass, insulating glass, tempered glass, laminated glass, and various transparent or translucent

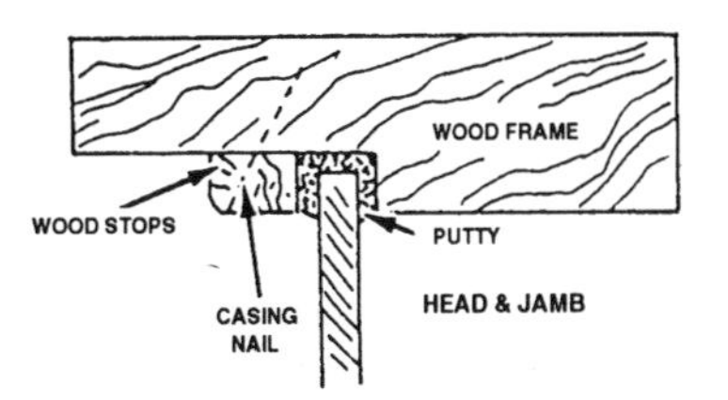

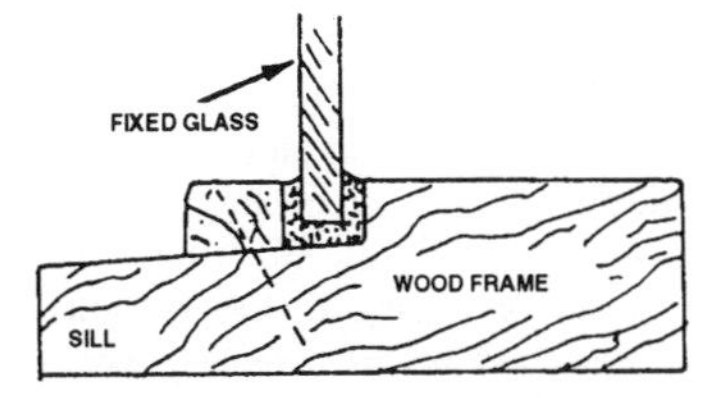

Figure 4-27.—Fixed glass in wood stops.

plastics. Also included may be ceramic-coated, corrugated, figured, and silvered and other decorative glass. Additional materials may include glazier's points, setting pads, glazing compounds, and other installation materials.

Sheet/Window

Sheet or window glass is manufactured by the flat or vertically drawn process. Because of the manufacturing process, a wave or draw distortion runs in one direction through the sheet. The degree of distortion controls the usefulness of this type of glass. For best appearance, window glass should be drawn horizontally or parallel with the ground. To ensure this, the width dimension is given first when you are ordering.

Plate

Plate glass is similar to window and heavy-sheet glass. The surface, rather than the composition or thickness, is the distinguishing feature. Plate glass is manufactured in a continuous ribbon and then cut into large sheets. Both sides of the sheet are ground and polished to a perfectly flat plane. Polished plate glass is furnished in thicknesses or from 1/8 inch to 1 1/4 inches. Thicknesses 5/16 inch and over are termed "heavy polished plate." Regular polished plate is available in three qualities: silvering, mirror glazing, and glazing. The glazing quality is generally used where ordinary glazing is required. Heavy polished plate is generally available in commercial quality only.

Heat Absorbing

Heat-absorbing glass contains controlled quantities of a ferrous iron admixture that absorbs much of the energy of the sun. Heat-absorbing glass is available in plate, heavy plate, sheet, patterned, tempered, wired, and laminated types. Heat-absorbing glass dissipates much of the heat it absorbs, but some of the heat is retained. Thus, heat-absorbing glass may become much hotter than ordinary plate glass.

Because of its higher rate of expansion, heat-absorbing glass requires careful cutting, handling, and glazing. Sudden heating or cooling may induce edge stresses, which can result in failure if edges are improperly cut or damaged. Large lights made of heat-absorbing glass that are partially shaded or heavily draped are subject to higher working stresses and require special design consideration.

Glare Reducing

Glare-reducing glass is available in two types. The first type is transparent with a neutral gray or other color tint, which lowers light transmission but preserves true color vision. The second type is translucent, usually white, which gives wide light diffusion and reduces glare. Both types absorb some of the sun's radiant energy and therefore have heat-absorbing qualities. The physical characteristics of glare-reducing glass are quite similar to those of plate glass. Although glare-reducing glass absorbs heat, it does not require the special precautions that heat-absorbing glass does.

Insulating

Insulating glass units consist of two or more sheets of glass separated by either 3/16-, 7/32-, or 1/4-inch air space. These units are factory-sealed. The captive air is dehydrated at atmospheric pressure. The edge seal can be made either by fusing the edges together or with metal spacing strips. A mastic seal and metal edge support the glass.

Insulating glass requires special installation precautions. Openings into which insulating glass is installed must be plumb and square. Glazing must be free of paint and paper because they can cause a heat trap that may result in breakage. There must be no direct contact between insulating glass and the frame into which it is installed. The glazing compound must be a nonhardening type that does not contain any materials that will attack the metal-to-glass seal of the insulating glass. Never use putty. Resilient setting blocks and spacers should be provided for uniform clearances on all units set with face stops. Use metal glazing strips for 1/2-inch-thick sash without face stops. Use a full bed of glazing compound in the edge clearance on the bottom of the sash and enough at the sides and top to make a weathertight seal. It is essential that the metal channel at the perimeter of each unit be covered by at least 1/8 inch of compound. This ensures a lasting seal.

Tempered

Tempered glass is plate or patterned glass that has been reheated to just below its melting point and then cooled very quickly by subjecting both sides to jets of air. This leaves the outside surfaces, which cool faster, in a state of compression. The inner portions of the glass are in tension. As a result, fully tempered glass has three to five times the strength against impact forces and temperature changes than untempered glass has. Tempered glass chipped or punctured on any edge or surface will shatter and disintegrate into small blunt pieces. Because of this, it cannot be cut or drilled.

Heat Strengthened

Heat-strengthened glass is plate glass or patterned glass with a ceramic glaze fused to one side. Preheating the glass to apply the ceramic glaze strengthens the glass considerably, giving it characteristics similar to tempered glass. Heat-strengthened glass is about twice as strong as plate glass. Like tempered glass, it cannot be cut or drilled.

Heat-strengthened glass is available in thicknesses of 1/4 and 5/16 inch and in limited standard sizes. It is opaque and is most often used for spandrel glazing in curtain wall systems. Framing members must be sturdy and rigid enough to support the perimeter of the tempered glass panels. Each panel should rest on resilient setting blocks. When used in operating doors and windows, it must not be handled or opened until the glazing compound has set.

Wired

Wired glass is produced by feeding wire mesh into the center of molten glass as it is passed through a pair of rollers. A hexagonal, diamond-shaped square, or rectangular pattern weld or twisted wire mesh may be used. To be given a fire rating, the mesh must be at least 25 gauge, with openings no larger than 1 1/8 inches. Also, the glass must be no less than 1/4 inch thick. Wired glass may be etched or sandblasted on one or both sides to soften the light or provide privacy. It may be obtained with a pattern on one or both sides.

Patterned

Patterned glass has the same composition as window and plate glass. It is semitransparent with distinctive geometric or linear designs on one or both sides. The pattern can be impressed during the rolling process or sandblasted or etched later. Some patterns are also available as wired glass. Pattern glass allows entry of light while maintaining privacy. It is also used for decorative screens and windows. Patterned glass must be installed with the smooth side to the face of the putty.

Laminated

Laminated glass is composed of two or more layers of either sheet or polished plate glass with one or more layers of transparent or pigmented plastic sandwiched between the layers. A vinyl plastic, such as plasticized polyvinyl resin butyl 0.015 to 0.025 inch thick, is generally used. Only the highest quality sheet or polished plate glass is used in making laminated glass. When this type of glass breaks, the plastic holds the pieces of glass and prevents the sharp fragments from shattering. When four or more layers of glass are laminated with three or more layers of plastic, the product is known as bullet-resisting glass. Safety glass has only two layers of glass and one of plastic.

Safety

Safety glass is available with clear or pigmented plastic, and either clear or heat-absorbing and glare-reducing glass. Safety glass is used where strong impact may be encountered and the hazard of flying glass must be avoided. Exterior doors with a pane area greater than 6 square feet and shower tubs and enclosures are typical applications.

Glazing compounds must be compatible with the layers of laminated plastic. Some compounds cause deterioration of the plastic in safety glass.

Mirrors

Mirrors are made with polished plate, window, sheet, and picture glass. The reflecting surface is a thin coat of metal, generally silver, gold, copper, bronze, or chromium, applied to one side of the glass. For special mirrors, lead, aluminum, platinum, rhodium, or other metals may be used. The metal film can be semi-transparent or opaque and can be left unprotected or protected with a coat of shellac, varnish, paint, or metal (usually copper). Mirrors used in building construction are usually either polished plate glass or tempered plate glass.

Proper installation requires that the weight of the mirror be supported at the bottom. Mastic installation is not recommended because it may cause silver spoilage.

Plexiglas®

Sheets made of thermoplastic acrylic resin (Plexiglas® and Lucite®, both trade names) are available in flat and corrugated sheets. This material is readily formed into curved shapes and, therefore, is often used in place of glass. Compared with glass, its surface is more readily scratched; hence, it should be installed in out-of-reach locations. This acrylic plastic is obtainable in transparent, translucent, or opaque sheets and in a wide variety of colors.

GLAZING MATERIALS

In this section, we'll discuss the various types of sealers you'll need to install, hold fast, and seal a window in its setting.

Wood-Sash Putty

Wood-sash putty is a cement composed of fine powdered chalk (whiting) or lead oxide (white lead) mixed with boiled or raw linseed oil. Putty may contain other drying oils, such as soybean or perilla. As the oil oxides, the putty hardens. Litharge (an oxide of lead) or special driers may be added if rapid hardening is required. Putty is used in glazing to set sheets of glass into frames. Special putty mixtures are available for interior and exterior glazing of aluminum and steel window sash.

A good grade of wood-sash putty resists sticking to the putty knife or glazier's hands, yet it should not be too dry to apply to the sash. In wood sash, apply a suitable primer, such as priming paints or boiled linseed oil.

Putty should not be painted until it has thoroughly set. Painting forms an airtight film, which slows the drying. This may cause the surface of the paint to crack. All putty should be painted for proper protection.

Metal-Sash Putty

Metal-sash putty differs from wood putty in that it is formulated to adhere to nonporous surfaces. It is used for glazing aluminum and steel sash either inside or outside. It should be applied as recommended by the manufacturer. Metal-sash putty should be painted within 2 weeks after application, but should be thoroughly set and hard before painting begins.

There are two grades of metal-sash putty: one for interior and one for exterior glazing. Both wood-sash putty and metal-sash putty are known as oleoresinous caulking compounds. The advantage of these materials is their low cost; their disadvantages include high shrinkage, little adhesion, and an exposed life expectancy of less than 5 years.

Elastic Compounds

Elastic glazing compounds are specially formulated from selected processed oils and pigments, which remain plastic and resilient over a longer period than the common hard putties. Butyl and acrylic compounds are the most common elastics. Butyl compounds tend to stain masonry and have a high shrinkage factor. Acrylic-based materials require heating to 110°F before application. Some shrinkage occurs during curing. At high temperatures, these materials sag considerably in vertical joints. At low temperatures, acrylic-based materials become hard and brittle. They are available in a wide range of colors and have good adhesion qualities.

Polybutane Tape

Polybutane tape is a nondrying mastic, which is available in extruded ribbon shapes. It has good adhesion qualities, but should not be used as a substitute or replacement for spacers. It can be used as a continuous bed material in conjunction with a polysulfide sealer compound. This tape must be pressure applied for proper adhesion.

Polysulfide Compounds

Polysulfide-base products are two-part synthetic rubber compounds based on a polysulfide polymer. The consistency of these compounds after mixing is similar to that of a caulking compound. The activator must be thoroughly mixed with the base compound at the job. The mixed compound is applied with either a caulking gun or spatula. The sealing surfaces must be extremely clean. Surrounding areas of glass should be protected before glazing. Excess and spilled material must be removed and the surfaces cleaned promptly. Once polysulfide elastomer glazing compound has cured, it is very difficult to remove. Any excess material left on the surfaces after glazing should be cleaned during the working time of the material (2 to 3 hours). Toluene and xylene are good solvents for this purpose.

Rubber Materials

Rubber compression materials are molded in various shapes. They are used as continuous gaskets and as intermittent spacer shims. A weathertight joint requires that the gasket be compressed at least 15 percent. Preformed materials reduce costs because careful cleaning of the glass is not necessary, and there is no waste of material.

MEASURING AND CUTTING GLASS

Always measure the length and width of the opening in which the glass is to fit at more than one place. Windows are often not absolutely square. If there is a difference between two measurements, use the smaller and then deduct 1/8 inch from the width and length to allow for expansion and contraction. Otherwise, the glass may crack with changes of temperature. This is especially true with steel casement windows.

Cutting glass is a matter of confidence—and experience. You can gain both by practicing on scrap glass before trying to cut window glass to size. Equipment required for glass cutting consists of a glass cutter, a flat, solid table, a tape measure, and a wood or metal T-square or straightedge. Look at figure 4-28. You should lightly oil the cutting wheel (view A) with a thin machine oil or lubricating fluid. Hold the cutter by resting your index finger on the flat part of the handle, as shown in view B.

To cut a piece of glass, lay a straightedge along the proposed cut, as shown in view C. Hold it down firmly with one hand and, with the glass cutter in the other, make one continuous smooth stroke along the surface of the glass with the side of the cutter pressed against the straightedge (view D). The objective is to score the glass, not cut through it. You should be able to hear the cutter bite into the glass as it moves along. Make sure the cut is continuous and that you have not skipped any section. Going over a cut is a poor practice as the glass is sure to break away at that point. Snap the glass immediately after cutting by placing a pencil or long dowel under the score line and pressing with your hands on each side of the cut (view E). Frosted or patterned glass should be cut on the smooth side. Wire-reinforced glass can be cut the same as ordinary glass, except that you will have to separate the wires by flexing the two pieces up and down until the wire breaks or by cutting the wires with side-cutting pliers.

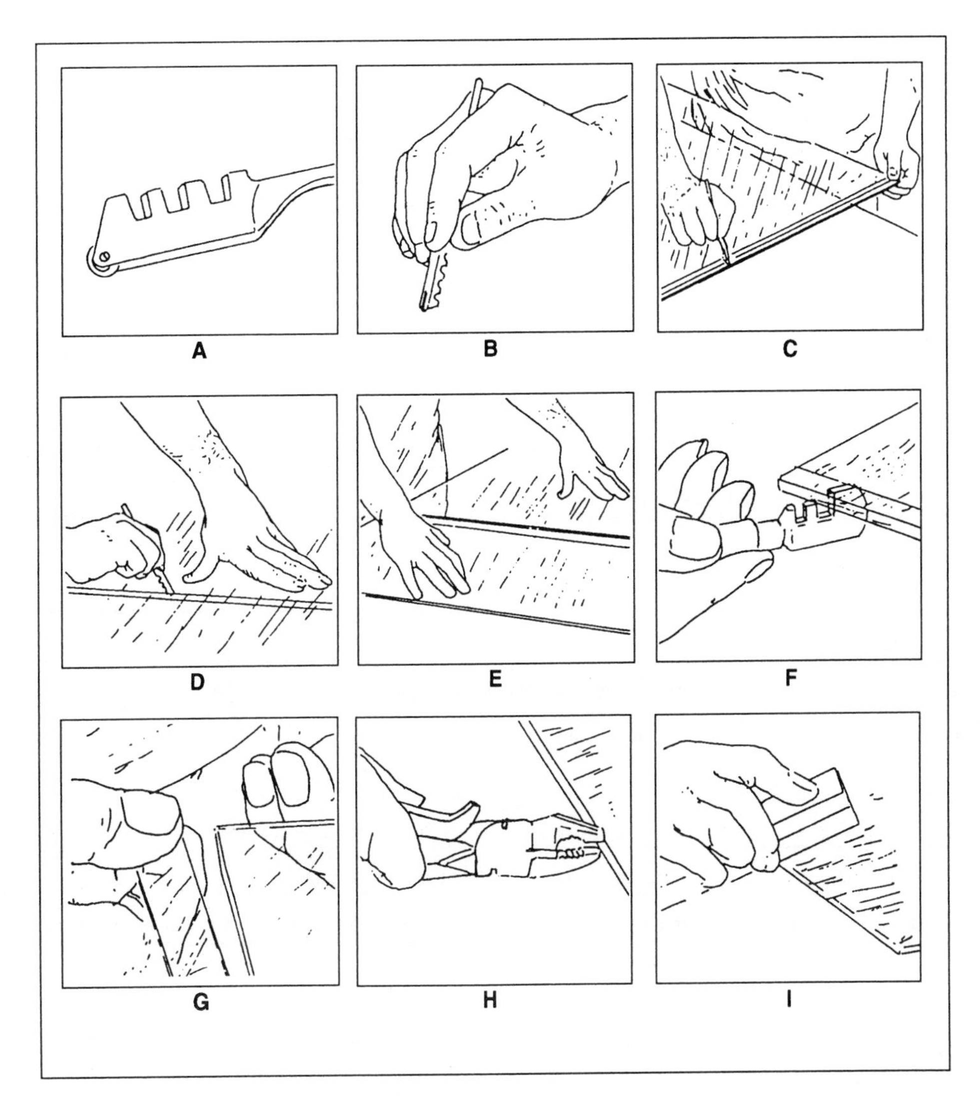

Figure 4-28.—Glass cutting.

To cut a narrow strip from a large piece of glass, score a line and then tap gently underneath the score line with the cutter to open up an inch or so of the score line (view F). Next, grasp the glass on each side of the line and gently snap off the waste piece (view G). Press downward, away from the score mark. If the strip does not break off cleanly, nibble it off with the pliers (view H) or the notches in the cutter. Slivers less than 1/2 inch wide are cut off by scoring the line and then nibbling off the waste. Do not nibble without first scoring a line. You can smooth off the edges of glass intended for shelving or tabletops with an oilstone dipped in water, as shown in view I. Rub the stone back and forth from end to end with the stone at a 45° angle to the glass. Rub the stone side to side only, not up and down.

No attempt should be made to change the size of heat-strengthened, tempered, or doubled-glazed units, since any such effort will result in permanent damage.

Table 4-1.—Weight and Maximum Sizes of Sheet Glass

THICKNESS (in.)	THICKNESS (mm)	WEIGHT (oz/sq ft)	WEIGHT (kg/m^2)	MAX. SIZE (in.)	MAX. SIZE (mm)
Window Glass					
SS 3/32	2.4	19	5.8	40 × 50	1,020 × 1,270
DS 1/8	3.2	26	7.9	60 × 80	1,520 × 2,030
Thick Sheet Glass					
3/16	4.8	40	12.2	120 × 84	3,050 × 2,130
7/32	5.6	45	13.7	120 × 84	3,050 × 2,130
1/4	6.3	52	15.9	120 × 84	3,050 × 2,130
3/8	9.5	77	23.5	160 × 84	4,060 × 2,130
7/16*	11.1	86	26.2	60 × 84	1,520 × 2,130

*Used for glass shelving and table tops

Table 4-2.—Grades of Sheet Glass

GRADE	USE
AA	For uses where superior quality is required
A	For selected glazing
B	For general glazing
Silvering quality A	For silvering mirror applications; seldom used for glazing
Silvering quality B	For mirror applications; seldom used for glazing
Greenhouse quality	For greenhouse glazing or similar applications where appearance is not critical

All heat-absorbing glass must be clean cut. Nibbling to remove flares or to reduce oversized dimensions of heat-absorbing glass is not permitted.

SHEET GLASS SIZES AND GRADES

Sheet glass is produced in a number of thicknesses, but only 3/32- and 1/8-inch sheets are commonly used as a window glass. These thicknesses are designated, respectively, as single strength (SS) and double strength (DS). Thick sheet glass, manufactured by the same method as window glass, is used in openings that exceed window-glass-size recommendations. Table 4-1 lists the thicknesses, weights, and recommended maximum sizes. Sheet glass comes in six grades (table 4-2).

The maximum size glass that may be used in a particular location is governed to a great extent by wind load. Wind velocities, and consequently wind pressures, increase with height above the ground. Various building codes or project specifications determine the maximum allowable glass area for wind load.

SASH PREPARATION

Attach the sash so that it will withstand the design load and comply with the specifications. Adjust, plumb, and square the sash to within 1/8 inch of nominal dimensions on shop drawings. Remove all rivets, screws, bolts, nail heads, welding fillets, and other projections from specified clearances. Seal all sash corners and fabrication intersections to make the sash watertight. Put a coat of primer paint on all sealing surfaces of wood sash and carbon steel sash. Use appropriate solvents to remove grease, lacquers, and other organic-protecting finishes from sealing surfaces of aluminum sash.

Wood

On old wood sashes, you must clean all putty runs of broken glass fragments and glazier's points—triangular pieces of zinc or galvanized steel driven into the rabbet. Remove loose paint and putty by scraping. Wipe the surface clean with a cloth saturated in mineral spirits or turpentine; prime the putty runs and allow them to dry.

On new wood sashes, you should remove the dust, prime the putty runs, and allow them to dry. All new wood sashes should be pressure treated for decay protection.

Metal

On old metal sashes, you must remove loose paint or putty by scraping. Use steel wool or sandpaper to remove rust. Clean the surfaces thoroughly with a cloth saturated in mineral spirits or turpentine. Prime bare metal and allow it to dry thoroughly.

On new metal sashes, you should wipe the sash thoroughly with a cloth saturated in mineral spirits or turpentine to remove dust, dirt, oil, or grease. Remove any rust with steel wool or sandpaper. If the sash is not already factory-primed, prime it with rust-inhibitive paint and allow it to dry thoroughly.

GLAZING

"Glazing" refers to the installation of glass in prepared openings of windows, doors, partitions, and curtain walls. Glass may be held in place with glazier's points, spring clips, or flexible glazing beads. Glass is kept from contact with the frame with various types of shims. Putty, sealants, or various types of caulking compounds are applied to make a weathertight joint between the glass and the frame.

Wood Sash

Most wood sash is face-glazed. The glass is installed in rabbets, consisting of L-shaped recesses cut into the sash or frame to receive and support panes of glass. The glass is held tightly against the frame by glazier's points. The rabbet is then filled with putty. The putty is pressed firmly against the glass and beveled back against the wood frame with a putty knife. A priming paint is essential in glazing wood sash. The priming seals the pores of the wood; preventing the loss of oil from the putty. Wood frames are usually glazed from the outside (fig. 4-29).

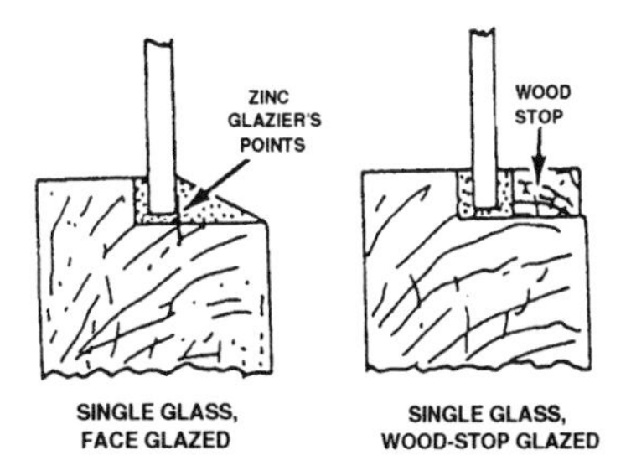

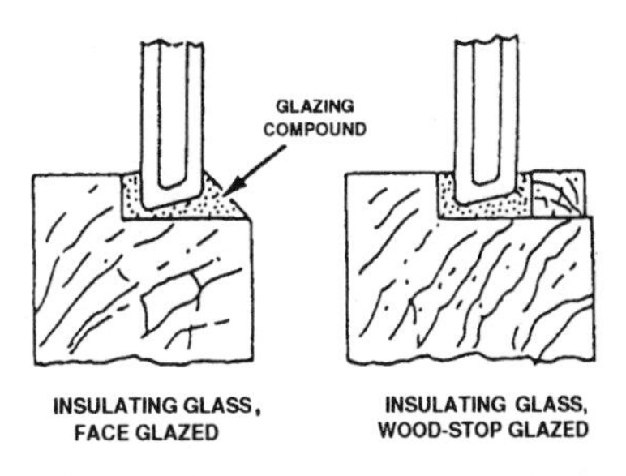

Figure 4-29.—Types of wood-sash glazing.

As we noted earlier, wood-sash putty is generally made with linseed oil and a pigment. Some putties contain soybean oil as a drying agent. Putty should not be painted until it is thoroughly set. A bead of putty or glazing compound is applied between the glass and the frame as a bedding. The bedding is usually applied to the frame before the glass is set. Back puttying is then used to force putty into spaces that may have been left between the frame and the glass.

Metal Windows and Doors

Glass set in metal frames must be prevented from making contact with metal. This may be accomplished by first applying a setting bed of metal-sash putty or glazing compound. Metal-sash putty differs from wood-sash putty in that it is formulated to adhere to a

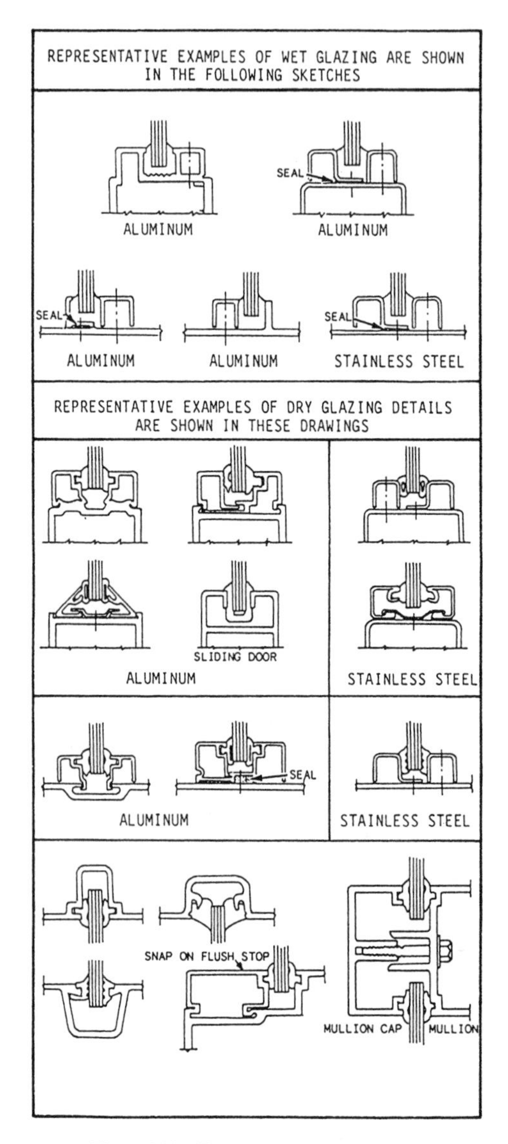

Figure 4-30.—Types of metal-sash glazing.

nonporous surface. Figure 4-30 shows examples of the types of metal-sash putty. Elastic glazing compounds may be used in place of putty. These compounds are produced from processed oils and pigments and will remain plastic and resilient over a longer period than will putty. A skin quickly forms over the outside of the compound after it is placed, while the interior remains soft. This type of glazing compound is used in windows or doors subject to twisting or vibration. It may be painted as soon as the surface has formed.

For large panes of glass, setting blocks may be placed between the glass edges and the frame to maintain proper spacing of the glass in the openings. The blocks may be of wood, lead, neoprene, or some flexible material. For large openings, flexible shims must be set between the face of the glass and the glazing channel to allow for movement. Plastics and heat-absorbing or reflective glass require more clearance to allow for greater expansion. The shims may be in the form of a continuous tape of a butyl-rubber-based compound, which has been extruded into soft, tacky, ready-to-use tape that adheres to any clean, dry surface. The tape is applied to the frame and the glass-holding stop before the glass is placed in a frame. Under compression, the tape also serves as a sealant.

Glass may be held in place in the frame by spring clips inserted in holes in the metal frame or by continuous angles or stops attached to the frame with screws or snap-on spring clips. The frames of metal windows are shaped either for outside or inside glazing.

SETTING GLASS IN WOOD AND METAL SASHES

Do not glaze or reglaze exterior sash when the temperature is 40°F or lower unless absolutely necessary. Sash and door members must be thoroughly cleaned of dust with a brush or cloth dampened with turpentine or mineral spirits. Lay a continuous 1/6-inch-thick bed of putty or compound in the putty run (fig. 4-31). The glazed face of the sash can be recognized as the size on which the glass was cut. If the glass has a bowed surface, it should be set with the concave side in. Wire glass is set with the twist vertical. Press the glass firmly into place so that the bed putty will fill all irregularities.

When glazing wood sash, insert two glazier's points per side for small lights and about 8 inches apart on all sides for large lights. When glazing metal sash, use wire clips or metal glazing beads.

After the glass has been bedded, lay a continuous bead of putty against the perimeter of the glass-face putty run. Press the putty with a putty knife or glazing tool with sufficient pressure to ensure its complete adhesion to the glass and sash. Finish with full, smooth, accurately formed bevels with clean-cut miters. Trim up

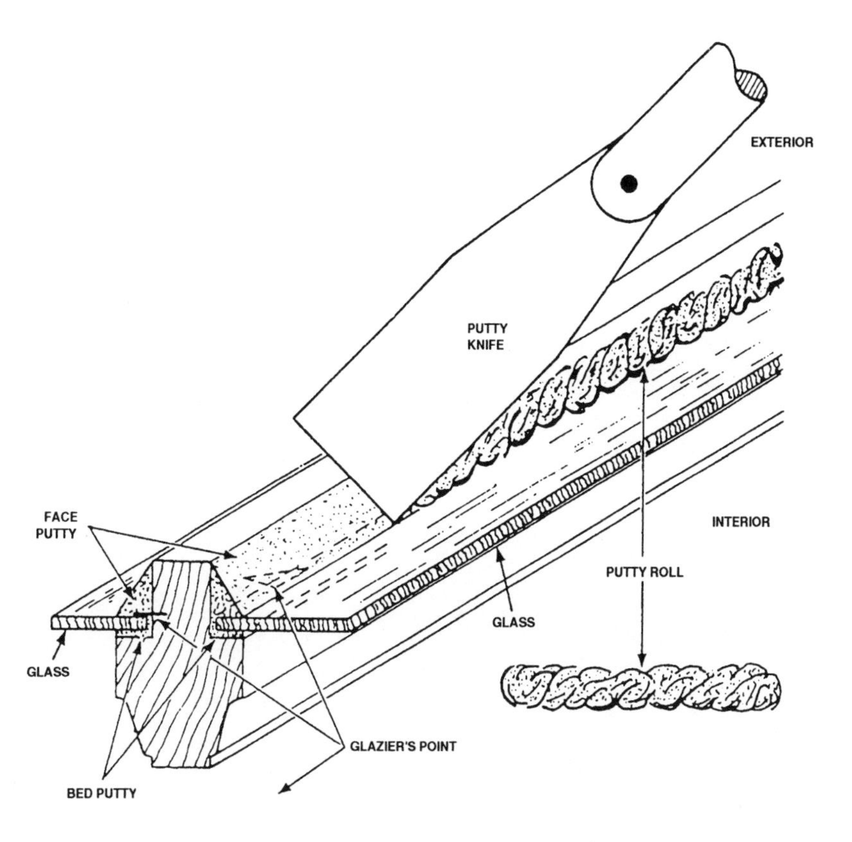

Figure 4-31.—Setting glass with glazier's points and putty.

the bed putty on the reverse side of the glass. When glazing or reglazing interior sash and transoms and interior doors, you should use wood or metal glazing beads. Exterior doors and hinged transoms should have glass secured in place with inside wood or metal glazing beads bedded in putty. In setting wired glass for security purposes, set wood or metal glazing beads, and secure with screws on the side facing the area to be protected.

Wood-sash putty should be painted as soon as it has surface-hardened. Do not wait longer than 2 months after glazing. When painting the glazing compound, overlap the glass 1/16 inch as a seal against moisture.

For metal sashes, use type 1 metal sash elastic compound. Metal-sash putty should be painted immediately after a firm skin has formed on the surface. Depending on weather conditions, the time for skinning over may be 2 to 10 days. Type II metal-sash putty can usually be painted within 2 weeks after placing. This putty should not be painted before it has hardened because early painting may retard the set.

Clean the glass on both sides after painting. A cloth moistened with mineral spirits will remove putty stains. When scrapers are used, care should be exercised to avoid breaking the paint seal at the putty edge.

After installing large glass units in buildings under construction, it is considered good practice to place a large "X" on the glass. Use masking tape or washable paint. This will alert workers so they will not walk into the glass or damage it with tools and materials.

RECOMMENDED READING LIST

NOTE

Although the following references were current when this TRAMAN was published, their continued currency cannot be assured. You therefore need to ensure that you are studying the latest revisions.

Basic Roof Framing, Benjamin Barnow, Tab Books, Inc., Blue Ridge Summit, Pa., 1986.

Design of Wood Frame Structures for Permanence, National Forest Products Association, Washington, D.C., 1988.

Exterior and Interior Trim, John E. Ball, Delmar Pub., Albany, N.Y., 1975.

Modern Carpentry, Willis H. Wagner, Goodheart-Wilcox Co., South Holland, Ill., 1983.

CHAPTER 5

INTERIOR FINISH OF WALLS AND CEILINGS

Builders are responsible for finishing the interior of the buildings of a construction project. Interior finish consists mainly of the coverings of the rough walls, ceilings, and floors, and installing doors and windows with trim and hardware. In this chapter, we'll discuss wall and ceiling coverings, including the closely related topics of insulation and ventilation. In the next chapter, we'll look at floor coverings, stairway construction, and interior door and wood trim installation.

DRYWALL AND OTHER COVERINGS

LEARNING OBJECTIVE: Upon completing this section, you should be able to describe drywall installation and finishing procedures, and identify various types of wall and ceiling coverings and the tools, fasteners, and accessories used in installation.

Though lath-and-plaster finish is still used in building construction today, drywall finish has become the most popular. Drywall finish saves time in construction, whereas plaster finish requires drying time before other interior work can be started. Drywall finish requires only short drying time since little, if any, water is required for application. However, a gypsum drywall demands a moderately low moisture content of the framing members to prevent "nail-pops." Nail-pops result when frame members dry out to moisture equilibrium, causing the nailhead to form small "humps" on the surface of the board. Stud alignment is also important for single-layer gypsum finish to prevent a wavy, uneven appearance. Thus, there are advantages to both plaster and gypsum drywall finishes and each should be considered along with the initial cost and maintenance.

DRYWALL

There are many types of drywall. One of the most widely used is gypsum board in 4- by 8-foot sheets. Gypsum board is also available in lengths up to 16 feet. These lengths are used in horizontal application. Plywood, hardboard, fiberboard, particleboard, wood paneling, and similar types are also used. Many of these drywall finishes come prefinished.

The use of thin sheet materials, such as gypsum board or plywood, requires that studs and ceiling joists have good alignment to provide a smooth, even surface. Wood sheathing often corrects misaligned studs on exterior walls. A strongback (fig. 5-1) provides for alignment of ceiling joists of unfinished attics. It can also be used at the center of a span when ceiling joists are uneven.

Gypsum wallboard is the most commonly used wall and ceiling covering in construction today. Because gypsum is nonflammable and durable, it is appropriate for application in most building types. Sheets of drywall are nailed or screwed into place, and nail indentions or "dimples" are filled with joint compound. Joints between adjoining sheets are built up with special tape and several layers (usually three) of joint compound. Drywall is easily installed, though joint work can be tedious.

Drywall varies in composition, thickness, and edge shape. The most common sizes with tapered edges are 1/2 inch by 4 feet by 8 feet and 1/2 inch by 4 feet by 12 feet.

Regular gypsum board is commonly used on walls and ceilings and is available in various thicknesses. The most common thicknesses are 1/2 inch and 5/8 inch. Type X gypsum board has special additives that make it fire resistant.

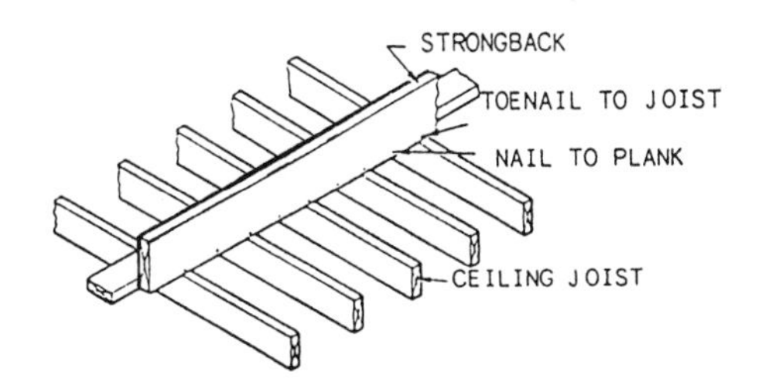

Figure 5-1.—Strongback for alignment of ceiling joists.

Types

MR (moisture resistant) or WR (water resistant) board is also called greenboard and blueboard. Being water resistant, this board is appropriate for bathrooms, laundries, and similar areas with high moisture. It also provides a suitable base for embedding tiles in mastic. MR or WR board is commonly 1/2 inch thick.

Sound-deadening board is a sublayer used with other layers of drywall (usually type X); this board is often 1/4 inch thick.

Backing board has a gray paper lining on both sides. It is used as a base sheet on multilayer applications. Backing board is not suited for finishing and decorating.

Foil-backed board serves as a vapor barrier on exterior walls. This board is available in various thicknesses.

Vinyl-surfaced board is available in a variety of colors. It is attached with special drywall finish nails and is left exposed with no joint treatment.

Plasterboard or gypsum lath is used for plaster base. It is available in thickness starting at 3/8 inch, widths 16 and 24 inches, and length is usually 48 inches. Because it comes in manageable sizes, it's widely used as a plaster base instead of metal or wood lath for both new construction and renovation. This material is not compatible with portland cement plaster.

The varying lengths of drywall allow you to lay out sheets so that the number of seams is kept to a minimum. End points can be a problem, however, since the ends of the sheets aren't shaped (only the sides are). As sheet length increases, so does weight, unwieldiness, and the need for helpers. Standard lengths are 8, 9, 10, 12, and 14 feet. Sixteen-foot lengths are also available. Use the thickness that is right for the job. One-half-inch drywall is the dimension most commonly used. That thickness, which is more than adequate for studs 16 inches on center (OC), is also considered adequate where studs are 24 inches OC. Where ceiling joists are 16 inches OC, use 1/2-inch drywall, whether it runs parallel or perpendicular to joists. Where ceiling joists are 24 inches OC, though, use 1/2-inch drywall only if the sheets are perpendicular to joists.

Drywall of 1/4- and 3/8-inch thicknesses is used effectively in renovation to cover existing finish walls with minor irregularities. Neither is adequate as a single layer for walls or ceiling, however. Two 1/4-inch-thick plies are also used to wrap curving walls.

Drywall of 5/8-inch thickness is favored for quality single-layer walls, especially where studs are 24 inches OC. Use 5/8-inch drywall for ceiling joists 24 inches OC, where sheets run parallel to joists. This thickness is widely used in multiple, fire-resistant combinations.

There are several types of edging in common use. Tapered allows joint tape to be bedded and built up to a flat surface. This is the most common edge used. Tapered round is a variation on the first type. Tapered round edges allow better joints. These edges are more easily damaged, however. Square makes an acceptable exposed edge. Beveled has an edge that, when left untapped, gives a paneled look.

Tools

Commonly used tools in drywall application include a tape measure, chalk line, level, utility and drywall knives, straightedge, and a 48-inch T square (drywall square) or framing square. Other basic tools include a keyhole saw, drywall hammer (or convex head hammer), screw gun, drywall trowel, corner trowel, and a foot lift. Some of these tools are shown in figure 5-2.

The tape measure, chalk line, and level are used for layout work. The utility and drywall knives, straightedge, and squares are used for scoring and breaking drywall. The keyhole saw is used for cutting irregular shapes and openings, such as outlet box openings. A convex head, or drywall, hammer used for drywall nails will "dimple" the material without tearing the paper. The screw gun quickly sinks drywall screws to the adjusted depth and then automatically disengages.

Drywall knives have a variety of uses. The 6-inch knife is used to bed the tape in the first layer of joint compound and for filling nail or screw dimples. The 12-inch finishing knife "feathers out" the second layer of joint compound and is usually adequate for the third or "topping" layer. Knives 16 inches and wider are used for applying the topping coat. Clean and dry drywall knives after use. Use only the drywall knives for the purpose intended—to finish drywall.

The drywall trowel resembles a concrete finishing trowel and is manufactured with a 3/16-inch concave bow. This trowel, also referred to as a "flaring," "feathering," or "bow" trowel, is used when applying the finish layer of joint compound. A corner trowel is almost indispensable for making clean interior corners.

For sanding dried joint compound smooth, use 220 grit sandpaper. Sandpaper should be wrapped around a sanding block or can be used on an orbital sander. When

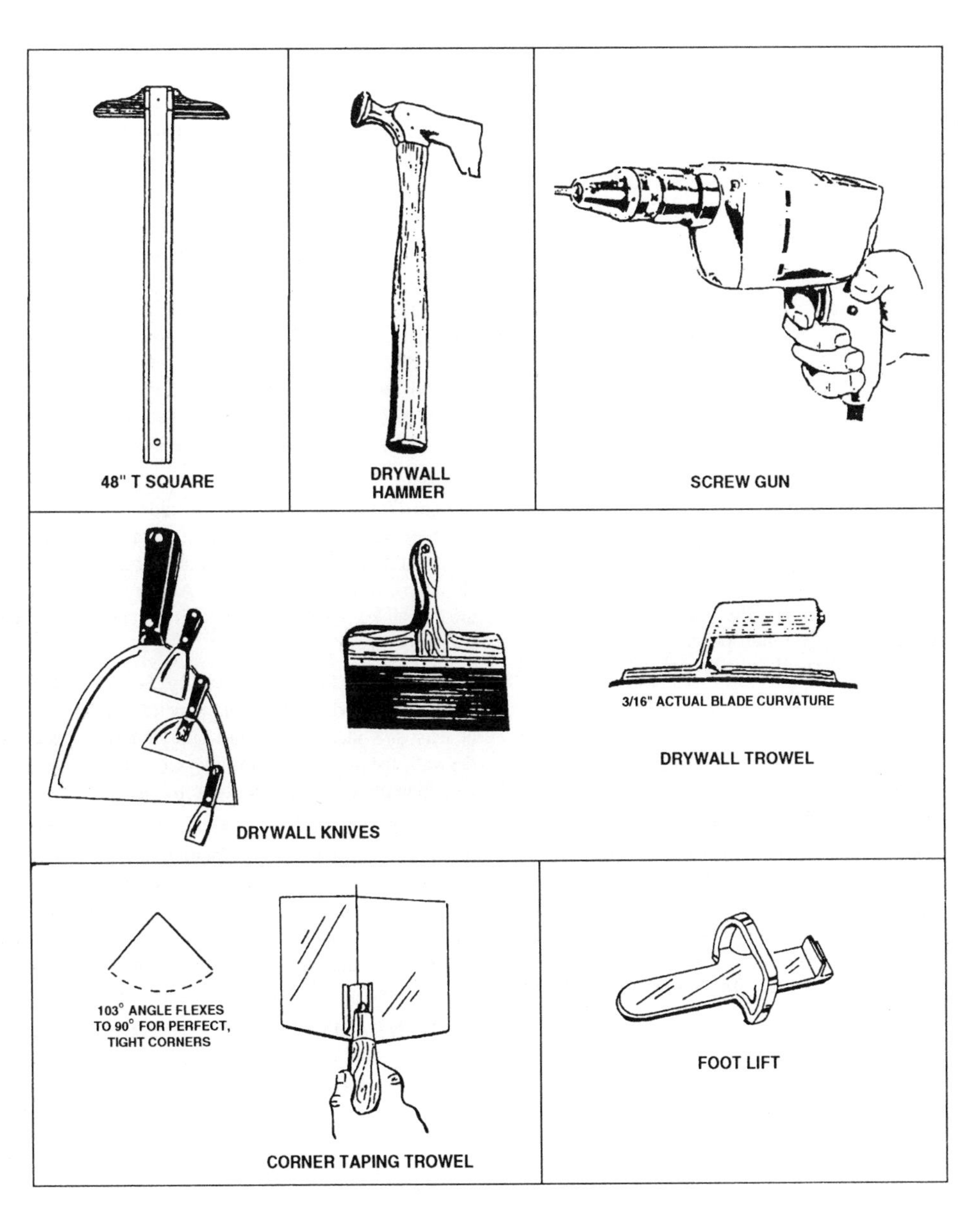

Figure 5-2.—Common tools for drywall installation.

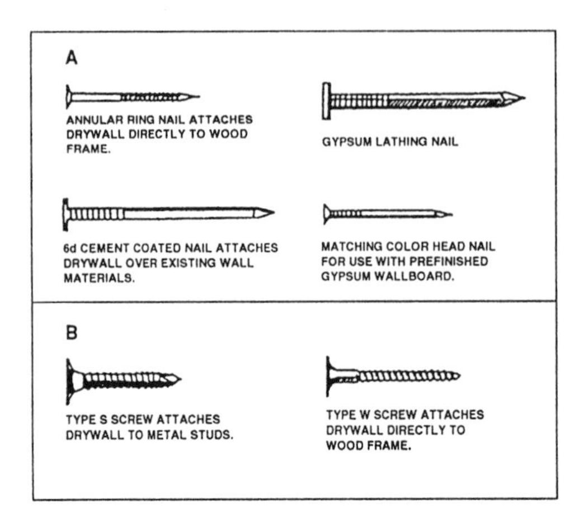

Figure 5-3.—Drywall fasteners.

sanding, ensure you're wearing the required personnel protective gear to prevent dust inhalation.

A foot lift helps you raise and lower drywall sheets while you plumb the edges. Be careful when using the foot lift—applying too much pressure to the lift can easily damage the drywall.

Fasteners

Which fasteners you use depends in part upon the material underneath. The framing is usually wood or metal studs, although gypsum is occasionally used as a base. Adhesives are normally used in tandem with screws or nails. This allows the installer to use fewer screws or nails, leaving fewer holes that require filling. For reasons noted shortly, you'll find the drywall screw the most versatile fastener for attaching drywall to framing members.

NAILS.—Drywall nails (fig. 5-3, view A) are specially designed, with oversized heads, for greater holding power. Casing or common nailheads are too small. Further, untreated nails can rust and stain a finish. The drywall nail most frequently used is the annular ring nail. This nail fastens securely into wood studs and joists. When purchasing such nails, consider the thickness of the layer or layers of drywall, and allow additional length for the nail to penetrate the underlying wood 3/4 inch. Example: 1/2-inch drywall plus 3/4-inch penetration requires a 1 1/4-inch nail. A longer nail does not fasten more securely than one properly sized, and the longer nail is subject to the expansion and contraction of a greater depth of wood.

Smooth-shank, diamond-head nails are commonly used to attach two layers of drywall; for example, when fireproofing a wall. Again, the nail length should be selected carefully. Smooth-shank nails should penetrate the base wood 1 inch. Predecorated drywall nails, which may be left exposed, have smaller heads and are color-matched to the drywall.

SCREWS.—Drywall screws (fig. 5-3, view B) are the preferred method of fastening among professional builders, cabinetmakers, and renovators. These screws are made of high-quality steel and are superior to conventional wood screws. Use a power screw gun or an electric drill to drive in the screws. Because this method requires no impact, there is little danger of jarring loose earlier connections. There are two types of drywall screws commonly used: type S and type W.

Type S.—Type S screws (fig. 5-3, view B) are designed for attachment to metal studs. The screws are self-tapping and very sharp, since metal studs can flex away. At least 3/8 inch of the threaded part of the screw should pass through a metal stud. Although other lengths are available, 1-inch type S screws are commonly used for single-ply drywall.

Type W.—Type W screws (fig. 5-3, view B) hold drywall to wood. They should penetrate studs or joists at least 5/8 inch. If you are applying two layers of drywall, the screws holding the second sheet need to penetrate the wood beneath only 1/2 inch.

TAPE.—Joint tape varies little. The major difference between tapes is whether they are perforated or not. Perforated types are somewhat easier to bed and cover. New self-sticking fiber-mesh types (resembling window screen) are becoming popular. Having the mesh design and being self-sticking eliminates the need for the first layer of bedding joint compound.

JOINT COMPOUND.—Joint compound comes ready-mixed or in powder form. The powder form must be mixed with water to a putty consistency. Ready-mixed compound is easier to work with, though its shelf life is shorter than the powdered form. Joint compounds vary according to the additive they contain. Always read and follow the manufacturer's specifications.

ADHESIVES.—Adhesives are used to bond single-ply drywall directly to the framing members, furring strips, masonry surfaces, insulation board, or other drywall. They must be used with nails or screws.

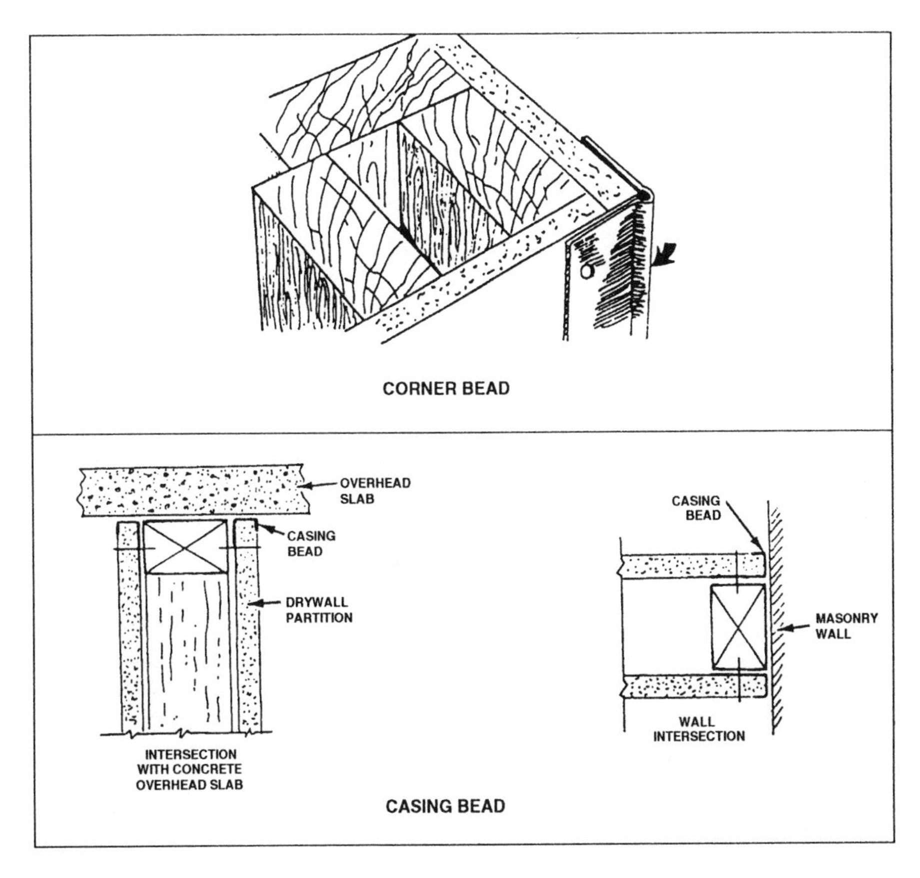

Figure 5-4.—Corner and casing beads.

Because adhesives are matched with specific materials, be sure to select the correct adhesive for the job. Read and follow the manufacturer's directions.

Accessories

A number of metal accessories have been developed to finish off or protect drywall. Corner beads (fig. 5-4) are used on all exposed corners to ensure a clean finish and to protect the drywall from edge damage. Corner bead is nailed or screwed every 5 inches through the drywall and into the framing members. Be sure the corner bead stays plumb as you fasten it in place. Casing beads (fig. 5-4), also called stop beads, are used where drywall sheets abut at wall intersections, wall and exposed ceiling intersections, or where otherwise specified. Casing beads are matched to the thickness of the drywall used.

Layout

When laying out a drywall job, keep in mind that each joint will require taping and sanding. You therefore should arrange the sheets so that there will be a minimum of joint work. Choose drywall boards of the maximum practical length.

Drywall can be hung with its length either parallel or perpendicular to joists or studs. Although both arrangements work, sheets running perpendicular afford better attachment. In double-ply installation, run base sheets parallel and top sheets perpendicular. For walls, the height of the ceiling is an important factor. Where

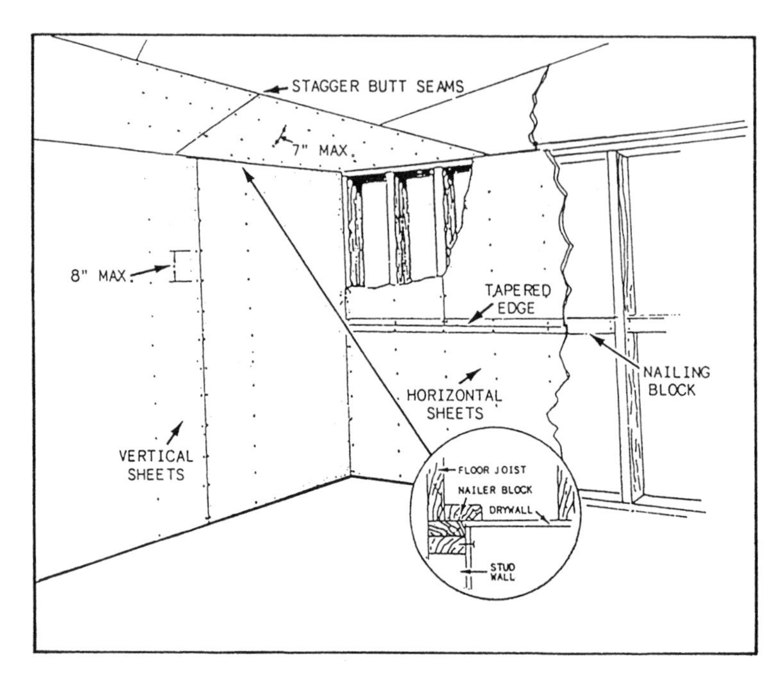

Figure 5-5.—Single-layer application of drywall.

ceilings are 8 feet 1 inch high or less, run wall sheets horizontally. Where they are higher, run wall sheets vertically, as shown in figure 5-5.

The sides of drywall taper, but the ends don't, so there are some layout constraints. End joints must be staggered where they occur. Such joints are difficult to feather out correctly. Where drywall is hung vertically, avoid side joints within 6 inches of the outside edges of doors or windows. In the case of windows, the bevel on the side of the drywall interferes with the finish trim, and the bevel may be visible. To avoid this difficulty, lay out vertical joints so they meet over a cripple (shortened) stud toward the middle of a door or window opening.

When installing drywall horizontally and an impact-resistant joint is required, you should use nailing blocks (fig. 5-5).

Handling

There are several things you can do to make working with drywall easier.

First, don't order drywall too far in advance. Drywall must be stored flat to prevent damage to the edges, and it takes up a lot of space.

Second, to cut drywall (fig. 5-6), you only need to cut through the fine-paper surface (view A). Then, grasp the smaller section and snap it sharply (view B). The gypsum core breaks along the scored line. Cut through the paper on the back (view C).

Third, when cutting a piece to length, never cut too closely. One-half-inch gaps are acceptable at the top and the bottom of a wall because molding covers these gaps. If you cut too closely, you may have difficulty getting the piece into place. Also, where walls aren't square, you may have to trim anyway.

Fourth, snap chalk lines on the drywall to indicate joists or stud centers underneath; attachment is much quicker. Remember: Drywall edges must be aligned over stud, joist, or rafter centers.

Fifth, when cutting out holes for outlet boxes, fixtures, and so on, measure from the nearest fixed point(s); for example, from the floor or edge of the next piece of drywall. Take two measurements from each

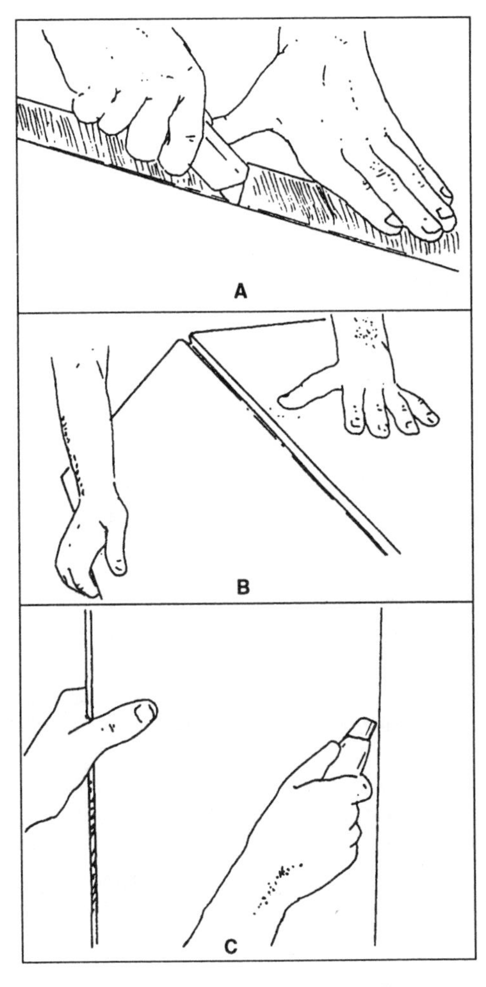

Figure 5-6.—Cutting gypsum drywall.

point, so you get the true height and width of the cutout. Locate the cutout on the finish side of the drywall. To start the cut, either drill holes at the corners or start cuts by stabbing the sharp point of the keyhole saw through the drywall and then finishing the cutting with a keyhole or compass saw. It is more difficult to cut a hole with just a utility knife, but it can be done.

Installation

When attaching drywall, hold it firmly against the framing to avoid nail-pops and other weak spots. Nails or screws must fasten securely in a framing member. If a nail misses the framing, pull it out, dimple the hole, and fill it in with compound; then try again. If you drive a nail in so deep that the drywall is crushed, drive in another reinforcing nail within 2 inches of the first.

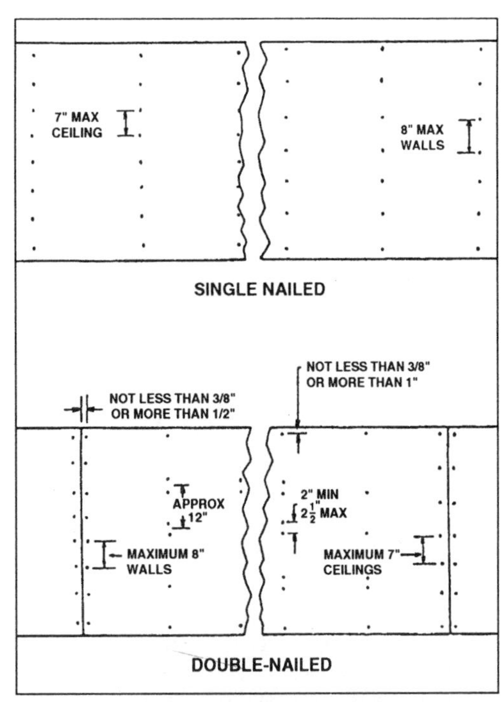

Figure 5-7.—Spacing for single and double nailing of gypsum drywall.

When attaching drywall sheets, nail (or screw) from the center of the sheet outward. Where you double-nail sheets, single nail the entire sheet first and then add the second (double) nails, again beginning in the middle of the sheet and working outward.

SINGLE AND DOUBLE NAILING.—Sheets are single- or double-nailed. Single nails are spaced a maximum of 8 inches apart on walls and 7 inches apart on ceilings. Where sheets are double-nailed, the centers of nail pairs should be approximately 12 inches apart. Space each pair of nails 2 to 2 1/2 inches apart. Do not double-nail around the perimeter of a sheet. Instead, nail as shown in figure 5-7. As you nail, it is important that you dimple each nail; that is, drive each nail in slightly below the surface of the drywall without breaking the surface of the material. Dimpling creates a pocket that can be filled with joint compound. Although special

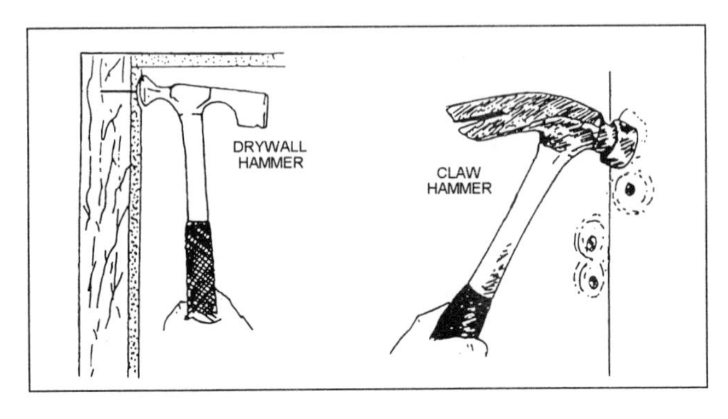

Figure 5-8.—Dimpling of gypsum drywall.

convex-headed drywall hammers are available for this operation, a conventional claw hammer also works (fig. 5-8).

SECURING WITH SCREWS.—Because screws attach more securely, fewer are needed. Screws are usually spaced 12 inches OC regardless of drywall thickness. On walls, screws may be placed 16 inches OC for greater economy, without loss of strength. Don't double up screws except where the first screw seats poorly. Space screws around the edges the same as nails.

SECURING WITH ADHESIVES.—Adhesive applied to wood studs allows you to bridge minor irregularities along the studs and to use about half the number of nails. When using adhesives, you can space the nails 12 inches apart (without doubling up). Don't alter nail spacing along end seams, however. To attach sheets to studs, use a caulking gun and run a 3/8-inch bead down the middle of the stud. Where sheets meet over a framing member, run two parallel beads. Don't make serpentine beads, as the adhesive could ooze out onto the drywall surface. If you are laminating a second sheet of drywall over a first, roll a liquid contact cement with a short-snap roller on the face of the sheet already in place. To keep adhesive out of your eyes, wear goggles. When the adhesive turns dark (usually within 30 minutes), it is ready to receive the second piece of drywall. Screw on the second sheet as described above.

CEILINGS.—Begin attaching sheets on the ceiling, first checking to be sure extra blocking (that will receive nails or screws) is in place above the top plates of the walls. By doing the ceiling first, you have maximum exposure of blocking to nail or screw into. If there are gaps along the intersection of the ceiling and wall, it is much easier to adjust wall pieces.

Ceilings can be covered by one person using two tees made from 2 by 4s. This practice is acceptable when dealing with sheets that are 8 foot in length. Sheets over this length will require a third tee, which is very awkward for one individual to handle. Two people should be involved with the installation of drywall on ceilings.

WALLS.—Walls are easier to hang than ceilings, and it's something one person working alone can do effectively, although the job goes faster if two people work together. As you did with the ceiling, be sure the walls have sufficient blocking in corners before you begin.

Make sure the first sheet on a wall is plumb and its leading edge is centered over a stud. Then, all you have to do is align successive sheets with the first sheet. The foot lift shown earlier in figure 5-2 is useful for raising or lowering a sheet while you level its edge. After you've sunk two or three screws or nails, the sheet will stay in place. A gap of 1/2 inch or so along the bottom of a sheet is not critical; it is easily covered by finish flooring, baseboards, and so on. If you favor a clean, modern line without trim, manufactured metal or vinyl edges (casing beads) are available for finishing the edges.

During renovation, you may find that hanging sheets horizontally makes sense. Because studs in older buildings often are not on regular centers, the joints of vertical sheets frequently do not align with the studs. Again, using the foot lift, level the top edge of the bottom sheet. Where studs are irregular, it's even more important that you note positions and chalk line stud centers onto the drywall face before hanging the sheet.

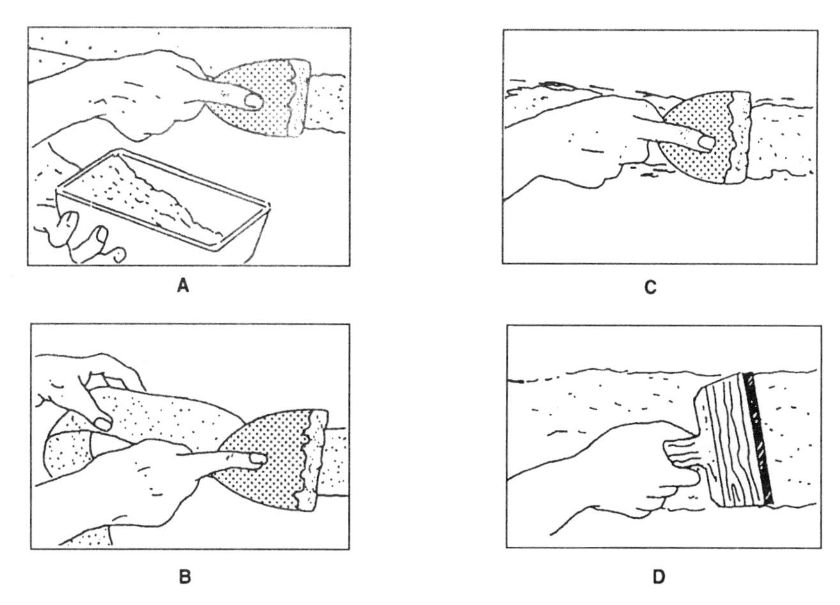

Figure 5-9.—Finishing drywall joints.

Applying drywall in older buildings yields a lot of waste because framing is not always standardized. Use the cutoffs in such out-of-the-way places as closets. Don't piece together small sections in areas where you'll notice seams. Never assume that ceilings are square with walls. Always measure from at least two points, and cut accordingly.

Drywall is quite good for creating or covering curved walls. For the best results, use two layers or 1/4-inch drywall, hung horizontally. The framing members of the curve should be placed at intervals of no more than 16 inches OC; 12 inches is better. For an 8-foot sheet applied horizontally, an arc depth of 2 to 3 feet should be no problem, but do check the manufacturer's specifications. Sharper curves may require backcutting (scoring slots into the back so that the sheet can be bent easily) or wetting (wet-sponging the front and back of the sheet to soften the gypsum). Results are not always predictable, though. When applying the second layer of 1/4-inch drywall, stagger the vertical butt joints.

Finishing

The finishing of gypsum board drywall is generally a three-coat application. Attention to drying times between coats prevents rework that has a cost involved as well as extra time.

Where sheets of drywall join, the joints are covered with joint tape and compound (fig. 5-9). The procedure is straightforward.

1. Spread a swath of bedding compound about 4 inches wide down the center of the joint (fig. 5-9, view A). Press the tape into the center of the joint with a 6-inch finish knife (fig. 5-9, view B). Apply another coat of compound over the first to bury the tape (fig. 5-9, view C). As you apply the compound over the tape, bear down so you take up any excess. Scrape clean any excess, however, as sanding it off can be tedious.
2. When the first coat is dry, sand the edges with fine-grit sand paper while wearing personal-protective equipment. Using a 12-inch knife, apply a topping of compound 2 to 4 inches wider than the first applications (view D).
3. Sand the second coat of compound when it is dry. Apply the third and final coat, feathering it out another 2 to 3 inches on each side of the joint. You should be able to do this with a 12-inch knife. Otherwise, you should use a 16-inch "feathering trowel."

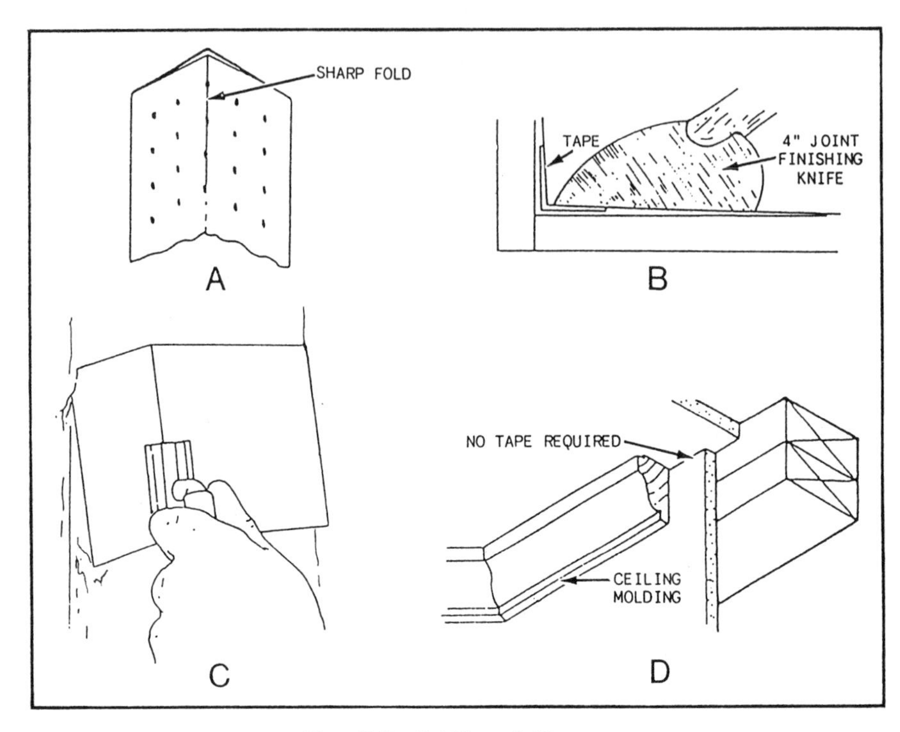

Figure 5-10.—Finishing an inside corner.

When finishing an inside corner (fig. 5-10), cut your tape the length of the corner angle you are going to finish. Apply the joint compound with a 4-inch knife evenly about 2 inches on each side of the angle. Use sufficient compound to embed the tape. Fold the tape along the center crease (view A) and firmly press it into the corner. Use enough pressure to squeeze some compound under the edges. Feather the compound 2 inches from the edge of the tape (view B). When the first coat is dry, apply a second coat. A corner trowel (view C) is almost indispensable for taping corners. Feather the edges of the compound 1 1/2 inches beyond the first coat. Apply a third coat if necessary, let it dry, and sand it to a smooth surface. Use as little compound as possible at the apex of the angle to prevent hairline cracking. When molding is installed between the wall and ceiling intersection, it is not necessary to tape the joint (view D).

When finishing an outside corner (fig. 5-11), be sure the corner bead is attached firmly. Using a 4-inch finishing knife, spread the joint compound 3 to 4 inches wide from the nose of the bead, covering the metal edges. When the compound is completely dry, sand lightly and apply a second coat, feathering edges 2 to 3 inches beyond the first coat. A third coat may be needed, depending on your coverage. Feather the edges of each coat 2 or 3 inches beyond each preceding coat. Corner beads are no problem if you apply compound with care and scrape the excess clean. Nail holes and screw holes usually can be covered in two passes, though shrinkage sometimes necessitates three. A tool that works well for sanding hard-to-reach places is a sanding block on an extension pole; the block has a swivel-head joint.

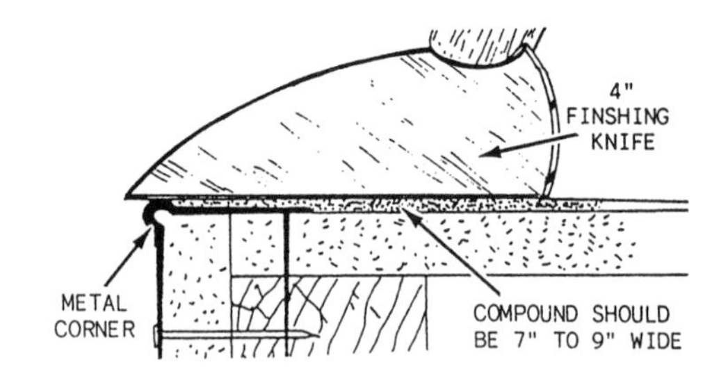

Figure 5-11.—Finishing an outside corner.

To give yourself the greatest number of decorating options in the future, paint the finished drywall surface with a coat of flat oil-base primer. Whether you intend to wallpaper or paint with latex, oil-base primer adheres best to the facing of the paper and seals it.

Renovation and Repair

For the best results, drywall should be flat against the surface to which it is being attached. How flat the nailing surface must be depends upon the desired finish effect. Smooth painted surfaces with spotlights on them require as nearly flawless a finish as you can attain. Similarly, delicate wall coverings—particularly those with close, regular patterns—accentuate pocks and lumps underneath. Textured surfaces are much more forgiving. In general, if adjacent nailing elements (studs, and so forth) vary by more than 1/4 inch, build up low spots. Essentially, there are three ways to create a flat nailing surface:

- Frame out a new wall—a radical solution. If the studs of partition walls are buckled and warped, it's often easier to rip the walls out and replace them. Where the irregular surface is a load-bearing wall, it may be easier to build a new wall within the old.
- Cover imperfections with a layer of 3/8-inch drywall. This thickness is flexible yet strong. Drywall of 1/4-inch thickness may suffice. Single-ply cover-up is a common renovation strategy where existing walls are ungainly but basically flat. Locate studs beforehand and use screws long enough to penetrate studs and joists at least 5/8 inch.
- Build up the surface by "furring out." In the "furring out" procedure, furring strips 1 by 2 inches are used. Some drywall manufacturers, however, consider that size too light for attachment, favoring instead a nominal size of 2 by 2 inches. Whatever size strips you use, make sure they (and the shims underneath) are anchored solidly to the wall behind.

By stretching strings taut between diagonal corners, you can get a quick idea of any irregularities in a wall. If studs are exposed, further assess the situation with a level held against a straight 2 by 4. Hold the straightedge plumb in front of each stud and mark low spots every 12 inches or so. Using a builder's crayon, write the depth of each low spot, relative to the straightedge, on the stud. If studs aren't exposed, locate each stud by test drilling and inserting a bent coat hanger into the hole. Chalk line

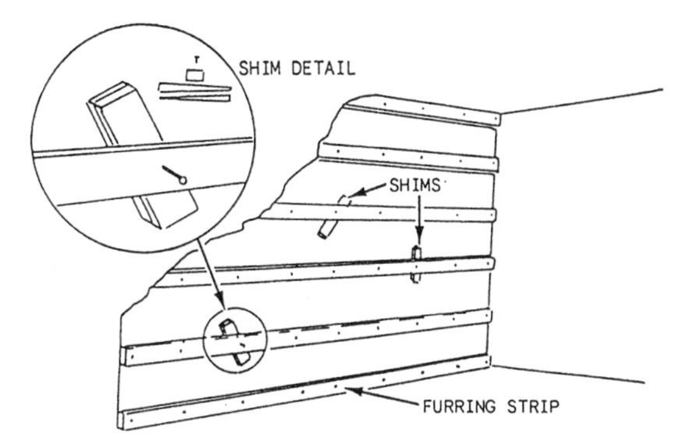

Figure 5-12.—Furring strips backed with shims.

the center of each stud on the existing surface. Here too, mark the depth of low spots.

The objective of this process is a flat plane of furring strips over existing studs. Tack the strips in place and add shims (wood shingles are best) at each low spot marked (see fig. 5-12). To make sure a furring strip doesn't skew, use two shims, with their thin ends reversed, at each point. Tack the shims in place and plumb the furring strips again. When you are satisfied, drive the nails or screws all the way in.

When attaching the finish sheets, use screws or nails long enough to penetrate through furring strips and into the studs behind. Strips directly over studs ensure the strongest attachment. Where finish materials are not sheets—for example, single-board vertical paneling—furring should run perpendicular to the studs.

Regardless of type, finish material must be backed firmly at all nailing points, corners, and seams. Where you cover existing finish surfaces or otherwise alter the thickness of walls, it's usually necessary to build up existing trim. Figure 5-13 shows how this might be done

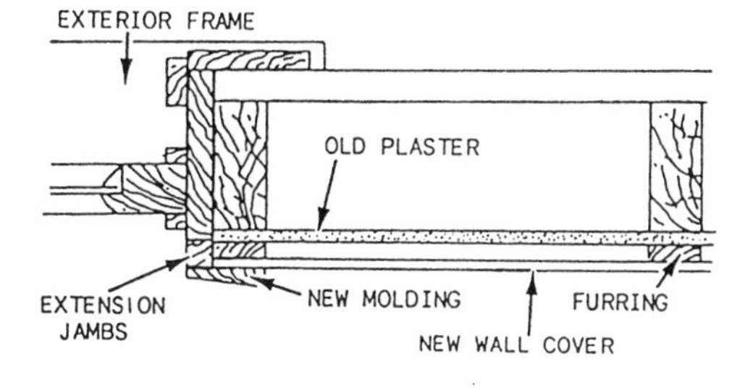

Figure 5-13.—Building up an interior window casing.

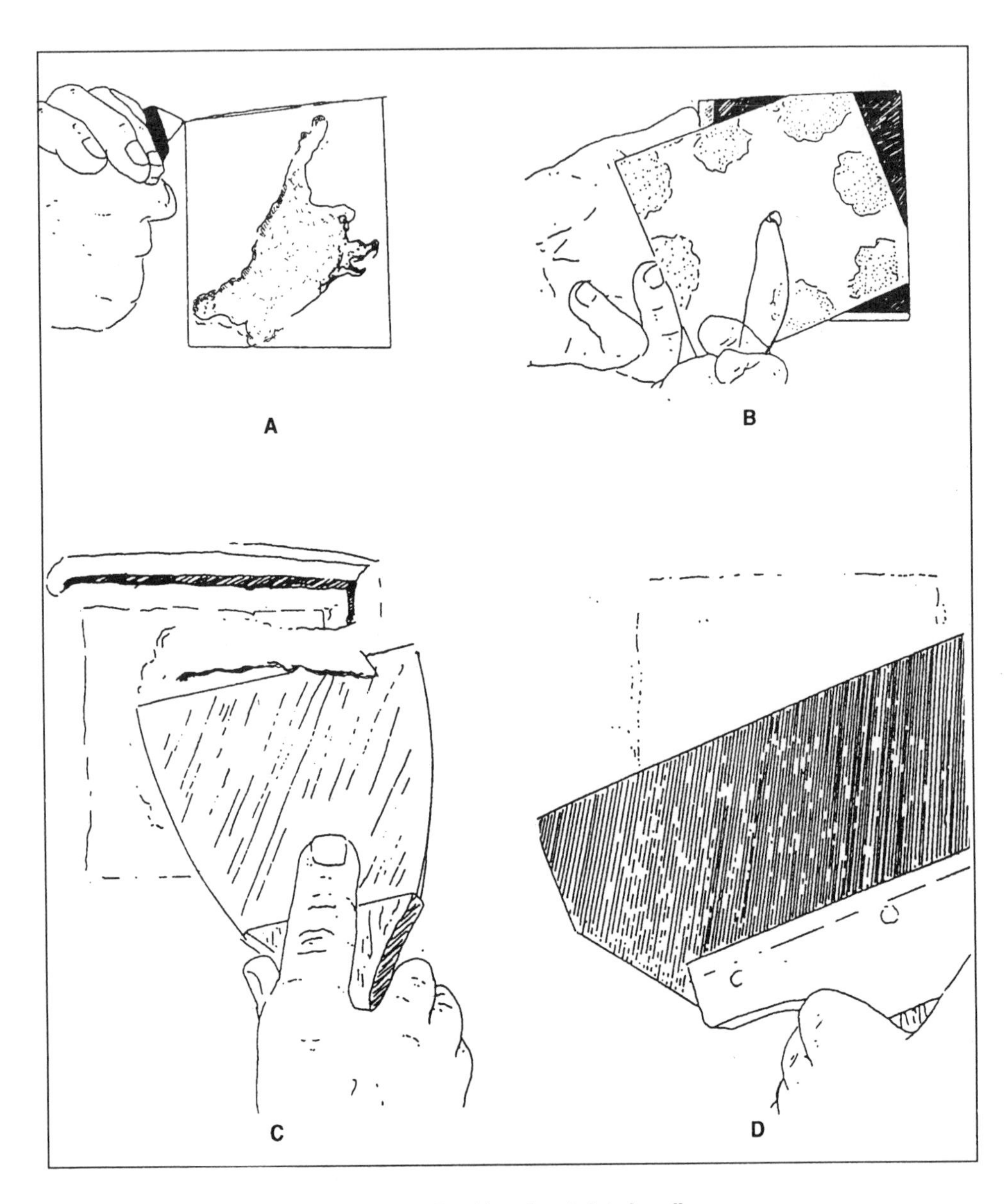

Figure 5-14.—Repairing a large hole in drywall.

for a window casing. Electrical boxes must also be extended with box extensions or plaster rings.

Masonry surfaces must be smooth, clean, and dry. Where the walls are below grade, apply a vapor barrier of polyethylene (use mastic to attach it) and install the furring strips. Use a power-actuated nail gun to attach strips to the masonry. Follow all safety procedures. If you hand nail, drive case-hardened nails into the mortar joints. Wear goggles; these nails can fragment.

Most drywall blemishes are caused by structural shifting or water damage. Correct any underlying problems before attacking the symptoms.

Popped-up nails are easily fixed by pulling them out or by dimpling them with a hammer. Test the entire wall for springiness and add nails or screws where needed. Within 2 inches of a popped-up nail, drive in another nail. Spackle both when the spots are dry, then sand and prime.

To repair cracks in drywall, cut back the edges of the crack slightly to remove any crumbly gypsum and to provide a good depression for a new filling of joint compound. Feather the edges of the compound. When dry, sand and prime them.

When a piece of drywall tape lifts, gently pull until the piece rips free from the part that's still well stuck. Sand the area affected and apply a new bed of compound for a replacement piece of tape. The self-sticking tape mentioned earlier works well here. Feather all edges.

If a sharp object has dented the drywall, merely sand around the cavity and fill it with spackling compound. A larger hole (bigger than your fist) should have a backing. One repair method is shown in figure 5-14. First, cut the edges of the hole clean with a utility knife (view A). The piece of backing should be somewhat larger than the hole itself. Drill a small hole into the middle of the backing piece and thread a piece of wire into the hole. This wire allows you to hold the piece of backing in place. Spread mastic around the edges of the backing. When the adhesive is tacky, fit the backing diagonally into the hole (view B) and, holding onto the wire, pull the piece against the back side of the hole. When the mastic is dry, push the wire back into the wall cavity. The backing stays in place. Now, fill the hole with plaster or joint compound (view C) and finish (view D). (**Note**: This is just one of several options available for repairing large surface damage to gypsum board.)

Compound sags in holes that are too big. If it happens, mastic a replacement piece of drywall to the backing piece. To avoid a bulge around the filled-in hole, feather the compound approximately 16 inches, or more. If the original drywall is 1/2 inch thick, use 3/8-inch plasterboard as a replacement on the backing piece.

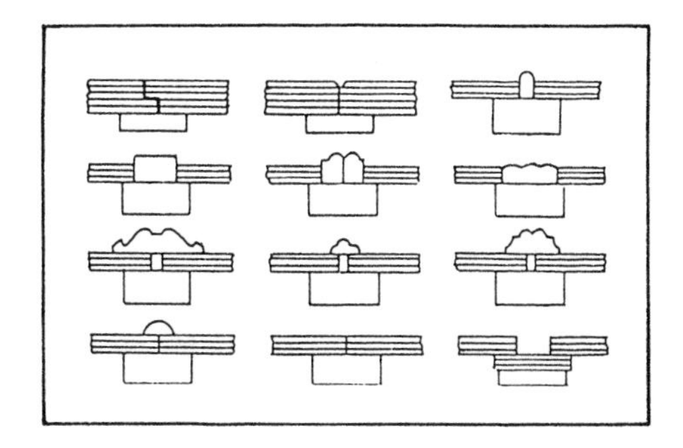

Figure 5-15.—Battens used for paneling joints.

Holes larger than 8 inches should be cut back to the centers of the nearest studs. Although you should have no problem nailing a replacement piece to the studs, the top and the bottom of the new piece must be backed. The best way to install backing is to screw drywall gussets (supports) to the back of the existing drywall. Then, put the replacement piece in the hole and screw it to the gussets.

PLYWOOD

Most of the plywood used for interior walls has a factory-applied finish that is tough and durable. Manufacturers can furnish prefinished matching trim and molding that is also easy to apply. Color-coordinated putty sticks are used to conceal nail holes.

Joints between plywood sheets can be treated in a number of ways. Some panels are fabricated with machine-shaped edges that permit almost perfect joint concealment. Usually, it is easier to accentuate the joints with grooves or use battens and strips. Some of the many different styles of battens are shown in figure 5-15.

Before installation, the panels should become adjusted (conditioned) to the temperature and humidity of the room. Carefully remove prefinished plywood from cartons and stack it horizontally. Place 1-inch spacer strips between each pair of face-to-face panels. Do this at least 48 hours before application.

Plan the layout carefully to reduce the amount of cutting and the number of joints. It is important to align panels with openings whenever possible. If finished panels are to have a grain, stand the panels around the walls and shift them until you have the most pleasing effect in color and grain patterns. To avoid mix-ups, number the panels in sequence after their position has been established.

When cutting plywood panels with a portable saw, mark the layout on the back side. Support the panel carefully and check for clearance below. Cut with the saw blade upward against the panel face. This minimizes splintering. This procedure is even more important when working with prefinished panels.

Plywood can be attached directly to the wall studs with nails or special adhesives. Use 3/8-inch plywood for this type of installation. When studs are poorly aligned or when the installation is made over an existing surface in poor condition, it is usually advisable to use furring. Nail 1- by 3- or 1- by 4-inch furring strips horizontally across the studs. Start at the floor line and continue up the wall. Spacing depends on the panel thickness. Thin panels need more support. Install vertical strips every 4 feet to support panel edges. Level uneven areas by shimmying behind the furring strips. Prefinished plywood panels can be installed with special adhesive. The adhesive is applied and the panels are simply pressed into place; no sustained pressure is required.

Begin installing panels at a corner. Scribe and trim the edges of the first panel so it is plumb. Fasten it in place before fitting the next panel. Allow approximately 1/4-inch clearance at the top and bottom. After all panels are in place, use molding to cover the space along the ceiling. Use baseboards to conceal the space at the floor line. If the molding strips, baseboards, and strips used to conceal panel joints are not prefinished, they should be spray painted or stained a color close to the tones in the paneling before installation.

On some jobs, 1/4-inch plywood is installed over a base of 1/2-inch gypsum wallboard. This backing is recommended for several reasons. It tends to bring the studs into alignment. It provides a rigid finished surface. And, it improves the fire-resistant qualities of the wall. (The plywood is bonded to the gypsum board with a compatible adhesive.)

HARDBOARD

Through special processing, hardboard (also called fiberboard) can be fabricated with a very low moisture absorption rate. This type is often scored to form a tile pattern. Panels for wall application are usually 1/4 inch thick.

Since hardboard is made from wood fibers, the panels expand and contract slightly with changes in humidity. They should be installed when they are at their maximum size. The panels tend to buckle between the studs or attachment points if installed when moisture content is low. Manufacturers of prefinished hardboard panels recommend that they be unwrapped and placed separately around the room for at least 48 hours before application.

Procedures and attachment methods for hardboard are similar to those for plywood. Special adhesives are available as well as metal or plastic molding in matching colors. You should probably drill nail holes for the harder types.

PLASTIC LAMINATES

Plastic laminates are sheets of synthetic material that are hard, smooth, and highly resistant to scratching and wear. Although basically designed for table and countertops, they are also used for wainscoting and wall paneling in buildings.

Since plastic laminate material is thin (1/32 to 1/16 inch), it must be bonded to other supporting panels. Contact bond cement is commonly used for this purpose. Manufacturers have recently developed prefabricated panels with the plastic laminate already bonded to a base or backer material. This base consists of a 1/32-inch plastic laminate mounted on 3/8-inch particleboard. Edges are tongue and grooved so that units can be blind-nailed into place. Various matching corner and trim moldings are available.

SOLID LUMBER PANELING

Solid wood paneling makes a durable and attractive interior wall surface and may be appropriately used in nearly any type of room. Several species of hardwood and softwood are available. Sometimes, grades with numerous knots are used to obtain a special appearance. Defects, such as the deep fissures in pecky cypress, can also provide a dramatic effect.

The softwood species most commonly used include pine, spruce, hemlock, and western red cedar. Boards range in widths from 4 to 12 inches (nominal size) and are dressed to 3/4 inch. Board and batten or shiplap joints are sometimes used, but tongue-and-groove (T&G) joints combined with shaped edges and surfaces are more popular.

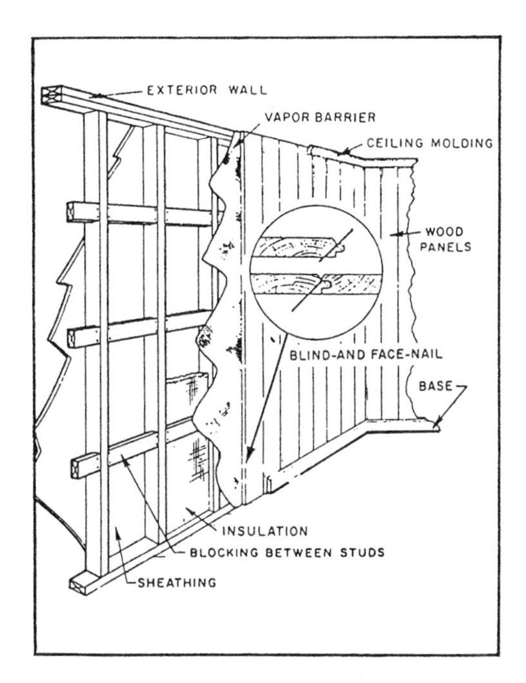

Figure 5-16.—Vertical wood paneling.

When solid wood paneling is applied horizontally, furring strips are not required—the boards are nailed directly to the studs. Inside corners are formed by butting the paneling units flush with the other walls. If random widths are used, boards on adjacent walls must match and be accurately aligned.

Vertical installations require furring strips at the top and bottom of the wall and at various intermediate spaces. Sometimes, 2- by 4-inch blocking is installed between the studs to serve as a nailing base (see fig. 5-16). Even when heavy T&G boards are used, these nailing members should not be spaced more than 24 inches apart.

Narrow widths (4 to 6 inches) of T&G paneling are blind-nailed (see insert in fig. 5-16). The nailheads do not appear on finished surfaces, and you eliminate the need for countersinking and filling nail holes. This nailing method also provides a smooth, blemish-free surface. This is especially important when clear finishes are used. Drive 6d finishing nails at a 45° angle into the base of the tongue and on into the bearing point. Carefully plumb the first piece installed and check for the plumbness at regular intervals. For lumber paneling (not tongue and grooved), use 6d casing or finishing nails. Use two nails at each nailing member for panels 6 inches or less in width and three nails for wider panels.

Exterior wall constructions, where the interior surface consists of solid wood paneling, should include a tight application of building paper located close to the backside of the boards. This prevents the infiltration of wind and dust through the joints. In cold climates, insulation and vapor barriers are important. Base, corner, and ceiling trim can be used for decorative purposes or to conceal irregularities in joints.

SUSPENDED ACOUSTIC CEILING SYSTEMS

LEARNING OBJECTIVE: Upon completing this section, you should be able to identify the materials used to install a suspended acoustical ceiling and explain the methods of installation.

Suspended acoustical ceiling systems can be installed to lower a ceiling, finish off exposed joints, cover damaged plaster, or make any room quieter and brighter. The majority of the systems available are primarily designed for acoustical control. However, many manufacturers offer systems that integrate the functions of lighting, air distribution, fire protection, and acoustical control. Individual characteristics of acoustical tiles, including sound-absorption co-efficients, noise-reduction coefficients, light-reflection values, flame resistance, and architectural applications, are available from the manufacturer.

Tiles are available in 12- to 30-inch widths, 12- to 60-inch lengths, and 3/16- to 3/4-inch thicknesses. The larger sizes are referred to as "panels." The most commonly used panels in suspended ceiling systems are the standard 2- by 2-foot and 2- by 4-foot acoustic panels composed of mineral or cellulose fibers.

It is beyond the scope of this training manual to acquaint you with each of the suspended acoustical ceiling systems in use today. Just as the components of these systems vary according to manufacturers, so do the procedures involved in their installation. With this in mind, the following discussion is designed to acquaint you with the principles involved in the installation of a typical suspended acoustical ceiling system.

PREPARATION FOR INSTALLATION

The success of a suspended ceiling project, as with any other construction project, is as dependent on planning as it is on construction methods and procedures. Planning, in this case, involves the selection of a grid system (either steel or aluminum), the selection and layout of a grid pattern, and the determination of

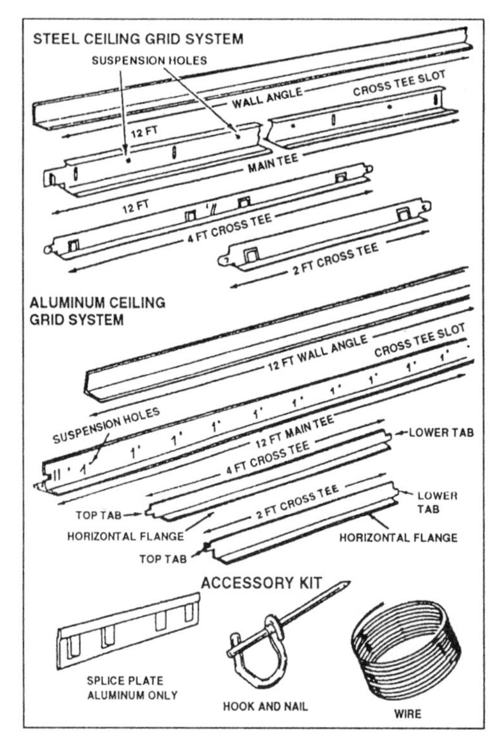

Figure 5-17.—Grid system components.

material requirements. Figure 5-17 shows the major components of a steel and aluminum ceiling grid system used for the 2- by 2-foot or 2- by 4-foot grid patterns shown in figure 5-18.

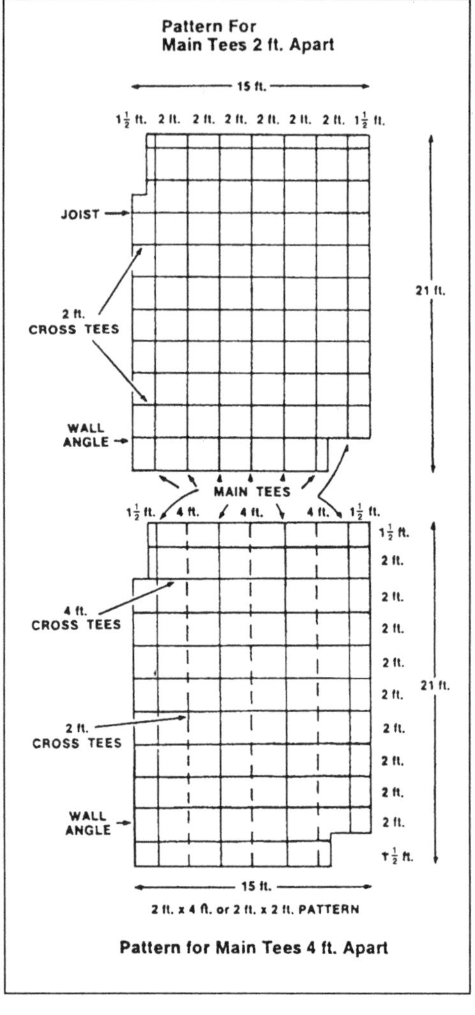

Figure 5-18.—Grid layout for main tees.

Pattern Layout

The layout of a grid pattern and the material requirements are based on the ceiling measurements and the length and width of the room at the new ceiling height. If the ceiling length or width is not divisible by 2 (that is, 2 feet), increase those dimensions to the next higher dimension divisible by 2. For example, if a ceiling measures 13 feet 7 inches by 10 feet 4 inches, the dimensions should be increased to 14 by 12 feet for material and layout purposes. Next, draw a layout on graph paper. Make sure the main tees run perpendicular to the joists. Position the main tees on your drawing so the border panels at room edges are equal and as large as possible. Try several layouts to see which looks best with the main tees. Draw in cross tees so the border panels at the room ends are equal and as large as possible. Try several combinations to determine the best. For 2- by 4-foot patterns, space cross tees 4 feet apart. For 2- by 2-foot patterns, space cross tees 2 feet apart. For smaller areas, the 2- by 2-foot pattern is recommended.

Material Requirements

As indicated in figure 5-17, wall angles and main tees come in 12-foot pieces. Using the perimeter of a room at suspended ceiling height, you can determine the number of pieces of wall angle by dividing the perimeter by 12 and adding 1 additional piece for any fraction. Determine the number of 12-foot main tees and 2-foot or 4-foot cross tees by counting them on the grid pattern

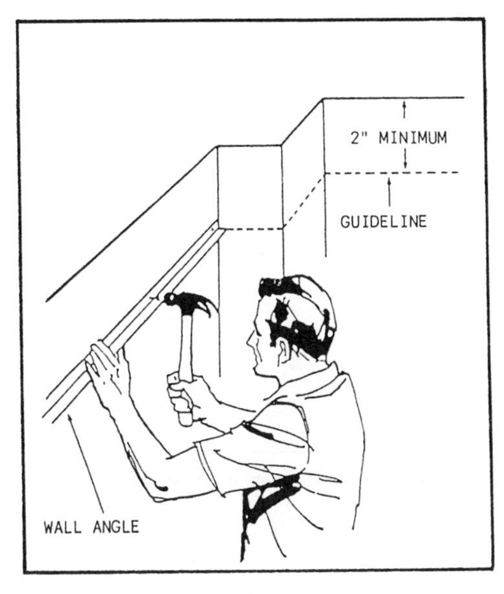

Figure 5-19.—Wall angle installation.

layout. In determining the number of 2-foot or 4-foot cross tees for border panels, you must remember that no more than 2 border tees can be cut from one cross tee.

INSTALLATION

The tools normally used to install a grid system include a hammer, chalk or pencil, pliers, tape measure, screwdriver, hacksaw, knife, and tin snips. With these, you begin by installing the wall angles, then the suspension wires, followed by the main tees, cross tees, and acoustical panels.

Wall Angles

The first step is to install the wall angles at the new ceiling height. This can be as close as 2 inches below the existing ceiling. Begin by marking a line around the entire room to indicate wall angle height and to serve as a level reference. Mark continuously to ensure that the lines at intersecting walls meet. On gypsum board, plaster, or paneled walls, install wall angles (fig. 5-19) with nails, screws, or toggle bolts. On masonry walls, use anchors or concrete nails spaced 24 inches apart. **Make sure the wall angle is level.** Overlap or miter the wall angle at corners (fig. 5-20). After the wall angles are installed, the next step is to attach the suspension wires.

Suspension Wires

Suspension wires are required every 4 feet along main tees and on each side of all splices (see fig. 5-21). Attach wires to the existing ceiling with nails or screw

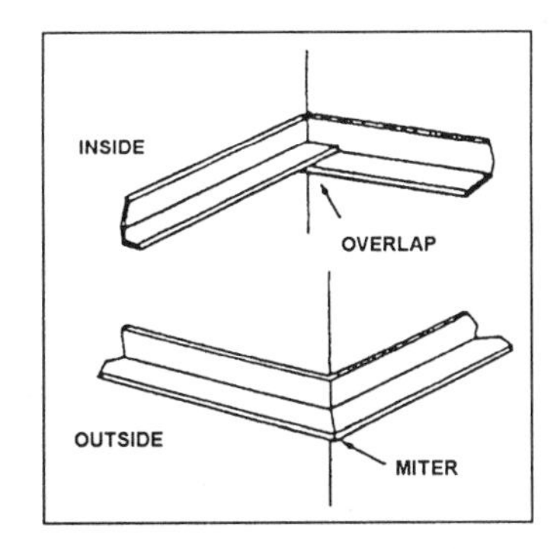

Figure 5-20.—Corner treatment.

eyelets. Before attaching the first wire, measure the distance from the wall to the first main tee. Then, stretch a guideline from an opposite wall angle to show the correct position of the first nail tee. Position suspension wires for the first tee along the guide. Wires should be cut to proper length, at least 2 inches longer than the distance between the old and new ceiling. Attach additional wires at 4-foot intervals. Pull wires to remove kinks and make 90° bends in the wires where they intersect the guideline. Move the guideline, as required, for each row. After the suspension wires are attached, the next step is to install the main tees.

Tees

In an acoustical ceiling, the panels rest on metal members called tees. The tees are suspended by wires.

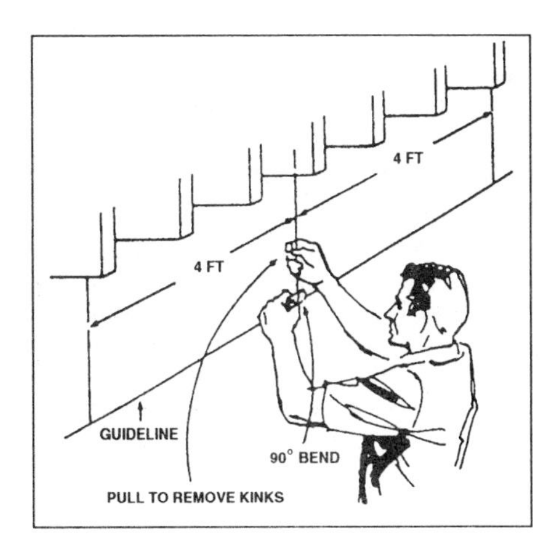

Figure 5-21.—Suspension wire installation.

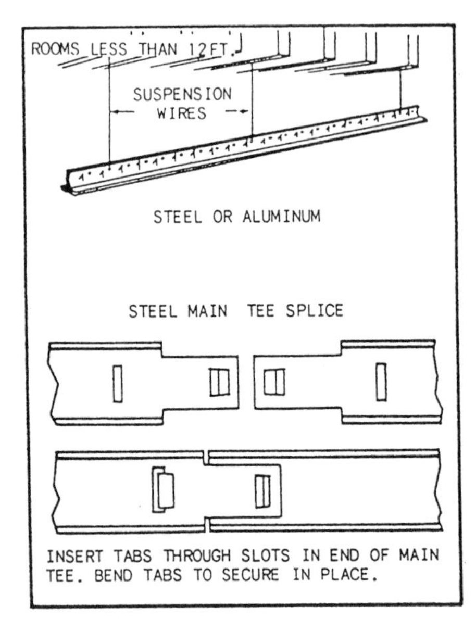

Figure 5-22.—Main tee suspension and steel splice.

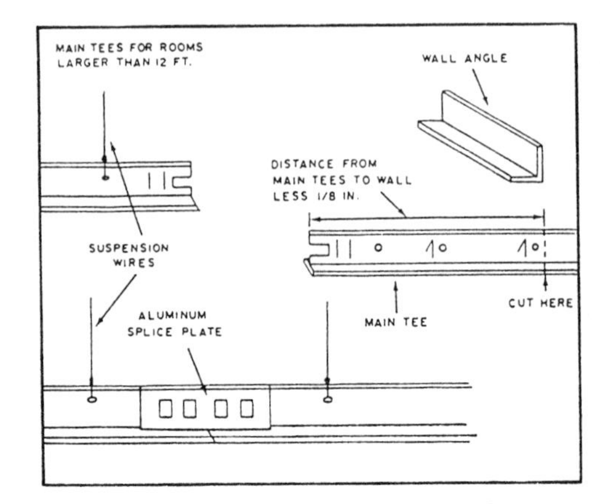

Figure 5-23.—Main tee and aluminum tee splice.

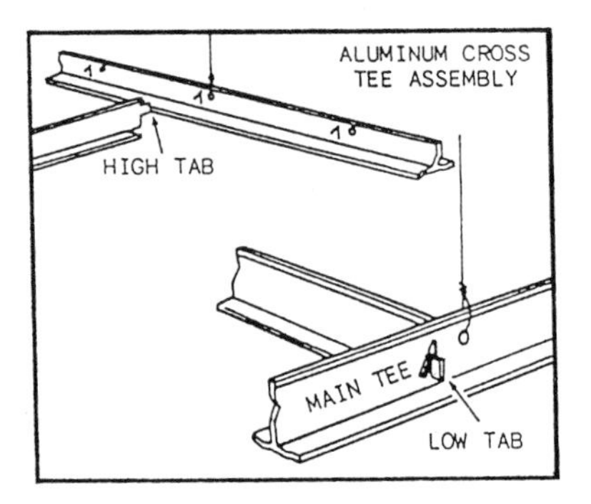

Figure 5-24.—Aluminum cross tee assembly.

MAIN TEES.—Install main tees of 12 feet or less by resting the ends on opposite wall angles and inserting the suspension wires (top view of fig. 5-22). Hang one wire near the middle of the main tee, level and adjust the wire length, then secure all wires by making the necessary turns in the wire.

For main tees over 12 feet, cut them so the cross tees do not intersect the main tee at a splice joint. Begin the installation by resting the cut end on the wall angle and attaching the suspension wire closest to the opposite end. Attach the remaining suspension wires, making sure the main tee is level before securing. The remaining tees are installed by making the necessary splices (steel splices are shown in fig. 5-22 and those for aluminum in 5-23) and resting the end on the opposite wall angle. After the main tees are installed, leveled, and secured, install the cross tees.

CROSS TEES.—Aluminum cross tees have "high" and "low" tab ends that provide easy positive installation without tools. Installation begins by cutting border tees (when necessary) to fit between the first main tee and the wall angle. Cut off the high tab end and rest this end in the main tee slot. Repeat this procedure until all border tees are installed on one side of the room. Continue across the room, installing the remaining cross tees according to your grid pattern layout. An aluminum cross tee assembly is shown in figure 5-24. At the opposite wall angle, cut off the low tab of the border tee and rest the cut end on the wall angle. If the border edge is less than half the length of the cross tee, use the remaining portion of the border of the previously cut tee.

Steel cross tees have the same tab on both ends and, like the aluminum tees, do not require tools for installation. The procedures used in their installation are the same as those just described for aluminum. A steel cross tee assembly is shown in figure 5-25. The final step after completion of the grid system is the installation of the acoustical panels.

Acoustical Panels

Panel installation is started by inserting all full ceiling panels. Border panels should be installed last, after they have been cut to proper size. To cut a panel, turn the finish side up, scribe with a sharp utility knife, and saw with a 12- or 14-point handsaw.

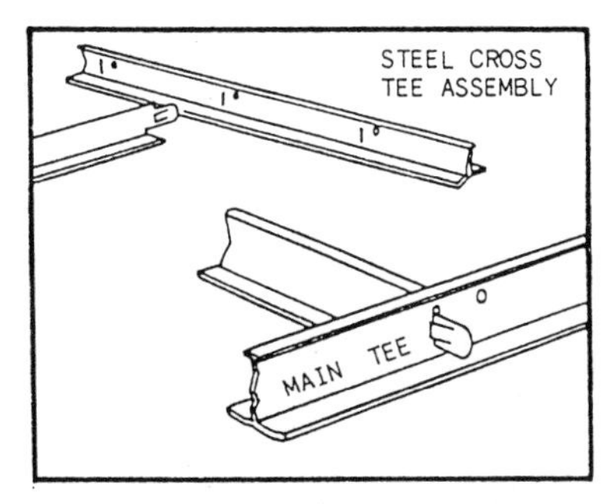

Figure 5-25.—Steel cross tee assembly.

Most ceiling panel patterns are random and do not require orientation. However, some fissured panels are designed to be installed in a specific direction and are so marked on the back with directional arrows. When installing panels on a large project, you should work from several cartons. The reason for this is that the color, pattern, or texture might vary slightly; and by working from several cartons, you avoid a noticeable change in uniformity.

Since ceiling panels are prefinished, handle them with care. Keep their surfaces clean by using talcum powder on your hands or by wearing clean canvas gloves. If panels do become soiled, use an art gum eraser to remove spots, smudges, and fingerprints. Some panels can be lightly washed with a sponge dampened with a mild detergent solution. However, before washing or performing other maintenance services, such as painting, refer to the manufacturer's instructions.

Ceiling Tile

Ceiling tile can be installed in several ways, depending on the type of ceiling or roof construction. When a flat-surfaced backing is present, such as between beams of a beamed ceiling in a low-slope roof, tiles are fastened with adhesive as recommended by the manufacturer. A small spot of a mastic type of construction adhesive at each corner of a 12- by 12-inch tile is usually sufficient. When tile is edge-matched, stapling is also satisfactory.

Perhaps the most common method of installing ceiling tile uses wood strips nailed across the ceiling joists or roof trusses (fig. 5-26, view A). These are

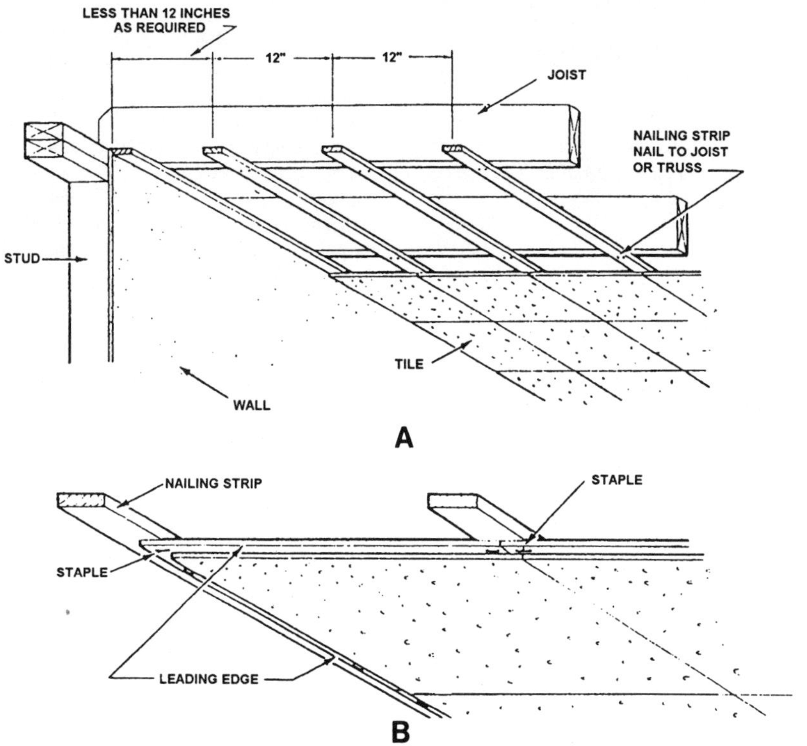

Figure 5-26.—Ceiling tile assembly.

spaced a minimum of 12 inches OC. A nominal 1- by 3-inch or 1- by 4-inch wood member can be used for roof or ceiling members spaced not more than 24 inches OC. A nominal 2- by 2-inch or 2- by 3-inch member should be satisfactory for truss or ceiling joist spacing of up to 48 inches.

In locating the strips, first measure the width of the room (the distance parallel to the direction of the ceiling joists). If, for example, this is 11 feet 6 inches, use ten 12-inch-square tiles and 9-inch-wide tile at each side edge. The second wood strips from each side are located so that they center the first row of tiles, that can now be ripped to a width of 9 inches. The last row will also be 9 inches, but do not rip these tiles until the last row is reached so that they fit tightly. The tile can be fitted and arranged the same way for the ends of the room.

Ceiling tiles normally have a tongue on two adjacent sides and a groove on the opposite adjacent sides. Start with the leading edge ahead and to the open side so that it can be stapled to the nailing strips. A small finish nail or adhesive should be used at the edge of the tiles in the first row against the wall. Stapling is done at the leading edge and the side edge of each tile (fig. 5-26, view B). Use one staple at each wood strip at the leading edge and two at the open side edge. At the opposite wall, a small finish nail or adhesive must again be used to hold the tile in place.

Most ceiling tile of this type has a factory finish; painting or finishing is not required after it is placed. Take care not to mar or soil the surface.

INSULATION

LEARNING OBJECTIVE: Upon completing this section, you should be able to identify the types of insulation and describe the methods of installation.

The inflow of heat through outside walls and roofs in hot weather or its outflow during cold weather is a major source of occupant discomfort. Providing heating or cooling to maintain temperatures at acceptable limits for occupancy is expensive. During hot or cold weather, insulation with high resistance to heat flow helps save energy. Also, you can use smaller capacity units to achieve the same heating or cooling result, an additional savings.

Most materials used in construction have some insulating value. Even air spaces between studs resist the passage of heat. However, when these stud spaces are filled or partially filled with material having a high insulating value, the stud space has many times the insulating ability of the air alone.

TYPES

Commercial insulation is manufactured in a variety of forms and types, each with advantages for specific uses. Materials commonly used for insulation can be grouped in the following general classes: (1) flexible insulation (blanket and batt); (2) loose-fill insulation; (3) reflective insulation; (4) rigid insulation (structural and nonstructural); and (5) miscellaneous types.

The insulating value of a wall varies with different types of construction, kinds of materials used in construction, and types and thicknesses of insulation. As we just mentioned, air spaces add to the total resistance of a wall section to heat transmission, but an air space is not as effective as the same space filled with an insulating material.

Flexible

Flexible insulation is manufactured in two types: blanket and batt. Blanket insulation (fig. 5-27, view A) is furnished in rolls or packages in widths to fit between studs and joists spaced 16 and 24 inches OC. It comes in thicknesses of 3/4 inch to 12 inches. The body of the blanket is made of felted mats of mineral or vegetable fibers, such as rock or glass wool, wood fiber, and cotton. Organic insulations are treated to make them resistant to fire, decay, insects, and vermin. Most blanket insulation is covered with paper or other sheet material with tabs on the sides for fastening to studs or joists. One covering sheet serves as a vapor barrier to resist movement of water vapor and should always face the warm side of the wall. Aluminum foil, asphalt, or plastic laminated paper is commonly used as barrier materials.

Batt insulation (fig. 5-27, view B) is also made of fibrous material preformed to thicknesses of 3 1/2 to 12 inches for 16- and 24-inch joist spacing. It is supplied with or without a vapor barrier. One friction type of fibrous glass batt is supplied without a covering and is designed to remain in place without the normal fastening methods.

Loose Fill

Loose-fill insulation (fig. 5-27, view C) is usually composed of materials used in bulk form, supplied in bags or bales, and placed by pouring, blowing, or packing by hand. These materials include rock or glass

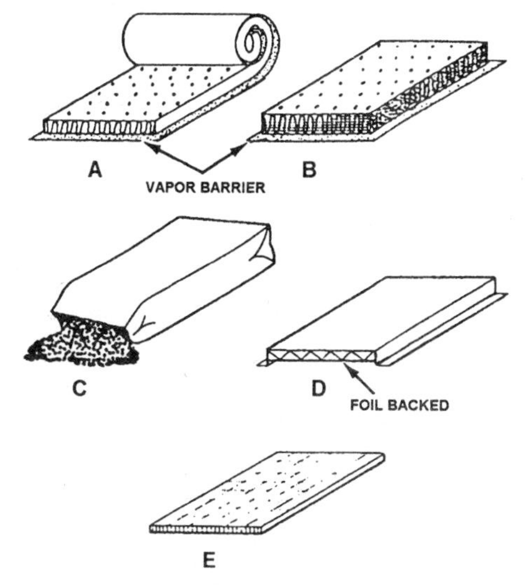

Figure 5-27.—Types of insulation.

wool, wood fibers, shredded redwood bark, cork, wood pulp products, vermiculite, sawdust, and shavings.

Fill insulation is suited for use between first-floor ceiling joists in unheated attics. It is also used in sidewalls of existing houses that were not insulated during construction. Where no vapor barrier was installed during construction, suitable paint coatings, as described later in this chapter, should be used for vapor barriers when blow insulation is added to an existing house.

Reflective

Most materials have the property of reflecting radiant heat, and some materials have this property to a very high degree. Materials high in reflective properties include aluminum foil, copper, and paper products coated with a reflective oxide. Such materials can be used in enclosed stud spaces, attics, and similar locations to retard heat transfer by radiation. Reflective insulation is effective only where the reflective surface faces an air space at least 3/4 inch deep. Where this surface contacts another material, the reflective properties are lost and the material has little or no insulating value. Proper installation is the key to obtaining the best results from the reflective insulation. Reflective insulation is equally effective whether the reflective surface faces the warm or cold side.

Reflective insulation used in conjunction with foil-backed gypsum drywall makes an excellent vapor barrier. The type of reflective insulation shown in figure 5-27, view D, includes a reflective surface. When properly installed, it provides an air space between other surfaces.

Rigid

Rigid insulation (fig. 5-27, view E) is usually a fiberboard material manufactured in sheet form. It is made from processed wood, sugar cane, or other vegetable products. Structural insulating boards, in densities ranging from 15 to 31 pounds per cubic foot, are fabricated as building boards, roof decking, sheathing, and wallboard. Although these boards have moderately good insulating properties, their primary purpose is structural.

Roof insulation is nonstructural and serves mainly to provide thermal resistance to heat flow in roofs. It is called slab or block insulation and is manufactured in rigid units 1/2 inch to 3 inches thick and usually 2- by 4-foot sizes.

In building construction, perhaps the most common forms of rigid insulation are sheathing and decorative covering in sheets or in tile squares. Sheathing board is made in thicknesses of 1/2 and 25/32 inch. It is coated or impregnated with an asphalt compound to provide water resistance. Sheets are made in 2- by 8-foot sizes for horizontal application and 4- by 8-foot (or longer) sizes for vertical application.

Miscellaneous

Some insulations are not easily classified, such a insulation blankets made up of multiple layers of corrugated paper. Other types, such as lightweight vermiculite and perlite aggregates, are sometimes used in plaster as a means of reducing heat transmission. Other materials in this category are foamed-in-place insulations, including sprayed and plastic foam types. Sprayed insulation is usually inorganic fibrous material blown against a clean surface that has been primed with an adhesive coating. It is often left exposed for acoustical as well as insulating properties.

Expanded polystyrene and urethane plastic forms can be molded or foamed in place. Urethane insulation can also be applied by spraying. Polystyrene and urethane in board form can be obtained in thicknesses from 1/2 to 2 inches.

LOCATION OF INSULATION

In most climates, all walls, ceilings, roofs, and floors that separate heated spaces from unheated spaces should be insulated. This reduces heat loss from the structure during cold weather and minimizes air conditioning during hot weather. The insulation should be placed on all outside walls and in the ceiling. In structures that have unheated crawl spaces, insulation should be placed between the floor joists or around the wall perimeter.

If a blanket or batt insulation is used, it should be well supported between joists by slats and a galvanized wire mesh, or by a rigid board. The vapor barrier should be installed toward the subflooring. Press-fit or friction insulations fit tightly between joists and require only a small amount of support to hold them in place.

Reflective insulation is often used for crawl spaces, but only dead air space should be assumed in calculating heat loss when the crawl space is ventilated. A ground cover of roll roofing or plastic film, such as polyethylene, should be placed on the soil of crawl spaces to decrease the moisture content of the space as well as of the wood members.

Insulation should be placed along all walls, floors, and ceilings that are adjacent to unheated areas. These include stairways, dwarf (knee) walls, and dormers of 1 1/2 story structures. Provisions should be made for ventilating the unheated areas.

Where attic space is unheated and a stairway is included, insulation should be used around the stairway as well as in the first-floor ceiling. The door leading to the attic should be weather stripped to prevent heat loss. Walls adjoining an unheated garage or porch should also be insulated. In structures with flat or low-pitched roofs, insulation should be used in the ceiling area with sufficient space allowed above for cleared unobstructed ventilation between the joists. Insulation should be used along the perimeter of houses built on slabs. A vapor barrier should be included under the slab.

In the summer, outside surfaces exposed to the direct rays of the sun may attain temperatures of 50°F or more above shade temperatures and tend to transfer this heat into the house. Insulation in the walls and in the attic areas retards the flow of heat and improves summer comfort conditions.

Where air conditioning is used, insulation should be placed in all exposed ceilings and walls in the same manner as insulating against cold-weather heat loss. Shading of glass against direct rays of the sun and the use of insulated glass helps reduce the air-conditioning load.

Ventilation of attic and roof spaces is an important adjunct to insulation. Without ventilation, an attic space may become very hot and hold the heat for many hours. Ventilation methods suggested for protection against cold-weather condensation apply equally well to protection against excessive hot-weather roof temperatures.

The use of storm windows or insulated glass greatly reduces heat loss. Almost twice as much heat loss occurs through a single glass as through a window glazed with insulated glass or protected by a storm sash. Double glass normally prevents surface condensation and frost forming on inner glass surfaces in winter. When excessive condensation persists, paint failures and decay of the sash rail can occur.

CAUTION

Prior to the actual installation of the insulation, consult the manufacturer's specifications and guidelines for personal-protection items required. Installing insulation is not particularly hazardous; however, there are some health safeguards to be observed when working with fiberglass.

INSTALLATION

Blanket insulation and batt insulation with a vapor barrier should be placed between framing members so that the tabs of the barrier lap the edge of the studs as well as the top and bottom plates. This method is not popular with contractors because it is more difficult to apply the drywall or rock lath (plaster base). However, it assures a minimum of vapor loss compared to the loss when the tabs are stapled to the sides of the studs. To protect the top and soleplates, as well as the headers over openings, use narrow strips of vapor barrier material along the top and bottom of the wall (fig. 5-28, view A). Ordinarily, these areas are not well covered by the vapor barrier on the blanket or batt. A hand stapler is commonly used to fasten the insulation and the vapor barriers in place.

For insulation without a vapor barrier (batt), a plastic film vapor barrier, such as 4-mil polyethylene, is commonly used to envelop the entire exposed wall and ceilings (fig. 5-28, views B and C). It covers the openings as well as the window and doorheaders and edge studs. This system is one of the best from the standpoint of resistance to vapor movement. Furthermore, it does not have the installation inconveniences encountered when tabs of the insulation are stapled over

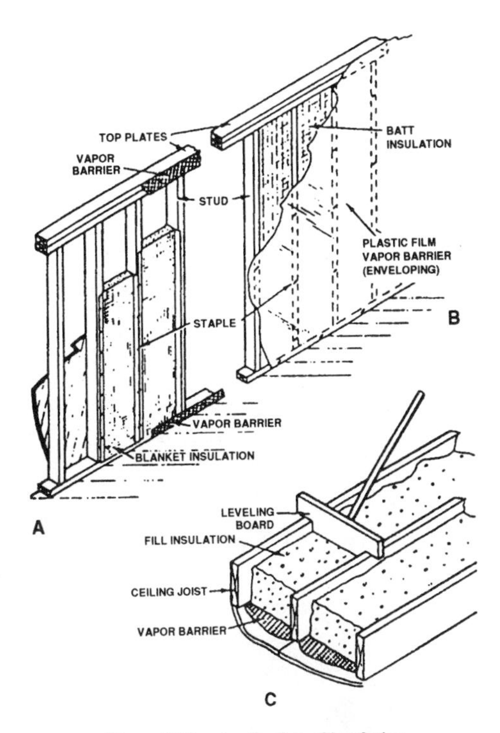

Figure 5-28.—Application of insulation.

the edges of the studs. After the drywall is installed or plastering is completed, the film is trimmed around the window and door openings.

Reflective insulation, in a single-sheet form with two reflective surfaces, should be placed to divide the space formed by the framing members into two approximately equal spaces. Some reflective insulations include air spaces and are furnished with nailing tabs. This type is fastened to the studs to provide at least a 3/4-inch space on each side of the reflective surfaces.

Fill insulation is commonly used in ceiling areas and is poured or blown into place (fig. 5-28, view C). A vapor barrier should be used on the warm side (the bottom, in case of ceiling joists) before insulation is placed. A leveling board (as shown) gives a constant insulation thickness. Thick batt insulation might also be combined to obtain the desired thickness with the vapor barrier against the back face of the ceiling finish. Ceiling insulation 6 or more inches thick greatly reduces heat loss in the winter and also provides summertime protection.

Areas around doorframes and window frames between the jambs and rough framing members also require insulation. Carefully fill the areas with

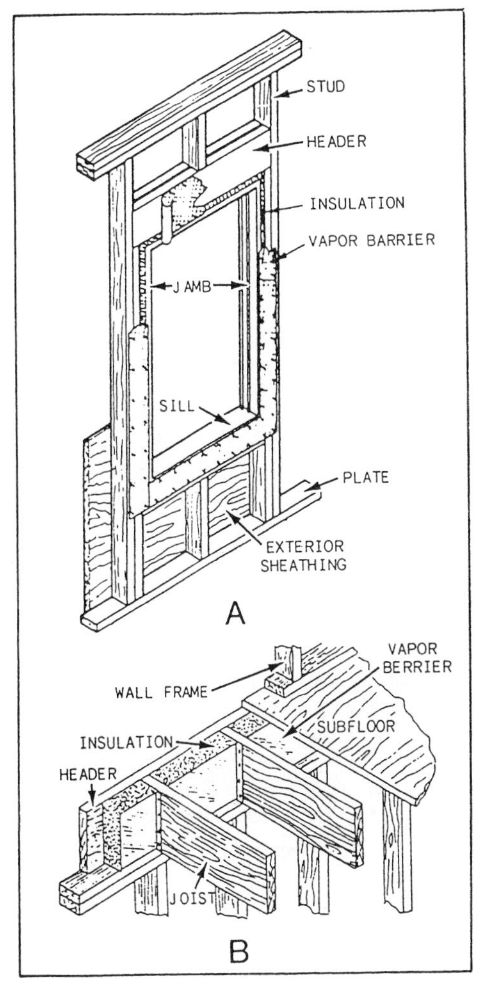

Figure 5-29.—Precautions in insulating.

insulation. Try not to compress the material, which may cause it to lose some of its insulating qualities. Because these areas are filled with small sections of insulation, a vapor barrier must be used around the openings as well as over the header above the openings (fig. 5-29, view A). Enveloping the entire wall eliminates the need for this type of vapor-barrier installation.

In 1 1/2- and 2-story structures and in basements, the area at the joist header at the outside walls should be insulated and protected with a vapor barrier (fig. 5-29, view B). Insulation should be placed behind electrical outlet boxes and other utility connections in exposed walls to minimize condensation on cold surfaces.

VAPOR BARRIER

Most building materials are permeable to water vapor. This presents problems because considerable water vapor can be generated inside structures. In cold climates during cold weather, this vapor may pass through wall and ceiling materials and condense in the wall or attic space. In severe cases, it may damage the exterior paint and interior finish, or even result in structural member decay. For protection, a material highly resistive to vapor transmission, called a vapor barrier, should be used on the warm side of a wall and below the insulation in an attic space.

Types

Effective vapor-barrier materials include asphalt laminated papers, aluminum foil, and plastic films. Most blanket and batt insulations include a vapor barrier on one side, and some of them with paper-backed aluminum foil. Foil-backed gypsum lath or gypsum boards are also available and serve as excellent vapor barriers.

Some types of flexible blanket and batt insulations have barrier material on one side. Such flexible insulations should be attached with the tabs at their sides fastened on the inside (narrow) edges of the studs, and the blanket should be cut long enough so that the cover sheet can lap over the face of the soleplate at the bottom and over the plate at the top of the stud space. However, such a method of attachment is not the common practice of most installers.

When a positive seal is desired, wall-height rolls of plastic-film vapor barriers should be applied over studs, plates, and window and doorheaders. This system, called "enveloping," is used over insulation having no vapor barrier or to ensure excellent protection when used over any type of insulation. The barrier should be fitted tightly around outlet boxes and sealed if necessary. A ribbon of sealing compound around an outlet or switch box minimizes vapor loss at this area. Cold-air returns, located in outside walls, should be made of metal to prevent vapor loss and subsequent paint problems.

Paint Coatings

Paint coatings cannot substitute for the membrane types of vapor barriers, but they do provide some protection for structures where other types of vapor barriers were not installed during construction. Of the various types of paint, one coat of aluminum primer followed by two decorative coats of flat wall oil base paint is quite effective. For rough plaster or for buildings in very cold climates, two coats of aluminum primer may be necessary. A pigmented primer and sealer, followed by decorative finish coats or two coats of rubber-base paint, are also effective in retarding vapor transmission.

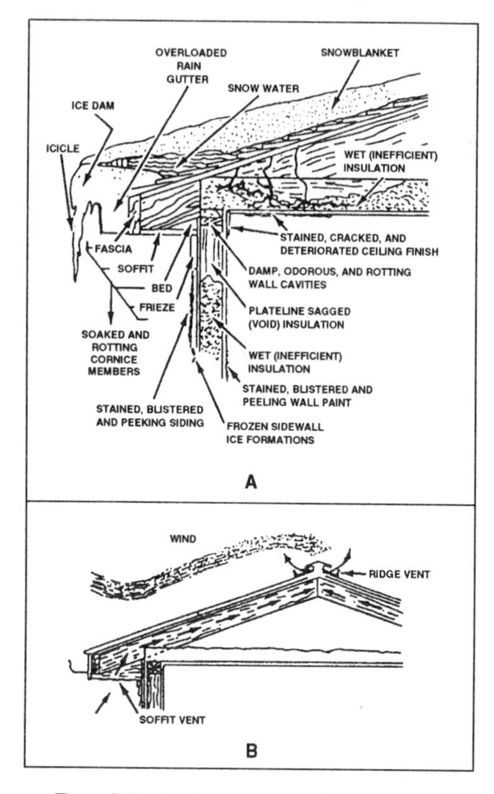

Figure 5-30.—Ice dams and protective ventilation.

VENTILATION

Condensation of moisture vapor may occur in attic spaces and under flat roofs during cold weather. Even where vapor barriers are used, some vapor will probably work into these spaces around pipes and other inadequately protected areas and through the vapor barrier itself. Although the amount might be unimportant if equally distributed, it may be sufficiently concentrated in some cold spots to cause damage. While wood shingle and wood shake roofs do not resist vapor movement, such roofings as asphalt shingles and built-up roofs are highly resistant. The most practical method of removing the moisture is by adequate ventilation of roof spaces.

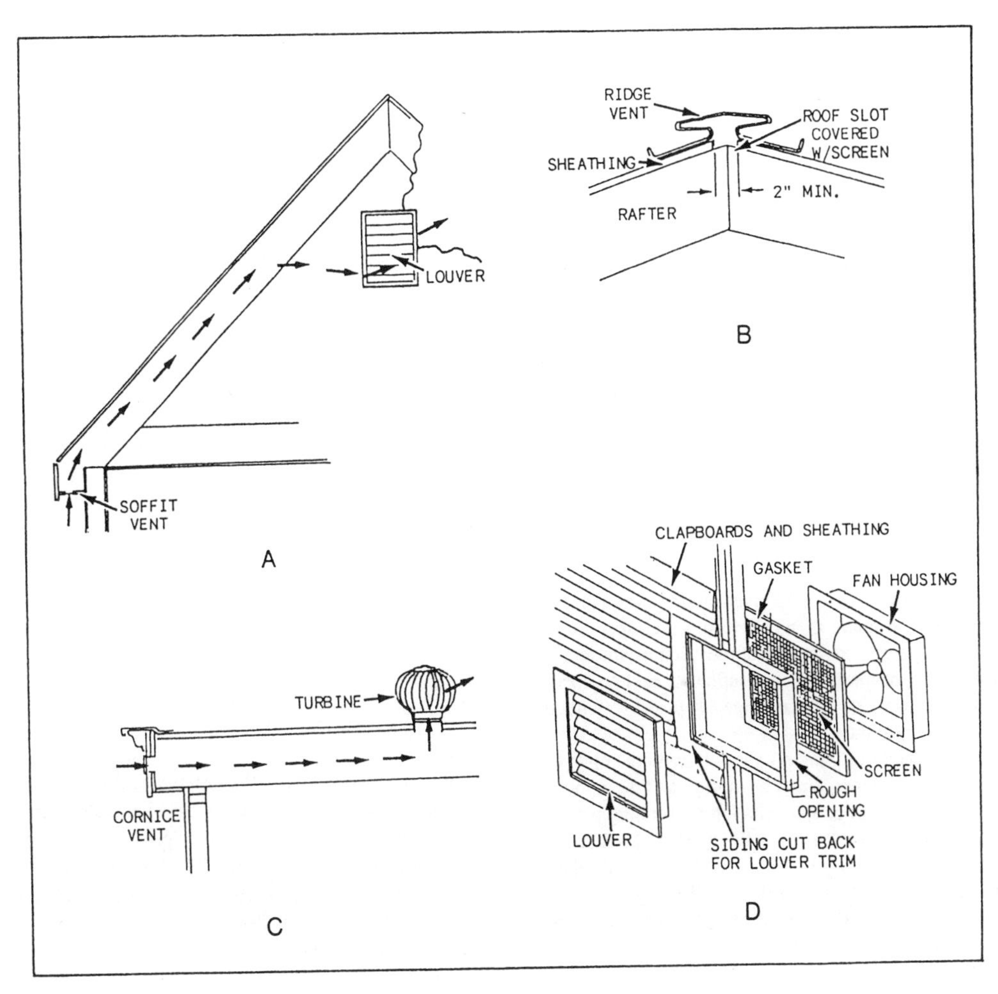

Figure 5-31.—Attic outlet vents.

A warm attic that is inadequately ventilated and insulated may cause formation of ice dams at the cornice (fig. 5-30, view A). During cold weather after a heavy snowfall, heat causes the snow next to the roof to melt. Water running down the roof freezes on the colder surface of the cornice, often forming an ice dam at the gutter that may cause water to back up at the eaves and into the wall and ceiling. Similar dams often form in roof valleys. Ventilation provides part of the solution to these problems. With a well-insulated ceiling and adequate ventilation (fig. 5-30 view B), attic temperatures are low and melting of snow over the attic space greatly reduced.

In hot weather, ventilation of attic and roof spaces offers an effective means of removing hot air and lowering the temperature in these spaces. Insulation should be used between ceiling joists below the attic or roof space to further retard heat flow into the rooms below and materially improve comfort conditions.

It is common practice to install louvered openings in the end walls of gable roofs for ventilation. Air movement through such openings depends primarily on wind direction and velocity. No appreciable movement can be expected when there is no wind. Positive air movement can be obtained by providing additional openings (vents) in the soffit areas of the roof overhang (fig. 5-31, view A) or ridge (view B). Hip-roof structures are best ventilated by soffit vents and by outlet ventilators along the ridge. The differences in temperature between the attic and the outside create an air movement independent of the wind, and also a more positive movement when there is wind. Turbine-type ventilators are also used to vent attic spaces (view C).

Where there is a crawl space under the house or porch, ventilation is necessary to remove the moisture vapor rising from the soil. Such vapor may otherwise condense on the wood below the floor and cause decay. As mentioned earlier, a permanent vapor barrier on the soil of the crawl space greatly reduces the amount of ventilation required.

Tight construction (including storm windows and storm doors) and the use of humidifiers have created potential moisture problems that must be resolved by adequate ventilation and the proper use of vapor barriers. Blocking of soffit vents with insulation, for example, must be avoided because this can prevent proper ventilation of attic spaces. Inadequate ventilation often leads to moisture problems, resulting in unnecessary maintenance costs.

Various styles of gable-end ventilators are available. Many are made with metal louvers and frames, whereas others may be made of wood to more closely fit the structural design. However, the most important factors are to have properly sized ventilators and to locate ventilators as close to the ridge as possible without affecting appearance.

Ridge vents require no special framing, only the disruption of the top course of roofing and the removal of strips of sheathing. Snap chalk lines running parallel to the ridge, down at least 2 inches from the peak. Using a linoleum cutter or a utility knife with a very stiff blade, cut through the roofing along the lines. Remove the roofing material and any roofing nails that remain. Set your power saw to cut through just the sheathing (not into the rafters) along the same lines. A carbide-tipped blade is best for this operation. Remove the sheathing. Nail the ridge vent over the slot you have created, using gasketed roofing nails. Remember to use compatible materials. For example, aluminum nails should be used with aluminum vent material. Because the ridge vent also covers the top of the roofing, be sure the nails are long enough to penetrate into the rafters. Caulk the underside of the vent before nailing.

The openings for louvers and in-the-wall fans (fig. 5-31, view D) are quite similar. In fact, fans are usually covered with louvers. Louver slats should have a downward pitch of 45° to minimize water blowing in. As with soffit vents, a backing of corrosion-resistant screen is needed to keep insects out. Ventilation fans may be manual or thermostatically controlled.

When installing a louver in an existing gable-end wall, disturb the siding, sheathing, or framing members as little as possible. Locate the opening by drilling small holes through the wall at each corner. Snap chalk lines to establish the cuts made with a reciprocating saw. Cut back the siding to the width of the trim housing the louver (or the louver-with-fan), but cut back the sheathing only to the dimensions of the fan housing. Box in the rough opening itself with 2 by 4s and nail or screw the sheathing to them. Flash and caulk a gable-end louver as you would a door or a window.

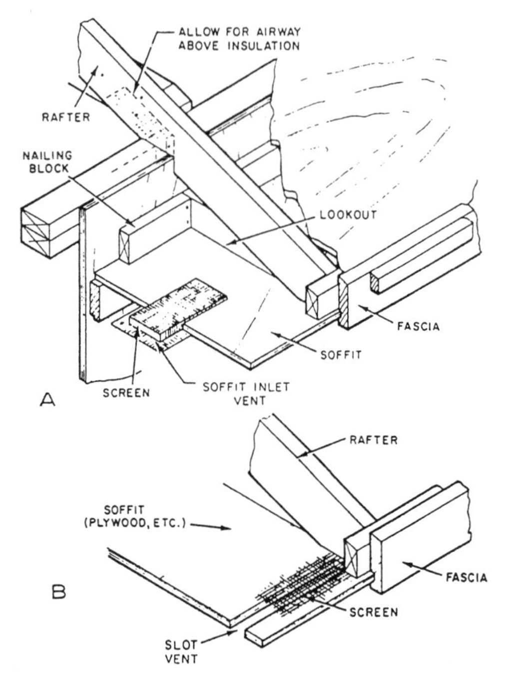

Figure 5-32.—Inlet vents.

Small, well-distributed vents or continuous slots in the soffit provide good inlet ventilation. These small louvered and screened vents (see fig. 5-32, view A) are easily obtained and simple to install. Only small sections need to be cut out of the soffit to install these vents, which can be sawed out before the soffit is installed. It is better to use several small, well-distributed vents than a few large ones. Any blocking that might be required between rafters at the wall line should be installed to provide an airway into the attic area.

A continuous screened slot vent, which is often desirable, should be located near the outer edge of the soffit near the fascia (fig. 5-32, view B). This location minimizes the chance of snow entering. This type of vent is also used on the overhang of flat roofs.

RECOMMENDED READING LIST

NOTE

Although the following references were current when this TRAMAN was published, their continued currency cannot be assured. You therefore need to ensure you are studying the latest revisions.

Drywall: Installation and Applications, W. Robert Harris, American Publishers, Inc., Homewood, Ill., 1979.

Modern Carpentry, Willis H. Wagner, Goodheart-Wilcox Co., South Holland, Ill., 1983.

Wood-Frame House Construction, L.O. Anderson, Forest Products Laboratory, U.S. Forest Service, U.S. Department of Agriculture, Washington, D.C., 1975.

CHAPTER 6

INTERIOR FINISH OF FLOORS, STAIRS, DOORS, AND TRIM

This chapter continues our discussion of interior finishing. In the previous chapter, we looked at the interior finishing of walls and ceilings, and related aspects of insulation and ventilation. Now, we'll examine the common types of flooring and the construction procedures for a stairway and interior doorframing. We'll also discuss the types of wood trim and the associated installation procedures.

FLOOR COVERINGS

LEARNING OBJECTIVE: Upon completing this section, you should be able to identify the common types of floor coverings and describe procedures for their placement.

Numerous flooring materials now available may be used over a variety of floor systems. Each has a property that adapts it to a particular usage. Of the practical properties, perhaps durability and ease of maintenance are the most important. However, initial cost, comfort, and appearance must also be considered. Specific service requirements may call for special properties, such as resistance to hard wear in warehouses and on loading platforms, or comfort to users in offices and shops.

There is a wide selection of wood materials used for flooring. Hardwoods and softwoods are available as strip flooring in a variety of widths and thicknesses, and as random-width planks and block flooring. Other materials include linoleum, asphalt, rubber, cork, vinyl, and tile and sheet forms. Tile flooring is also available in a particleboard, which is manufactured with small wood particles combined with resin and formed under high pressure. In many areas, ceramic tile and carpeting are used in ways not thought practical a few years ago. Plastic floor coverings used over concrete or a stable wood subfloor are another variation in the types of finishes available.

WOOD-STRIP FLOORING

Softwood finish flooring costs less than most hardwood species and is often used to good advantage in bedroom and closet areas where traffic is light. However, it is less dense than the hardwoods, less wear-resistant, and shows surface abrasions more readily. Softwoods most commonly used for flooring are southern pine, Douglas fir, redwood, and western hemlock.

Softwood flooring has tongue-and-groove edges and may be hollow-backed or grooved. Some types are also end-matched. Vertical-grain flooring generally has better wearing qualities than flat-grain flooring under hard usage.

Hardwoods most commonly used for flooring are red and white oak, beech, birch, maple, and pecan, any of which can be prefinished or unfinished.

Hardwood strip flooring is available in widths ranging from 1 1/2 to 3 1/4 inches. Standard thicknesses include 3/8, 1/2, and 3/4 inch. A useful feature of hardwood strip flooring is the undercut. There is a wide groove on the bottom of each piece that enables it to lay flat and stable, even when the subfloor surface is slightly uneven.

These strips are laid lengthwise in a room and normally at right angles to the floor joists. A subfloor of diagonal boards or plywood is normally used under the finish floor. The strips are tongue and groove and end-matched (fig. 6-1, view A). Strips are random length

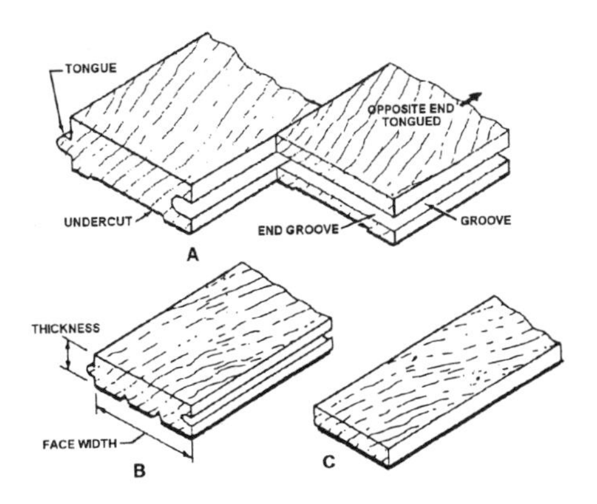

Figure 6-1.—Types of strip flooring.

and may vary from 2 to more than 16 feet. The top is slightly wider than the bottom so that tight joints result when flooring is laid. The tongue fits tightly into the groove to prevent movement and floor squeaks.

Thin strip flooring (fig. 6-1. view B) is made of 3/8- by 2-inch strips. This flooring is commonly used for remodeling work or when the subfloor is edge-blocked or thick enough to provide very little deflection under loads.

Square-edged strip flooring (fig. 6-1, view C) is also occasionally used. The strips are usually 3/8 inch by 2 inches and laid over a substantial subfloor. Face-nailing is required for this type of flooring.

Plank floors are usually laid in random widths. The pieces are bored and plugged to simulate wooden pegs originally used to fasten them in place. Today, this type of floor has tongue-and-groove edges. It is laid similar to regular strip flooring. Solid planks are usually 3/4 inch thick. Widths range from 3 to 9 inches in multiples of 1 inch.

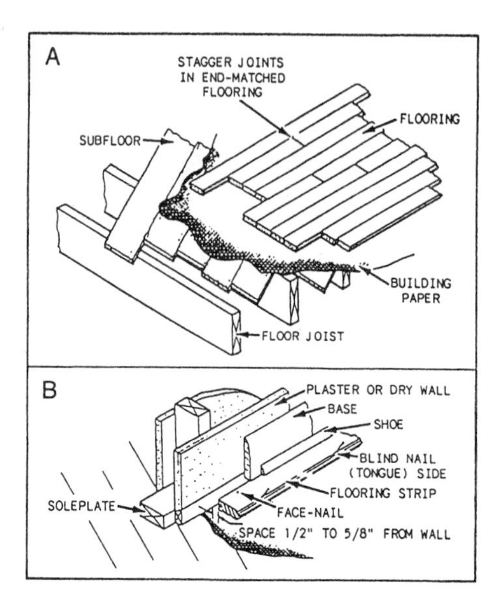

Figure 6-2.—Application of strip flooring.

Installation

Flooring should be laid after drywall, plastering, or other interior wall and ceiling finish is completed and dried out. Windows and exterior doors should be in place, and most of the interior trim, except base, casing, and jambs, should be installed to prevent damage by wetting or construction activity.

Board subfloors should be clean and level and covered with felt or heavy building paper. The felt or paper stops a certain amount of dust, somewhat deadens sound, and, where a crawl space is used, increases the warmth of the floor by preventing air infiltration. As a guide to provide nailing into the joists, wherever possible, mark with a chalk line the location of the joists on the paper. Plywood subflooring does not normally require building paper.

Strip flooring should normally be laid crosswise to the floor joists (fig. 6-2, view A). In conventional structures, the floor joists span the width of the building over a center-supporting beam or wall. Thus, the finish flooring of the entire floor areas of a rectangular structure will be laid in the same direction. Flooring with "L"- or "I"-shaped plans will usually have a direction change, depending on joist direction. As joists usually span the short way in a room, the flooring will be laid lengthwise to the room. This layout has a pleasing appearance and also reduces shrinkage and swelling of the flooring during seasonal changes.

Storing

When the flooring is delivered, store it in the warmest and driest place available in the building. Moisture absorbed after delivery to the building site is the most common cause of open joints between flooring strips that appear after several months of the heating season.

Floor Squeaks

Floor squeaks are usually caused by the movement of one board against another. Such movement can occur for a number of reasons: floor joists too light, causing excessive deflection; sleepers over concrete slabs not held down tightly; loose fitting tongues; or poor nailing. Adequate nailing is an important means of minimizing squeaks. Another is to apply the finish floors only after the joists have dried to 12-percent moisture content or less. A much better job results when it is possible to nail through the finish floor, through the subfloor, and into the joists than if the finish floor is nailed only to the subfloor.

Nailing

Various types of nails are used in nailing different thicknesses of flooring. Before using any type of nail, you should check with the floor manufacturer's

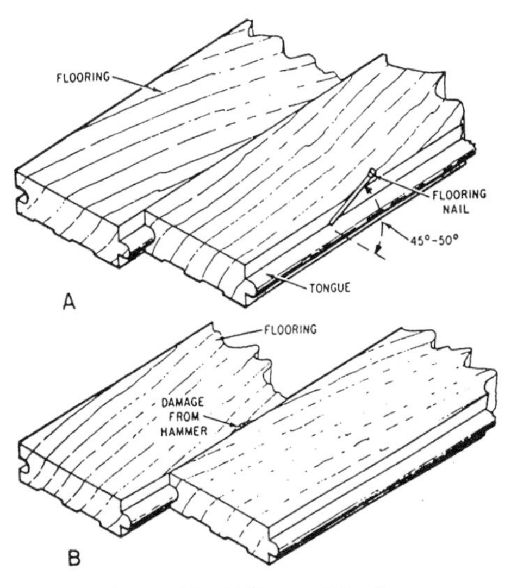

Figure 6-3.—Nailing wood flooring.

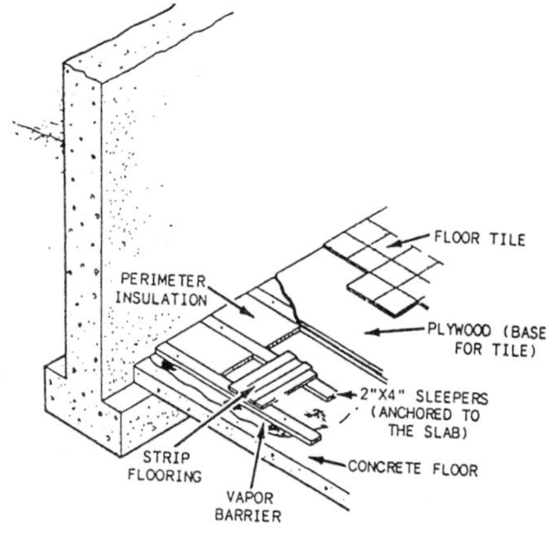

Figure 6-4.—Floor detail for existing concrete construction.

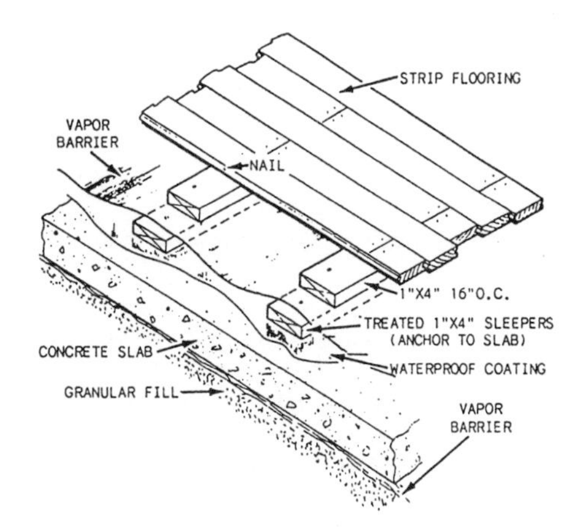

Figure 6-5.—Base for wood flooring on a slab with vapor barrier.

recommendations as to size and diameter for specific uses. Flooring brads are also available with blunted points to prevent splitting the tongue.

Figure 6-2, view B, shows how to nail the first strip of flooring. This strip should be placed 1/2 to 5/8 inch away from the wall. The space is to allow for expansion of the flooring when moisture content increases. The first nails should be driven straight down, through the board at the groove edge. The nails should be driven into the joist and near enough to the edge so that they will be covered by the base or shoe molding. The first strip of flooring can also be nailed through the tongue (fig. 6-3, view A). This figure shows in detail how nails should be driven into the tongue of the flooring at an angle of 45° to 50°. Don't drive the nails flush; this prevents damaging the edge by the hammerhead (fig. 6-3 view B). These nails should be set with a nail set.

To prevent splitting the flooring, predrill through the tongue, especially at the ends of the strip. For the second course of flooring from the wall, select pieces so that the butt joints are well separated from those in the first course. Under normal conditions, each board should be driven up tightly against the previous board. Crooked pieces may require wedging to force them into alignment or may be cut and used at the ends of the course or in closets. In completing the flooring, you should provide a 1/2- to 5/8-inch space between the wall and the last flooring strip. This strip should be face-nailed just like the first strip so that the base or shoe covers the set nailheads (fig. 6-2, view B).

Installation over Concrete

One of the most critical factors in applying wood flooring over concrete is the use of a good vapor barrier under the slab to resist ground moisture. The vapor barrier should be placed under the slab during construction. However, an alternate method must be used when the concrete is already in place (shown in fig. 6-4).

A system of preparing a base for wood flooring when there is a vapor barrier under the slab is shown in figure 6-5. Treated 1- by 4-inch furring strips should be

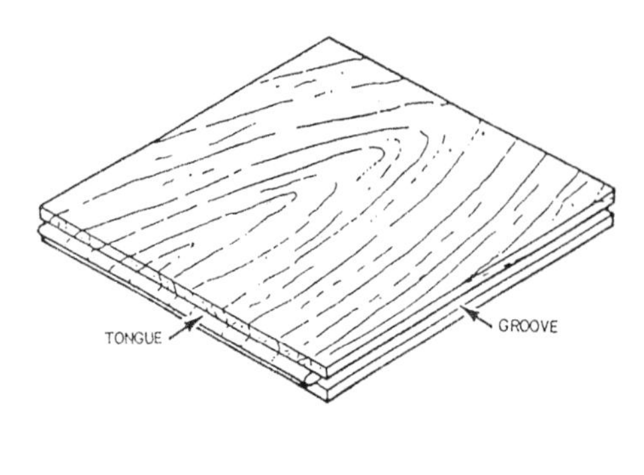

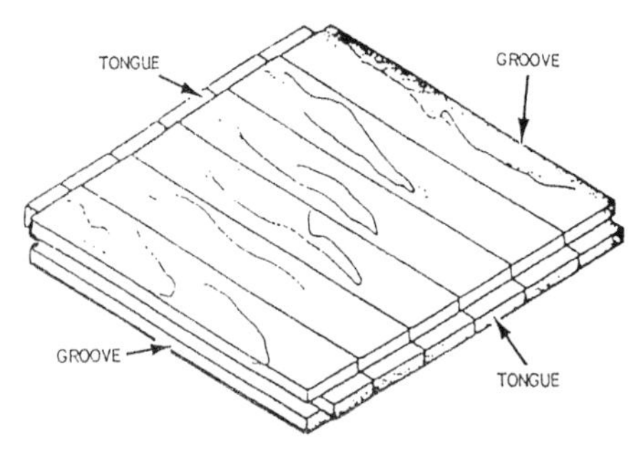

Figure 6-6.—Wood block (parquet) laminated flooring.

anchored to the existing slab. Shims can be used, when necessary, to provide a level base. Strips should be spaced no more than 16 inches on center (OC). A good waterproof or water-vapor resistant coating on the concrete before the treated strips are installed is usually recommended to aid in further reducing moisture movement. A vapor barrier, such as a 4-mil polyethylene or similar membrane, is then laid over the anchored 1- by 4-inch wood strips and a second set of 1 by 4s nailed to the first. Use 1 1/2-inch-long nails spaced 12 to 16 inches apart in a staggered pattern. The moisture content of these second members should be approximately the same as that of the strip flooring to be applied. Strip flooring can then be installed as previously described.

When other types of finish floor, such as a resilient tile, are used, plywood underlayment is placed over the 1 by 4s as a base.

WOOD BLOCK FLOORING

Wood block (parquet) flooring (fig. 6-6) is used to produce a variety of elaborate designs formed by small wood block units. A block unit consists of short lengths of flooring, held together with glue, metal splines, or other fasteners. Square and rectangular units are produced. Generally, each block is laid with its grain at right angles to the surrounding units.

Blocks, called laminated units, are produced by gluing together several layers of wood. Unit blocks are commonly produced in 3/4-inch thicknesses. Dimensions (length and width) are in multiples of the widths of the strips from which they are made. For example, squares assembled from 2 1/4-inch strips are 6 3/4 by 6 3/4 inches, 9 by 9 inches, or 11 1/4 by 11 1/4 inches. Wood block flooring is usually tongue and groove.

UNDERLAYMENT

Flooring materials, such as asphalt, vinyl, linoleum, and rubber, usually reveal rough or irregular surfaces in the flooring structure upon which they are laid. Conventional subflooring does not provide a satisfactory surface. An underlayment of plywood or hardboard is required. On concrete floors, a special mastic material is sometimes used when the existing surface is not suitable as a base for the finish flooring.

An underlayment also prevents the finish flooring materials from checking or cracking when slight movements take place in a wood subfloor. When used for carpeting and resilient materials, the underlayment is usually installed as soon as wall and ceiling surfaces are complete.

Hardboard and Particleboard

Hardboard and particleboard both meet the requirements of an underlayment board. The standard thickness for hardboard is 1/4 inch. Particleboard thicknesses range from 1/4 to 3/4 inch.

This type of underlayment material will bridge small cups, gaps, and cracks. Larger irregularities should be repaired before the underlayment is applied. High spots should be sanded down and low areas filled. Panels should be unwrapped and placed separately around the room for at least 24 hours before they are installed. This equalizes the moisture content of the panels before they are installed.

INSTALLATION.—To install hardboard or particleboard, start at one corner and fasten each panel securely before laying the next. Some manufacturers print a nailing pattern on the face of the panel. Allow at least a 1/8- to 3/8-inch space next to a wall or any other vertical surface for panel expansion.

Stagger the joints of the underlayment panel. The direction of the continuous joints should be at right

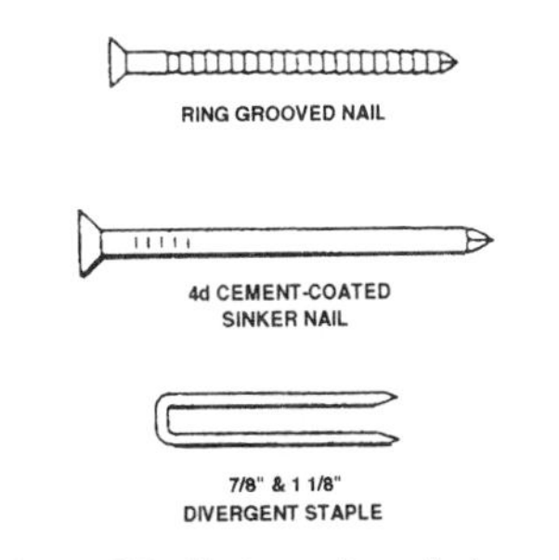

Figure 6-7.—Fasteners for underlayment.

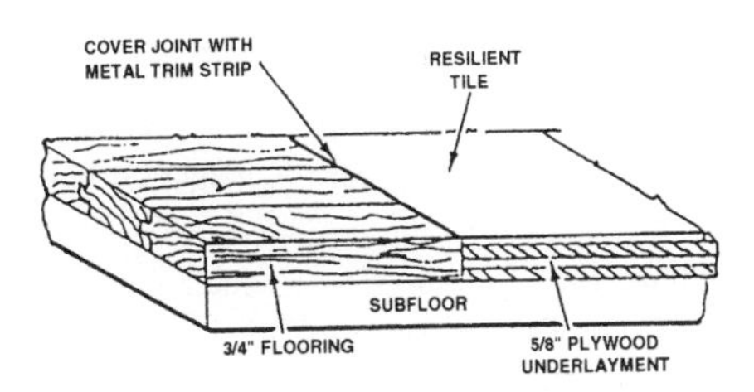

Figure 6-8.—Alignment of finish flooring materials.

angles to those in the subfloor. Be especially careful to avoid aligning any joints in the underlayment with those in the subfloor. Leave a 1/32-inch space at the joints between hardboard panels. Particleboard panels should be butted lightly.

FASTENERS.—Underlayment panels should be attached to the subfloor with approved fasteners. Examples are shown in figure 6-7. For hardboard, space the fasteners 3/8 inch from the edge.

Spacing for particleboard varies for different thicknesses. Be sure to drive nailheads flush. When fastening underlayment with staples, use a type that is etched or galvanized and at least 7/8 inch long. Staples should not be spaced over 4 inches apart along panel edges.

Special adhesives can also be used to bond underlayment to subfloors. They eliminate the possibility of nail-popping under resilient floors.

Plywood

Plywood is preferred by many for underlayment. It is dimensionally stable, and spacing between joints is not critical. Since a range of thicknesses is available, alignment of the surfaces of various finish flooring materials is easy. An example of aligning resilient flooring with wood strip flooring is shown in figure 6-8.

To install plywood underlayment, follow the same general procedures described for hardboard. Turn the grain of the face-ply at right angles to the framing supports. Stagger the end joints. Nails may be spaced farther apart for plywood but should not exceed a field spacing of 10 inches (8 inches for 1/4- and 3/8-inch thicknesses) and an edge spacing of 6 inches OC. You should use ring-grooved or cement-coated nails to install plywood underlayment.

RESILIENT FLOOR TILE

After the underlayment is securely fastened, sweep and vacuum the surface carefully. Check to see that surfaces are smooth and joints level. Rough edges should be removed with sandpaper or a block plane.

The smoothness of the surface is extremely important, especially under the more pliable materials (vinyl, rubber, linoleum). Over a period of time, these materials will "telegraph" (show on the surface) even the slightest irregularities or rough surfaces. Linoleum is especially susceptible. For this reason, a base layer of felt is often applied over the underlayment when linoleum, either in tile or sheet form, is installed.

Because of the many resilient flooring materials on the market, it is essential that each application be made according to the recommendations and instructions furnished by the manufacturer of the product.

Installing Resilient Tile

Start a floor tile layout by locating the center of the end walls of the room. Disregard any breaks or irregularities in the contour. Establish a main center line by snapping a chalk line between these two points. When snapping long lines, remember to hold the line at various intervals and snap only short sections.

Next, lay out another center line at right angles to the main center line. This line should be established by using a framing square or set up a right triangle (fig. 6-9)

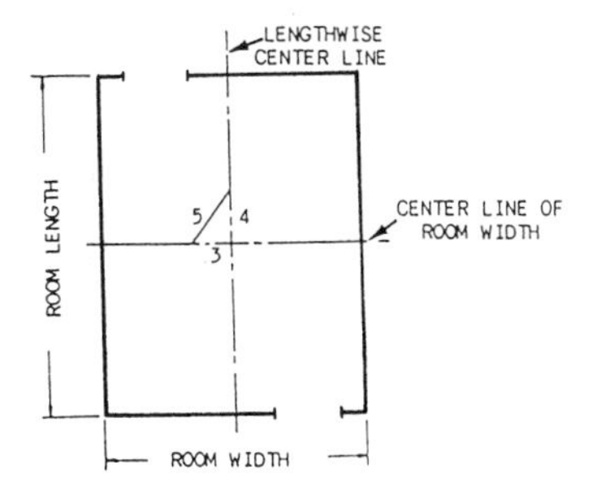

Figure 6-9.—Establishing center for laying floor tile.

with length 3 feet, height 4 feet, and hypotenuse 5 feet. In a large room, a 6:8:10-foot triangle can be used. To establish this triangle, you can either use a chalk line or draw the line along a straightedge.

With the center lines established, make a trial layout of tile along the center lines. Measure the distance between the wall and last tile. If the distance is less than 2 inches or more than 8 inches, move the center line half the width of the tile (4 1/2 inches for a 9 by 9 tile) closer to the wall. This adjustment eliminates the need to install border tiles that are too narrow. (As you will learn shortly, border tiles are installed as a separate operation—after the main area has been tiled.) Check the layout along the other center line in the same way. Since the original center line is moved exactly half the tile size, the border tile will remain uniform on opposite sides of the room. After establishing the layout, you are now ready to spread the adhesive.

SPREADING ADHESIVE.—Before you spread the adhesive, reclean the floor surface. Using a notched trowel, spread the adhesive over one-quarter of the total area bringing the spread up to the chalk line but not covering it. Be sure the depth of the adhesive is the depth recommended by the manufacturer.

The spread of adhesive is very important. If it is too thin, the tile will not adhere properly. If too heavy, the adhesive will bleed between the joints.

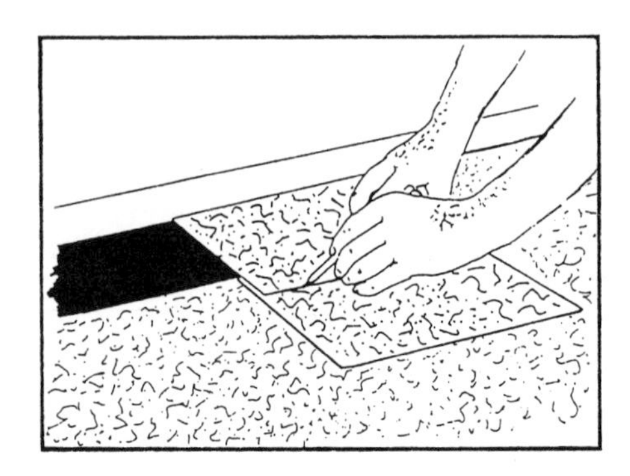

Figure 6-10.—Layout of a border tile.

Allow the adhesive to take an initial set before a single tile is laid. The time required will vary from a minimum of 15 minutes to a much longer time, depending on the type of adhesive used. Test the surface with your thumb. It should feel slightly tacky but should not stick to your thumb.

LAYING THE TILE.—Start laying the tile at the center of the room. Make sure the edges of the tile align with the chalk line. If the chalk line is partially covered with the adhesive, snap a new one or tack down a thin, straight strip of wood to act as a guide in placing the tile.

Table 6-1.—Estimating Adhesive for Floor Tile

ADHESIVE FOR FLOOR TILE	
Type and Use	**Approximate Coverage in sq. ft. Per Gallon**
Primer—For treating on or below grade concrete subfloors before installing asphalt tile	250 to 350
Asphalt cement—For installing asphalt tile over primed concrete subfloors in direct contact with the ground	200
Emulsion adhesive—For installing asphalt tile over lining felt	130 to 150
Lining paste—For cementing lining felt to wood subfloor	160
Floor and wall sealer—For priming chalky or dusty suspended concrete subfloors before installing resilient tile other than asphalt	200 to 300
Waterproof cement—Recommended for installing linoleum, rubber, and cork tile over any type of suspended subfloor in areas where surface moisture is a problem	130 to 150

Butt each tile squarely to the adjoining tile, with the corners in line. Carefully lay each tile in place. Do not slide the tile; this causes the adhesive to work up between the joints and prevents a tight fit. Take sufficient time to position each tile correctly. There is usually no hurry since most adhesives can be "worked" over a period of several hours.

To remove air bubbles, rubber, vinyl, and linoleum are usually rolled after a section of the floor is laid. Be sure to follow the manufacturer's recommendations. Asphalt tile does not need to be rolled.

After the main area is complete, set the border tile as a separate operation. To lay out a border tile, place a loose tile (the one that will be cut and used) over the last tile in the outside row. Now, take another tile and place it in position against the wall and mark a sharp pencil line on the first tile (fig. 6-10).

Cut the tile along the marked line, using heavy-duty shears or tin snips. Some types of tile require a special cutter or they may be scribed and broken. Asphalt tile, if heated, can be easily cut with snips.

After all sections of the floor have been completed, install the cove along the wall and around fixtures. A special adhesive is available for this operation. Cut the proper lengths and make a trial fit. Apply the adhesive to the cove base and press it into place.

Check the completed installation carefully. Remove any spots of adhesive. Work carefully using cleaners and procedures approved by the manufacturer.

SELF-ADHERING TILE.—Before installing self-adhering tile, you must first ensure that the floors are dry, smooth, and completely free of wax, grease, and dirt. Generally, tiles can be laid over smooth-faced resilient floors. Embossed floors, urethane floors, or cushioned floors should be removed.

Self-adhering tile is installed in basically the same way as previously mentioned types of tile. Remove the paper from the back of the tile, place the tile in position on the floor, and press it down.

Table 6-2.—Estimating Floor Tile

SQ. FT. OF FLOOR	NUMBER OF TILES	
	9" × 9"	12" × 12"
1	2	1
2	4	2
3	6	3
4	8	4
5	9	5
6	11	6
7	13	7
8	15	8
9	16	9
10	18	10
20	36	20
30	54	30
40	72	40
50	89	50
60	107	60
70	125	70
80	143	80
90	160	90
100	178	100
200	356	200
300	534	300
400	712	400
500	890	500
Waste Tile		
1 to 50 sq. ft		14%
50 to 100 sq. ft		10%
100 to 200 sq. ft		8%
200 to 300 sq. ft		7%
300 to 1,000 sq. ft		5%
Over 1,000 sq. ft		3%

Estimating Floor Tile Materials

Use table 6-1 when estimating resilient floor tile materials. This table gives you approximate square feet coverage per gallon of different types of primer and adhesives. Be sure to read and follow the manufacturer's directions. Table 6-2 provides figures for estimating the two sizes of tile most commonly used. After calculating the square feet of the area to be tiled, refer to the table to find the number of tiles needed, then add the waste factor.

To find the number of tiles required for an area not shown in this table, such as the number of 9- by 9-inch tiles required for an area of 550 square feet, add the number of tiles needed for 50 square feet to the number of tile needed for 500 square feet. The result will be 979 tiles, to which you must add 5 percent for waste. The total number of tiles required is 1,028.

When tiling large areas, work from several different boxes of tile. This will avoid concentrating one color shade variation in one area of the floor.

SHEET VINYL FLOORING

Because of its flexibility, vinyl flooring is very easy to install. Since sheets are available in 6- to 12-foot widths, many installations can be made free of seams. Flexible vinyl flooring is fastened down only around the edges and at seams. It can be installed over concrete, plywood, or old linoleum.

To install, spread the sheet smoothly over the floor. Let excess material turn up around the edges of the room. Where there are seams, carefully match the pattern. Fasten the two sections to the floor with adhesive. Trim the edges to size by creasing the vinyl sheet material at intersections of the floor and walls and cutting it with a utility knife drawn along a straightedge. Be sure the straightedge is parallel to the wall.

After the edges are trimmed and fitted, secure them with a staple gun, or use a band of double-faced adhesive tape. Always study the manufacturer's directions carefully before starting the work.

WALL-TO-WALL CARPETING

Wall-to-wall carpeting can make a small room look larger, insulate against drafty floors, and do a certain amount of soundproofing. Carpeting is not difficult to install.

All carpets consist of a surface pile and backing. The surface pile may be nylon, polyester, polypropylene, acrylic, wool, or cotton. Each has its advantages and disadvantages. The type you select depends on your needs. Carpeting can be purchased in 9-, 12-, and 15-foot widths.

Measuring and Estimating

Measure the room in the direction in which the carpet will be laid. To broaden long, narrow rooms, lay patterned or striped carpeting across the width. For conventionally rectangular rooms, measure the room lengthwise. Include the full width of doorframes so the carpet will extend slightly into the adjoining room. When measuring a room with alcoves or numerous wall projections, calculate on the basis of the widest and longest points. This will result in some waste material, but is safer than ordering less than what you need.

Most wall-to-wall carpeting is priced by the square yard. To determine how many square yards you need, multiply the length by the width of the room in feet and divide the result by 9.

Underlayment

Except for so-called "one-piece" and cushion-backed carpeting, underlayment or padding is essential to a good carpet installation. It prolongs the life of the carpeting, increases its soundproofing qualities, and adds to underfoot comfort.

The most common types of carpet padding are latex (rubber), sponge-rubber foams, soft-and-hardback vinyl foams, and felted cushions made of animal hair or of a combination of hair and jute. Of all types, the latex and vinyl foams are generally considered the most practical. Their waffled surface tends to hold the carpet in place. Most carpet padding comes in a standard 4 1/2-foot width.

Cushion-backed carpeting is increasing in popularity, especially with do-it-yourself homeowners. The high-density latex backing is permanently fastened to the carpet, which eliminates the need for a separate underpadding. It is nonskid and heavy enough to hold the carpet in place without the use of tacks. In addition, the foam rubber backing keeps the edges of the carpet from unraveling so that it need not be bound. Foam rubber is mildewproof and unaffected by water, so the carpet can be used in basements and other below-grade installations. It can even be laid directly over unfinished concrete.

The key feature of this backing, however, is the dimensional stability it imparts to the carpet. This added characteristic means the carpet will not stretch, nor will it expand and contract from temperature or humidity changes. Because of this, these carpets can be loose-laid, with no need for adhesive or tacks to give them stability.

Preparing the Floor

To lay carpets successfully on wood floors, you must ensure that the surface is free of warps, and that all nails and tacks are either removed or hammered flush. Nail down any loose floorboards and plane down the ridges of warped boards. Fill wide cracks between floorboards with strips of wood or wood putty. Cover floors that are warped and cracked beyond reasonable repair with hardboard or plywood.

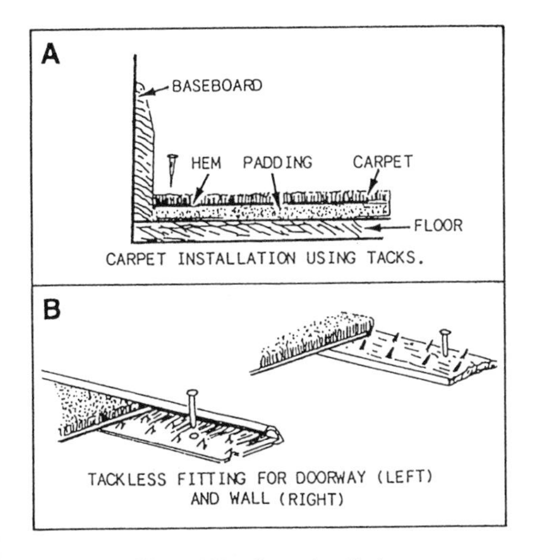

Figure 6-11.—Carpet installation.

Stone or concrete floors that have surface ridges or cracks should be treated beforehand with a floor-leveling compound to reduce carpet wear. These liquid compounds are also useful for sealing the surface of dusty or powdery floors. A thin layer of the compound, which is floated over the floor, will keep dust from working its way up through the underlayment and into the carpet pile.

The best carpeting for concrete and hard tile surfaces is the indoor-outdoor type. The backing of this carpet is made of a closed-pore type of either latex or vinyl foam, which keeps out most moisture. It is not wise to lay any of the standard paddings on top of floor tiles unless the room is well ventilated and free of condensation. Vinyl and asbestos floor tiles accumulate moisture when carpeting is laid over them. This condensation soaks through into the carpet and eventually causes a musty odor. It can also produce mildew stains.

Fastening Carpets

The standard fastening methods are with tacks or by means of tackless fittings. Carpets can also be loose-laid with only a few tacks at entrances. Carpet tack lengths are 3/4 and 1 inch. The first is long enough to go through a folded carpet hem and anchor it firmly to the floor (fig. 6-11, view A). The 1-inch tacks are used in corners where the folds of the hem make three thicknesses.

Tackless fittings (fig. 6-11, view B) are a convenient fastening method. They consist of a 4-foot wooden batten with a number of spikes projecting at a 60° angle. The battens are nailed to the floor around the entire room, end to end and 1/4 inch from the baseboard, with the spikes facing toward the wall. The spikes grip the backing of the carpet to hold it in place. On stone or concrete floors, the battens are glued in place with special adhesives.

Though cushion-backed carpeting can stay in place without fastening, securing with double-face tape is the preferred method. Carpets can also be attached to the floor with Velcro™ tape where the frequent removability of the carpet for cleaning and maintenance is a factor.

Carpet Installation

To install a carpet, you will need a hammer, large scissors, a sharp knife, a 3-foot rule, needle and carpet thread, chalk and chalk line, latex adhesive, and carpet tape. The only specialized tool you will need is a carpet stretcher, often called a knee-kicker.

Before starting the job, remove all furniture and any doors that swing into the room. When cutting the carpet, spread it out on a suitable floor space and chalk the exact pattern of the room on the pile surface; then cut along the chalk line with the scissors or sharp knife.

Join unseamed carpet by placing the two pieces so the pile surfaces meet edge to edge. Match patterned carpets carefully. With plain carpets, lay each piece so the piles run the same way. Join the pieces with carpet thread, taking stitches at 18-inch intervals along the seam. Pull the carpet tight after each stitch to take up slack. Sew along the seam between stitches. Tuck any protruding fibers back into the pile. Carpet can also be seamed by cementing carpet tape to the backing threads with latex adhesive.

Open the carpet to room length and position it before starting to put down the padding. The pile should fall away from windows to avoid uneven shading in daylight. Fold one end of the carpet back halfway and put the padding down on the exposed part of the floor. Do the same at the other end. This avoids wrinkles caused by movement of the padding.

To tack, start at the corner of the room that is formed by the two walls with the fewest obstructions. Butt the carpet up against the wall, leaving about 1 1/2 inches up the baseboard for hemming. Attach the carpet temporarily with tacks about 6 inches from the baseboard along these two walls. Use the knee-kicker to stretch the carpet, first along the length, then the width. Start from the middle of the wall, stretching alternately toward opposite corners. When it is smooth, tack down the stretched area temporarily.

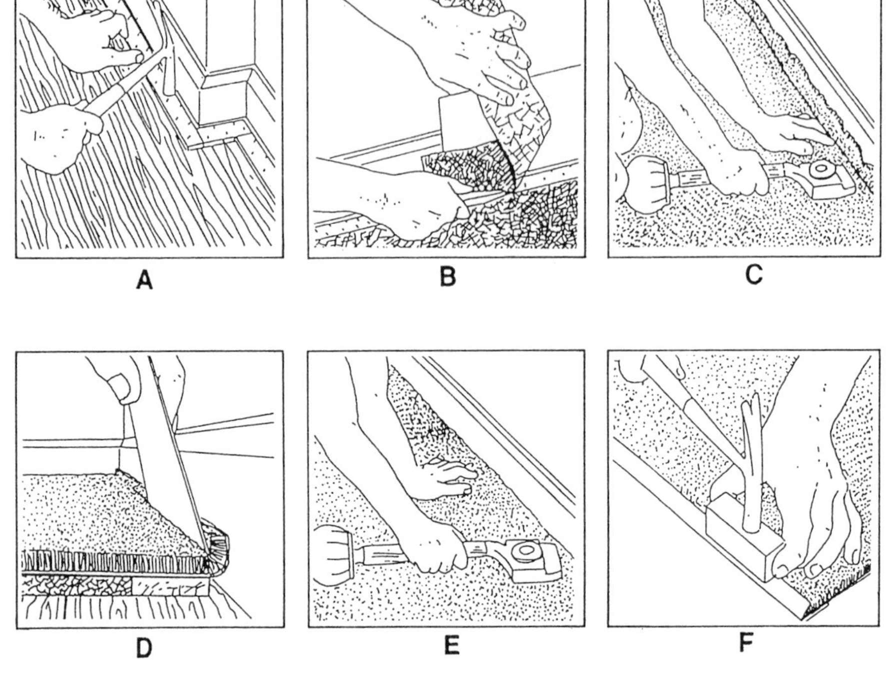

Figure 6-12.—Carpet installation using tackless fastenings.

Cut slots for pipes, fireplace protrusions, and radiators. Trim back the padding to about 2 inches from the wall to leave a channel for the carpet hem. Fold the hem under and tack the carpet in place with a tack every 5 inches. Be sure the tacks go through the fold.

When installing carpet, use tackless fastening strips, as shown in figure 6-12, view A. Position and trim the padding (view B) so that it meets the strip at the wall, but does not overlap the strip. Tack it down so it does not move. Lay out the carpeting and, using a knee-kicker, stretch the carpet over the nails projecting out of the tackless strip (view C). Trim the carpet, leaving a 3/8-inch overlap, which is tucked into place between the wall and the tackless strip (view D). (If you trim too much carpeting, lift the carpeting off the spikes of the tackless strip and use the knee-kicker to restretch the carpet [view E]). Protect the exposed edge of the carpet at doorways with a special metal binder strip or bar (view F). The strip is nailed to the floor at the doorway and the carpet slipped under a metal lip, which is then hammered down to grip the carpet edge.

Tacks can be used as an alternative to a binder strip. Before tacking, tape the exposed edge of woven carpet to prevent fraying if the salvage has been trimmed off. Cement carpet tape to the backing threads with latex adhesive. Nonwoven or latex-backed carpet will not fray, but tape is still advisable to protect exposed edges. Any door that drags should be removed and trimmed.

When installing cushion-backed carpeting, you can eliminate several steps. For instance, you don't need to use tack strips or a separate padding. Although these instructions apply to most such carpeting, read the manufacturer's instructions for any deviation in technique or use of material.

To install a cushioned carpet, apply 2-inch-wide double-face tape flush with the wall around the entire room (fig. 6-13, view A). Roll out and place the carpet. Fold back the carpet and remove the protective paper from the tape. Press the carpet down firmly over the tape and trim away excess (view B). A metal binder strip or an aluminum saddle is generally installed in doorways (view C). If your room is wider than the carpet, you will have to seam two pieces together. Follow the manufacturer's recommendations.

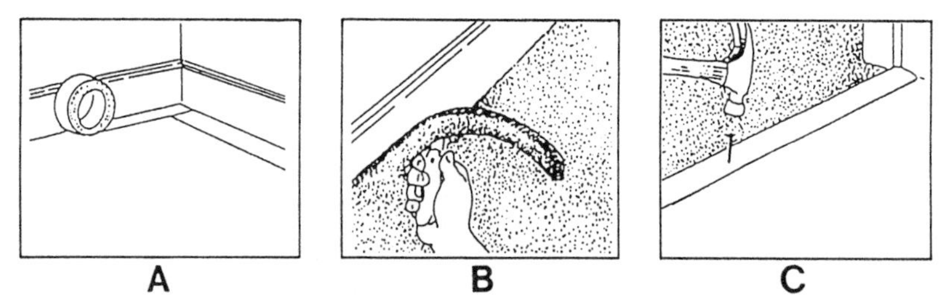

Figure 6-13.—Installing cushion-backed carpeting.

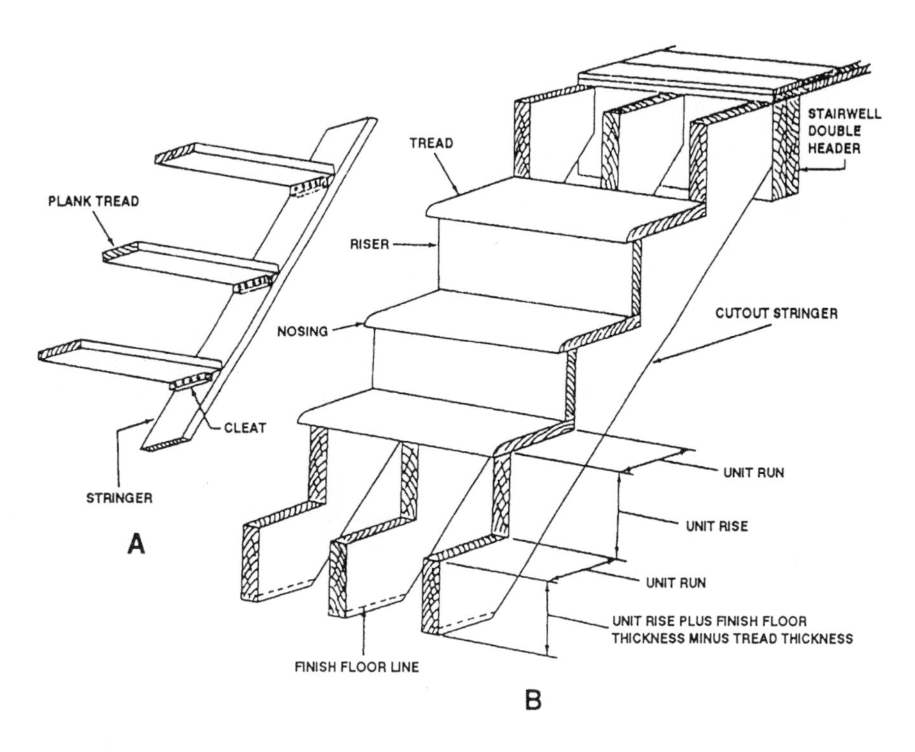

Figure 6-14.—Stairways.

STAIRS

LEARNING OBJECTIVE: Upon completing this section, you should be able to describe a stairway layout and how to frame stairs according to drawings and specifications.

There are many different kinds of stairs (interior and exterior), each serving the same purpose—the movement of personnel and products from one floor to another. All stairs have two main parts, called treads and stringers. The underside of a simple stairway, consisting only of stringers and treads, is shown in figure 6-14, view A. Treads of the type shown are called plank treads. This simple type of stairway is called a cleat stairway because of the cleats attached to the stringers to support the treads.

A more finished type of stairway has the treads mounted on two or more sawtooth-edged stringers, and includes risers (fig. 6-14, view B). The stringers shown

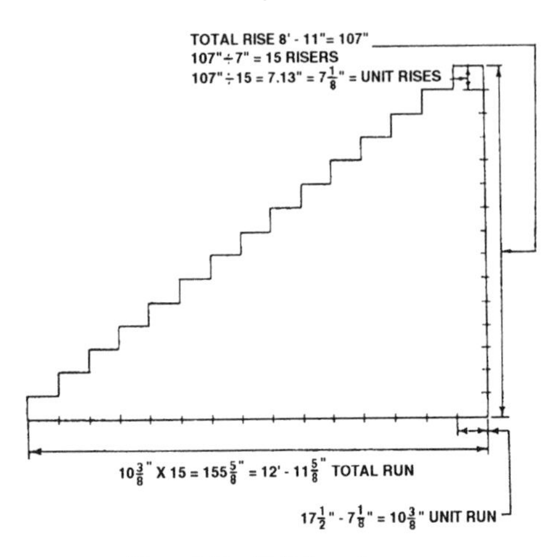

Figure 6-15.—Unit rise and run.

are cut from solid pieces of dimensional lumber (usually 2 by 12s) and are called cutout, or sawed, stringers.

STAIRWAY LAYOUT

The first step in stairway layout is to determine the unit rise and unit run (fig. 6-14, view B). The unit rise is calculated on the basis of the total rise of the stairway, and the fact that the customary unit rise for stairs is 7 inches.

The total rise is the vertical distance between the lower finish-floor level and the upper finish-floor level. This may be shown in the elevations. However, since the actual vertical distance as constructed may vary slightly from that shown in the plans, the distance should be measured.

At the time stairs are laid out, only the subflooring is installed. If both the lower and the upper floors are to be covered with finish flooring of the same thickness, the measured vertical distance from the lower subfloor surface to the upper subfloor surface will be the same as the eventual distance between the finish floor surfaces. The distance is, therefore, equal to the total rise of the stairway. But if you are measuring up from a finish floor, such as a concrete basement floor, then you must add to the measured distance the thickness of the upper finish flooring to get the total rise of the stairway. If the upper and lower finish floors will be of different thickness, then you must add the difference in thickness to the measured distance between subfloor surfaces to get the rise of the stairway. To measure the vertical distance, use a straight piece of lumber plumbed in the stair opening with a spirit level.

Let's assume that the total rise measures 8 feet 11 inches, as shown in figure 6-15. Knowing this, you can determine the unit rise as follows. First, reduce the total rise to inches—in this case it comes to 107 inches. Next, divide the total rise in inches by the average unit rise, which is 7 inches. The result, disregarding any fraction, is the number of risers the stairway will have—in this case, 107/7 or 15. Now, divide the total rise in inches by the number of risers—in this case, 107/15, or nearly 7 1/8 inches. This is the unit rise, as shown in figure 6-15.

The unit run is calculated on the basis of the unit rise and a general architect's rule that the sum of the unit run and unit rise should be 17 1/2 inches. Then, by this rule, the unit run is 17 1/2 inches minus 7 1/8 inches or 10 3/8 inches.

You can now calculate the total run of the stairway. The total run is the unit run multiplied by the total number of treads in the stairway. However, the total number of treads depends upon the manner in which the upper end of the stairway will be anchored to the header.

In figure 6-16, three methods of anchoring the upper end of a stairway are shown. In view A, there is a complete tread at the top of the stairway. This means the number of complete treads is the same as the number of

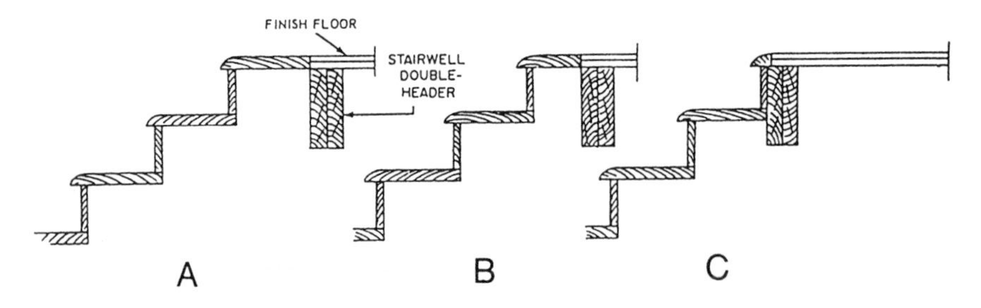

Figure 6-16.—Method for anchoring upper end of a stairway.

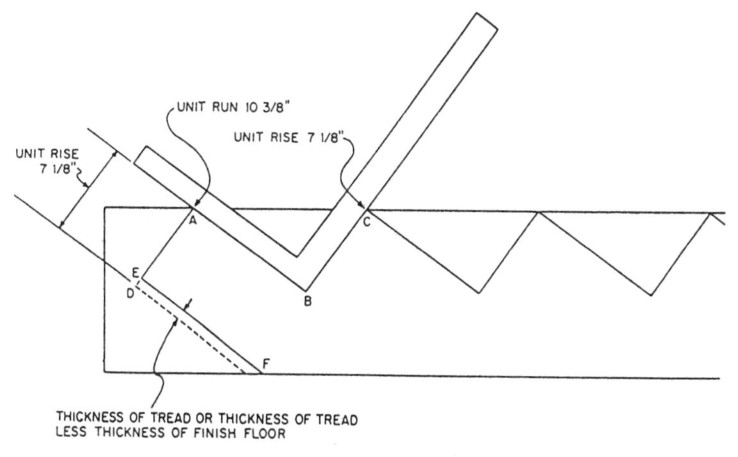

Figure 6-17.—Layout of lower end of cutout stringer.

risers. For the stairway shown in figure 6-15, there are 15 risers and 15 complete treads. Therefore, the total run of the stairway is equal to the unit run times 15, or 12 feet 11 5/8 inches.

In view B, only part of a tread is at the top of the stairway. If this method were used for the stairway shown in figure 6-15, the number of complete treads would be one less than the number of risers, or 14. The total run of the stairway would be the product of 14 multiplied by 10 3/8, plus the run of the partial tread at the top. Where this run is 7 inches, for example, the total run equals 152 1/4 inches, or 12 feet 8 1/4 inches.

In view C, there is no tread at all at the top of the stairway. The upper finish flooring serves as the top tread. In this case, the total number of complete treads is again 14, but since there is no additional partial tread, the total run of the stairway is 14 times 10 3/8 inches, or 145 1/4 inches, or 12 feet 1 1/4 inches.

When you have calculated the total run of the stairway, drop a plumb bob from the header to the floor below and measure off the total run from the plumb bob. This locates the anchoring point for the lower end of the stairway.

As mentioned earlier, cutout stringers for main stairways are usually made from 2 by 12 stock. Before cutting the stringer, you will first need to solve for the length of stock you need.

Assume that you are to use the method of upper-end anchorage shown in view A of figure 6-16 to lay out a stringer for the stairway shown in figure 6-15. This stairway has a total rise of 8 feet 11 inches and a total run of 12 feet 11 5/8 inches. The stringer must be long enough to form the hypotenuse of a triangle with sides of those two lengths. For an approximate length estimate, call the sides 9 and 13 feet long. Then, the length of the hypotenuse will equal the square root of 9^2 plus 13^2. This is the square root of 250, about 15.8 feet or 15 feet 9 1/2 inches.

Extreme accuracy is required in laying out the stringers. Be sure to use a sharp pencil or awl and make the lines meet on the edge of the stringer material.

Figure 6-17 shows the layout at the lower end of the stringer. Set the framing square to the unit run on the tongue and the unit rise on the blade, and draw the line AB. This line represents the bottom tread. Then, draw AD perpendicular to AB. Its length should be equal to the unit rise. This line represents the bottom riser in the stairway. You may have noticed that the thickness of a tread in the stairway has been ignored. This thickness is now about to be accounted for by making an allowance in the height of this first riser. This process is called "dropping the stringer."

As you can see in figure 6-14, view B, the unit rise is measured from the top of one tread to the top of the next for all risers except the bottom one. For the bottom riser, unit rise is measured from the finished floor surface to the surface of the first tread. If AD were cut to the unit rise, the actual rise of the first step would be the sum of the unit rise plus the thickness of a tread. Therefore, the length of AD is shortened by the thickness of a tread, as shown in figure 6-17, by the

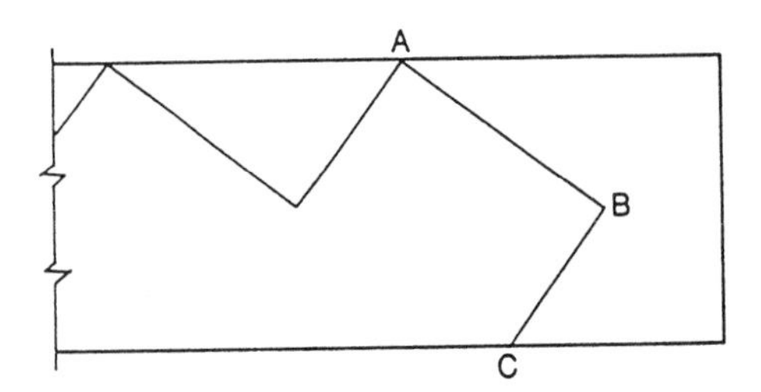

Figure 6-18.—Layout of upper end of cutout stringer.

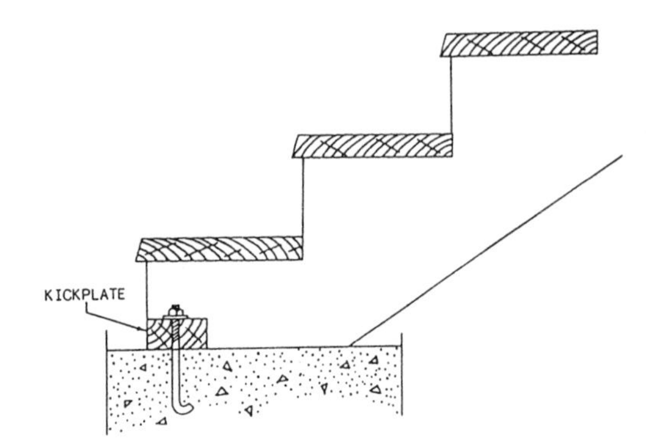

Figure 6-19.—Kickplate for anchoring stairs to concrete.

thickness of a tread less the thickness of the finish flooring. The first is done if the stringer rests on a finish floor, such as a concrete basement floor. The second is done where the stringer rests on subflooring.

When you have shortened AD to AE, draw EF parallel to AB. This line represents the bottom horizontal anchor edge of the stringer. Then, proceed to lay off the remaining risers and treads to the unit rise and unit run until you have laid off 15 risers and 15 treads. Figure 6-18 shows the layout at the upper end of the stringer. The line AB represents the top, the 15th tread. BC, drawn perpendicular to AB, represents the upper vertical anchor edge of the stringer. This edge butts against the stairwell header.

In a given run of stairs, be sure to make all the risers the same height and treads the same width. An unequal riser, especially one that is too high, is dangerous.

STAIRWAY CONSTRUCTION

We have been dealing with a common straight-flight stairway—meaning one which follows the same direction throughout. When floor space is not extensive enough to permit construction of a straight-flight stairway, a change stairway is installed—meaning one which changes direction one or more times. The most common types of these are a 90° change and a 180° change. These are usually platform stairways, successive straight-flight lengths, connecting platforms at which the direction changes 90° or doubles back 180°. Such a stairway is laid out simply as a succession of straight-flight stairways.

The stairs in a structure are broadly divided into principal stairs and service stairs. Service stairs are porch, basement, and attic stairs. Some of these may be simple cleat stairways; others may be open-riser stairways. An open-riser stairway has treads anchored on cutout stringers or stair-block stringers, but no risers. The lower ends of the stringers on porch, basement, and other stairs anchored on concrete are fastened with a kickplate (shown in fig. 6-19).

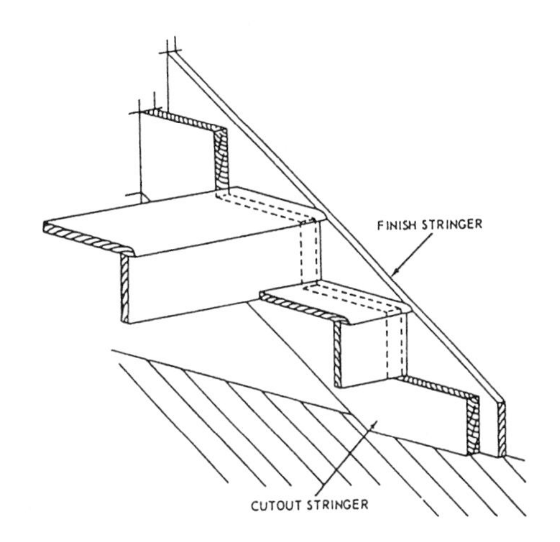

Figure 6-20.—Finish stringer.

When dealing with stairs, it is vitally important to remember the allowable head room. Head room is defined as the minimum vertical clearance required from any tread on the stairway to any part of the ceiling structure above the stairway. In most areas, the local building codes specify a height of 6 feet 8 inches for main stairs, and 6 feet 4 inches for basement stairs.

A principal stairway usually has a finished appearance. Rough cutout stringers are concealed by

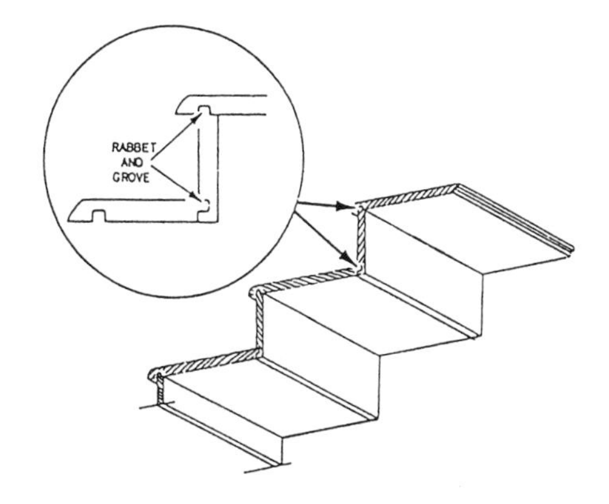

Figure 6-21.—Rabbet-and-groove-jointed treads and risers.

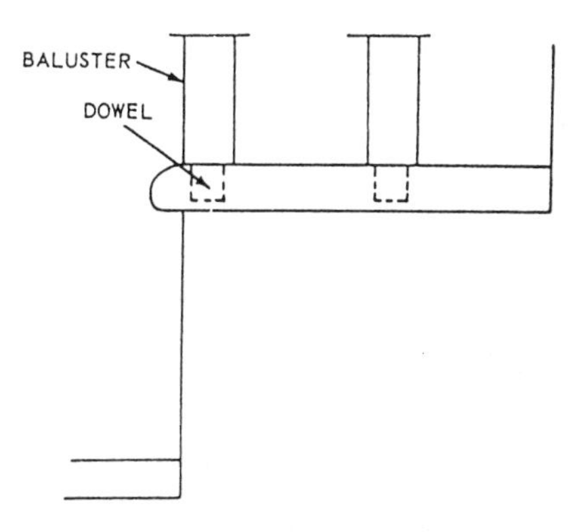

Figure 6-22.—Joining a baluster to the tread.

finish stringers (see fig. 6-20). Treads and risers are often rabbet-jointed as in figure 6-21.

Vertical members that support a stairway handrail are called balusters. Figure 6-22 shows a method of joining balusters to treads. Here, dowels, shaped on the lower ends of the balusters, are glued into holes bored in the treads.

Stringers should be toenailed to stairwell double headers with 10d nails, three to each side of the stringer. Those which face against trimmer joists should each be nailed to the joists with at least three 16d nails. At the bottom, a stringer should be toenailed with 10d nails, four to each side, driven into the subflooring and, if possible, into a joist below.

Treads and risers should be nailed to stringers with 6d, 8d, or 10d finish nails, depending on the thickness of the stock.

INTERIOR DOORFRAMING

LEARNING OBJECTIVE: Upon completing this section, you should be able to describe the procedures for laying out and installing interior doorframes, doors, and the hardware used.

Rough openings for interior doors are usually framed to be 3 inches higher than the door height and 2 1/2 inches wider than the door width. This provides for the frame and its plumbing and leveling in the opening. Interior doorframes are made up of two side jambs, a head jamb, and the stop moldings upon which the door closes. The most common of these jambs is the one-piece type (shown in fig. 6-23, view A). Jambs can be obtained in standard 5 1/4 inch widths for plaster walls and 4 5/8 inch widths for walls with 1/2-inch drywall finish. The two- and three-piece adjustable jambs (views B and C) are also standard types. Their principal advantage is in being adaptable to a variety of wall thicknesses.

Some manufacturers produce interior doorframes with the doors fitted and prehung, ready for installing. Installation of the casing completes the job. When used with two- or three-piece jambs, casings can even be installed at the factory.

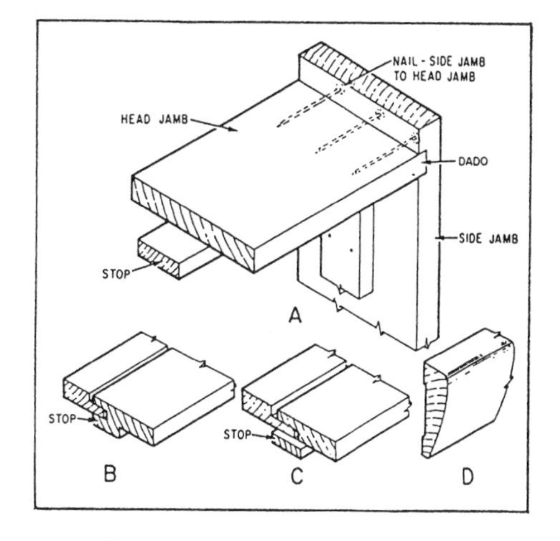

Figure 6-23.—Interior door framing parts.

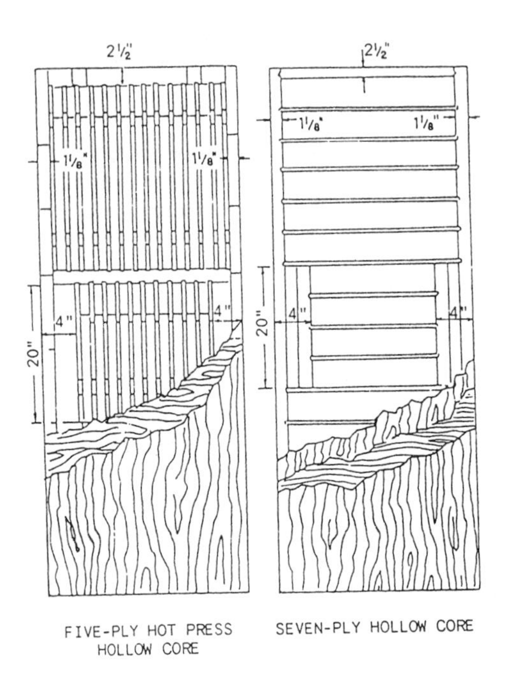

Figure 6-24.—Hollow-core construction of flushed doors.

Common minimum widths for single interior doors are as follows: bedrooms and other habitable rooms, 2 feet 6 inches; bathrooms, 2 feet 4 inches; and small closets and linen closets, 2 feet. These sizes vary a great deal, and sliding doors, folding door units, and similar types are often used for wardrobes and may be 6 feet or more in width. However, in most cases, the jamb stop and casing parts are used in some manner to frame and finish the opening.

CASING

Casing is the edge trim around interior door openings and is also used to finish the room side of windows and exterior doorframes. Casing usually varies in widths from 2 1/4 to 3 1/2 inches, depending on the style. Casing is available in thicknesses from 1/2 to 3/4 inch, although 11/16 inch is standard in many of the narrow-line patterns. A common casing pattern is shown in figure 6-23, view D.

The two general types of interior doors are the flush and the panel. Flush interior doors usually have a hollow core of light framework and are faced with thin plywood or hardboard (shown in fig. 6-24). Plywood-faced flush

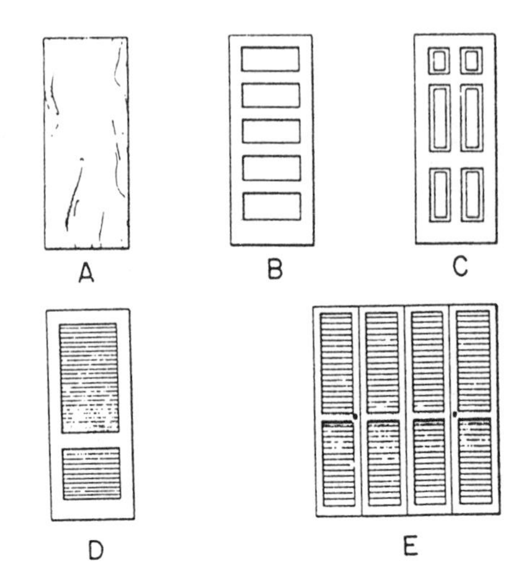

Figure 6-25.—Interior door types.

doors (fig. 6-25, view A) may be obtained in gum, birch, oak, mahogany, and several other wood species, most of which are suitable for natural finish. Nonselected grades are usually painted as hardboard-faced doors.

The panel door consists of solid stiles (vertical side members), rails (cross pieces), and panels of various types. The five-cross panel and the colonial-type panel doors are perhaps the most common of this style (fig. 6-25, views B and C). The louvered door (view D) is also popular and is commonly used for closets because it provides some ventilation. Large openings for wardrobes are finished with sliding or folding doors, or with flush or louvered doors (view E). Such doors are usually 1 1/8 inches thick.

Hinged doors should open or swing in the direction of natural entry, against a blank wall whenever possible. They should not be obstructed by other swinging doors. Doors should never be hinged to swing into a hallway.

FRAME AND TRIM INSTALLATION

When the frame and doors are not assembled and prefitted, the side jambs should be fabricated by nailing through the dado into the head jamb with three 7d or 8d coated nails (fig. 6-23 view A). The assembled frames are then fastened in the rough openings by shingle wedges used between the side jamb and the stud (fig. 6-26, view A). One jamb is plumbed and leveled using four or five sets of shingle wedges for the height of the frame. Two 8d finishing nails should be used at each

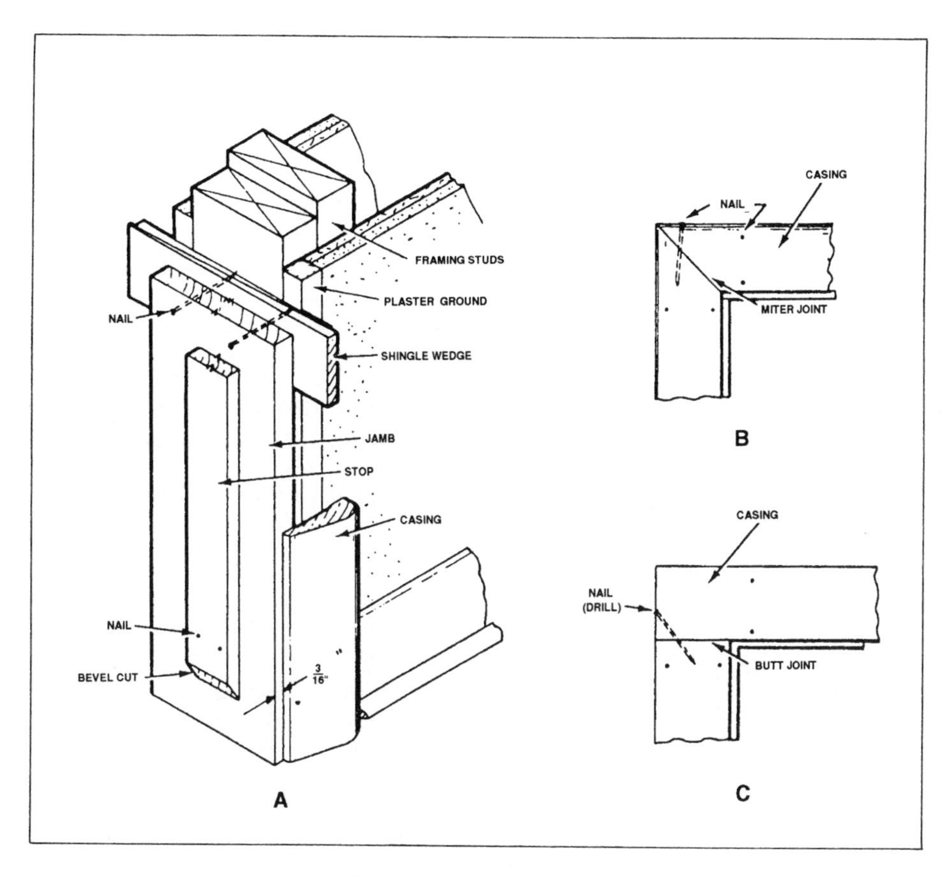

Figure 6-26.—Doorframe and trim.

wedged area, one driven so that the doorstop covers it. The opposite side jamb is then fastened in place with shingle wedges and finishing nails, using the first jamb as a guide in keeping a uniform width.

Casings should be nailed to both the jamb and the framing members. You should allow about a 3/16-inch edge distance from the face of the jamb. Use 6d or 7d finish or casing nails, depending on the thickness of the casing. To nail into the stud, use 4d or 5d finish nails or 1 1/2-inch brads to fasten the thinner edge of the casing to the jamb. For hardwood casing, it is advisable to predrill to prevent splitting. Nails in the casing should be located in pairs and spaced about 16 inches apart along the full height of the opening at the head jamb.

Casing with any form of molded shape must have a mitered joint at the corners (fig. 6-26, view B). When casing is square-edged, a butt joint may be made at the junction of the side and head casing (fig. 6-26, view C). If the moisture content of the casing material is high, a mitered joint may open slightly at the outer edge as the material dries. This can be minimized by using a small glued spline at the corner of the mitered joint. Actually, use of a spline joint under any moisture condition is considered good practice, and some prefitted jamb, door, and casing units are provided with splined joints. Nailing into the joint after drilling helps retain a close fit.

The door opening is now complete except for fitting and securing the hardware and nailing the stops in proper position. Interior doors are normally hung with two 3 1/2- by 3 1/2-inch loose-pin butt hinges. The door is fitted into the opening with the clearances shown in

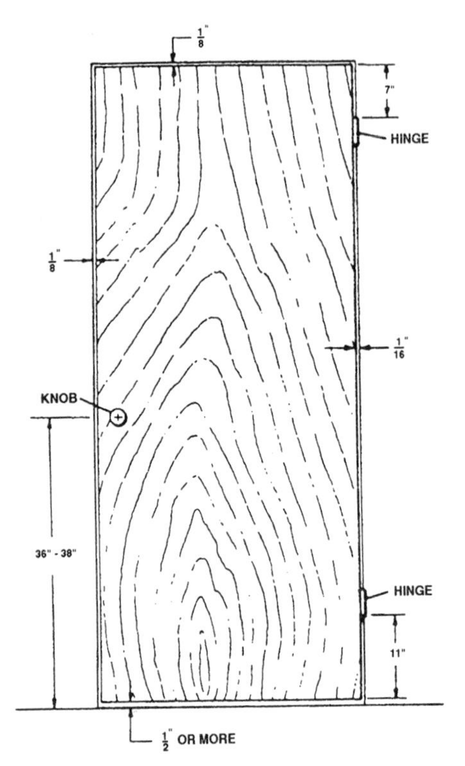

Figure 6-27.—Door clearances.

figure 6-27. The clearance and location of hinges, lockset, and doorknob may vary somewhat, but they are generally accepted by craftsmen and conform to most millwork standards. The edge of the lock stile should be beveled slightly to permit the door to clear the jamb when swung open. If the door is to swing across heavy carpeting, the bottom clearance may need to be slightly more.

When fitting doors, you should temporarily nail the stop in place; this stop will be nailed in permanently when the door has been hung. Stops for doors in single-piece jambs are generally 7/16 inch thick and may be 3/4 inch to 2 1/4 inches wide. They are installed with a mitered joint at the junction of the side and head jambs. A 45° bevel cut at the bottom of the stop, about 1 to 1 1/2 inches above the finish floor, eliminates a dirt pocket and makes cleaning or refinishing of the floor easier (fig. 6-26, view A).

Some manufacturers supply prefitted doorjambs and doors with the hinge slots routed and ready for installation. A similar door buck (jamb) of sheet metal with formed stops and casing is also available.

DOOR HARDWARE INSTALLATION

Hardware for doors is available in a number of finishes, with brass, bronze, and nickel being the most common. Door sets are usually classified as entry lock for interior doors; bathroom set (inside lock control with safety slot for opening from the outside); bedroom lock (keyed lock); and passage set (without lock).

As mentioned earlier, doors should be hinged so that they open in the direction of natural entry. They should also swing against a blank wall whenever possible and never into a hallway. The door swing directions and sizes are usually shown on the working drawings. The "hand of the door" (fig. 6-28) is the expression used to describe the direction in which a door is to swing (normal or reverse) and the side from which it is to hang (left or right).

When ordering hardware for a door, be sure to specify whether it is a left-hand door, a right-hand door, a left-hand reverse door, or a right-hand reverse door.

Hinges

You should use three hinges for hanging 1 3/4-inch exterior doors and two hinges for the lighter interior doors. The difference in exposure on the opposite sides of exterior doors causes a tendency to warp during the winter. Three hinges reduce this tendency. Three hinges are also useful on doors that lead to unheated attics and for wider and heavier doors that may be used within the structure. If a third hinge is required, center it between the top and bottom hinges.

Loose-pin butt hinges should be used and must be of the proper size for the door they support. For 1 3/4-inch-thick doors, use 4- by 4-inch butts; for 1 3/8-inch doors, you should use 3 1/2- by 3 1/2-inch butts. After the door is fitted to the framed opening with the proper clearances, hinge halves are fitted to the door. They are routed into the door edge with about a 3/16-inch back distance (fig. 6-29, view A). One hinge half should be set flush with the surface and must be fastened square with the edge of the door. Screws are included with each pair of hinges.

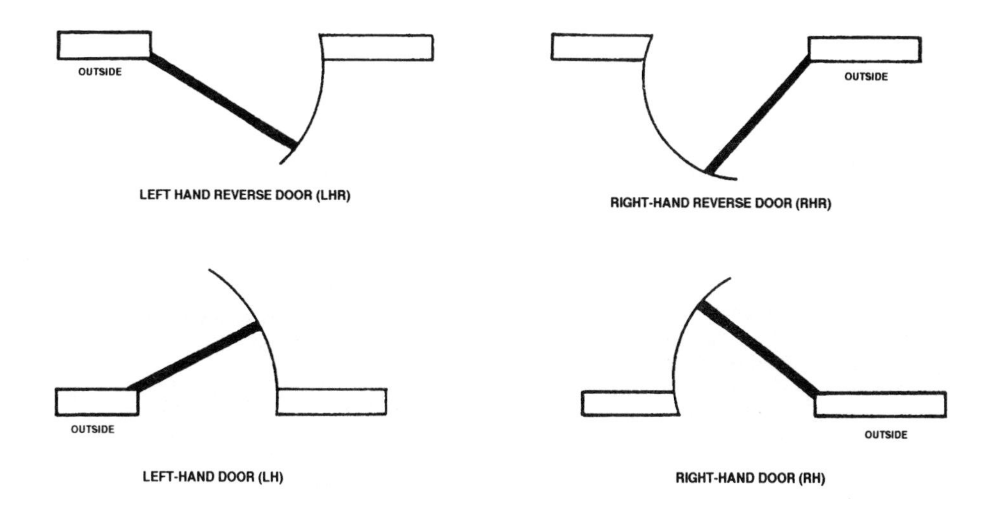

Figure 6-28.—"Hands" of doors.

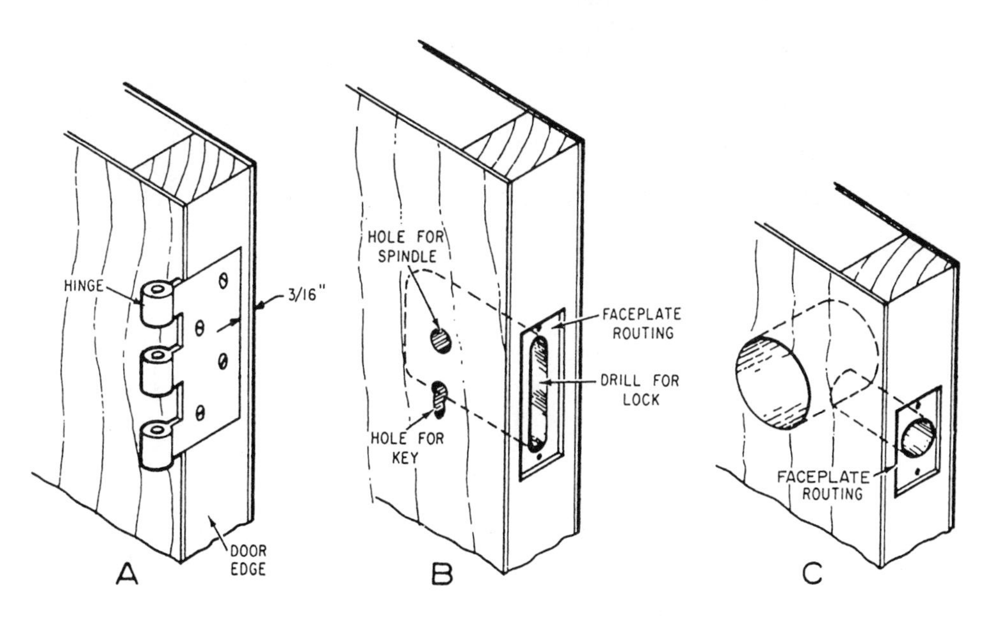

Figure 6-29.—Installation of door hardware.

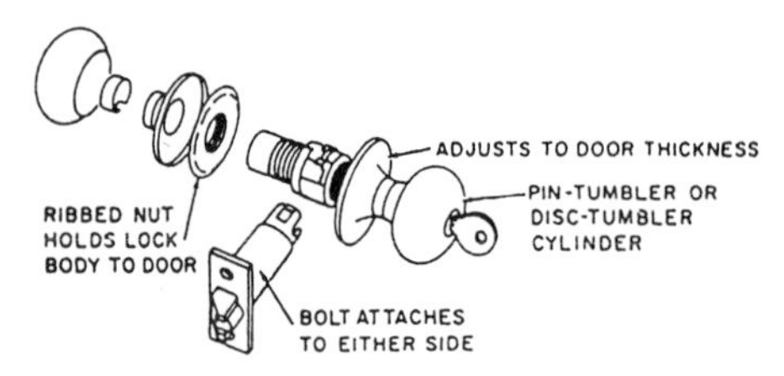

Figure 6-30.—Parts of a cylinder lock.

The door should now be placed in the opening and blocked up at the bottom for proper clearance. The jamb should be marked at the hinge locations, and the remaining hinge half routed and fastened in place. The door should then be positioned in the opening and the pins slipped in place. If you have installed the hinges correctly and the jambs are plumb, the door should swing freely.

Locks

The types of door locks differ with regard to installation, cost, and the amount of labor required to set them. Some types, such as mortise locks, combination dead bolts, and latch locksets, require drilling of the edge and face of the door and then routing of the edge to accommodate the lockset and faceplate (fig. 6-29, view B). A bored lockset (view C) is easy to install since it requires only one hole drilled in the edge and one in the face of the door. Boring jigs and faceplate markers are available to ensure accurate installation.

The lock should be installed so that the doorknob is 36 to 38 inches above the floor line. Most sets come with paper templates, marking the location of the lock and size of the holes to be drilled. Be sure to read the manufacturer's installation instructions carefully. Recheck your layout measurements before you drill any holes.

The parts of an ordinary cylinder lock for a door are shown in figure 6-30. The procedure for installing a lock of this type is as follows:

1. Open the door to a convenient working position and check it in place with wedges under the bottom near the outer edge.
2. Measure up 36 inches from the floor (the usual knob height), and square a line across the face and edge of the lock stile.
3. Place the template, which is usually supplied with a cylinder lock, on the face of the door at the proper height and alignment with layout lines and mark the centers of the holes to be drilled. (A typical template is shown in fig. 6-31.)

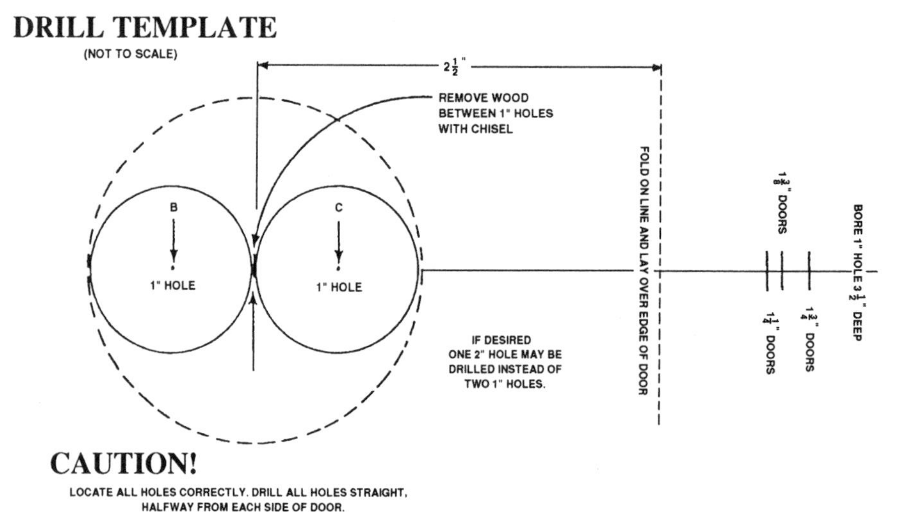

Figure 6-31.—Drill template for locksets.

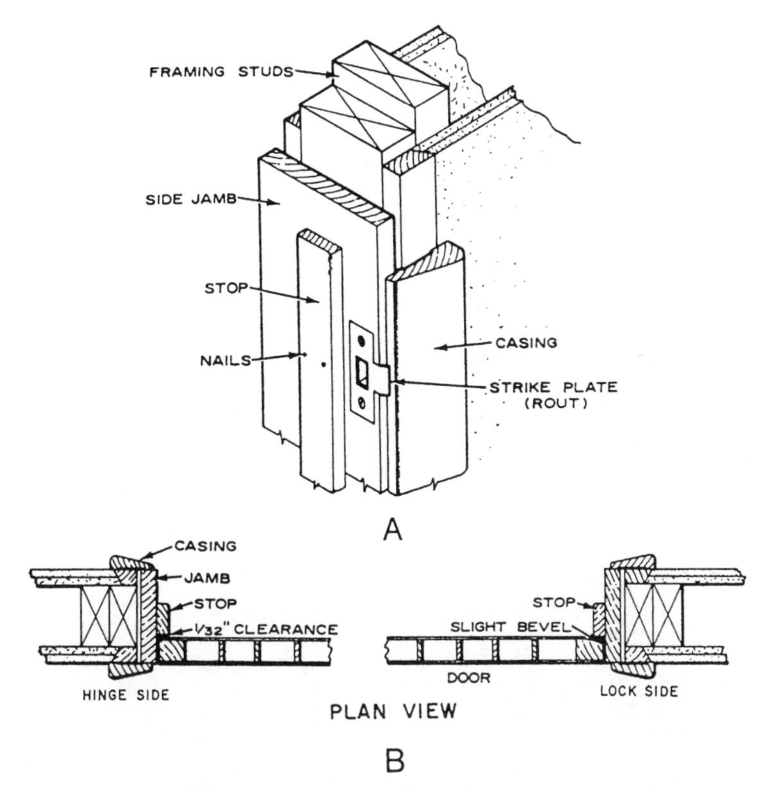

Figure 6-32.—Door details.

4. Drill the holes through the face of the door and then the hole through the edge to receive the latch bolt. It should be slightly deeper than the length of the bolt.
5. Cut a gain for the latch-bolt mounting plate, and install the latch unit.
6. Install the interior and exterior knobs.
7. Find the position of the strike plate and install it in the jamb.

Strike Plates

The strike plate, which is routed into the doorjamb, holds the door in place by contact with the latch. To install, mark the location of the latch on the doorjamb and locate the position of the strike plate by outlining it. Rout out the marked outline with a chisel and also rout for the latch (fig. 6-32, view A). The strike plate should be flush with or slightly below the face of the doorjamb. When the door is latched, its face should be flush with the edge of the jamb.

Doorstops

The stops that have been temporarily set during the fitting of the door and the hardware may now be nailed in permanently. You should use finish nails or brads, 1 1/2 inches long. The stop at the lock side (fig. 6-32, view B) should be nailed first, setting it tight against the door face when the door is latched. Space the nails in pairs 16 inches apart.

The stop behind the hinge side should be nailed next, and a 1/32-inch clearance from the door face should be allowed to prevent scraping as the door is opened. The head-jamb stop should then be nailed in place. Remember that when the door and trim are painted, some of the clearance will be taken up.

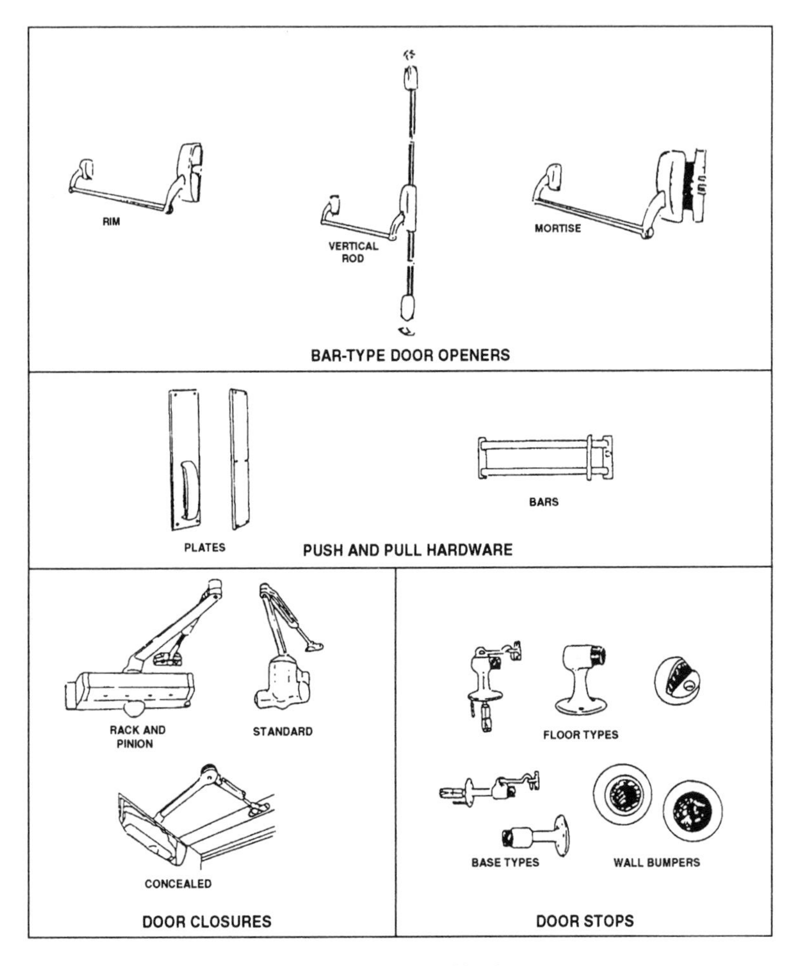

Figure 6-33.—Commercial hardware.

COMMERCIAL/INDUSTRIAL HARDWARE

The items of commercial/industrial door hardware shown in figure 6-33 are usually installed in commercial or industrial buildings, not residential housing. These items are used where applicable, in new construction or in alterations or repairs of existing facilities. Most of these items are made for use in or on metal doors, but some items are made for wood doors. Follow the manufacturer's installation instructions. Recommended door hardware locations for standard steel doors are shown in figure 6-34. Standard 7-foot doors are usually used in commercial construction.

INTERIOR WOOD TRIM

LEARNING OBJECTIVE: Upon completing this section, you should be able to identify the types of interior wood trim and the associated installation procedures.

The casing around the window frames on the interior of a structure should be the same pattern as that used around the interior doorframes. Other trim used for a double-hung window frame includes the sash, stops, stool, and apron (fig. 6-35, view A). Another method of using trim around windows has the entire opening

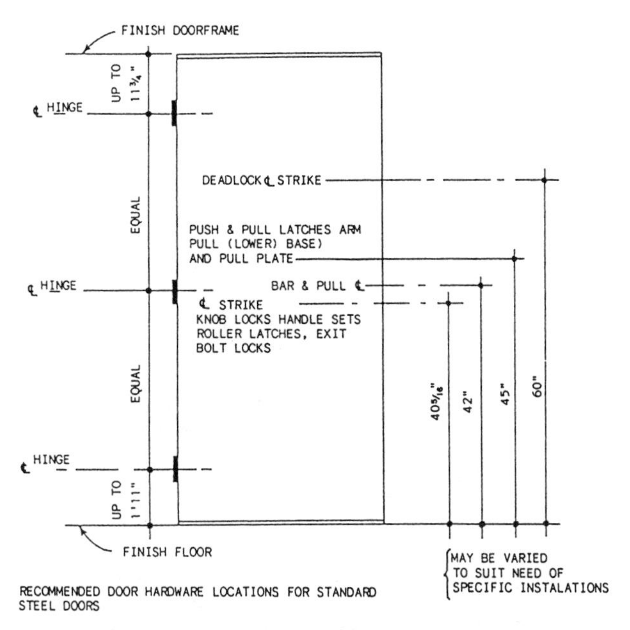

Figure 6-34.—Location of hardware for steel doors.

enclosed with casing (fig. 6-35, view B). The stool serves as a filler trim member between the bottom sash rail and the bottom casing.

The stool is the horizontal trim member that laps the windowsill and extends beyond the casing at the sides, with each end notched against the plastered wall. The apron serves as a finish member below the stool. The window stool is the first piece of window trim to be installed and is notched and fitted against the edge of the jamb and plaster line, with the outside edge being flush against the bottom rail of the window sash. The stool is blind-nailed at the ends so that the casing and the stop cover the nailheads. Predrilling is usually necessary to prevent splitting. The stool should also be nailed at the midpoint of the sill and to the apron with finishing nails. Face-nailing to the sill is sometimes substituted or supplemented with toenailing of the outer edge to the sill.

The window casing should be installed and nailed as described for doorframes (fig. 6-26, view A) except for the inner edge. This edge should be flush with the inner face of the jambs so that the stop covers the joint between the jamb and casing. The window stops are then nailed to the jambs so that the window sash slides smoothly. Channel-type weather stripping often

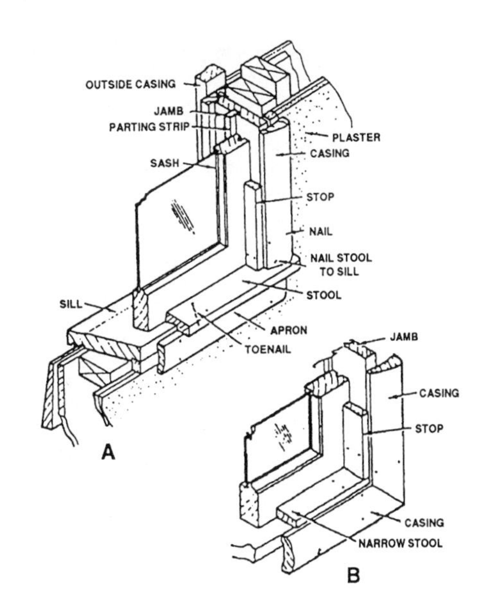

Figure 6-35.—Installation of window trim.

includes full-width metal subjambs into which the upper and lower sash slide, replacing the parting strip. Stops are located against these instead of the sash to provide a small amount of pressure. The apron is cut to a length equal to the outer width of the casing line (fig. 6-35, view A). It should be nailed to the windowsill and to the 2- by 4-inch framing sill below.

When casing is used to finish the bottom of the window frame, as well as the sides and top, the narrow stool butts against the side window jamb. Casing should then be mitered at the bottom corners (fig. 6-35, view B) and nailed as previously described.

BASE MOLDING

Base molding serves as a finish between the finished wall and floor. It is available in several widths and forms. Two-piece base consists of a baseboard topped with a small base cap (fig. 6-36, view A). When plaster is not straight and true, the small base molding will conform more closely to the variations than will the wider base alone. A common size for this type of baseboard is 5/8 inch by 3 1/4 inches or wider. One-piece base varies in size from 7/16 inch by 2 1/4 inches to 1/2 inch by 3 1/4 inches and wider (fig. 6-36, views B and C). Although a baseboard is desirable at the junction of the wall and carpeting to serve as a protective bumper, wood trim is sometimes eliminated entirely.

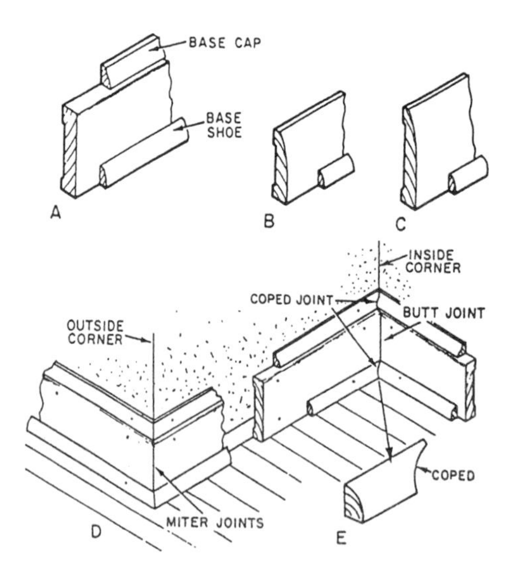

Figure 6-36.—Base moldings.

Most baseboards are finished with a 1/2- by 3/4-inch base shoe (fig. 6-36, view A). A single base molding without the shoe is sometimes placed at the wall-floor junction, especially where carpeting might be used.

Square-edged baseboard should be installed with a butt joint at the inside corners and a mitered joint at the outside corners (fig. 6-36, view D). It should be nailed to each stud with two 8d finishing nails. Molded single-piece base, base moldings, and base shoe should have a coped joint at the inside corners and a mitered joint at the outside corners. In a coped joint, the first piece is square-cut against the plaster or base and the second piece of molding coped. This is done by sawing a 45° miter cut and using a coping saw to trim the molding along the inner line of the miter (fig. 6-36, view E). The base shoe should be nailed into the baseboard itself. Then, if there is a small amount of shrinkage of the joists, no opening will occur under the shoe.

To butt-join a piece of baseboard to another piece already in place at an inside corner, set the piece to be joined in position on the floor, bring the end against or near the face of the other piece, and take off the line of the face with a scriber (fig. 6-37). Use the same procedure when butting ends of the baseboard against the side casings of the doors.

For miter-joining at an outside corner, proceed as shown in figure 6-38. First, set a marker piece of baseboard across the wall corner, as shown view A, and mark the floor along the edge of the piece. Then set the piece to be mitered in place. Mark the point where the wall corner intersects the top edge and the point where

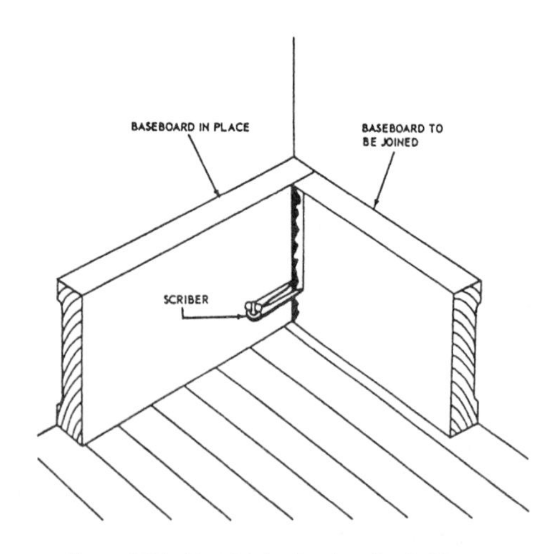

Figure 6-37.—Butt-joining baseboard at inside corners.

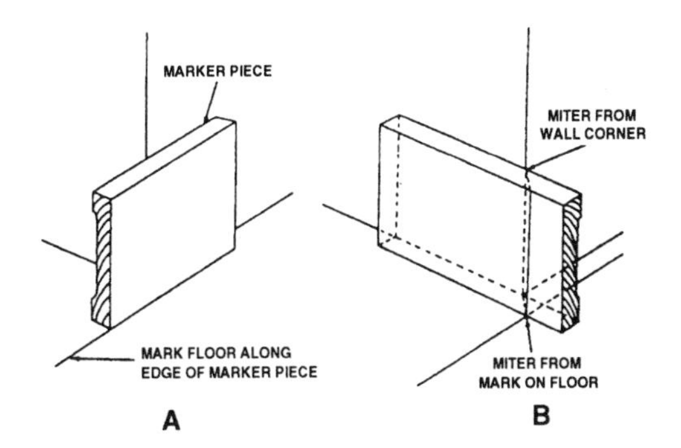

Figure 6-38.—Miter-joining at inside corners.

the mark on the floor intersects the bottom edge. Lay 45° lines across the edge from these points to make a 90° corner. Connect these lines with a line across the face (view B), and miter to the lines as indicated.

The most economical, and sometimes the quickest, method of installing baseboard is to use vinyl. In addition to its flexibility, it comes with premolded inside and outside corners. When installing vinyl base, follow the manufacturer's recommended installation procedures for both the base and adhesive.

CEILING MOLDING

Ceiling moldings (fig. 6-39) are sometimes used at the junction of the wall and ceiling for an architectural effect or to terminate drywall paneling of gypsum board or wood. As with base moldings, inside corners should be cope-jointed (fig. 6-39, view A). This ensures a tight joint and retains a good fit if there are minor moisture changes.

A cutback edge at the outside of the molding (view B) partially conceals any unevenness of the plaster and makes painting easier where there are color changes. For gypsum drywall construction, a small, simple molding (view C) might be desirable. Finish nails should be driven into the upper wall plates and also into the ceiling joists for large molding when possible.

DECORATIVE TREATMENT

The decorative treatment for interior doors, trim, and other millwork may be painted or given a natural finish with stain, varnish, or other nonpigmented material. The paint or natural finish desired for the woodwork in various rooms often determines the species of wood to be used.

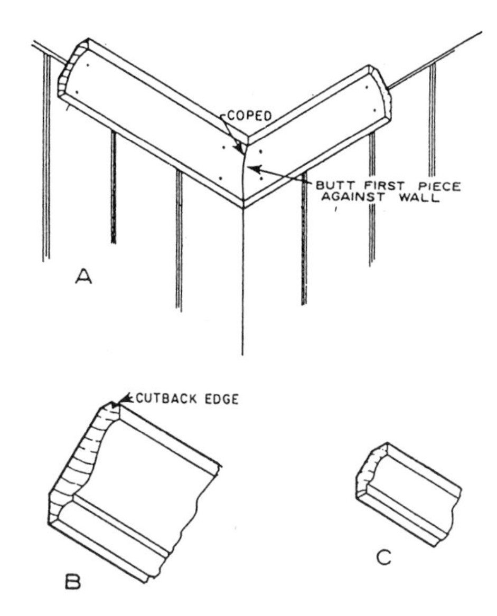

Figure 6-39.—Ceiling moldings.

Interior finish to be painted should be smooth, close-grained, and free from pitch streaks. Species meeting these requirements include ponderosa pine, northern white pine, redwood, and spruce. Birch, gum, and yellow poplar are recommended for their hardness and resistance to hard usage. Ash, birch, cherry, maple, oak, and walnut provide a beautiful natural finish decorative treatment. Some require staining to improve appearance.

RECOMMENDED READING LIST

NOTE

Although the following references were current when this TRAMAN was published, their continued currency cannot be assured. You therefore need to ensure that you are studying the latest revisions.

Carpentry, Leonard Koel, American Technical Publishers, Inc., Alsip, Ill., 1985.

Exterior and Interior Trim, John E. Ball, Delmar Publishers, Inc., Albany, N.Y., 1975.

Wood Frame House Construction, L.O.Anderson, Forest Products Laboratory, U.S. Forest Service, U.S. Department of Agriculture, Washington, D.C., 1975.

CHAPTER 7

PLASTERING, STUCCOING, AND CERAMIC TILE

Plaster and stucco are like concrete in that they are construction materials applied in a plastic condition that harden in place. They are also basically the same material. The fundamental difference between the two is location. If used internally, the material is called plaster; if used externally, it is called stucco. Ceramic tile is generally used to partially or entirely cover interior walls, such as those in bathrooms, showers, galleys, and corridors. The tile is made of clay, pressed into shape, and baked in an oven.

This chapter provides information on the procedures, methods, and techniques used in plastering, stuccoing, and tile setting. Also described are various tools, equipment, and materials the Builder uses when working with these materials.

PLASTER

LEARNING OBJECTIVE: Upon completing this section, you should be able to identify plaster ingredients, state the principles of mix design, and describe common types and uses of gypsum plaster.

A plaster mix, like a concrete mix, is made plastic by the addition of water to dry ingredients (binders and aggregates). Also, like concrete, a chemical reaction of the binder and the water, called hydration, causes the mix to harden.

The binders most commonly used in plaster are gypsum, lime, and portland cement. Because gypsum plaster should not be exposed to water or severe moisture conditions, it is usually restricted to interior use. Lime and portland cement plaster may be used both internally and externally. The most commonly used aggregates are sand, vermiculite, and perlite.

GYPSUM PLASTER

Gypsum is a naturally occurring sedimentary gray, white, or pink rock. The natural rock is crushed, then heated to a high temperature. This process (known as calcining) drives off about three-quarters of the water of crystallization, which forms about 20 percent of the weight of the rock in its natural state. The calcined material is then ground to a fine powder. Additives are used to control set, stabilization, and other physical or chemical characteristics.

For a type of gypsum plaster called Keene's cement, the crushed gypsum rock is heated until nearly all the crystallization water is removed. The resulting material, called Keene's cement, produces a very hard, fine-textured finish coat.

The removal of crystallization water from natural gypsum is a dehydration process. In the course of setting, mixing water (water of hydration) added to the mix rehydrates with the gypsum, causing recrystallization. Recrystallization results in hardening of the plaster.

Base Coats

There are four common types of gypsum base coat plasters. Gypsum neat plaster is gypsum plaster without aggregate, intended for mixing with aggregate and water on the job. Gypsum ready-mixed plaster consists of gypsum and ordinary mineral aggregate. On the job, you just add water. Gypsum wood-fibered plaster consists of calcined gypsum combined with at least 0.75 percent by weight of nonstaining wood fibers. It may be used as is or mixed with one part sand to produce base coats of superior strength and hardness. Gypsum bond plaster is designed to bond to properly prepared monolithic concrete. This type of plaster is basically calcined gypsum mixed with from 2- to 5-percent lime by weight.

Finish Coats

There are five common types of gypsum-finish coat plasters.

Ready-mix gypsum-finish plasters are designed for use over gypsum-plaster base coats. They consist of finely ground calcined gypsum, some with aggregate and others without. On the job, just add water.

Gypsum acoustical plasters are designed to reduce sound reverberation. Gypsum gauging plasters contain lime putty. The putty provides desirable setting properties, increases dimensional stability during drying, and provides initial surface hardness.

Gauging plasters are obtainable in slow-set, quick-set, and special high-strength mixtures.

Gypsum molding plaster is used primarily in casting and ornamental plasterwork. It is available neat (that is, without admixtures) or with lime. As with portland cement mortar, the addition of lime to a plaster mix makes the mix more "buttery."

Keene's cement is a fine, high-density plaster capable of a highly polished surface. It is customarily used with fine sand, which provides crack resistance.

LIME PLASTER

Lime is obtained principally from the calcining of limestone, a very common mineral. Chemical changes occur that transform the limestone into quicklime, a very caustic material. When it comes in contact with water, a violent reaction, hot enough to boil the water, occurs.

Today, the lime manufacturers slake the lime as part of the process of producing lime for mortar. Slaking is done in large tanks where water is added to convert the quicklime to hydrated lime without saturating it with water. The hydrated lime is a dry powder with just enough water added to supply the chemical reaction. Hydration is usually a continuous process and is done in equipment similar to that used in calcining. After the hydrating process, the lime is pulverized and bagged. When received by the plasterer, hydrated lime still requires soaking with water.

In mixing medium-slaking and slow-slaking limes, you should add the water to the lime. Slow-slaking lime must be mixed under ideal conditions. It is necessary to heat the water in cold weather. Magnesium lime is easily drowned, so be careful you don't add too much water to quick-slaking calcium lime. When too little water is added to calcium and magnesium limes, they can be burned. Whenever lime is burned or drowned, a part of it is spoiled. It will not harden and the paste will not be as viscous and plastic as it should be. To produce plastic lime putty, soak the quicklime for an extended period, as much as 21 days.

Because of the delays involved in the slaking process of quicklime, most building lime is the hydrated type. Normal hydrated lime is converted into lime putty by soaking it for at least 16 hours. Special hydrated lime develops immediate plasticity when mixed with water and may be used right after mixing. Like calcined gypsum, lime plaster tends to return to its original rock-like state after application.

For interior base coat work, lime plaster has been largely replaced by gypsum plaster. Lime plaster is now used mainly for interior finish coats. Because lime putty is the most plastic and workable of the cementitious materials used in plaster, it is often added to other less workable plaster materials to improve plasticity. For lime plaster, lime (in the form of either dry hydrate or lime putty) is mixed with sand, water, and a gauging material. The gauging material is intended to produce early strength and to counteract shrinkage tendencies. It can be either gypsum gauging plaster or Keene's cement for interior work or portland cement for exterior work. When using gauging plaster or Keene's cement, mix only the amount you can apply within the initial set time of the material.

PORTLAND CEMENT PLASTER

Portland cement plaster is similar to the portland cement mortar used in masonry. Although it may contain only cement, sand, and water, lime or some other plasterizing material is usually added for "butteriness."

Portland cement plaster can be applied directly to exterior and interior masonry walls and over metal lath. Never apply portland cement plaster over gypsum plasterboard or over gypsum tile. Portland cement plaster is recommended for use in plastering walls and ceilings of large walk-in refrigerators and cold-storage spaces, basements, toilets, showers, and similar areas where an extra hard or highly water-resistant surface is required.

AGGREGATES

As we mentioned earlier, there are three main aggregates used in plaster: sand, vermiculite, and perlite. Less frequently used aggregates are wood fiber and pumice.

Sand

Sand for plaster, like sand for concrete, must contain no more than specified amounts of organic impurities and harmful chemicals. Tests for these impurities and chemicals are conducted by Engineering Aids.

Proper aggregate gradation influences plaster strength and workability. It also has an effect on the tendency of the material to shrink or expand while setting. Plaster strength is reduced if excessive fine aggregate material is present in a mix. The greater quantity of mixing water required raises the water-cement ratio, thereby reducing the dry-set density. The cementitious material becomes over-extended since it must coat a relatively larger overall aggregate surface. An excess of coarse aggregate adversely affects workability—the mix becomes harsh working and difficult to apply.

Plaster shrinkage during drying can be caused by an excess of either fine or coarse aggregate. You can minimize this problem by properly proportioning the raw material, and using good, sharp, properly size-graded sand.

Generally, any sand retained on a No. 4 sieve is too coarse to use in plaster. Only a small percentage of the material (about 5 percent) should pass the No. 200 sieve.

Vermiculite

Vermiculite is a micaceous mineral (that is, each particle is laminated or made up of adjoining layers). When vermiculite particles are exposed to intense heat, steam forms between the layers, forcing them apart. Each particle increases from 6 to 20 times in volume. The expanded material is soft and pliable with a color varying between silver and gold.

For ordinary plasterwork, vermiculite is used only with gypsum plaster; therefore, its use is generally restricted to interior applications. For acoustical plaster, vermiculite is combined with a special acoustical binder.

The approximate dry weight of a cubic foot of 1:2 gypsum-vermiculite plaster is 50 to 55 pounds. The dry weight of a cubic foot of comparable sand plaster is 104 to 120 pounds.

Perlite

Raw perlite is a volcanic glass that, when flash-roasted, expands to form irregularly shaped frothy particles containing innumerable minute air cells. The mass is 4 to 20 times the volume of the raw particles. The color of expanded perlite ranges from pearly white to grayish white.

Perlite is used with calcined gypsum or portland cement for interior plastering. It is also used with special binders for acoustical plaster. The approximate dry weight of a cubic foot of 1:2 gypsum-perlite plaster is 50 to 55 pounds, or about half the weight or a cubic foot of sand plaster.

Wood Fiber and Pumice

Although sand, vermiculite, and perlite make up the great majority of plaster aggregate, other materials, such as wood fiber and pumice, are also used. Wood fiber may be added to neat gypsum plaster, at the time of manufacture, to improve its working qualities. Pumice is a naturally formed volcanic glass similar to perlite, but heavier (28 to 32 pounds per cubic foot versus 7.5 to 15 pounds for perlite). The weight differential gives perlite an economic advantage and limits the use of pumice to localities near where it is produced.

WATER

In plaster, mixing water performs two functions. First, it transforms the dry ingredients into a plastic, workable mass. Second, it combines with the binder to induce hardening. As with concrete, there is a maximum quantity of water per unit of binder required for complete hydration; an excess over this amount reduces the plaster strength.

In all plaster mixing, though, more water is added than is necessary for complete hydration of the binder. The excess is necessary to bring the mix to workable consistency. The amount to be added for workability depends on several factors: the characteristics and age of the binder, application method, drying conditions, and the tendency of the base to absorb water. A porous masonry base, for example, draws a good deal of water out of a plaster mix. If this reduces the water content of the mix below the maximum required for hydration, incomplete curing will result.

As a general rule, only the amount of water required to attain workability is added to a mix. The water should be potable and contain no dissolved chemicals that might accelerate or retard the set. **Never use water previously used to wash plastering tools for mixing plaster.** It may contain particles of set plaster that may accelerate setting. Also avoid stagnant water; it may contain organic material that can retard setting and possibly cause staining.

PLASTER BASE INSTALLATION

LEARNING OBJECTIVE: Upon completing this section, you should be able to associate the names and purposes of each type of lath used as a plaster base. You should also be able to describe the procedures used in plastering, including estimating materials and the procedures for mixing and applying plaster bases.

For plastering, there must be a continuous surface to which the plaster can be applied and to which it will cling—the plaster base. A continuous concrete or masonry surface may serve as a base without further treatment.

BASES

For plaster bases, such as those defined by the inner edges of the studs or the lower edges of the joists, a base material, called lath, must be installed to form a continuous surface spanning the spaces between the structural members.

Wood Lath

Wood lath is made of white pine, spruce, fir, redwood, and other soft, straight-grained woods. The standard size of wood lath is 5/16 inch by 1 1/2 inches by 4 feet. Each lath is nailed to the studs or joists with 3-penny (3d) blued lathing nails.

Laths are nailed six in a row, one above the other. The next six rows of lath are set over two stud places. The joints of the lath are staggered in this way so cracks will not occur at the joinings. Lath ends should be spaced 1/4 inch apart to allow movement and prevent buckling. Figure 7-1 shows the proper layout of wood lath. To obtain a good key (space for mortar), space the laths not less than 3/8 inch apart. Figure 7-2 shows good spacing with strong keys.

Wood laths come 50 to 100 to the bundle and are sold by the thousand. The wood should be straight-grained, and free of knots and excessive pitch. Don't use old lath; dry or dirty lath offers a poor bonding surface. Lath must be damp when the mortar is applied. Dry lath pulls the moisture out of the mortar, preventing proper setting. The best method to prevent dry lath is to wet it thoroughly the day before plastering. This lets the wood swell and reach a stable condition ideal for plaster application.

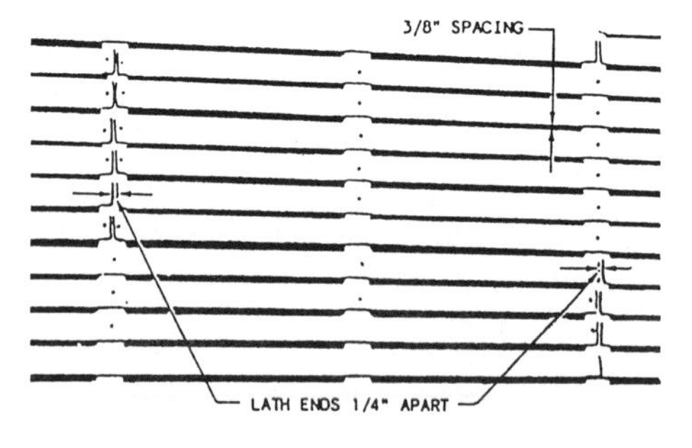

Figure 7-1.—Wood lath with joints staggered every sixth course.

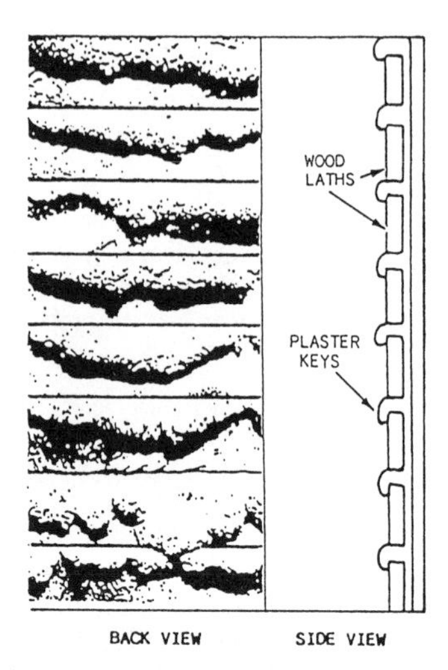

Figure 7-2.—Wood lath, showing proper keys.

Board Lath

Of the many kinds of lathing materials available, board lath is the most widely used today. Board lath is manufactured from mineral and vegetable products. It is produced in board form, and in sizes generally standardized for each application to studs, joists, and various types of wood and metal furring.

Board lath has a number of advantages. It is rigid, strong, stable, and reduces the possibility of dirt filtering through the mortar to stain the surface. It is insulating and strengthens the framework structure. Gypsum board lath is fire resistant. Board lath also requires the least amount of mortar to cover the surface.

Board laths are divided into two main groups: gypsum board and insulation board. Gypsum lath is made in a number of sizes, thicknesses, and types. Each type is used for a specific purpose or condition. **Note:** Only gypsum mortar can be used over gypsum lath. Never apply lime mortar, portland cement, or any other binding agent to gypsum lath.

The most commonly used size gypsum board lath is the 3/8 inch by 16 inches by 48 inches, either solid or perforated. This lath will not burn or transmit temperatures much in excess of 212°F until the gypsum is completely calcined. The strength of the bond of plaster to gypsum lath is great. It requires a pull of 864 pounds per square foot to separate gypsum plaster from gypsum lath (based on a 2:1 mix of sand and plaster mortar).

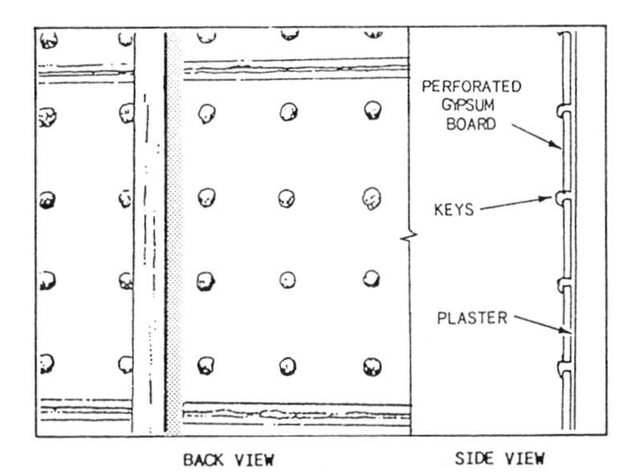

Figure 7-3.—Keys formed with perforated gypsum board.

There is also a special fire-retardant gypsum lath, called type X. It has a specially formulated core, containing minerals giving it additional fire protection.

Use only one manufacturer's materials for a specified job or area. This ensures compatibility. Always strictly follow the manufacturer's specifications for materials and conditions of application.

Plain gypsum lath plaster base is used in several situations: for applying nails and staples to wood and nailable steel framing; for attaching clips to wood framing, steel studs, and suspended metal grillage; and for attaching screws to metal studs and furring channels. Common sizes include 16 by 48 inches, 3/8 or 1/2 inch thick, and 16 by 96 inches, 3/8 inch thick.

Perforated gypsum lath plaster base is the same as plain gypsum lath except that 3/4-inch round holes are punched through the lath 4 inches on center (OC) in each direction. This gives one 3/4-inch hole for each 16 square inches of lath area. This provides mechanical keys in addition to the natural plaster bond and obtains higher fire ratings. Figure 7-3 shows back and side views of a completed application.

Insulating gypsum lath plaster base is the same as plain gypsum lath, but with bright aluminum foil laminated to the back. This creates an effective vapor barrier at no additional labor cost. In addition, it provides positive insulation when installed with the foil facing a 3/4-inch minimum air space. When insulating gypsum lath plaster is used as a ceiling, and under winter heating conditions, its heat-resistance value is approximately the same as that for 1/2-inch insulation board.

Long lengths of gypsum lath are primarily used for furring the interior side of exterior masonry walls. It is available in sizes 24 inches wide, 3/8 inch thick, and up to 12 feet in length.

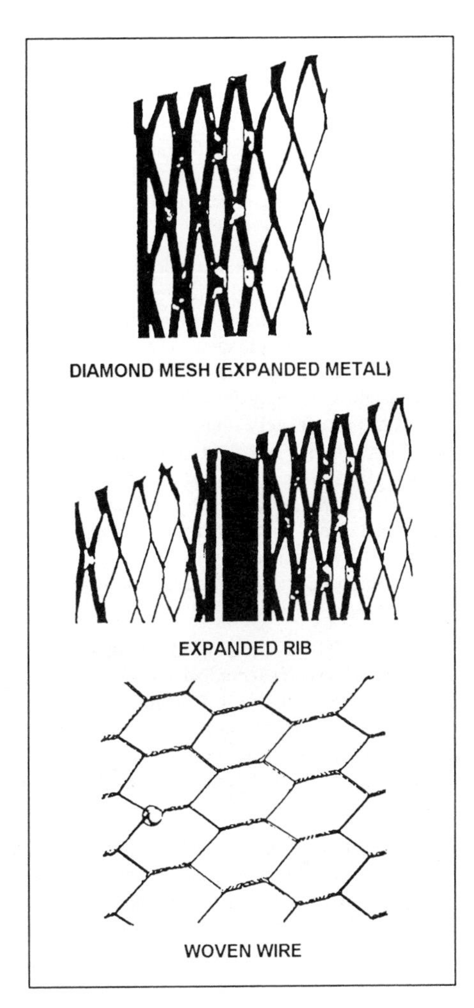

Figure 7-4.—Types of metal lath.

Gypsum lath is easily cut by scoring one or both sides with a utility knife. Break the lath along the scored line. Be sure to make neatly fitted cutouts for utility openings, such as plumbing pipes and electrical outlets.

Metal Lath

Metal lath is perhaps the most versatile of all plaster bases. Essentially a metal screen, the bond is created by keys formed by plaster forced through the openings. As the plaster hardens, it becomes rigidly interlocked with the metal lath.

Three types of metal lath are commonly used: diamond mesh (expanded metal), expanded rib, and wire mesh (woven wire). These are shown in figure 7-4.

DIAMOND MESH.—The terms "diamond mesh" and "expanded metal" refer to the same type of lath (fig. 7-4). It is manufactured by first cutting staggered slits in a sheet and then expanding or stretching the sheet to form the screen openings. The standard diamond mesh lath has a mesh size of 5/16 by 9/16 inch. Lath is made in sheets of 27 by 96 inches and is packed 10 sheets to a bundle (20 square yards).

Diamond mesh lath is also made in a large mesh. This is used for stucco work, concrete reinforcement, and support for rock wool and similar insulating materials. Sheet sizes are the same as for the small mesh. The small diamond mesh lath is also made into a self-furring lath by forming dimples into the surface that hold the lath approximately 1/4 inch away from the wall surface. This lath may be nailed to smooth concrete or masonry surfaces. It is widely used when replastering old walls and ceilings when the removal of the old plaster is not desired. Another lath form is paper-backed where the lath has a waterproof or kraft paper glued to the back of the sheet. The paper acts as a moisture barrier and plaster saver.

EXPANDED RIB.—Expanded rib lath (fig. 7-4) is like diamond mesh lath except that various size ribs are formed in the lath to stiffen it. Ribs run lengthwise of the lath and are made for plastering use in 1/8-, 3/8-, and 3/4-inch rib height. The sheet sizes are 27 to 96 inches in width, and 5-, 10-, and 12-foot lengths for the 3/4-inch rib lath.

WIRE MESH.—Woven wire lath (fig. 7-4) is made of galvanized wire of various gauges woven or twisted together to form either squares or hexagons. It is commonly used as a stucco mesh where it is placed over tar paper on open-stud construction or over various sheathing.

INSTALLATION

Let's now look at the basic installation procedures for plaster bases and accessories.

Gypsum Lath

Gypsum lath is applied horizontally with staggered end joints, as shown in figure 7-5. Vertical end joints should be made over the center of studs or joists. Lath joints over openings should not occur at the jamb line. Do not force the boards tightly together; let them butt loosely so the board is not under compression before the plaster is applied. Use small pieces only where necessary. The most common method of attaching the boards has been the lath nail. More recently, though, staples have gained wider use (due mainly to the ready availability of power guns).

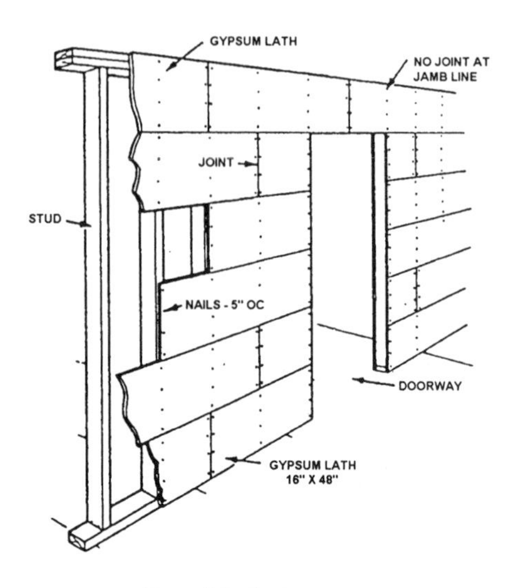

Figure 7-5.—Lath joints.

The nails used are 1 1/8 inches by 13 gauge, flat headed, blued gypsum lath nails for 3/8-inch-thick boards and 1 1/4 inches for 1/2-inch boards. There are also resin-coated nails, barbed-shaft nails, and screw-type nails in use. Staples should be No. 16 U.S. gauge flattened galvanized wire formed with a 7/16-inch-wide crown and 7/8-inch legs with divergent points for 3/8-inch lath. For 1/2-inch lath, use 1-inch-long staples.

Four nails or staples are used on each support for 16-inch-wide lath and five for 2-foot-wide lath. Some special fire ratings, however, require five nails or staples per 16-inch board. Five nails or staples are also recommended when the framing members are spaced 24 inches apart.

Start nailing or stapling 1/2 inch from the edges of the board. Nail on the framing members falling on the center of the board first, then work to either end. This should prevent buckling.

Insulating lath should be installed much the same as gypsum lath except that slightly longer blued nails are used. A special waterproof facing is provided on one type of gypsum board for use as a ceramic tile base when the tile is applied with an adhesive.

Metal Lath

All metal lath is installed with the sides and ends lapped over each other. The laps between supports should be securely tied, using 18-gauge tie wire. In general, metal lath is applied with the long length at right angles to the supports. Rib lath is placed with the ribs against the supports and the ribs nested where the lath overlaps. Generally, metal lath and wire lath are lapped at least 1 inch at the ends and 1/2 inch at the sides. Some wire lath manufacturers specify up to 4 1/2-inch end lapping and 2-inch side laps. This is done to mesh the wires and the paper backing.

Lath is either nailed, stapled, or hog-tied (heavy wire ring installed with a special gun) to the supports at 6-inch intervals. Use 1 1/2-inch barbed roofing nails with 7/16-inch heads or 1 inch 4-gauge staples for the flat lath on wood supports. For ribbed lath, heavy wire lath, and sheet lath, nails or staples must penetrate the wood 1 3/8 inches for horizontal application and at least 3/4 inch for vertical application. When common nails are used, they must be bent across at least three lath strands.

On channel iron supports, the lath is tied with No. 18-gauge tie wire at 4-inch intervals using lathers' nippers. For wire lath, the hog tie gun can be used. Lath must be stretched tight as it is applied so that no sags or buckles occur. Start tying or nailing at the center of the sheet and work toward the ends. Rib lath should have ties looped around each rib at all supports, as the main supporting power for rib lath is the rib.

When you install metal laths at both inside and outside corners, bend the lath to form a corner and carry it at least 4 inches in or around the corner. This provides the proper reinforcement for the angle or corner.

Lath Accessories

A wide variety of metal accessories is produced for use with gypsum and metal lathing. Lathing accessories are usually installed before plastering to form true corners, act as screeds for the plasterer, reinforce possible weak points, provide control joints, and provide structural support.

Lathing accessories consist of structural components and miscellaneous accessories. The principal use of structural components is in the construction of hollow partitions. A hollow partition is one containing no building framing members, such as studs and plates. Structural components are lathing accessories that take the place of the missing framing members supporting

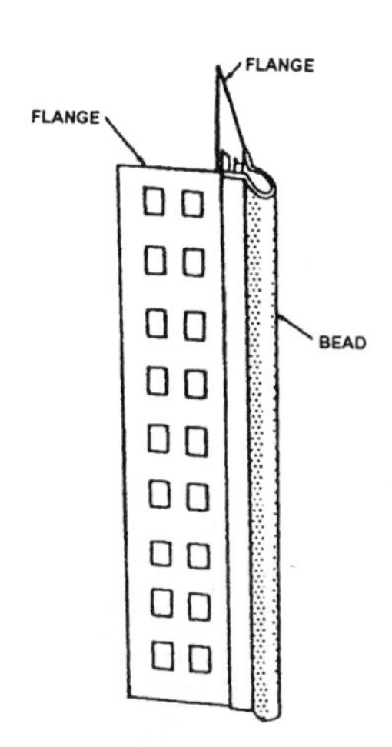

Figure 7-6.—Perforated flanged corner bead.

the lath. These include prefabricated metal studs and floor and ceiling runner tracks. The runner tracks take the place of missing stud top and bottom plates. They usually consist of metal channels. Channels are also used for furring and bracing.

Miscellaneous accessories consist of components attached to the lath at various locations. They serve to define and reinforce corners, provide dividing strips between plaster and the edges of baseboard or other trim, and define plaster edges at unframed openings.

Corner beads fit over gypsum lath outside corners to provide a true, reinforced corner. They are available in either small-nose or bullnose types, with flanges of either solid or perforated (fig. 7-6) metal. They are available with expanded metal flanges.

Casing beads are similar to corner beads and are used both as finish casings around openings in plaster walls and as screeds to obtain true surfaces around doors and windows. They are also used as stops between a plaster surface and another material, such as masonry or wood paneling. Casing beads are available as square sections, modified-square sections, and quarter-rounds.

Base or parting screeds are used to separate plaster from other flush surfaces, such as concrete. Ventilating expansion screed is used on the underside of closed soffits and in protected vertical surfaces for ventilation of enclosed attic spaces. Drip screeds act as terminators of exterior portland cement plaster at concrete foundation walls. They are also used on external horizontal corners of plaster soffits to prevent drip stains

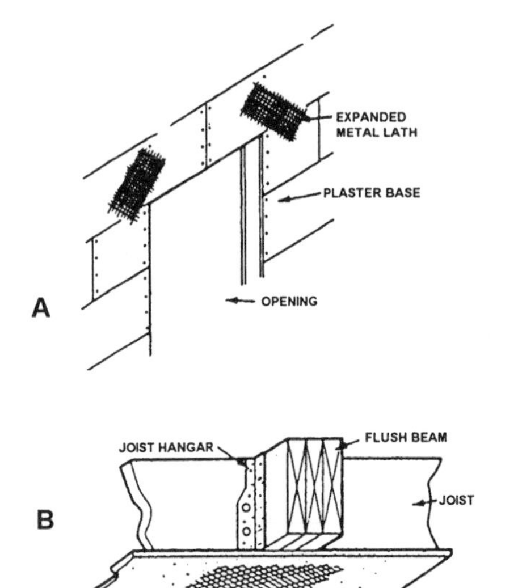

Figure 7-7.—Metal lath used to minimize cracking.

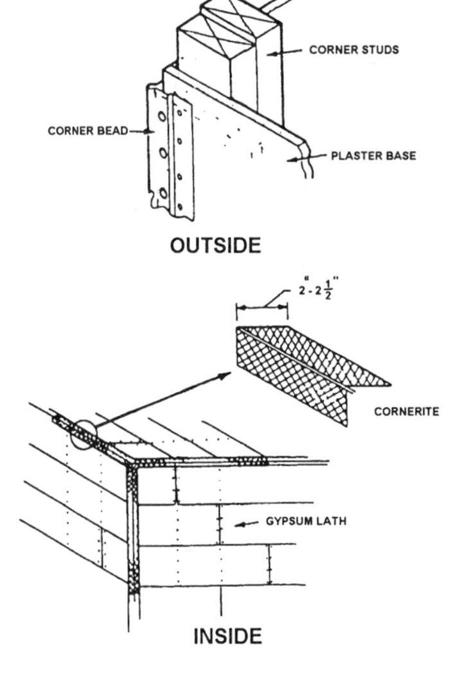

Figure 7-8.—Plaster reinforcing at corners.

on the underside of the soffit. A metal base acts as a flush base at the bottom of a plaster wall. It also serves as a plaster screed.

Joint Reinforcing

Because some drying usually takes place in the wood framing members after a structure is completed, some shrinkage is expected. This, in turn, may cause plaster cracks to develop around openings and in the corners. To minimize, if not eliminate, these cracks, use expanded metal lath in key positions over the plaster-base material as reinforcements. Strip reinforcement (strips of expanded metal lath) can be used over door and window openings (fig. 7-7, view A). A 10- to 20-inch strip is placed diagonally across each upper corner of the opening and tacked in place.

Strip reinforcement should also be used under flush ceiling beams (fig. 7-7, view B) to prevent plaster cracks. On wood drop beams extending below the ceiling line, the metal lath is applied with furring nails to provide space for keying the plaster.

Corner beads of expanded metal lath or of perforated metal (fig. 7-8) should be installed on all outside corners. They should be applied plumb and level. Each bead acts as a leveling edge when walls are plastered and reinforces the corner against mechanical damage. To minimize plaster cracks, reinforce the inside corners at the juncture of walls and ceilings. Metal lath, or wire fabric, is tacked lightly in place in these corners.

Control joints (an example of which is shown in fig. 7-9) are formed metal strips used to relieve stresses and strains in large plaster areas or at junctures of dissimilar materials on walls and ceilings. Cracks can develop in plaster or stucco from a single cause or a combination of causes, such as foundation settlement, material shrinkage, building movement, and so forth. The control joint minimizes plaster cracking and assures proper plaster thickness. The use of control joints is extremely important when portland cement plaster is used.

Plastering Grounds

Plastering grounds are strips of wood used as plastering guides or strike-off edges and are located around window and door openings and at the base of the walls. Grounds around interior door openings (such as fig. 7-10, view A) are full-width pieces nailed to the sides over the studs and to the underside of the header. They are 5 1/4 inches wide, which coincides with the standard jamb width for interior walls with a plaster

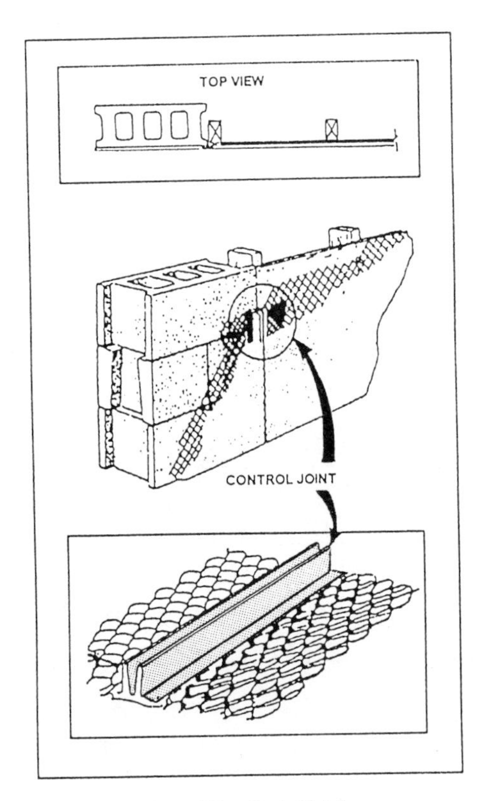

Figure 7-9.—Control joint.

finish. They are removed after the plaster has dried. Narrow strip grounds (fig. 7-10, view B) can also be used around interior openings.

In window and exterior door openings, the frames are normally in place before the plaster is applied. Thus, the inside edges of the side and head jamb can, and often do, serve as grounds. The edge of the window might also be used as a ground, or you can use a narrow 7/8-inch-thick ground strip nailed to the edge of the 2- by 4-inch sill (fig. 7-10, view C). These are normally left in place and covered by the casing.

A similar narrow ground or screed is used at the bottom of the wall to control the thickness of the gypsum plaster and to provide an even surface for the baseboard and molding. This screed is also left in place after the plaster has been applied.

Mixing

Some plaster comes ready-mixed, requiring only the addition of enough water to attain minimum required workability. For job mixing, tables are available giving recommended ingredient proportions for gypsum, lime, lime-portland cement, and portland cement plaster for base coats on lath or on various types of concrete or masonry surfaces, and for finish coats of various types. In this chapter, we'll cover recommended proportions for only the more common types of plastering situations.

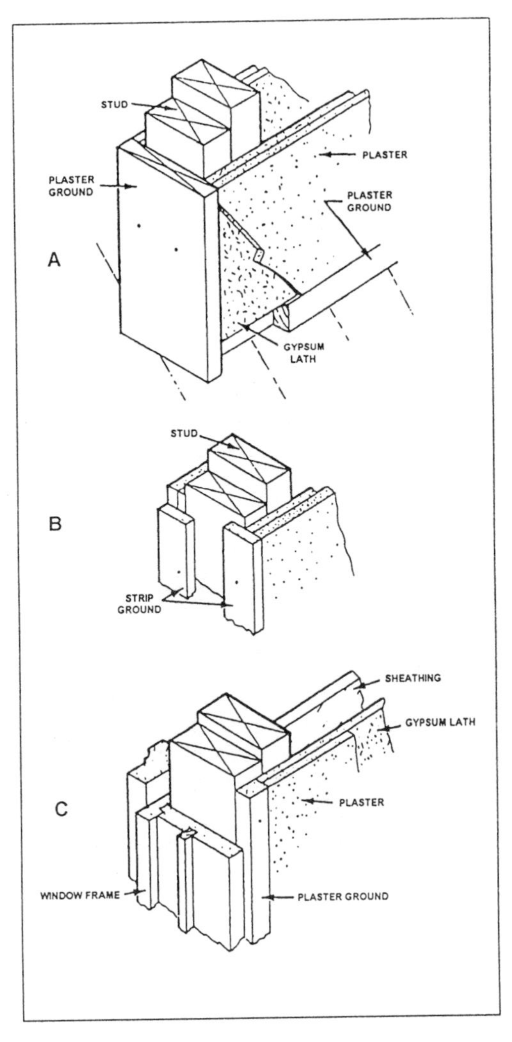

Figure 7-10.—Plaster grounds.

In the following discussion, one part of cementitious material means 100 pounds (one sack) of gypsum, 100 pounds (two sacks) of hydrated lime, 1 cubic foot

Table 7-1.—Base Coat Proportions for Different Types of Work

INGREDIENT	PROPORTION
Two-Coat Masonry Work[1]	
Gypsum plaster	1:3
Lime plaster using hydrated lime	1:7.5
Lime plaster using lime putty	1:3.5
Three-Coat Work on a Masonry Base[2]	
Gypsum plaster	Both coats 1:3
Portland cement plaster	Both coats 1:3 to 1:5
Work on a Metal Lath	
Gypsum plaster	Same as for three-coat work on gypsum lath
Lime plaster using hydrated lime	Scratch 1:6.75, brown 1:9
Lime plaster using lime putty	Scratch 1:3, brown 1:4
Portland cement plaster	Both coats 1:3 to 1:5

[1]Portland cement plaster is **not** used for two-coat work, and two-coat work is **not** usually done on metal lath

[2]Lime plaster is not usually used for three-coat work on a masonry base

of lime putty, or 94 pounds (one sack) of portland cement. One part of aggregate means 100 pounds of sand or 1 cubic foot of vermiculite or perlite. **Note**: Vermiculite and perlite are **not** used with lime plaster. While aggregate parts given for gypsum or portland cement plaster may be presumed to refer to either sand or vermiculite/perlite, the aggregate part given for lime plaster means **sand only**.

BASE COAT PROPORTIONS.—Two-coat plasterwork consists of a single base coat and a finish coat. Three-coat plasterwork consists of two base coats (the scratch coat and the brown coat) and a finish coat.

Portland cement plaster **cannot** be applied to a gypsum base. Lime plaster can, but, in practice, only gypsum plaster is applied to gypsum lath as a base coat. For two-coat work on gypsum lath, the recommended base coat proportions for gypsum plaster are 1:2.5. For two-coat work on a masonry (either monolithic concrete or masonry) base, the recommended base coat proportions are shown in table 7-1. Also shown in table 7-1 are proportions for three-coat work on a masonry base and proportions for work on metal lath.

For three-coat work on gypsum lath, the recommended base coat proportions for gypsum plaster are shown in table 7-2.

FINISH COAT PROPORTIONS.—A lime finish can be applied over a lime, gypsum, or portland cement base coat. Other finishes should be applied only to base coats containing the same cementitious material. A gypsum-vermiculite finish should be applied only to a gypsum-vermiculite base coat.

Finish coat proportions vary according to whether the surface is to be finished with a trowel or with a float. (These tools are described later.) The trowel attains a smooth finish; the float produces a textured finish.

For a trowel-finish coat using gypsum plaster, the recommended proportions are 200 pounds of hydrated lime or 5 cubic feet of lime putty to 100 pounds of gypsum gauging plaster.

Table 7-2.—Recommended Base Coat Proportions for Gypsum Plaster

COAT	PROPORTION
Scratch coat	1:2
Brown coat	1:3
Both coats	1:2.5

For a trowel-finish coat using lime-Keene's cement plaster, the recommended proportions are, for a medium-hard finish, 50 pounds of hydrated lime or 100 pounds of lime putty to 100 pounds of Keene's cement. For a hard finish, the recommended proportions are 25 pounds of hydrated lime or 50 pounds of lime putty to 100 pounds of Keene's cement.

For a trowel-finish coat using lime-portland cement plaster, the recommended proportions are 200 pounds of hydrated lime or 5 cubic feet of lime putty to 94 pounds of portland cement.

For a finish coat using portland cement-sand plaster, the recommended proportions are 300 pounds of sand to 94 pounds of portland cement. This plaster may be either troweled or floated. Hydrated lime up to 10 percent by weight of the portland cement, or lime putty up to 24 percent of the volume of the portland cement, may be added as a plasticizer.

For a trowel-finish coat using gypsum gauging or gypsum neat plaster and vermiculite aggregate, the recommended proportions are 1 cubic foot of vermiculite to 100 pounds of plaster.

Estimates

The total volume of plaster required for a job is the product of the thickness of the plaster times the net area to be covered. Plaster specifications state a minimum thickness, which you must not go under. Also, you should exceed the specs as little as possible due to the increased tendency of plaster to crack with increased thickness.

Mixing Plaster

The two basic operations in mixing plaster are determining the correct proportions and the actual mixing methods used.

PROPORTIONS.—The proper proportions of the raw ingredients required for any plastering job are found in the job specifications. The specs also list the types of materials to use and the type of finish required for each area. Hardness and durability of the plaster surface depend upon how accurately you follow the correct proportions. Too much water gives you a fluid plaster that is hard to apply. It also causes small holes to develop in the finish mortar coat. Too much aggregate in the mix, without sufficient binder to unite the mixture, causes aggregate particles to crumble off. **Without exception, consult the specifications prior to the commencement of any plaster job.**

MIXING METHODS.—As a Builder, you will be mixing plaster either by hand or using a machine.

Hand Mixing.—To hand-mix plaster, you will need a flat, shallow mixing box and a hoe. The hoe usually has one or more holes in the blade. Mixed plaster is transferred from the mixing box to a mortar board, similar to that used in bricklaying. Personnel applying the plaster pick it up from the mortar board.

In hand mixing, first place the dry ingredients in a mixing box and thoroughly mix until a uniform color is obtained. After thoroughly blending the dry ingredients, you then cone the pile and add water to the mix. Begin mixing by pulling the dry material into the water with short strokes. Mixing is continued until the materials have been thoroughly blended and proper consistency has been attained. With experience, a person acquires a **feel** for proper consistency. Mixing should not be continued for more than 10 to 15 minutes after the materials have been thoroughly blended. Excessive agitation may hasten the rate of solution of the cementitious material and reduce initial set time.

Finish-coat lime plaster is usually hand-mixed on a 5- by 5-foot mortar board called a finishing board. Hydrated lime is first converted to lime putty by soaking in an equal amount of water for 16 hours. In mixing the plaster, you first form the lime putty into a ring on the finishing board. Next, pour water into the ring and sift the gypsum or Keene's cement into the water to avoid lumping. Last, allow the mix to stand for 1 minute, then thoroughly blend the materials. Sand, if used, is then added and mixed in.

Machine Mixing.—For a quicker, more thorough mix, use a plaster mixing machine. A typical plaster mixing machine (shown in fig. 7-11) consists primarily of a metal drum containing mixing blades, mounted on

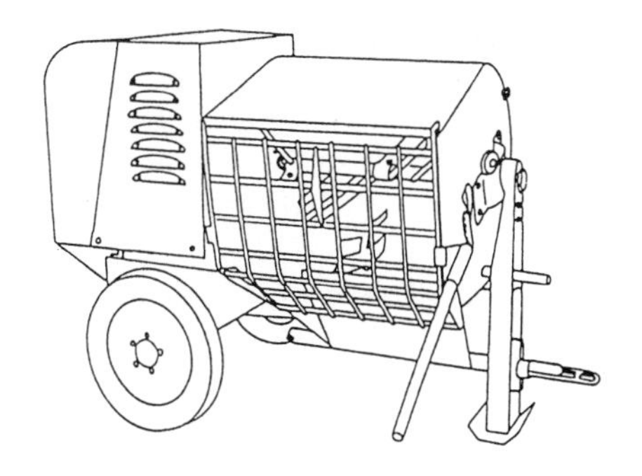

Figure 7-11.—Plaster mixing machine.

a chassis equipped with wheels for road towing. Sizes range from 3 1/2 to 10 cubic feet and can be powered by an electric or a gasoline motor. Mixing takes place either by rotation of the drum or by rotation of the blades inside the drum. Tilt the drum to discharge plaster into a wheelbarrow or other receptacle.

When using a plaster mixer, add the water first, then add about half the sand. Next, add the cement and any admixture desired. Last, add the rest of the sand. Mix until the batch is uniform and has the proper consistency—3 to 4 minutes is usually sufficient. Note that excessive agitation of mortar speeds up the setting time. Most mixers operate at top capacity when the mortar is about 2 inches, at most, above the blades. When the mixer is charged higher than this, proper mixing fails to take place. Instead of blending the materials, the mixer simply folds the material over and over, resulting in excessively dry mix on top and too wet mix underneath—a bad mix. Eliminate this situation by not overloading the machine.

Handling Materials

Personnel handling cement or lime bags should wear recommended personnel protective gear. Always practice personal cleanliness. Never wear clothing that is hard and stiff with cement. Such clothing irritates the skin and may cause serious infection. Any susceptibility of skin to cement and lime burns should be immediately reported to your supervisor.

Don't pile bags of cement or lime more than 10 bags high on a pallet except when stored in bins or enclosures built for such purposes. Place the bags around the outside of the pallet with the tops of the bags facing the center. To prevent piled bags from falling outward, crosspile the first five tiers of bags, each way from any corner, and make a setback starting with the sixth tier. If you have to pile above the 10th tier, make another setback. The back tier, when not resting against a wall of sufficient strength to withstand the pressure, should be set back one bag every five tiers, the same as the end tiers.

During unpiling, the entire top of the pile should be kept level and the setbacks maintained for every five tiers.

Lime and cement must be stored in a dry place to help prevent the lime from crumbling and the cement from hydrating before it is used.

PLASTER APPLICATION TOOLS AND TECHNIQUES

LEARNING OBJECTIVE: Upon completing this section, you should be able to state the uses of plastering tools, and describe the techniques of plastering.

A plaster layer must have uniform thickness to attain complete structural integrity. Also, a plane plaster surface must be flat enough to appear flat to the eye and receive surface-applied materials, such as casings and other trim, without the appearance of noticeable spaces. Specified flatness tolerance is usually 1/8 inch in 10 feet.

TOOLS

Plastering requires the use of a number of tools, some specialized, including trowels, hawk, float, straight and feather edges, darby, scarifier, and plastering machines.

Trowels

Steel trowels are used to apply, spread, and smooth plaster. The shape and size of the trowel blade are determined by the purpose for which the tool is used and the manner of using it.

The four common types of plastering trowels are shown in figure 7-12. The rectangular trowel, with a blade approximately 4 1/2 inches wide by 11 inches long, serves as the principle conveyor and manipulator

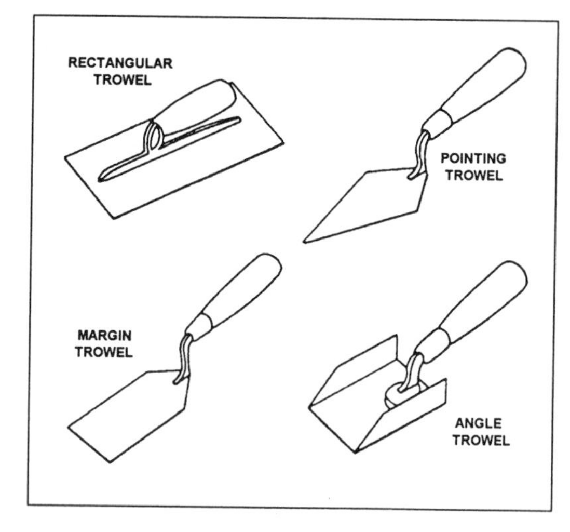

Figure 7-12.—Plastering trowels.

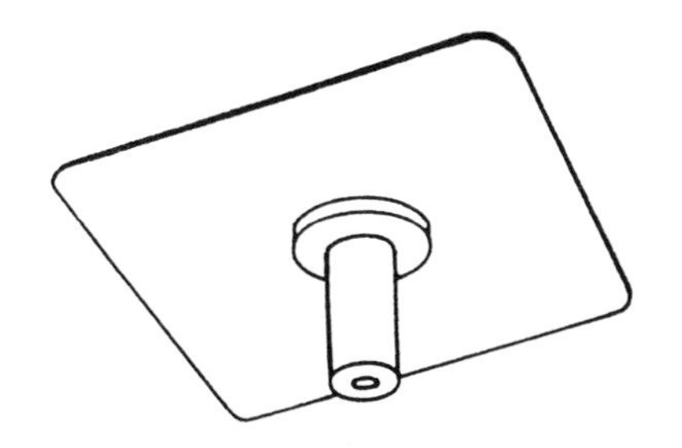

Figure 7-13.—Plastering hawk.

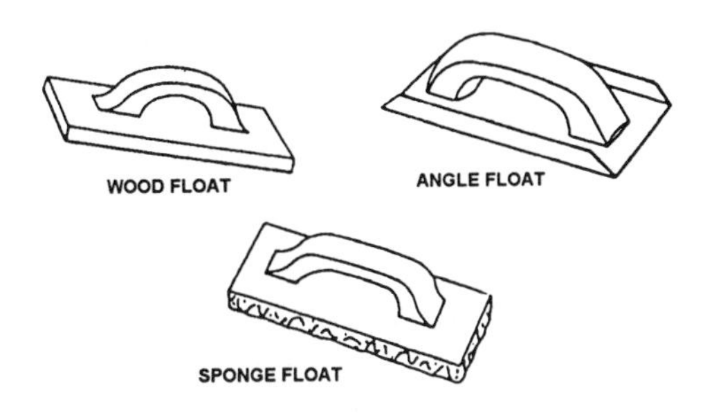

Figure 7-14.—Plastering floats.

of plaster. The pointing trowel, 2 inches wide and about 10 inches long, is used in places where the rectangular trowel doesn't fit. The margin trowel is a smaller trowel, similar to the pointing trowel, but with a square, rather than a pointed, end. The angle trowel is used for finishing corner angles formed by adjoining right-angle plaster surfaces.

Hawk

The hawk (fig. 7-13) is a square, lightweight sheet-metal platform with a vertical central handle, used for carrying mortar from the mortar board to the place where it is to be applied. The plaster is then removed from the hawk with the trowel. The size of a hawk varies from a 10- to a 14-inch square. A hawk can be made in the field from many different available materials.

Float

A float is glided over the surface of the plaster to fill voids and hollows, to level bumps left by previous operations, and to impart a texture to the surface. The most common types of float are shown in figure 7-14. The wood float has a wood blade 4 to 5 inches wide and about 10 inches long. The angle float has a stainless steel

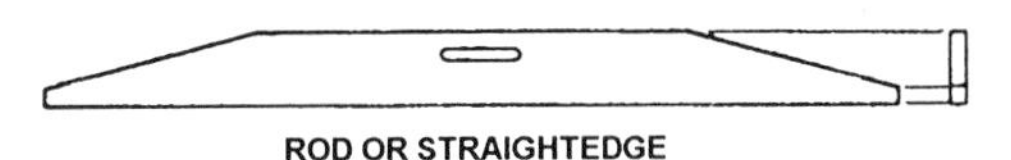

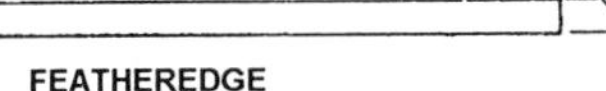

Figure 7-15.—Straightedge and featheredge.

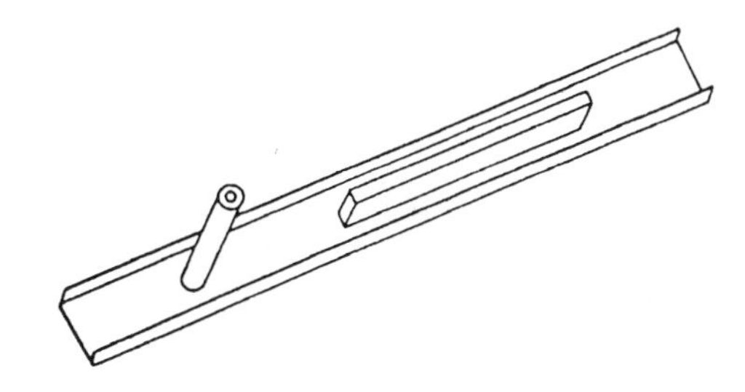

Figure 7-16.—Darby.

or aluminum blade. The sponge float is faced with foam rubber or plastic, intended to attain a certain surface texture.

In addition to the floats just mentioned, other floats are also used in plaster work. A carpet float is similar to a sponge float, but faced with a layer of carpet material. A cork float is faced with cork.

Straight and Feather Edges

The rod or straightedge consists of a wood or lightweight metal blade 6 inches wide and 4 to 8 feet long (see fig. 7-15). This is the first tool used in leveling and straightening applied plaster between the grounds. A wood rod has a slot for a handle cut near the center of the blade. A metal rod usually has a shaped handle running the length of the blade.

The featheredge (fig. 7-15) is similar to the rod except that the blade tapers to a sharp edge. It is used to cut in inside corners and to shape sharp, straight lines at outside corners where walls intersect.

Darby

The darby (fig. 7-16) is, in effect, a float with an extra long (3 1/2 to 4 foot) blade, equipped with handles for two-handed manipulation. It is used for further straightening of the base coat, after rodding is

Figure 7-17.—Scarifier.

completed, to level plaster screeds and to level finish coats. The blade of the darby is held nearly flat against the plaster surface, and in such a way that the line of the edge makes an angle 45° with the line of direction of the stroke.

When a plaster surface is being leveled, the leveling tool must move over the plaster smoothly. If the surface is too dry, lubrication must be provided by moistening. In base coat operations, dash or brush on water with a water-carrying brush called a browning brush. This is a fine-bristled brush about 4 to 5 inches wide and 2 inches thick, with bristles about 6 inches long. For finish coat operations, a finishing brush with softer, more pliable bristles is used.

Scarifier

The scarifier (fig. 7-17) is a raking tool that leaves furrows approximately 1/8 inch deep, 1/8 inch wide, and 1/2 inch to 3/4 inch apart. The furrows are intended to improve the bond between the scratch coat and the brown coat.

Plastering Machines

There are two types of plastering machines: wet mix and dry mix. The wet-mix pump type carries mixed plaster from the mixing machine to a hose nozzle. The dry-mix machine carries dry ingredients to a mixing nozzle where water under pressure combines with the mix and provides spraying force. Most plastering machines are of the wet-mix pump variety.

A wet-mix pump may be of the worm-drive, piston-pump, or hand-hopper type. In a worm-drive machine, mixed plaster is fed into a hopper and forced through the hose to the nozzle by the screw action of a rotor and stator assembly in the neck of the machine. A machine of this type has a hopper capacity of from 3 to 5 cubic feet and can deliver from 0.5 to 2 cubic feet of plaster per minute. On a piston-pump machine, a hydraulic, air-operated, or mechanically operated piston supplies the force for moving the wet plaster. On a hand-hopper machine, the dry ingredients are placed in a hand-held hopper just above the nozzle. Hopper capacity is usually around 1/10 cubic foot. These machines are mainly used for applying finish plaster.

Machine application reduces the use of the hawk and trowel in initial plaster application. However, the use of straightening and finishing hand tools remains about the same for machine-applied plaster.

CREWS

A typical plastering crew for hand application consists of a crew leader, two to four plasterers, and two to four tenders. The plasterers, under the crew leader's supervision, set all levels and lines and apply and finish the plaster. The tenders mix the plaster, deliver it to the plasterers, construct scaffolds, handle materials, and do cleanup tasks.

For a machine application, a typical crew consists of a nozzle operator who applies the material, two or three plasterers leveling and finishing, and two to three tenders.

BASE COAT APPLICATION

Lack of uniformity in the thickness of a plaster coat detracts from the structural performance of the plaster, and the thinner the coat, the smaller the permissible variation from uniformity. Specifications usually require that plaster be finished "true and even, within 1/8-inch tolerance in 10 feet, without waves, cracks, or imperfections." The standard of 1/8 inch appears to be the closest practical tolerance to which a plasterer can work by the methods commonly in use.

The importance of adhering to the recommended minimum thickness for the plaster cannot be overstressed. A plaster wall becomes more rigid as thickness over the minimum recommended increases. As a result, the tendency to crack increases as thickness increases. However, tests have shown that a reduction of thickness from a recommended minimum of 1/2 inch to 3/8 inch, with certain plasters, decreases resistance by as much as 60 percent, while reduction to 1/4 inch decreases it as much as 82 percent.

Gypsum

The sequence of operations in three-coat gypsum plastering is as follows:

1. Install the plaster base.
2. Attach the grounds.
3. Apply the scratch coat approximately 3/16 inch thick.
4. Before the scratch coat sets, rake and cross rake.
5. Allow the scratch coat to set firm and hard.
6. Apply plaster screeds (if required).
7. Apply the brown coat to a depth of the screeds.
8. Using the screeds as guides, straighten the surface with a rod.
9. Fill in any hollows and rod again.
10. Level and compact the surface with a darby; then rake and cross rake to receive the finish coat.
11. Define angles sharply with an angle float and a featheredge. Trim back the plaster around the grounds so the finish coat can be applied flush with the grounds.

Lime

The steps for lime base coat work are similar to those for gypsum work except that, for lime, an additional floating is required the day after the brown coat is applied. This extra floating is required to increase the density of the slab and to fill in any cracks that may have developed because of shrinkage of the plaster. A wood float with one or two nails protruding 1/8 inch from the sole (called a devil's float) is used for this purpose.

Portland Cement

Portland cement plaster is actually cement mortar. It is usually applied in three coats, the steps being the same as those described for gypsum plaster. Minimum recommended thicknesses are usually 3/8 inch for the scratch coat and brown coat, and 1/8 inch for the finish coat.

Portland cement plaster should be moist-cured, similar to concrete. The best procedure is fog-spray curing. The scratch coat and the brown coat should both be fog-sprayed cured for 48 hours. The finish coat should not be applied for at least 7 days after the brown coat. It too should be spray-cured for 48 hours.

FINISH COAT APPLICATION

Interior plaster can be finished by troweling, floating, or spraying. Troweling makes a smooth finish; floating or spraying makes a finish of a desired surface texture.

Smooth Finish

Finish plaster made of gypsum gauging plaster and lime putty (called white coat or putty coat) is the most widely used material for smooth finish coats. A putty coat is usually applied by a team of two or more persons. The steps are as follows:

1. One person applies plaster at the angles.
2. Another person follows immediately, straightening the angles with a rod or featheredge.
3. The remaining surface is covered with a skim coat of plaster. Pressure on the trowel must be sufficient to force the material into the rough surface of the base coat to ensure a good bond.
4. The surface is immediately doubled back to bring the finish coat to final thickness.
5. All angles are floated, with additional plaster added if required to fill hollows.
6. The remaining surface is floated, and all hollows filled. This operation is called drawing up. The hollows being filled are called cat faces.
7. The surface is allowed to draw for a few minutes. As the plaster begins to set, the surface-water glaze disappears and the surface becomes dull. At this point, troweling should begin. The plasterer holds the water brush in one hand and the trowel in the other, so troweling can be done immediately after water is brushed on.
8. Water is brushed on lightly, and the entire surface is rapidly troweled with enough pressure to compact the finish coat fully. The troweling operation is repeated until the plaster has set.

The sequence of steps for trowel finishes for other types of finish plasters is about the same. Gypsum-finish plaster requires less troweling than white-coat plaster. Regular Keene's cement requires longer troweling, but quick-setting Keene's cement requires less. Preliminary finishing of portland cement-sand is done with a wood float, after which the steel trowel is used. To avoid excessive drawing of fines to the surface, delay troweling of the portland cement-sand as long as possible. For the same reason, the surface must not be troweled too long.

The steps in float finishing are about the same as those described for trowel finishing except, of course, that the final finish is obtained with the float. A surface is usually floated twice: a rough floating with a wooden float first, then a final floating with a rubber or carpet float. With one hand the plasterer applies with the brush, while moving the float in the other hand in a circular motion immediately behind the brush.

Special Textures

Some special interior-finish textures are obtained by methods other than or in addition to floating. A few of these are listed below.

STIPPLED.—After the finish coat has been applied, additional plaster is daubed over the surface with a stippling brush or roller.

SPONGE.—By pressing a sponge against the surface of the finish coat, you get a very soft, irregular texture.

DASH.—The dash texture is obtained by throwing plaster onto the surface from a brush. It produces a fairly coarse finish that can be modified by brushing the plaster with water before it sets.

TRAVERTINE.—The plaster is jabbed at random with a whisk broom, wire brush, or other tool that will form a dimpled surface. As the plaster begins to set, it is troweled intermittently to form a pattern of rough and smooth areas.

PEGGLE.—A rough finish, called peggle, is obtained by throwing small pebbles or crushed stone against a newly plastered surface. If necessary, a trowel is used to press the stones lightly into the plaster.

STUCCO

LEARNING OBJECTIVE: Upon completing this section, you should be able to identify the composition of stucco, and state the procedures for mixing, applying, and curing.

"Stucco" is the term applied to plaster whenever it is applied to the exterior of a building or structure. Stucco can be applied over wood frames or masonry structures. A stucco finish lends warmth and interest to projects.

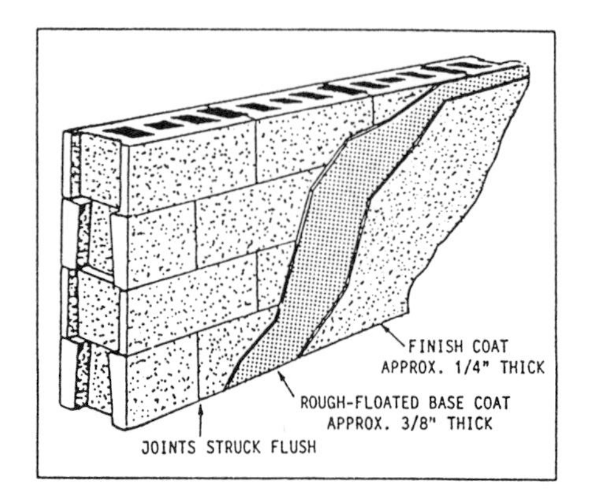

Figure 7-18.—Masonry (two-coat work directly applied).

COMPOSITION

Stucco is a combination of cement or masonry cement, sand, and water, and frequently a plasticizing material. Color pigments are often used in the finish coat, which is usually a factory-prepared mix. The end product has all the desirable properties of concrete. Stucco is hard, strong, fire resistant, weather resistant, does not deteriorate after repeated wetting and drying, resists rot and fungus, and retains colors.

The material used in a stucco mix should be free of contaminants and unsound particles. Type I normal portland cement is generally used for stucco, although type II, type III, and air-entraining may be used. The plasticizing material added to the mix is hydrated lime. Mixing water must be potable. The aggregate used in cement stucco can greatly affect the quality and performance of the finished product. It should be well graded, clean, and free from loam, clay, or vegetable matter, which can prevent the cement paste from properly binding the aggregate particles together. Follow the project specifications as to the type of cement, lime, and aggregate to be used.

APPLICATION

Metal reinforcement should be used whenever stucco is applied on wood frame, steel frame, flashing, masonry, or any surface not providing a good bond. Stucco may be applied directly on masonry.

The rough-floated base coat is approximately 3/8 inch thick. The finish coat is approximately 1/4 inch thick. Both are shown in figure 7-18 applied to a masonry surface. On open-frame construction

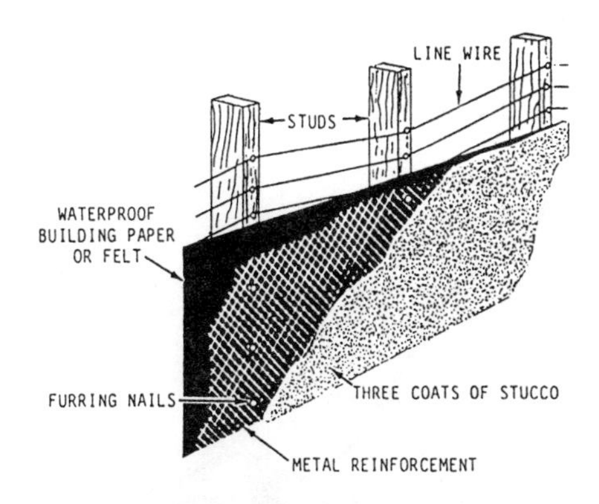

Figure 7-19.—Open-frame construction.

(fig. 7-19), nails are driven one-half their length into the wood. Spacing should be 5 to 6 inches OC from the bottom. Nails should be placed at all corners and openings throughout the entire structure on the exterior.

The next step is to place wire on the nails. This is called installing the line wire. Next, a layer of waterproof paper is applied over the line wire. Laps should be 3 to 4 inches and nailed with roofing nails. Install wire mesh (stucco netting), which is used as the reinforcement for the stucco.

Furring nails (fig. 7-20) are used to hold the wire away from the paper to a thickness of three- eighths of an inch. Stucco or sheathed frame construction is the same as open frame except no line wire is required. The open and sheathed frame construction requires three coats of 3/8-inch scratch coat horizontally scored or scratched, a 3/8-inch brown coat, and a 1/8-inch finish coat.

Stucco is usually applied in three coats. The first coat is the scratch coat; the second the brown coat; and the final coat the finish coat. On masonry where no reinforcement is used, two coats may be sufficient. Start at the top and work down the wall. This prevents mortar from falling on the completed work. The first, or scratch, coat should be pushed through the mesh to ensure the metal reinforcement is completely embedded for mechanical bond. The second, or brown, coat should be applied as soon as the scratch coat has set up enough to carry the weight of both coats (usually 4 or 5 hours). The brown coat should be moist-cured for about 48 hours and then allowed to dry for about 5 days. Just before the application of the finish coat, the brown coat should be uniformly dampened. The third, or finish, coat is frequently pigmented to obtain decorative colors. Although the colors may be job-mixed, a factory-prepared mix is recommended. The finish coat may be applied by hand or machine. Stucco finishes are available in a variety of textures, patterns, and colors.

Surface Preparation

Before the various coats of stucco can be applied, the surfaces have to be prepared. Roughen the surfaces of masonry units enough to provide good mechanical key, and clean off paint, oil, dust, soot, or any other material that may prevent a tight bond. Joints may be

Figure 7-20.—Several types of furring nails.

struck off flush or slightly raked. Old walls softened and disintegrated by weather action, surfaces that cannot be cleaned thoroughly, such as painted brickwork, and all masonry chimneys should be covered with galvanized metal reinforcement before applying the stucco. When masonry surfaces are not rough enough to provide good mechanical key, one or more of the following actions may be taken:

- Old cast-in-place concrete or other masonry may be roughened with bush hammers or other suitable hand tools. Roughen at least 70 percent of the surface with the hammer marks uniformly distributed. Wash the roughened surface free of chips and dust. Let the wall dry thoroughly.
- Concrete surfaces may be roughened with an acid wash. Use a solution of 1 part muriatic acid to 6 parts water. **Note**: Add muriatic acid to the water; **never add water to the acid**. First, wet the wall so the acid will act on the surface only. More than one application may be necessary. After the acid treatment, wash the wall thoroughly to remove all acid. Allow the washed wall to dry thoroughly.

CAUTION

When your crewmembers are using muriatic acid, make sure they wear goggles, rubber gloves, and other protective clothing and equipment.

- You can quickly rough masonry surfaces using a power-driven roughing machine (such as that shown in figure 7-21) equipped with a cylindrical cage fitted with a series of hardened steel cutters. The cutters should be mounted to provide a flailing action that results in a scored pattern. After roughing, wash the wall clean of all chips and dust and let it dry.

Suction is absolutely necessary to attain a proper bond of stucco on concrete and masonry surfaces. It is also necessary in first and second coats so the following coats bond properly. Uniform suction also helps obtain a uniform color. If one part of the wall draws more moisture from the stucco than another, the finish coat may be spotty. Obtain uniform suction by dampening the wall evenly, but not soaking, before applying the stucco. The same applies to the scratch and brown coats. If the surface becomes dry in spots, dampen those areas again to restore suction. Use a fog spray for dampening.

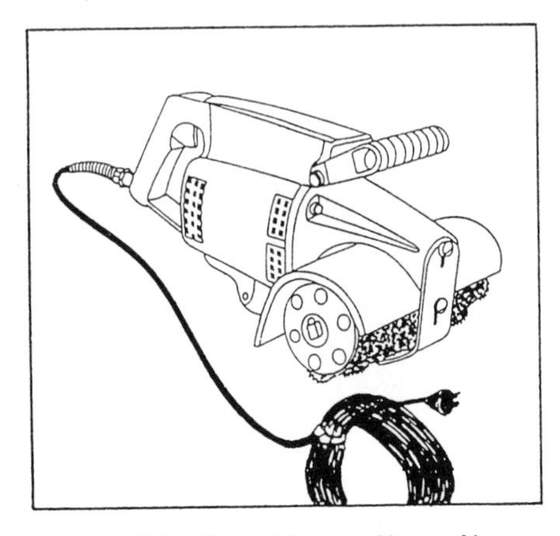

Figure 7-21.—Power-driven roughing machine.

When the masonry surface is not rough enough to ensure an adequate bond for a trowel-applied scratch coat, use the dash method. Acid-treated surfaces usually require a dashed scratch coat. Dashing on the scratch coat aids in getting a good bond by excluding air that might get trapped behind a trowel-applied coat. Apply the dash coat with a fiber brush or whisk broom, using a strong whipping motion at right angles to the wall. A cement gun or other machine that can apply the dash coat with considerable force also produces a suitable bond. Keep the dash coat damp for at least 2 days immediately following its application and then allow it to dry.

Protect the finish coat against exposure to sun and wind for at least 6 days after application. During this time, keep the stucco moist by frequent fog-spraying.

Mixing

Mixing procedures for stucco are similar to those for plaster. Three things you need to consider before mixing begins are the type of material you are going to use, the backing to which the material will be applied, and the method used to mix the material (hand or machine). As with plaster, addition of too much of one raw ingredient or the deletion of a raw material gives you a bad mix. Prevent this by allowing **only** the required amount of ingredients in the specified mix.

Applying

Stucco can be applied by hand or machine. Machine application allows application of material over a large area without joinings (joinings are a problem for hand-applied finishes). To apply stucco, begin at the top of the wall and work down. Make sure the crew has sufficient personnel to finish the total wall surface without joinings (laps or interruptions).

Curing

The curing of stucco depends on the surface to which it is applied, the thickness if the material, and the weather. Admixtures can be used to increase workability, prevent freezing, and to waterproof the mortar. Using high-early cement reduces the curing time required for the cement to reach its initial strength (3 days instead of 7). Air-entraining cement is used to resist freezing action.

COMMON FAULTS

There are times when the finish you get is not what you expected. Some of the most common reasons for discoloration and stains are listed below:

- Failure to have uniform suction in either of the base coats;
- Improper mixing of the finish coat materials;
- Changes in materials or proportions during the work;
- Variations in the amount of mixing water;
- Use of additional water to retemper mortar; and
- Corrosion and rust from flashing or other metal attachments and failure to provide drips and washes on sills and projecting trim.

CERAMIC TILE

LEARNING OBJECTIVE: Upon completing this section, you should be able to identify the different types of ceramic tile and associated mortars, adhesives, and grouts, and state the procedures for setting tiles.

Ceramic tile is used extensively where sanitation, stain resistance, ease in cleaning, and low maintenance are desired. Ceramic tiles are commonly used for walls and floors in bathrooms, laundry rooms, showers, kitchens, laboratories, swimming pools, and locker rooms. The tremendous range of colors, patterns, and designs available in ceramic tile even includes three-dimensional sculptured tiles. Extensive use has been made of ceramic tile for decorative effects throughout buildings, both inside and outside.

CLASSIFICATIONS

Tile is usually classified by exposure (interior or exterior) and location (walls or floors), although many tiles may be used in all locations. Since exterior tile must be frostproof, the tiles are kiln fired to a point where they have a very low absorption. Tiles vary considerably in quality among manufacturers. This may affect their use in various exposures and locations.

SIZES

Tile is generally available in the following square sizes: 4 1/4 by 4 1/4, 6 by 6, 3 by 3, and 1 3/8 by 1 3/8 inches. Rectangular sizes available include 8 1/2 by 4 1/4, 6 by 4 1/4, and 1 3/8 by 4 1/4 inches. Tile often comes mounted into sheets (usually between 1 and 2 square feet) with some type of backing on the sheet or between the tiles to hold them together.

Tiles with less than 6 square inches of face area and about 1/4 inch thick are called ceramic mosaics. Ceramic mosaic tile sizes range from 3/8 by 3/8 inch to about 2 by 2 inches, and they are available from the manufactures in both sheet and roll form. Often, large tile is scored by the manufacturer to resemble small tiles.

FINISHES

Tile finishes include glazed, unglazed, textured (matte) glazed, porcelain, and abrasive. Glazed and matte glazed finishes may be used for light-duty floors but should not be used in areas of heavy traffic where the glazed surface may be worn away. Glazed ceramic wall tiles usually have a natural clay body (nonvitreous, 7- to 9-percent absorption), and a vitreous glaze is fused to the face of the tile. This type of tile is not recommended for exterior use. Glazed tile should never be cleaned with acid, which mars the finish. Use only soap and water. Unglazed ceramic mosaics have dense, nonvitreous bodies uniformly distributed through the tile. Certain glazed mosaics are recommended for interior use only, others for wall use only. Porcelain tiles have a smoother surface than mosaics and are denser, with an impervious body of less than one-half of 1-percent absorption. This type of tile may be used throughout the interior and exterior of a building. An abrasive finish is available as an aggregate embedded in the surface or an irregular surface texture.

Tiles are available with self-spacing lugs, square edges, and cushioned edges (slightly rounded) (see

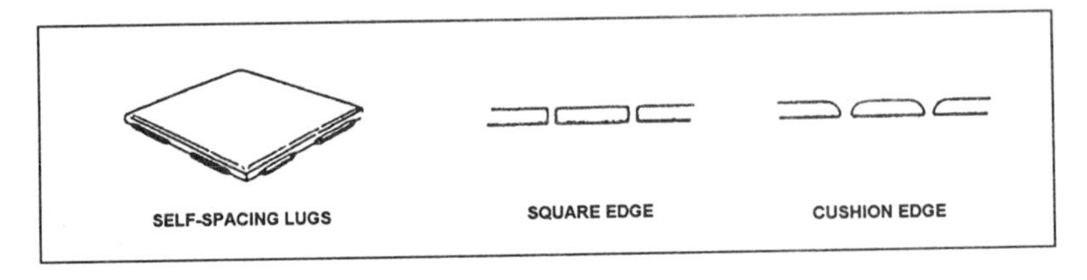

Figure 7-22.—Tile edges.

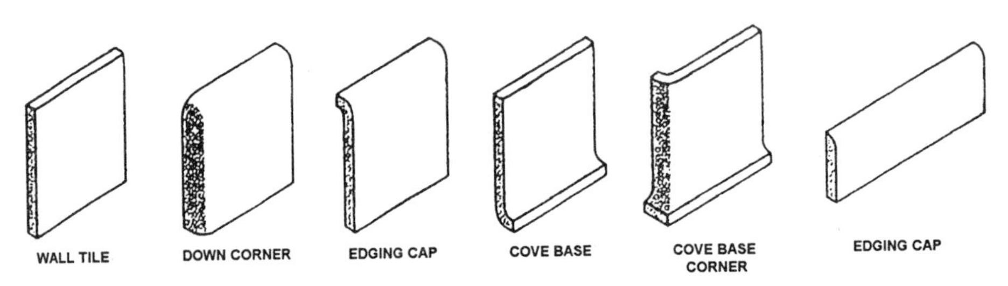

Figure 7-23.—Trimmer shapes.

fig. 7-22). The lugs assure easy setting and uniform joints. The edges available vary with the size of the tile and the manufacturer.

Margins, corners, and base lines are finished with trimmers of various shapes (fig. 7-23). A complete line of shaped ceramic trim is available from manufacturers. Other accessories include towel bars, shelf supports, paper holders, grab rails, soap holders, tumbler holders, and combination toothbrush and tumbler holders, to list a few of the more popular units.

MORTARS AND ADHESIVES

The resistance of ceramic tile to traffic depends primarily on base and bonding material rigidity, grout strength, hardness, and the accurate leveling and smoothness of the individual tiles in the installation. The four basic installation methods are cement mortar (the only thick bed method), dry-set mortar, epoxy mortar, and organic adhesives (mastic).

Cement Mortar

Cement mortar for setting ceramic tiles is composed of a mixture of portland cement and sand. The mix proportions for floors may vary from 1:3 to 1:6 by volume. For walls, a portland cement, sand, and hydrated lime mix may vary from 1:3:1 to 1:5 1/2:1. These proportion ratios are dictated by the project specifications. The mortar is placed on the surface 3/4 to 1 inch thick on walls and 3/4 inch to 1 1/4 inches thick on floors. A neat cement bond coat is applied over it while the cement mortar is fresh and plastic. After soaking in water for at least 30 minutes, the tiles are installed over the neat cement bond coat. This type of installation, with its thick mortar bed, permits wall and floor surfaces to be sloped. This installation provides a bond strength of 100 to 200 pounds per square inch. A waterproof backing is sometimes required, and the mortar must be damp-cured.

Dry-Set Mortar

Dry-set mortar is a thin-bed mortar of premixed portland cement, sand, and admixtures that control the setting (hardening) time of the mortar. It may be used over concrete, block, brick, cellular foamed glass, gypsum wallboard, and unpainted dry cement plaster, as well as other surfaces. A sealer coat is often required when the base is gypsum plaster. It is not recommended for use over wood or wood products. Dry-set mortar can be applied in one layer 3/32 inch thick, and it provides a bond strength of 500 pounds per square inch. This method has excellent water and impact resistance and may be used on exteriors. The tiles do not have to be presoaked, but the mortar must be damp-cured.

Epoxy Mortar

Epoxy mortar can be applied in a bed as thin as 1/8 inch. When the epoxy resin and hardener are mixed on the job, the resulting mixture hardens into an extremely strong, dense setting bed. Pot life, once the parts are mixed, is about 1 hour if the temperature is 82°F or higher. This mortar has excellent resistance to the

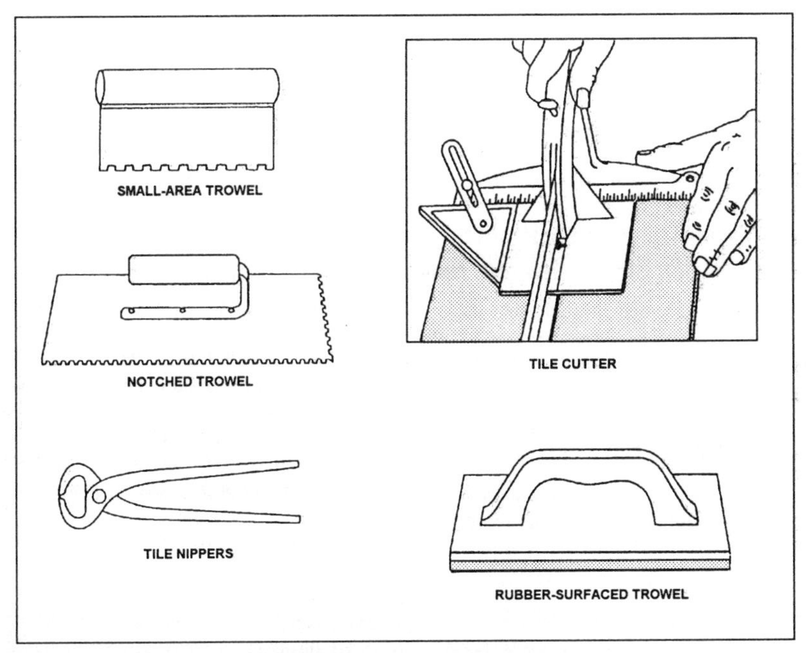

Figure 7-24.—Special tile-setting tools.

corrosive conditions often encountered in industrial and commercial installations. It may be applied over bases of wood, plywood, concrete, or masonry. This type of mortar is nonshrinking and nonporous. A bond strength of over 1,000 pounds per square inch is obtained with this installation method.

Organic Adhesives

Organic adhesives (mastics) are applied in a thin layer with a notched trowel. They are solvent-base, rubber material. Porous materials should be primed before mastic is applied to prevent some of the plasticizers and oils from soaking into the backing. Suitable surfaces include wood, concrete, masonry, gypsum wallboard, and plaster. The bond strength available varies considerably among manufacturers, but the average is about 100 pounds per square inch.

GROUTS

The joints between the tiles must be filled with a grout selected to meet the tile requirements and exposure. Tile grouts may be portland cement base, epoxy base, furans, or latex.

Cement grout consists of portland cement and admixtures. This is better in terms of waterproofing, uniform color, whiteness, shrink resistance, and fine texture than a plain cement. It may be colored and used in all areas subject to ordinary use. When the grout is placed, the tiles should be wet. Moisture is required for proper curing.

Drywall grout has the same characteristics as dry-set mortar and is suitable for areas of ordinary use. Tiles to be set in drywall grout do not require wetting except during very dry conditions.

Epoxy grout consists of an epoxy resin and hardener. It produces a joint that is stainproof, resistant to chemicals, hard, smooth, impermeable, and easy to clean. It is used extensively in counters that must be kept sanitary for foods and chemicals. It has the same basic characteristics as epoxy mortars.

Furan resin grout is used in industrial areas requiring high resistance to acids and weak alkalies. Special installation techniques are required with this type of grouting.

Latex grout is used for a more flexible and less permeable finish than cement grout. It is made by introducing a latex additive into the portland cement grout mix.

TOOLS

A selection of special tools, shown in figure 7-24, should be available when doing tile installation work.

A primary tool is a notched trowel with the notches of the depth recommended by the adhesive manufacturers. A trowel with notches on one side and smooth on the other is preferred. Different sized trowels are available.

A tile cutter is the most efficient tool for cutting ceramic tile. The scribe on the cutter has a tungsten carbide tip. A glass cutter can be used but quickly dulls.

Use tile nippers when trimming irregular shapes. Nip off very small pieces of the tile you are cutting. Attempting to take big chunks at one time can crack the tile.

A rubber-surfaced trowel is used to force grout into the joints of the tile.

APPLICATION TECHNIQUES

There are three primary steps in tile installation: applying a mortar bed, applying adhesive, and setting tiles in place.

Mortar

Before applying a mortar bed to a wall having wooden studs, you first tack a layer of waterproof paper to the studs. You then nail metal lath over the paper. The first coat of mortar applied to a wall for setting tiles is a scratch coat; the second is a float, leveling, or brown coat.

A scratch coat for application as a foundation coat must be at least 1/4 inch thick and composed of 1 part cement to 3 parts sand, with the addition of 10-percent hydrated lime by volume of the cement used. While still plastic, the scratch coat is deeply scored or scratched and cross scratched. Keep the scratch coat protected and reasonably moist during the hydration period. All mortar for scratch and float coats should be used within 1 hour after mixing. Do not retemper partially hardened mortar. Apply the scratch coat not more than 48 hours, nor less than 24 hours, before setting the tile.

The float coat should be composed of 1 part cement, 1 part of hydrated lime, and 3 1/2 parts sand. It should be brought flush with screeds or temporary guide strips, placed to give a true and even surface at the proper distance from the finished face of the tile.

Wall tiles should be thoroughly soaked for a minimum of 30 minutes in clean water before being set. Set tiles by troweling a skim coat of neat portland cement mortar on the float coat, or applying a skim coat to the back of each tile unit and immediately floating the tiles into place. Joints must be straight, level, perpendicular, of even width, and not exceeding 1/16 inch. Wainscots are built of full courses. These may extend to a greater or lesser height, but in no case more than 1 1/2-inch from the specified or figured height. Vertical joints must be maintained plumb for the entire height of the tile work.

All joints in wall tile should be grouted full with a plastic mix of neat white cement or commercial tile grout immediately after a suitable area of the tile has been set. Tool the joints slightly concave; cut off and wipe excess mortar from the face of tiles. Any spaces, crevices, cracks, or depressions in the mortar joints after the grout has been cleaned from the surface should be roughened at once and filled to the line of the cushioned edge (if applicable) before the mortar begins to harden. Tile bases or coves should be solidly backed with mortar. Make all joints between wall tiles and plumbing or other built-up fixtures with a light-colored caulking compound. Immediately after the grout has set, apply a protective coat of noncorrosive soap or other approved protection to the tile wall surfaces.

The installation of wall tile over existing and patched or new plaster surfaces in an existing building is completed as previously described, except that an adhesive is used as the bonding agent. Where wall tile is to be installed in areas subject to intermittent or continual wetting, prime the wall areas with adhesive following the manufacturer's recommendations.

Adhesive

Wall tiles may be installed either by floating or buttering the adhesive. In floating, apply the adhesive uniformly over the prepared wall surface using quantities recommended by the manufacturer. Use a notched trowel held at the proper angle to spread adhesive to the required uniform thickness. Touch up thin or bare spots with an additional coating of adhesive. The area coated at one time should not be any larger than that recommended by the manufacturer. In the buttering method, daub the adhesive on the back of each tile. Use enough so that, when compressed, the adhesive forms a coating not less than 1/16 inch thick over 60 percent of the back of each tile.

Laying Tile

The key to a professional-looking ceramic tile job is to start working with a squared-off area. Most rooms do not have perfectly square corners. As a result, the first step is to mark off a square area in such a way that fractional tiles at the corners (edges) are approximately the same size. Begin by finding the lowest point of the wall you are tiling. From this corner draw a horizontal line one full tile height above the low point and extend

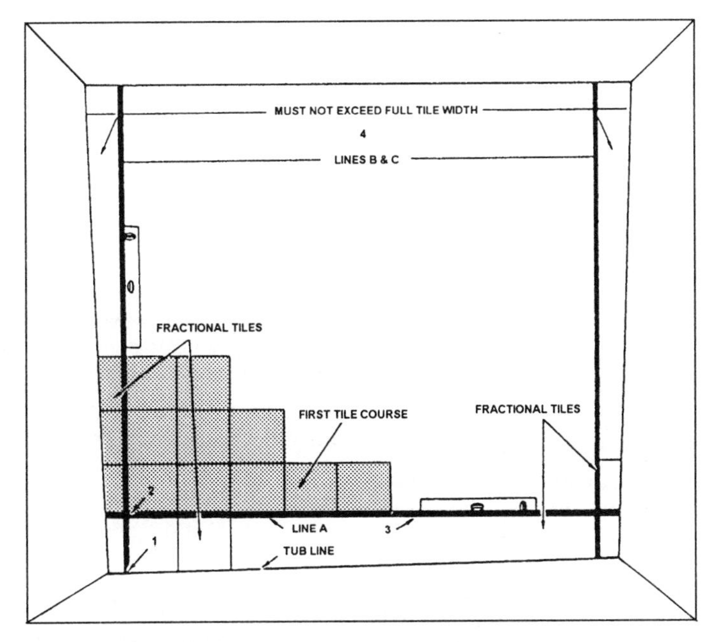

Figure 7-25.—Steps used for squaring a wall.

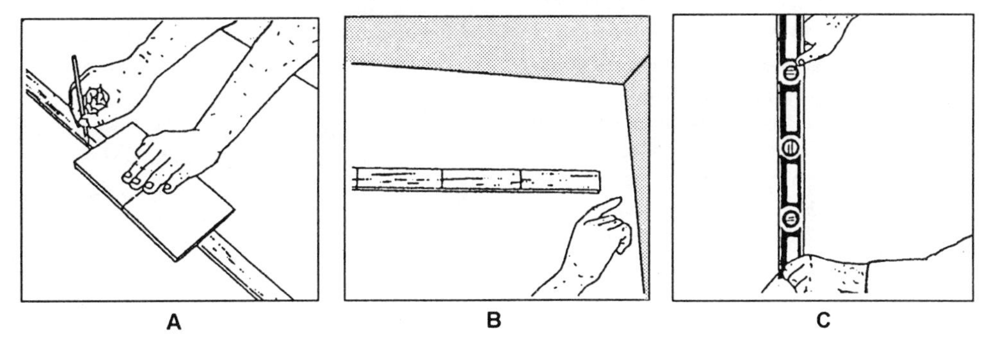

Figure 7-26.—Layout for installing ceramic wall tile.

this line level across the entire width of the room. Refer to the bathroom wall example in figure 7-25 as you study the following steps:

1. Find the low point of the tub.
2. Measure up the height of one full tile at the low point. Draw a horizontal line A. It must be level.
3. Use a tile-measuring stick (fig. 7-26) to determine the position of full-width tiles in such a way that fractional tiles at each corner or edge are equal.
4. Draw vertical lines B and C perpendicular to line A (fig. 7-25). Apply tiles to the squared-off area first. Then cut and apply fractional tiles.

Another method for figuring fractional tiles (edges) is to employ the "half-tile rule." (The stick method is good for short walls, but the half-tile rule is needed for long walls.) Take the number of full-size tiles required for one course, multiply this by the tile size, subtract this answer from the wall length in inches, add one full tile size and divide by 2. The result is the size of end tiles.

After determining fractional tiles, use a piece of scrap wood from 36 inches to 48 inches in length to mark up a tile-measuring stick (fig. 7-26, view A). Mark off a series of lines equal to the width of a tile. Lay this stick on the wall and shift it back and forth to determine the starting point for laying the tiles. Make sure the fractional tiles at the end of each row are of equal widths (fig. 7-26, view B).

Use a level to establish a line perpendicular to the horizontal starting line (fig. 7-26, view C). At both ends of the horizontal line, draw vertical lines to form the squared-off area. To make the tile application easier, you can fasten battens to the wall on the outside of the drawn lines.

Use a trowel to spread the mastic over approximately a 3- by 3-foot area of the wall. Use the notched side to form ridges in the mastic, pressing hard against the surface so that the ridges are the same height as the notches on the tool. Allow the mastic to set for 24 hours before applying grout. Follow the manufacturer's mixing instructions closely and use a rubber-surfaced trowel to spread the grout over the tile surface. Work the trowel in an arc, holding it at a slight angle so that grout is forced into the spaces between the tiles.

Start tiling at either of the vertical lines and tile half the wall at a time, working in horizontal rows. Press each tile into the mastic, but do not slide them—the mastic may be forced up the edges onto the tile surface. After each course of tile is applied, check with the level before spreading more mastic. If a line is crooked, remove all tiles in that line and apply fresh ones. Do not use the removed tiles until the mastic has been cleaned off. Finish tiling the main area before fitting edge tiles.

When the grout begins to dry, wipe the excess from the tiles with a damp rag. After the grout is thoroughly dry, rinse the wall and wipe it with a clean towel.

Nonstaining caulking compound should be used at all joints between built-in fixtures and tile work, and at the top of ceramic tile bases to ensure complete waterproofing. Inside corners should be caulked before a corner bead is applied.

Promptly replace cracked and broken tiles. This protects the edges of adjacent tiles and helps maintain waterproofing and appearance. Timely pointing of displaced joint material and spalled areas in joints is necessary to keep tiles in place.

A new tile surface should be cleaned according to the tile manufacturer's recommendations to avoid damage to the glazed surfaces.

RECOMMENDED READING LIST

NOTE

Although the following references were current when this TRAMAN was published, their continued currency cannot be assured. You therefore need to ensure that you are studying the latest revisions.

Gypsum Construction Handbook, United States Gypsum Company, Chicago, Ill., 1987.

Materials and Methods of Architectural Construction, John Wiley & Sons, New York, 1958.

Plastering Skills, American Technical Publishers, Inc., Alsip, Ill., 1984.

CHAPTER 8

STRUCTURAL COATINGS AND PRESERVATIVES

The final stage of most construction projects is the application of protective coatings, or "painting." As with all projects, you should follow the plans and specifications for surface preparation and application of the finish coat. The specifications give all the information you need to complete the tasks. But, to have a better understanding of structural coatings, you need to know their purposes, methods of surface preparation, and application techniques.

PURPOSES OF STRUCTURAL COATINGS

LEARNING OBJECTIVE: Upon completing this section, you should be able to state the purposes of the different types of structural coatings and how each is employed.

The protection of surfaces is the most important consideration in determining the maintenance cost of structures. Structural coatings serve as protective shields between the base construction materials and elements that attack and deteriorate them. Regularly programmed structural coatings offer long-range protection, extending the useful life of a structure.

PREVENTIVE MAINTENANCE

The primary purpose of a structural coating is protection. This is provided initially with new construction and maintained by a sound and progressive preventive maintenance program. Programmed painting enforces inspection and scheduling. A viable preventive maintenance program will help ensure that minor problems are detected at an early stage—before they become major failures later. An added advantage derived from preventive maintenance is the detection of faulty structural conditions or problems caused by leakage or moisture.

Resistance to moisture from rain, snow, ice, and condensation constitutes perhaps the greatest single protective characteristic of paint, the most common type of structural coating. Moisture causes metal to corrode and wood to swell, warp, or rot. Interior wall finishes of buildings can be ruined by moisture entering through neglected exterior surfaces. Porous masonry is attacked and destroyed by moisture. Therefore, paint films must be as impervious to moisture as possible to provide a protective, water- proof film over the surface to which they are applied. Paint also acts as a protective film against acids, alkalies, material organisms, and other damaging elements.

SANITATION AND CLEANLINESS

Painting is an essential part of general maintenance programs for hospitals, kitchens, mess halls, offices, warehouses, and living quarters. Paint coatings provide smooth, nonabsorptive surfaces that are easily washed and kept free of dirt and foodstuffs. Adhering foodstuffs harbor germs and cause disease. Coating rough or porous areas seals out dust and grease that would otherwise be difficult to remove.

Odorless paints are used in these areas because conventional paint solvent odors are obnoxious to personnel. In food preparation areas, the odors may be picked up by nearby food.

FIRE RETARDANCE

Certain types of structural coatings delay the spread of fire and assist in confining a fire to its area of origin. Fire-retardant coatings should not be considered substitutes for conventional paints. The use of fire-retardant coatings is restricted to areas of highly combustible surfaces, and must be justified and governed by the specific agency's criteria. Fire-retardant coatings are not used in buildings containing automatic sprinkler systems.

CAMOUFLAGE

Camouflage paints have special properties, making them different from conventional paints. Their uses are limited to special applications. Do not use camouflage paints as substitutes for conventional paints. Use this paint only on exterior surfaces to render buildings and structures inconspicuous by blending them in with the surrounding environment.

ILLUMINATION AND VISIBILITY

White and light-tinted coatings applied to ceilings and walls reflect both natural and artificial light and help brighten rooms and increase visibility. On the other hand, darker colors reduce the amount of reflected light. Flat coatings diffuse, soften, and evenly distribute illumination, whereas gloss finishes reflect more like mirrors and may create glare. Color contrasts improve visibility of the painted surface, especially when paint is applied in distinctive patterns. For example, white on black, white on orange, or yellow on black can be seen at greater distances than single colors or other combinations of colors.

IDENTIFICATION AND SAFETY

Certain colors are used as standard means of identifying objects and promoting safety. For example, fire protection equipment is painted red. Containers for kerosene, gasoline, solvents, and other flammable liquids should be painted a brilliant yellow and marked with large black letters to identify their contents. The colors of signal lights and painted signs help control traffic safely by providing directions and other travel information.

TYPES OF COATINGS

LEARNING OBJECTIVE: Upon completing this section, you should be able to identify the types of structural coatings and finishes, and the general characteristics of each.

As a Builder, you must consider many factors when selecting a coating for a particular job. One important factor is the type of coating, which depends on the composition and properties of the ingredients. Paint is composed of various ingredients, such as pigment, nonvolatile vehicle, or binder, and solvent, or thinner. Other coatings may contain only a single ingredient.

PAINT

In this section, we'll cover the basic components of paint—pigment, vehicles, and solvents—and explain the characteristics of different types of paint.

Composition

Paint is composed of two basic ingredients: pigment and a vehicle. A thinner may be added to change the application characteristics of the liquid.

PIGMENT.—Pigments are insoluble solids, ground finely enough to remain suspended in the vehicle for a considerable time after thorough stirring or shaking. Opaque pigments give the paint its hiding, or covering, capacity and contribute other properties (white lead, zinc oxide, and titanium dioxide are examples). Color pigments give the paint its color. These may be inorganic, such as chrome green, chrome yellow, and iron oxide, or organic, such as toluidine red and phthalocyanine blue. Transparent or extender pigments contribute bulk and also control the application properties, durability, and resistance to abrasion of the coating. There are other special-purpose pigments, such as those enabling paint to resist heat, control corrosion, or reflect light.

VEHICLES, OR BINDERS.—The vehicle, or binder, of paint is the material holding the pigment together and causing paint to adhere to a surface. In general, paint durability is determined by the resistance of the binder to the exposure conditions. Linseed oil, once the most common binder, has been replaced, mainly by the synthetic alkyd resins. These result from the reaction of glycerol phthalate and an oil and may be made with almost any property desired. Other synthetic resins, used either by themselves or mixed with oil, include phenolic resin, vinyl, epoxy, urethane, polyester, and chlorinated rubber. Each has its own advantages and disadvantages. When using these materials, it is particularly important that you exactly follow the manufacturers' instructions.

SOLVENTS, OR THINNERS.—The only purpose of a solvent, or thinner, is to adjust the consistency of the material so that it can be applied readily to the surface. The solvent then evaporates, contributing nothing further to the film. For this reason, the cheapest suitable solvent should be used. This solvent is likely to be naphtha or mineral spirits. Although turpentine is sometimes used, it contributes little that other solvents do not and costs much more.

NOTE

Synthetic resins usually require a special solvent. It is important the correct one be used; otherwise, the paint may be spoiled entirely.

Types

Paints, by far, comprise the largest family of structural coatings you will be using to finish products, both interior and exterior. In the following section, we'll cover some of the most commonly encountered types.

OIL-BASED PAINTS.—Oil-based paints consist mainly of a drying oil (usually linseed) mixed with one or more pigments. The pigments and quantities of oil in oil paints are usually selected on the basis of cost and their ability to impart to the paint the desired properties, such as durability, economy, and color. An oil-based paint is characterized by easy application and slow drying. It normally chalks in such a manner as to permit recoating without costly surface preparation. Adding small amounts of varnish tends to decrease the time it takes an oil-based paint to dry and to increase the paint's resistance to water. Oil-based paints are not recommended for surfaces submerged in water.

ENAMEL.—Enamels are generally harder, tougher, and more resistant to abrasion and moisture penetration than oil-based paints. Enamels are obtainable in flat, semigloss, and gloss. The extent of pigmentation in the paint or enamel determines its gloss. Generally, gloss is reduced by adding lower cost pigments called extenders. Typical extenders are calcium carbonate (whiting), magnesium silicate (talc), aluminum silicate (clay), and silica. The level of gloss depends on the ratio of pigment to binder.

EPOXY.—Epoxy paints are a combined resin and a polyamide hardener that are mixed before use. When mixed, the two ingredients react to form the end product. Epoxy paints have a limited working, or pot, life, usually 1 working day. They are outstanding in hardness, adhesion, and flexibility—plus, they resist corrosion, abrasion, alkali, and solvents. The major uses of epoxy paints are as tile-like glaze coatings for concrete or masonry, and for structural steel in corrosive environments. Epoxy paints tend to chalk on exterior exposure; low-gloss levels and fading can be anticipated. Otherwise, their durability is excellent.

LATEX.—Latex paints contain a synthetic chemical, called latex, dispersed in water. The kinds of latex usually found in paints are styrene-butadiene (so-called synthetic rubber), polyvinyl acetate (PVA or vinyl), and acrylic. Latex paints differ from other paints in that the vehicle is an emulsion of binder and water. Being water-based, latex paints have the advantage of being easy to apply. They dry through evaporation of the water. Many latex paints have excellent durability. This makes them particularly useful for coating plaster and masonry surfaces. Careful surface preparation is required for their use.

RUBBER-BASED.—Rubber-based paints are solvent thinned and should not be confused with latex binders (often called rubber-based emulsions). Rubber-based paints are lacquer-type products and dry rapidly to form finishes highly resistant to water and mild chemicals. They are used for coating exterior masonry and areas that are wet, humid, or subject to frequent washing, such as laundry rooms, showers, washrooms, and kitchens.

PORTLAND CEMENT.—Portland cement mixed with several ingredients acts as a paint binder when it reacts with water. The paints are supplied as a powder to which the water is added before being used. Cement paints are used on rough surfaces, such as concrete, masonry, and stucco. They dry to form hard, flat, porous films that permit water vapor to pass through readily. When properly cured, cement paints of good quality are quite durable. When improperly cured, they chalk excessively on exposure and may present problems in repainting.

ALUMINUM.—Aluminum paints are available in two forms: ready mixed and ready to mix. Ready-mixed aluminum paints are supplied in one package and are ready for use after normal mixing. They are made with vehicles that will retain metallic brilliance after moderate periods of storage. They are more convenient to use and allow for less error in mixing than the ready-to-mix form.

Ready-to-mix aluminum paints are supplied in two packages: one containing clear varnish and the other, the required amount of aluminum paste (usually two-thirds aluminum flake and one-third solvent). You mix just before using by slowly adding the varnish to the aluminum paste and stirring. Ready-to-mix aluminum paints allow a wider choice of vehicles and present less of a problem with storage stability. A potential problem with aluminum paints is moisture in the closed container. When present, **moisture may react with the aluminum flake to form hydrogen gas that pressurizes the container**. Pressure can cause the container to bulge or even pop the cover off the container. Check the containers of ready-mixed paints for bulging. If they do, puncture the covers carefully before opening to relieve the pressure. Be sure to use dry containers when mixing aluminum paints.

VARNISHES

In contrast to paints, varnishes contain little or no pigment and do not obscure the surface to which applied. Usually a liquid, varnish dries to a hard, transparent coating when spread in a thin film over a surface, affording protection and decoration.

Of the common types of varnishes, the most important are the oils, including spar, flat, rubbing, and color types. These are extensively used to finish and refinish interior and exterior wood surfaces, such as floors, furniture, and cabinets. Spar varnish is intended for exterior use in normal or marine environments, although its durability is limited. To increase durability, exterior varnishes are especially formulated to resist weathering.

Varnishes produce a durable, elastic, and tough surface that normally dries to a high-gloss finish and does not easily mar. Often, a lower gloss may be obtained by rubbing the surface with a very fine steel wool. However, it is simpler to use a flat varnish with the gloss reduced by adding transparent-flatting pigments, such as certain synthetic silicas. These pigments are dispersed in the varnish to produce a clear finish that dries to a low gloss, but still does not obscure the surface underneath (that is, you can still see the grain of the wood).

SHELLAC

Shellac is purified lac formed into thin flakes and widely used as a binder in varnishes, paints, and stains. (Lac is a resinous substance secreted by certain insects.) The vehicle is wood alcohol. The natural color of shellac is orange, although it can be obtained in white. Shellac is used extensively as a finishing material and a sealant. Applied over knots in wood, it prevents bleeding.

LACQUERS

Lacquers may be clear or pigmented and can be lusterless, semigloss, or glossy. Lacquers dry or harden quickly, producing a firm oil- and water-resistant film. But many coats are required to achieve adequate dry-film thickness. It generally costs more to use lacquers than most paints.

STAINS

Stains are obtainable in four different kinds: oil, water, spirit, and chemical. Oil stains have an oil vehicle; mineral spirits can be added to increase penetration. Water stains are solutions of aniline dyes and water. Spirit stains contain alcohol. Chemical stains work by means of a chemical reaction when dissolved by water. The type of stain to use depends largely on the purpose, the location, and the type of wood being covered.

SURFACE PREPARATION

LEARNING OBJECTIVE: Upon completing this section, you should be able to describe the procedures used in preparing surfaces for painting.

The most essential part of any painting job is proper surface preparation and repair. Each type of surface requires specific cleaning procedures. Paint will not adhere well, provide the protection necessary, or have the desired appearance unless the surface is in proper condition for painting. Exterior surface preparation is especially important because hostile environments can accelerate deterioration.

METALS

As a Builder, you are most likely to paint three types of metals: ferrous, nonferrous, and galvanized. Improper protection of metals is likely to cause fatigue in the metal itself and may result in costly repairs or even replacement. Correct surface preparation, prior to painting, is essential.

Ferrous

Cleaning ferrous metals, such as iron and steel, involves the removal of oil, grease, previous coatings, and dirt. Keep in mind that once you prepare a metal surface for painting, it will start to rust immediately unless you use a primer or pretreatment to protect the surface.

Nonferrous

The nonferrous metals are brass, bronze, copper, tin, zinc, aluminum, nickel, and others not derived from iron ore. Nonferrous metals are generally cleaned with a solvent type of cleaner. After cleaning, you should apply a primer coat or a pretreatment.

Galvanized

Galvanized iron is one of the most difficult metals to prime properly. The galvanizing process forms a hard, dense surface that paint cannot penetrate. Too often, galvanized surfaces are not prepared properly, resulting in paint failure. Three steps must be taken to develop a sound paint system.

1. Wash the galvanized surface with a solvent to remove grease, waxes, or silicones. Manufacturers sometimes apply these to resist "white rust" that may form on galvanized sheets stored

under humid conditions. Mineral spirits or acid washes should definitely not be used at this stage.

2. Etch the surface with a mild phosphoric acid wash. Etching increases paint adhesion and helps overcome the stress forces generated by expansion and contraction of the galvanized coating. After acid washing the surface, rinse it with clean water and allow to dry. When using acid, remember the situation can represent actual or potential danger to yourself and other employees in the area. Continuous and automatic precautionary measures minimize safety problems and improve both efficiency and morale of the crew.

3. Apply a specially formulated primer. Two basic types of primer are in common use: zinc-bound and cementitious-resin. The zinc-bound type is used for normal exposure. Most types of finish can be used over this type of primer. Latex emulsion paints provide a satisfactory finish. Oil-based products should not be used over cementitious-resin primers. A minimum of two coats of finish is recommended over each type of primer.

CONCRETE AND MASONRY

In Navy construction, concrete and masonry are normally not painted unless painting is required for damp-proofing. Cleaning concrete and masonry involves the removal of dirt, mildew, and efflorescence (a white, powdery crystalline deposit that often forms on concrete and masonry surfaces).

Dirt and Fungus

Dirt and fungus are removed by washing with a solution of trisodium phosphate. The strength of the solution may vary from 2 to 8 ounces per gallon of water, depending upon the amount of dirt or mildew on the surface. Immediately after washing, rinse off all the trisodium phosphate with clear water. If using oil paint, allow the surface to dry thoroughly before painting.

Efflorescence

For efflorescence, first remove as much of the deposit as possible by dry brushing with a wire brush or a stiff fiber brush. Next, wet the surface thoroughly with clear water; then, scrub with a stiff brush dipped in a 5-percent solution (by weight) of muriatic acid. Allow the acid solution to remain on the surface about 3 minutes before scrubbing, but rinse thoroughly with clear water immediately after scrubbing. Work on small areas not larger than 4 square feet. Wear rubber gloves, a rubber apron, and goggles when mixing and applying the acid solution. In mixing the acid, always add acid to water. **Do not add water to acid; this can cause the mixture to explode**. For a very heavy deposit, the acid solution may be increased to 10 percent and allowed to remain on the surface for 5 minutes before it is scrubbed.

Repairing Defects

All defects in a concrete or masonry surface must be repaired before painting. To repair a large crack, cut the crack out to an inverted-V shape and plug it with grout (a mixture of two or three parts of mortar sand, one part of portland cement, and enough water to make it putty-like in consistency). After the grout sets, damp cure it by keeping it wet for 48 hours. If oil paint is to be used, allow at least 90 days for weathering before painting over a grout-filled crack.

PLASTER AND WALLBOARD

Whenever possible, allow new plaster to age at least 30 days before painting if oil-based paint is being applied. Latex paint can be applied after 48 hours, although a 30-day wait is generally recommended. Before painting, fill all holes and cracks with spackling compound or patching plaster. Cut out the material along the crack or hole in an inverted-V shape. To avoid excessive absorption of water from the patching material, wet the edges and bottom of the crack or hole before applying the material. Fill the opening to within 1/4 inch of the surface and allow the material to set partially before bringing the level up flush with the surface. After the material has thoroughly set (depending on the type of filler used), use fine sandpaper to smooth out the rough spots. Plaster and wallboard should have a sealer or a prime coat applied before painting. When working with old work, remove all loose or scaling paint, sand lightly, and wash off all dirt, oil, and stains. Allow the surface to dry thoroughly before applying the new finish coat.

WOOD

Before being painted, a wood surface should be closely inspected for loose boards, defective lumber, protruding nail heads, and other defects or irregularities. Loose boards should be nailed tight, defective lumber should be replaced, and all nail heads should be countersunk.

A dirty wood surface is cleaned for painting by sweeping, dusting, and washing with solvent or soap and water. In washing wood, take care to avoid excessive wetting, which tends to raise the grain. Wash a small area at a time, then rinse and dry it immediately.

Wood that is to receive a natural finish (meaning not concealed by an opaque coating) may require bleaching to a uniform or light color. To bleach, apply a solution of 1 pound of oxalic acid to 1 gallon of hot water. More than one application may be required. After the solution has dried, smooth the surface with fine sandpaper.

Rough wood surfaces must be sanded smooth for painting. Mechanical sanders are used for large areas, hand sanding for small areas. For hand sanding, you should wrap sandpaper around a rubber, wood, or metal sanding block. For a very rough surface, start with a coarse paper, about No. 2 or 2 1/2. Follow this with a No. 1/2, No. 1, or No. 1 1/2. You should finish with about a No. 2/0 grit. For fine work, such as furniture sanding, you should finish with a finer grit.

Sap or resin in wood can stain through a coat, or even several coats, of paint. Remove sap or resin by scraping or sanding. Knots in resinous wood should be treated with knot sealer.

Green lumber contains a considerable amount of water, most of which must be removed before use. This not only prevents shrinkage after installation, but prevents blistering, cracking, and loss of adhesion after applied paint. Be sure all lumber used has been properly dried and kept dry before painting.

CONDITIONERS

Conditioners are often applied on masonry to seal a chalky surface to improve adhesion of water-based

Table 8-1.—Treatments of Various Substrates

MECHANICAL	WOOD	METAL		CONCRETE AND MASONRY	PLASTER AND WALLBOARD
		Steel	Other		
Hand Cleaning	S	S	S	S	S
Power Tool Cleaning	S*	S	S	S	
Flame Cleaning		S			
Blast Cleaning			S		
Brush-Off		S		S	
All Other		S			
Chemical and Solvent			S		
Solvent Cleaning	S	S			
Alkali Cleaning		S		S	
Steam Cleaning		S		S	
Acid Cleaning		S		S	
Pickling		S			
Pretreatments					
Hot Phosphate		S			
Cold Phosphate		S	S		
Wash Primers		S			
Conditioners, Sealers, and Fillers					
Conditioners				S	
Sealers	S				
Fillers	S			S	

S—Satisfactory for use as indicated
*—Sanding only

topcoats. Sealers are used on wood to prevent resin running or bleeding. Fillers are used to produce a smooth finish on open-grained wood and rough masonry. Table 8-1 presents the satisfactory treatments of the various surfaces.

Since water-thinned latex paints do not adhere well to chalky masonry surfaces, an oil-based conditioner is applied to the chalky substrate before latex paint is applied. The entire surface should be vigorously wire brushed by hand or power tools, then dusted to remove all loose particles and chalk residue. The conditioner is then brushed on freely to assure effective penetration and allowed to dry. Conditioner is not intended for use as a finish coat.

SEALERS

Sealers are applied to bare wood like coats of paint. Freshly exuded resin, while still soft, may be scraped off with a putty knife and the area cleaned with alcohol. Remove hardened resin by scraping or sanding. Since sealer is not intended as a prime coat, it should be used only when necessary and applied only over the affected area. When previous paint becomes discolored over knots on pine lumber, the sealer should be applied over the old paint before the new paint is applied.

FILLERS

Fillers are used on porous wood, concrete, and masonry to provide a smoother finish coat.

Wood

Wood fillers are used on open-grained hardwoods. In general, hardwoods with pores larger than those found in birch should be filled. Table 8-2 lists the characteristics of various woods and which ones require fillers. The table also contains notes on finishing. Filling is done after staining. Stain should be allowed to dry for 24 hours before the filler is

Table 8-2.—Characteristics of Wood

NAME OF WOOD	TYPE OF GRAIN			NOTES ON FINISHING
	SOFT	HARD		
	Closed	Open	Closed	
Ash		X		Requires filler
Alder	X			Stains well
Aspen			X	Paints well
Basswood			X	Paints well
Beech			X	Paints poorly; varnishes well
Birch			X	Paints and varnishes well
Cedar	X			Paints and varnishes well
Cherry			X	Varnishes well
Chestnut		X		Requires filler; paints poorly
Cottonwood			X	Paints well
Cypress			X	Paints and varnishes well
Elm		X		Requires filler; paints poorly
Fir	X			Paints poorly
Gum			X	Varnishes well
Hemlock	X			Paints fairly well
Hickory		X		Requires filler
Mahogany		X		Requires filler
Maple			X	Varnishes well
Oak		X		Requires filler
Pine	X			Variable depending on grain
Teak		X		Requires filler
Walnut		X		Requires filler
Redwood	X			Paints well
Note: Any type of finish may be applied unless otherwise specified				

applied. If staining is not warranted, natural (uncolored) filler is applied directly to the bare wood. The filler may be colored with some of the stain to accentuate the grain pattern of the wood.

To apply, you first thin the filler with mineral spirits to a creamy consistency, then liberally brush it across the grain, followed by a light brushing along the grain. Allow it to stand 5 to 10 minutes until most of the thinner has evaporated. At this time, the finish will have lost its glossy appearance. Before it has a chance to set and harden, wipe the filler off across the grain using burlap or other coarse cloth, rubbing the filler into the pores of the wood while removing the excess. Finish by stroking along the grain with clean rags..All excess filler must be removed.

Knowing when to start wiping is important. Wipng too soon pulls the filler out of the pores. Allowing the filler to set too long makes it hard to wipe off. A simple test for dryness consists of rubbing a finger across the surface. If a ball is formed, it's time to wipe. If the filler slips under the pressure of the finger, it is still too wet for wiping. Allow the filler to dry for 24 hours before applying finish coats.

Masonry

Masonry fillers are applied by brush to bare and previously prepared (all loose, powdery, flaking material removed) rough concrete, concrete block, stucco, or other masonry surfaces. The purpose is to fill the open pores in the surface, producing a fairly smooth finish. If the voids on the surface are large, you should apply two coats of filler, rather than one heavy coat. This avoids mud cracking. Allow 1 to 2 hours drying time between coats. Allow the final coat to dry 24 hours before painting.

PAINT MIXING AND CONDITIONING

LEARNING OBJECTIVE: Upon completing this section, you should be able to describe the techniques used in mixing and applying paint.

Most paints used in the Navy are ready-mixed, meaning the ingredients are already combined in the proper proportions. When oil paint is left in storage for long periods of time, the pigments settle to the bottom. These must be remixed into the vehicle before the paint is used. The paint is then strained, if necessary. All paint should be placed in the paint shop at least 24 hours before use. This is to bring the paint to a temperature between 65°F and 85°F.

There are three main reasons to condition and mix paint. First, you need to redisperse, or reblend, settled pigment with the vehicle. Second, lumps, skins, or other impediments to proper application need to be

Table 8-3.—Mixing Procedures

COATING	EQUIPMENT	REMARKS
Enamel, semigloss, or flat paints (oil type)	Manual, propeller, or shaker	Mix until homogeneous
Water-based paints (latex type)	Manual or propeller	Use extreme care to avoid air entrapment
Clear finishes	Manual, propeller, or shaker	Generally require little or no mixing
Extremely viscous finishes; for example, coal tar paints	Drum-type mixer	Use extreme care to avoid air entrapment
Two-package metallic paints; for example, aluminum paints	Propeller	Add small amount of liquid to paste; mix well. Slowly add remainder of vehicle, while stirring, until coating is homogeneous. With metallic powder, first make into a paste with solvent, and then proceed as above
Two-Component Systems	Propeller, shaker, or drum-type mixer	Mix until homogeneous. Check label for special instructions

eliminated. And, third, the paint must be brought to its proper application temperature.

MIXING

Paints should be mixed, or blended, in the paint shop just before they are issued. Mixing procedures vary among different types of paints. Regardless of the procedure used, try not to overmix; this introduces too much air into the mixture. Table 8-3 outlines the types of equipment and remarks for various coatings. Mixing is done by either a manual or mechanical method. The latter is definitely preferred to ensure maximum uniformity. Manual mixing is less efficient than mechanical in terms of time, effort, and results. It should be done only when absolutely necessary and be limited to containers no larger than 1 gallon. Nevertheless, it is possible to mix 1-gallon and 5-gallon containers by hand. To do so, first pour half of the paint vehicle into a clean, empty container. Stir the paint pigment that has settled to the bottom of the container into the remaining paint vehicle. Continue to stir the paint as you return the other half slowly to its original container. Stir and pour the paint from can to can. This process of mixing is called boxing paint. The mixed paint must have a completely blended appearance with no evidence of varicolored swirls at the top. Neither should there be lumps of undispersed solids or foreign matter. Figure 8-1 illustrates the basic steps for boxing paint.

There are only three primary true-pigmented colors: red, blue, and yellow. Shades, tints, and hues are derived by mixing these colors in various proportions. Figure 8-2 shows a color triangle with one primary color at each of its points. The lettering

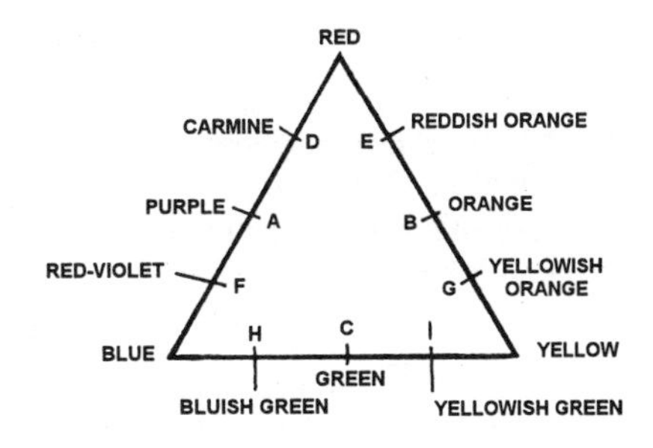

Figure 8-2.—A color triangle.

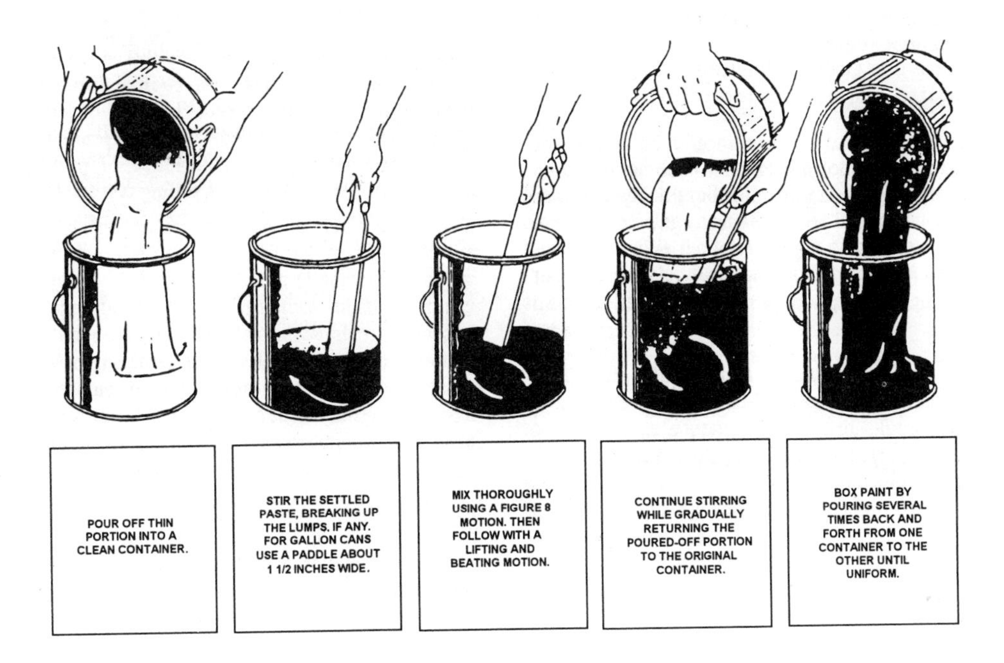

Figure 8-1.—Manual mixing and boxing.

in the triangle indicates the hues that result when colors are mixed.

A—Equal proportions of red and blue produce purple.

B—Equal proportions of red and yellow produce orange.

C—Equal proportions of blue and yellow produce green.

D—Three parts of red to one part of blue produce carmine.

E—Three parts of red to one part of yellow produce reddish orange.

F—Three parts of blue to one part of red produce red-violet.

G—Three parts of yellow to one part of red produce yellowish orange.

H—Three parts of blue to one part of yellow produce bluish green.

I—Three parts of yellow to one part of blue produce yellowish green.

Hues are known as chromatic colors, whereas black, white, and gray are achromatic (neutral colors). Gray can be produced by mixing black and white in different proportions.

Thinning

When received, paints should be ready for application by brush or roller. Thinner can be added for either method of application, but the supervisor or inspector must give prior approval. Thinning is often required for spray application. Unnecessary or excessive thinning causes an inadequate thickness of the applied coating and adversely affects coating longevity and protective qualities. When necessary, thinning is done by competent personnel using only the thinning agents named by the specifications or label instructions. Thinning is not done to make it easier to brush or roll cold paint materials. They should be preconditioned (warmed) to bring them up to 65°F to 85°F.

Straining

Normally, paint in freshly opened containers does not require straining. But in cases where lumps, color flecks, or foreign matter are evident, paints should be strained after mixing. When paint is to be sprayed, it must be strained to avoid clogging the spray gun.

Skins should be removed from the paint before mixing. If necessary, the next step is thinning. Finally, the paint is strained through a fine sieve or commercial paint strainer.

Tinting

Try not to tint paint. This will reduce waste and eliminate the problem of matching special colors at a later date. Tinting also affects the properties of the paint, often reducing performances to some extent. One exception is the tinting of an intermediate coat to differentiate between that coat and a topcoat; this helps assure you don't miss any areas. In this case, use only colorants of known compatibility. Try not to add more than 4 ounces of tint per gallon of paint. If more is added, the paint may not dry well or otherwise perform poorly.

When necessary, tinting should be done in the paint shop by experienced personnel. The paint must be at application viscosity before tinting. Colorants must be compatible, fresh, and fluid to mix readily. Mechanical agitation helps distribute the colorants uniformly throughout the paint.

APPLICATION

The common methods of applying paint are brushing, rolling, and spraying. The choice of method is based on several factors, such as speed of application, environment, type and amount of surface, type of coating to be applied, desired appearance of finish, and training and experience of painters. Brushing is the slowest method, rolling is much faster, and spraying is usually the fastest by far. Brushing is ideal for small surfaces and odd shapes or for cutting in corners and edges. Rolling and spraying are efficient on large, flat surfaces. Spraying can also be used for round or irregular shapes.

Local surroundings may prohibit the spraying of paint because of fire hazards or potential damage from over-spraying (accidentally getting paint on adjacent surfaces). When necessary, adjacent areas not to be coated must be covered when spraying is performed. This results in loss of time and, if extensive, may offset the speed advantage of spraying.

Brushing may leave brush marks after the paint is dry. Rolling leaves a stippled effect. Spraying yields the smoothest finish, if done properly. Lacquer products, such as vinyls, dry rapidly and should be sprayed. Applying them by brush or roller may be difficult, especially in warm weather or outdoors on breezy days. The painting method requiring the most training is spraying. Rolling requires the least training.

PAINT FAILURES

LEARNING OBJECTIVE: Upon completing this section, you should be able to identify the common types of coating failures and recognize the reasons for each.

A coating that prematurely reaches the end of its useful life is said to have failed. Even protective coatings properly selected and applied on well-prepared surfaces gradually deteriorate and eventually fail. The speed of deterioration under such conditions is less than when improper painting procedures are carried out. Inspectors and personnel responsible for maintenance painting must recognize signs of deterioration to establish an effective and efficient system of inspection and programmed painting. Repainting at the proper time avoids the problems resulting from painting either too soon or too late. Applying coatings ahead of schedule is costly and eventually results in a heavy buildup that tends to quicken deterioration of the coating. Applying a coating after it is scheduled results in costly surface preparation and may be responsible for damage to the structure, which may then require expensive repairs.

In the following sections, we'll look at some of the more common types of paint failures, the reasons for such failures, methods of prevention, and cures.

SURFACE PREPARATION FAULTS

Paint failures can result from many causes. Here, we'll look at some of the most common caused by faults in surface preparation.

Alligatoring

Figure 8-3.—Alligatoring.

Alligatoring (fig. 8-3) refers to a coating pattern that looks like the hide of an alligator. It is caused by uneven expansion and contraction of the undercoat. Alligatoring can have several causes: applying an enamel over an oil primer; painting over bituminous paint, asphalt, pitch, or shellac; and painting over grease or wax.

Peeling

Figure 8-4.—Peeling.

Peeling (fig. 8-4) results from inadequate bonding of the topcoat with the undercoat or the underlying surface. It is nearly always caused by inadequate surface preparation. A topcoat peels when applied to a wet, dirty, oily or waxy, or glossy surface. All glossy surfaces must be sanded before painting. Also, the use of incompatible paints can cause the loss of adhesion. The stresses in the hardening film can then cause the two coatings to separate and the topcoat to flake and peel.

Blistering

Blistering is caused by the development of gas or liquid pressure under the paint. Examples are shown in figure 8-5. The root cause of most blistering, other than that caused by excessive heat, is inadequate ventilation plus some structural defect allowing moisture to accumulate under the paint. A prime source of this problem, therefore, is the use of essentially porous major construction materials that allow moisture to pass through. Insufficient drying time between coats is another prime reason for blistering. All blisters should be scraped off, the paint edges feathered with sandpaper, and the bare places primed before the blistered area is repainted.

Prolonged Tackiness

A coat of paint is dry when it ceases to be "tacky" to the touch. Prolonged tackiness indicates excessively slow drying. This may be caused by insufficient drier in the paint, a low-quality vehicle in the paint, applying the paint too thickly, painting over an undercoat that is not thoroughly dry, painting over a waxy, oily, or greasy surface, or painting in damp weather.

Inadequate Gloss

Sometimes a glossy paint fails to attain the normal amount of gloss. This may be caused by inadequate surface preparation, application over an undercoat that is not thoroughly dry, or application in cold or damp weather.

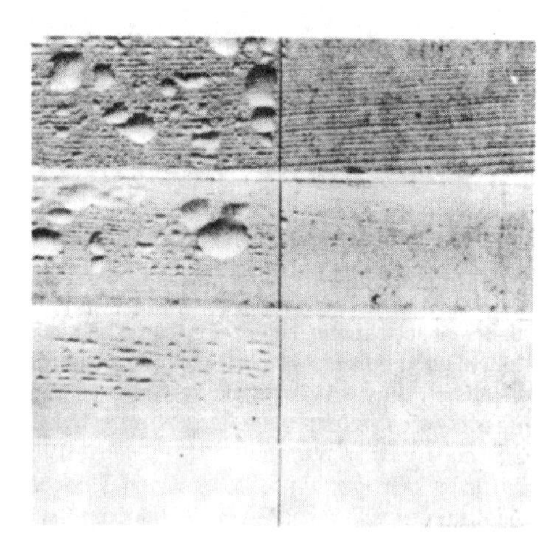

Figure 8-5.—Blistering.

IMPROPER APPLICATION

One particular area you, as a Builder, have direct control over is application. It takes a lot of practice, but you should be able to eliminate the two most common types of application defects: crawling and wrinkling.

Crawling

Crawling (fig. 8-6) is the failure of a new coat of paint to wet and form a continuous film over the preceding coat. This often happens when latex paint is applied over high-gloss enamel or when paints are applied on concrete or masonry treated with a silicone water repellent.

Wrinkling

When coatings are applied too thickly, especially in cold weather, the surface of the coat dries to a skin over a layer of undried paint underneath. This usually causes wrinkling (fig. 8-7). Wrinkling can be avoided in brush painting or roller painting by brushing or rolling each coat of paint as thinly as possible. In spray painting, you can avoid wrinkling by keeping the gun in constant motion over the surface whenever the trigger is down.

PAINT DEFECTS

Not all painting defects are caused by the individual doing the job. It sometimes happens that the coating itself is at fault. Chalking, checking, and cracking are the most common types of product defects you will notice in your work as a Builder.

Figure 8-6.—Crawling.

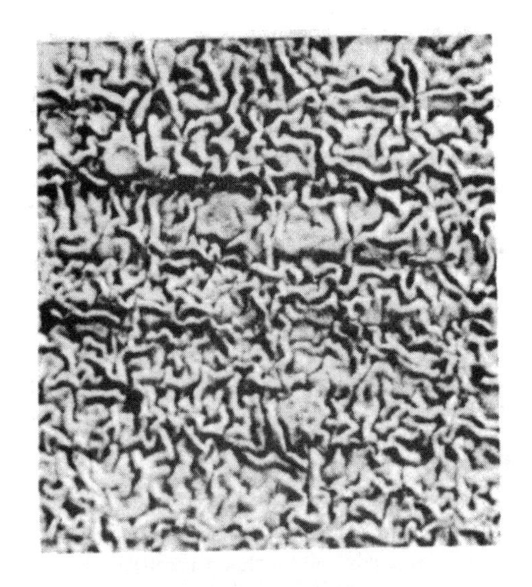

Figure 8-7.—Wrinkling.

Chalking

Chalking (fig. 8-8) is the result of paint weathering at the surface of the coating. The vehicle is broken down by sunlight and other destructive forces, leaving behind loose, powdery pigment that can easily be rubbed off with the finger. Chalking takes place rapidly with soft paints, such as those based on linseed oil. Chalking is most rapid in areas exposed to sunshine. In the Northern Hemisphere, for example, chalking is most rapid on the south side of a building. On the other hand, little chalking takes place in areas protected from sunshine and rain, such as under eaves or overhangs. Controlled chalking can be an asset, especially in white paints where it acts as a self-cleaning process and helps to keep the surface clean and white. The gradual wearing away reduces the thickness of the coating, thus allowing continuous repainting without making the coating too thick for satisfactory service.

Do not use a chalking or self-cleaning paint above natural brick or other porous masonry surfaces. The chalking will wash down and stain or discolor these areas.

Chalked paints are generally easier to repaint since the underlying paint is in good condition and requires little surface preparation. But, this is not the case with water-thinned paints; they adhere poorly to chalky surfaces.

Checking and Cracking

Checking and cracking are breaks in a coating formed as the paint becomes hard and brittle. Temperature changes cause the substrate and overlying paint to expand and contract. As the paint becomes hard, it gradually loses its ability to expand without breaking. Checking (fig. 8-9) consists of tiny breaks in only the upper coat or coats of the paint film

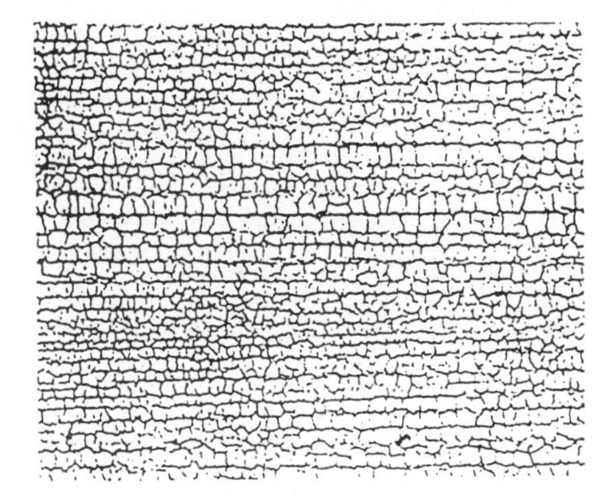

Figure 8-9.—Severe checking.

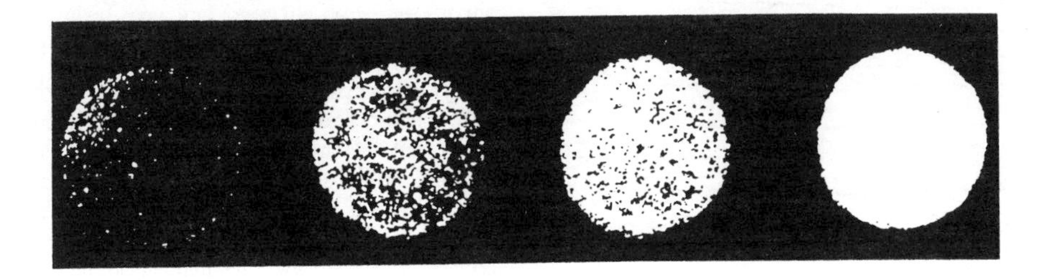

Figure 8-8.—Degrees of chalk.

without penetrating to the substrate. The pattern is usually similar to that of a crow's foot. Cracking is larger with longer breaks extending through to the substrate (fig. 8-10). Both result from stresses exceeding the strength of the coating. But, whereas checking arises from stress within the paint film, cracking is caused by stresses between the film and the substrate.

Cracking generally takes place to a greater extent on wood, due to its grain, than on other substrates. The stress in the coating is greatest across the grain, causing cracks to form parallel to the grain of the wood. Checking and cracking are aggravated by excessively thick coatings that have reduced elasticity. Temperature variations, humidity, and rainfall are also concerns for checking or cracking.

WOOD PRESERVATIVES

LEARNING OBJECTIVE: Upon completing this section, you should be able to describe how to treat wood for protection against dry rot, termites, and decay.

There are three destructive forces against which most wood protective measures are directed: biological deterioration (wood is attacked by a number of organisms), fire, and physical damage. In this section, we'll deal with protecting wood products against biological deterioration.

Damage to wood buildings and other structures by termites, wood bores, and fungi is a needless waste. The ability of wood to resist such damage can be greatly increased by proper treatment and continued maintenance. Wood defects are also caused by improper care after preservation treatment. All surfaces of treated wood that are cut or drilled to expose the untreated interior must be treated with a wood preservative.

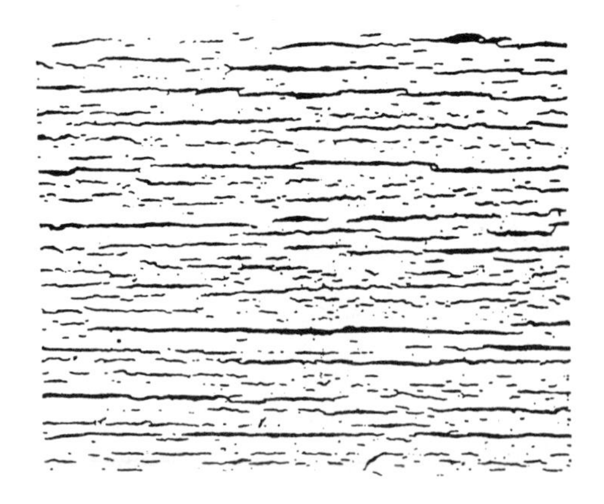

Figure 8-10.—Severe cracking.

APPLICATION METHODS

There are two basic methods for treating wood: pressure and nonpressure. Pressure treatment is superior to nonpressure, but costly and time consuming. Building specifications dictate which method to use.

Pressure

The capacity of any wood to resist dry rot, termites, and decay can be greatly increased by impregnating the wood with a general-purpose wood preservative or fungicide. It's important to remember that good **pressure** treatment adds to the service life of wood in contact with damp ground. It does not, however, **guarantee** the wood will remain serviceable throughout the life of the building it supports.

Woods of different timber species do not treat with equal ease. Different woods have different capacities for absorbing preservatives or other liquids. In any given wood, sapwood is more absorbent than heartwood. Hardwoods are, in general, less absorbent than softwoods. Naturally, the extent to which a preservative protects increases directly with the depth it penetrates below the surface of the wood. As we just mentioned, the best penetration is obtained by a pressure method. Table 8-4 shows the ease of preservative penetration into various woods. In the table, use E for easy, M for moderate, and D for difficult.

Nonpressure

Nonpressure methods of applying preservatives to a surface include dipping, brushing, and spraying. Figure 8-11 shows how you can improvise long tanks for the dipping method. Absorption is rapid at first, then much slower. A rule of thumb holds that in 3 minutes wood absorbs half the total amount of preservative it will absorb in 2 hours. However, the extent of the penetration depends upon the type of wood, its moisture content, and the length of time it remains immersed.

Surface application by brush or spray is the least satisfactory method of treating wood from the

Table 8-4.—Preservative Penetration

SPECIES	RELATIVE EASE OF GETTING PENETRATION INTO	
	Sapwood	Heartwood
Pines (most species)	E	M to D
Ponderosa pine	E	M
White fir	E	M
Most other true firs	E to M	D
Eastern hemlock	M	D
Western hemlock	E	E to M
Redwood	M to D	M to D
Douglas fir (Coast)	E to M	M to D
Douglas fir (Rocky Mountain)	D	D
Western larch	E	D
Sitka spruce	M	M
Most other spruces	D	D
Western red cedar	D	D
White oak	E	D
Selected red oaks	E	E

Notes: E—Easy; M—Moderate; D—Difficult

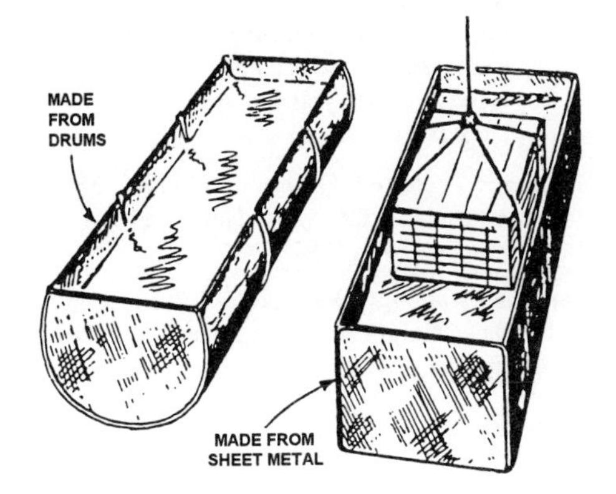

Figure 8-11.—Improvised tanks for dip treating lumber.

standpoint of maximum penetration. However, it is more or less unavoidable in the case of already installed wood, as well as treated wood that has been cut or drilled to expose the untreated interior.

FIELD-MIXED PRESERVATIVES

Pentachlorophenol and creosote coal tar are likely to be the only field-mixed preservatives used by the Builder. The type of treatment or preservative depends on the severity of exposure and the desired life of the end product.

Preservatives can be harmful to personnel if improperly handled. When applying preservatives, you should take the following precautions:

- Avoid undue skin contact;
- Avoid touching the face or rubbing the eyes when handling pretreated material;
- Avoid inhalation of toxic (poisonous) material;
- Work only in a properly ventilated space and use approved respirators; and
- Wash with soap and water after contact.

PAINTING SAFETY

LEARNING OBJECTIVE: Upon completing this section, you should be able to state the principal fire and health hazards associated with painting operations.

Every painting assignment exposes Builders to conditions and situations representing actual or potential danger. Toxic and flammable materials, pressurized equipment, ladders, scaffolding, and rigging always make painting a hazardous job. Hazards may also be inherent in the very nature of the environment or result from ignorance or carelessness by the painter.

The main causes of painting accidents are unsafe working conditions or equipment, and careless personnel. The proper setting up and dismantling of equipment, the required safety checks, and the proper care of equipment may require more time than is spent using it. Nevertheless, safety measures must be taken.

FIRE HAZARDS

Certain general rules regarding fire and explosion hazards apply to all situations. All paint materials should have complete label instructions stipulating the potential fire hazards and precautions to be taken. Painters must be advised and reminded of the fire hazards that exist under the particular conditions of each job. They need to be aware of the dangers involved and the need to work safely. Proper fire-fighting equipment must always be readily available in the paint shop, spray room, and other work areas where potential fire hazards exist. Electric wiring and equipment installed or used in the paint shop, including the storage room and spray room, must conform to the applicable requirements of the National Electrical Code (NEC) for hazardous areas.

HEALTH HAZARDS

Many poisons, classified as toxic and skin-irritating, are used in the manufacture of paint. Although your body can withstand small quantities of poisons for short periods, overexposure can have harmful effects. Continued exposure to even small amounts may cause the body to become sensitized; subsequent contact, even in small amounts, may cause an aggravated reaction. The poisons in paint are definite threats to normally healthy individuals and serious dangers to persons having chronic illnesses or disorders. Nevertheless, health hazards can be avoided by a common-sense approach of avoiding unnecessary contact with toxic or skin-irritating materials.

As with all tasks the Builder undertakes, safety must be a primary concern from the earliest planning stages to the final cleanup. Shortcuts, from personnel protection to equipment-related safety devices, should not be permitted. Follow the project safety plan, and consult all applicable safety manuals when involved with any paint operation. Remember, work safe, stay safe.

RECOMMENDED READING LIST

NOTE

Although the following references were current when this TRAMAN was published, their continued currency cannot be assured. You therefore need to ensure that you are studying the latest revisions.

Paints and Protective Coatings, NAVFAC MO-110, Departments of the Army, Navy, and Air Force, Washington, D.C., 1991.

Wood Preservation, NAVFAC MO-312, Department of the Navy, Naval Facilities Engineering Command, Washington, D.C., 1968.

CHAPTER 9

ADVANCED BASE FIELD STRUCTURES AND EMBARKATION

The primary responsibility of the Seabees is the construction of advanced bases during the early phases of crises and other emergency situations. As Builders, it is our job to move swiftly to hostility areas and build temporary facilities and structures to support U.S. military operations. We are expected to react expediently.

The most widely used structure for expediency and as a temporary facility is the preengineered building. This chapter covers the process involved with the erection of such buildings, as well as wood-frame tents, latrines, and the process of embarkation.

PREENGINEERED BUILDINGS

LEARNING OBJECTIVE: Upon completing this section, you should be able to explain the principles and procedures involved in the preparation and erection of preengineered metal buildings.

The preengineered building (PEB) discussed here is a commercially designed structure, fabricated by civilian industry to conform to armed forces specifications. A preengineered structure offers an advantage in that it is designed for erection in the shortest possible time. Each PEB is shipped as a complete building kit. All necessary materials and instructions for erection are included. Preengineered structures are available from various manufacturers.

The typical PEB is a 40- by 100-foot structure. The 20- by 48-foot PEB is a smaller version of the 40- by 100-foot PEB using the same erection principles. Layout and erection of either size PEB is normally assigned to Builders.

PREPLANNING

A preplan of the erection procedures should be made based on a study of the working drawings or manufacturers' instructions. Preplanning should include the establishment of the most logical and expeditious construction sequence. Consideration should be given to the manpower, equipment, rigging, and tools required. Everything necessary for erection should then be procured. The advantages of constructing and using jigs and templates for assembling parts of similar trusses, frames, and so on, should also be evaluated.

Although Builders must be familiar with the layout and erection procedures for both the 40- by 100-foot and the 20- by 48-foot PEBs, we will use the 20- by 48-foot rigid-frame, straight-walled PEB as the model for our discussion. This building is prefabricated and shipped in compact crates ready for erection. Each structure is shipped as a complete kit, including all materials and an instruction manual. It is extremely important to follow the manual; you can easily install a part incorrectly.

The 20- by 48-foot rigid-frame building is designed for erection with basic hand tools and a minimum number of people. The instruction manual may suggest the PEB can be erected by seven persons. For military construction, though, two teams or work crews supervised by an E-6 are recommended. The building is designed for erection on a floor system of piers, concrete blocks, or a concrete slab.

When completed, a single rigid-frame building is easy to expand for additional space. Buildings can be erected end to end, as in figure 9-1, or side by side "in multiple." As this type of building uses only bolted connections, it can be disassembled easily, moved to a new location, and re-erected without waste or damage.

Component Parts

The component parts of a prefabricated structure are shipped knocked down (KD). A manufacturer's instruction manual accompanies each shipment. The manual contains working drawings and detailed instructions on how the parts should be assembled. Directions vary with different types of structures, but there are certain basic erection procedures that should be followed in all cases.

Working Drawings

The working drawings show which items are **not** prefabricated or included in the shipment. These must be constructed in the field. Make plans in advance for the procurement of necessary materials for these items. Foundations, for example, are often designated "to be constructed in the field."

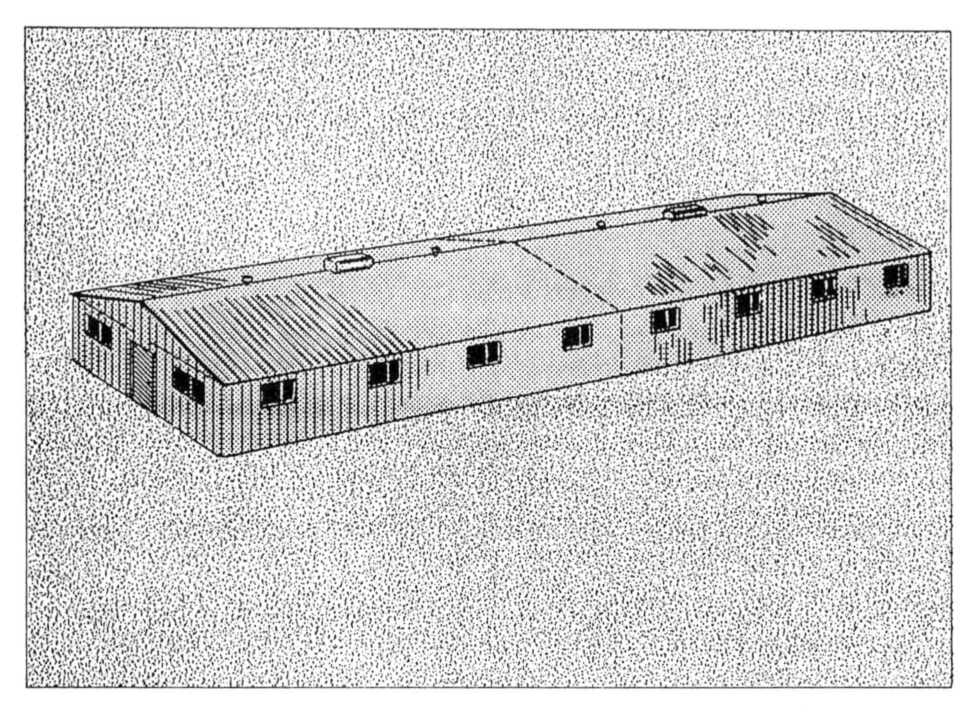

Figure 9-1.—Two 20- by 48-foot rigid-frame, preengineered buildings erected end to end.

FOUNDATIONS

In addition to the usual reasons for stressing the importance of a square and level foundation, there is another reason peculiar to the erection of a prefabricated structure. Prefabricated parts are designed to fit together without forcing. If the foundation is even slightly out of square or not perfectly level, many of the parts will not fit together as designed. Continuously check the alignment of the anchor bolts prior to, during, and after concrete is poured.

PREERECTION WORK

A lot of preliminary work is necessary before the erection of a PEB can begin. After the building site is selected, prepare to pour the foundation and slab.

Before concrete for the foundation piers can be poured, templates for the anchor bolts are placed on the forms, and anchor bolts are inserted in the holes. The threads of the bolts are greased and nuts are placed on them to protect the threads from the concrete. After a last-minute check to ensure all forms are level and the anchor bolts are properly aligned, concrete is placed in the forms and carefully worked around the bolts so they remain vertical and true.

While the foundation is being prepared, other crewmembers are assigned various kinds of preliminary work. This work includes uncrating material and inventory, bolting up rigid-frame assemblies, assembling door leaves, and glazing windows. When all preliminary work is properly completed, assembly and erection of the entire building are quicker and you have fewer problems.

All materials, except paneling, should be uncrated and laid out in an orderly manner so parts can be easily found. Don't uncrate paneling until it is ready to install. When the crates are opened, don't damage the lumber; you can use it for scaffolding, props, and sawhorses.

After the building foundation has been prepared, and where practical, building materials should be placed on the building site near the place where they will be used. Girts, purlins, eave struts, and brace rods should be equally divided along each side of the foundation. Panels and miscellaneous parts that will not be used immediately should be placed on boards on each side of the foundation and covered with tarpaulins or similar covering until needed. Parts making up the rigid-frame assemblies should be laid out ready for assembly and in position for raising.

Always exercise care when unloading materials. **Remember**: Damaged parts can delay getting the job done in the shortest possible time. To avoid damage, lower the materials to the ground—don't drop them.

ERECTION

Figure 9-2, view A, will help you identify the various structural members of a PEB and their locations. View B shows the placement of liner panels. Each part serves a specific purpose and must be installed in the proper location to ensure a sound

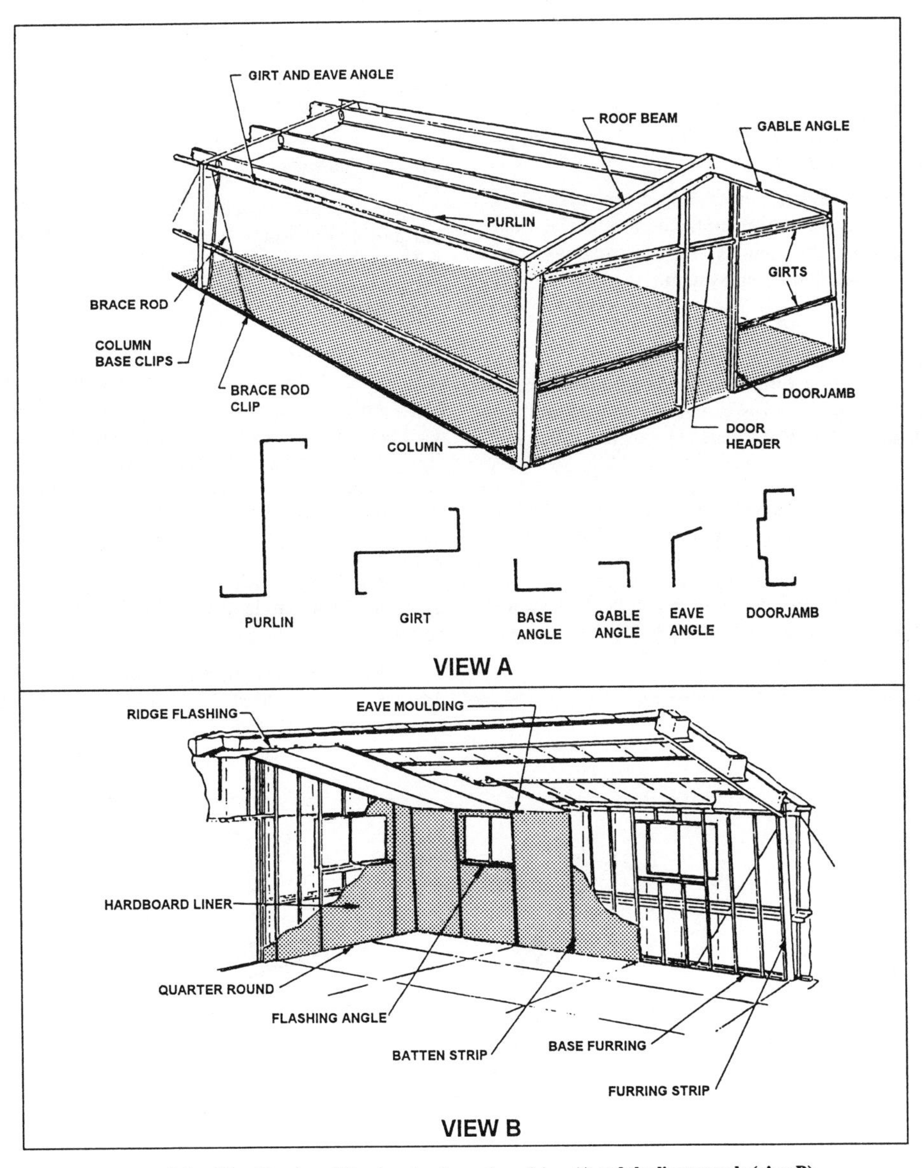

Figure 9-2.—Identification of the structural members (view A) and the liner panels (view B).

structure. Never omit any part called for on the detailed erection drawings. Each of the members, parts, and accessories is labeled by stencil; it is not necessary to guess which part goes where. Refer to the erection plans and find the particular members you need as you work.

High-strength steel bolts are used at rigid-frame connections: roof-beam splice and roof beam to column. These high-strength bolts are identified by a **Y** stamped into the head, as shown in figure 9-3. All high-strength steel bolts and nuts should be tightened to give at least the required minimum bolt tension values. The bolts may be tightened with a torque wrench, an impact wrench, or an open-end wrench.

When a PEB is not erected on a concrete slab, a floor system by the same manufacturer should be used. Read and follow the manufacturer's instructions when you are installing the floor system.

Layout

After the floor system or concrete slab has been prepared, the next step is to uncrate and lay out the structural parts. Lay the parts out in the following manner:

- Parts making up the frame assembly should be laid out, ready for assembly and in position for raising.
- Girts, purlins, and base angles should be divided (equally) along each side of the foundation.
- End-wall parts should be divided equally between the two ends.
- All miscellaneous parts should be centrally located.
- Panels and other parts not used immediately should be placed on boards and protected from the environment and jobsite debris.
- Lay out the column and roof beams for assembly, using crate lumber to block up the frames. Erect the center frame first. Use the minimum number of high-strength bolts to bring the frame members together. Install the remaining bolts to get the proper tightness.

Exterior Assembly

Use galvanized machine bolts to assemble the girt and purlin clips to the frame. Keep in mind that the end frames have girt and purlin clips on one side only. The center frame has girt and purlin clips on each side of the frame.

The eave girts should be attached to the eave angles with 5/16-inch left-hand nuts and shoulder bolts. An example is shown in figure 9-4. You will need two eave angles for each eave girt. In fastening these together, remember the short section of the eave angle is always fastened to the left side of the eave girt. The long section of the eave angle is fastened to the right side of the eave girt.

Use 3/8- by 1-inch galvanized machine bolts to attach the gable angle and doorjamb top clips to the bottom flange of the end frame roof beams.

A-FRAME.—To erect the frame, place A-frame props in position—one 8-foot frame at each side of the building and a 10-foot frame in the center of the

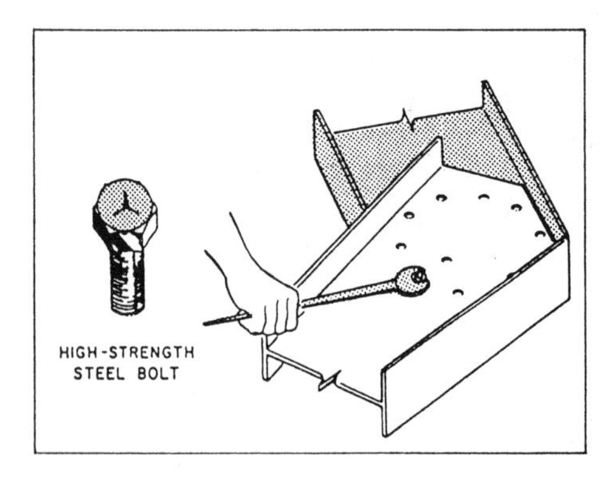

Figure 9-3.—High-strength steel bolt.

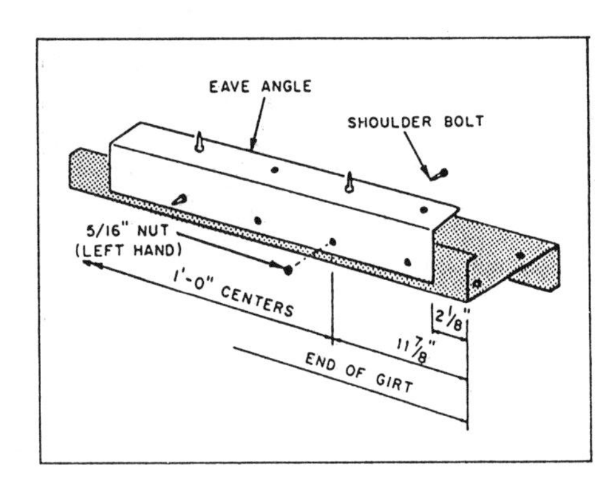

Figure 9-4.—Attaching eave angle to girt.

building (see fig. 9-5). Prop the frame on two sawhorses and attach tag lines to assist in raising the ridge. Raise the frame and brace it up with the A-frames. The end frames are erected in a similar manner, except that they are held in position by installing purlins and girts.

BRACE RODS.—After all sidewall girts, eave girts, and base angles have been installed, install the brace rods. Look at figure 9-6. First, attach brace rod clips to the floor. Then, insert the end of the brace rod down through the hole in the sidewall girt. Connect the top end through the eave girt and the eave girt clip. Finally, connect the bottom end through the clip on the floor.

As soon as the four brace rods are in position, use them to plumb the building. To plumb the rigid frame, tighten or loosen the rod nuts at the brace rod clips to adjust the column to plumb condition. Don't forget: When you tighten one side, the other side must be loosened.

To make sure you are installing the end-wall members correctly, snap a chalk line across the building, using one edge of the columns for positioning the line. Mark the center of the building on this line. Then, drop a plumb bob from the center of the joint of the roof beams at the ridge, with the line over the same side of the roof beam as the chalk line. Adjust the frame so

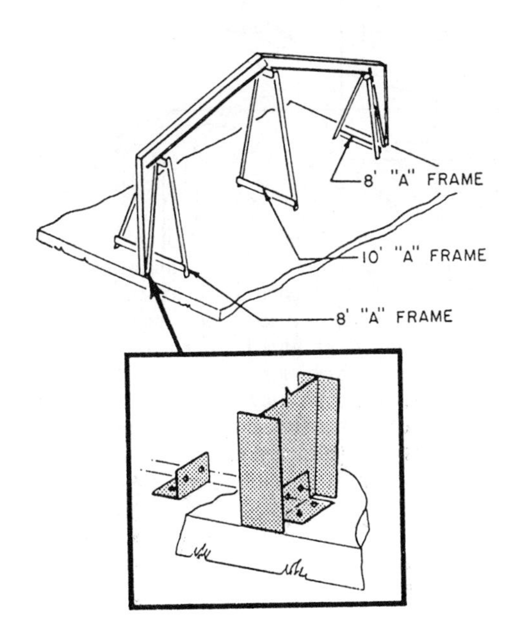

Figure 9-5.—Frame erection with A-frame props.

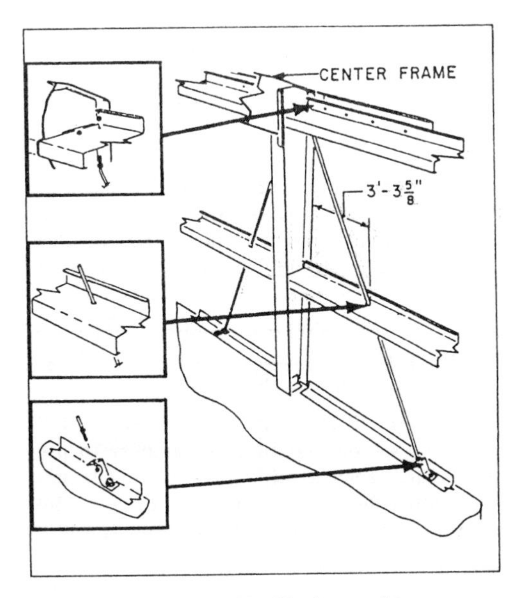

Figure 9-6.—Installing brace rods.

the plumb bob is directly over the center mark. Brace the roof beam in this position until the roof panels are in place.

EXTERIOR WALLS.—Uncrate exterior panels and distribute them near where they will be used. First, separate and place panels for each end wall. Place full-length wall panels for each corner. Centrally locate lower and upper sidewall panels and above and below window panels, along each side of the building. Place roof and ridge panels in stacks of eight each on the floor. Make sure you fashion all joints properly. Next, tighten all fasteners using metal-backed neoprene washers with all roof fasteners and with all shoulder bolts in the sidewalls. Then, properly apply black mastic or sealant to all roof panel side laps and end laps.

Start paneling the end wall at one corner and work across to the other corner. Install the corner panel, locating the bottom of the panel over the first two shoulder bolts in the base angle. Use a level to plumb this panel with the other shoulder bolts located at the center of the corrugations. Locate the "below window panel" over the base angle shoulder bolts, and impale over the shoulder bolts. Remove the panel and reinstall it so that it underlaps the first panel by pulling out on the corrugated edge of the first panel.

For installing paneling and windows, follow the same general instructions as those given for the end wall. However, be sure that the girts are in a straight line before impaling panels onto shoulder bolts. It is important to block the girts in a straight line with crating lumber cut to the correct length. The drawings should be checked for proper location of shoulder bolts. The first shoulder bolt should be 12 inches from the center of the column, then 12 inches on center (OC).

Recheck the plumb of the center frame. Adjust the brace rods to plumb if necessary. Check the drawings for the location of base angles.

The upper wall panels must overlap the lower wall panels for weathertightness. Remember to use metal-backed neoprene washers and No. 10 hex nuts on shoulder bolts. Use machine screws (1/4 by 3/4 inch) for panel-to-panel connections at side laps.

ROOF ASSEMBLIES.—Since the roof panels are factory-punched for panel-to-purlin connection, the purlins must be accurately aligned. Spacer boards constructed from crating lumber can be used to align purlins, as shown in figure 9-7. Move the spacer boards ahead to the next bay as the paneling progresses. Before you actually start paneling the roof, place the spacer board over the shoulder bolts and insert nails in the 5/16-inch holes in the ridge purlins.

Start roof paneling at one end of the building. Place the panels so the holes in the corrugation line up with the shoulder bolts in the roof beam, eave angles, and ridge purlins. Install one eave panel to each side of the building. The eave panels should be installed one row ahead of the ridge panels (see fig. 9-8). Before proceeding with the work, make sure you are applying enough black mastic or sealant. Roof paneling should continue in this order to ensure a weathertight joint at the corner laps. However, you should check the drawings for the location and installation of the smoke stacks and ridge ventilation.

DOORS AND WINDOWS.—Doorjambs can be hung anytime after the end-wall structural parts are completed. But, they must be hung before installing the interior lining. A helpful hint is to hang the doors before installing the exterior end-wall paneling. This makes adjustments on the door frame easier.

Hinges are factory-welded to the doorjamb, and the entrance doors are supposed to swing to the inside of the building. Remove the hinge leaf from the doorjamb and attach it to the door with 1-inch No. 10 flathead wood screws. Hang the door and make adjustments to get the proper clearances at the top and sides of the door. Install the lockset in the door and attach the faceplate to the door with 3/4-inch No. 8 flathead wood screws. Attach the strike plate to the doorjamb with 1/2-inch No. 8 flathead machine screws.

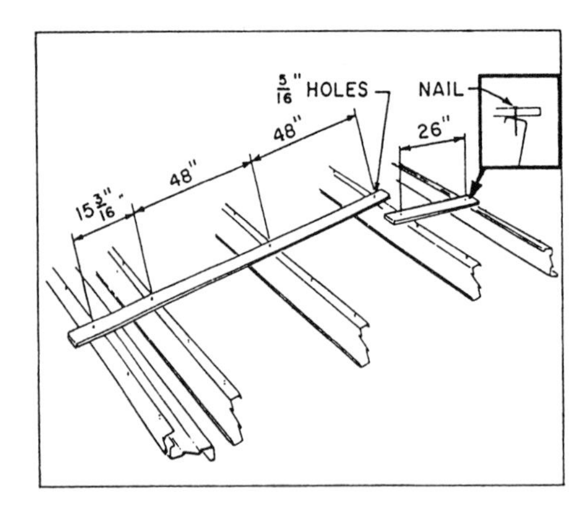

Figure 9-7.—Aligning purlins with spacer board.

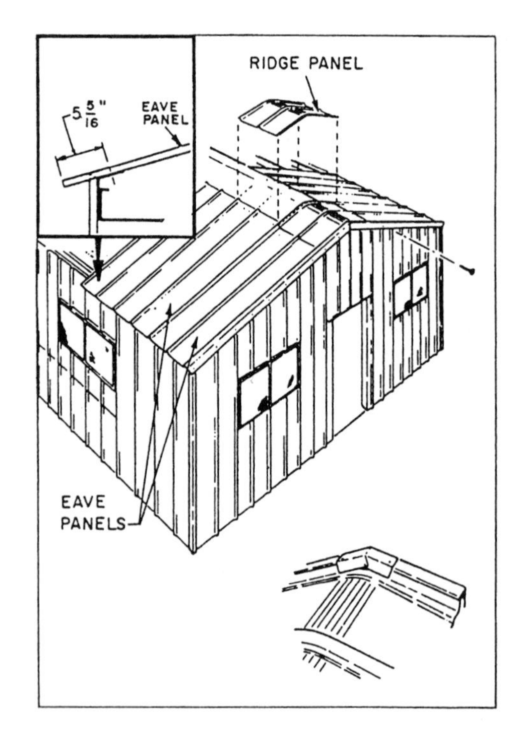

Figure 9-8.—Installing ridge and eave panels.

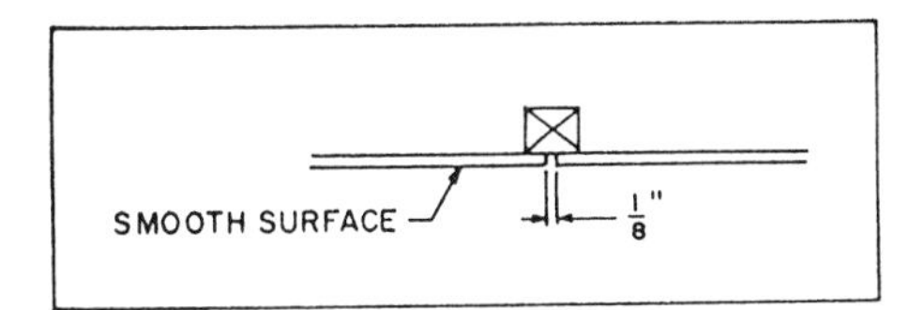

Figure 9-9.—Installing hardboard.

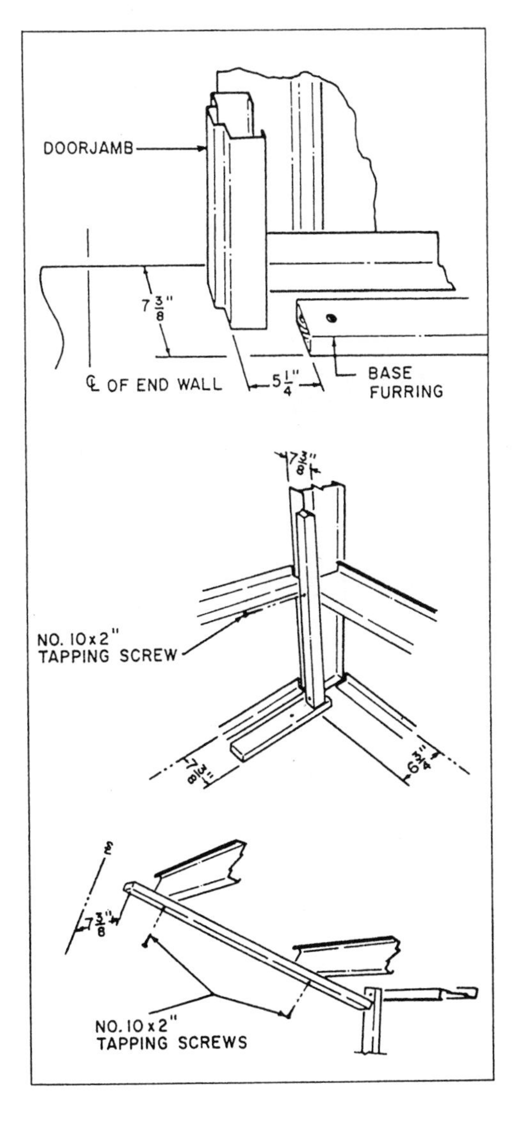

Figure 9-10.—Installing furring for the end-wall liners.

Hinges are also factory-welded to screen doors, which swing to the outside of the building. The method used in hanging screen doors is similar to hanging entrance doors. A spring, however, is needed to hold the screen door closed.

Interior Assembly

After the exterior members have been erected, work can begin on installing the interior assemblies. These include the liner panels, furring, hardboard flashing, and the trim.

LINER PANELS.—Installation of the liner panels consists of installing furring strips, hardboard liner panels, and trim and battens. Various liner panel parts were shown earlier in view B of figure 9-2.

To install end walls properly, precut the liner panels according to the cutting diagrams. The hardboard must be installed with the smooth surface exposed and with a 1/8-inch gap between panels to allow for expansion (fig. 9-9). A scrap piece of hardboard or batten can be used as a shim or spacer to maintain the proper gap.

BASE FURRING.—Nail the base furring to the floor 3 inches from each end and 2 feet 8 inches OC, with the inside edge 7 3/8 inches from the building structural line. You can get a better of this by referring to figure 9-10. When base furring is to be used on a wood floor, use 8d box nails. Use 1 1/4-inch No. 9 concrete nails for a concrete floor. Drill the 2 by 2s and girts with a 5/32-inch bit so furring can be attached to the sidewall and eave girt with 2-inch panhead No. 10 sheet-metal screws. Attach the hardboard to the furring strips with 1 1/4-inch aluminum shingle nails on 4-inch centers at the sides and ends (see fig. 9-11). Use 8-inch centers at the intermediate furring.

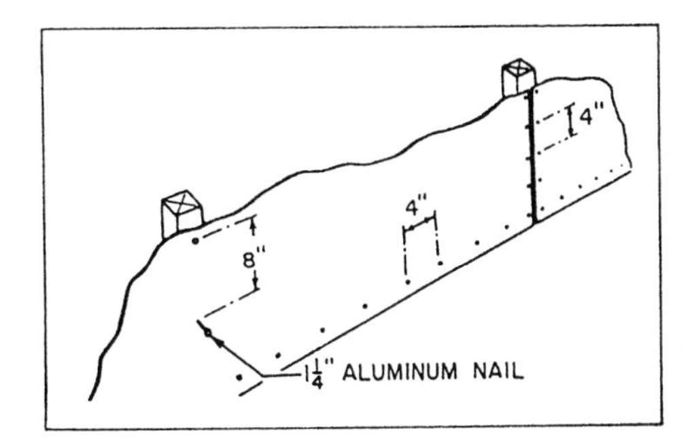

Figure 9-11.—Nailing pattern for attaching hardboard to furring.

VERTICAL FURRING.—The vertical furring (fig. 9-12) should be installed immediately after the base, corner, and gable furrings are in place. The center line of the furring on each side of the window should be in line with the center line of the end-wall panel corrugations (shown in the inserts). After the end-wall hardboard has been installed, attach side and top flashing to the door. Attach flashing to the furring with 4d aluminum nails and to the door frames with 1/2-inch No. 10 sheet-metal screws, as shown in figure 9-13.

SIDEWALL AND CEILING FURRING.—After installing the end-wall liner, install furring for the sidewall and ceiling. Cut the base so the end just clears the inside flange of the center-frame column. Nail the furring in the same manner as the end walls.

HARDBOARD FLASHING.—With the furring in place, you can now install the hardboard liner. Install the top and bottom hardboard flashing, as shown in figure 9-14. Insert the outside edge into the retaining grooves in the window. Nail metal flashing angle and hardboard to the horizontal furring with 4d aluminum nails 1 foot 8 inches OC. Install side hard- board flashing and metal flashing angles using the same procedures discussed above.

The installed ceiling furring should intersect sidewall furring. When all the ceiling furring has been installed, the hardboard liner can be installed. Remember the 1/8-inch gap between panels. The smoke stack assembly should be attached to the block- ing and furring with 4d aluminum nails. Hand trim the hardboard flashing for the ends of the ventilator opening and attach the metal ventilator flashing (see fig. 9-15).

TRIM.—Install the eave molding with the beveled edge against the ceiling panels. Attach each sidewall furring strip with 4d aluminum nails. Use

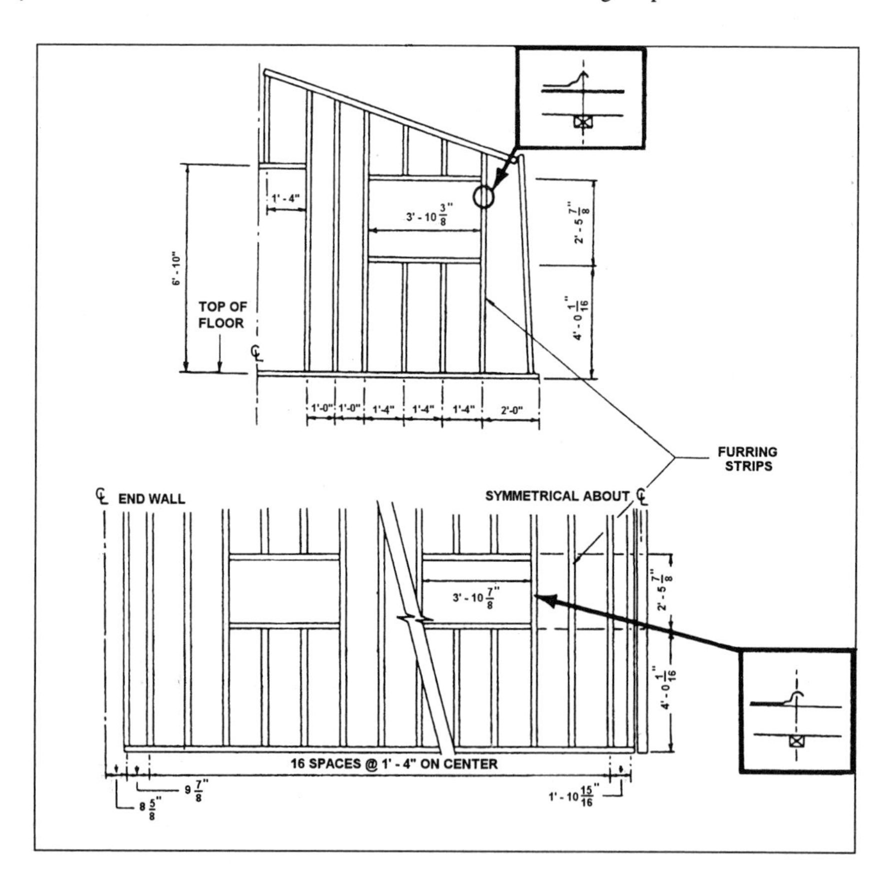

Figure 9-12.—Placing furring for liners.

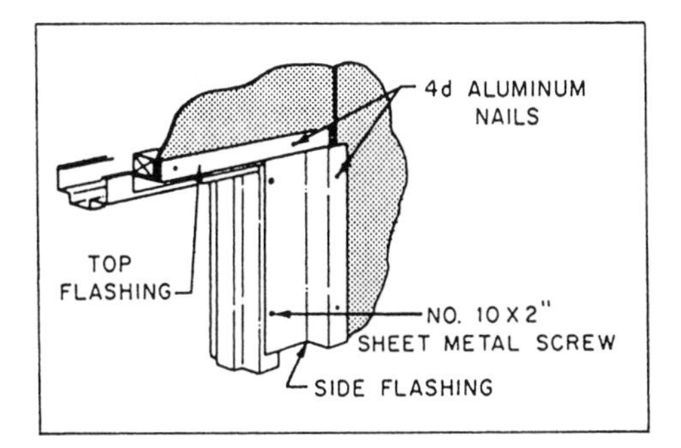

Figure 9-13.—Side and top flashing for the doors.

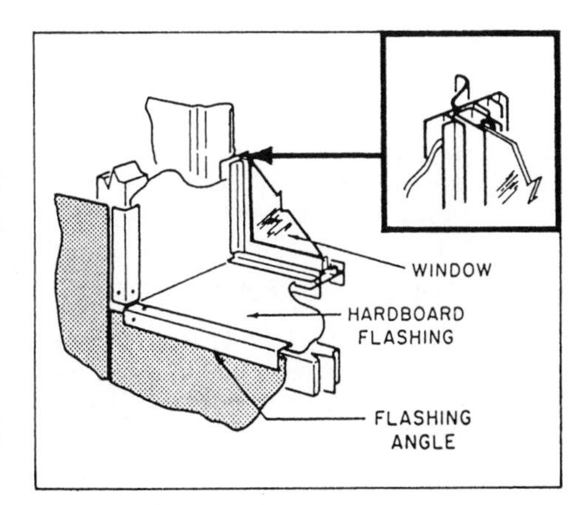

Figure 9-14.—Top and bottom hardboard flashing.

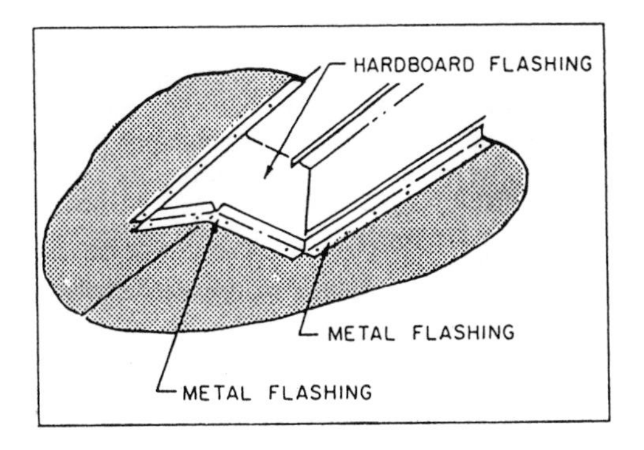

Figure 9-15.—Metal ventilator flashing.

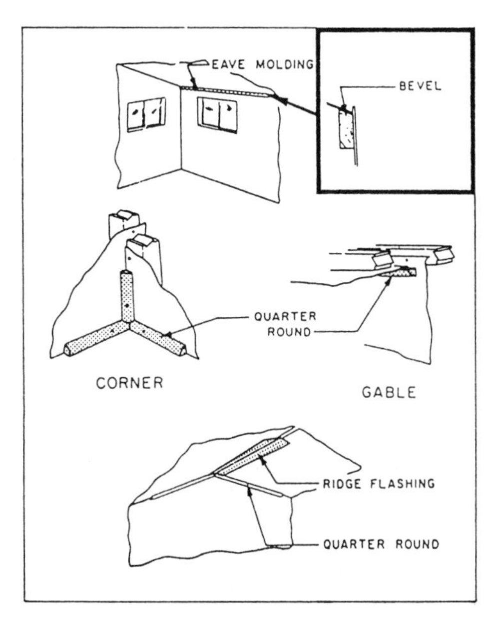

Figure 9-16.—Interior trim.

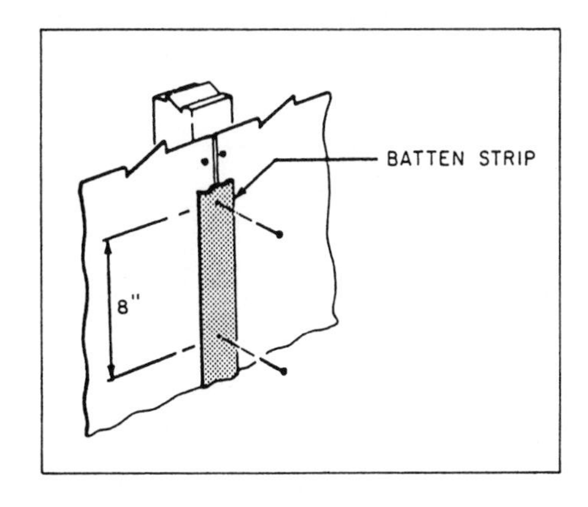

Figure 9-17.—Batten strip.

quarter-round molding to trim the ceiling to the end wall, end wall to sidewall, and walls to floor. Use metal ridge flashing, as shown in figure 9-16, to trim the ridge of the ceiling liner. It can be attached to the ceiling furrings with 4d aluminum nails. Check the drawings to make sure you are installing it correctly. Next, cut battens to the required length and attach them to the furring with 4d aluminum nails, 8 inches OC. An example of this is shown in figure 9-17.

General Comments

Don't be careless with bolts, nuts, and miscellaneous fasteners just because they are furnished in quantities greater than actual requirements. Be careful when using these fasteners to prevent scattering them on the ground. Each evening, empty your pockets of fasteners and other small parts before leaving the erection site.

An extra amount of mastic or sealant is also furnished with each PEB. Here too, reasonable care in applying mastic to roof panels and roof accessories ensures an adequate supply.

Crating lumber can be used to construct an entrance platform and stairs at each end of a PEB. Figure 9-18 shows one way this might be done.

DISASSEMBLY PROCEDURES

Disassembly of a preengineered building should not be difficult once you are familiar with the erection procedures. Basically, it involves accurately marking the parts and following some basic steps.

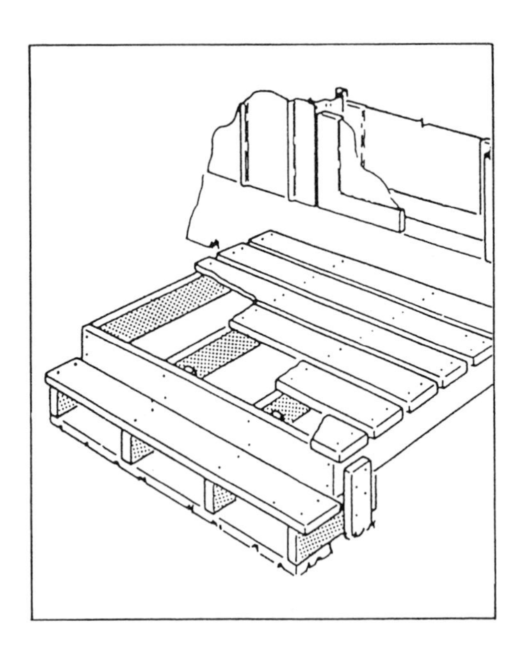

Figure 9-18.—Crate platform.

Marking

It's obvious but worth repeating: In disassembling a building, be sure to clearly mark or number all parts. You will then know where the parts go when reassembling the building.

Steps

There are five main steps in disassembling a PEB:

1. Remove hardboard liner panels.

2. Remove windows, door leaves, and end wall.

3. Remove diagonal brace angles and sag rods.

4. Remove braces, girts, and purlins.

5. Let down frames.

Handling of the building components during disassembly is very important. You may have to reuse these same components again at another location. As you complete disassembly, protect those components from damage. Any damaged components will have to be replaced, and time might not be on your side.

WOOD-FRAME CONSTRUCTION

LEARNING OBJECTIVE: Upon completing this section, you should be able to identify the characteristics of wood-frame tents, SEA huts, and field-type latrines.

There are three basic types of wood-frame construction of concern to Builders: tents with wood frames for support; SEA huts (developed in Southeast Asia during the Vietnam war); and field latrines.

WOOD-FRAME TENTS

Figure 9-19 shows working drawings for framing and flooring of a 16- by 32-foot wood-frame tent. Tents of this type are used for temporary housing, storage, showers, washrooms, latrines, and utility spaces at an advanced base.

Tent floors consist of floor joists (16-foot lengths of 2 by 4s) and sheathing (4- by 8-foot sheets of 1/2-inch plywood). The supports for the floor framing are doubled 2 by 4 posts anchored on 2 by 12 by 12 mudsills. The wall-framing members are 2 by 4 studs, spaced 4-feet OC. The roof-framing members are 2 by 4 rafters, spaced 4-feet OC. The plates (2 by 4s) and the bracing members (1 by 6s) are fabricated in the field. A representative floor-framing plan for a

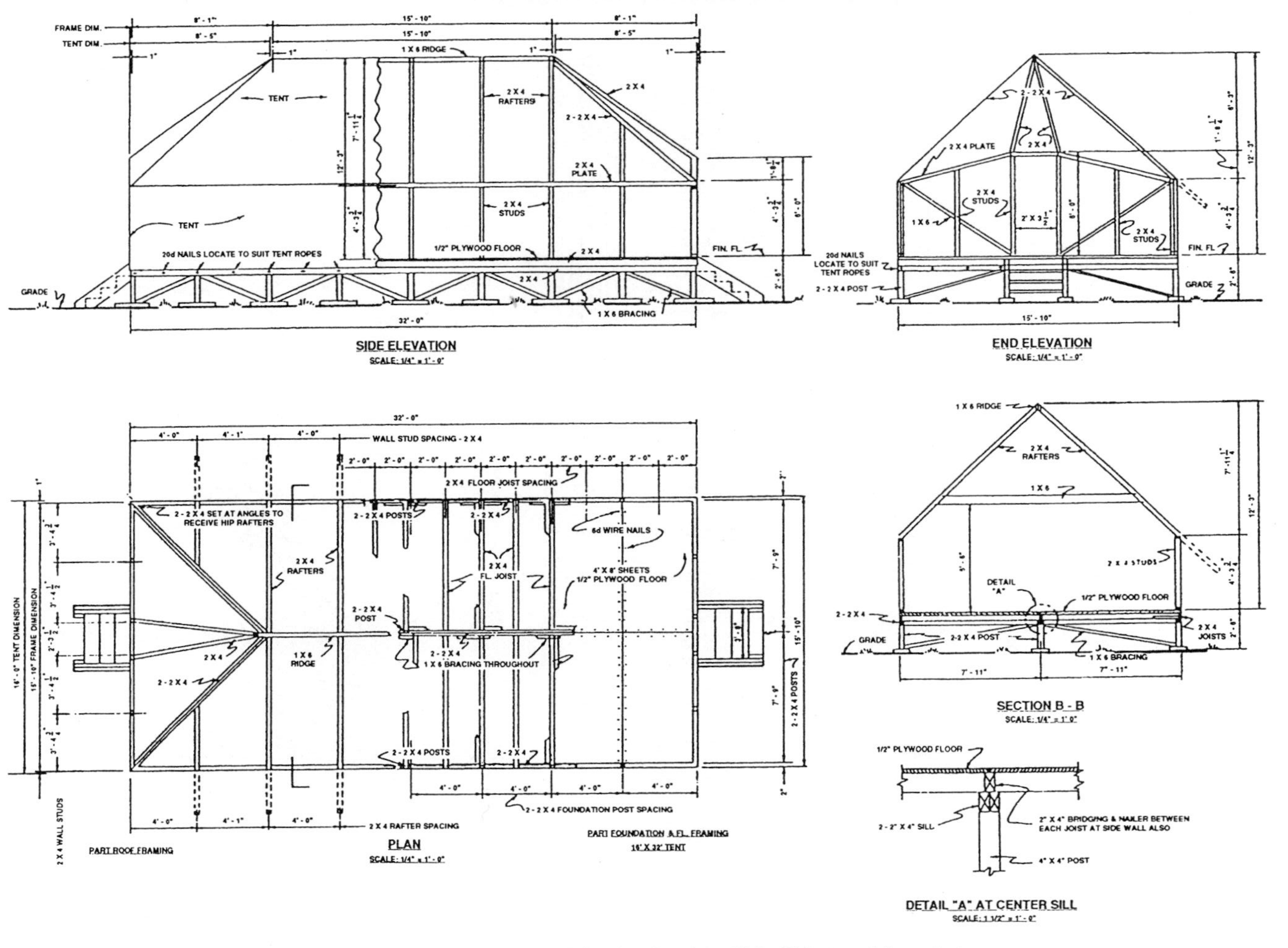

Figure 9-19.—Framing and flooring plans for a 16- by 32-foot wood-frame tent.

field-type shower and a washroom is shown in figure 9-20.

Basically, all field structures are derived from the 16- by 32-foot wood-frame tent. However, if more tent space is needed, a 40- by 100-foot model is available. This tent is not difficult to assemble because it is put together without a floor. It can be erected without a strongback frame since it comes complete with ridge pieces, poles, stakes, and line, and does not require framing. But no matter how easy erection may seem, **always read and follow the instructions**.

SEA HUT

When the 16- by 32-foot wood-frame tent is modified with a metal roof, extended rafters, and screened-in areas, it is called a Southeast Asia (SEA) hut. An example of the completed product is shown in figure 9-21. The SEA hut was originally developed in Vietnam for use in tropical areas by U.S. troops for berthing; but, it can readily be adapted for any use in any situation. It is also known as a strongback because of the roof and sidewall materials.

The SEA hut is usually a standard prefabricated unit, but the design can be easily changed to fit local

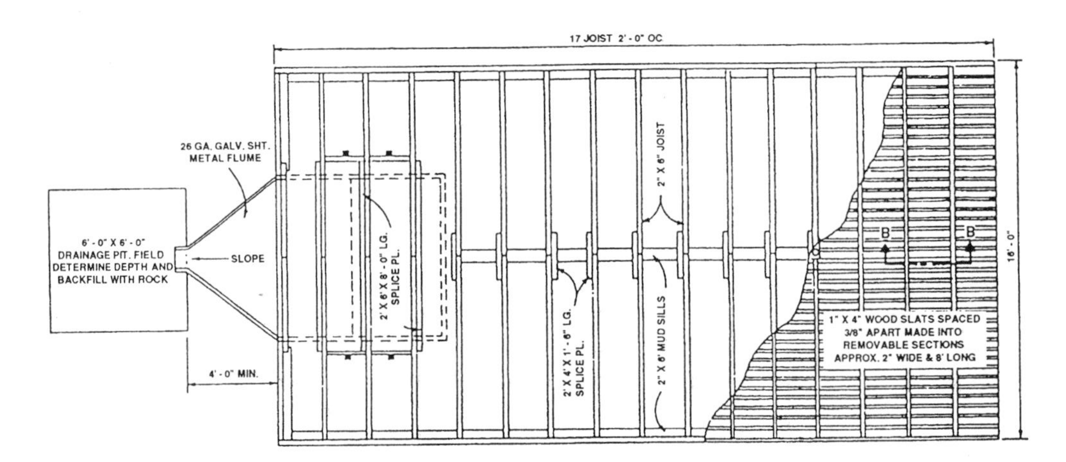

Figure 9-20.—Floor-framing plan.

Figure 9-21.—Completed SEA huts.

requirements, such as lengthening the floor or making the roof higher. The standard prefabrication of a SEA hut permits disassembly for movement to other locations when structures are needed rapidly. As with all disassembly of buildings, ensure it is not damaged in the process.

FIELD-TYPE LATRINES

Temporary facilities for disposal of human waste are one of the first things to be constructed at an advance base. A number of field-type latrines are designed for this purpose; a 16- by 32-foot wood-frame tent may be used to shelter the latrine.

Four Seat

A prefabricated four-seat latrine box is shown in figure 9-22. It can be collapsed for shipment, as shown in figure 9-23.

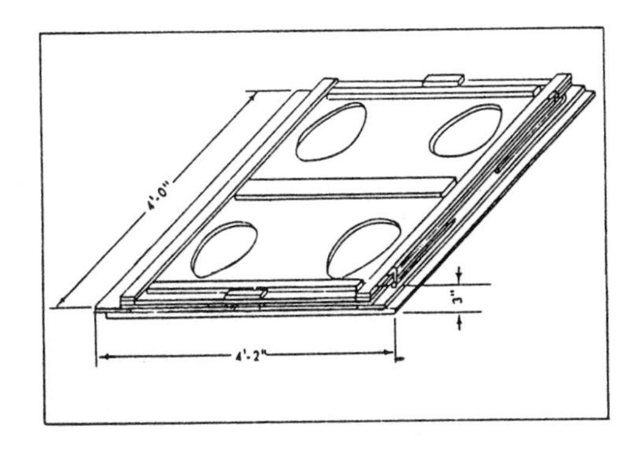

Figure 9-23.—Latrine box collapsed for shipment.

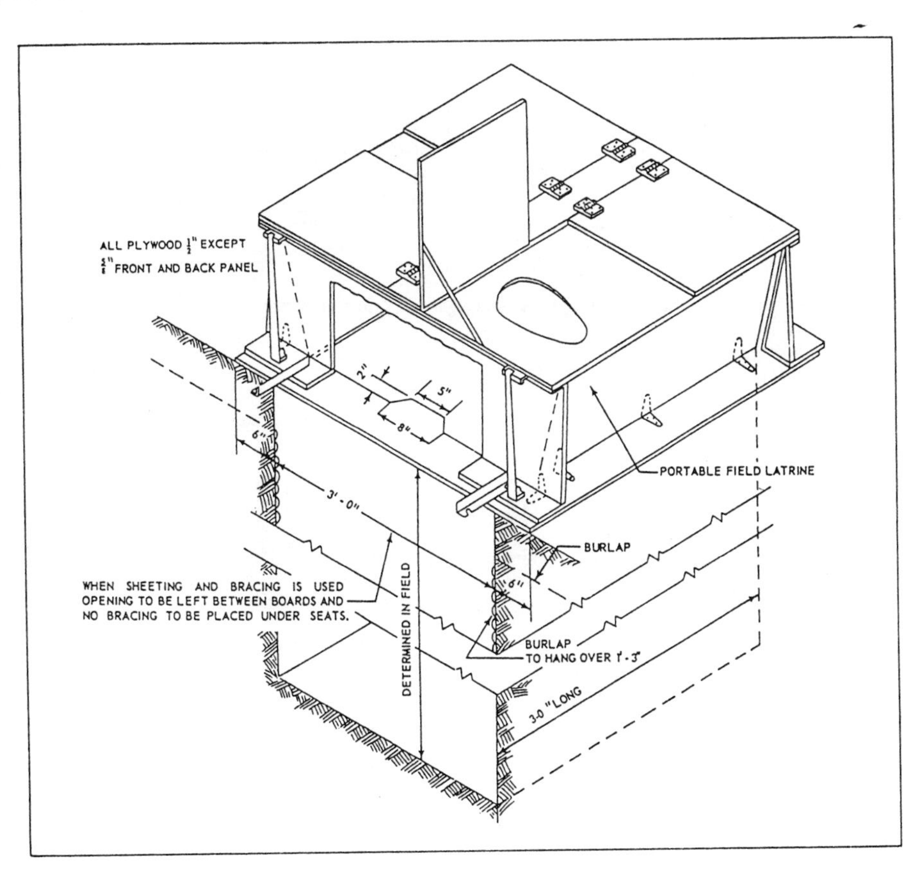

Figure 9-22.—Prefabricated four-seat latrine box.

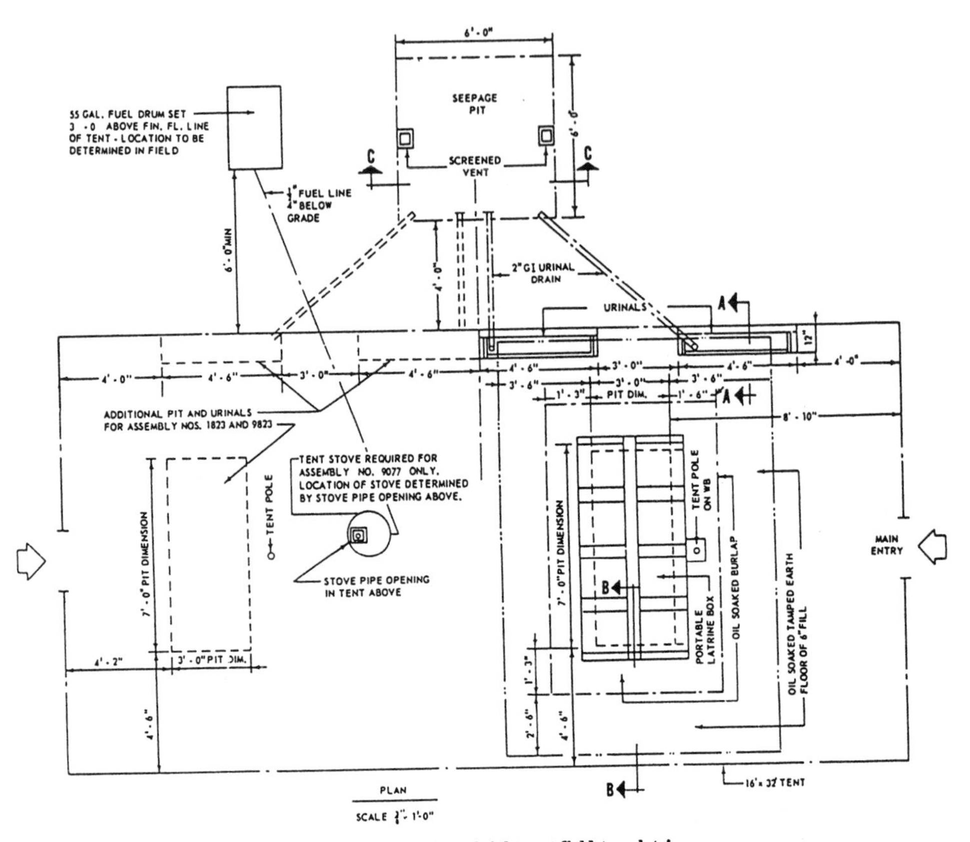

Figure 9-24.—Plan view of eight-seat field-type latrine.

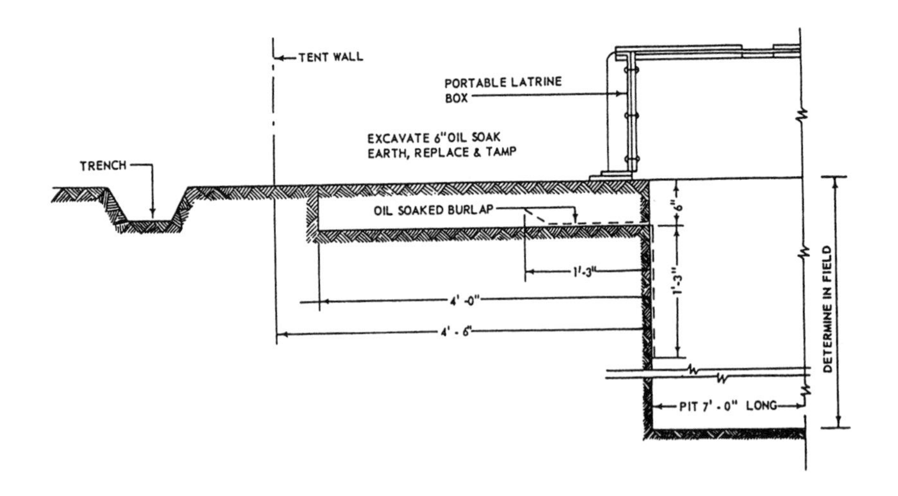

Figure 9-25.—Margin of oil-soaked earth around latrine boxes.

Eight Seat

A plan view of an eight-seat field-type latrine is shown in figure 9-24. Two four-seat boxes straddle a 3- by 7-foot pit. After the pit is dug, but before the boxes are placed, a 4-foot-wide margin around the pit is excavated to a depth of 6 inches, as shown in figure 9-25. A layer of oil-soaked burlap is laid in this excavation. Then, the excavated earth is soaked with oil, replaced, and tamped down to keep surface water out.

Two 4-foot 6-inch trough-type urinals are furnished with the eight-seat latrine. Each is mounted in a frame constructed as shown in figure 9-26. A 2-inch urinal drainpipe leads from the downpipe on each urinal to a 6- by 6-foot urinal seepage pit. The seepage pit is constructed as shown in figure 9-27.

As indicated in figure 9-24, the eight-seat field-type latrine can be expanded to a 16-seat field-type latrine.

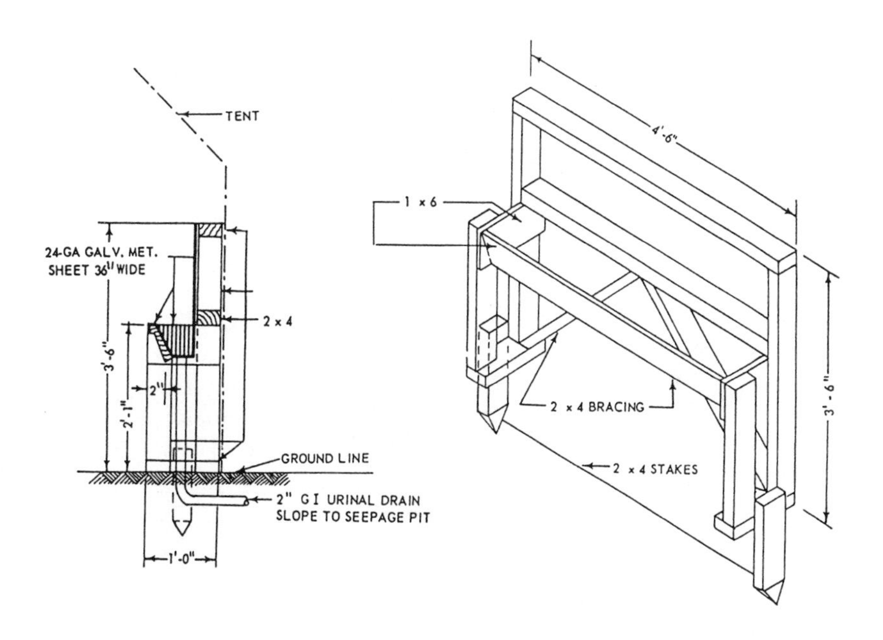

Figure 9-26.—Frame for urinal trough.

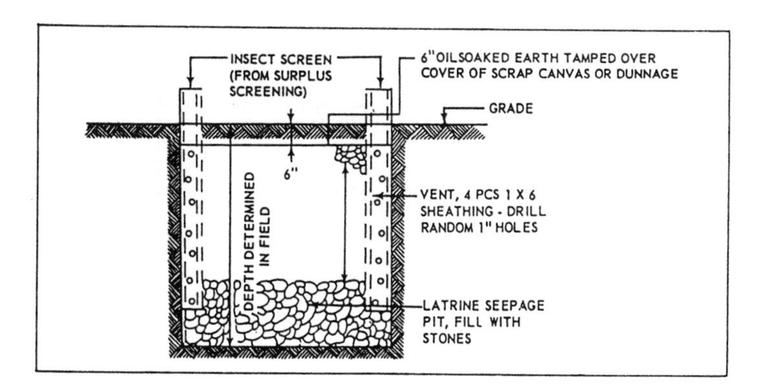

Figure 9-27.—Urinal seepage pit.

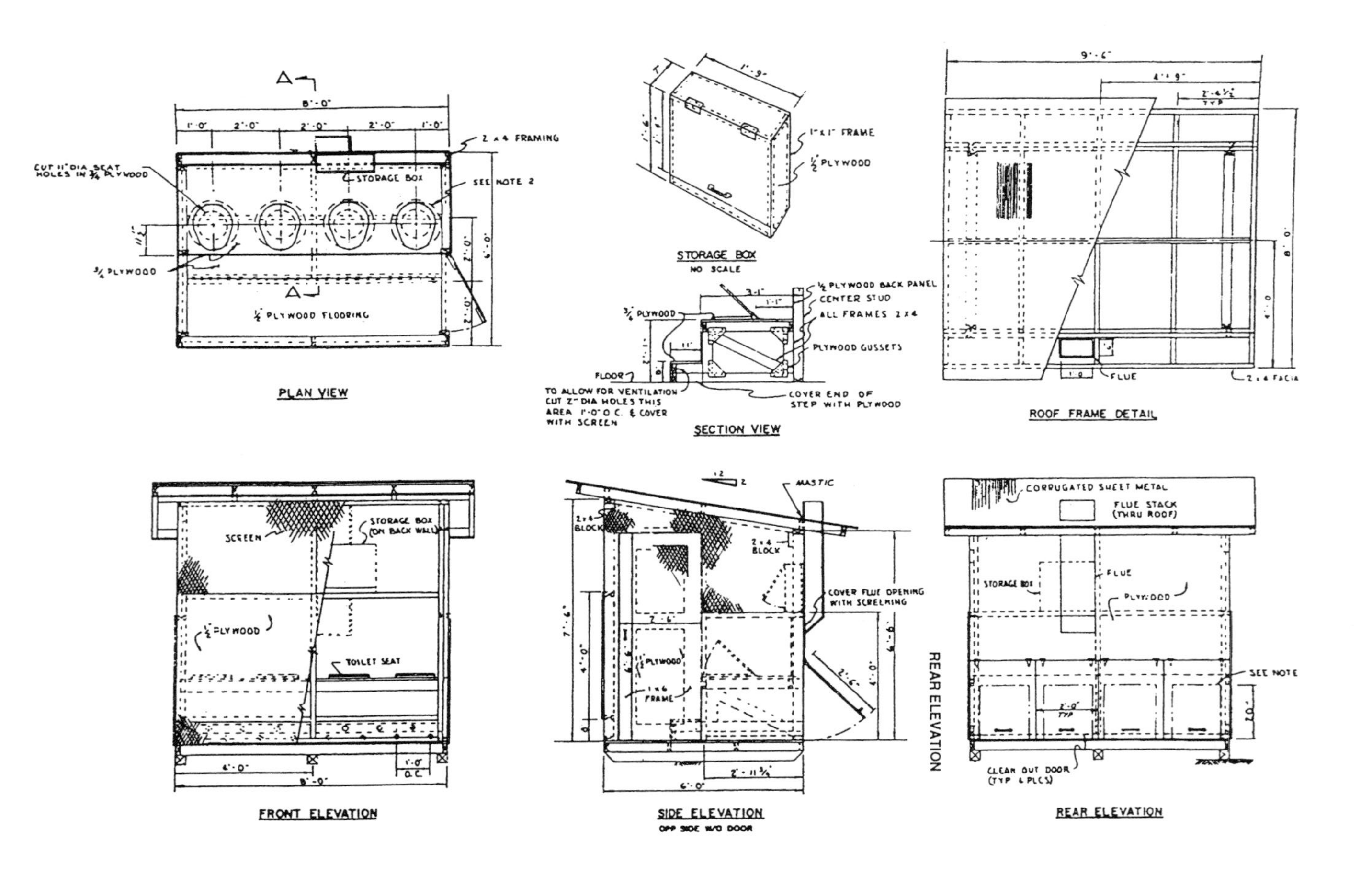

Figure 9-28.—Burnout type four-hole latrine.

Burnout Type

A complete plan view of a four-hole burnout field-type latrine is shown in figure 9-28. The waste goes into removable barrels. The waste is then disposed of at another location. This type of latrine is used at most advanced or temporary bases. The burnout latrine is kept in an orderly condition (daily) by the camp maintenance personnel or the assigned sanitation crew. It can be easily maintained by spreading lime over the waste material or using diesel fuel to burn the waste material.

MAINTENANCE

Once wood frame facilities are completed and occupied by the tenants, maintenance becomes the major priority. The life span of a facility is greatly increased with proper maintenance. Even though the majority of these buildings are temporary in nature, most can be dismantled and reassembled at another site. Establishment of a regularly scheduled maintenance program ensures the buildings are in a consistent state of readiness.

K-SPAN BUILDING

LEARNING OBJECTIVE: Upon completing this section, you should be able to identify the components of, preparation procedures for, and procedures used in the erection of a K-span building.

The K-span building is a relatively new form of construction in the Seabee community. The intended uses of these buildings are as flexible as the SEA huts discussed earlier. Training key personnel in the operation of the related equipment associated with the K-span is essential. These same personnel, once trained, can instruct other members of the crew in the safe erection of a K-span. The following section gives you some, but not all, of the key elements associated with K-span construction. As with other equipment, always refer to the manufacturers' manuals.

OPERATING INSTRUCTIONS

The main component of the K-span system is the trailer-mounted machinery shown in figure 9-29. This

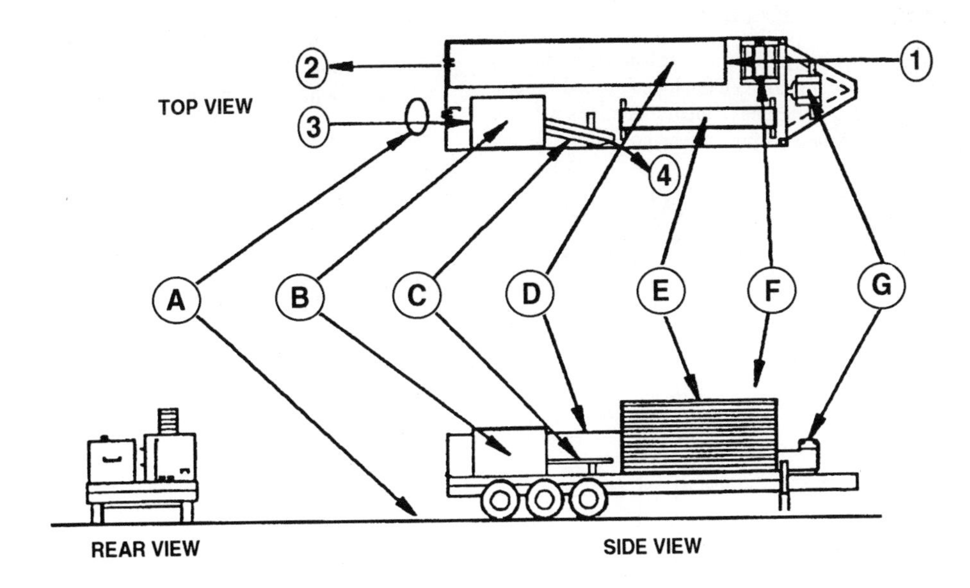

A. Operator's station
B. Panel curving machine
C. Primary curving run-out table
D. Straight panel roll forming machine
E. Run-out tables in transporting rack
F. Coil stock pay-out reel and stand
G. Engine/hydraulic pump primary power unit

General Operation (Top View)
1. Coil stock enters panel forming machine
2. Formed straight panel
3. Straight panel entry to panel curving machine
4. Curved building panel

Figure 9-29.—Trailer-mounted machinery.

figure shows the primary components of the trailer as well as general operations. The key element is the operator's station at the rear of the trailer (shown in fig. 9-30). The individual selected for this station must be able to understand the machine operations and manuals. From here, the operator controls all the elements required to form the panels. The operator must remain at the controls at all times. The forming of the panels is a complex operation that becomes easier with a thorough understanding of the manuals. From the placement of the trailer on site, to the completion of the curved panel, attention to detail is paramount.

As you operate the panel, you will be adjusting the various machine-operating components. Make adjustments for thickness, radius, and the curving machine according to the manuals. **Do not permit short cuts in adjustments**. Any deviations in adjustments, or disregard for the instructions found in the operating manuals, will leave you with a pile of useless material and an inconsistent building.

MACHINERY PLACEMENT

To avoid setup problems, preplanning of the site layout is important. Uneven or sloped ground is not a concern as long as the bed of the trailer aligns with the general lay of the existing surface conditions. Using figure 9-31 as a guide, consider the following items when placing the machinery:

- Maneuvering room for the towing of the trailer, or leave it attached to vehicle (as shown at A);
- Length of unit is 27 feet 8 inches long by 7 feet 4 inches wide (B);
- Allow enough room for run-out stands to hold straight panels. Stands have a net length of 9 feet 6 inches each (C);
- Find point X: From center of curve, measure distance equal to radius in line with front of curved frame. From point X, scribe an arc equal to radius. This arc will define path of curved panel. Add 10 feet for run-out stands and legs (D);
- Storage area required to store coil stock and access for equipment to load onto machine (E);
- Direction curved panels must be carried after being formed (F);
- Level area required to lay panels on ground for seaming. Building will not be consistent if panels are not straight when seaming (G); and
- Space required for crane operations (H).

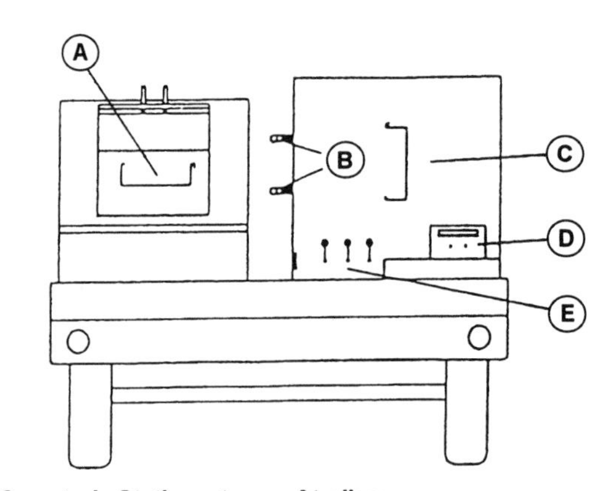

Operator's Station, at rear of trailer:

A. Straight panel exit and cut-off
B. Curving machine radius adjustments and dials
C. Straight panel entrance to curving machine
D. Engine remote control panel
E. Hydraulic controls

Figure 9-30.—Rear of K-span trailer.

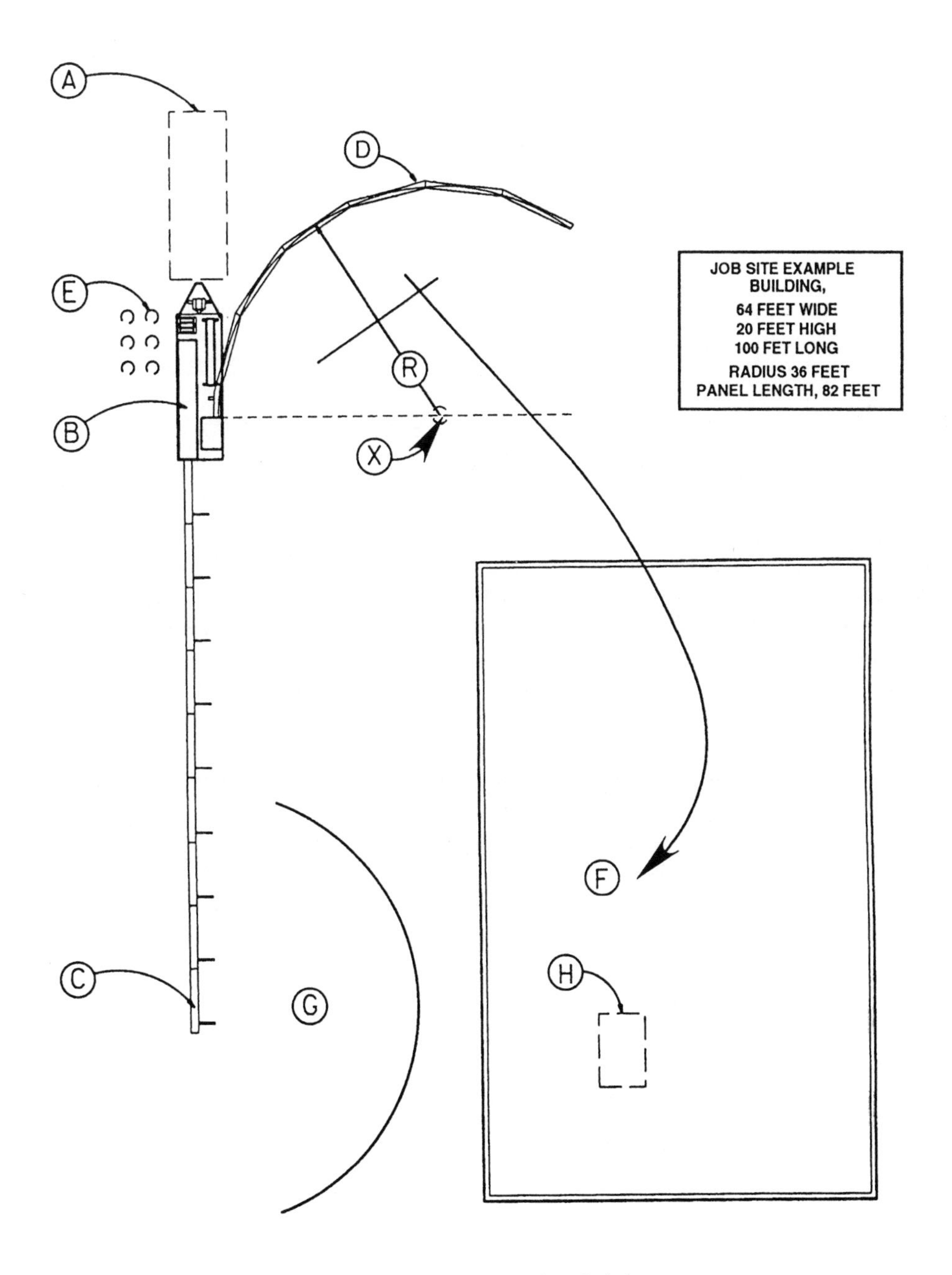

Figure 9-31.—Machinery placement calculations.

FOUNDATIONS

The design of the foundation for a K-span building depends on the building's size, existing soil conditions, and wind load. The foundations for the buildings are simple and easy to construct. With the even distribution of the load in a standard arch building, the size of the continuous strip footing is smaller and more economical than the foundations for conventional buildings.

The concrete forms and accessories provided are sufficient to form the foundations for a building 100 feet long by 50 feet wide. When a different configuration is required, forms are available upon request from the manufacturer.

The actual footing construction is based, as all project are, on the building plans and specifications. The location of the forms, placement of steel, and the psi (pounds per square inch) of the concrete are critical. Since the building is welded to the angle in the footer prior to the concrete placement, all aspects of the footer construction must be thoroughly checked for alignment and square. Once concrete is placed, there is no way to correct mistakes.

As mentioned above, forms are provided for the foundation. Using table 9-1 as a guide, figure 9-32 gives you a simple foundation layout by parts designation. As noted in figure 9-32, the cross pipes are not provided in the kit. They are provided by the contractor.

With the foundation forms in place, and the building panels welded to the attaching angle (fig. 9-33) at 12 inches OC, you are ready to place the concrete. When placing the concrete, remember it is extremely important that it be well-vibrated. This helps eliminate voids under all embedded items. As the concrete begins to set, slope the top exterior portion of the concrete cap about 5 inches (fig. 9-34) to allow water to drain away from the building. The elevation and type of interior floor are not relevant as long as the finish of the interior floor is not higher than the top of the concrete cap.

BUILDING ERECTION

With the placement of the machinery and forming of the building panels in progress, your next considerations are the placement and the weight-lifting capabilities of the crane. Check the crane's weight-lifting chart for its maximum weight capacity. This dictates the number of panels you can safely lift at the operating distance. As with all crane operations, attempting to lift more than the rated capacity can cause the crane to turn over.

Table 9-1.—Concrete Forms Included in Kit

DESCRIPTION (Each set of forms is sufficient to erect a building 100 feet long by 50 feet wide)	PART NUMBER
Side form panels, 1′ × 10′, 12-gauge steel	F-1
Transition panels, 1′ × 12″, 12-gauge steel	F-2
Transition panels, 1′ × 28″, 12-gauge steel	F-3
End-wall caps, 1′ × 15″, 12-gauge steel	F-4
Side-wall caps, 1′ × 19″, 12-gauge steel	F-5
Filler form, 1′ × 12′, 12-gauge steel	F-6
Sidewall inside stop, 1′ × 12″, 12-gauge steel	F-7
End wall inside stop, 1′ × 12″, 12-gauge steel	F-8
Stakes, 1/4″ diameter, bar steel	F-9
All-thread rod, 1/2-13 × 18″	F-10
Hex nuts, 1/2-13	F-11
Hex bolts, 1/8-16 × 1-1/2″	F-12
Hex nuts, 3/8-16	F-13
Flat washers, 1/8″ SAE	F-14
Corner angles, 2″ × 2″ × 12″, steel angle	F-15

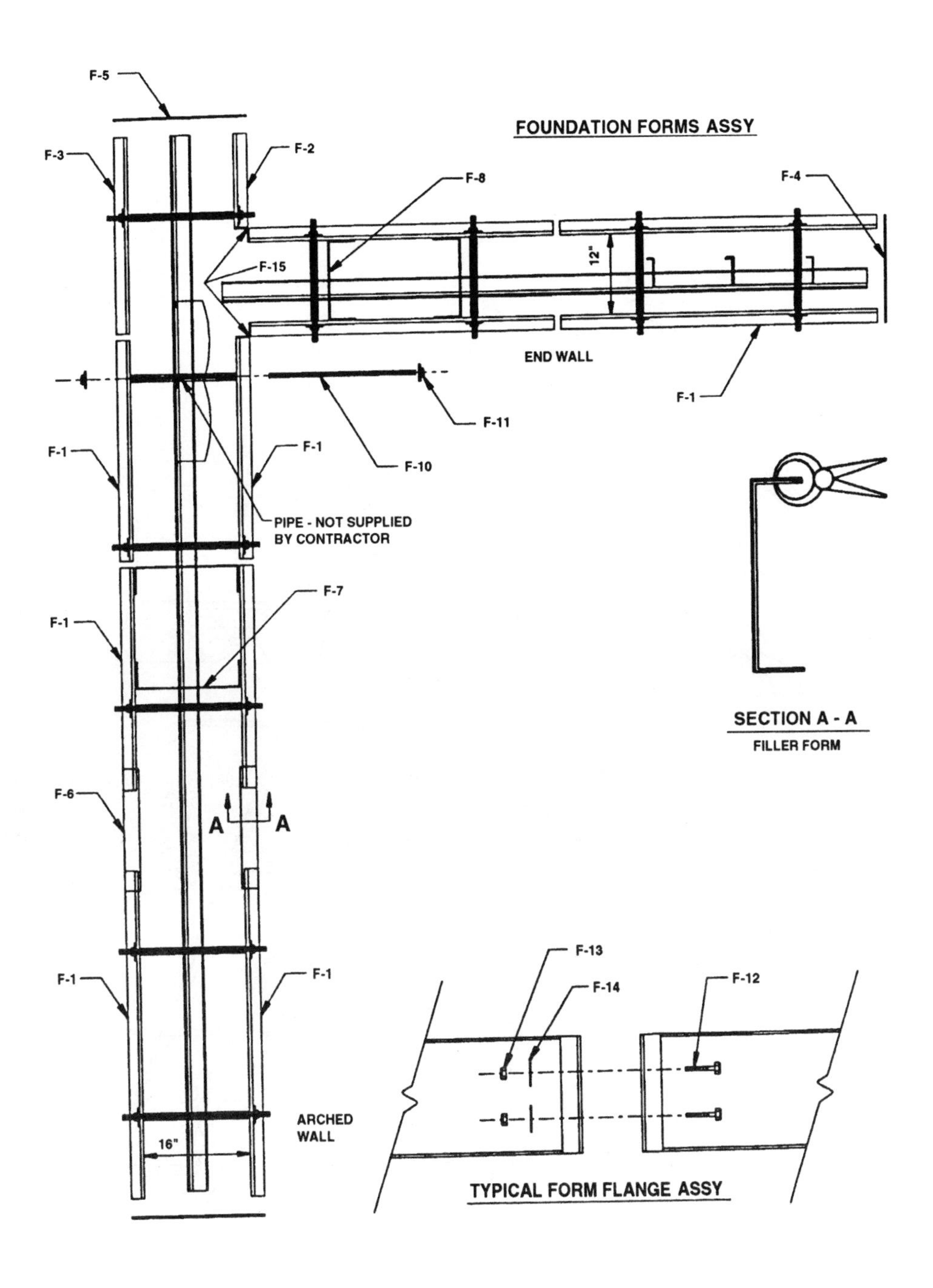

Figure 9-32.—Simple form assembly.

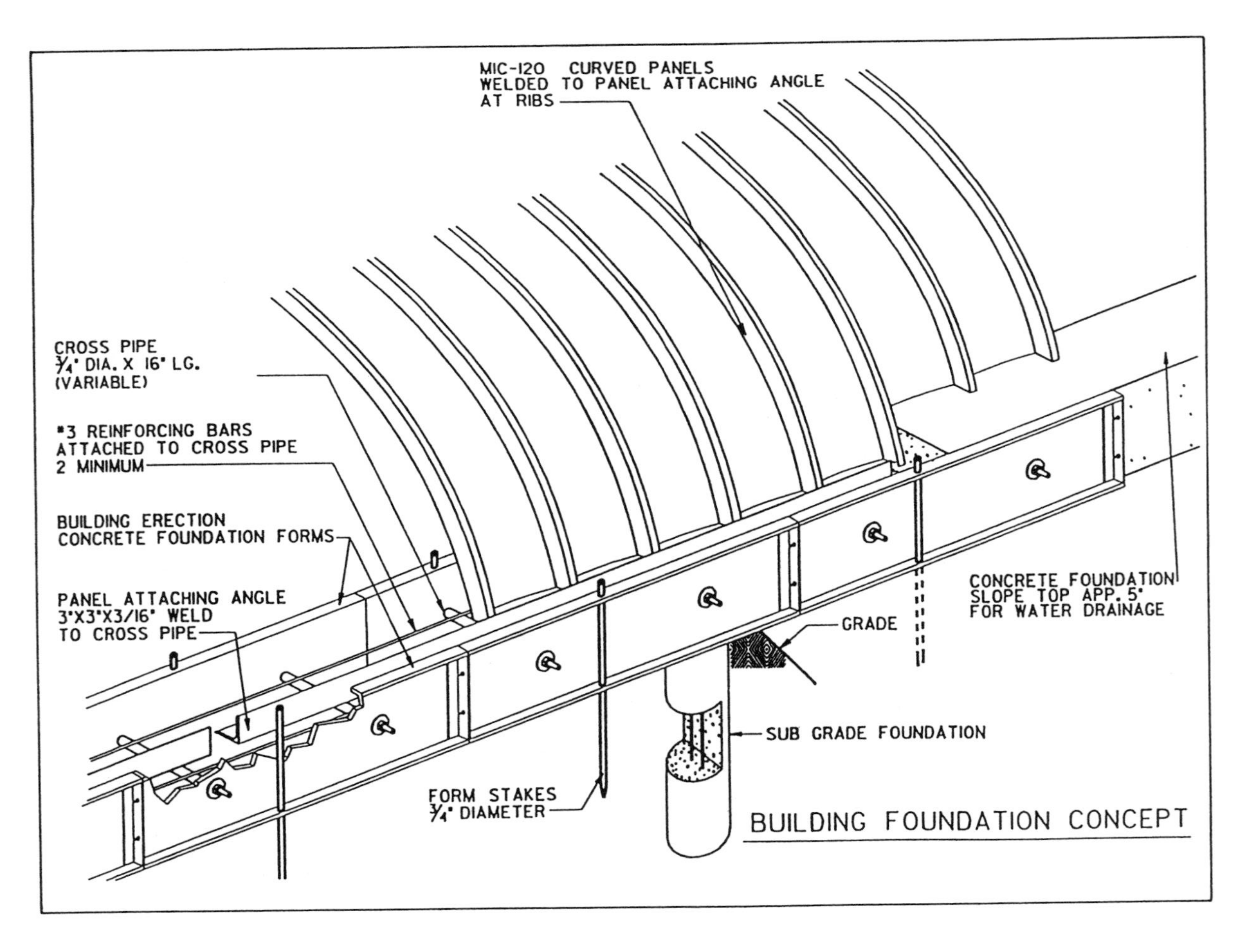

Figure 9-33.—Building foundation concept.

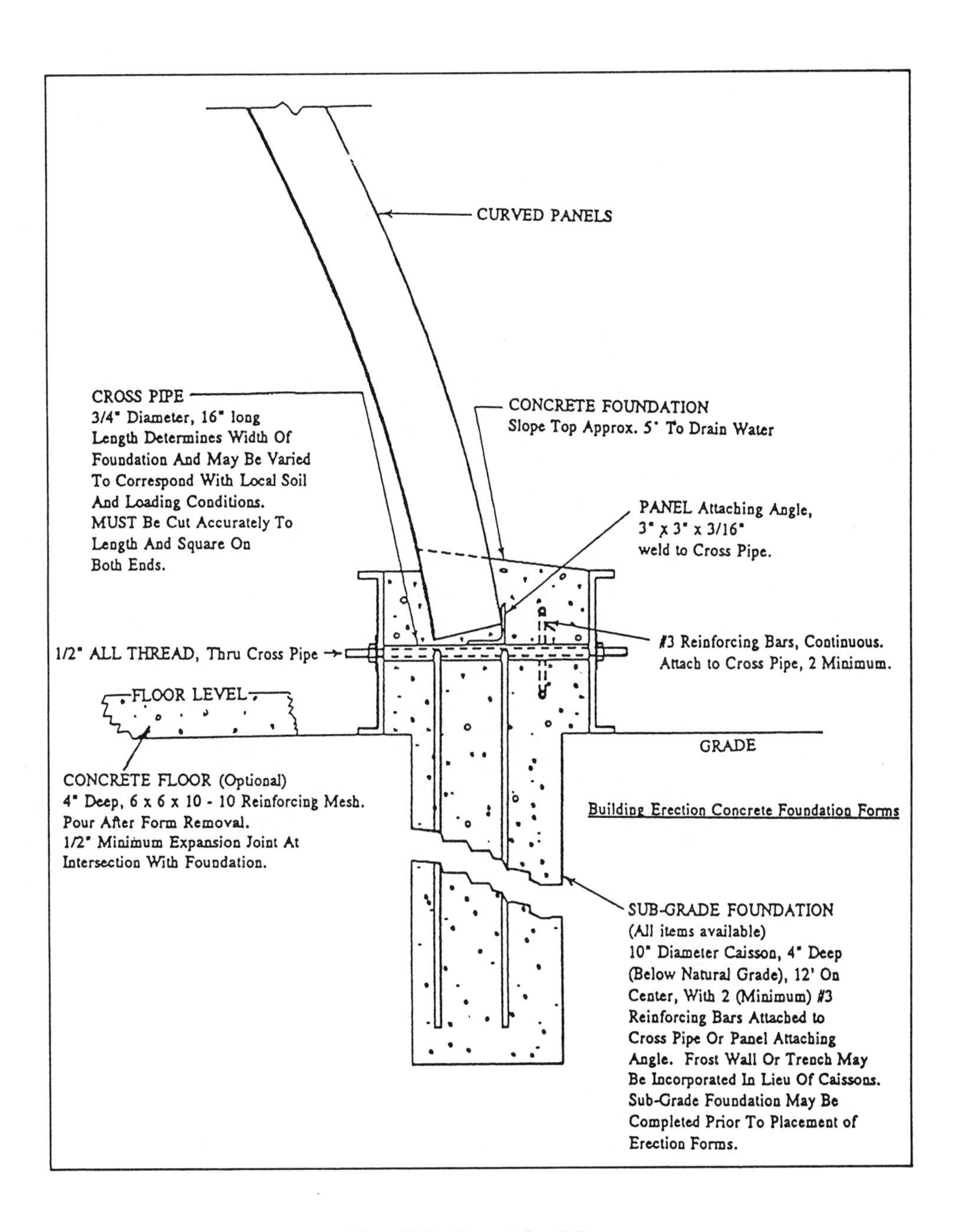

Figure 9-34.—Concrete foundation.

Attaching the spreader bar (fig. 9-35) to the curved formed panels is a critical step; failure to tightly clamp the panel can cause the panels to slip and fall with potential harm to personnel and damage to the panel. With guide ropes attached (fig. 9-36) and personnel manning these ropes, lift panels for placement. When lifting, lift only as high as necessary, position two men at each free end to guide in place, and remind

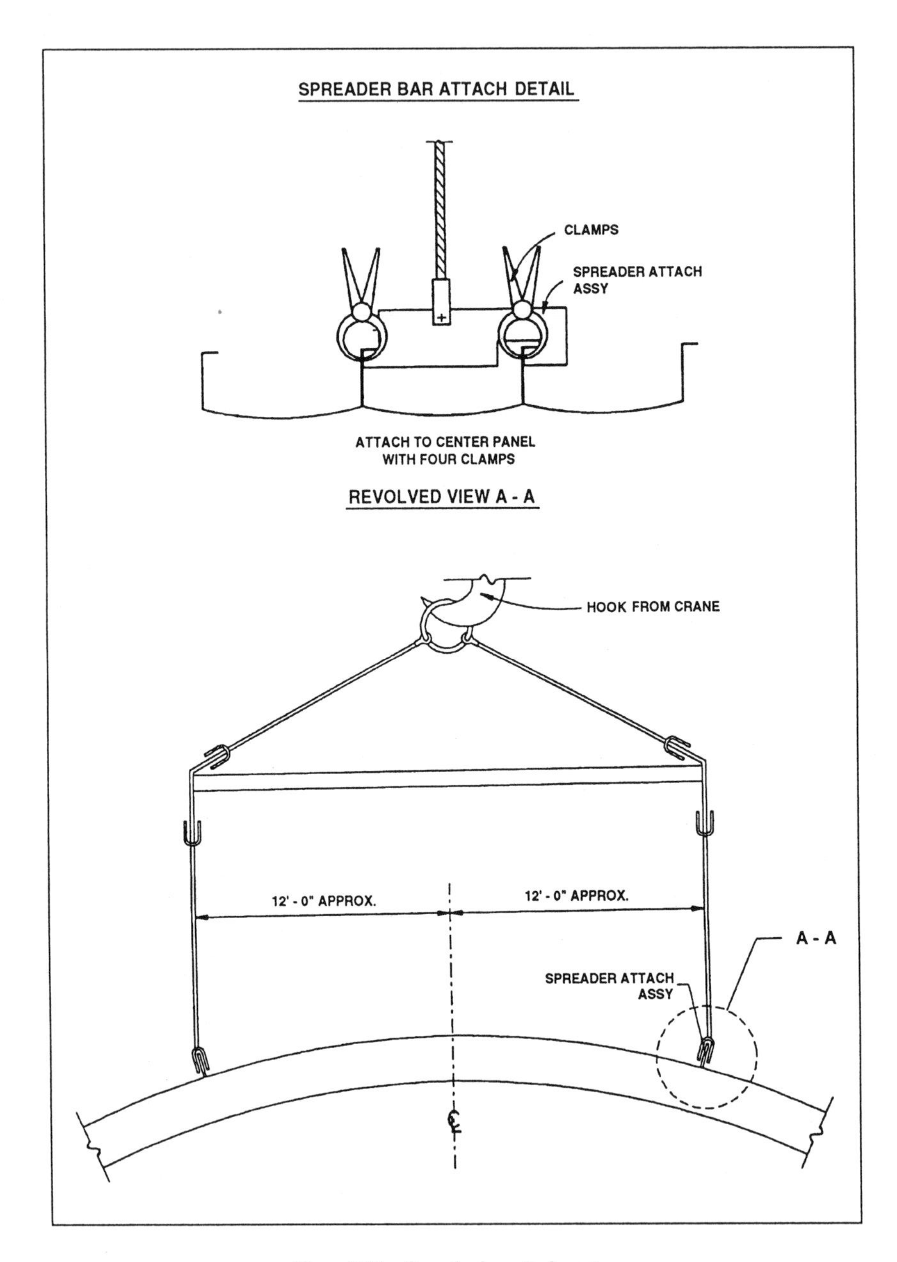

Figure 9-35.—Spreader bar attachment.

crewmembers to keep their feet from under the ends of the arches. Never attempt lifting any sets of panels in high winds.

Place the first set of panels on the attaching angle of foundation and position so there will be room for the end-wall panels. After positioning the first set of panels, clamp them to the angle, plumb with guide ropes, and secure the ropes to previously anchored stakes. Detach the spreader bar and continue to place panel sets. Seam each set to standing panels before detaching spreader bar.

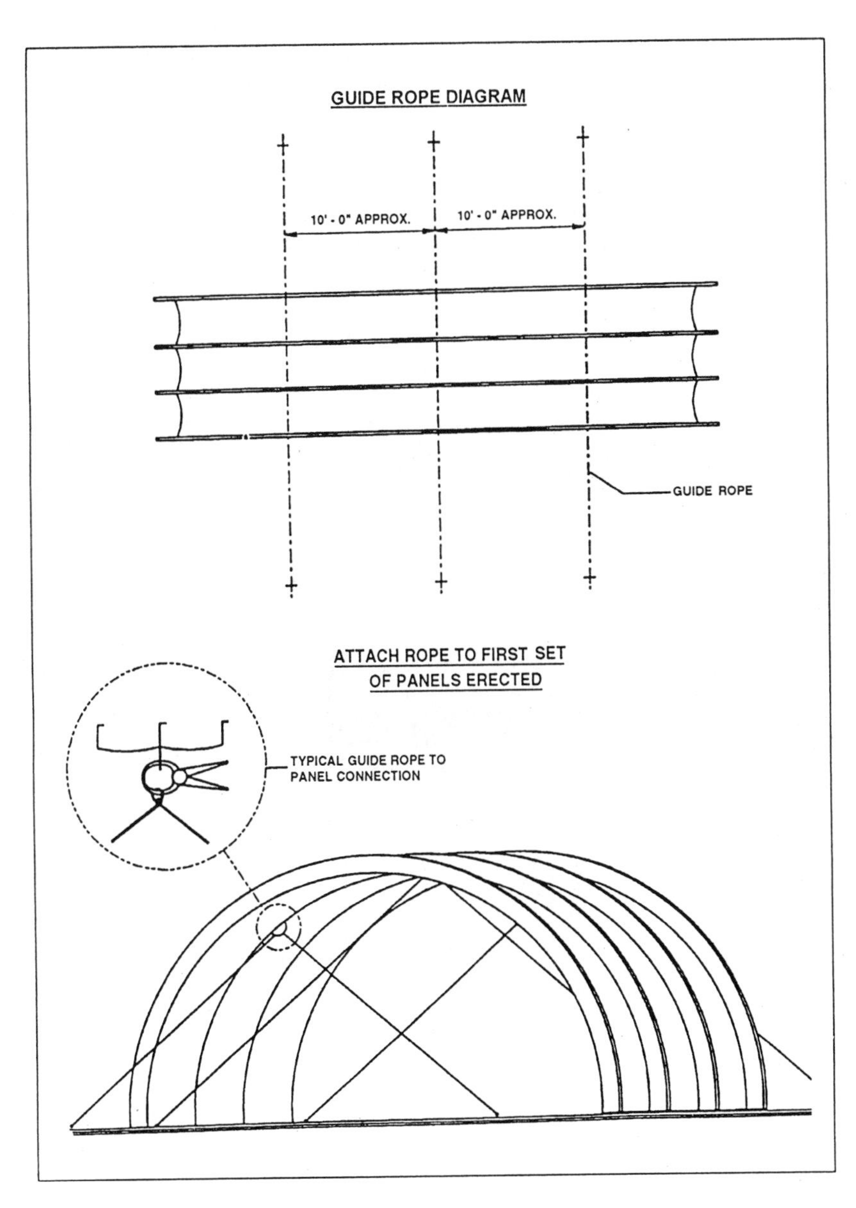

Figure 9-36.—Guide rope diagram.

After about 15 panels (3 sets) are in place, measure the building length at both ends (just above forms) and at the center of the arch. This measurement will seldom be exactly one foot per panel (usually slightly more), but should be equal for each panel. Adjust the ends to equal the center measure. Panels are flexible enough to adjust slightly. Check these measurements periodically during building construction. Since exact building lengths are difficult to predict, the end wall attaching angle on the finishing end of building should not be put in place until all panels are set.

After arches are in place, set the longest end-wall panel in the form, plumb, and clamp it in place. Work from the longest panel outward and be careful to maintain plumb.

CONSTRUCTION DETAILS

The K-span building system is similar to other types of preengineered or prefabricated buildings in that windows, doors, and roll-up doors can be installed only when erection is completed. When insulation of the building is required, insulation boards (usually 4 by 8 feet) may be of any semirigid material that can be bent to match the radius of the building. The insulation is installed using clips, as shown in figure 9-37.

When the integrity of the end-wall panels is continuous from ground to roof line, the end walls become self-supporting. The installation of windows (fig. 9-38), and aluminum and wood doors (figs. 9-39 and 9-40, respectively), presents no problem since the integrity of the of the wall system is not interrupted.

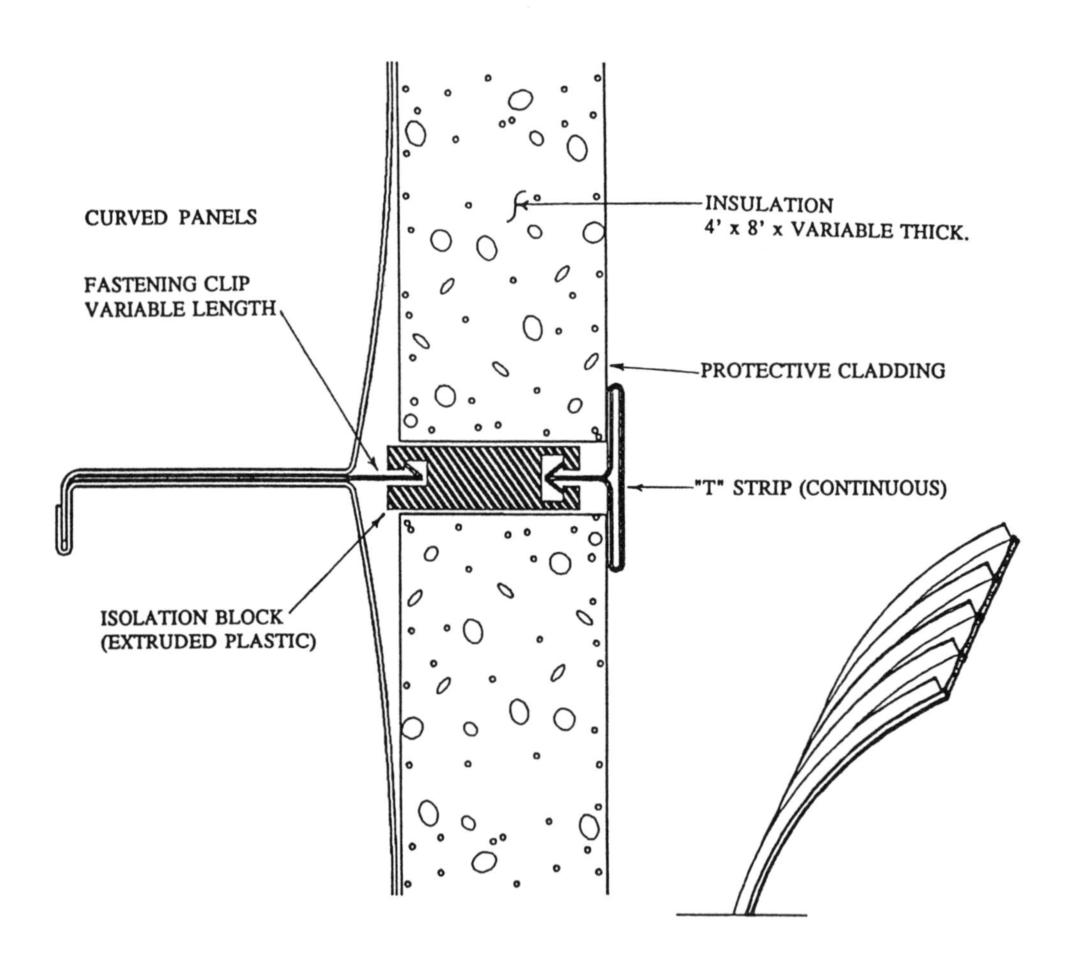

Figure 9-37.—Insulation.

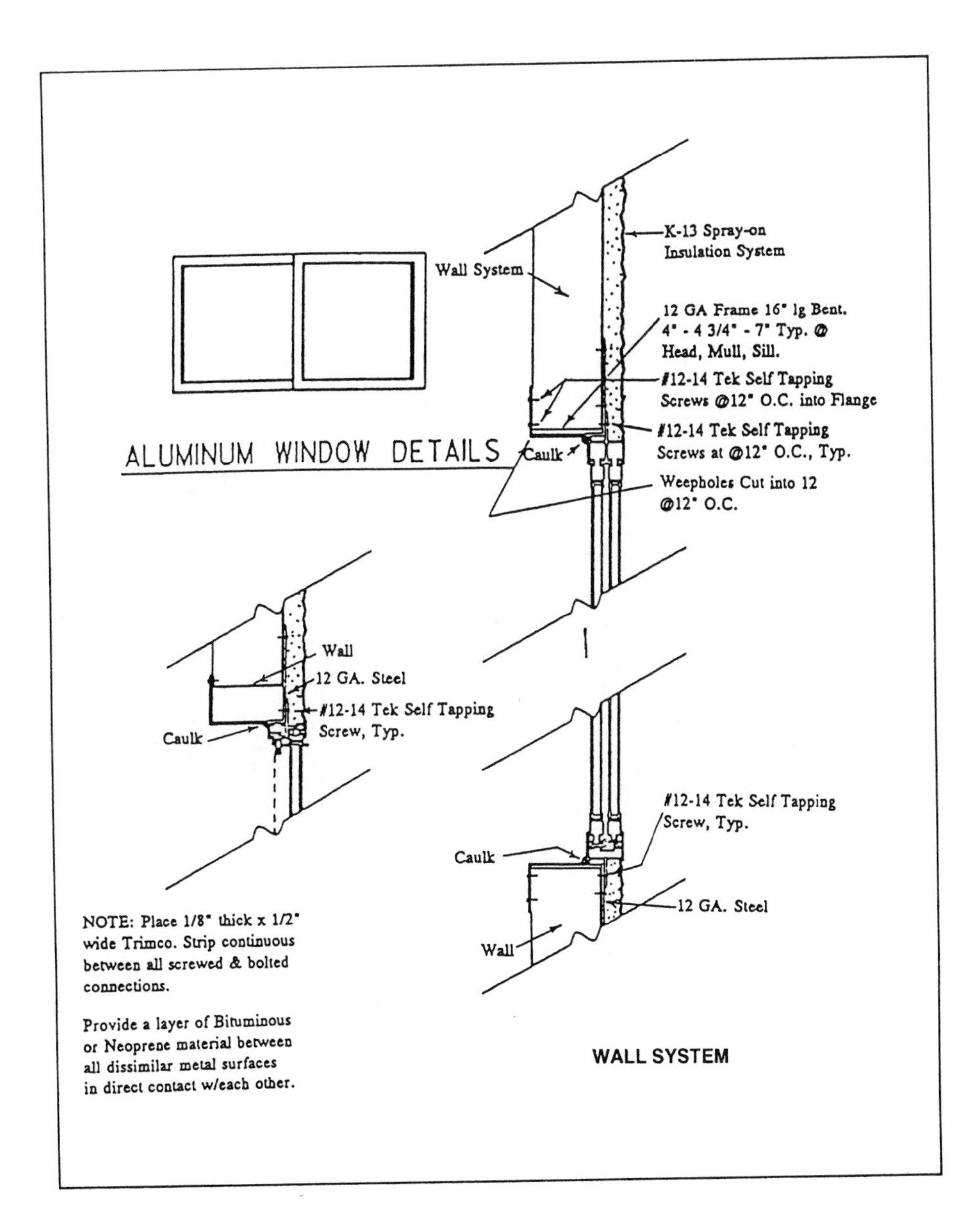

Figure 9-38.—Aluminum window installation.

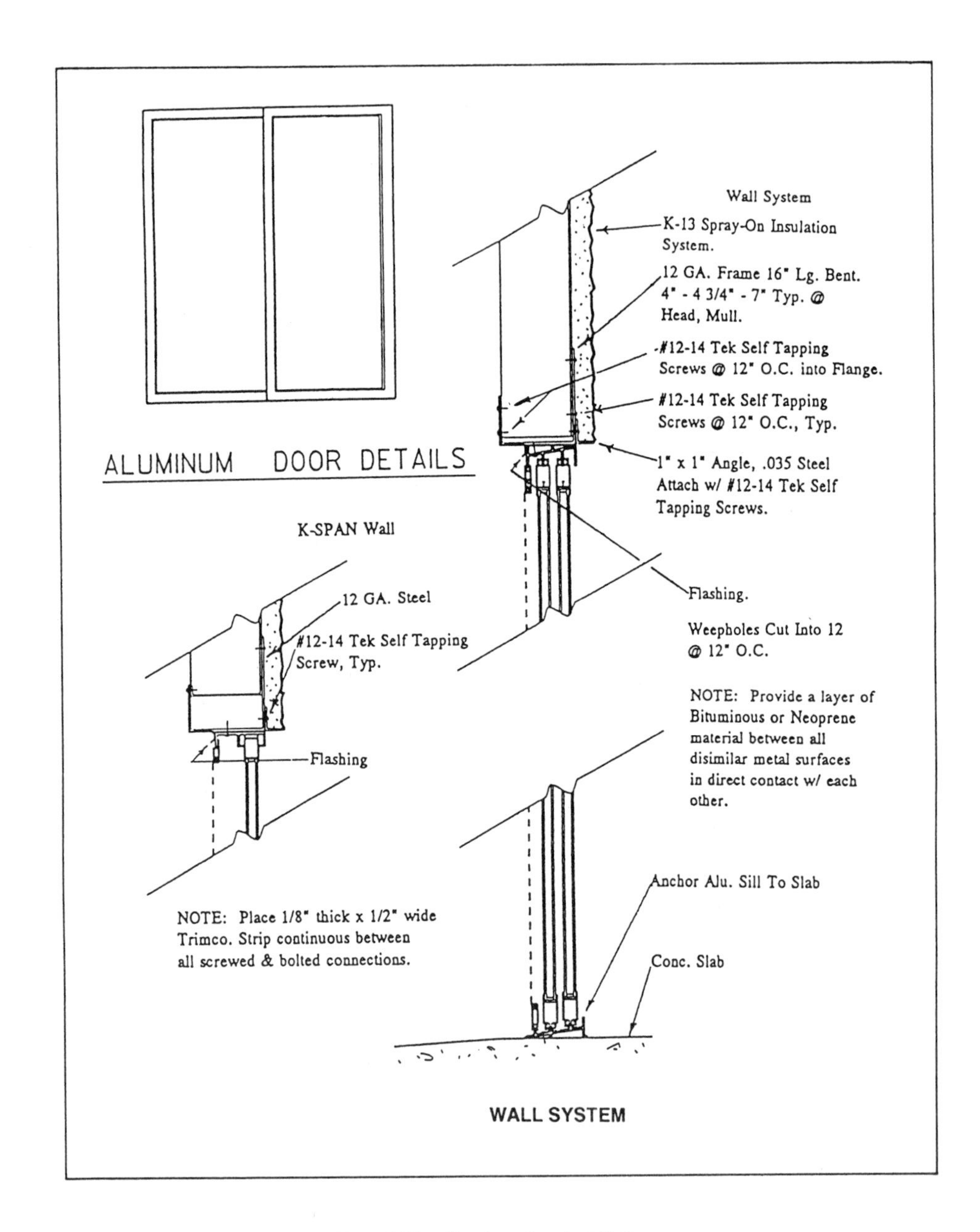

Figure 9-39.—Aluminum door installation.

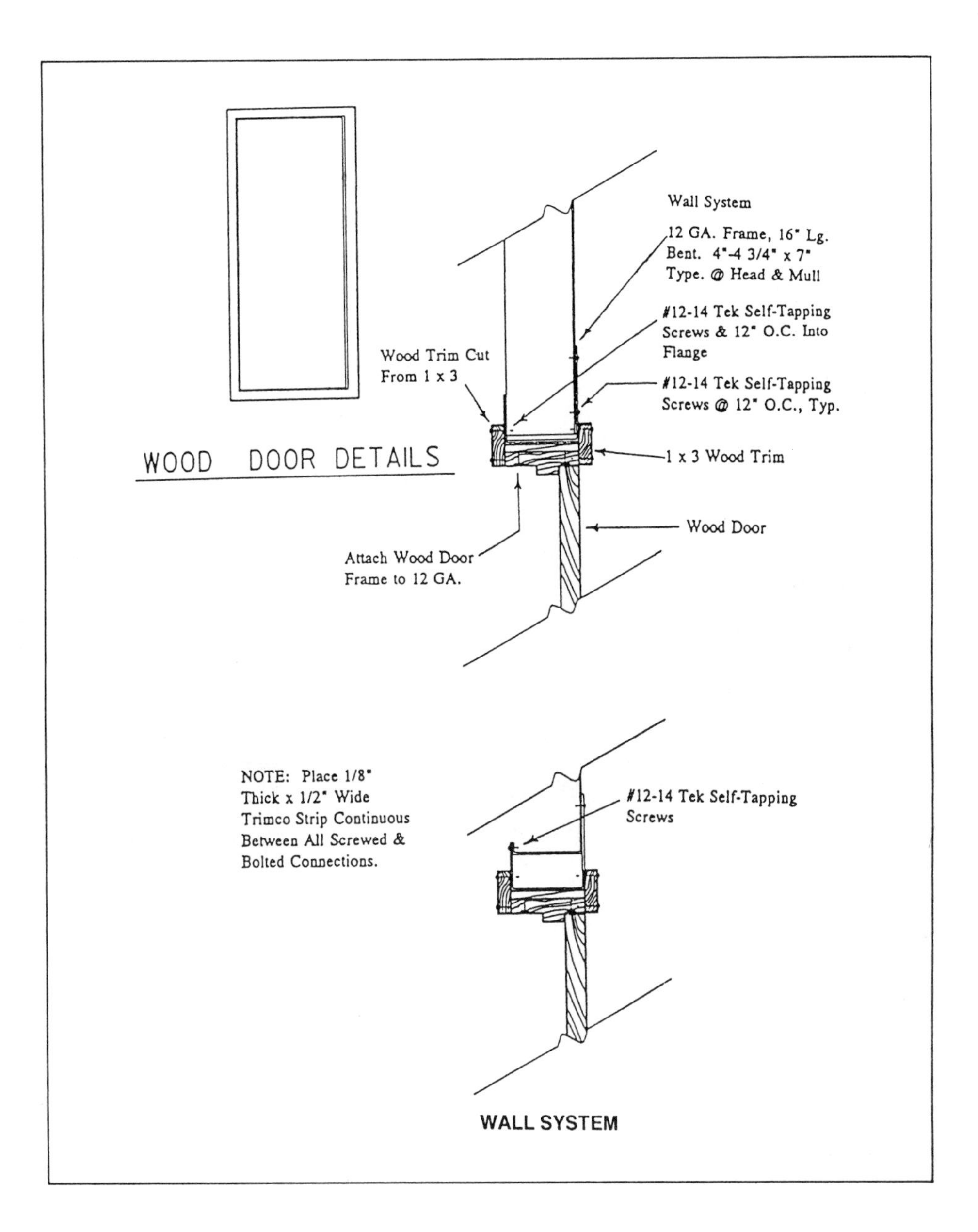

Figure 9-40.—Wood door installation.

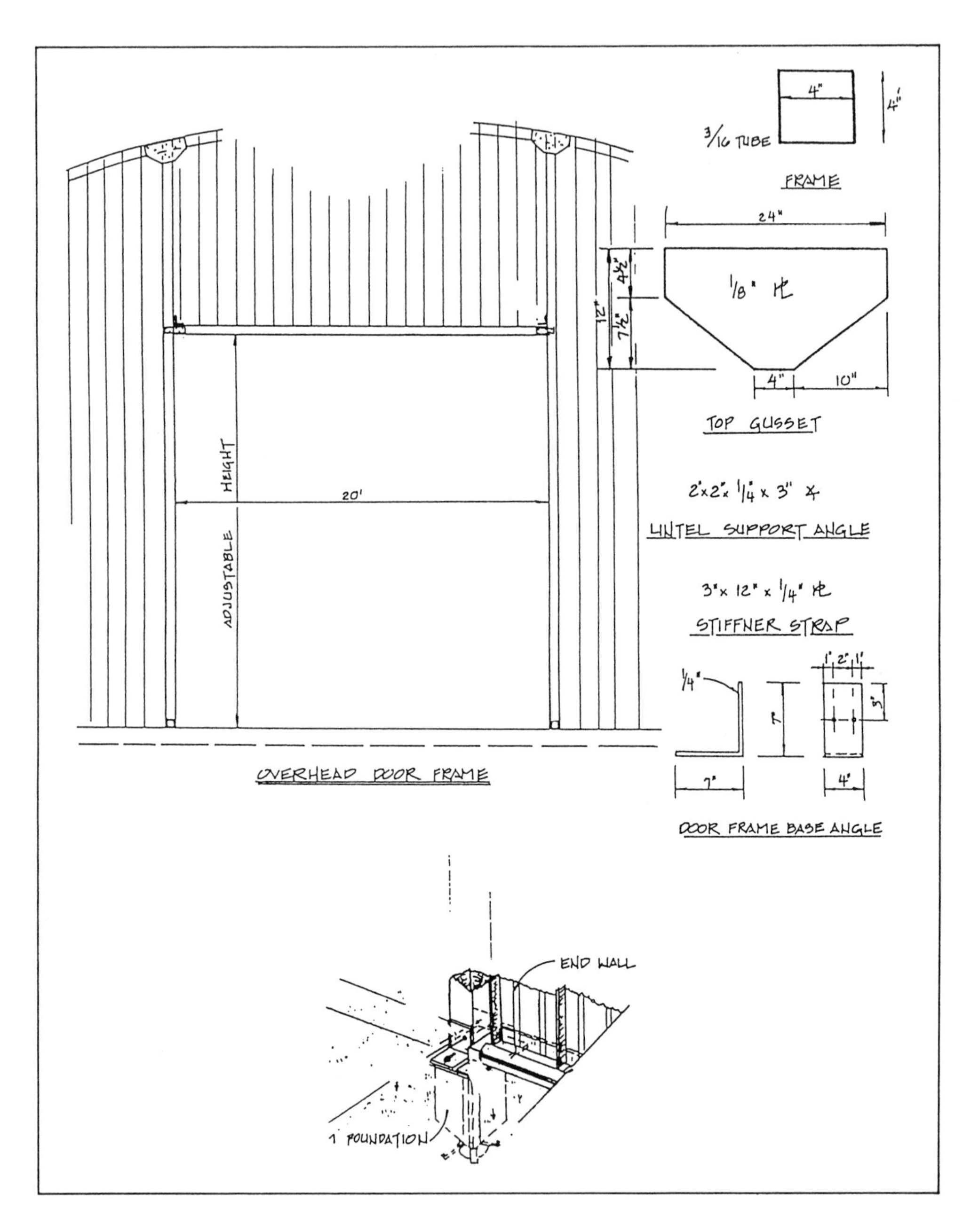

Figure 9-41.—Overhead door frame.

The installation of the overhead door (fig. 9-41) does present a problem in that it does interrupt the integrity of the wall system. This situation is quickly overcome by the easily installed and adjustable (height and width) door frame package that supports both the door and end wall. This door frame package is offered by the manufacturer.

Keep in mind that the information provided in this section on the K-span building is minimal. During the actual construction of this building, **you must consult the manufacturer's complete set of manuals**.

EMBARKATION

LEARNING OBJECTIVE: Upon completing this section, you should be able to identify the procedures and techniques used in preparing material for embarkation.

For a smooth, expedient mount-out, careful pre-planning and organizing are required. Embarkation, whether by air, land, sea, or any combinations thereof, is an all-hands evolution. A successful move requires 100-percent support.

Flexibility is extremely important. Proper embarkation depends to a large extent on the mutual understanding of objectives and capabilities, and full cooperation in planning and execution by both the unit mobilizing and the organization providing the lift. Whenever possible, early communication and coordination between the two is extremely important.

PLANNING

Embarkation planning involves all measures necessary to assure timely and effective out-loading of the amphibious task force and portions thereof. Planning for embarkation also applies to all unit moves, regardless of the method used for movement. These measures are determined by the availability of transportation and the transportation requirements of the unit moving. In amphibious embarkation, the OPNAV level in the chain of command determines overall shipping requirements and the embarkation schedules. This enables subordinate units to prepare detailed loading plans for individual ships. Planning requires constant coordination between commanders in the Navy and the Air Force; they must have a mutual understanding of the problems of each support group. However, in the final analysis, the embarkation plan must support the tactical deployment plan of the unit. In the case of an amphibious landing, it must support the tactical plan for landing and the scheme of maneuvers ashore.

Embarkation planning requires detailed knowledge of the characteristics, capabilities, and limitations of ships, aircraft, and amphibious vehicles, and their relationships to the troops, supplies, and equipment to be embarked. The planner must be familiar with transport types of amphibious ships, Military Sealift Command (MSC) ships, merchant ships, and cargo aircraft. MSC ships and merchant ships pose certain problems; basically, they are not designed, equipped, or have a crew large enough for amphibious operations. But, their use must be anticipated. The additional requirements of hatch crews, winchmen, cargo-handling equipment, cargo nets, assault craft, and other facilities must be provided by the user.

Principles

Whether by ship during amphibious operations or by aircraft for assault force support operations, you must observe certain principles to ensure proper embarkation.

First, embarkation plans must support the plan for landing and the scheme of maneuvers ashore. Personnel, equipment, and supplies must be loaded so they can be unloaded at the time and in the sequence required to support operations ashore.

Second, embarkation plans must provide for the highest possible degree of unit self-sufficiency. Troops should not be separated from their combat equipment and supplies. Weapons crews should be embarked on the same ship or aircraft with their weapons; radio operators with their radios; and equipment operators with their equipment. In addition, each unit should embark with sufficient combat supplies, such as ammunition, gasoline, and radio batteries, to sustain its combat operations during the initial period in the operational area. All personnel should have sufficient water and rations to sustain themselves for 24 hours.

Third, plans must provide for rapid unloading in the objective area. This can be achieved by a balanced distribution of equipment and supplies.

Fourth, and last, plans must provide for dispersion of critical units and supplies among several ships or aircraft. The danger of not doing so is obvious. If critical units and supplies are not dispersed, loss of one ship, or a relatively few ships or aircraft, could result in a major loss of combat capability. Accomplishment of the mission can be seriously jeopardized.

Team Planning

Effective embarkation planning by the embarkation team is dependent upon the early receipt of information from higher authority. Detailed planning begins with the determination of team composition and the assignment of shipping. The following information should be included in the team's embarkation planning:

- Designation of the team embarkation officer(s);
- Preparation and submission of basic loading forms by troop units of the embarkation team;
- Preparation of the detailed loading plan;
- Designation of the ship's platoon, billeting, messing, and duty officers during the period of the embarkation;
- Designation and movement of advance parties and advance details to the embarkation area;
- Establishment of liaison with the embarkation control office in the embarkation area;
- Preparation for the schedule for movement of troops, vehicles, equipment, and supplies to the embarkation area; and
- Preparation of plans for the security of cargo in the embarkation area.

Plans

Three basic embarkation plans are normally prepared by the various command levels within the landing force: the landing force embarkation plan, the group embarkation plan, and the unit embarkation plan.

LANDING FORCE EMBARKATION PLAN.—The landing force embarkation plan includes the organization for embarkation; supplies and equipment to be embarked; embarkation points and cargo assembly areas; control, movement and embarkation of personnel; and miscellaneous infor- mation. The landing force embarkation plan contains information from which the embarkation group commander prepares a more detailed plan.

GROUP EMBARKATION PLAN.—The group embarkation plan, prepared by the embarkation group commander, establishes the formation for embarkation units and assigns shipping to each embarkation unit. It contains the same information as the landing force embarkation plan, but in much greater detail. The group embarkation plan has attached to it or included within the embarkation organization a shipping assignment table.

UNIT EMBARKATION PLAN.—The unit embarkation plan prepared by the embarkation unit commander establishes the formation of embarkation teams and assigns each embarkation team to a ship. It contains, generally, the same information as the group embarkation plan, but in greater detail. Attached to the unit embarkation plan is the unit embarkation organization and shipping assignment table. Naval construction force (NCF) units embarking alone outside of the landing force, either by amphibious means or by air, should prepare an embarkation plan incorporating all of the information necessary for proper embarkation by the unit.

PACKAGING

Standard boxing procedures are required to minimize shipping, packing, and repacking of allowance items and to establish uniformity among the NCF units. Present mobility requirements necessitate being partially packed for redeployment at all times. The best method of obtaining this state of readiness is to use packing boxes for day-to-day storage and for dispensing all types of battalion allowance items. Each NCF unit must fabricate mount-out boxes according to the *Embarkation Manual*, COMCBPAC/COMCBLANTINST 3120.1, for all authorized allowance items within the unit's TOA that can be boxed. Existing boxes may be used if the color and marking codes conform with standard box markings.

Packing Lists

Packing lists must be prepared for each box. One copy is placed inside the box; one copy is mounted in a protective packet on the outside of the box; one copy is kept on file in the embarkation mount-out control center; and, one copy is retained by the department to which the supplies or equipment belong. Packing lists must be sufficiently detailed to locate needed items without having to open and search several boxes.

Mount-Out

When constructing mount-out boxes, observe the following considerations:

- Screw nails (or flathead screws) and glue must be used to assemble the boxes.
- Covers must be bolted to tapped metal inserts, as shown in COMCBPAC/COMCBLANT-INST 3120.1, or an equivalent bolting method.
- Box interiors may be compartmented to suit the contents.
- Gross weight of the boxes should be limited to 250 pounds each for easy handling without material-handling equipment.
- Boxes must be fabricated of 3/4-inch exterior-grade plywood, reinforced with 2 by 4 ends.
- Special boxes for large items are authorized, but should conform to the criteria set forth in COMCBPAC/COMCBLANATINST 3120.1.
- Metal corners or other protection may be installed to prevent shipping damage.

Pre-positioned Stocks and Supplies

Because of the mobile nature of the NCF, it is necessary to pre-position certain supplies and equipment in anticipation of use in contingency mount-outs. These stocks include oil, gasoline, lubricants, rations, and ammunition, plus a full allowance of equipment. During a contingency mount-out, all or part of these pre-positioned stocks may be used. As part of the planning phase, NCF units should check the plan to be supported to determine the exact amount and types of supplies to be embarked and the location of the supplies.

RECOMMENDED READING LIST

NOTE

Although the following references were current when this TRAMAN was published, their continued currency cannot be assured. You therefore need to ensure that you are studying the latest revisions.

Automatic Building Machine Type K-Span Operating Manual, MIC-120 ABM, M.I.C. Industries, Inc.

Naval Construction Force/Seabee Chief Petty Officer, NAVEDTRA 10600, Naval Education and Training Command, Pensacola, Fla., June 1989.

Naval Construction Force/Seabee Petty Officer First Class, NAVEDTRA 10601, Naval Education and Training Command, Pensacola, Fla., December 1989.

CHAPTER 10

HEAVY CONSTRUCTION

Heavy construction includes structures made of steel, timber, concrete, or a combination of these materials. Examples include trestles, timber piers, and waterfront structures. The requirement for heavy construction today is not as important as in earlier years; however, the need to understand this type of construction still remains.

In this chapter, we'll examine the materials used in building heavy structures. We'll also discuss the methods and techniques of heavy construction, including shoring and excavation. In addition, we'll look at the procedures used in maintaining the structures.

TIMBER TRESTLES

LEARNING OBJECTIVE: Upon completing this section, you should be able to identify the parts of a trestle, and describe the procedures for erecting bents and superstructures.

A trestle is a braced framework of timbers, piles, or steel members. It is typically built to carry a roadway across a depression, such as a gully, a canyon, or the valley of a stream. The two main parts of a trestle are the substructure, consisting of the supporting members, and the superstructure, consisting of the decking and the stringers on which the decking is laid.

The substructure of a timber trestle is a series of transverse frameworks called bents. Trestle bents are used on solid, dry ground, or in shallow water with a solid bottom. Pile bents are used in soft or marshy ground, or where the water is so deep or the current so swift that the use of trestle bents is impossible. The posts of a pile bent are bearing piles or vertical members driven into the ground.

NOMENCLATURE

The following terms are common to timber trestle construction:

Abutment—The ground support at each of the extreme ends of a trestle superstructure. Examples are shown in figures 10-1 and 10-2.

Bracing—The timbers used to brace a trestle bent, called transverse bracing, or the timbers used to brace bents to each other, called longitudinal

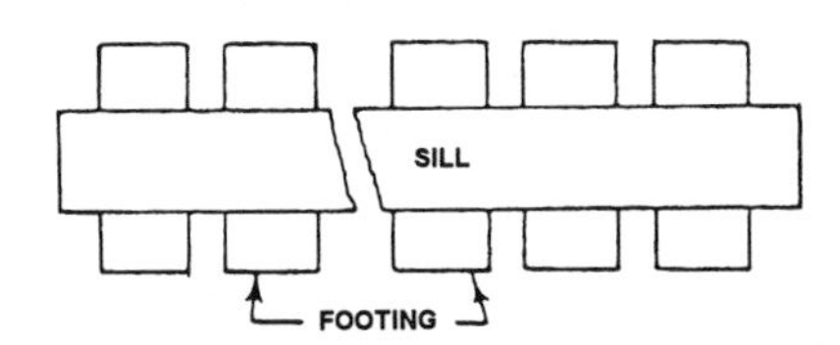

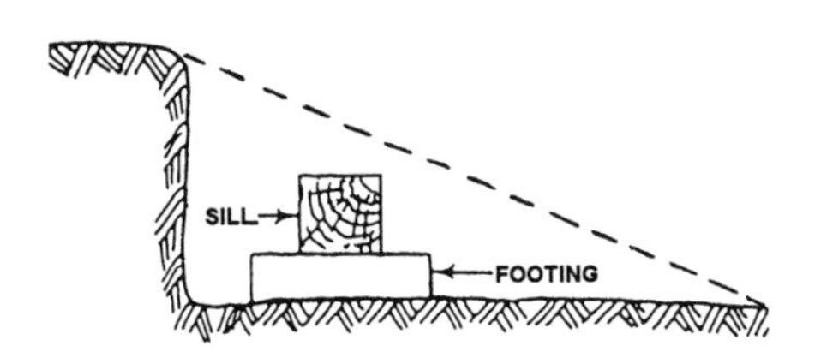

Figure 10-1.—Abutment sill and footing and abutment excavation.

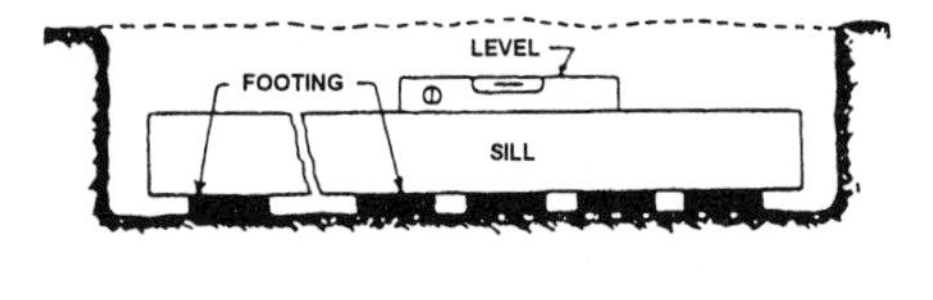

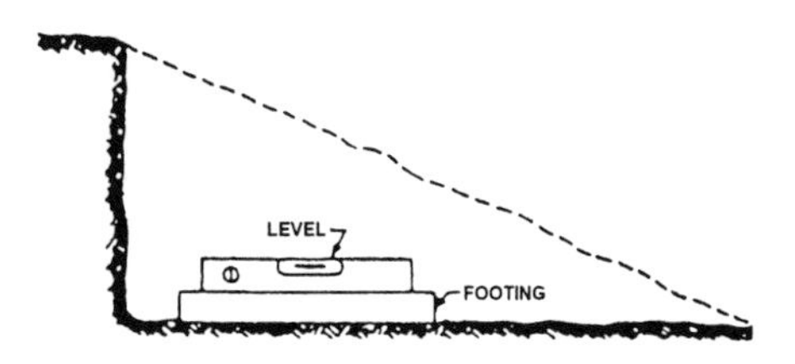

Figure 10-2.—Placing and leveling abutment footings and abutment sill.

bracing. Figure 10-3 shows both types for a two-story trestle bent.

Cap—The uppermost transverse horizontal structural member of a bent. It is laid across the tops of the posts.

Decking—The structure laid on the girders to form the roadway across the trestle. It consists of a lower layer of timbers (flooring) and an upper layer of timbers (treadway).

Footings—The supports placed under the sills. In an all-timber trestle, the footings consist of a series of short lengths of plank. Whenever possible, however, the footings are made of concrete.

Girder—One of a series of longitudinal supports for the deck, which is laid on the caps. Also called a stringer.

Post—One of the vertical structural members.

Sill—The bottom transverse horizontal structural member of a trestle bent, on which the posts are anchored, or transverse horizontal member, which supports the ends of the girders at an abutment.

Substructure—The supporting structure of braced trestle bents, as distinguished from the superstructure.

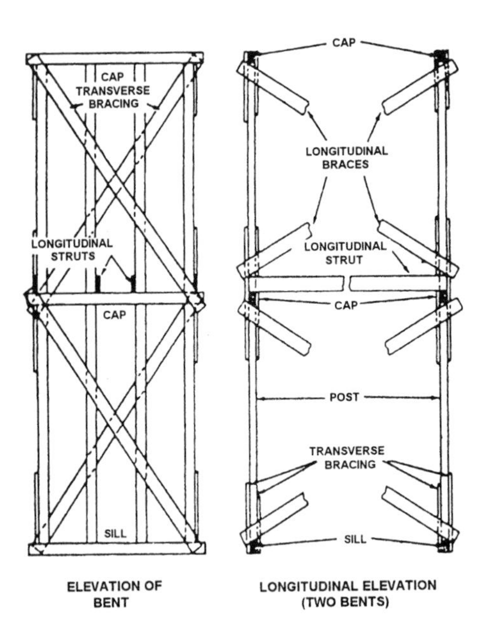

Figure 10-3.—Two-story trestle bent.

Superstructure—The spanning structure of girders and decking, as distinguished from the substructure.

Trestle Bent—A single-story bent or a multistory bent and the support framework or substructure of a trestle. The parts of a single-story bent are shown in figure 10-4. A two-story bent is shown above in figure 10-3.

CONSTRUCTION

After the center line of a trestle has been determined, the next step is to locate the abutment on each bank at the desired or prescribed elevation. The abutments are then excavated to a depth equal to the combined depths of the decking and the stringers, less an allowance for settlement. The abutment footings and the abutment sills are then cut, placed, and leveled (as in fig. 10-2).

The horizontal distance from an abutment sill to the first bent and from one bent to the next is controlled by the length of the girder stock. It is usually equal to the length of the stock, minus about 2 feet for overlap. Girder stock is usually in 14-foot lengths. The center-to-center horizontal distance between bents is usually 14 minus 2, or 12 feet.

To determine the locations of the seats for the trestle bents and the heights of the bents (fig. 10-5), first stretch a tape from the abutment along the center line. Use a builder's level or a line level to level the tape. Drop a plumb bob from the 12-foot mark on the tape to the ground. The position of the plumb bob on the ground will be the location of the first bent. The vertical distance from the location of the bob to the horizontal tape,

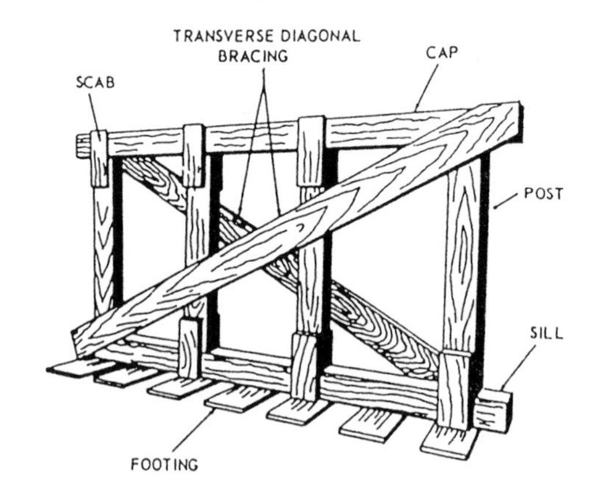

Figure 10-4.—Components of a single-story trestle bent.

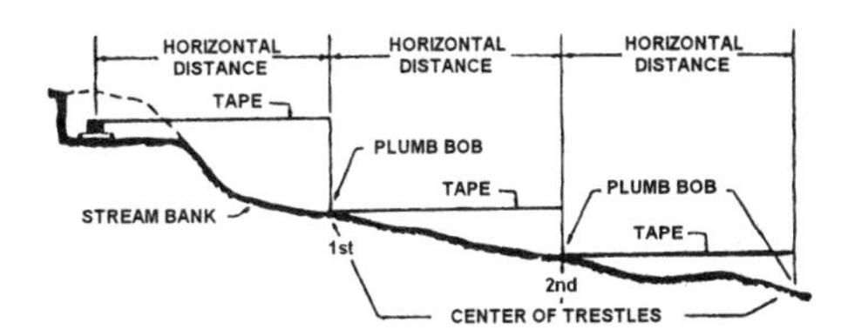

Figure 10-5.—Locating seats for trestle bents.

less the thickness of a footing, will be the height of the first bent.

Next, stretch the tape from the location of the first bent, level it as before, and again drop a plumb bob from the 12-foot mark. The position of the plumb bob will be the location of the section bent. The vertical distance from the location of the bob to the horizontal tape, plus the height of the first bent, less the thickness of the footing, will be the height of the second bent.

Finally, stretch the tape from the location of the second bent and proceed as before. The vertical distance from the location of the bob to the horizontal tape, plus the height of the second bent, less the thickness of a footing, will be the height of the third bent, and so on.

CONSTRUCTING A TRESTLE BENT

When a trestle bent is laid out and constructed, the length of the posts is equal to the height of the bent, less the combined depths of the cap and sill. In a four-post bent, the centers of the two outside posts are located from 1 to 2 1/2 feet inboard of the ends of the sill, and the centers of the two inner posts are spaced equally distant between the other two.

Sills, caps, and posts are commonly made of stock that ranges in size from 12 by 12s to 14 by 16s. If a sill or cap is not square in a cross section, the larger dimension should be placed against the ends of the posts. The usual length for a sill or cap is 2 feet more than the width of the roadway on the trestle. The minimum width for a single-lane trestle is 14 feet; for a two-lane trestle, 18 feet.

Layout

Part of the terrain at an assembly site may be graded flat and used as a framing yard, or a low platform may be constructed for use as a framing platform. To assemble a bent, lay the posts out parallel and properly spaced, and set the cap and sill in position against the ends. Bore the holes for the driftpins through the cap and the sill into the ends of the posts, and drive in the driftpins. Cut a pair of 2- by 8- by 18-inch scabs for each joint and then spike, lag-screw, or bolt the scabs to the joints.

Finally, measure the diagonals to determine the lengths of the transverse diagonal braces. Cut the braces to length and spike, lag-screw, or bolt them to the sills, caps, and posts. Transverse diagonal bracing is usually made of 2 by 8 stock.

Trestle Bent Erection

After assembly, the trestle bent is moved to the abutment, and set in place on the footings at the seat. Carefully plumb the bent and temporarily brace it with timbers running from the top of the bent to stakes driven at the abutment. Lay the superstructure (girders and decking) from the abutment out to the top of the first bent. The second bent is then brought out to the end of the superstructure and set in place. Plumb the second bent and measure the diagonals to determine the lengths of the longitudinal diagonal braces between the first and second bents. Then, cut the braces and spike, lag-screw, or bolt them in place.

The superstructure is then carried out to the second bent, after which the third bent is brought to the end of the superstructure. This procedure is repeated, usually by parties working out from both abutments, until the entire span is completed.

SUPERSTRUCTURE

Timber girders are usually 10 by 16s, 14 feet long, spaced 3 feet 3 1/2 inches on center (OC). Various methods of fastening **timber stringers** to timber caps are shown in figure 10-6, view A. Various methods of fastening **steel girders** to timber caps are shown in view B. This view also shows three ways of fastening a

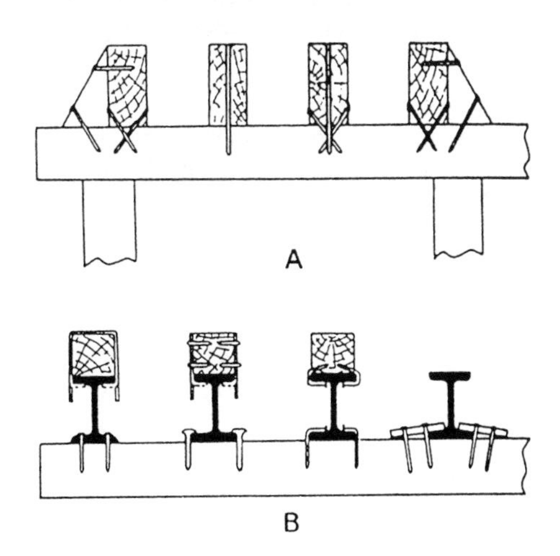

Figure 10-6.—Methods of fastening timber stringers and steel girders to timber caps.

timber-nailing anchorage for flooring to the top of a steel girder.

Timber decking consists of two layers of 3-inch planks. The lower layer, called the flooring, is laid at right angles to the stringers and nailed with two 60d spikes to each stringer crossing. The upper layer, called the tread (fig. 10-7), is laid securely and nailed at a 90° angle to the flooring.

Most of the flooring planks and all of the tread planks are cut to lengths that will bring the ends of the planks flush with the outer faces of the outside stringers. However, at 5-foot intervals along the superstructure, a flooring plank is left long enough to extend 2 feet 8 inches beyond the outer faces of the outside stringers. The extension serves as support for the curb risers, the curb, and the handrail posts, as shown in figure 10-7. The curb risers consist of 3-foot lengths of 6 by 6 timbers, one of which is set in front of each handrail post as shown. A continuous 2 by 6 handrail is nailed to 4 by 4 handrail posts. Each handrail post is supported by a 2 by 4 knee brace, as shown.

An end dam, such as that shown in figure 10-8, is set at each end of the superstructure. This prevents the approach of the road to the trestle from washing out or eroding between the abutment and the girders.

PILE DRIVING TERMINOLOGY AND TECHNIQUES

LEARNING OBJECTIVE: Upon completing this section, you should be able to identify the types of piles used in heavy construction and state the procedures for constructing a timber pier.

The principal structural members in many waterfront structures are piles. There are different types of and uses for piles. The common terms used with piles and pile driving are explained below.

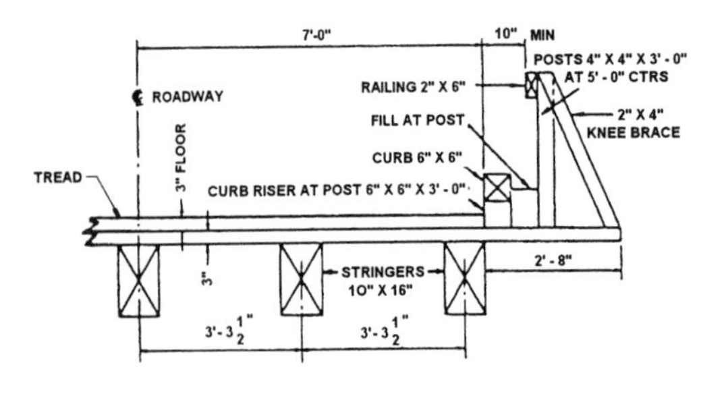

Figure 10-7.—Details of superstructure of a timber trestle.

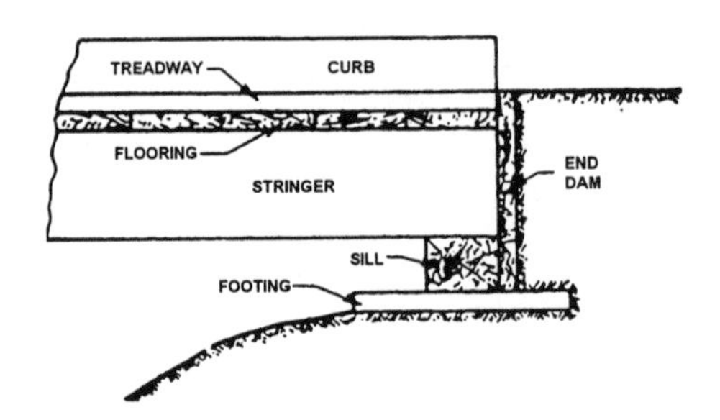

Figure 10-8.—End dam.

TYPES OF PILES

A pile is a load-bearing member made of timber, steel, concrete, or a combination of these materials. It is usually forced into the ground to transfer the load to underlying soil or rock layers when the surface soils at a proposed site are too weak or compressible to provide enough support.

Timber Bearing

Timber bearing piles are usually straight tree trunks cut off above ground swell with the branches closely trimmed and the bark removed. Occasionally, sawed timbers may be used as bearing piles.

CHARACTERISTICS.—A good timber pile has the following characteristics:

- It is free of sharp bends, large or loose knots, shakes, splits, and decay.
- It is uniformly tapered from butt to tip.
- The centers of the butt and tip are end points of a straight line that lies within the body of the pile.

Cross-section dimensions for timber piles should be as follows:

- Piles shorter than 40 feet, tip diameters between 8 and 11 inches, and butt diameters between 12 and 18 inches.
- Piles longer than 40 feet, tip diameters between 6 and 8 inches, and butt diameters between 13 and 20 inches. The butt diameter must not be greater than the distance between the pile leads.

PREPARATION FOR DRIVING.—Timber piles can be damaged while being driven, particularly under hard-driving conditions. To protect a pile against damage, cut the butt of the pile squarely (so the pile hammer will strike it evenly) and chamfer it. When a driving cap is used, the chamfered butt must fit the cap. When a cap is not used, the top end of the pile is wrapped with 10 or 12 turns of wire rope at a distance of about one diameter below the head of the pile (fig. 10-9, views A and B). When a hole is bored in the butt of the pile, double wrappings are used (view C). The pile can also be wrapped or clamped if the butt is crushed or split. As an alternative to wrapping, two half-rings of 3/8-inch steel are clamped around the butt (view D).

The tip of the pile is cut off perpendicular to its axis. When driven into soft or moderately compressible soil, the tip of the pile may be left unpointed. A blunt-end pile provides a larger bearing surface than a pointed-end pile when used as an end-bearing pile. When driven, a blunt-end pile that strikes a root or small obstruction may break through it.

Where soil is only slightly compressible and must be displaced, the tip of the pile is usually sharpened to the shape of an inverted truncated pyramid (fig. 10-9, view A). The blunt end is about 4 to 6 inches square; the length of the point is 1 1/2 to 2 times the diameter of the pile at its foot. A crooked pile may be pointed for driving, as shown in view B.

For hard driving, steel shoes are used to protect the pile tips. A manufactured shoe is shown in view C, and an improvised steel shoe is shown in view D.

Steel Bearing

Steel ranks next to timber in importance, especially where the construction must accommodate heavy loads or the foundations are expected to be used over a long period of time. Steel is best suited for use as bearing piles where piles must be driven under any of the following conditions:

- Piles are longer than 80 feet.
- Column strength exceeds the compressive strength of timber.
- To reach bedrock for maximum bearing surface through overlying layers of partially decomposed rock.
- To penetrate layers of coarse gravel or soft rock, such as coral.
- To attain great depth of penetration for stability (for example, driving piles in a rock-bedded, swiftly flowing stream where timber piles cannot be driven deeply enough for stability).

One of the most common types of steel bearing piles is the pipe pile. An open-end pipe pile is open at the bottom. A closed-end pipe pile is closed at the bottom. Another common type of steel pile is the H-type, often seen as HP. When driving HPs, a special driving cap (shown in fig. 10-10) is used.

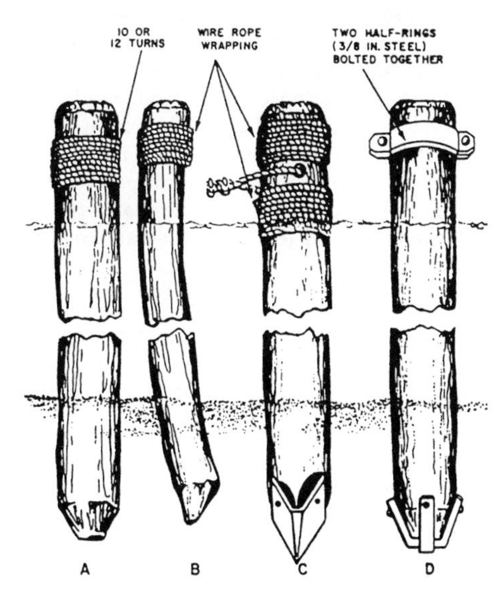

Figure 10-9.—Preparation of timber piles for driving.

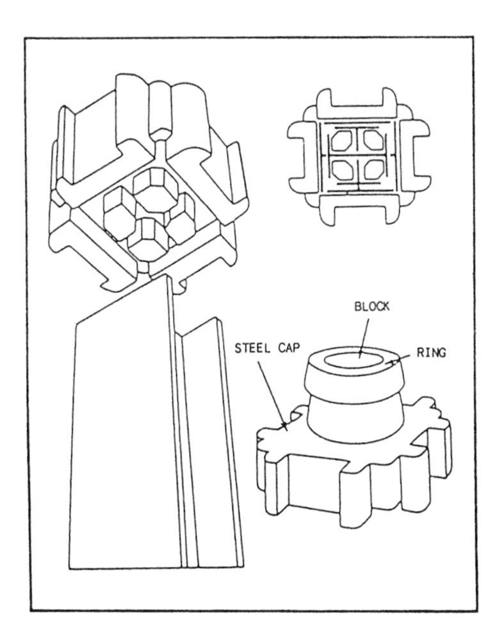

Figure 10-10.—HP-bearing pile and special cap for driving.

Concrete

A concrete bearing pile may be cast in-place or precast. A cast-in-place concrete pile may be a shell type or a shell-less type.

A shell type of cast-in-place pile is constructed as shown in figure 10-11. A steel core, called a mandrel, is used to drive a hollow steel shell into the ground. The mandrel is then withdrawn, and the shell is filled with concrete. If the shell is strong enough, it may be driven without a mandrel.

A shell-less cast-in-place concrete pile is made by placing the concrete in direct contact with the earth. The hole for the pile may be made by driving a shell or a mandrel and shell, or it may be simply bored with an earth auger. If a mandrel and shell are used, the mandrel, and usually also the shell, are removed before the concrete is poured. In one method, however, a cylindrical mandrel and shell are used, and only the mandrel is removed before the concrete is poured. The concrete is poured into the shell, after which the shell is extracted. This sequence of events is shown in figure 10-12.

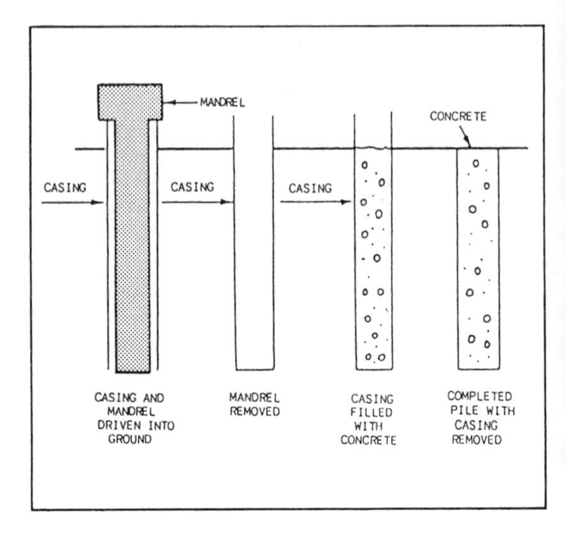

Figure 10-12.—Procedure for cast-in-ground concrete piles.

Casting in place is not usually feasible for concrete piles used in waterfront structures. Concrete piles for waterfront structures are usually precast. The cross section of precast concrete piles is usually either square or octagonal (eight-sided). Square-section piles run from 6 to 24 inches square. Concrete piles more than 100 feet long can be cast, but are usually too heavy for handling without special equipment.

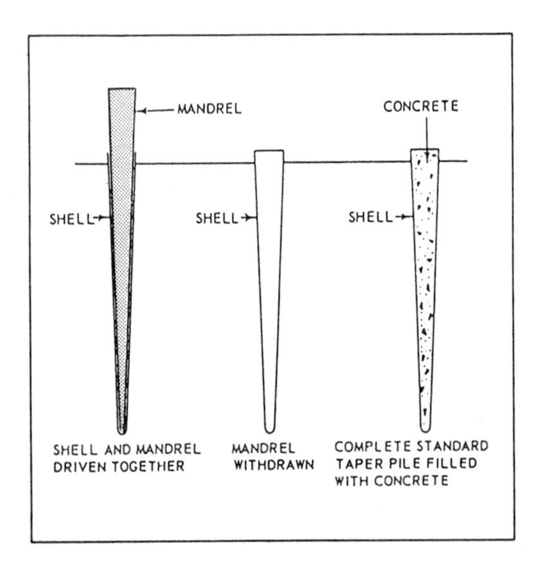

Figure 10-11.—Shell type cast-in-place concrete pile.

Sheet

Sheet piles are special shapes of interlocking piles made of steel, wood, or formed concrete. They are widely used to form a continuous wall to resist horizontal pressures resulting from earth or water loads. Examples include retaining walls, cutoff walls, trench sheathing, cofferdams, and bulkheads in wharves, docks, or other waterfront structures. Cofferdams exclude water and earth from an excavation so that construction can proceed easily. Cutoff walls are built beneath water-retaining structures to retard the flow of water through the foundation.

Sheet piles may also be used in the construction of piers for bridges and left in place. Here, steel piles are driven to form a square or rectangular enclosure. The material inside is then excavated to the desired depth and replaced with concrete.

Timber Pier Piles

Working drawings for advanced base timber piers are contained in *Facilities Planning Guide*, Volume I, NAVFAC P-437. Figure 10-13 shows a general plan; figure 10-14, a part plan; and

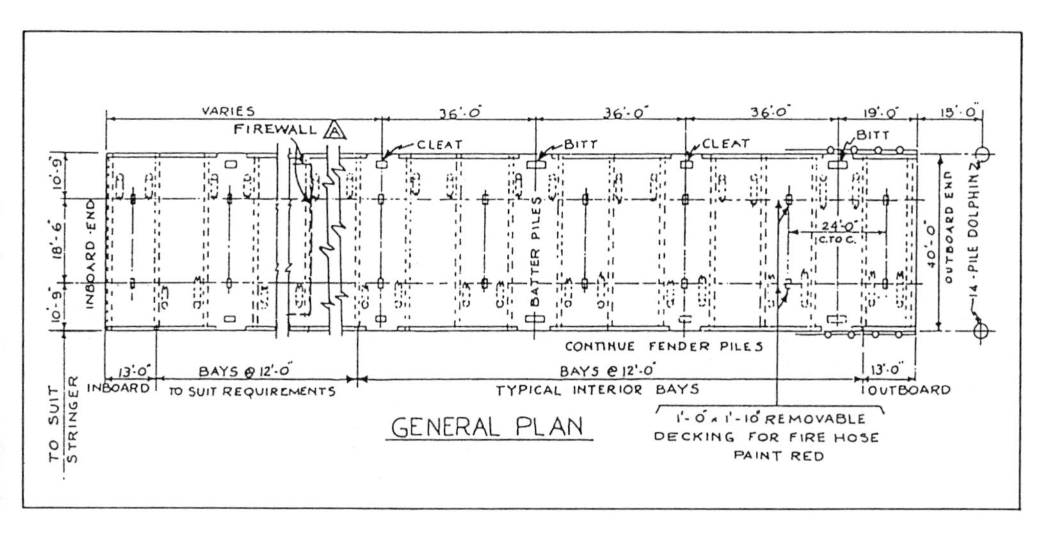

Figure 10-13.—General plan of an advanced base 40-foot-timber pier.

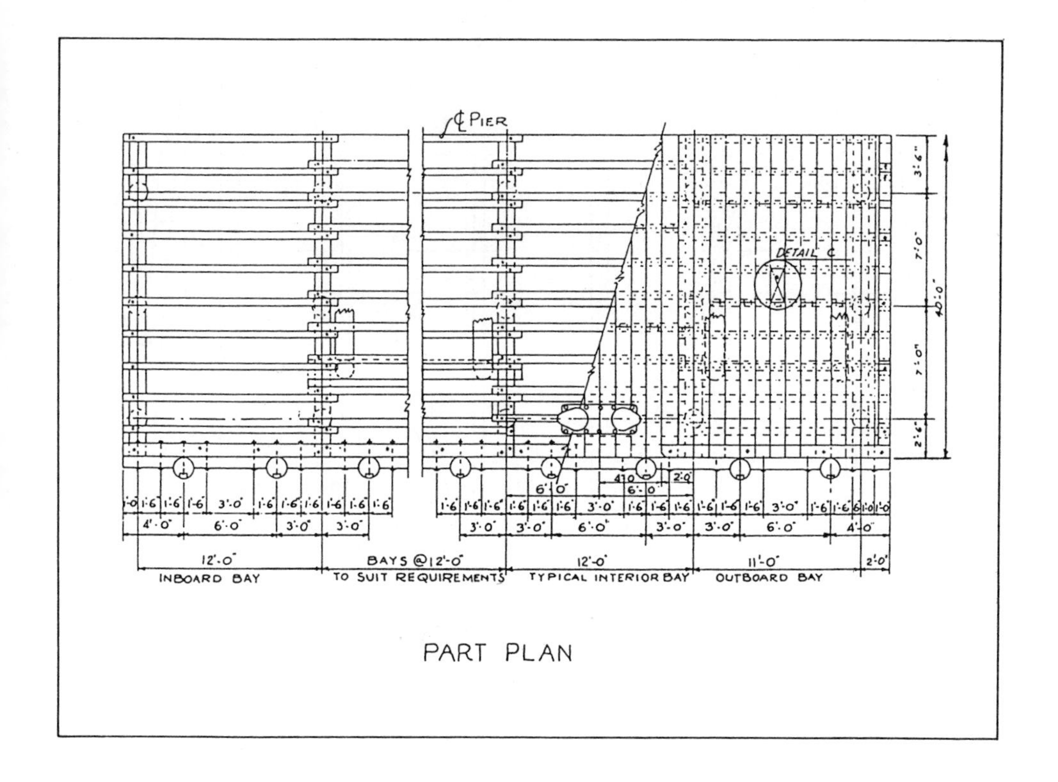

Figure 10-14.—Part plan of an advanced base timber pier.

figure 10-15, a cross section for a 40-foot pier. Thedrawings (examples are shown in figs. 10-13, 10-14, and 10-15) include a bill of materials, showing the dimensions and location of all structural members, driftpins, bolts, and hardware. Figures 10-13 and 10-14 are parts of NAVFAC Drawing No. 6028173; figure 10-15 is a part off NAVFAC Drawing 6028174.

The size of the pier is designated by its width. The width is equal to the length of a bearing-pile cap.

Each part of a pier lying between adjacent pile bents is called a bay, and the length of a bay is equal to the OC spacing of the bents. The general plan (fig. 10-13) shows that the advanced base 40-foot timber pier consists of one 13-foot outboard bay, one 13-foot inboard bay, and as many 12-foot interior bays as needed to meet requirements.

The cross section (fig. 10-15) shows that each bent consists of six bearing piles. The bearing piles are braced transversely by diagonal braces. Additional transverse bracing for each bent is provided by a pair of batter piles. The batter angle is specified as 5 in 12. One pile of each pair is driven on either side of the bent, as shown in the general plan. The butts of the batter piles are joined to 12-inch by 12-inch by 14-foot longitudinal batter-pile caps. Each of these is bolted to the undersides of two adjacent bearing-pile caps with bolts in the positions shown in the part plan (fig. 10-14). The batter-pile caps are placed 3 feet inboard of the center lines of the outside bearing piles in the bent. They are backed by 6- by 14-inch batter-pile cap blocks, each of which is bolted to a bearing-pile cap. Longitudinal bracing between bents consists of 14-foot lengths of 3 by 10 planks, bolted to the bearing piles.

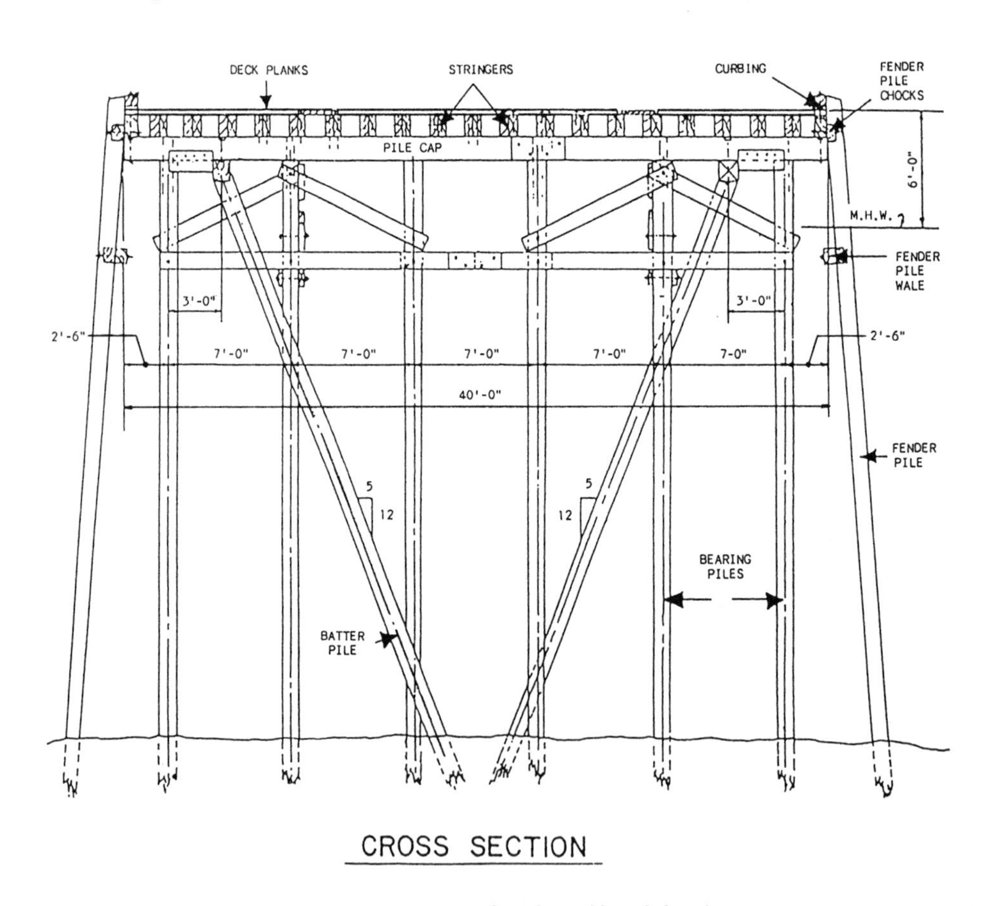

Figure 10-15.—Cross section of an advanced base timber pier.

The superstructure (fig. 10-15) consists of a single layer of 4 by 12 planks laid on 19 inside stringers measuring 6 inches by 14 inches by 14 feet. The inside stringers are fastened to the pile caps with driftbolts. The outside stringers are fastened to the pile caps with through-bolts. The deck planks are fastened to the stringers with 3/8- by 8-inch spikes. After the deck is laid, 12-foot lengths of 8 by 10s are laid over the outside stringers to form the curbing. The lengths of curbing are distributed as shown in the general plan. The curbing is bolted to the outside stringers to form the curbing. The lengths of curbing are distributed as shown in the general plan. The curbing is bolted to the outside stringers with bolts.

The pier is equipped with a fender system for protection against shock, caused by contact with vessels coming or lying alongside. Fender piles, spaced as shown in the part plan, are driven along both sides of the pier and bolted to the outside stringers with bolts. The heads of these bolts are countersunk below the surfaces of the piles. An 8 by 10 fender wale is bolted to the backs of the fender piles with bolts. Lengths of 8 by 10 fender pile chocks are cut to fit between the piles and bolted to the outside stringers and the fender wales. The spacing for these bolts is shown in the part plan. As indicated in the general plan, the fender system also includes two 14-pile dolphins, located 15 feet beyond the end of the pier. A dolphin is an isolated cluster of piles, constructed as shown in figure 10-16. A similar cluster attached to a pier is called a pile cluster.

PILE DRIVING TECHNIQUES

When driving piles of any type, always watch both the pile and equipment. Care must be taken to avoid damaging the pile or the driving hammer. Watch the piles carefully for any indications of splitting or breaking below ground. The next section covers some of the more common problems you might encounter.

Springing and Bouncing

Springing means that the pile vibrates too much laterally. Springing may occur when a pile is crooked, when the butt has not been squared off properly, or

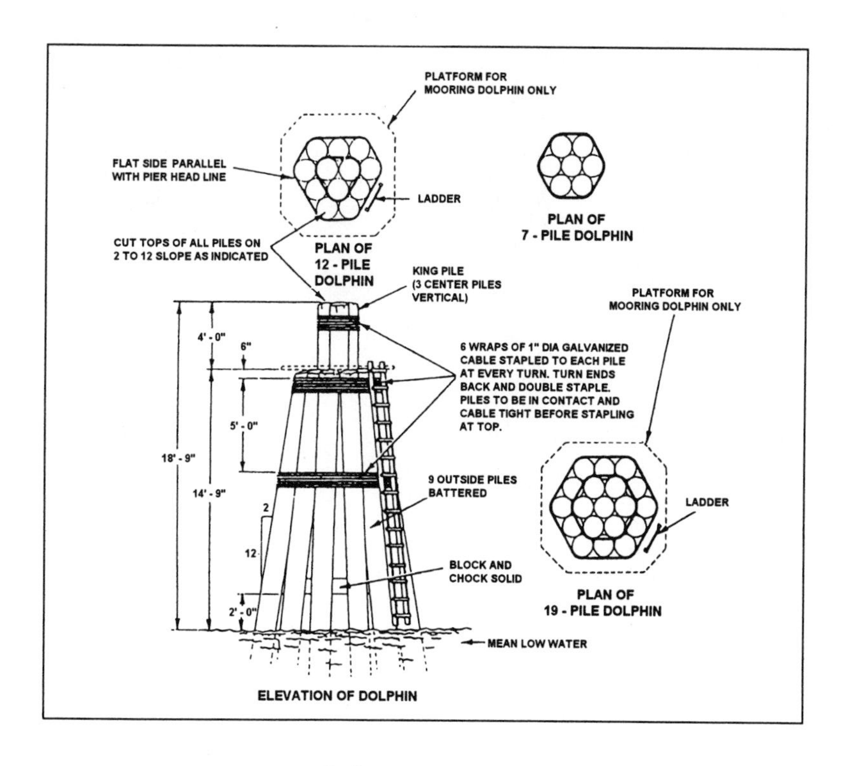

Figure 10-16.—Dolphins.

when the pile is not in line with the fall of the hammer. Always make sure the fall of the hammer is in line with the pile axis. Otherwise, the head of the pile and the hammer may be severely damaged and much of the energy of the hammer blow lost.

Excessive bouncing may be caused by a hammer that is too light. However, it usually occurs when the butt of the pile becomes crushed or broomed, as when the pile meets an obstruction or penetrates to a solid footing. When a double-acting hammer is being used, bouncing may result from too much steam or air pressure. With a closed-end diesel hammer, if the hammer lifts on the upstroke of the ram piston, the throttle setting is probably too high. Back off on the throttle control just enough to avoid this lifting. If the butt of the timber pile has been crushed or broomed more than an inch or so, it should be cut back to sound wood before you drive it any more.

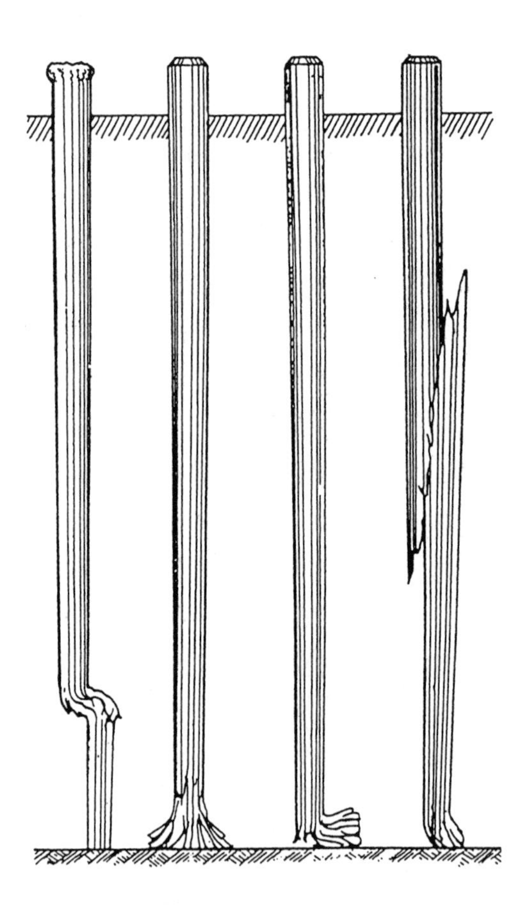

Figure 10-17.—Types of pile damage caused by overdriving timber piles.

Obstruction and Refusal

When a pile reaches a level where 6 blows of a drop hammer or 20 blows of a steam or air hammer do not drive it more than an average of 1/8 inch per blow, the pile has either hit an obstruction or has been driven to refusal. In either case, further driving is likely to break or split the pile. Examples of typical damage are shown if figure 10-17.

If the lack of penetration seems to be caused by an obstruction, 10 or 15 blows of less than maximum force may be tried. This may cause the pile to displace or penetrate the obstruction. For obstructions that cannot be disposed of in this manner, it is often necessary to pull (extract) the pile and clear the obstruction.

When a pile has been driven to a depth where deeper penetration is prevented by friction, the pile has been driven to refusal. It is not always necessary to drive a friction pile to refusal. Such a pile needs to be driven only to the depth where friction develops the required load-bearing capacity.

Straightening

Piles should be straightened when any misalignment is noticed during driving. The accuracy of

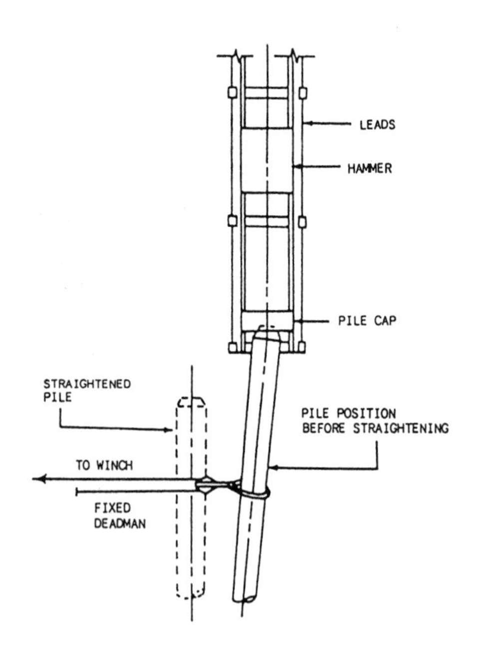

Figure 10-18.—Realigning pile by pull on a line to a winch.

alignment desirable for a finished job depends on various factors. Generally, though, a pile more than a few inches out of its plumb line should be trued. The greater the penetration along the wrong line, the harder to get the pile back into plumb. There are several methods of realigning a pile.

One method of realignment is to use pull from a block and tackle, with the impact of the hammer jarring the pile back into line (fig. 10-18). The straightening of steel bearing piles must include twisting the individual piles to bring the webs of the piles parallel to the center line of the bent.

Another method of realignment is to use a jet (fig. 10-19), either alone or with either of the other two methods. Jetting a pile can be done with either water or air.

When all piles in a bent have been driven, they can be pulled into proper spacing and alignment with a block and tackle and an aligning frame, as shown in figures 10-20 and 10-21.

Pulling

A pile that has hit an obstruction, has been driven in the wrong place, has been split or broken in driving, or is to be salvaged (steel sheet piles are frequently salvaged for reuse) is usually pulled (extracted). Pulling should be done as soon as possible after driving. The longer the pile stays in the soil, the more compact the soil becomes, and the greater the resistance to pulling will be.

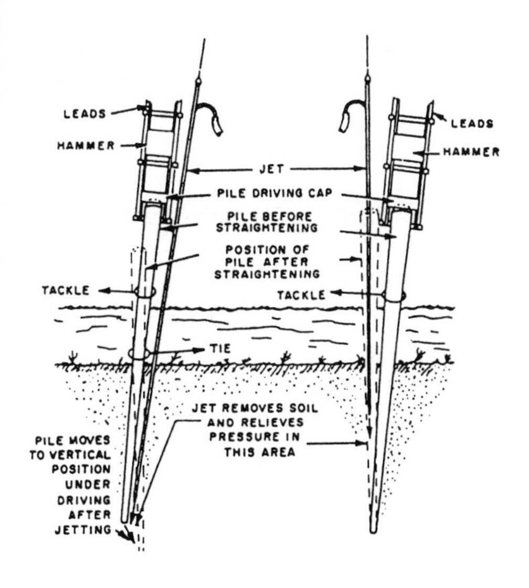

Figure 10-19.—Realigning pile by jetting.

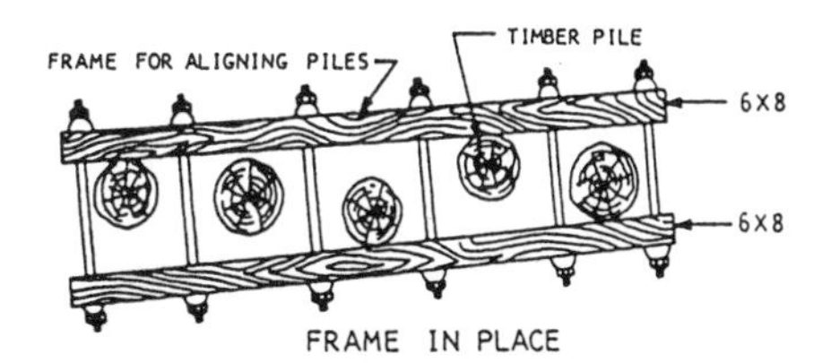

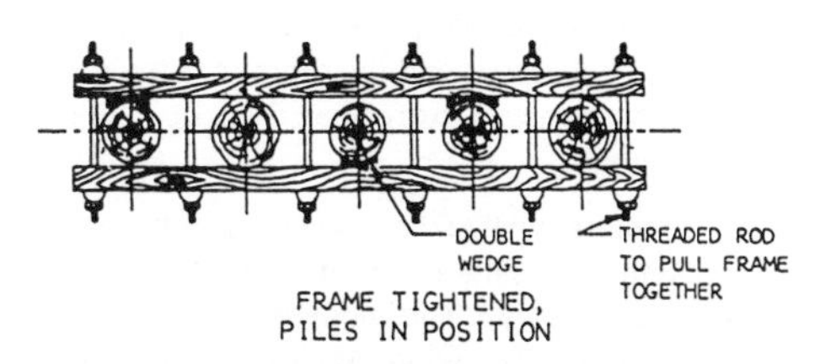

Figure 10-20.—Aligning framing used for timber pile bent.

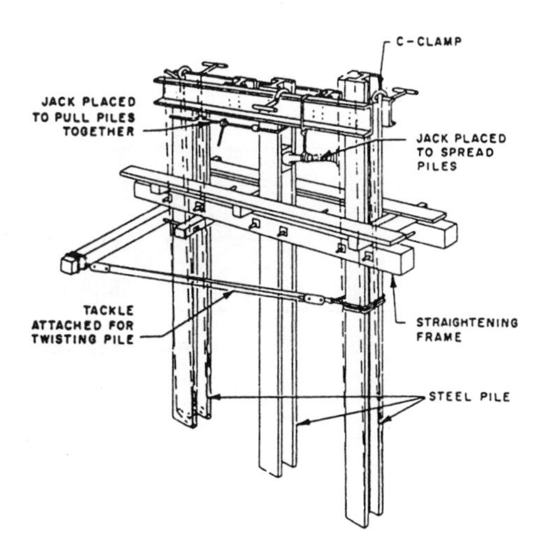

Figure 10-21.—Aligning and capping steel pile bents.

WATERFRONT STRUCTURES

LEARNING OBJECTIVE: Upon completing this section, you should be able to describe the uses of and construction methods for offshore, alongshore, wharfage, and below the water table construction.

Waterfront structures are broadly divided into three main categories: offshore structures creating a sheltered harbor; alongshore structures establishing and maintaining a stable shoreline; and wharfage structures allowing vessels to lie alongside for loading or unloading.

OFFSHORE

Offshore structures include breakwaters and jetties. They are alike in construction and differ mainly in function.

Breakwaters and Jetties

In an offshore barrier, the breakwater interrupts the action of the waves of open water to create an area of calm water between it and the shore. A jetty works to direct and confine a current or tidal flow into a selected channel.

The simplest type of breakwater or jetty is the rubble mound (also called rock mound). An example is shown in figure 10-22. The width of its cap may vary from 15 to 70 feet. The width of its base depends on the width of the cap, height of the structure, and slope of the inner and outer faces.

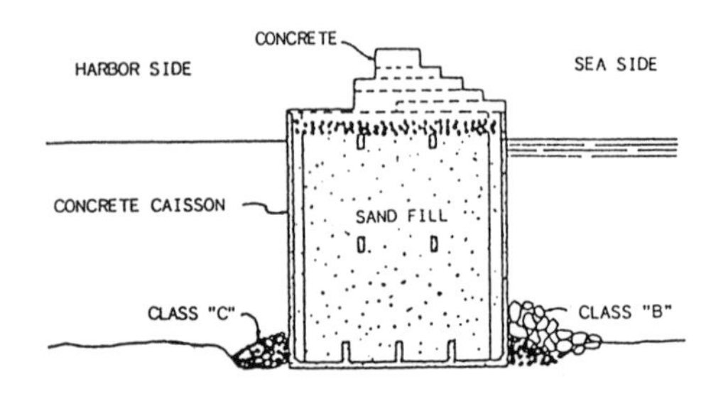

Figure 10-24.—Caisson breakwater/jetty.

Rubble-mound breakwaters or jetties are constructed by dumping rock from either barges or railcars (running on temporary pile-bent structures) and by placing upper rock and cap rock with floating cranes.

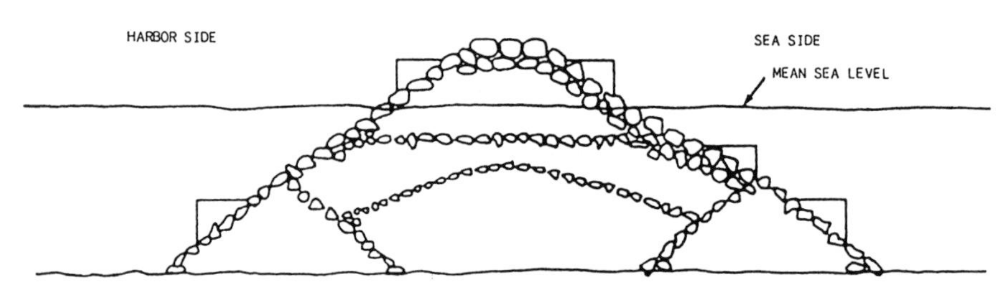

Figure 10-22.—Rubble-mound breakwater/jetty.

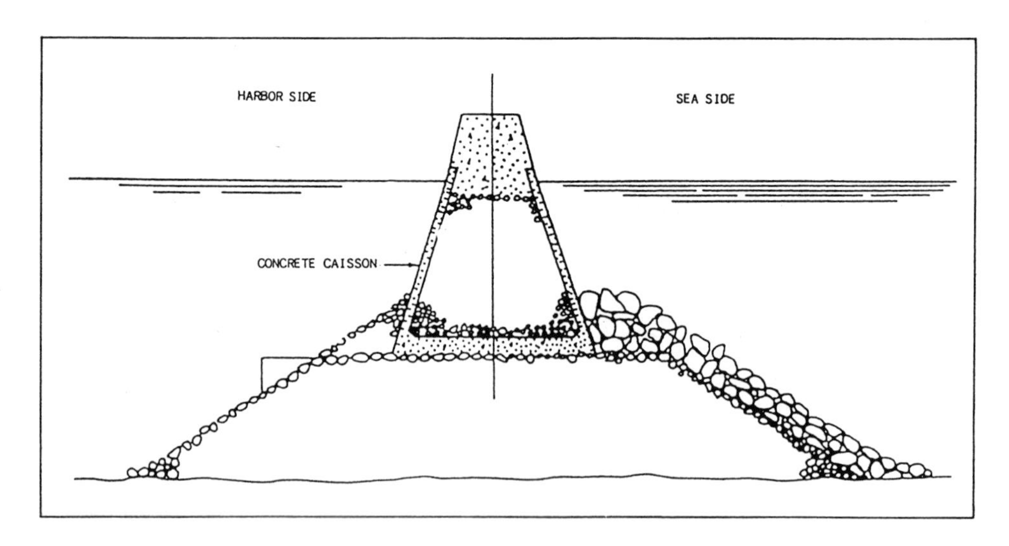

Figure 10-23.—Composite breakwater/jetty.

For a deepwater site or one with an extreme range between high and low tides, a rubble-mound breakwater or jetty may by topped with a cap structure to form the composite type shown in figure 10-23. In this case, the cap structure consists of a series of precast concrete boxes called caissons, each of which is floated over its final location and sunk into place by filling with rock. A single-piece concrete cap is then cast in place on the top of each caisson. Breakwaters and jetties are sometimes built entirely of caissons. A typical caisson breakwater/jetty is shown in figure 10-24. A jetty may also be constructed to serve as a wharfage structure. If so, it is still called a jetty.

ALONGSHORE

Alongshore structures include seawalls, groins, and bulkheads. Their main purpose is to stabilize a shoreline.

Seawalls

Seawalls vary widely in details of design and materials, depending on the severity of the exposure, the value of the property to be protected, and other considerations. Basically, though, they consist of some form of barrier designed to break up or reflect the waves and a deep, tight cutoff wall to preclude washing out of the sand or soil behind and under the barrier. The cutoff wall is generally constructed of timber, steel, or concrete sheet piling. Figure 10-25

Figure 10-25.—Riprap seawall.

shows a rubble-mound seawall. The stone protecting the shoreline against erosion is called riprap. Therefore, a rubble-stone seawall is called a riprap seawall.

Various types of cast-in-place concrete seawalls are the vertical-face, inclined-face, curved-face, stepped-face, and combination curved-face and stepped-face. The sea or harbor bottom along the toe (bottom of the outside face) of a seawall is usually protected against erosion (caused by the backpull of receding waves) by riprap piles against the toe.

Groins

Groins, built like breakwaters or jetties, extend outward from the shore. Again, they differ mainly in function. A groin is used where a shoreline is in danger of erosion caused by a current or wave action running obliquely against or parallel to the shoreline. It is placed to arrest the current or wave action or to deflect it away from the shoreline.

Groins generally consist of tight sheet piling of creosoted timber, steel, or concrete, braced with wales and with round piles of considerable length. Groins are usually built with their tops a few feet above the sloping beach surface that is to be maintained or restored.

Bulkheads

A bulkhead has the same general purpose as a seawall: to establish and maintain a stable shoreline. But, whereas a seawall is self-contained, relatively thick, and supported by its own weight, a bulkhead is a relatively thin wall supported by a series of tie wires or tie rods, running back to a buried anchorage (deadman). A timber bulkhead for a bridge abutment is shown in figure 10-26. It is made of wood sheathing (square-edged, single-layer planks), laid horizontally.

Most bulkheads, however, are made of steel sheet piles, an example of which is shown in figure 10-27. The outer ends of the tie rods are anchored to a steel wale running horizontally along the outer face of the bulkhead.

This wale is usually made up of pairs of steel channels bolted together, back to back. A channel is a structural steel member with a U-shaped section. Sometimes the wale is placed on the inner face of the bulkhead, and the piles are bolted to it.

The anchorage shown in figure 10-27 is covered by backfill. In stable soil above the groundwater level, the anchorage may consist simply of a buried timber, a concrete deadman, or a row of driven and buried sheet piles. A more substantial anchorage for each tie rod is used below the groundwater level. Two common types of anchorages are shown in figure 10-28. In view A, the anchorage for each tie rod consists of a timber cap, supported by a batter pile. In view B, the anchorage consists of a reinforced concrete cap, supported by a pair of batter piles. As indicated in the figure, tie rods are supported by piles located midway between the anchorage and the bulkhead.

Bulkheads are constructed from working drawings like those shown in figure 10-29. The detail plan for the bulkhead shows that the anchorage consists of a row of sheet piles to which the inner ends

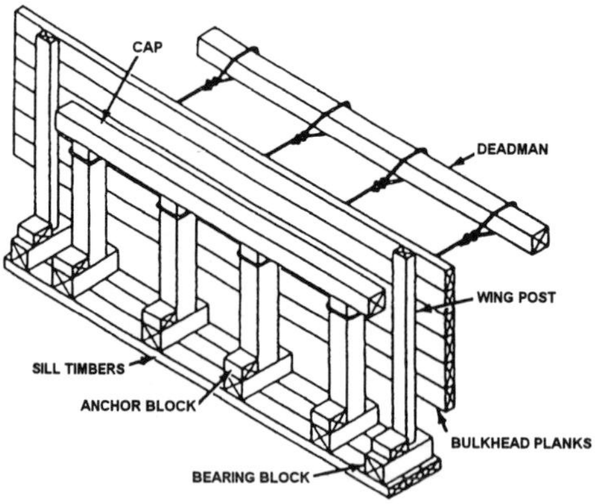

CAP
DEADMAN
WING POST
SILL TIMBERS
ANCHOR BLOCK
BULKHEAD PLANKS
BEARING BLOCK

Figure 10-26.—Timber bulkhead for bridge abutment.

Figure 10-27.—Constructed steel sheet pile bulkhead.

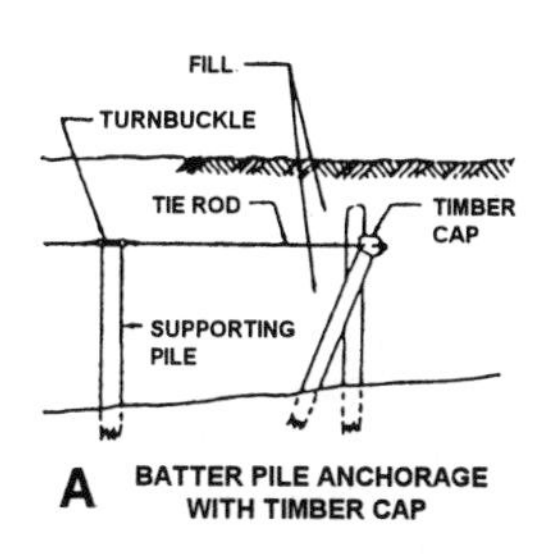

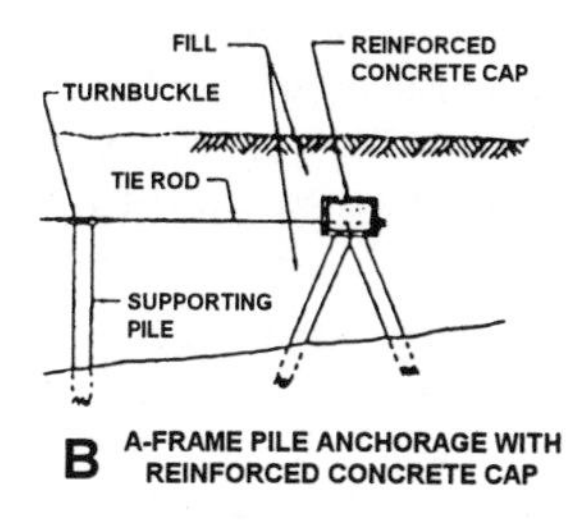

Figure 10-28.—Two types of tie-rod anchorages for bulkheads.

of the tie rods are anchored by means of a channel wale.

In the figure, the construction sequence begins when the shore and bottom are first excavated to the level of the long, sloping dotted line. The sheet piles for the bulkhead and the anchorage are then driven. The supporting piles for the tie rods are driven next, after which the tie rods between the bulk and the anchorage are set in place and the wales are bolted on. The tie rods are prestressed lightly and uniformly, and the backfilling then begins.

The first backfilling operation consists of placing fill over the anchorage, out to the dotted line shown in the plan. The turnbuckles on the tie rods are then set to bring the bulkhead plumb, and the rest of the backfill is worked out to the bulkhead. After the backfilling is completed, the bottom outside the bulkhead is dredged to the desired depth.

WHARFAGE

As mentioned earlier, wharfage structures allow vessels to lie alongside for loading or unloading. Moles and jetties are the most typical forms.

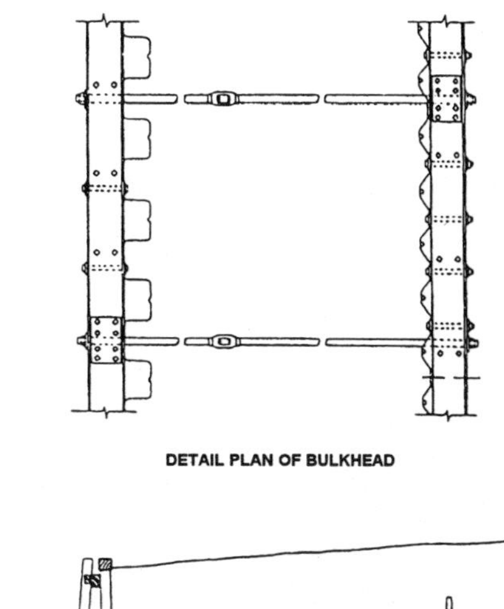

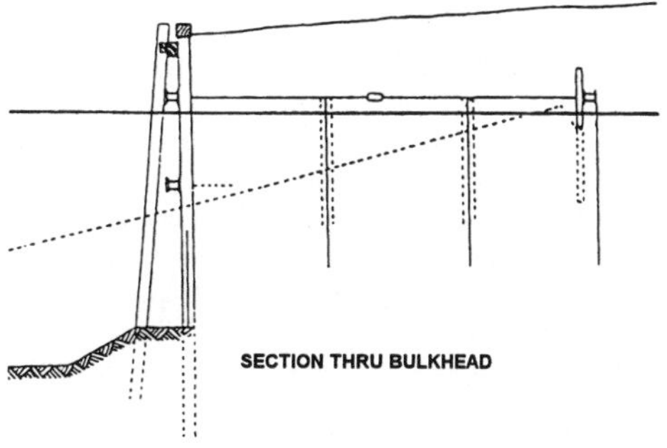

Figure 10-29.—Working drawings for a steel sheet pile bulkhead.

Moles and Jetties

A mole is simply a breakwater that serves as a wharfage structure. The only difference is that its inner or harbor face must be vertical and its top must function as a deck. In a similar way, jetties also serve as wharfage structures.

BELOW THE WATER TABLE

When construction is carried on below the groundwater level, or when underwater structures like seawalls, bridge piers, and the like, are erected, it is usually necessary to temporarily keep the water out of the construction area. This is typically done with well points, cofferdams, or caissons.

Well Points

Well points are long pipes thrust into the ground down to the level at which the water must be excluded.

They are connected to each other by a pipeline system that heads up at a water pump. Well point engineers determine the groundwater level and the direction of flow of the groundwater, and the well point system is placed so as to cut off the flow into the construction area. Well pointing requires highly specialized personnel and expensive equipment.

Cofferdams

The cofferdam is a temporary structure, usually built in place, and tight enough so that the water can be pumped out of the structure and kept out while construction on the foundations is in progress. Common cofferdam types are earthen, steel sheeting, wooden sheathing, and crib. Figure 10-30 shows a cofferdam under construction.

An earthen cofferdam is built by dumping earth fill into the water, shaped to surround the construction area without encroaching upon it. Because swiftly moving currents can carry the material away, earthen cofferdams are limited to sluggish waterways where the velocities do not exceed 5 feet per second. Use is also limited to shallow waters; the quantities of material required in deep waters would be excessive due to the flat slopes to which the earth settles when deposited in the water. For this reason, the earthen type is commonly combined with another type, such as sheathing or cribbing, to reduce the quantities of earthwork.

Steel is commonly used for cofferdam construction. Sheet piling is manufactured in many interlocking designs and in many weights and shapes for varying load conditions. The piling is driven as sheeting in a row to form a relatively tight structure surrounding the construction area. This pile wall is supported in several ways. It may be supported by a framework of stringers and struts. A cofferdam wall can consist of a double row of piles tied together with heavy steel ties and filled with earth. This can square, rectangular, circular, or oval shape for stability around the construction area.

Wooden sheathing, instead of steel, is similarly used in cofferdam constructions. Interlocking timber sheathing is driven as a single wall and supported by stringers and cross struts between walls, or it is driven in double rows as a wall. The sheathing in each row is connected and tied with braces.

Wooden or concrete cribbing may be used in cofferdam construction. The cribbing offers stability

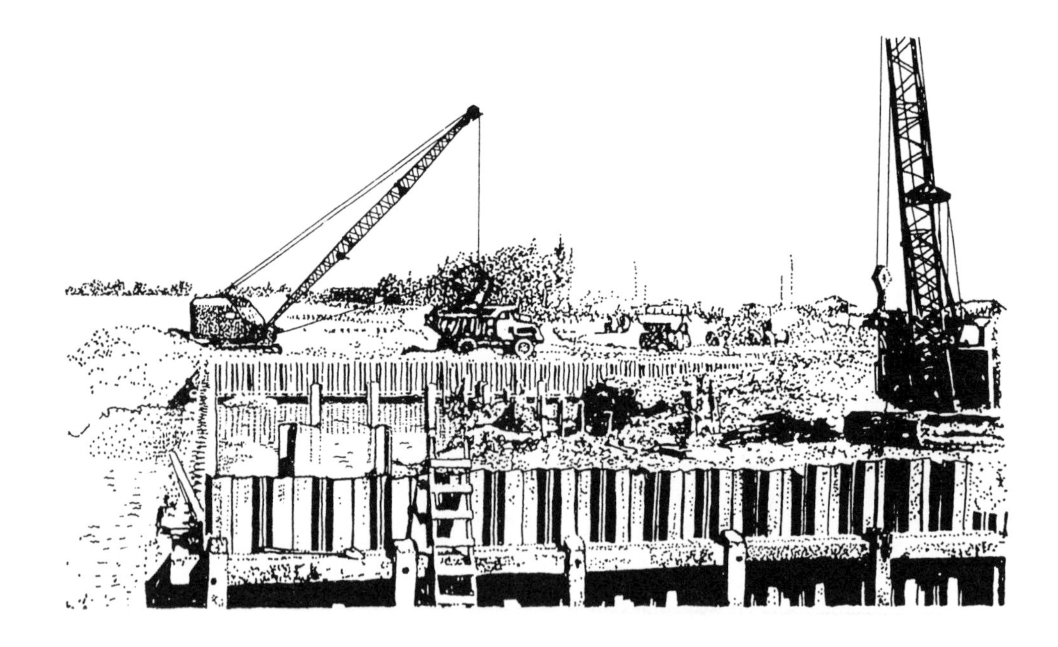

Figure 10-30.—Cofferdam under construction.

to the cofferdam wall. It also provides watertightness when filled with earth and rock.

Movable cofferdams of timber, steel, or concrete have been built, but their uses and designs are very similar to those discussed under boxes and open caissons, below.

Caissons

Caissons are boxes or chambers used for construction work underwater. There are three forms of caissons used in constructing foundations underwater: box, open, and pneumatic caisson. If the structure is open at the top and closed at the bottom, it is called a box caisson. If it is open both at the top and the bottom, it is an open caisson. If it is open at the bottom and closed at the top, and compressed air is used, it is a pneumatic caisson.

It is sometimes difficult to distinguish between a cofferdam and caisson. In general, if the structure is self-contained and does not depend upon the surrounding material for support, it is a caisson. However, if the structure requires such support as sheathing or sheet piling, it is a cofferdam. Retaining walls and piers may be built of boxes of wood, steel, or reinforced concrete, floated into place and then filled with various materials. These are known as floating caissons. Open caissons may be constructed of wood or steel sheet piling.

The preceding information provides only a basic understanding of heavy construction. As with other phases of construction, specialized tools and equipment will be required. The Table of Allowance (TOA) at your command will have these items. Follow all safety rules and manufacturers' recommendations for operations and maintenance.

RECOMMENDED READING LIST

NOTE

Although the following reference was current when this TRAMAN was published, its continued currency cannot be assured. You therefore need to ensure that you are studying the latest revision.

Pile Construction, Field Manual 5-134, Headquarters, Department of the Army, Washington, D.C., 1985.

APPENDIX I

GLOSSARY

ABUTMENT—Masonry, timber, or timber and earth structures supporting the end of a bridge or an arch.

ACOUSTICAL TILE—Any tile composed of materials that absorb sound waves.

AGGREGATE—Crushed rock or gravel screened to size for use in road surfaces, concrete, or bituminous mixes.

AIR-ENTRAINED CONCRETE—Concrete containing millions of trapped air bubbles.

ALLIGATORING—A defect in a painted surface, resulting from the application of a hard finish coat over a soft primer. The checked pattern is caused by the slipping of the new coat over the old coat. The old coat can be seen through the cracks.

ANCHOR BOLTS—Bolts used to fasten columns, girders, soleplates, or other members to concrete or masonry.

ANCHORS—Devices giving stability to one part of a structure by securing it to another part, such as toggle bolts holding structural wood members to a masonry block wall.

AS-BUILT DRAWINGS—Drawings made during or after construction, illustrating how various elements of the project were actually installed.

ASPHALT SHINGLE—A type of composition shingle made of felt and saturated with asphalt or tar pitch.

ASTRAGAL—A closure between the two leaves of a double-swing or double-slide door to close the joint. This can also be a piece of molding.

AUGER—A boring bit.

BATCH—The amount of concrete mixed at one time regardless of quantity.

BATTER BOARDS—Two boards nailed at right angles to posts set up near the proposed corner of an excavation for a building and used for transferring building lines.

BEARING PILE—A pile carrying a superimposed vertical load.

BERM—An artificial ridge of earth.

BINDER—Hot melted pitch (or asphalt) applied between the layers of a built-up roof to bind the layers of felt together.

BIRD'S-MOUTH—A notch cut in the lower edge of a rafter, to fit over the top wall plate. Formed by a level line and a plumb cut.

BOX NAILS—Lightweight nails with large heads.

BRAD—A slender nail with a small head.

BREAKWATER—A barrier constructed to shield the interior waters of a harbor from wave forces.

BRICK—Solid blocks of fine clay.

BRIDGING—Crossed or solid supports installed between joists (floor or ceiling) to help evenly distribute the load and brace the joists against side sway.

BULKHEAD—A retaining wall, generally vertical.

BUTTERING—Putting mortar on a brick or block with a trowel before laying.

CAISSON—A watertight box structure surrounding work below water.

CANTILEVER—A projecting beam supported only at one end.

CARRIAGE BOLT—A partially threaded bolt with a head that is flat on the underside and rounded on top.

CASING—The trim around doors and windows.

CASING NAILS—Twopenny (2d) to fortypenny (40d) nails with flaring heads.

CEMENT—Fuzed and pulverized limestone and clay.

CHASE—A vertical recess in a wall for pipes.

COFFERDAM—A watertight enclosure.

COMMON BOND—Five stretcher courses with the sixth as an all header course.

COMMON NAILS—Twopenny (2d) to sixtypenny (60d) strong nails.

COMPOSITE PILES—Piles formed of one material in the lower section and another in the upper.

CONCRETE—Artificial stone made of cement, water, sand, and aggregate.

CONCRETE BUGGY—Two-wheeled buggy for transporting concrete, nicknamed "Georgia Buggy."

CONCRETE PILES—Piles made of concrete, either cast in place or precast.

CONSTRUCTION JOINT—A joint that runs through concrete. made by pouring sections of a structure at different times.

CORNICE—The area under the eaves where the roof and sidewalls meet.

COURSE—A single layer of bricks, stone, or other masonry.

CREOSOTE—A coal tar distillate used for preserving wood.

CRIPPLE—Any frame member shorter than a regular member.

CROWN—The outside curve of a twisted, bowed, or cupped board.

CURING—The process of keeping concrete damp and at favorable temperatures to ensure complete hardening.

DOLPHIN—A group of piles in water driven close (clustered) together and tied so that the group is capable of withstanding large lateral forces from vessels and other floating objects.

DRESSING—Trimming or planing; a term usually applied to lumber.

DRY ROT—Fungus growth making wood soft or brittle.

DRYWALL—A system of interior wall finish using sheets of gypsum board and taped joints.

EAVE—The part of a roof projecting over the sidewall.

EFFLORESCENCE—A white powdery substance forming on masonry surfaces. It is caused by calcium carbide in the mortar.

END-BEARING PILE—A bearing pile deriving practically all its support from a firm underlying stratum.

ESSEX BOARD MEASURE—A method for rapidly calculating board feet.

EXPANSION JOINT—Construction joint with expandable material at the contact points.

FASCIA—The flat outside horizontal member of a cornice placed in a vertical position.

FERROUS—Any metal containing a high percentage of iron.

FINISHING NAILS—Twopenny (2d) to twentypenny (20d) sizes with small barrel-shaped heads.

FOOTING—An enlargement at the lower end of a wall to distribute the load to a wider area of supporting soil.

FURRING—Any extra material added to another piece or member to bring an uneven surface to a true plane and to prove additional nailing surface.

GAIN—An area removed by chiseling where hinges and locks can be mounted flush with a surface.

GIN POLE—An upright guy pole with hoisting tackle and foot-mounted snatch block used for vertical lifts.

GIRDER—A supporting beam laid crosswise of the building; a long truss.

GIRT—A horizontal brace used on outside walls covered with vertical sliding.

GLAZE—The process of installing glass panes in window frames and doorframes and applying putty to hold the glass in place.

GROIN—A bulkhead, generally made of piling, built out from the shoreline perpendicular to the direction of the current or drift to cut off and prevent the carrying of beach materials along the shore.

GROUT—A mixture of sand, cement, and water that can be poured.

GUNITE—A patent name for spray concrete.

GUSSET—A plate connecting members of a truss together.

HONEYCOMBING—Sections of weak, porous concrete.

HYDRATION—The chemical reaction between cement and water causing the cement paste to harden and to bind the aggregates together to form mortar or concrete.

JETTING—A method of forcing water around and under a pile to displace and lubricate the surrounding soil.

JETTY—A term designating various types of small wharf structures, such as a small boat jetty or a refueling jetty. In harbor-protection works, a rock mound or other structure extending into a body of water to direct and confine the stream or tidal flow to a selected channel.

JOIST—A member that makes up the body of the floor and ceiling frames.

LAG SCREW—A screw with a wrench head and wood screw threads.

LEADS—Points at which block and brick are laid up a few courses and used as guides.

LINE—Strands of natural or synthetic fiber twisted together, sometimes referred to as "rope."

LINTEL—A support beam placed over an opening in a wall.

MILLWORK—In woodworking, any material that has been machined, finished, and partly assembled at the mill.

MITER—A butt joint of two members at equal angles.

MOLE—A massive stone or masonry breakwater constructed of concrete or steel sheet pile and constructed on the inner side of a jetty for unloading and loading ships.

MONOLITHIC POUR—Concrete cast in a single pour.

MORTAR—Sand, water, and cementing material in proper proportions.

MOUSING—Turns of cordage around the opening of a block hook.

MULLION—The division between multiple windows or screens.

MUNTIN—The small members dividing glass panes in a window frame; vertical separators between panels in a panel door.

PARAPET—The part of a wall above the roof line.

PEB—Preengineered building.

PERLITE—Lightweight concrete aggregate.

PIGMENT—An insoluble coloring substance, usually in powder form, mixed with oil or water to color paints.

PILE—Load-bearing member made of timber, steel, concrete, or a combination of these materials; usually forced into the ground.

PILE BENT—Two or more piles driven in a row transverse to the long dimension of the structure and fastened together by capping and (sometimes) bracing.

PILE BUTT—The larger end of a tapered pile; usually the upper end as driven.

PILE CAPS—A structural member placed on top of a pile to distribute loads from the structure to the pile.

PUMP CREATE—A method of placing small aggregate concrete by means of a pump.

PURLIN—Horizontal members of a roof supporting common rafters. Also, members between trusses supporting sheathing.

PUTLOG—Horizontal boards set perpendicular to scaffold lengths that directly support the platform planks.

QUAY—A margin wharf adjacent to the shore and generally of solid filled construction.

RAFTER—A sloping roof member supporting the roof covering and extending from the ridge or the hip of the roof to the eaves.

RAKE—The inclined position of a cornice; also the angle of slope of a roof rafter.

REEVING—Threading or placement of a working line.

RIDGE—The long joining members placed at the angle where two slopes of a roof meet at the peak.

RIDGEBOARD—The horizontal timber at the upper end of the common rafters to which the rafters are nailed.

RISE—In a roof, the vertical distance between the plate and the ridge. In a stair, the total height of the stair.

SASH—The movable part of a window.

SHEAVE—A grooved wheel used to support cable or change its direction of travel (pronounced "shiv").

SHRINKAGE—Concrete contraction due to curing and excess water in mix.

SILLS—The first members of a frame set in place.

SLUMP TEST—A means of sample testing concrete for consistency; a measure of the plasticity of a concrete mix.

SLURRY—Thin watery mixture of water and cement.

SOFFIT—The underside of a subordinate member of a building.

SPAN—The shortest distance between a pair of rafter seats.

SPECIFICATIONS—Written instructions containing information about the materials, style, workmanship, and finish for the job.

STRIPPING—The removal of mold forms from hardened concrete.

STUDS—The vertical members of walls, wooden forms, and frames.

TERRAZZO—A concrete surface of Portland cement, fine sand, and marble chips.

TIES—Metal strips used to tie the outer wall of brick or masonry to the inner wall. Also used to tie concrete forms together.

TRUSS—A combination of members, such as beams, bars, and ties, usually arranged in triangular units to form a rigid framework for supporting loads over a span, usually a roof or bridge.

WAINSCOT—A wall covering for the lower part of an interior wall; can be wood, glass, or tile.

WIRE ROPE—A rope formed of wires wrapped around a central core; a steel cable.

APPENDIX II

REFERENCES USED TO DEVELOP THE TRAMAN

Volume 1

Although the following references were current when this TRAMAN was published, their continued currency cannot be assured. You therefore need to ensure that you are studying the latest revision.

Chapter 1

Naval Construction Force Manual, NAVFAC P-315, Naval Facilities Engineering Command, Washington, D.C., 1985.

Naval Construction Force Occupational Safety and Health Program Manual, COMCBPAC/COMCBLANTINST 5100.1F CH-2, Commander, Naval Construction Battalions U.S. Pacific Fleet Pearl Harbor, Hawaii, and Commander, Naval Construction Battalions U.S. Atlantic Fleet, Naval Amphibious Base Little Creek, Norfolk, Va., 1991.

Seabee Planner's and Estimators Handbook, NAVFAC P-405, Naval Facilities Engineering Command, Alexandria, Va., 1983.

Chapter 2

Blueprint Reading and Sketching, NAVEDTRA 10077-F1, Naval Education and Training Program Management Support Activity, Pensacola, Fl., 1988.

Engineering and Design Criteria for Navy Facilities, Military Bulletin, Naval Construction Battalion Center, Port Hueneme, Calif., 1991.

Chapter 3

Cabinetmaking, Patternmaking, and Millwork, Gasper J. Lewis, Delmar Publishers Inc., Albany, N.Y., 1981.

Carpentry I, EN5155, U.S. Army Engineer School, Fort Belvoir, Va., 1987.

Carpentry lll, ENO533, U.S. Army Engineer School, Fort Belvoir, Va., 1987.

Chapter 4

Safety and Health Requirements Manual, EM 385-1-1, U.S. Army Corps of Engineers, Washington, D.C., 1987.

Chapter 5

Engineering Aid: Intermediate and Advanced, NAVEDTRA 12540, Naval Education and Training Program Management Support Activity, Pensacola, Fla., 1994.

Chapter 6

Concrete and Masonry, Headquarters, Department of the Army, Washington, D.C., 1985.

Design and Control of Concrete Mixtures, Portland Cement Association, 5420 Old Orchard Road, Skokie, Ill., 1988.

Chapter 7

Concrete Formwork, Leonard Koel, American Technical Publishers, Inc., Homewood, Ill., 1988.

Concrete and Masonry, Headquarters, Department of the Army, Washington, D.C., 1985.

Chapter 8

Concrete and Masonry, Headquarters, Department of the Army, Washington, D.C., 1985.

Chapter 9

Naval Construction Force/Seabee Petty Officer First Class, NAVEDTRA 10601, Naval Education and Training Program Management Support Activity, Pensacola, Fla., 1989.

Seabee Planner's and Estimator's Handbook, NAVFAC P-405, Naval Facilities Engineering Command, Alexandria, Va., 1983.

APPENDIX II

Volume 2

Chapter 1

Carpentry, Leonard Koel, American Technical Publishers, Inc., Alsip, Ill., 1985.

Design of Wood Frame Structures for Permanence, National Products Association, Washington, D.C., 1988.

Exterior and Interior Trim, John E. Ball, Delmar Publishers, Inc., Albany, N.Y., 1975.

Chapter 2

Basic Roof Framing, Benjamin Barnow, Tab Books, Inc., Blue Ridge Summit, Pa., 1986.

Chapter 3

Basic Roof Framing, Benjamin Barnow, Tab Books, Inc., Blue Ridge Summit, Pa., 1986.

Design of Wood Frame Structures for Permanence, National Forest Products Association, Washington, D.C., 1988.

Exterior and Interior Trim, John E. Ball, Delmar Publishers, Inc., Albany, N.Y., 1975.

Manual of Built-up Roof Systems, Charles William Griffin, McGraw-Hill Book Co., New York, N.Y., 1982.

Modern Carpentry, Willis H. Wagner, Goodheart-Wilcox Co., South Holland, Ill., 1983.

Chapter 4

Basic Roof Framing, Benjamin Barnow, Tab Books, Inc., Blue Ridge Summit, Pa., 1986.

Design of Wood Frame Structures for Permanence, National Forest Products Association, Washington, D.C., 1988.

Exterior and Interior Trim, John E. Ball, Delmar Publishers, Inc., Albany, N.Y., 1975.

Modern Carpentry, Willis H. Wagner, Goodheart-Wilcox Co., South Holland, Ill., 1983.

Chapter 5

Drywall: Installation and Application, W. Robert Harris, American Technical Publishers, Inc., Homewood, Ill., 1979.

Modern Carpentry, Willis H. Wagner, Goodheart-Wilcox Co., South Holland, Ill., 1983.

Wood Frame House Construction, L.O. Anderson, Forest Products Laboratory, U.S. Forest Service, U.S. Department of Agriculture, Washington, D.C., 1975.

Chapter 6

Carpentry, Leonard Koel, American Technical Publishers, Inc., Alsip, Ill., 1985.

Exterior and Interior Trim, John E. Ball, Delmar Publishers, Inc., Albany, N.Y., 1975.

Wood Frame House Construction, L.O. Anderson, Forest Products Laboratory, U.S. Forest Service, U.S. Department of Agriculture, Washington, D.C., 1975.

Chapter 7

Handbook of Ceramic Tile Installation, Tile Council of America, Inc., Princeton, N.J., 1990.

Plastering Skills, F. Van Den Branden and Thomas L. Hartsell, American Technical Publishers, Inc., Alsip, Ill., 1984.

Chapter 8

Paints and Protective Coatings, NAVFAC-MO-110, Departments of the Army, Navy, and Air Force, Washington, D.C., 1981.

Wood Preservation, NAVFAC-MO-312, Naval Facilities Engineering Command, Department of the Navy, Washington, D.C., 1968.

Chapter 9

Facilities Planning Guide, NAVFAC P-437 (Revised), Naval Facilities Engineering Command, Department of the Navy, Alexandria, Va., 1989.

Chapter 10

Pile Construction, Field Manual 5-134, Headquarters, Department of the Army, Washington, D.C., 1985.

APPENDIX III

HAND SIGNALS

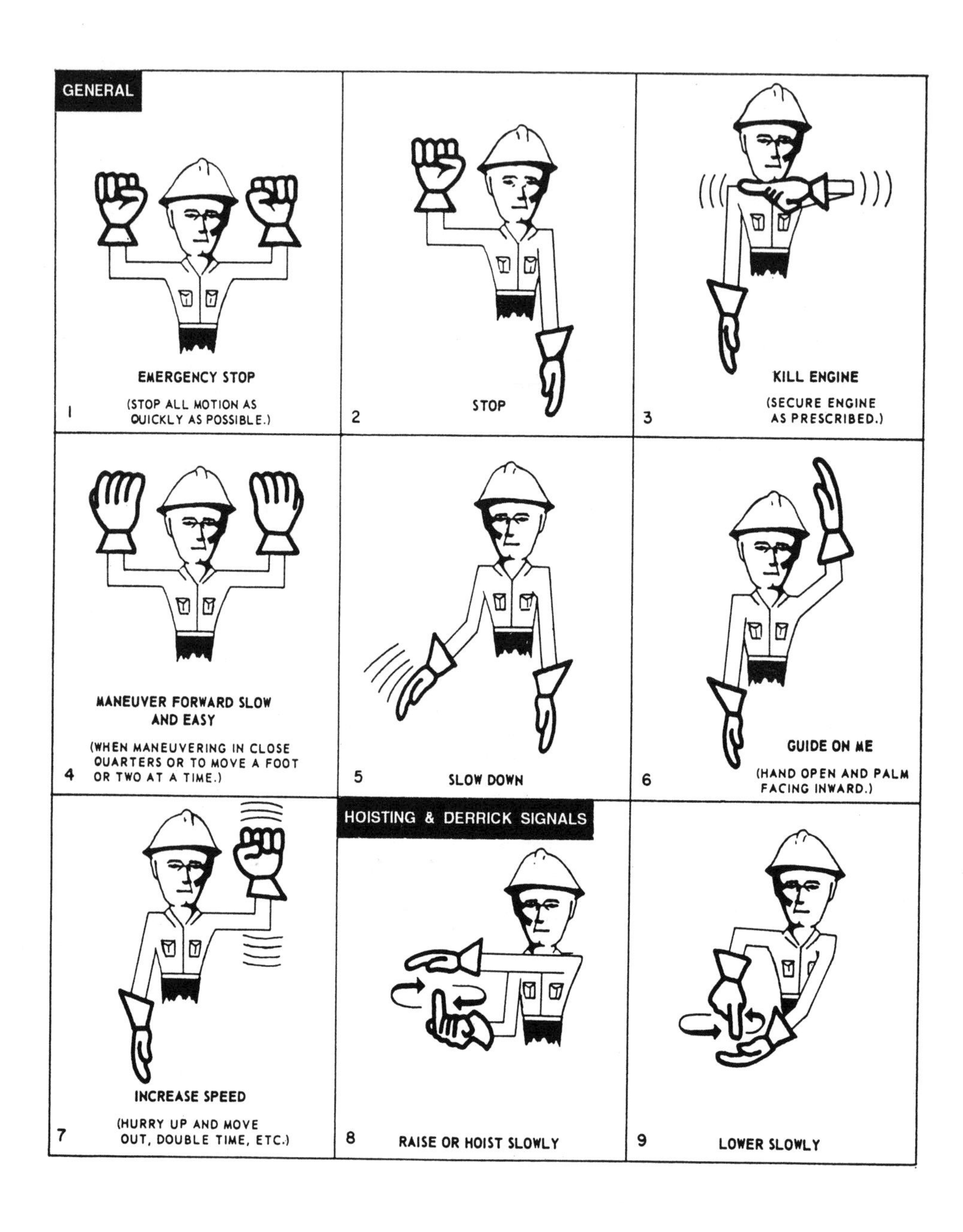
GENERAL
EMERGENCY STOP
1 (STOP ALL MOTION AS QUICKLY AS POSSIBLE.)
2 STOP
KILL ENGINE
3 (SECURE ENGINE AS PRESCRIBED.)
MANEUVER FORWARD SLOW AND EASY
4 (WHEN MANEUVERING IN CLOSE QUARTERS OR TO MOVE A FOOT OR TWO AT A TIME.)
5 SLOW DOWN
GUIDE ON ME
6 (HAND OPEN AND PALM FACING INWARD.)
INCREASE SPEED
7 (HURRY UP AND MOVE OUT, DOUBLE TIME, ETC.)
HOISTING & DERRICK SIGNALS
8 RAISE OR HOIST SLOWLY
9 LOWER SLOWLY

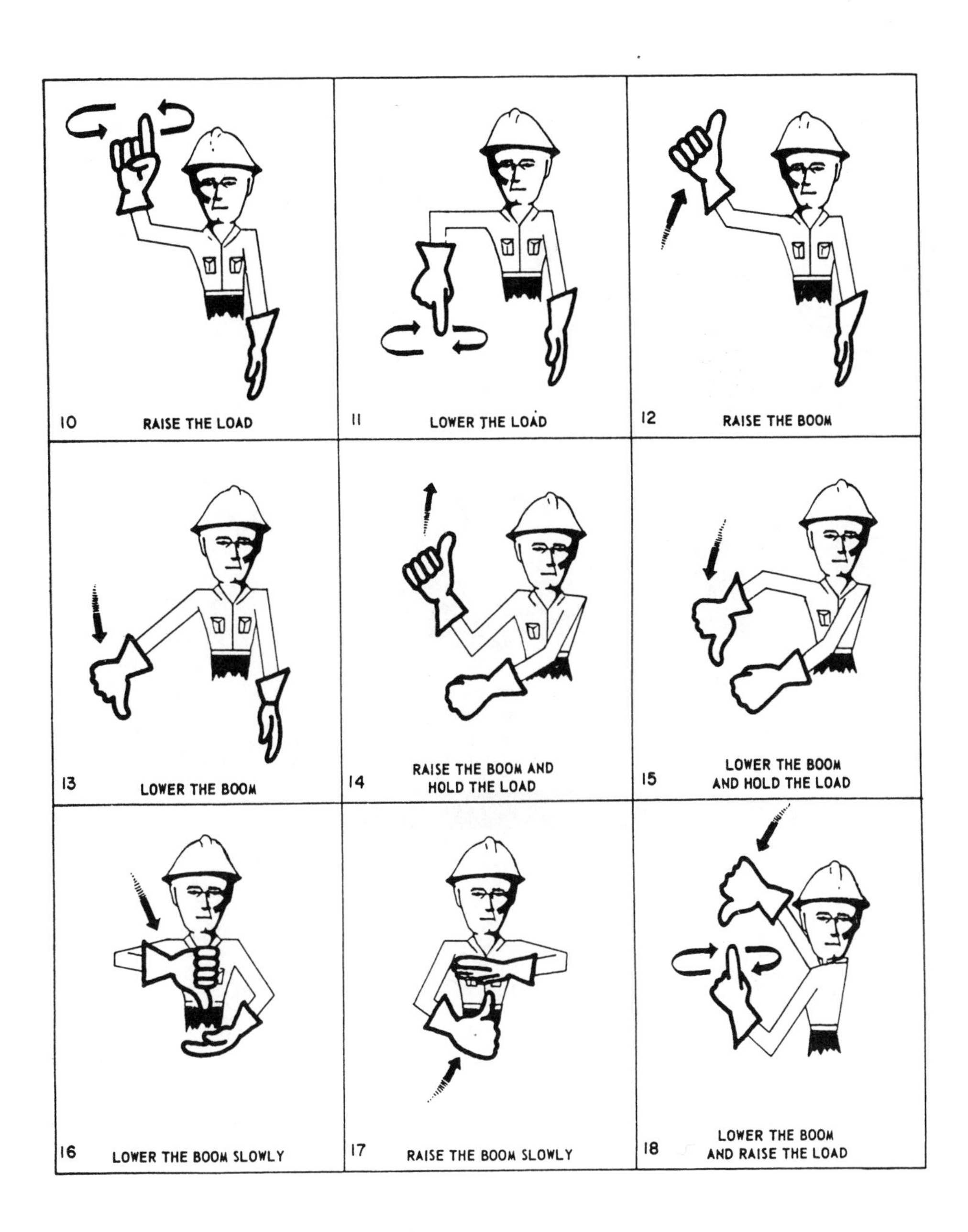
10
RAISE THE LOAD
11
LOWER THE LOAD
12
RAISE THE BOOM
13
LOWER THE BOOM
14
RAISE THE BOOM AND HOLD THE LOAD
15
LOWER THE BOOM AND HOLD THE LOAD
16
LOWER THE BOOM SLOWLY
17
RAISE THE BOOM SLOWLY
18
LOWER THE BOOM AND RAISE THE LOAD

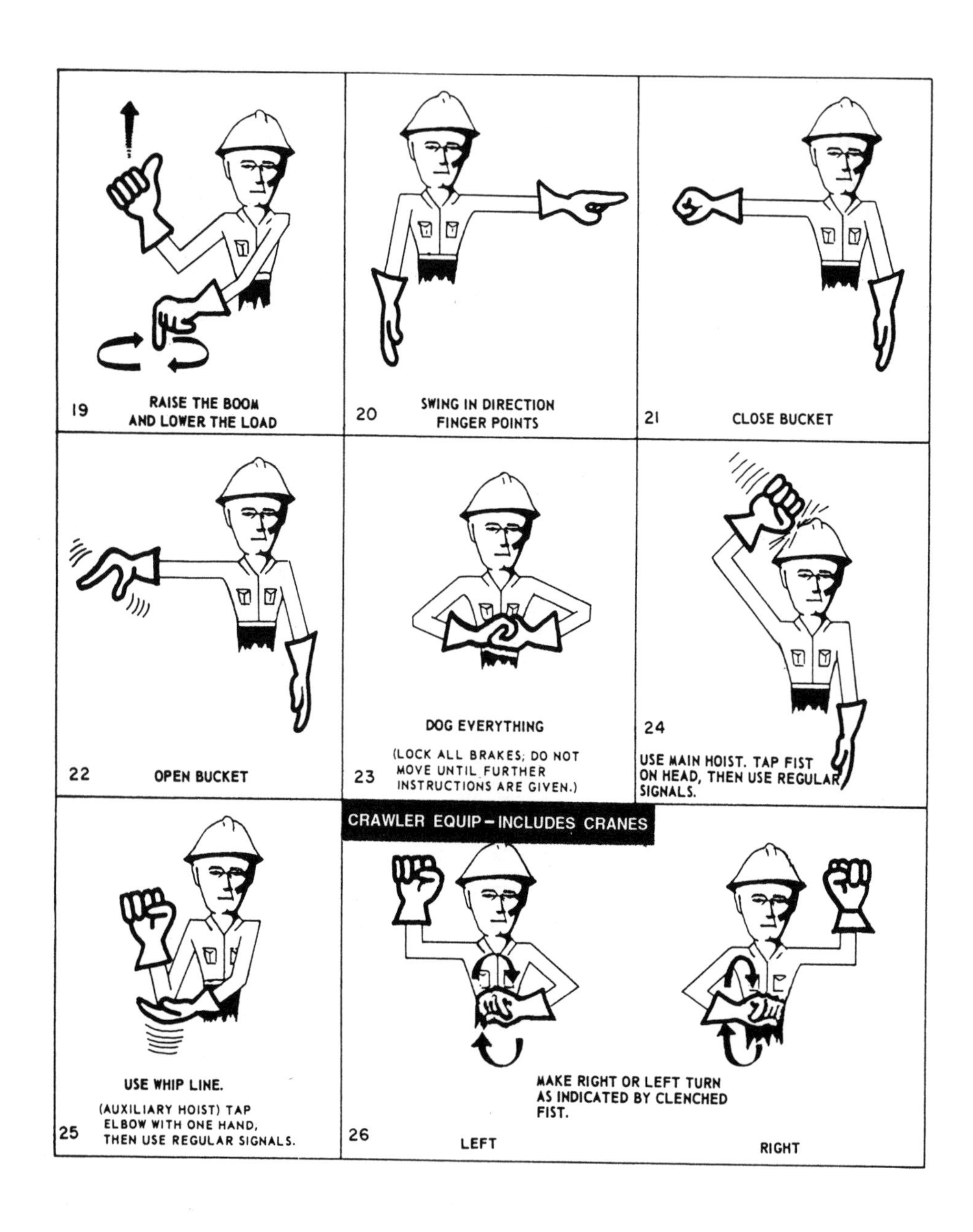
19
RAISE THE BOOM
AND LOWER THE LOAD
20
SWING IN DIRECTION
FINGER POINTS
21
CLOSE BUCKET
22
OPEN BUCKET
DOG EVERYTHING
(LOCK ALL BRAKES; DO NOT
MOVE UNTIL FURTHER
INSTRUCTIONS ARE GIVEN.)
23
24
USE MAIN HOIST. TAP FIST
ON HEAD, THEN USE REGULAR
SIGNALS.
CRAWLER EQUIP – INCLUDES CRANES
USE WHIP LINE.
(AUXILIARY HOIST) TAP
ELBOW WITH ONE HAND,
THEN USE REGULAR SIGNALS.
25
MAKE RIGHT OR LEFT TURN
AS INDICATED BY CLENCHED
FIST.
26
LEFT
RIGHT

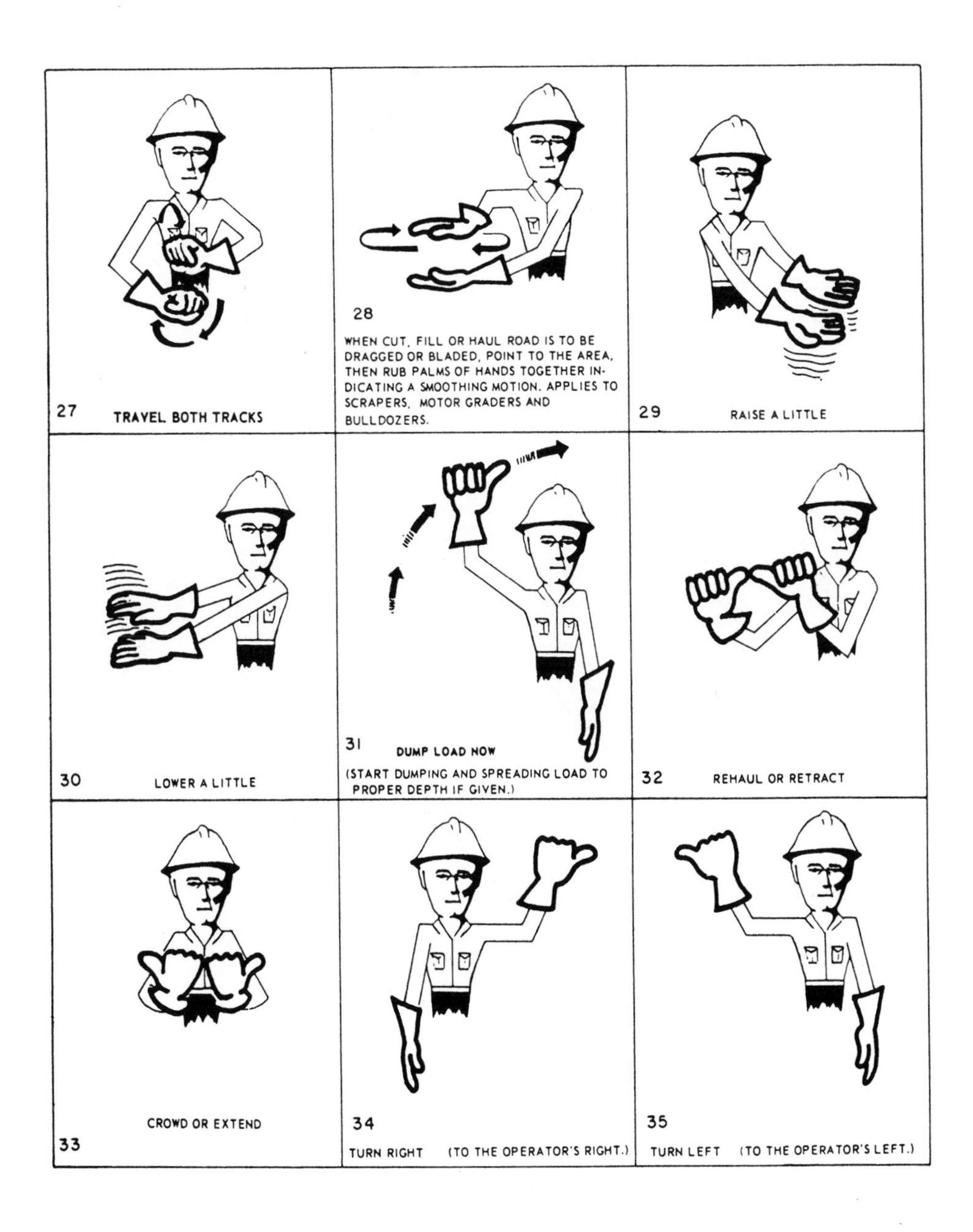
27 TRAVEL BOTH TRACKS
28
WHEN CUT, FILL OR HAUL ROAD IS TO BE DRAGGED OR BLADED, POINT TO THE AREA, THEN RUB PALMS OF HANDS TOGETHER INDICATING A SMOOTHING MOTION. APPLIES TO SCRAPERS, MOTOR GRADERS AND BULLDOZERS.
29 RAISE A LITTLE
30 LOWER A LITTLE
31 DUMP LOAD NOW
(START DUMPING AND SPREADING LOAD TO PROPER DEPTH IF GIVEN.)
32 REHAUL OR RETRACT
33 CROWD OR EXTEND
34
TURN RIGHT (TO THE OPERATOR'S RIGHT.)
35
TURN LEFT (TO THE OPERATOR'S LEFT.)

INDEX